American Prisoners of War
Held at Plymouth
During the War of 1812

Heritage Books by the Society of the War of 1812
in the State of Ohio:

Transcribed by Harrison Scott Baker

*American Prisoners of War Held at Bermuda,
Cape of Good Hope and Jamaica During the War of 1812*

*American Prisoners of War Held at Barbados,
Newfoundland and New Providence During the War of 1812*

*American Prisoners of War Held at Halifax
During the War of 1812, Volume I and II*

Transcribed by Eric Eugene Johnson

American Prisoners of War Held at Dartmoor During the War of 1812

*American Prisoners of War Held in Montreal
and Quebec During the War of 1812*

*American Prisoners of War Held at Plymouth
During the War of 1812*

*American Prisoners of War Held at Quebec
During the War of 1812, 8 June 1813–11 December 1814*

*American Prisoners of War Paroled at Dartmouth,
Halifax, Jamaica and Odiham During the War of 1812*

*American Sea Fencibles in the War of 1812:
United States Sea Fencibles, State Sea Fencibles*

Black Regulars in the War of 1812

Black Regulars and Militiamen in the War of 1812

Forgotten Americans Who Served in the War of 1812

Ohio and the War of 1812: A Collection of Lists, Musters and Essays

Ohio's Regulars in the War of 1812

Heritage Books by the Society of the War of 1812
in the State of Maryland:

Maryland Regulars in the War of 1812
Transcribed by Eric Eugene Johnson; Foreword by Christos Christou

American Prisoners of War

Held at
Plymouth during the War of 1812

Transcribed by
Eric Eugene Johnson

Society of the War of 1812
in the
State of Ohio

HERITAGE BOOKS
2018

HERITAGE BOOKS
AN IMPRINT OF HERITAGE BOOKS, INC.

Books, CDs, and more—Worldwide

For our listing of thousands of titles see our website
at
www.HeritageBooks.com

Published 2018 by
HERITAGE BOOKS, INC.
Publishing Division
5810 Ruatan Street
Berwyn Heights, Md. 20740

International Standard Book Number
Paperbound: 978-0-7884-5825-5

- Table of Contents -

Introduction

This is a transcription of American prisoner of war records from the U.S. Navy, privateers and merchant vessels (plus some civilians) who were captured and then interned by the British Empire at Plymouth, England during the War of 1812.

This volume was compiled from copies of the *General Entry Book of American Prisoners of War* ledgers of the British Admiralty made by the Public Records Office in London, Great Britain (ADM 103 series). The *General Entry Book* (GEB) records are composed of lines for the recording of names and personal information of those incarcerated. The record of each prisoner is found on two facing pages. The clerk making the entries wrote the page number on the upper right-side corner of each page. The names and information of ten men can be found on each double page for the Americans interned at the Plymouth.

Microfilm	POW facilities	Dates	Number of prisoners
ADM 103/268	Plymouth	Oct 1812 – May 1813	1,207
ADM 103/269	Plymouth	May 1813 – Jan 1814	1,247
ADM 103/270	Plymouth	Jan 1814 – Jun 1814	251
"	"	Feb 1815 – Jun 1815	863
Total			3,568

Below are the column headers from the GEB ledgers used at Plymouth. Titles in brackets, "[" and "]", indicates that the transcriber has changed the column headers from the original to a more meaningful header. Some of the columns have been eliminated in this book while other columns have been combined.

Column 1 – Number
    Each prisoner of war arriving at a prisoner facility was assigned a number.

Column 2 – By what Ship or how taken [How taken]
    This column lists the Royal Naval ship or privateer which captured the prisoner. Also included in this column are the men who gave themselves up as a prisoner, who were impressed by the Royal Navy, taken ashore or captured by the Royal Army.

Column 3 – Time when [When Taken]
    The date the prisoner was taken into custody.

Column 4 – Where Taken
    This column lists the location of capture which could indicate latitudes and longitudes if at sea, a port, or a geographic region or location.

Column 5 – Name of Prize [Prize Name]
    The name of the ship or vessel of the prisoner of war, or if from land forces, the name of the regiment or service.

Column 6 - Whether Man of War, Privateer, or Merchant Vessel [Ship Type]
    The type of ship or vessel, that is: man of war (warship), revenue cutter, privateer, letter of marque, merchant vessel, or prize of a privateer. Land forces indicate regular army, volunteers or militia.

Column 7 – Prisoner's Names [title not used]
    The prisoner names given in the GEB are first name then last name, e.g. John Smith, but in this book the last name is given first then the first name.

Column 8 – Quality [Rank]
    This column gives the rank of the prisoner. In addition to naval, privateer and merchant vessel ranks, there are also civilians, merchants, supercargoes and passengers found in these records.

Column 9 – Time when received into custody [Date Received]
The date when a prisoner arrived at a prisoner facility.

Column 10 – From what ship, or whence received [From what ship]
The location of the prisoner before being received at the current prisoner of war facility. This could be the capturing ship, the ship that transported the prisoner to England, a hospital, another prison facility or parole location.

Column 11 – Place of Nativity [Born]
Lists the birth place of a prisoner.

Column 12 – Age
The age of the prisoner in years, most likely, from his last birthday.

Columns 13 through 18 are not used in this book [Race]
These columns indicate the height, hair color, color of eyes, type of complexion and any body marks (including tattoos) or wounds of a prisoner. Included in these physical descriptions are the races of non-Caucasians, which are Black, Negro, Mulatto, Creole or Chinese. The author has created a new column entitled Race to indicate non-Caucasians.

Columns 19 through 32 are not used in this book
The personal items of a prisoner were inventoried upon arrival at a prisoner of war facility and the required missing items were replaced. These items include hammocks, beds, straw mattresses, cushions, blankets, hats, jackets, waistcoats, trousers, shirts, shoes, stockings and handkerchiefs.

Columns 33 through 35 have been combined into a single entry [titles not used]
This new column indicates when a prisoner died at a prisoner facility, when he escaped from a prisoner facility, when he was discharged to another prison facility, or when he was released and sent back to the United States.

Column 33 – Exchanged, Discharged, Died or Escaped [field not used]
"E" or "R" indicates that the prisoner escaped from the parole location while "D" indicates that he was discharged. "DD" indicates that he died while assigned to a parole location.

Column 34 – Time When [field not used]
Contains the date of the event from column 33.

Column 35 - Whither, and by what order if discharged. [field not used]
This column shows the place or ship that the prisoner was sent when he was discharged from the prison facility. Orders were given by the His Majesty's Transport Board.

The Royal Navy's Plymouth Naval Base was the home of one of the three prisoner of war prison ship facilities in England which were used during the War of 1812 to house American prisoners of war. The facility had been used since 1796 to intern French prisoners of war during the Napoleonic Wars. The other two prison ship facilities were located at Portsmouth and at Chatham.

The naval base was used as the entry point for receiving American POWs before they were re-assigned to the other prison depots and prison ships in England. Between 22 October 1812 and 24 June 1814, 2,707 Americans were processed at Plymouth spending no more than thirty days on the prison ships. The men were captured at the beginning of the war off American ships in English ports or men who were in port awaiting to sign on to a new ship as crew men. Many men were taken off Royal Naval warships and English merchant vessels as enemy aliens. Other men were captured on American privateers and merchant vessels which were operating in the Irish Sea, the English Channel and the Bay of Biscay.

A total of 256 senior officers, warrant officers and captured civilian were sent to the Ashburton on honor parole. These individuals lived freely in this village and did not have to live in a prison or prison ship. 364 men at Plymouth were transferred to Portsmouth, 1,284 to Dartmoor, 265 to Chatham, and 400 to Stapleton. Fifteen men died on the prison ship *Caton* or at the Mill Prison Hospital in Plymouth. There were 281 African-Americans and one Chinese among the Americans assigned to Plymouth.

On 3 February 1815 the prison ships were once again opened to receive American POWs. A total of 863 Americans, including 111 African-Americans and one Indian, spent no more than three months on the prison ships. 315 were sent to Dartmoor and two to Ashburton. The rest of the men were released in June 1815 and sent back to the United States. No Americans died during the second phase that the prison ships were used.

A new GEB ledger was started during the second phase at the prison facility and the prisoner numbers restarted at "1". The first 863 men in each phase have the same prisoner numbers. The book treats the two phases as two separate prison facilities to avoid this confusion.

The penmanship in these ledgers was very good. The spelling of non-familiar names was done phonetically.

Any errors or omissions are regretted and are the fault of the transcriber.

Eric Eugene Johnson

President (2008-2011)
Society of the War of 1812 in the State of Ohio

Registrar General (2017-   )
General Society of the War of 1812

Abbreviations

HM – His Majesty
HMS – His Majesty's Ship
HMT – His Majesty's Transport
US – United States
War – Warship
MV – Merchant Vessel
P – Privateer
LM – Letter of Marque

Also, standard state name abbreviations were used

The locations in Maine are listed as MA and not ME. During the War of 1812, Maine was still a part of Massachusetts.

(1), (2), etc. – Different men with the same names assigned to a single ship or a number of ships with the same name.

- In memory of those who did not return -

The Honored Dead

Allen, William Henry
Baxter, Charles
Coutie, Thomas
Delphy, Richard
Eggert, Francis
English, Edward
Gallo, Joseph
Glass, Reuben
Jordan, Joseph
Manuel, Joseph
Moneys, Robert
Montel, Pierre
Randol, Thomas
White, James (1)
White, James (2)

- Those who die in service to the United States should not be forgotten –

Alphabetical listing of names
October 1812 through June 1814

Abbey, Obadiah - Seaman - Number: 2269 - Prize name: Fanny - Ship type: MV - How taken: HM Frigate Eurotas - When taken: 25 Dec 1813 - Where taken: at sea - Date received: 20 Jan 1814 - From what ship: HMS Eurotas - Born: Henfield - Age: 31 - Discharged on 31 Jan 1814 and sent to Dartmoor.

Abbott, Ephraim - Seaman - Number: 2462 - Prize name: Fair American - Ship type: MV - How taken: HM Frigate Andromache - When taken: 19 Jan 1814 - Where taken: Bay of Biscay - Date received: 22 Feb 1814 - From what ship: HMS York - Born: Boston - Age: 25 - Discharged on 10 May 1814 and sent to Dartmoor.

Abbott, Timothy - Seaman - Number: 1781 - Prize name: Orders in Council - Ship type: LM - How taken: HM Frigate Surveillante - When taken: 1 Jun 1813 - Where taken: Bay of Biscay - Date received: 7 Aug 1813 - From what ship: HMS Gleaner - Born: Hampshire - Age: 25 - Discharged on 8 Sep 1813 and sent to Dartmoor.

Abraham, William - 2nd Mate - Number: 333 - Prize name: Mariner - Ship type: MV - How taken: HM Brig Lyra - When taken: 15 Dec 1812 - Where taken: Bay of Biscay - Date received: 22 Jan 1813 - From what ship: HMS Lyra - Born: Nantucket - Age: 28 - Sent to Chatham on 29 Mar 1813 on the HMS Braham.

Acher, Joseph - Chief Mate - Number: 900 - Prize name: Dick - Ship type: MV - How taken: HM Brig Dispatch - When taken: 17 Mar 1813 - Where taken: Bay of Biscay - Date received: 2 Apr 1813 - From what ship: HMS Plymouth - Born: Boston - Age: 38 - Sent on 4 Apr 1813 to Ashburton on parole.

Adam, Joseph - Passenger - Number: 1595 - Prize name: Governor Gerry - Ship type: MV - How taken: HM Brig Royalist - When taken: 31 May 1813 - Where taken: Bay of Biscay - Date received: 14 Jun 1813 - From what ship: HMS Royalist - Born: Paris - Age: 55 - Discharged on 7 Jul 1813 and sent to Ashburton on parole.

Adams, James - Chief Mate - Number: 1931 - Prize name: Ellen & Emeline - Ship type: MV - How taken: HM Schooner Telegraph - When taken: 13 Aug 1813 - Where taken: off St. Andrews - Date received: 25 Aug 1813 - From what ship: Ellen & Emeline - Born: New York - Age: 34 - Discharged on 30 Aug 1813 and sent to Ashburton on parole.

Adams, John - Seaman - Number: 1142 - Prize name: Magdalen - Ship type: MV - How taken: HM Ship-of-the-Line Superb - When taken: 15 Apr 1813 - Where taken: off Belle Isle - Date received: 22 Apr 1813 - From what ship: HMS Superb - Born: Pennsylvania - Age: 16 - Sent to Dartmoor on 1 Jul 1813.

Adams, John - Seaman - Number: 2588 - Prize name: Rambler - Ship type: MV - How taken: Transport May - When taken: Jan 1813 - Where taken: Isle de France - Date received: 10 May 1814 - From what ship: HMS Lion - Born: Boston - Age: 25 - Discharged on 14 Jun 1814 and sent to Dartmoor.

Adams, John - Seaman - Number: 2153 - Prize name: Agnes, prize of the Privateer Rambler - Ship type: LM - How taken: Cutter Jane - When taken: 28 Nov 1813 - Where taken: Bay of Biscay - Date received: 10 Dec 1813 - From what ship: Cutter Jane - Born: Boston - Age: 17 - Discharged on 31 Jan 1814 and sent to Dartmoor.

Adams, Peter William - Cook - Number: 2574 - How taken: Delivered himself up from the MV Hebrus - Date received: 8 May 1814 - From what ship: MV Hebrus - Born: Norfolk - Age: 24 - Race: Black - Discharged on 7 May 1814 and sent to Ashburton on parole.

Adams, Samuel Reed - Mate - Number: 995 - Prize name: Polly - Ship type: MV - How taken: HM Frigate Surveillante - When taken: 27 Mar 1813 - Where taken: Bay of Biscay - Date received: 16 Apr 1813 - From what ship: HMS Fairy - Born: Marblehead - Age: 23 - Sent on 22 Apr 1813 to Ashburton on parole.

Adams, Theophilus - Chief Mate - Number: 215 - Prize name: Vengeance - Ship type: LM - How taken: HM Frigate Phoebe - When taken: 1 Jan 1813 - Where taken: Lat 44.4 Long 23 - Date received: 9 Jan 1813 - From what ship: HMS Phoebe - Born: Barnstable - Age: 27 - Sent on 13 Jan 1813 to Ashburton on parole.

Addigo, Henry - Marine - Number: 1871 - Prize name: US Brig Argus - Ship type: War - How taken: HM Brig Pelican - When taken: 14 Aug 1813 - Where taken: Irish Channel - Date received: 23 Aug 1813 - From what ship: HMS Pelican - Born: New York - Age: 41 - Discharged on 3 Nov 1813 and sent to Dartmoor.

Adgate, William - Supercargo - Number: 944 - Prize name: Good Friends - Ship type: MV - How taken: HM Frigate

Andromache - When taken: 2 Apr 1813 - Where taken: Bay of Biscay - Date received: 7 Apr 1813 - From what ship: HMS Sea Lark - Born: Connecticut - Age: 36 - Sent on 10 Apr 1813 to Ashburton on parole.

Akerman, William - Seaman - Number: 2596 - How taken: HM Frigate Havannah - Date received: 16 May 1814 - From what ship: HMS Repulse - Born: New York - Age: 28 - Discharged on 14 Jun 1814 and sent to Dartmoor.

Akins, William - Seaman - Number: 751 - Prize name: Charlotte - Ship type: MV - How taken: HM Ship-of-the-Line Warspite - When taken: 3 Mar 1813 - Where taken: Bay of Biscay - Date received: 19 Mar 1813 - From what ship: HMS Warspite - Born: New Orleans - Age: 23 - Sent to Dartmoor on 2 Apr 1813.

Albert, Hezekiah - Seaman - Number: 368 - Prize name: Orbit - Ship type: MV - How taken: HM Brig Achates - When taken: 29 Jan 1813 - Where taken: Lat 49 N Long 13 W - Date received: 4 Feb 1813 - From what ship: HMS Achates - Born: Rhode Island - Age: 20 - Sent to Chatham on 29 Mar 1813 on the HMS Braham.

Albertson, John Nathaniel - Seaman - Number: 1275 - Prize name: Caroline - Ship type: MV - How taken: HM Frigate Medusa - When taken: 12 Apr 1813 - Where taken: Bay of Biscay - Date received: 10 May 1813 - From what ship: HMS Medusa - Born: Boston - Age: 27 - Discharged on 3 Jul 1813 and sent to Stapleton Prison.

Alexander, Richard - Seaman - Number: 2174 - Prize name: Wolf, prize of the Privateer Grand Turk - Ship type: P - How taken: HM Frigate Briton - When taken: 1 Dec 1813 - Where taken: Bay of Biscay - Date received: 17 Dec 1813 - From what ship: HMS Briton - Born: Marblehead - Age: 21 - Discharged on 17 Dec 1813 and released to HMS Britton.

Allen, Eleazer - 1st Mate - Number: 471 - Prize name: Orbit - Ship type: MV - How taken: HM Brig Achates - When taken: 29 Jan 1813 - Where taken: Lat 49 N Long 13 W - Date received: 9 Feb 1813 - From what ship: Orbit - Born: New Bedford - Age: 25 - Sent on 10 Feb 1813 to Ashburton on parole.

Allen, Elihu - Seaman - Number: 31 - Prize name: Catharine - Ship type: MV - How taken: HM Frigate Leonidas - When taken: 31 Jul 1812 - Where taken: off Ireland - Date received: 23 Nov 1812 - From what ship: HMS Stork - Born: New Bedford - Age: 18 - Sent to Portsmouth on 29 Dec 1812 on the HMS Northumberland.

Allen, Elissia - Prize Master - Number: 2082 - Prize name: Colin West, prize of Privateer True Blooded Yankee - Ship type: P - How taken: HM Schooner Helicon & HM Schooner Whiting - When taken: 26 Oct 1813 - Where taken: Channel - Date received: 2 Nov 1813 - From what ship: HMS Whiting - Born: Massachusetts - Age: 27 - Discharged on 3 Nov 1813 and sent to Dartmoor.

Allen, Henry - Seaman - Number: 729 - Prize name: Mars - Ship type: MV - How taken: HM Ship-of-the-Line Warspite - When taken: 26 Feb 1813 - Where taken: Bay of Biscay - Date received: 19 Mar 1813 - From what ship: HMS Warspite - Born: Vermont - Age: 21 - Sent to Dartmoor on 2 Apr 1813.

Allen, Henry - Seaman - Number: 1658 - Prize name: Mars - Ship type: MV - How taken: HM Ship-of-the-Line Warspite - When taken: 26 Feb 1813 - Where taken: Bay of Biscay - Date received: 15 Jun 1813 - From what ship: Dartmoor Prison - Born: Vermont - Age: 21 - Discharged on 16 Jun 1813 and released to HMS Salvador del Mundo.

Allen, Jacob - Seaman - Number: 405 - Prize name: Union - Ship type: MV - How taken: HM Frigate Iris - When taken: 17 Jan 1813 - Where taken: Lat 44 N Long 2.3 W - Date received: 5 Feb 1813 - From what ship: HMS San Josef - Born: America - Age: 32 - Sent to Chatham on 29 Mar 1813 on the HMS Braham.

Allen, James - Seaman - Number: 1952 - Prize name: Joel Barlow - Ship type: LM - How taken: HM Frigate Briton - When taken: 3 Jul 1813 - Where taken: off Bordeaux - Date received: 31 Aug 1813 - From what ship: HMS Clarence - Born: New London - Age: 44 - Discharged on 8 Sep 1813 and sent to Dartmoor.

Allen, John - Seaman - Number: 1470 - Prize name: Paul Jones - Ship type: P - How taken: HM Frigate Leonidas - When taken: 23 May 1813 - Where taken: off Cape Clear - Date received: 26 May 1813 - From what ship: HMS Leonidas - Born: Seabrook, MA - Age: 37 - Discharged on 30 Jun 1813 and sent to Stapleton.

Allen, John - Boy - Number: 924 - Prize name: Prompt - Ship type: MV - How taken: Chance, privateer - When taken: 22 Mar 1813 - Where taken: Bay of Biscay - Date received: 6 Apr 1813 - From what ship: Fonvey - Born: Boston - Age: 14 - Sent on 21 Jun 1813 to Portsmouth on HMS Prometheus.

Allen, John - Cook - Number: 1718 - Prize name: Orders in Council - Ship type: LM - How taken: HM Frigate Surveillante - When taken: 1 Jun 1813 - Where taken: Bay of Biscay - Date received: 4 Jul 1813 - From what ship:

HMS Iris - Born: New York - Age: 26 - Race: Mulatto - Discharged on 22 Aug 1813 and released to HM Brig Redpole.

Allen, John - Seaman - Number: 671 - How taken: Impressed at Liverpool - When taken: 10 Jan 1813 - Date received: 15 Mar 1813 - From what ship: HMS Bittern - Born: Richmond - Age: 23 - Sent to Dartmoor on 1 Jul 1813.

Allen, John L. - Seaman - Number: 591 - Prize name: St. Martin's Plantation, prize of the Privateer Paul Jones - Ship type: P - How taken: HM Ship-of-the-Line Dublin - When taken: 9 Feb 1815 - Where taken: Lat 43 N Long 33.5 W - Date received: 25 Feb 1813 - From what ship: HMS Dublin - Born: Gloucester - Age: 21 - Sent to Dartmoor on 2 Apr 1813.

Allen, Joseph - Seaman - Number: 1881 - Prize name: US Brig Argus - Ship type: War - How taken: HM Brig Pelican - When taken: 14 Aug 1813 - Where taken: Irish Channel - Date received: 23 Aug 1813 - From what ship: HMS Pelican - Born: Boston - Age: 27 - Discharged on 8 Sep 1813 and sent to Dartmoor.

Allen, Philip - Seaman - Number: 1936 - Prize name: Marmion - Ship type: MV - How taken: HM Frigate President - When taken: 14 Aug 1813 - Where taken: off Nantes - Date received: 27 Aug 1813 - From what ship: HMS Urgent - Born: Maryland - Age: 28 - Race: Negro - Discharged on 8 Sep 1813 and sent to Dartmoor.

Allen, Thomas - Seaman - Number: 1104 - Prize name: Viper - Ship type: MV - How taken: HM Ship-of-the-Line Superb - When taken: 15 Apr 1813 - Where taken: Bay of Biscay - Date received: 22 Apr 1813 - From what ship: HMS Superb - Born: Alexandria - Age: 23 - Sent to Dartmoor on 1 Jul 1813.

Allen, William - Chief Mate - Number: 1572 - Prize name: Governor Gerry - Ship type: MV - How taken: HM Brig Royalist - When taken: 31 May 1813 - Where taken: Bay of Biscay - Date received: 10 May 1813 - From what ship: Governor Gerry - Born: New Bedford - Age: 20 - Discharged on 30 Jun 1813 and sent to Ashburton on parole.

Allen, William - Seaman - Number: 678 - Prize name: Brizeland - Ship type: MV - How taken: Olive Branch, letter of marque - When taken: 28 Dec 1812 - Date received: 15 Mar 1813 - From what ship: HMS Bittern - Born: Boston - Age: 35 - Sent to Dartmoor on 2 Apr 1813.

Allen, William Henry - Captain - Number: 1847 - Prize name: US Brig Argus - Ship type: War - How taken: HM Brig Pelican - When taken: 14 Aug 1813 - Where taken: Irish Channel - Date received: 18 Aug 1813 - From what ship: USS Argus - Born: Rhode Island - Age: 27 - Died on 18 Aug 1813 in the Mill Prison Hospital.

Allen, William Howard - 2nd Lieutenant - Number: 1858 - Prize name: US Brig Argus - Ship type: War - How taken: HM Brig Pelican - When taken: 14 Aug 1813 - Where taken: Irish Channel - Date received: 23 Aug 1813 - From what ship: HMS Pelican - Born: Hudson, NY - Age: Kristiansand 23 - Discharged on 1 Sep 1813 and sent to Ashburton on parole.

Alley, Samuel - Seaman - Number: 1005 - Prize name: Eliza - Ship type: MV - How taken: HM Frigate Surveillante - When taken: 27 Mar 1813 - Where taken: Bay of Biscay - Date received: 16 Apr 1813 - From what ship: HMS Fairy - Born: New Havens - Age: 35 - Sent to Dartmoor on 8 Sep 1813.

Allister, Isaac - Seaman - Number: 1829 - Prize name: US Brig Argus - Ship type: War - How taken: HM Brig Pelican - When taken: 14 Aug 1813 - Where taken: Irish Channel - Date received: 17 Aug 1813 - From what ship: USS Argus - Born: Habanos, America - Age: 26 - Discharged on 8 Sep 1813 and sent to Dartmoor.

Alman, John - Seaman - Number: 1426 - Prize name: Paul Jones - Ship type: P - How taken: HM Frigate Leonidas - When taken: 23 May 1813 - Where taken: off Cape Clear - Date received: 26 May 1813 - From what ship: HMS Leonidas - Born: Baltimore - Age: 25 - Discharged on 3 Jul 1813 and sent to Stapleton Prison.

Alton, Peter - Marine - Number: 2652 - How taken: Delivered himself up as a prisoner of war - Date received: 31 May 1814 - From what ship: HMS Gronville - Born: Ballston - Age: 21 - Discharged on 14 Jun 1814 and sent to Dartmoor.

Ammerson, Charles - Seaman - Number: 1227 - Prize name: Essex - Ship type: MV - How taken: HM Frigate Pyramus - When taken: 2 Apr 1813 - Where taken: Bay of Biscay - Date received: 9 May 1813 - From what ship: HMS Andromache - Born: Massachusetts - Age: 18 - Discharged on 3 Jul 1813 and sent to Stapleton Prison.

Amos, Cheney - Seaman - Number: 1782 - Prize name: Orders in Council - Ship type: LM - How taken: HM Frigate Surveillante - When taken: 1 Jun 1813 - Where taken: Bay of Biscay - Date received: 7 Aug 1813 - From what ship: HMS Gleaner - Born: Connecticut - Age: 21 - Discharged on 8 Sep 1813 and sent to Dartmoor.

Anderson, Daniel - Seaman - Number: 1633 - Prize name: Leo - Ship type: LM - How taken: HM Frigate Magiciene - When taken: 4 Jun 1813 - Where taken: Lat 45 Long 14 - Date received: 14 Jun 1813 - From what ship: HMS Orestes - Born: Massachusetts - Age: 22 - Discharged on 30 Jun 1813 and sent to Stapleton.

Anderson, David - Seaman - Number: 1256 - Prize name: Caroline - Ship type: MV - How taken: HM Frigate Medusa - When taken: 12 Apr 1813 - Where taken: Bay of Biscay - Date received: 10 May 1813 - From what ship: HMS Medusa - Born: Maryland - Age: 30 - Race: Negro - Discharged on 3 Jul 1813 and sent to Stapleton Prison.

Anderson, Edward - Seaman - Number: 2340 - Prize name: Apparencen, prize of the Privateer Bunker Hill - Ship type: P - How taken: Cartdian - When taken: 27 Jan 1814 - Where taken: off Ushant - Date received: 30 Jan 1814 - From what ship: Cartdian - Born: Baltimore - Age: 26 - Discharged on 31 Jan 1814 and sent to Dartmoor.

Anderson, George - Seaman - Number: 2433 - Prize name: US Brig Argus - Ship type: War - How taken: HM Brig Pelican - When taken: 14 Aug 1813 - Where taken: Irish Channel - Date received: 12 Feb 1814 - From what ship: HMS Salvador del Mundo - Born: Liddy - Age: 24 - Discharged on 10 May 1814 and sent to Dartmoor.

Anderson, James - Cook - Number: 1452 - Prize name: Paul Jones - Ship type: P - How taken: HM Frigate Leonidas - When taken: 23 May 1813 - Where taken: off Cape Clear - Date received: 26 May 1813 - From what ship: HMS Leonidas - Born: Maryland - Age: 25 - Race: Black - Discharged on 30 Jun 1813 and sent to Stapleton.

Anderson, James - Seaman - Number: 196 - Prize name: Hunter - Ship type: P - How taken: HM Frigate Phoebe - When taken: 24 Dec 1812 - Where taken: off Western Islands - Date received: 9 Jan 1813 - From what ship: HMS Phoebe - Born: Kristiansand, Norway - Age: 18 - Sent to Portsmouth on 8 Feb 1813 on the HMS Colossus.

Anderson, James - Seaman - Number: 992 - Prize name: Lightning - Ship type: MV - How taken: HM Frigate Medusa - When taken: 2 Apr 1813 - Where taken: Bay of Biscay - Date received: 16 Apr 1813 - From what ship: HMS Fairy - Born: Hampton - Age: 24 - Sent on 7 Jul 1813 to HMS Salvador del Mundo.

Anderson, James - Seaman - Number: 2022 - How taken: Impressed at Plymouth - Date received: 23 Oct 1813 - From what ship: Plymouth - Born: Boston - Age: 27 - Discharged on 3 Nov 1813 and sent to Dartmoor.

Anderson, Mathew - Seaman - Number: 2010 - Prize name: Ned - Ship type: LM - How taken: HM Brig Royalist - When taken: 6 Sep 1812 - Where taken: coast of France - Date received: 24 Sep 181 - From what ship: HMS Rippon - Born: Virginia - Age: 19 - Discharged on 27 Sep 1813 and sent to Dartmoor.

Anderson, Robert - Seaman - Number: 606 - How taken: Delivered himself up from HM Brig Foxhound - Date received: 27 Feb 1813 - From what ship: HMS Salvador del Mundo - Born: Rhode Island - Age: 26 - Sent to Dartmoor on 2 Apr 1813.

Anderson, Robert - Seaman - Number: 1492 - How taken: Delivered himself up from HM Brig Foxhound - Date received: 27 May 1813 - From what ship: Dartmoor prison - Born: Rhode Island - Age: 26 - Discharged on 8 Jul 1813 and sent to Chatham on HM Tender Neptune.

Anderson, Samuel - Mate - Number: 1073 - Prize name: John & Frances - Ship type: MV - How taken: HM Frigate Belle Poule - When taken: 19 Mar 1813 - Where taken: Bay of Biscay - Date received: 22 Apr 1813 - From what ship: HMS Superb - Born: New York - Age: 27 - Sent on 25 Apr 1813 to Ashburton on parole.

Anderson, William - Seaman - Number: 278 - Prize name: Leader - Ship type: MV - How taken: HM Frigate Briton - When taken: 10 Dec 1812 - Where taken: off Bordeaux - Date received: 21 Jan 1813 - From what ship: HMS Abercrombie - Born: Norway - Age: 20 - Sent to Portsmouth on 8 Feb 1813 on the HMS Colossus.

Anderson, William - Seaman - Number: 257 - Prize name: Ocean, prized to the Privateer Diligent - Ship type: P - How taken: HM Frigate Surveillante - When taken: 20 Dec 1812 - Where taken: Lat 44 N, Long 6 W - Date received: 21 Jan 1813 - From what ship: HMS Ocean - Born: New York - Age: 29 - Race: Negro - Sent to Portsmouth on 8 Feb 1813 on the HMS Colossus.

Andres, David - Seaman - Number: 1348 - Prize name: Fox - Ship type: LM - How taken: HM Sloop Pheasant - When taken: 23 Apr 1813 - Where taken: Bay of Biscay - Date received: 14 May 1813 - From what ship: HMS Pleasant - Born: Philadelphia - Age: 25 - Discharged on 3 Jul 1813 and sent to Stapleton Prison.

Andrews, Benjamin - Seaman - Number: 2204 - Prize name: General Kempt, prize of the Privateer Grand Turk - Ship type: P - How taken: HM Brig Foxhound - When taken: 18 Dec 1813 - Where taken: Lat 48.4 Long 6 - Date received: 21 Dec 1813 - From what ship: HMS Foxhound - Born: Massachusetts - Age: 23 - Discharged on 31 Jan 1814 and sent to Dartmoor.

Andrews, Charles - Seaman - Number: 1057 - Prize name: Virginia Planter - Ship type: MV - How taken: HM Frigate Pyramus - When taken: 17 Mar 1813 - Where taken: off Nantes - Date received: 22 Apr 1813 - From what ship: HMS Superb - Born: Newport - Age: 36 - Sent to Dartmoor on 1 Jul 1813.

Andrews, Charles - Seaman - Number: 1777 - Prize name: Virginia Planter - Ship type: MV - How taken: HM Frigate Pyramus - When taken: 17 Mar 1813 - Where taken: off Nantes - Date received: 3 Aug 1813 - From what ship: Dartmoor Prison - Born: Newport - Age: 36 - Discharged on 14 Sep 1813 and sent to Dartmoor.

Andrews, George - Seaman - Number: 2088 - Prize name: Sybille - Ship type: MV - How taken: HM Brig Zenobia - When taken: 17 Jun 1813 - Where taken: off Cape Henry - Date received: 2 Nov 1813 - From what ship: HMS Bacchus - Born: Philadelphia - Age: 30 - Discharged on 3 Nov 1813 and sent to Dartmoor.

Andrews, Joseph - Seaman - Number: 2170 - Prize name: Wolf, prize of the Privateer Grand Turk - Ship type: P - How taken: HM Frigate Briton - When taken: 1 Dec 1813 - Where taken: Bay of Biscay - Date received: 17 Dec 1813 - From what ship: HMS Briton - Born: Marblehead - Age: 25 - Discharged on 17 Dec 1813 and released to HMS Britton.

Andrews, Thomas - Seaman - Number: 627 - Prize name: Good Intent, prize of the Privateer Thrasher - Ship type: P - How taken: HM Frigate Pyramus - When taken: 26 Jun 1813 - Where taken: off Bordeaux - Date received: 2 Mar 1813 - From what ship: HMS Insolent - Born: Marblehead - Age: 21 - Sent to Dartmoor on 2 Apr 1813.

Anthony, Abraham - Seaman - Number: 172 - Prize name: Hunter - Ship type: P - How taken: HM Frigate Phoebe - When taken: 24 Dec 1812 - Where taken: off Western Islands - Date received: 9 Jan 1813 - From what ship: HMS Phoebe - Born: New Haven - Age: 24 - Sent to Portsmouth on 8 Feb 1813 on the HMS Colossus.

Anthony, John - Seaman - Number: 92 - Prize name: Experiment - Ship type: MV - How taken: HM Brig Rover - When taken: 21 Oct 1812 - Where taken: off Bordeaux - Date received: 25 Dec 1812 - From what ship: HMS Northumberland - Born: New Orleans - Age: 28 - Sent to Portsmouth on 29 Dec 1812 on the HMS Northumberland.

Anthony, Joseph - Seaman - Number: 2256 - Prize name: US Frigate Chesapeake - Ship type: War - How taken: HM Frigate Shannon - When taken: 1 Jul 1813 - Where taken: off Boston - Date received: 15 Jan 1814 - From what ship: HMS Teazer - Born: America - Age: 20 - Discharged on 31 Jan 1814 and sent to Dartmoor.

Anthony, Stephen - Seaman - Number: 1561 - Prize name: Courier - Ship type: LM - How taken: HM Brig Rover - When taken: 14 Mar 1813 - Where taken: Bay of Biscay - Date received: 29 May 1813 - From what ship: HMS Hannibal - Born: Maryland - Age: 28 - Race: Black - Discharged on 30 Jun 1813 and sent to Stapleton.

Appine - Chinese Servant to captain - Number: 1856 - Prize name: US Brig Argus - Ship type: War - How taken: HM Brig Pelican - When taken: 14 Aug 1813 - Where taken: Irish Channel - Date received: 19 Aug 1813 - From what ship: USS Argus - Born: Canton, China - Age: 21 - Race: Chinese - Discharged on 8 Sep 1813 and sent to Dartmoor.

Archer, Joseph - Seaman - Number: 2192 - Prize name: General Kempt, prize of the Privateer Grand Turk - Ship type: P - How taken: HM Brig Foxhound - When taken: 18 Dec 1813 - Where taken: Lat 48.4 Long 6 - Date received: 21 Dec 1813 - From what ship: HMS Foxhound - Born: Salem - Age: 28 - Discharged on 31 Jan 1814 and sent to Dartmoor.

Aris, James - Seaman - Number: 1205 - Prize name: Zebra - Ship type: LM - How taken: HM Frigate Pyramus - When taken: 20 Apr 1813 - Where taken: Bay of Biscay - Date received: 9 May 1813 - From what ship: HMS Andromache - Born: New York - Age: 19 - Sent on 3 Jul 1813 to Stapleton prison.

Armistead, Edward - Seaman - Number: 2503 - Prize name: Diamond, prize to the Privateer True Blooded Yankee - Ship type: P - How taken: HM Ship-of-the-Line Vengeur - When taken: 6 Mar 1814 - Where taken: Lat 47.4N Long 5.4W - Date received: 3 Apr 1814 - From what ship: HMS Rippon - Born: Norfolk - Age: 21 - Discharged on 10 May 1814 and sent to Dartmoor.

Armstrong, Daniel - Seaman - Number: 2703 - Prize name: Margaret, prize to the Privateer Surprize - Ship type: P - How taken: HM Brig Foxhound - When taken: 27 May 1814 - Where taken: off Scylly - Date received: 16 Jun 1814 - From what ship: HMS Foxhound - Born: Baltimore - Age: 19 - Discharged on 20 Jun 1814 and sent to

Dartmoor.

Armstrong, James - Seaman - Number: 1667 - How taken: Taken at Liverpool - When taken: 19 Mar 1813 - Date received: 17 Jun 1813 - From what ship: HMS Bittern - Born: Alexandria - Age: 26 - Discharged on 30 Jun 1813 and sent to Stapleton.

Armstrong, Joseph - Seaman - Number: 2653 - How taken: Sent into custody from HM Transport Agolent - Date received: 3 Jun 1814 - From what ship: HMS Gronville - Born: Richmond - Age: 45 - Race: Black - Discharged on 14 Jun 1814 and sent to Dartmoor.

Armstrong, Nicholas - Seaman - Number: 417 - Prize name: Union - Ship type: MV - How taken: HM Frigate Iris - When taken: 17 Jan 1813 - Where taken: Lat 44 N Long 2.3 W - Date received: 5 Feb 1813 - From what ship: HMS San Josef - Born: Philadelphia - Age: 17 - Sent to Chatham on 29 Mar 1813 on the HMS Braham.

Armstrong, William - Seaman - Number: 569 - Prize name: Rolla - Ship type: MV - How taken: HM Frigate Surveillante - When taken: 11 Feb 1813 - Where taken: Bay of Biscay - Date received: 23 Feb 1813 - From what ship: HMS Surveillante - Born: New York - Age: 27 - Sent to Dartmoor on 2 Apr 1813.

Armstrong, William - Seaman - Number: 1698 - Prize name: Governor Gerry - Ship type: MV - How taken: HM Brig Royalist - When taken: 31 May 1813 - Where taken: Bay of Biscay - Date received: 26 Jun 1813 - From what ship: HMS Duncan - Born: Philadelphia - Age: 36 - Race: Black - Discharged on 30 Jun 1813 and sent to Stapleton.

Armstrong, William - Seaman - Number: 1659 - Prize name: Rolla - Ship type: MV - How taken: HM Frigate Surveillante - When taken: 11 Feb 1813 - Where taken: Bay of Biscay - Date received: 15 Jun 1813 - From what ship: Dartmoor Prison - Born: New York - Age: 27 - Discharged on 16 Jun 1813 and released to HMS Salvador del Mundo.

Arnold, Alfred - Seaman - Number: 1122 - Prize name: Viper - Ship type: MV - How taken: HM Ship-of-the-Line Superb - When taken: 15 Apr 1813 - Where taken: Bay of Biscay - Date received: 22 Apr 1813 - From what ship: HMS Superb - Born: New Orleans - Age: 23 - Sent to Dartmoor on 1 Jul 1813.

Arnold, James - Master - Number: 251 - How taken: Delivered into custody by the mayor of Plymouth - Date received: 17 Jan 1813 - From what ship: HMS Plymouth - Born: Weymouth - Age: 27 - Sent to Chatham on 8 Jul 1813 on HM Tender Neptune.

Arnold, Thomas - Passenger & Chief Mate of a MV - Number: 2593 - Prize name: Favorite - Ship type: MV - How taken: HM Brig Racehorse - When taken: 16 Nov 1812 - Where taken: Cape of Good Hope - Date received: 10 May 1814 - From what ship: HMS Lion - Born: New York - Age: 40 - Discharged on 12 May 1814 and sent to Ashburton on parole.

Arnold, Welcome - 1st Mate - Number: 2591 - Prize name: Valentine - Ship type: MV - How taken: HM Brig Racehorse - When taken: 16 Nov 1812 - Where taken: Cape of Good Hope - Date received: 10 May 1814 - From what ship: HMS Lion - Born: Providence - Age: 35 - Discharged on 12 May 1814 and sent to Ashburton on parole.

Arthur, Alexander - Boy - Number: 86 - Prize name: Argus - Ship type: MV - How taken: Fancy, Cutter - When taken: 19 Dec 1812 - Where taken: Bay of Biscay - Date received: 24 Dec 1812 - From what ship: Fancy, Cutter - Born: New York - Age: 15 - Sent to Portsmouth on 29 Dec 1812 on the HMS Northumberland.

Ash, Samson - Boy - Number: 1923 - Prize name: US Brig Argus - Ship type: War - How taken: HM Brig Pelican - When taken: 14 Aug 1813 - Where taken: Irish Channel - Date received: 23 Aug 1813 - From what ship: HMS Pelican - Born: New York - Age: 18 - Race: Black - Discharged on 8 Sep 1813 and sent to Dartmoor.

Ashfield, Henry - Seaman - Number: 929 - Prize name: Weasel - Ship type: MV - How taken: HM Brig Foxhound - When taken: 25 Mar 1813 - Where taken: Bay of Biscay - Date received: 6 Apr 1813 - From what ship: HMS Foxhound - Born: New York - Age: 19 - Sent on 21 Jun 1813 to Portsmouth on HMS Prometheus.

Ashmore, Edward - Seaman - Number: 1997 - Prize name: Ned - Ship type: LM - How taken: HM Brig Royalist - When taken: 6 Sep 1812 - Where taken: coast of France - Date received: 22 Sep 1813 - From what ship: HMS Royalist - Born: Laberton - Age: 26 - Discharged on 27 Sep 1813 and sent to Dartmoor.

Ashton, Joseph - Mate - Number: 383 - Prize name: Union - Ship type: MV - How taken: HM Frigate Iris - When taken: 17 Jan 1813 - Where taken: Lat 44 N Long 2.3 W - Date received: 5 Feb 1813 - From what ship: HMS San Josef - Born: Philadelphia - Age: 26 - Sent on 18 Jan 1813 to Ashburton on parole.

Askwith, William Vickery - Sailing Master - Number: 1499 - Prize name: Paul Jones - Ship type: P - How taken: HM Frigate Leonidas - When taken: 23 May 1813 - Where taken: off Cape Clear - Date received: 27 May 1813 - From what ship: HMS Leonidas - Born: Hudson, NY - Age: 29 - Discharged on 30 Jun 1813 and sent to Stapleton.

Astrop, Hans Christian - Seaman - Number: 2137 - Prize name: Amiable - Ship type: LM - How taken: HM Ship-of-the-Line Magnificant - When taken: 30 Oct 1813 - Where taken: off Lorient - Date received: 29 Nov 1813 - From what ship: HMS Dublin - Born: Bergen, Norway - Age: 29 - Discharged on 4 Dec 1813 and sent to Dartmoor.

Athroun, Samuel - 2nd Mate - Number: 604 - Prize name: Rolla - Ship type: MV - How taken: HM Frigate Surveillante - When taken: 11 Feb 1813 - Where taken: Bay of Biscay - Date received: 25 Feb 1813 - From what ship: HMS Plymouth - Born: Philadelphia - Age: 30 - Sent to Dartmoor on 2 Apr 1813.

Atkens, Martyn - Passenger - Number: 1944 - Prize name: Marmion - Ship type: MV - How taken: HM Frigate President - When taken: 14 Aug 1813 - Where taken: off Nantes - Date received: 27 Aug 1813 - From what ship: HMS Urgent - Born: Charlestown - Age: 48 - Discharged on 8 Sep 1813 and sent to Ashburton on parole.

Atkins, James - Seaman - Number: 1522 - Prize name: Courier - Ship type: LM - How taken: HM Brig Rover - When taken: 14 Mar 1813 - Where taken: Bay of Biscay - Date received: 29 May 1813 - From what ship: HMS Hannibal - Born: Massachusetts - Age: 21 - Discharged on 30 Jun 1813 and sent to Stapleton.

Atkins, Uriah - Seaman - Number: 1650 - Prize name: Tyger - Ship type: MV - How taken: Detained at Gibraltar - When taken: 8 Aug 1813 - Date received: 15 Jun 1813 - From what ship: Dartmoor Prison - Born: Penobscot - Discharged on 16 Jun 1813 and released to HMS Salvador del Mundo.

Atkins, Uriah - Seaman - Number: 511 - Prize name: Tyger - Ship type: MV - How taken: Detained at Gibraltar - When taken: 8 Aug 1812 - Date received: 15 Feb 1813 - From what ship: HMS Andromeda - Born: Buxton - Age: 25 - Sent to Dartmoor on 2 Apr 1813.

Atkinson, William - Seaman - Number: 1025 - Prize name: Two Brothers - Ship type: MV - How taken: Bootle of Liverpool, letter of marque - When taken: 18 Mar 1813 - Where taken: Western Islands - Date received: 21 Apr 1813 - From what ship: HMS Bittern - Born: Harford County - Age: 36 - Sent to Dartmoor on 1 Jul 1813.

Atwick, Thomas C. - Paymaster - Number: 2501 - Prize name: Diamond, prize to the Privateer True Blooded Yankee - Ship type: P - How taken: HM Ship-of-the-Line Vengeur - When taken: 6 Mar 1814 - Where taken: Lat 47.4N Long 5.4W - Date received: 3 Apr 1814 - From what ship: HMS Rippon - Born: Portland - Age: 35 - Discharged on 14 Jun 1814 and sent to Dartmoor.

Atwood, John - Seaman - Number: 2645 - How taken: Delivered himself up from MV Young William - Date received: 20 May 1814 - From what ship: Cygnet - Born: New Jersey - Age: 28 - Discharged on 14 Jun 1814 and sent to Dartmoor.

Atwood, Thomas - Seaman - Number: 909 - Prize name: Prompt - Ship type: MV - How taken: Chance, privateer - When taken: 22 Mar 1813 - Where taken: Bay of Biscay - Date received: 3 Apr 1813 - From what ship: Mary - Born: Arlington - Age: 29 - Sent on 21 Jun 1813 to Portsmouth on HMS Prometheus.

Audibert, Joseph - Passenger - Number: 822 - Prize name: Decornau - Ship type: MV - How taken: HM Sloop Pheasant - When taken: 15 Mar 1813 - Where taken: Bay of Biscay - Date received: 20 Mar 1813 - From what ship: HMS Pheasant - Entered on 9 Apr 1813 into the French General Entry Book as prisoner number 39426.

Augustin, Anthony - Seaman - Number: 573 - Prize name: Rolla - Ship type: MV - How taken: HM Frigate Surveillante - When taken: 11 Feb 1813 - Where taken: Bay of Biscay - Date received: 23 Feb 1813 - From what ship: HMS Surveillante - Born: New Orleans - Age: 48 - Sent to Dartmoor on 2 Apr 1813.

Augustin, Anthony - Seaman - Number: 1738 - Prize name: Rolla - Ship type: MV - How taken: HM Frigate Surveillante - When taken: 11 Feb 1813 - Where taken: Bay of Biscay - Date received: 10 Jul 1813 - From what ship: Dartmoor Prison - Born: New Orleans - Age: 47 - Discharged on 10 Jul 1813 and released to HMS Salvador del Mundo.

Augustus, Amos - Seaman - Number: 623 - Prize name: Governor McKean - Ship type: LM - How taken: HM Brig Rover - When taken: 26 Jun 1813 - Where taken: off Bordeaux - Date received: 2 Mar 1813 - From what ship: HMS Insolent - Born: Wilmington - Age: 17 - Race: Black - Sent to Dartmoor on 2 Apr 1813.

Aulajo, Thomas - Seaman - Number: 222 - Prize name: Vengeance - Ship type: LM - How taken: HM Frigate Phoebe

- When taken: 1 Jan 1813 - Where taken: Lat 44.4 Long 23 - Date received: 9 Jan 1813 - From what ship: HMS Phoebe - Born: Baltimore - Age: 23 - Sent to Portsmouth on 8 Feb 1813 on the HMS Colossus.

Aurel, Leonard - Seaman - Number: 971 - Prize name: Ferox - Ship type: MV - How taken: HM Frigate Medusa & HM Brig Lyra - When taken: 28 Mar 1813 - Where taken: off Cape Ortagle - Date received: 9 Apr 1813 - From what ship: HMS Lyra - Born: Guttenberg - Age: 29 - Sent to Chatham on 8 Jul 1813 on HM Tender Neptune.

Austin, Jonathan - Seaman - Number: 1311 - How taken: Delivered himself up from HM Ship-of-the-Line Clarence - Date received: 10 May 1813 - From what ship: HMS Clarence - Born: Massachusetts - Age: 29 - Discharged on 8 Jul 1813 and sent to Chatham on tender Neptune.

Autest, Jean - Seaman - Number: 2504 - Prize name: Diamond, prize to the Privateer True Blooded Yankee - Ship type: P - How taken: HM Ship-of-the-Line Vengeur - When taken: 6 Mar 1814 - Where taken: Lat 47.4N Long 5.4W - Date received: 3 Apr 1814 - From what ship: HMS Rippon - Born: Brest - Age: 30 - Discharged on 10 May 1814 and sent to Dartmoor.

Averell, Loring - Seaman - Number: 756 - Prize name: William Bayard - Ship type: MV - How taken: HM Ship-of-the-Line Warspite - When taken: 3 Mar 1813 - Where taken: Bay of Biscay - Date received: 19 Mar 1813 - From what ship: HMS Warspite - Born: Connecticut - Age: 19 - Sent to Dartmoor on 2 Apr 1813.

Ayres, Robert - Seaman - Number: 1912 - Prize name: US Brig Argus - Ship type: War - How taken: HM Brig Pelican - When taken: 14 Aug 1813 - Where taken: Irish Channel - Date received: 23 Aug 1813 - From what ship: HMS Pelican - Born: New Jersey - Age: 28 - Discharged on 8 Sep 1813 and sent to Dartmoor.

Ayers, John - Seaman - Number: 2611 - How taken: MV Alemena - When taken: 25 Jan 1813 - Date received: 16 May 1814 - From what ship: HMS Repulse - Born: New Jersey - Age: 26 - Discharged on 14 Jun 1814 and sent to Dartmoor.

Baas, John Thomas - Supercargo - Number: 1076 - Prize name: John & Frances - Ship type: MV - How taken: HM Frigate Belle Poule - When taken: 19 Mar 1813 - Where taken: Bay of Biscay - Date received: 22 Apr 1813 - From what ship: HMS Superb - Born: Charlestown - Age: 24 - Sent on 25 Apr 1813 to Ashburton on parole.

Babi, Joseph - Seaman - Number: 2104 - Prize name: Chesapeake - Ship type: LM - How taken: HM Frigate Hotspur & HM Frigate Pyramus - When taken: 26 Oct 1813 - Where taken: Bay of Biscay - Date received: 22 Nov 1813 - From what ship: HMS Pyramus - Born: Massachusetts - Age: 22 - Discharged on 29 Nov 1813 and sent to Dartmoor.

Bacus, David - Captain's Steward - Number: 1850 - Prize name: US Brig Argus - Ship type: War - How taken: HM Brig Pelican - When taken: 14 Aug 1813 - Where taken: Irish Channel - Date received: 18 Aug 1813 - From what ship: USS Argus - Born: Boston - Age: 15 - Race: Black - Discharged on 8 Sep 1813 and sent to Dartmoor.

Baday, John - Seaman - Number: 1725 - Prize name: Hannah & Eliza - Ship type: MV - How taken: HM Brig Lyra - When taken: 29 May 1813 - Where taken: off north coast of Spain - Date received: 4 Jul 1813 - From what ship: HMS Iris - Born: New York - Age: 24 - Discharged on 3 Nov 1813 and sent to Dartmoor.

Badson, Jacob - Seaman - Number: 2532 - Prize name: Young Dixon - Ship type: MV - How taken: HM Transport Hydra - When taken: 3 Apr 1814 - Where taken: at sea - Date received: 7 Apr 1814 - From what ship: HMS Fly - Born: Boston - Age: 31 - Discharged on 10 May 1814 and sent to Dartmoor.

Bagley, William - Seaman - Number: 1033 - How taken: Impressed at Liverpool - When taken: 4 Apr 1813 - Date received: 21 Apr 1813 - From what ship: HMS Bittern - Born: Portland - Age: 34 - Sent to Dartmoor on 1 Jul 1813.

Bagot, Lewis - Carpenter & Passenger - Number: 348 - Prize name: Vengeance - Ship type: LM - How taken: HM Frigate Phoebe - When taken: 1 Jan 1813 - Where taken: Lat 44.4 Long 23 - Date received: 29 Jan 1813 - From what ship: HMS Phoebe - Sent to France on 1 Feb 1813 on a cartel.

Bailey, John - Seaman - Number: 934 - Prize name: Weasel - Ship type: MV - How taken: HM Brig Foxhound - When taken: 25 Mar 1813 - Where taken: Bay of Biscay - Date received: 6 Apr 1813 - From what ship: HMS Foxhound - Born: Gloucester - Age: 29 - Sent on 21 Jun 1813 to Portsmouth on HMS Prometheus.

Bailey, Joseph - Seaman - Number: 302 - Prize name: Stephen - Ship type: MV - How taken: HM Frigate Briton & HM Frigate Andromache - When taken: 17 Dec 1812 - Where taken: off Bordeaux - Date received: 21 Jan 1813 -

From what ship: HMS Abercrombie - Born: Pennsylvania - Age: 21 - Race: Black - Sent to Portsmouth on 8 Feb 1813 on the HMS Colossus.

Bailey, Samuel - Seaman - Number: 962 - Prize name: Ferox - Ship type: MV - How taken: HM Frigate Medusa & HM Brig Lyra - When taken: 28 Mar 1813 - Where taken: off Cape Ortagle - Date received: 9 Apr 1813 - From what ship: HMS Lyra - Born: Portland - Age: 32 - Sent to Chatham on 8 Jul 1813 on HM Tender Neptune.

Baird, David - Seaman - Number: 2363 - Prize name: Hannah - Ship type: MV - How taken: HM Ship-of-the-Line Conquestador - When taken: 14 Jan 1814 - Where taken: Lat 47.3 Long 7 - Date received: 31 Jan 1814 - From what ship: HMS Surveillante - Born: Pelham - Age: 29 - Discharged on 27 Feb 1814 and sent to Chatham on HMS Haleyon.

Baker, Israel - Seaman - Number: 2013 - Prize name: Ned - Ship type: LM - How taken: HM Brig Royalist - When taken: 6 Sep 1812 - Where taken: coast of France - Date received: 24 Sep 181 - From what ship: HMS Rippon - Born: New York - Age: 24 - Discharged on 27 Sep 1813 and sent to Dartmoor.

Baker, John - Seaman - Number: 726 - Prize name: Mars - Ship type: MV - How taken: HM Ship-of-the-Line Warspite - When taken: 26 Feb 1813 - Where taken: Bay of Biscay - Date received: 19 Mar 1813 - From what ship: HMS Warspite - Born: Baltimore - Age: 25 - Sent to Dartmoor on 2 Apr 1813.

Baker, John - Seaman - Number: 1117 - Prize name: Viper - Ship type: MV - How taken: HM Ship-of-the-Line Superb - When taken: 15 Apr 1813 - Where taken: Bay of Biscay - Date received: 22 Apr 1813 - From what ship: HMS Superb - Born: Pennsylvania - Age: 19 - Sent to Dartmoor on 1 Jul 1813.

Baldwin, John - Seaman - Number: 1342 - Prize name: Fox - Ship type: LM - How taken: HM Sloop Pheasant - When taken: 23 Apr 1813 - Where taken: Bay of Biscay - Date received: 14 May 1813 - From what ship: HMS Pleasant - Born: Boston - Age: 23 - Discharged on 3 Jul 1813 and sent to Stapleton Prison.

Baldwin, John - Seaman - Number: 2463 - Prize name: Fair American - Ship type: MV - How taken: HM Frigate Andromache - When taken: 19 Jan 1814 - Where taken: Bay of Biscay - Date received: 22 Feb 1814 - From what ship: HMS York - Born: Boston - Age: 23 - Discharged on 10 May 1814 and sent to Dartmoor.

Baldwin, Pierson - Passenger - Number: 2234 - Prize name: Squirrel - Ship type: MV - How taken: HM Frigate Belle Poule - When taken: 14 Dec 1813 - Where taken: Bay of Biscay - Date received: 15 Jan 1814 - From what ship: HMS Bellona - Born: Connecticut - Age: 29 - Discharged on 16 Jan 1814 and sent to Ashburton on parole.

Bancroft, Samuel - Seaman - Number: 1155 - Prize name: Essex - Ship type: MV - How taken: HM Frigate Pyramus - When taken: 2 Apr 1813 - Where taken: Bay of Biscay - Date received: 2 Mar 1813 - From what ship: HMS Rota - Born: Marblehead - Age: 29 - Sent on 24 Apr 1813 to Ashburton on parole.

Banks, Perry - Seaman - Number: 2705 - Prize name: Margaret, prize to the Privateer Surprize - Ship type: P - How taken: HM Brig Foxhound - When taken: 27 May 1814 - Where taken: off Scylly - Date received: 16 Jun 1814 - From what ship: HMS Foxhound - Born: Baltimore - Age: 29 - Race: Black - Discharged on 20 Jun 1814 and sent to Dartmoor.

Bannister, George - Seaman - Number: 582 - Prize name: Cashiere - Ship type: LM - How taken: HM Brig Reindeer - When taken: 3 Feb 1813 - Where taken: Bay of Biscay - Date received: 23 Feb 1813 - From what ship: HMS Surveillante - Born: New England - Age: 23 - Race: Black - Sent to Dartmoor on 2 Apr 1813.

Bannister, Joshua - Seaman - Number: 2048 - How taken: Impressed at Liverpool - When taken: 16 Oct 1813 - Date received: 1 Nov 1813 - From what ship: HMS Bittern - Born: Philadelphia - Age: 26 - Race: Black - Discharged on 3 Nov 1813 and sent to Dartmoor.

Bansall, Lewis - Seaman - Number: 1179 - How taken: Delivered himself up from HM Ship-of-the-Line Clarence - Date received: 6 May 1813 - From what ship: HMS Stag - Born: Boston - Age: 27 - Sent to Chatham on 8 Jul 1813 on HM Tender Neptune.

Baptist, John - Cook - Number: 613 - Prize name: Governor McKean - Ship type: LM - How taken: HM Brig Rover - When taken: 26 Jun 1813 - Where taken: off Bordeaux - Date received: 2 Mar 1813 - From what ship: HMS Insolent - Born: New Orleans - Age: 21 - Race: Negro - Sent to Dartmoor on 2 Apr 1813.

Baransan, John - Seaman - Number: 1569 - Prize name: Miranda, prize of the Privateer Paul Jones - Ship type: P - How taken: HM Frigate Unicorn - When taken: 21 Mar 1813 - Where taken: off Ushant - Date received: 9 Jun

1813 - From what ship: HMS Conquestador - Born: Rochelle, France - Age: 22 - Discharged on 30 Jun 1813 and sent to Stapleton.

Barbadoes, Robert - Seaman - Number: 2342 - Prize name: Apparencen, prize of the Privateer Bunker Hill - Ship type: P - How taken: Cartdian - When taken: 27 Jan 1814 - Where taken: off Ushant - Date received: 30 Jan 1814 - From what ship: Cartdian - Born: Boston - Age: 29 - Discharged on 31 Jan 1814 and sent to Dartmoor.

Barber, William - Seaman - Number: 1544 - Prize name: Zebra - Ship type: LM - How taken: HM Frigate Pyramus - When taken: 20 Apr 1813 - Where taken: Bay of Biscay - Date received: 29 May 1813 - From what ship: HMS Hannibal - Born: Newport - Age: 40 - Race: Black - Discharged on 30 Jun 1813 and sent to Stapleton.

Barchman, John - Master's Mate - Number: 151 - Prize name: Hunter - Ship type: P - How taken: HM Frigate Phoebe - When taken: 24 Dec 1812 - Where taken: off Western Islands - Date received: 9 Jan 1813 - From what ship: HMS Phoebe - Born: Dover - Age: 26 - Sent to Portsmouth on 8 Feb 1813 on the HMS Colossus.

Barford, William - Seaman - Number: 2655 - How taken: Sent into custody from HM Frigate Nereus - Date received: 5 Jun 1814 - From what ship: HMS Gronville - Born: Philadelphia - Age: 26 - Discharged on 14 Jun 1814 and sent to Dartmoor.

Barker, Charles G. - Seaman - Number: 2254 - How taken: Impressed at Cove of Cork - When taken: 26 Dec 1813 - Date received: 15 Jan 1814 - From what ship: HMS Teazer - Born: Boston - Age: 23 - Discharged on 31 Jan 1814 and sent to Dartmoor.

Barker, Reuben - Seaman - Number: 1217 - Prize name: Zebra - Ship type: LM - How taken: HM Frigate Pyramus - When taken: 20 Apr 1813 - Where taken: Bay of Biscay - Date received: 9 May 1813 - From what ship: HMS Andromache - Born: New York - Age: 18 - Discharged on 3 Jul 1813 and sent to Stapleton Prison.

Barkman, Henry - 2nd Mate - Number: 1131 - Prize name: Magdalen - Ship type: MV - How taken: HM Ship-of-the-Line Superb - When taken: 15 Apr 1813 - Where taken: off Belle Isle - Date received: 22 Apr 1813 - From what ship: HMS Superb - Born: Pennsylvania - Age: 36 - Sent to Dartmoor on 1 Jul 1813.

Barlow, John - Seaman - Number: 1833 - Prize name: US Brig Argus - Ship type: War - How taken: HM Brig Pelican - When taken: 14 Aug 1813 - Where taken: Irish Channel - Date received: 17 Aug 1813 - From what ship: USS Argus - Born: Amsterdam, Holland - Age: 22 - Discharged on 8 Sep 1813 and sent to Dartmoor.

Barnard, John - Marine - Number: 1479 - Prize name: Paul Jones - Ship type: P - How taken: HM Frigate Leonidas - When taken: 23 May 1813 - Where taken: off Cape Clear - Date received: 26 May 1813 - From what ship: HMS Leonidas - Born: New Orleans - Age: 37 - Discharged on 30 Jun 1813 and sent to Stapleton.

Barnard, Timothy - Captain - Number: 470 - Prize name: Orbit - Ship type: MV - How taken: HM Brig Achates - When taken: 29 Jan 1813 - Where taken: Lat 49 N Long 13 W - Date received: 9 Feb 1813 - From what ship: Orbit - Born: Nantucket - Age: 28 - Sent on 10 Feb 1813 to Ashburton on parole.

Barnes, Nathaniel - Seaman - Number: 1501 - Prize name: Paul Jones - Ship type: P - How taken: HM Frigate Leonidas - When taken: 23 May 1813 - Where taken: off Cape Clear - Date received: 27 May 1813 - From what ship: HMS Leonidas - Born: New York - Age: 35 - Discharged on 30 Jun 1813 and sent to Stapleton.

Barnes, Robert - Seaman - Number: 528 - Prize name: Howard - Ship type: MV - How taken: Detained at Gibraltar - When taken: 8 Jul 1813 - Date received: 15 Feb 1813 - From what ship: HMS Andromeda - Born: Rehoboth - Age: 23 - Sent to Dartmoor on 2 Apr 1813.

Barnes, Robert (alias Murray) - Seaman - Number: 1657 - Prize name: Howard - Ship type: MV - How taken: Detained at Gibraltar - When taken: 8 Jul 1813 - Date received: 15 Jun 1813 - From what ship: Dartmoor Prison - Born: Rehoboth - Age: 23 - Discharged on 16 Jun 1813 and released to HMS Salvador del Mundo.

Barnes, William - Seaman - Number: 527 - Prize name: Howard - Ship type: MV - How taken: Detained at Gibraltar - When taken: 8 Jul 1813 - Date received: 15 Feb 1813 - From what ship: HMS Andromeda - Born: Rehoboth - Age: 26 - Sent to Dartmoor on 2 Apr 1813.

Baron, Thomas - Servant - Number: 1853 - Prize name: US Brig Argus - Ship type: War - How taken: HM Brig Pelican - When taken: 14 Aug 1813 - Where taken: Irish Channel - Date received: 19 Aug 1813 - From what ship: USS Argus - Born: Norfolk - Age: 21 - Race: Negro - Discharged on 8 Sep 1813 and sent to Dartmoor.

Barradaile, Thomas - Passenger - Number: 388 - Prize name: Union - Ship type: MV - How taken: HM Frigate Iris -

When taken: 17 Jan 1813 - Where taken: Lat 44 N Long 2.3 W - Date received: 5 Feb 1813 - From what ship: HMS San Josef - Born: New Jersey - Age: 21 - Sent on 10 Feb 1813 to Ashburton on parole.

Barret, Anthony - Seaman - Number: 725 - Prize name: Mars - Ship type: MV - How taken: HM Ship-of-the-Line Warspite - When taken: 26 Feb 1813 - Where taken: Bay of Biscay - Date received: 19 Mar 1813 - From what ship: HMS Warspite - Born: New Orleans - Age: 58 - Sent to Dartmoor on 2 Apr 1813.

Barrett, James - Seaman - Number: 1281 - Prize name: Price - Ship type: LM - How taken: HM Frigate Medusa - When taken: 13 Apr 1813 - Where taken: Bay of Biscay - Date received: 10 May 1813 - From what ship: HMS Medusa - Born: Massachusetts - Age: 33 - Discharged on 21 Jun 1813 and released to American ship Mount Hope.

Barron, William - Seaman - Number: 643 - Prize name: Criterion - Ship type: MV - How taken: HM Frigate Belle Poule - When taken: 14 Feb 1813 - Where taken: Bay of Biscay - Date received: 4 Mar 1813 - From what ship: HMS Strenuous - Born: New York - Age: 25 - Sent to Dartmoor on 2 Apr 1813.

Barry, James - Seaman - Number: 1875 - Prize name: US Brig Argus - Ship type: War - How taken: HM Brig Pelican - When taken: 14 Aug 1813 - Where taken: Irish Channel - Date received: 23 Aug 1813 - From what ship: HMS Pelican - Born: Annapolis - Age: 29 - Discharged on 8 Sep 1813 and sent to Dartmoor.

Barstow, Samuel - Master - Number: 1947 - Prize name: Marmion - Ship type: MV - How taken: HM Frigate President - When taken: 14 Aug 1813 - Where taken: off Nantes - Date received: 30 Aug 1813 - From what ship: Marmion - Born: Pembroke - Age: 27 - Discharged on 31 Aug 1813 and sent to Ashburton on parole.

Bartell, Robert - Master - Number: 876 - Prize name: Meteor - Ship type: MV - How taken: HM Frigate Briton - When taken: 12 Mar 1813 - Where taken: Bay of Biscay - Date received: 30 Mar 1813 - From what ship: HMS Plymouth - Born: Massachusetts - Age: 44 - Sent on 31 Mar 1813 to Ashburton on parole.

Bartell, William - Master - Number: 507 - Prize name: Tyger - Ship type: MV - How taken: Detained at Gibraltar - When taken: 8 Aug 1812 - Date received: 15 Feb 1813 - From what ship: HMS Andromeda - Born: Marblehead - Age: 46 - Sent on 17 Feb 1813 to Ashburton on parole.

Bartholf, Nicholas - Seaman - Number: 892 - Prize name: Tiger - Ship type: MV - How taken: HM Brig Scylla - When taken: 22 Mar 1813 - Where taken: Bay of Biscay - Date received: 1 Apr 1813 - From what ship: HMS Scylla - Born: America - Age: 20 - Sent on 21 Jun 1813 to Portsmouth on HMS Prometheus.

Bartis, John - Cook - Number: 907 - Prize name: Prompt - Ship type: MV - How taken: Chance, privateer - When taken: 22 Mar 1813 - Where taken: Bay of Biscay - Date received: 3 Apr 1813 - From what ship: Mary - Born: New Orleans - Age: 26 - Race: Black - Sent on 21 Jun 1813 to Portsmouth on HMS Prometheus.

Bartlett, Caleb - Seaman - Number: 1622 - Prize name: Leo - Ship type: LM - How taken: HM Frigate Magiciene - When taken: 4 Jun 1813 - Where taken: Lat 45 Long 14 - Date received: 14 Jun 1813 - From what ship: HMS Orestes - Born: Plymouth, MA - Age: 24 - Discharged on 30 Jun 1813 and sent to Stapleton.

Bartlett, John - Passenger - Number: 1706 - Prize name: Joseph - Ship type: MV - How taken: HM Frigate Iris - When taken: 8 Jun 1813 - Where taken: Bay of Biscay - Date received: 4 Jul 1813 - From what ship: HMS Iris - Born: Marblehead - Age: 25 - Race: Black - Discharged on 22 Aug 1813 and released to HM Brig Redpole.

Bartlett, Thomas - Seaman - Number: 545 - Prize name: Spitfire - Ship type: MV - How taken: HM Brig Achates - When taken: 14 Feb 1813 - Where taken: off Ushant - Date received: 16 Feb 1813 - From what ship: HMS Achates - Born: Marblehead - Age: 21 - Sent to Dartmoor on 2 Apr 1813.

Barton, Eseek - Seaman - Number: 2122 - Prize name: Chesapeake - Ship type: LM - How taken: HM Frigate Hotspur & HM Frigate Pyramus - When taken: 26 Oct 1813 - Where taken: Bay of Biscay - Date received: 22 Nov 1813 - From what ship: HMS Pyramus - Born: Rhode Island - Age: 54 - Discharged on 29 Nov 1813 and sent to Dartmoor.

Barton, Nathan - Seaman - Number: 292 - Prize name: Columbia - Ship type: MV - How taken: HM Frigate Briton - When taken: 17 Dec 1812 - Where taken: off Bordeaux - Date received: 21 Jan 1813 - From what ship: HMS Abercrombie - Born: Baltimore - Age: 26 - Sent to Portsmouth on 8 Feb 1813 on the HMS Colossus.

Barton, Robert - Seaman - Number: 758 - Prize name: William Bayard - Ship type: MV - How taken: HM Ship-of-the-Line Warspite - When taken: 3 Mar 1813 - Where taken: Bay of Biscay - Date received: 19 Mar 1813 - From what ship: HMS Warspite - Born: Pennsylvania - Age: 35 - Sent to Dartmoor on 2 Apr 1813.

Barttist, John - Seaman - Number: 1715 - Prize name: Orders in Council - Ship type: LM - How taken: HM Frigate

Surveillante - When taken: 1 Jun 1813 - Where taken: Bay of Biscay - Date received: 4 Jul 1813 - From what ship: HMS Iris - Born: New Orleans - Age: 37 - Discharged on 22 Aug 1813 and released to HM Brig Redpole.

Bass, Prince - Seaman - Number: 2172 - Prize name: Wolf, prize of the Privateer Grand Turk - Ship type: P - How taken: HM Frigate Briton - When taken: 1 Dec 1813 - Where taken: Bay of Biscay - Date received: 17 Dec 1813 - From what ship: HMS Briton - Born: Barnstable - Age: 23 - Discharged on 17 Dec 1813 and released to HMS Britton.

Basset, David - Seaman - Number: 1693 - Prize name: Governor Gerry - Ship type: MV - How taken: HM Brig Royalist - When taken: 31 May 1813 - Where taken: Bay of Biscay - Date received: 26 Jun 1813 - From what ship: HMS Duncan - Born: Baltimore - Age: 43 - Race: Black - Discharged on 30 Jun 1813 and sent to Stapleton.

Bassett, Gorham - Seaman - Number: 2660 - Prize name: Indian Lass, prize of the Privateer Grand Turk - Ship type: P - How taken: HM Transport Akbar - When taken: 29 Apr 1814 - Where taken: at sea - Date received: 5 Jun 1814 - From what ship: HMS Gronville - Born: Barnstable - Age: 22 - Discharged on 14 Jun 1814 and sent to Dartmoor.

Bassonet, Charles - Seaman - Number: 2109 - Prize name: Chesapeake - Ship type: LM - How taken: HM Frigate Hotspur & HM Frigate Pyramus - When taken: 26 Oct 1813 - Where taken: Bay of Biscay - Date received: 22 Nov 1813 - From what ship: HMS Pyramus - Born: Pennsylvania - Age: 25 - Discharged on 29 Nov 1813 and sent to Dartmoor.

Bateman, Charles - Mate - Number: 18 - Prize name: Jenny - Ship type: MV - How taken: Taken at Liverpool - When taken: 18 Oct 1812 - Date received: 21 Nov 1812 - From what ship: HMS Salvador del Mundo - Born: Baltimore - Age: 28 - Sent on 8 Dec 1812 to Ashburton on parole.

Batman, John - Seaman - Number: 1381 - Prize name: Tom - Ship type: LM - How taken: HM Frigate Surveillante - When taken: 27 Apr 1813 - Where taken: Bay of Biscay - Date received: 15 May 1813 - From what ship: HMS Foxhound - Born: Maryland - Age: 38 - Discharged on 3 Jul 1813 and sent to Stapleton Prison.

Baughton, Glover - Seaman - Number: 2050 - How taken: Pallas - Where taken: St. Johns, New Brunswick - Date received: 1 Nov 1813 - From what ship: HMS Bittern - Born: Marblehead - Age: 17 - Discharged on 3 Nov 1813 and sent to Dartmoor.

Bawnham, Francis Aortas - Supercargo - Number: 2350 - Prize name: Hannah - Ship type: MV - How taken: HM Ship-of-the-Line Conquestador - When taken: 14 Jan 1814 - Where taken: Lat 47.3 Long 7 - Date received: 31 Jan 1814 - From what ship: HMS Surveillante - Born: Marblehead - Age: 26 - Discharged on 31 Jan 1814 and sent to Ashburton on parole.

Baxter, Charles - Seaman - Number: 1806 - Prize name: US Brig Argus - Ship type: War - How taken: HM Brig Pelican - When taken: 14 Aug 1813 - Where taken: Irish Channel - Date received: 17 Aug 1813 - From what ship: USS Argus - Died on 2 Sep 1813 in the Mill Prison Hospital.

Bayard, Joseph - Seaman - Number: 1435 - Prize name: Paul Jones - Ship type: P - How taken: HM Frigate Leonidas - When taken: 23 May 1813 - Where taken: off Cape Clear - Date received: 26 May 1813 - From what ship: HMS Leonidas - Born: France - Age: 27 - Discharged on 30 Jun 1813 and sent to Stapleton.

Bazin, Francois Gaston - Merchant Agent & Passenger - Number: 1327 - Prize name: Fox - Ship type: LM - How taken: HM Sloop Pheasant - When taken: 23 Apr 1813 - Where taken: Bay of Biscay - Date received: 14 May 1813 - From what ship: HMS Pleasant - Born: Guadeloupe - Age: 23 - Discharged on 15 May 1813 and sent to Ashburton on parole.

Beachman, George - Carpenter - Number: 55 - Prize name: Independence - Ship type: MV - How taken: HM Frigate Medusa - When taken: 9 Nov 1812 - Where taken: San Sebastian - Date received: 27 Nov 1812 - From what ship: HMS Wasp - Born: Baltimore - Age: 32 - Sent to Portsmouth on 29 Dec 1812 on the HMS Northumberland.

Beard, Francis - Seaman - Number: 1347 - Prize name: Fox - Ship type: LM - How taken: HM Sloop Pheasant - When taken: 23 Apr 1813 - Where taken: Bay of Biscay - Date received: 14 May 1813 - From what ship: HMS Pleasant - Born: Saint Domingue (Haiti) - Age: 21 - Discharged on 3 Jul 1813 and sent to Stapleton Prison.

Bears, Moses - Passenger - Number: 52 - Prize name: Independence - Ship type: MV - How taken: HM Frigate Medusa - When taken: 9 Nov 1812 - Where taken: San Sebastian - Date received: 27 Nov 1812 - From what ship: HMS Wasp - Born: Newburgh - Age: 27 - Sent on 8 Dec 1812 to Ashburton on parole.

Beckett, William - Seaman - Number: 14 - Prize name: Hibernia - Ship type: MV - How taken: Taken at Liverpool - When taken: 18 Oct 1812 - Date received: 21 Nov 1812 - From what ship: HMS Salvador del Mundo - Born: Virginia - Age: 28 - Sent to Portsmouth on 29 Dec 1812 on the HMS Northumberland.

Beckworth, Benjamin - Seaman - Number: 1376 - Prize name: Tom - Ship type: LM - How taken: HM Frigate Surveillante - When taken: 27 Apr 1813 - Where taken: Bay of Biscay - Date received: 15 May 1813 - From what ship: HMS Foxhound - Born: Pennsylvania - Age: 25 - Discharged on 3 Jul 1813 and sent to Stapleton Prison.

Beecher, Thaddeus - Master - Number: 921 - Prize name: Prompt - Ship type: MV - How taken: Chance, privateer - When taken: 22 Mar 1813 - Where taken: Bay of Biscay - Date received: 6 Apr 1813 - From what ship: Fonvey - Born: New Haven - Age: 23 - Sent on 6 Apr 1813 to Ashburton on parole.

Beek, Steward - Seaman - Number: 1396 - Prize name: Henry Clements - Ship type: MV - How taken: HM Brig Orestes - When taken: 13 Apr 1813 - Where taken: Bay of Biscay - Date received: 15 May 1813 - From what ship: HMS Orestes - Born: Maryland - Age: 28 - Discharged on 3 Jul 1813 and sent to Stapleton Prison.

Been, James - Seaman - Number: 2428 - Prize name: US Brig Argus - Ship type: War - How taken: HM Brig Pelican - When taken: 14 Aug 1813 - Where taken: Irish Channel - Date received: 12 Feb 1814 - From what ship: HMS Salvador del Mundo - Born: Hampton - Age: 37 - Discharged on 10 May 1814 and sent to Dartmoor.

Belfast, Richard - Seaman - Number: 1130 - Prize name: Viper - Ship type: MV - How taken: HM Ship-of-the-Line Superb - When taken: 15 Apr 1813 - Where taken: Bay of Biscay - Date received: 22 Apr 1813 - From what ship: HMS Superb - Born: Connecticut - Age: 24 - Sent to Dartmoor on 1 Jul 1813.

Belford, Isaac - Seaman - Number: 294 - Prize name: Columbia - Ship type: MV - How taken: HM Frigate Briton - When taken: 17 Dec 1812 - Where taken: off Bordeaux - Date received: 21 Jan 1813 - From what ship: HMS Abercrombie - Born: Boston - Age: 19 - Sent to Portsmouth on 8 Feb 1813 on the HMS Colossus.

Bell, George - Seaman - Number: 274 - Prize name: Leader - Ship type: MV - How taken: HM Frigate Briton - When taken: 10 Dec 1812 - Where taken: off Bordeaux - Date received: 21 Jan 1813 - From what ship: HMS Abercrombie - Born: Boston - Age: 30 - Sent to Portsmouth on 8 Feb 1813 on the HMS Colossus.

Bell, James - Seaman - Number: 1768 - Prize name: Union, prize of the Privateer Brutus - Ship type: LM - How taken: HM Brig Goldfinch - When taken: 17 Jul 1813 - Where taken: Bay of Biscay - Date received: 25 Jul 1813 - From what ship: HMS Pyramus - Born: Virginia - Age: 27 - Discharged on 8 Sep 1813 and sent to Dartmoor.

Bell, James - Seaman - Number: 1956 - Prize name: Joel Barlow - Ship type: LM - How taken: HM Frigate Briton - When taken: 3 Jul 1813 - Where taken: off Bordeaux - Date received: 31 Aug 1813 - From what ship: HMS Clarence - Born: Pennsylvania - Age: 19 - Discharged on 8 Sep 1813 and sent to Dartmoor.

Bellinger, William - Seaman - Number: 738 - Prize name: Pert - Ship type: MV - How taken: HM Ship-of-the-Line Warspite - When taken: 1 Mar 1813 - Where taken: Bay of Biscay - Date received: 19 Mar 1813 - From what ship: HMS Warspite - Born: New Jersey - Age: 21 - Sent to Dartmoor on 28 Jun 1813.

Benedict, William - Marine - Number: 1864 - Prize name: US Brig Argus - Ship type: War - How taken: HM Brig Pelican - When taken: 14 Aug 1813 - Where taken: Irish Channel - Date received: 23 Aug 1813 - From what ship: HMS Pelican - Born: New York - Age: 21 - Discharged on 8 Sep 1813 and sent to Dartmoor.

Bennett, Charles - 2nd Mate - Number: 1933 - Prize name: Marmion - Ship type: MV - How taken: HM Frigate President - When taken: 14 Aug 1813 - Where taken: off Nantes - Date received: 27 Aug 1813 - From what ship: HMS Urgent - Born: New York - Age: 27 - Discharged on 8 Sep 1813 and sent to Dartmoor.

Benner, Lewis - Seaman - Number: 94 - Prize name: Experiment - Ship type: MV - How taken: HM Brig Rover - When taken: 21 Oct 1812 - Where taken: off Bordeaux - Date received: 25 Dec 1812 - From what ship: HMS Northumberland - Born: Baltimore - Age: 22 - Race: Mulatto - Sent to Portsmouth on 29 Dec 1812 on the HMS Northumberland.

Bennet, James - Seaman - Number: 808 - Prize name: Decornau - Ship type: MV - How taken: HM Sloop Pheasant - When taken: 15 Mar 1813 - Where taken: Bay of Biscay - Date received: 19 Mar 1813 - From what ship: HMS Pheasant - Born: Frankfort - Age: 26 - Sent to Dartmoor on 28 Jun 1813.

Bennett, Charles - Seaman - Number: 1285 - Prize name: Price - Ship type: LM - How taken: HM Frigate Medusa - When taken: 13 Apr 1813 - Where taken: Bay of Biscay - Date received: 10 May 1813 - From what ship: HMS

Medusa - Born: New York - Age: 24 - Discharged on 3 Jul 1813 and sent to Stapleton Prison.

Benny, David - Boy - Number: 2643 - Prize name: Ateline - Ship type: MV - How taken: HM Frigate Magiciene - When taken: 14 Mar 1814 - Where taken: off Cape Finisterre, Spain - Date received: 17 May 1814 - From what ship: HMS Tortois - Born: New York - Age: 17 - Discharged on 14 Jun 1814 and sent to Dartmoor.

Benny, Mallock - Seaman - Number: 301 - Prize name: Stephen - Ship type: MV - How taken: HM Frigate Briton & HM Frigate Andromache - When taken: 17 Dec 1812 - Where taken: off Bordeaux - Date received: 21 Jan 1813 - From what ship: HMS Abercrombie - Born: Norwich - Age: 25 - Race: Black - Sent to Portsmouth on 8 Feb 1813 on the HMS Colossus.

Benson, James - Seaman - Number: 1882 - Prize name: US Brig Argus - Ship type: War - How taken: HM Brig Pelican - When taken: 14 Aug 1813 - Where taken: Irish Channel - Date received: 23 Aug 1813 - From what ship: HMS Pelican - Born: Delaware - Age: 30 - Race: Black - Discharged on 8 Sep 1813 and sent to Dartmoor.

Bensted, John - Seaman - Number: 2449 - How taken: Taken out of a Russian ship - When taken: 28 Jan 1814 - Where taken: Cove of Cork - Date received: 15 Feb 1814 - From what ship: HMS Zealous - Born: Newbury - Age: 33 - Discharged on 10 May 1814 and sent to Dartmoor.

Bentsten, John - Seaman - Number: 111 - Prize name: Otter - Ship type: MV - How taken: HM Ship-Sloop Jalousie - When taken: 1 Dec 1812 - Where taken: off Cape Vincent - Date received: 30 Dec 1812 - From what ship: HMS Leonidas - Born: Virginia - Age: 20 - Sent to Portsmouth on 4 Jan 1813 on the HMS Revolutionnaire.

Berdick, Simon - Seaman - Number: 1961 - Prize name: Joel Barlow - Ship type: LM - How taken: HM Frigate Briton - When taken: 3 Jul 1813 - Where taken: off Bordeaux - Date received: 31 Aug 1813 - From what ship: HMS Clarence - Born: Rhode Island - Age: 27 - Discharged on 8 Sep 1813 and sent to Dartmoor.

Beriston, Peter - Seaman - Number: 1261 - Prize name: Caroline - Ship type: MV - How taken: HM Frigate Medusa - When taken: 12 Apr 1813 - Where taken: Bay of Biscay - Date received: 10 May 1813 - From what ship: HMS Medusa - Born: Philadelphia - Age: 32 - Discharged on 3 Jul 1813 and sent to Stapleton Prison.

Berry, Brook - Seaman - Number: 518 - Prize name: Allegany - Ship type: MV - How taken: Detained at Gibraltar - When taken: 8 Aug 1812 - Date received: 15 Feb 1813 - From what ship: HMS Andromeda - Born: Prince Georges County - Age: 20 - Sent to Dartmoor on 2 Apr 1813.

Bertine, John - Seaman - Number: 994 - Prize name: Lightning - Ship type: MV - How taken: HM Frigate Medusa - When taken: 2 Apr 1813 - Where taken: Bay of Biscay - Date received: 16 Apr 1813 - From what ship: HMS Fairy - Born: New York - Age: 20 - Sent to Chatham on 8 Jul 1813 on HM Tender Neptune.

Berto, John - Seaman - Number: 2438 - Prize name: Prince of Wales, prize of the Privateer Prince de Neufchatel - Ship type: P - How taken: Transport Nelson - When taken: 4 Feb 1814 - Where taken: at sea - Date received: 14 Feb 1814 - From what ship: HMS Halcyon - Born: New Orleans - Age: 40 - Discharged on 10 May 1814 and sent to Dartmoor.

Bertol, Samuel - Seaman - Number: 2632 - Prize name: Ateline - Ship type: MV - How taken: HM Frigate Magiciene - When taken: 14 Mar 1814 - Where taken: off Cape Finisterre, Spain - Date received: 17 May 1814 - From what ship: HMS Tortois - Born: Freeport - Age: 23 - Discharged on 14 Jun 1814 and sent to Dartmoor.

Besson, Phillipe - Seaman - Number: 2416 - Prize name: Joseph - Ship type: MV - How taken: HM Brig Royalist - When taken: 18 Jan 1814 - Where taken: Bay of Biscay - Date received: 9 Feb 1814 - From what ship: HMS Sparrow - Born: Marblehead - Age: 18 - Discharged on 10 May 1814 and sent to Dartmoor.

Best, Robert - Seaman - Number: 1644 - Prize name: Tickler - Ship type: LM - How taken: HM Frigate Magiciene - When taken: 5 Jun 1813 - Where taken: Lat 47 Long 13 - Date received: 14 Jun 1813 - From what ship: HMS Orestes - Born: New Jersey - Age: 22 - Discharged on 30 Jun 1813 and sent to Stapleton.

Bevers, Clement - Seaman - Number: 226 - Prize name: Vengeance - Ship type: LM - How taken: HM Frigate Phoebe - When taken: 1 Jan 1813 - Where taken: Lat 44.4 Long 23 - Date received: 9 Jan 1813 - From what ship: HMS Phoebe - Born: Baltimore - Age: 43 - Sent to Portsmouth on 8 Feb 1813 on the HMS Colossus.

Bevin, William - Seaman - Number: 1895 - Prize name: Betsey, prize to the US Brig Argus - Ship type: War - How taken: HM Frigate Leonidas - When taken: 12 Aug 1813 - Where taken: Channel - Date received: 23 Aug 1813 - From what ship: HMS Pelican - Born: Chatham - Age: 37 - Discharged on 8 Sep 1813 and sent to Dartmoor.

Bickford, Ebenezer - Master's Mate - Number: 153 - Prize name: Hunter - Ship type: P - How taken: HM Frigate Phoebe - When taken: 24 Dec 1812 - Where taken: off Western Islands - Date received: 9 Jan 1813 - From what ship: HMS Phoebe - Born: Salem, NJ - Age: 26 - Sent to Portsmouth on 8 Feb 1813 on the HMS Colossus.

Bidson, Thomas - Seaman - Number: 644 - Prize name: Criterion - Ship type: MV - How taken: HM Frigate Belle Poule - When taken: 14 Feb 1813 - Where taken: Bay of Biscay - Date received: 4 Mar 1813 - From what ship: HMS Strenuous - Born: New York - Age: 22 - Sent to Dartmoor on 2 Apr 1813.

Bienfaux, Allen - Seaman - Number: 2067 - Prize name: Betsy, prize to the Privateer True Blooded Yankee - Ship type: P - How taken: HM Frigate Eurotas - When taken: 26 Oct 1813 - Where taken: off Ushant - Date received: 1 Nov 1813 - From what ship: HMS Hannibal - Born: Brest - Age: 43 - Discharged on 3 Nov 1813 and sent to Dartmoor.

Billings, John - 2nd Mate - Number: 1721 - Prize name: Hannah & Eliza - Ship type: MV - How taken: HM Brig Lyra - When taken: 29 May 1813 - Where taken: off north coast of Spain - Date received: 4 Jul 1813 - From what ship: HMS Iris - Born: Providence - Age: 20 - Discharged on 22 Aug 1813 and released to HM Brig Redpole.

Birch, Andrew - Seaman - Number: 493 - Prize name: Cashiere - Ship type: LM - How taken: HM Brig Reindeer - When taken: 3 Feb 1813 - Where taken: Bay of Biscay - Date received: 12 Feb 1813 - From what ship: HMS Reindeer - Born: Baltimore - Age: 19 - Sent to Dartmoor on 2 Apr 1813.

Bird, Comfort - Seaman - Number: 2483 - How taken: Impressed at Liverpool - When taken: 1 Feb 1814 - Date received: 13 Mar 1814 - From what ship: HMS Bittern - Born: Dorchester - Age: 24 - Discharged on 10 May 1814 and sent to Dartmoor.

Birnent, Edward - Seaman - Number: 221 - Prize name: Vengeance - Ship type: LM - How taken: HM Frigate Phoebe - When taken: 1 Jan 1813 - Where taken: Lat 44.4 Long 23 - Date received: 9 Jan 1813 - From what ship: HMS Phoebe - Born: Salisbury - Age: 17 - Sent to Portsmouth on 8 Feb 1813 on the HMS Colossus.

Bishop, Edward - Seaman - Number: 902 - Prize name: Dick - Ship type: MV - How taken: HM Brig Dispatch - When taken: 17 Mar 1813 - Where taken: Bay of Biscay - Date received: 2 Apr 1813 - From what ship: HMS Warspite - Born: New York - Age: 27 - Sent on 21 Jun 1813 to Portsmouth on HMS Prometheus.

Bishop, William - Steward - Number: 543 - Prize name: Spitfire - Ship type: MV - How taken: HM Brig Achates - When taken: 14 Feb 1813 - Where taken: off Ushant - Date received: 16 Feb 1813 - From what ship: HMS Achates - Born: Danvers - Age: 17 - Sent to Dartmoor on 2 Apr 1813.

Bishop, William - Seaman - Number: 1740 - Prize name: Spitfire - Ship type: MV - How taken: HM Brig Achates - When taken: 14 Feb 1813 - Where taken: off Ushant - Date received: 10 Jul 1813 - From what ship: Dartmoor Prison - Born: Danvers - Age: 17 - Discharged on 10 Jul 1813 and released to HMS Salvador del Mundo.

Bisley, Horace - Seaman - Number: 712 - Prize name: Star - Ship type: MV - How taken: HM Ship-of-the-Line Superb - When taken: 9 Feb 1813 - Where taken: Bay of Biscay - Date received: 19 Mar 1813 - From what ship: HMS Warspite - Born: Rockhill - Age: 18 - Sent to Dartmoor on 2 Apr 1813.

Biss, Daniel W. - Seaman - Number: 1528 - Prize name: Courier - Ship type: LM - How taken: HM Brig Rover - When taken: 14 Mar 1813 - Where taken: Bay of Biscay - Date received: 29 May 1813 - From what ship: HMS Hannibal - Born: Massachusetts - Age: 25 - Discharged on 30 Jun 1813 and sent to Stapleton.

Bitters, John - Seaman - Number: 2672 - Prize name: John, prize to Amelia - Ship type: P - How taken: HM Ship-of-the-Line Sterling Castle - When taken: 10 May 1814 - Where taken: Lat 36 Long 37 - Date received: 5 Jun 1814 - From what ship: HMS Gronville - Born: Philadelphia - Age: 23 - Race: Black - Discharged on 14 Jun 1814 and sent to Dartmoor.

Black, George - Seaman - Number: 1964 - Prize name: Joel Barlow - Ship type: LM - How taken: HM Frigate Briton - When taken: 3 Jul 1813 - Where taken: off Bordeaux - Date received: 31 Aug 1813 - From what ship: HMS Clarence - Born: Not legible - Age: 29 - Discharged on 8 Sep 1813 and sent to Dartmoor.

Blacklee, Barnard - Seaman - Number: 784 - Prize name: Cannoniere - Ship type: P - How taken: HM Ship-of-the-Line Warspite - When taken: 14 Mar 1813 - Where taken: Bay of Biscay - Date received: 19 Mar 1813 - From what ship: HMS Warspite - Born: Newburgh - Age: 26 - Sent to Dartmoor on 28 Jun 1813.

Blackledge, John - Marine - Number: 1865 - Prize name: US Brig Argus - Ship type: War - How taken: HM Brig

Pelican - When taken: 14 Aug 1813 - Where taken: Irish Channel - Date received: 23 Aug 1813 - From what ship: HMS Pelican - Born: New Jersey - Age: 26 - Discharged on 8 Sep 1813 and sent to Dartmoor.

Blackman, Moses - Seaman - Number: 932 - Prize name: Weasel - Ship type: MV - How taken: HM Brig Foxhound - When taken: 25 Mar 1813 - Where taken: Bay of Biscay - Date received: 6 Apr 1813 - From what ship: HMS Foxhound - Born: Boston - Age: 40 - Sent on 21 Jun 1813 to Portsmouth on HMS Prometheus.

Blackston, Edward - Seaman - Number: 496 - Prize name: Cashiere - Ship type: LM - How taken: HM Brig Reindeer - When taken: 3 Feb 1813 - Where taken: Bay of Biscay - Date received: 12 Feb 1813 - From what ship: HMS Reindeer - Born: Pennsylvania - Age: 23 - Sent to Dartmoor on 2 Apr 1813.

Blackston, Edward - Seaman - Number: 1733 - Prize name: Cashiere - Ship type: LM - How taken: HM Brig Reindeer - When taken: 3 Feb 1813 - Where taken: Bay of Biscay - Date received: 10 Jul 1813 - From what ship: Dartmoor Prison - Born: New Sharon - Age: 23 - Discharged on 10 Jul 1813 and released to HMS Salvador del Mundo.

Blackwood, Bailey - Mate - Number: 1088 - Prize name: Young Holkar - Ship type: MV - How taken: HM Ship-of-the-Line Superb - When taken: 10 Apr 1813 - Where taken: off Belle Isle - Date received: 22 Apr 1813 - From what ship: HMS Superb - Born: New Jersey - Age: 28 - Sent on 27 Apr 1813 to Ashburton on parole.

Bladen, John - Seaman - Number: 1843 - Prize name: US Brig Argus - Ship type: War - How taken: HM Brig Pelican - When taken: 14 Aug 1813 - Where taken: Irish Channel - Date received: 17 Aug 1813 - From what ship: USS Argus - Born: Fairfax - Age: 23 - Discharged on 8 Sep 1813 and sent to Dartmoor.

Blaisdoll, Jonathan - Seaman - Number: 2041 - How taken: Impressed at Liverpool - When taken: 16 Oct 1813 - Date received: 1 Nov 1813 - From what ship: HMS Bittern - Born: Kennebunkport - Age: 20 - Discharged on 3 Nov 1813 and sent to Dartmoor.

Blake, Alexander - Seaman - Number: 1093 - Prize name: Young Holkar - Ship type: MV - How taken: HM Ship-of-the-Line Superb - When taken: 10 Apr 1813 - Where taken: off Belle Isle - Date received: 22 Apr 1813 - From what ship: HMS Superb - Born: Charlestown - Age: 19 - Sent to Dartmoor on 1 Jul 1813.

Blake, Philip - 2nd Mate - Number: 868 - Prize name: Charlotte - Ship type: MV - How taken: HM Ship-of-the-Line Warspite - When taken: 1 Mar 1813 - Where taken: Bay of Biscay - Date received: 28 Mar 1813 - From what ship: HMS Warspite - Born: Boston - Age: 21 - Sent to Dartmoor on 28 Jun 1813.

Blake, Thomas - Cook - Number: 2144 - Prize name: Amiable - Ship type: LM - How taken: HM Ship-of-the-Line Magnificant - When taken: 30 Oct 1813 - Where taken: off Lorient - Date received: 29 Nov 1813 - From what ship: HMS Dublin - Born: Virginia - Age: 27 - Race: Black - Discharged on 4 Dec 1813 and sent to Dartmoor.

Blanchard, Carvan - Seaman - Number: 514 - Prize name: Tyger - Ship type: MV - How taken: Detained at Gibraltar - When taken: 8 Aug 1812 - Date received: 15 Feb 1813 - From what ship: HMS Andromeda - Born: Milford - Age: 21 - Sent to Dartmoor on 2 Apr 1813.

Blanchard, George - Seaman - Number: 754 - Prize name: William Bayard - Ship type: MV - How taken: HM Ship-of-the-Line Warspite - When taken: 3 Mar 1813 - Where taken: Bay of Biscay - Date received: 19 Mar 1813 - From what ship: HMS Warspite - Born: Elizabethtown - Age: 21 - Sent to Dartmoor on 2 Apr 1813.

Blanchard, George - Seaman - Number: 1656 - Prize name: William Bayard - Ship type: MV - How taken: HM Ship-of-the-Line Warspite - When taken: 3 Mar 1813 - Where taken: Bay of Biscay - Date received: 15 Jun 1813 - From what ship: Dartmoor Prison - Born: Elizabethtown - Age: 21 - Discharged on 16 Jun 1813 and released to HMS Salvador del Mundo.

Blanchard, Simon - Seaman - Number: 1301 - Prize name: Price - Ship type: LM - How taken: HM Frigate Medusa - When taken: 13 Apr 1813 - Where taken: Bay of Biscay - Date received: 10 May 1813 - From what ship: HMS Medusa - Born: Charlestown - Age: 21 - Discharged on 3 Jul 1813 and sent to Stapleton Prison.

Blisset, James - Seaman - Number: 1885 - Prize name: US Brig Argus - Ship type: War - How taken: HM Brig Pelican - When taken: 14 Aug 1813 - Where taken: Irish Channel - Date received: 23 Aug 1813 - From what ship: HMS Pelican - Born: Gloucester - Age: 24 - Discharged on 8 Sep 1813 and sent to Dartmoor.

Blodget, Caleb - Seaman - Number: 1230 - Prize name: Essex - Ship type: MV - How taken: HM Frigate Pyramus - When taken: 2 Apr 1813 - Where taken: Bay of Biscay - Date received: 9 May 1813 - From what ship: HMS Andromache - Born: New Hampshire - Age: 23 - Discharged on 3 Jul 1813 and sent to Stapleton Prison.

Bloom, Joseph - Seaman - Number: 578 - Prize name: Cashiere - Ship type: LM - How taken: HM Brig Reindeer - When taken: 3 Feb 1813 - Where taken: Bay of Biscay - Date received: 23 Feb 1813 - From what ship: HMS Surveillante - Born: Pennsylvania - Age: 25 - Sent to Dartmoor on 2 Apr 1813.

Bloomdose, John - Seaman - Number: 2381 - Prize name: Devon, prize to the Privateer Bunker Hill - Ship type: P - How taken: HM Brig Fly - When taken: 21 Jan 1814 - Where taken: at sea - Date received: 31 Jan 1814 - From what ship: HMS Fly - Born: Albany - Age: 23 - Discharged on 27 Feb 1814 and sent to Chatham on HMS Haleyon.

Blue, Peter - Carpenter - Number: 683 - Prize name: Nope - Ship type: MV - How taken: Chance, privateer - When taken: 15 Feb 1813 - Where taken: off Bordeaux - Date received: 15 Mar 1813 - From what ship: Growler - Born: Lisbon - Age: 46 - Sent to Chatham on 8 Jul 1813 on HM Tender Neptune.

Bodfish, William - Seaman - Number: 2295 - Prize name: Siro - Ship type: LM - How taken: HM Brig Pelican - When taken: 13 Jan 1814 - Where taken: at sea - Date received: 23 Jan 1814 - From what ship: HMS Pelican - Born: Boston - Age: 26 - Discharged on 31 Jan 1814 and sent to Dartmoor.

Boffs, James - 2nd Mate - Number: 612 - Prize name: Governor McKean - Ship type: LM - How taken: HM Brig Rover - When taken: 26 Jun 1813 - Where taken: off Bordeaux - Date received: 2 Mar 1813 - From what ship: HMS Insolent - Born: Londonderry, Ireland - Age: 62 - Sent to Dartmoor on 2 Apr 1813.

Bogart, Killian William - Seaman - Number: 1839 - Prize name: US Brig Argus - Ship type: War - How taken: HM Brig Pelican - When taken: 14 Aug 1813 - Where taken: Irish Channel - Date received: 17 Aug 1813 - From what ship: USS Argus - Born: New York - Age: 21 - Discharged on 8 Sep 1813 and sent to Dartmoor.

Boggs, James - 2nd Mate - Number: 1728 - Prize name: Governor McKean - Ship type: LM - How taken: HM Brig Rover - When taken: 26 Jan 1813 - Where taken: off Bordeaux - Date received: 5 Jul 1813 - From what ship: Dartmoor Prison - Born: Londonderry Island - Age: 62 - Discharged on 22 Aug 1813 and released to HM Brig Redpole.

Bonie, James - Seaman - Number: 1046 - Prize name: Independence - Ship type: MV - How taken: HM Ship-of-the-Line Superb - When taken: 16 Mar 1813 - Where taken: Bay of Biscay - Date received: 22 Apr 1813 - From what ship: HMS Superb - Born: Maryland - Age: 26 - Sent to Dartmoor on 1 Jul 1813.

Bonnell, James - Seaman - Number: 2681 - How taken: Sent into custody from HM Ship-of-the-Line Minden - Date received: 5 Jun 1814 - From what ship: HMS Gronville - Born: New Jersey - Age: 33 - Discharged on 14 Jun 1814 and sent to Dartmoor.

Booth, Charles - Seaman - Number: 2185 - Prize name: Charlotte - Ship type: MV - How taken: Cutter Dwarf - When taken: 4 Nov 1812 - Where taken: off Bordeaux - Date received: 18 Dec 1813 - From what ship: HMS Conquistador - Born: New York - Age: 30 - Discharged on 31 Jan 1814 and sent to Dartmoor.

Booth, Stephen - Seaman - Number: 1979 - Prize name: Frederick & Augusta - Ship type: MV - How taken: Sir John Sherbrook, privateer - When taken: 12 Apr 1813 - Where taken: off Newport - Date received: 19 Sep 1813 - From what ship: HMS Bittern - Born: Connecticut - Age: 24 - Discharged on 27 Sep 1813 and sent to Dartmoor.

Booth, William - Captain - Number: 1946 - Prize name: Ellen & Emeline - Ship type: MV - How taken: HM Schooner Telegraph - When taken: 13 Aug 1813 - Where taken: off St. Andrews - Date received: 28 Aug 1813 - From what ship: HMS Telegraph - Born: Belville, NJ - Age: 36 - Discharged on 30 Aug 1813 and sent to Ashburton on parole.

Borddell, Justice - Seaman - Number: 2556 - Prize name: Bunker Hill - Ship type: P - How taken: HM Frigate Pomone & HM Frigate Cydnus - When taken: 4 Mar 1814 - Where taken: Bay of Biscay - Date received: 17 Apr 1814 - From what ship: HMS Teazer - Born: New York - Age: 17 - Discharged on 10 May 1814 and sent to Dartmoor.

Borgin, Gabriel - Seaman - Number: 1732 - Prize name: Star - Ship type: MV - How taken: HM Ship-of-the-Line Superb - When taken: 9 Feb 1813 - Where taken: Bay of Biscay - Date received: 10 Jul 1813 - From what ship: Dartmoor Prison - Born: Somerset County - Age: 17 - Discharged on 10 Jul 1813 and released to HMS Salvador del Mundo.

Borgin, Gabriel - Seaman - Number: 702 - Prize name: Star - Ship type: MV - How taken: HM Ship-of-the-Line Superb - When taken: 9 Feb 1813 - Where taken: Bay of Biscay - Date received: 19 Mar 1813 - From what ship: HMS Warspite - Born: Somerset County - Age: 17 - Sent to Dartmoor on 2 Apr 1813.

Boriesa, John - Seaman - Number: 1552 - Prize name: Good Friends - Ship type: MV - How taken: HM Frigate

Andromache - When taken: 2 Apr 1813 - Where taken: Bay of Biscay - Date received: 29 May 1813 - From what ship: HMS Hannibal - Born: Maryland - Age: 23 - Race: Negro - Discharged on 30 Jun 1813 and sent to Stapleton.

Boss, Thomas - Seaman - Number: 357 - Prize name: Louisa, prize of the Privateer Decatur - Ship type: P - How taken: HM Frigate Andromache - When taken: 11 Jan 1813 - Where taken: off Bordeaux - Date received: 4 Feb 1813 - From what ship: HMS Cornwall - Born: Hampton - Age: 31 - Sent to Chatham on 29 Mar 1813 on the HMS Braham.

Bottelio, Antony - Seaman - Number: 2306 - Prize name: Siro - Ship type: LM - How taken: HM Brig Pelican - When taken: 13 Jan 1814 - Where taken: at sea - Date received: 23 Jan 1814 - From what ship: HMS Pelican - Born: St. Michael - Age: 19 - Discharged on 31 Jan 1814 and sent to Dartmoor.

Bowdler, Thomas - Boy - Number: 207 - Prize name: Hunter - Ship type: P - How taken: HM Frigate Phoebe - When taken: 24 Dec 1812 - Where taken: off Western Islands - Date received: 9 Jan 1813 - From what ship: HMS Phoebe - Born: Charlestown - Age: 15 - Sent to Portsmouth on 8 Feb 1813 on the HMS Colossus.

Bowen, John - Seaman - Number: 1684 - Prize name: Revenge - Ship type: LM - How taken: HM Frigate Belle Poule - When taken: 11 May 1813 - Where taken: Bay of Biscay - Date received: 26 Jun 1813 - From what ship: HMS Duncan - Born: Charleston - Age: 24 - Discharged on 30 Jun 1813 and sent to Stapleton.

Bowen, Lewis - 2nd Mate - Number: 216 - Prize name: Vengeance - Ship type: LM - How taken: HM Frigate Phoebe - When taken: 1 Jan 1813 - Where taken: Lat 44.4 Long 23 - Date received: 9 Jan 1813 - From what ship: HMS Phoebe - Born: New York - Age: 20 - Sent to Portsmouth on 8 Feb 1813 on the HMS Colossus.

Bower, Joseph - Seaman - Number: 1538 - Prize name: Courier - Ship type: LM - How taken: HM Brig Rover - When taken: 14 Mar 1813 - Where taken: Bay of Biscay - Date received: 29 May 1813 - From what ship: HMS Hannibal - Born: Maryland - Age: 26 - Discharged on 30 Jun 1813 and sent to Stapleton.

Bowers, Jonathan - Master - Number: 2147 - Prize name: Charlotte - Ship type: MV - How taken: Cutter Dwarf - When taken: 4 Nov 1812 - Where taken: off Bordeaux - Date received: 7 Dec 1813 - From what ship: Cutter Dwarf - Born: Massachusetts - Age: 28 - Discharged on 8 Dec 1813 and sent to Ashburton on parole.

Bowers, Mrs. - Woman - Number: 2149 - Prize name: Charlotte - Ship type: MV - How taken: Cutter Dwarf - When taken: 4 Nov 1812 - Where taken: off Bordeaux - Date received: 7 Dec 1813 - From what ship: Cutter Dwarf - Discharged on 8 Dec 1813 and sent to Ashburton on parole.

Bowman, Benjamin - Seaman - Number: 983 - How taken: Impressed at Dublin - When taken: 12 Feb 1813 - Date received: 13 Apr 1813 - From what ship: HMS Frederick - Born: Connecticut - Age: 26 - Sent to Chatham on 8 Jul 1813 on HM Tender Neptune.

Bowman, Isaac - Seaman - Number: 2551 - Prize name: Bunker Hill - Ship type: P - How taken: HM Frigate Pomone & HM Frigate Cydnus - When taken: 4 Mar 1814 - Where taken: Bay of Biscay - Date received: 17 Apr 1814 - From what ship: HMS Teazer - Born: Philadelphia - Age: 26 - Discharged on 10 May 1814 and sent to Dartmoor.

Bowman, Leon - Steward - Number: 2157 - Prize name: Charlotte - Ship type: MV - How taken: Cutter Dwarf - When taken: 4 Dec 1813 - Where taken: off Bordeaux - Date received: 9 Dec 1813 - From what ship: Cutter Dwarf - Born: Switzerland - Age: 41 - Discharged on 7 Feb 1814 and sent to Harwich on Transport Luenter.

Bowne, Asher - Seaman - Number: 1903 - Prize name: US Brig Argus - Ship type: War - How taken: HM Brig Pelican - When taken: 14 Aug 1813 - Where taken: Irish Channel - Date received: 23 Aug 1813 - From what ship: HMS Pelican - Born: Middletown - Age: 23 - Discharged on 8 Sep 1813 and sent to Dartmoor.

Boyd, Edward - Seaman - Number: 186 - Prize name: Hunter - Ship type: P - How taken: HM Frigate Phoebe - When taken: 24 Dec 1812 - Where taken: off Western Islands - Date received: 9 Jan 1813 - From what ship: HMS Phoebe - Born: Salem - Age: 19 - Sent to Portsmouth on 8 Feb 1813 on the HMS Colossus.

Boyd, John - Master - Number: 1571 - Prize name: Hannah & Eliza - Ship type: MV - How taken: HM Brig Lyra - When taken: 29 May 1813 - Where taken: off north coast of Spain - Date received: 10 May 1813 - From what ship: Hannah & Eliza - Born: Havre de Grace - Age: 31 - Discharged on 11 May 1813 and sent to Ashburton on parole.

Boyd, William - Seaman - Number: 2448 - How taken: Taken out of a Russian ship - When taken: 28 Jan 1814 - Where taken: Cove of Cork - Date received: 15 Feb 1814 - From what ship: HMS Zealous - Born: Salem - Age:

36 - Discharged on 14 Jun 1814 and sent to Dartmoor.

Bradford, Charles - Seaman - Number: 1219 - Prize name: Grand Napoleon - Ship type: MV - How taken: HM Frigate Belle Poule - When taken: 3 Apr 1813 - Where taken: off Bordeaux - Date received: 9 May 1813 - From what ship: HMS Andromache - Born: Massachusetts - Age: 19 - Discharged on 3 Jul 1813 and sent to Stapleton Prison.

Bradford, Henry - Chief Mate - Number: 976 - Prize name: Grand Napoleon - Ship type: MV - How taken: HM Frigate Belle Poule - When taken: 3 Apr 1813 - Where taken: off Bordeaux - Date received: 11 Apr 1813 - From what ship: Napoleon - Born: Duxbury - Age: 27 - Sent on 12 Apr 1813 to Ashburton on parole.

Bradford, James - Cook - Number: 1978 - Prize name: Hero - Ship type: MV - How taken: Tenedos - When taken: 12 Sep 1813 - Where taken: Western ocean - Date received: 19 Sep 1813 - From what ship: HMS Bittern - Born: Maryland - Age: 22 - Race: Black - Discharged on 27 Sep 1813 and sent to Dartmoor.

Bradley, Hugh - Seaman - Number: 1119 - Prize name: Viper - Ship type: MV - How taken: HM Ship-of-the-Line Superb - When taken: 15 Apr 1813 - Where taken: Bay of Biscay - Date received: 22 Apr 1813 - From what ship: HMS Superb - Born: Philadelphia - Age: 31 - Sent to Dartmoor on 1 Jul 1813.

Bradt, Francis - Marine - Number: 1861 - Prize name: US Brig Argus - Ship type: War - How taken: HM Brig Pelican - When taken: 14 Aug 1813 - Where taken: Irish Channel - Date received: 23 Aug 1813 - From what ship: HMS Pelican - Born: New York - Age: 35 - Discharged on 8 Sep 1813 and sent to Dartmoor.

Braley, George - Captain - Number: 1590 - Prize name: Governor Gerry - Ship type: MV - How taken: HM Brig Royalist - When taken: 31 May 1813 - Where taken: Bay of Biscay - Date received: 14 Jun 1813 - From what ship: HMS Royalist - Born: Massachusetts - Age: 31 - Discharged on 30 Jun 1813 and sent to Ashburton on parole.

Bramblecome, David - Seaman - Number: 2 - Prize name: Cornelia - Ship type: MV - How taken: HM Brig Zenobia - When taken: 14 Aug 1812 - Where taken: Western ocean - Date received: 22 Oct 1812 - From what ship: HMS Frederick - Born: Marblehead - Age: 36 - Sent to Portsmouth on 29 Dec 1812 on the HMS Northumberland.

Brandage, John - Seaman - Number: 1640 - Prize name: Tickler - Ship type: LM - How taken: HM Frigate Magiciene - When taken: 5 Jun 1813 - Where taken: Lat 47 Long 13 - Date received: 14 Jun 1813 - From what ship: HMS Orestes - Born: Greece - Age: 22 - Discharged on 30 Jun 1813 and sent to Stapleton.

Brandy, Francis - Mate - Number: 235 - Prize name: Eunice - Ship type: MV - How taken: Impressed at Liverpool - When taken: 17 Nov 1812 - Date received: 12 Jan 1813 - From what ship: HMS Bittern - Born: New Orleans - Age: 25 - Sent to Portsmouth on 8 Feb 1813 on the HMS Colossus.

Branegen, John - Seaman - Number: 1976 - Prize name: Montgomery - Ship type: P - How taken: HM Frigate Nymphe - When taken: 5 May 1813 - Where taken: Boston Bay - Date received: 19 Sep 1813 - From what ship: HMS Bittern - Born: Gothenburg, Sweden - Age: 26 - Discharged on 27 Sep 1813 and sent to Dartmoor.

Brant, Thomas - Seaman - Number: 1210 - Prize name: Zebra - Ship type: LM - How taken: HM Frigate Pyramus - When taken: 20 Apr 1813 - Where taken: Bay of Biscay - Date received: 9 May 1813 - From what ship: HMS Andromache - Born: New York - Age: 19 - Discharged on 3 Jul 1813 and sent to Stapleton Prison.

Bredey, Jason - Seaman - Number: 2411 - How taken: Sent to Mill Prison by the major of Plymouth - Date received: 9 Feb 1814 - From what ship: Plymouth - Born: New York - Age: 23 - Discharged on 10 May 1814 and sent to Dartmoor.

Brereton, Benjamin - Boy - Number: 650 - Prize name: Criterion - Ship type: MV - How taken: HM Frigate Belle Poule - When taken: 14 Feb 1813 - Where taken: Bay of Biscay - Date received: 4 Mar 1813 - From what ship: HMS Strenuous - Age: 18 - Sent on 1 May 1813 to Ashburton on parole.

Brevoort, William - Master - Number: 940 - Prize name: Gleamer - Ship type: MV - How taken: Brothers, privateer - When taken: 26 Mar 1813 - Where taken: Bay of Biscay - Date received: 8 Apr 1813 - From what ship: Wasp, sloop from Guernsey - Born: New York - Age: 29 - Sent on 9 Apr 1813 to Ashburton on parole.

Brewer, James - Seaman - Number: 276 - Prize name: Leader - Ship type: MV - How taken: HM Frigate Briton - When taken: 10 Dec 1812 - Where taken: off Bordeaux - Date received: 21 Jan 1813 - From what ship: HMS Abercrombie - Born: Boston - Age: 19 - Sent to Portsmouth on 8 Feb 1813 on the HMS Colossus.

Bribion, Madam - 1st child - Passenger - Number: 253 - Prize name: Vengeance - Ship type: LM - How taken: HM Frigate Phoebe - When taken: 1 Jan 1813 - Where taken: Lat 44.4 Long 23 - Date received: 16 Jan 1813 - From what ship: HMS Phoebe - Sent to France on 1 Feb 1813 on a cartel.

Bribion, Madam - 2nd child - Passenger - Number: 254 - Prize name: Vengeance - Ship type: LM - How taken: HM Frigate Phoebe - When taken: 1 Jan 1813 - Where taken: Lat 44.4 Long 23 - Date received: 16 Jan 1813 - From what ship: HMS Phoebe - Sent to France on 1 Feb 1813 on a cartel.

Bribion, Madam - 3rd child - Passenger - Number: 255 - Prize name: Vengeance - Ship type: LM - How taken: HM Frigate Phoebe - When taken: 1 Jan 1813 - Where taken: Lat 44.4 Long 23 - Date received: 16 Jan 1813 - From what ship: HMS Phoebe - Sent to France on 1 Feb 1813 on a cartel.

Bribion, Madam (first name not listed) - Passenger - Number: 252 - Prize name: Vengeance - Ship type: LM - How taken: HM Frigate Phoebe - When taken: 1 Jan 1813 - Where taken: Lat 44.4 Long 23 - Date received: 16 Jan 1813 - From what ship: HMS Phoebe - Sent to France on 1 Feb 1813 on a cartel.

Bridge, Francis - Seaman - Number: 551 - Prize name: Spitfire - Ship type: MV - How taken: HM Brig Achates - When taken: 14 Feb 1813 - Where taken: off Ushant - Date received: 16 Feb 1813 - From what ship: HMS Achates - Born: Marblehead - Age: 32 - Sent to Dartmoor on 2 Apr 1813.

Bridges, John - Seaman - Number: 1027 - Prize name: Two Brothers - Ship type: MV - How taken: Bootle of Liverpool, letter of marque - When taken: 18 Mar 1813 - Where taken: Western Islands - Date received: 21 Apr 1813 - From what ship: HMS Bittern - Born: Castine - Age: 23 - Sent to Dartmoor on 1 Jul 1813.

Briggs, William - Seaman - Number: 2495 - How taken: Impressed at Cork - When taken: 6 Mar 1814 - Date received: 26 Mar 1814 - From what ship: Earl Spencer, cutter - Born: New York - Age: 23 - Discharged on 10 May 1814 and sent to Dartmoor.

Bright, George - Seaman - Number: 1360 - Prize name: Fox - Ship type: LM - How taken: HM Sloop Pheasant - When taken: 23 Apr 1813 - Where taken: Bay of Biscay - Date received: 15 May 1813 - From what ship: HMS Foxhound - Born: New Jersey - Age: 22 - Discharged on 3 Jul 1813 and sent to Stapleton Prison.

Brightman, John - Seaman - Number: 2518 - Prize name: Bunker Hill - Ship type: P - How taken: HM Frigate Pomone & HM Frigate Cydnus - When taken: 4 Mar 1814 - Where taken: Bay of Biscay - Date received: 4 Apr 1814 - From what ship: HMS Virago - Born: Portsmouth - Age: 24 - Race: Black - Discharged on 10 May 1814 and sent to Dartmoor.

Brightman, Joseph - Quartermaster - Number: 25 - How taken: Delivered himself up from HM Ship-of-the-Line San Josef - Date received: 22 Nov 1812 - From what ship: HMS San Josef - Born: Boston - Age: 36 - Sent to Portsmouth on 29 Dec 1812 on the HMS Northumberland.

Brill, John - Boy - Number: 937 - Prize name: Weasel - Ship type: MV - How taken: HM Brig Foxhound - When taken: 25 Mar 1813 - Where taken: Bay of Biscay - Date received: 6 Apr 1813 - From what ship: HMS Foxhound - Born: New York - Age: 13 - Sent on 21 Jun 1813 to Portsmouth on HMS Prometheus.

Bristol, Nehemiah - Chief Mate - Number: 1036 - Prize name: Amphitrite - Ship type: MV - How taken: HM Ketch Gleaner - When taken: 27 Feb 1813 - Where taken: Bay of Biscay - Date received: 21 Apr 1813 - From what ship: HMS Silvia - Born: Milford - Age: 27 - Sent on 23 Apr 1813 to Ashburton on parole.

Broadwater, Samuel - Seaman - Number: 1270 - Prize name: Caroline - Ship type: MV - How taken: HM Frigate Medusa - When taken: 12 Apr 1813 - Where taken: Bay of Biscay - Date received: 10 May 1813 - From what ship: HMS Medusa - Born: Maryland - Age: 21 - Discharged on 3 Jul 1813 and sent to Stapleton Prison.

Bronston, Job Ellis - Seaman - Number: 630 - Prize name: Mars - Ship type: MV - How taken: HM Ship-of-the-Line Warspite - When taken: 26 Feb 1813 - Where taken: Bay of Biscay - Date received: 4 Mar 1813 - From what ship: Mars - Born: Buckberry - Age: 22 - Sent to Dartmoor on 2 Apr 1813.

Brooks, Hayden Theophilus - Seaman - Number: 1203 - Prize name: Zebra - Ship type: LM - How taken: HM Frigate Pyramus - When taken: 20 Apr 1813 - Where taken: Bay of Biscay - Date received: 9 May 1813 - From what ship: HMS Andromache - Born: New York - Age: 22 - Sent on 3 Jul 1813 to Stapleton prison.

Brooks, Russell - Seaman - Number: 2553 - Prize name: Bunker Hill - Ship type: P - How taken: HM Frigate Pomone & HM Frigate Cydnus - When taken: 4 Mar 1814 - Where taken: Bay of Biscay - Date received: 17 Apr 1814 - From what ship: HMS Teazer - Born: East Haddam - Age: 33 - Discharged on 10 May 1814 and sent to Dartmoor.

Brooks, Thomas - Seaman - Number: 2492 - How taken: Impressed at Cork - When taken: Feb 1814 - Date received:

26 Mar 1814 - From what ship: Earl Spencer, cutter - Born: Virginia - Age: 23 - Discharged on 10 May 1814 and sent to Dartmoor.

Brown, Alexander - Marine - Number: 1872 - Prize name: US Brig Argus - Ship type: War - How taken: HM Brig Pelican - When taken: 14 Aug 1813 - Where taken: Irish Channel - Date received: 23 Aug 1813 - From what ship: HMS Pelican - Born: Norway - Age: 37 - Discharged on 8 Sep 1813 and sent to Dartmoor.

Brown, Ambrose James - Master - Number: 2410 - Prize name: Joseph - Ship type: MV - How taken: HM Brig Royalist - When taken: 18 Jan 1814 - Where taken: Bay of Biscay - Date received: 7 Feb 1814 - From what ship: Joseph -

Born: Marblehead - Age: 29 - Discharged on 7 Feb 1814 and sent to Ashburton on parole.

Brown, Arn - Seaman - Number: 1507 - Prize name: Grand Napoleon - Ship type: MV - How taken: HM Brig Goldfinch - When taken: 17 Apr 1813 - Where taken: Bay of Biscay - Date received: 29 May 1813 - From what ship: HMS Hannibal - Born: Rhode Island - Age: 38 - Discharged on 8 Sep 1813 and sent to Dartmoor.

Brown, Benjamin - Seaman - Number: 816 - Prize name: Decornau - Ship type: MV - How taken: HM Sloop Pheasant - When taken: 15 Mar 1813 - Where taken: Bay of Biscay - Date received: 19 Mar 1813 - From what ship: HMS Pheasant - Born: New Haven - Age: 29 - Sent to Dartmoor on 28 Jun 1813.

Brown, Benjamin - Seaman - Number: 1683 - Prize name: Revenge - Ship type: LM - How taken: HM Frigate Belle Poule - When taken: 11 May 1813 - Where taken: Bay of Biscay - Date received: 26 Jun 1813 - From what ship: HMS Duncan - Born: Westborough - Age: 21 - Discharged on 30 Jun 1813 and sent to Stapleton.

Brown, Charles - Seaman - Number: 1467 - Prize name: Paul Jones - Ship type: P - How taken: HM Frigate Leonidas - When taken: 23 May 1813 - Where taken: off Cape Clear - Date received: 26 May 1813 - From what ship: HMS Leonidas - Born: Virginia - Age: 23 - Race: Black - Discharged on 30 Jun 1813 and sent to Stapleton.

Brown, David - Seaman - Number: 2226 - Prize name: Zephyr - Ship type: MV - How taken: HM Frigate Pyramus - When taken: 30 Nov 1813 - Where taken: off Lorient - Date received: 3 Jan 1814 - From what ship: HMS Warspite - Born: Maryland - Age: 35 - Race: Negro - Discharged on 10 May 1814 and sent to Dartmoor.

Brown, Ebenezer - Seaman - Number: 1966 - Prize name: Joel Barlow - Ship type: LM - How taken: HM Frigate Briton - When taken: 3 Jul 1813 - Where taken: off Bordeaux - Date received: 31 Aug 1813 - From what ship: HMS Clarence - Born: Rhode Island - Age: 19 - Race: Black - Discharged on 8 Sep 1813 and sent to Dartmoor.

Brown, Francis - Chief Mate - Number: 1798 - Prize name: Godfrey & Mary - Ship type: MV - How taken: Robert Todd, privateer - When taken: 23 Jun 1813 - Where taken: Lat 21.3N Long 53W - Date received: 16 Aug 1813 - From what ship: HMS Bittern - Born: New London - Age: 22 - Discharged on 8 Sep 1813 and sent to Dartmoor.

Brown, George - Prize Master - Number: 1564 - Prize name: Margaret, prize of the Privateer True Blooded Yankee - Ship type: P - How taken: HM Brig Nimrod - When taken: 9 Mar 1813 - Where taken: off Morant Bay, Jamaica - Date received: 6 Jun 1813 - From what ship: Dartmoor prison - Born: Wellfleet - Age: 37 - Discharged on 29 Jun 1813 and sent to London.

Brown, George - Prize Master - Number: 668 - Prize name: Margaret, prize of the Privateer True Blooded Yankee - Ship type: P - How taken: HM Brig Nimrod - When taken: 9 Mar 1813 - Where taken: off Morant Bay, Jamaica - Date received: 14 Mar 1813 - From what ship: HMS Salvador del Mundo - Born: Wellfleet - Age: 37 - Sent to Dartmoor on 1 Jul 1813.

Brown, Henry - Seaman - Number: 1734 - Prize name: Criterion - Ship type: MV - How taken: HM Frigate Belle Poule - When taken: 14 Feb 1813 - Where taken: Bay of Biscay - Date received: 10 Jul 1813 - From what ship: Dartmoor Prison - Born: New York - Age: 17 - Discharged on 10 Jul 1813 and released to HMS Salvador del Mundo.

Brown, Henry - Boy - Number: 648 - Prize name: Criterion - Ship type: MV - How taken: HM Frigate Belle Poule - When taken: 14 Feb 1813 - Where taken: Bay of Biscay - Date received: 4 Mar 1813 - From what ship: HMS Strenuous - Born: New York - Age: 17 - Sent to Dartmoor on 2 Apr 1813.

Brown, James - Boatswain - Number: 841 - Prize name: Pallas - Ship type: MV - How taken: HM Brig Rebuff - When taken: 23 Dec 1812 - Where taken: off Cadiz - Date received: 23 May 1813 - From what ship: HMS Dauntless - Born: Wethersfield - Age: 31 - Sent to Dartmoor on 28 Jun 1813.

Brown, Jesse - 2nd Mate - Number: 872 - Prize name: Star - Ship type: MV - How taken: HM Ship-of-the-Line Superb - When taken: 9 Feb 1813 - Where taken: Bay of Biscay - Date received: 28 Mar 1813 - From what ship: HMS Warspite - Born: Connecticut - Age: 26 - Sent to Dartmoor on 28 Jun 1813.

Brown, Jesse - Boy - Number: 206 - Prize name: Hunter - Ship type: P - How taken: HM Frigate Phoebe - When taken: 24 Dec 1812 - Where taken: off Western Islands - Date received: 9 Jan 1813 - From what ship: HMS Phoebe - Born: Boston - Age: 17 - Sent to Portsmouth on 8 Feb 1813 on the HMS Colossus.

Brown, Jesse - Seaman - Number: 2600 - How taken: HM Ship-of-the-Line Trident - When taken: 28 Oct 1812 - Date received: 16 May 1814 - From what ship: HMS Repulse - Born: New York - Age: 27 - Race: Mulatto - Discharged on 14 Jun 1814 and sent to Dartmoor.

Brown, John - Seaman - Number: 1300 - Prize name: Price - Ship type: LM - How taken: HM Frigate Medusa - When taken: 13 Apr 1813 - Where taken: Bay of Biscay - Date received: 10 May 1813 - From what ship: HMS Medusa - Born: New Jersey - Age: 23 - Discharged on 3 Jul 1813 and sent to Stapleton Prison.

Brown, John - Seaman - Number: 68 - Prize name: Goldfinch - Ship type: MV - How taken: HM Brig Goldfinch - When taken: 27 Nov 1812 - Where taken: Sound - Date received: 29 Nov 1812 - From what ship: HMS Wasp - Born: Boston - Age: 34 - Sent to Portsmouth on 29 Dec 1812 on the HMS Northumberland.

Brown, John - 2nd Mate - Number: 108 - Prize name: Otter - Ship type: MV - How taken: HM Ship-Sloop Jalousie - When taken: 1 Dec 1812 - Where taken: off Cape Vincent - Date received: 30 Dec 1812 - From what ship: HMS Leonidas - Born: Marblehead - Age: 34 - Sent to Portsmouth on 4 Jan 1813 on the HMS Revolutionnaire.

Brown, John - Chief Mate - Number: 664 - Prize name: Nope - Ship type: MV - How taken: Chance, privateer - When taken: 15 Feb 1813 - Where taken: off Gorduan - Date received: 10 Mar 1813 - From what ship: Sloop Ann - Born: Rhode Island - Age: 27 - Sent on 12 Feb 1813 to Ashburton on parole.

Brown, John - Boy - Number: 205 - Prize name: Hunter - Ship type: P - How taken: HM Frigate Phoebe - When taken: 24 Dec 1812 - Where taken: off Western Islands - Date received: 9 Jan 1813 - From what ship: HMS Phoebe - Born: New York - Age: 15 - Sent to Portsmouth on 8 Feb 1813 on the HMS Colossus.

Brown, John - Seaman - Number: 598 - Prize name: St. Martin's Plantation, prize of the Privateer Paul Jones - Ship type: P - How taken: HM Ship-of-the-Line Dublin - When taken: 9 Feb 1815 - Where taken: Lat 43 N Long 33.5 W - Date received: 25 Feb 1813 - From what ship: HMS Dublin - Born: New York - Age: 19 - Sent to Dartmoor on 8 Sep 1813.

Brown, John - Seaman - Number: 2453 - Prize name: US Brig Argus - Ship type: War - How taken: HM Brig Pelican - When taken: 14 Aug 1813 - Where taken: Irish Channel - Date received: 15 Feb 1814 - From what ship: HMS Salvador del Mundo - Born: Ireland - Age: 23 - Discharged on 20 Jun 1814 and sent to HMS Salvador del Mundo.

Brown, John - Seaman - Number: 1902 - Prize name: US Brig Argus - Ship type: War - How taken: HM Brig Pelican - When taken: 14 Aug 1813 - Where taken: Irish Channel - Date received: 23 Aug 1813 - From what ship: HMS Pelican - Born: Ireland - Age: 23 - Discharged on 24 Aug 1813 and released to HMS Salvador del Mundo.

Brown, John William - Seaman - Number: 1369 - Prize name: Tom - Ship type: LM - How taken: HM Frigate Surveillante - When taken: 27 Apr 1813 - Where taken: Bay of Biscay - Date received: 15 May 1813 - From what ship: HMS Foxhound - Born: Albany - Age: 27 - Discharged on 3 Jul 1813 and sent to Stapleton Prison.

Brown, Lodwick - Seaman - Number: 818 - Prize name: Decornau - Ship type: MV - How taken: HM Sloop Pheasant - When taken: 15 Mar 1813 - Where taken: Bay of Biscay - Date received: 19 Mar 1813 - From what ship: HMS Pheasant - Born: Virginia - Age: 38 - Sent to Dartmoor on 28 Jun 1813.

Brown, Richard - Cook - Number: 2093 - Prize name: Indian - Ship type: MV - How taken: Impressed at Liverpool - When taken: 14 Sep 1813 - Date received: 3 Nov 1813 - From what ship: HMS Bittern - Born: Baltimore - Age: 38 - Race: Black - Discharged on 29 Nov 1813 and sent to Dartmoor.

Brown, Samuel - Seaman - Number: 1228 - Prize name: Essex - Ship type: MV - How taken: HM Frigate Pyramus - When taken: 2 Apr 1813 - Where taken: Bay of Biscay - Date received: 9 May 1813 - From what ship: HMS Andromache - Born: New Hampshire - Age: 19 - Discharged on 3 Jul 1813 and sent to Stapleton Prison.

Brown, Samuel - Seaman - Number: 820 - Prize name: Decornau - Ship type: MV - How taken: HM Sloop Pheasant - When taken: 15 Mar 1813 - Where taken: Bay of Biscay - Date received: 19 Mar 1813 - From what ship: HMS

Pheasant - Born: New Orleans - Age: 38 - Sent to Dartmoor on 28 Jun 1813.

Brown, Stephen - 4th Lieutenant - Number: 2274 - Prize name: Siro - Ship type: LM - How taken: HM Brig Pelican - When taken: 13 Jan 1814 - Where taken: at sea - Date received: 23 Jan 1814 - From what ship: HMS Pelican - Born: Laban - Age: 22 - Discharged on 31 Jan 1814 and sent to Dartmoor.

Brown, Sandy - Seaman - Number: 2302 - Prize name: Siro - Ship type: LM - How taken: HM Brig Pelican - When taken: 13 Jan 1814 - Where taken: at sea - Date received: 23 Jan 1814 - From what ship: HMS Pelican - Born: New York - Age: 25 - Discharged on 31 Jan 1814 and sent to Dartmoor.

Brown, Thomas - Master - Number: 106 - Prize name: Otter - Ship type: MV - How taken: HM Ship-Sloop Jalousie - When taken: 1 Dec 1812 - Where taken: off Cape Vincent - Date received: 30 Dec 1812 - From what ship: HMS Leonidas - Born: Ipswich - Age: 39 - Sent on 8 Dec 1812 to Ashburton on parole.

Brown, Thomas - Prize Master - Number: 1748 - Prize name: Fox Packet, prize of the Privateer Fox - Ship type: P - How taken: Superior, letter of marque - When taken: 25 Jun 1813 - Where taken: Lat 50N Long 21W - Date received: 19 Jul 1813 - From what ship: HMS Bittern - Born: Portsmouth - Age: 20 - Discharged on 10 Jul 1813 and released to HMS Salvador del Mundo.

Brown, Thomas - Seaman - Number: 2598 - How taken: HM Ship-of-the-Line Trident - When taken: 28 Oct 1812 - Date received: 16 May 1814 - From what ship: HMS Repulse - Born: Philadelphia - Age: 39 - Discharged on 14 Jun 1814 and sent to Dartmoor.

Brown, Wheeler - Carpenter - Number: 362 - Prize name: Orbit - Ship type: MV - How taken: HM Brig Achates - When taken: 29 Jan 1813 - Where taken: Lat 49 N Long 13 W - Date received: 4 Feb 1813 - From what ship: HMS Achates - Born: New London - Age: 43 - Sent to Chatham on 29 Mar 1813 on the HMS Braham.

Brown, William - Seaman - Number: 741 - Prize name: Charlotte - Ship type: MV - How taken: HM Ship-of-the-Line Warspite - When taken: 3 Mar 1813 - Where taken: Bay of Biscay - Date received: 19 Mar 1813 - From what ship: HMS Warspite - Born: New York - Age: 30 - Sent to Dartmoor on 2 Apr 1813.

Brown, William - Seaman - Number: 1171 - Prize name: Hebe - Ship type: MV - How taken: HM Frigate Stag - When taken: 18 Apr 1813 - Where taken: Bay of Biscay - Date received: 6 May 1813 - From what ship: HMS Stag - Born: Copenhagen - Age: 26 - Sent on 3 Jul 1813 to Stapleton prison.

Brown, William - Cook - Number: 1367 - Prize name: Tom - Ship type: LM - How taken: HM Frigate Surveillante - When taken: 27 Apr 1813 - Where taken: Bay of Biscay - Date received: 15 May 1813 - From what ship: HMS Foxhound - Born: New York - Age: 33 - Race: Black - Discharged on 3 Jul 1813 and sent to Stapleton Prison.

Brown, William - Seaman - Number: 1972 - How taken: Impressed at Liverpool - When taken: 8 Sep 1813 - Date received: 19 Sep 1813 - From what ship: HMS Bittern - Born: New Castle - Age: 25 - Discharged on 27 Sep 1813 and sent to Dartmoor.

Brown, William - Seaman - Number: 1974 - How taken: Out of an American licensed ship - When taken: 27 Aug 1813 - Date received: 19 Sep 1813 - From what ship: HMS Bittern - Born: Philadelphia - Age: 28 - Discharged on 27 Sep 1813 and sent to Dartmoor.

Brown, William - Seaman - Number: 1776 - Prize name: Matilda - Ship type: P - How taken: HM Frigate Revolutionnaire - When taken: 27 Jul 1813 - Where taken: off Larvine - Date received: 31 Jul 1813 - From what ship: Matilda, prize - Born: Virginia - Age: 20 - Discharged on 8 Sep 1813 and sent to Dartmoor.

Brownell, Richard - Seaman - Number: 812 - Prize name: Decornau - Ship type: MV - How taken: HM Sloop Pheasant - When taken: 15 Mar 1813 - Where taken: Bay of Biscay - Date received: 19 Mar 1813 - From what ship: HMS Pheasant - Born: Massachusetts - Age: 24 - Sent to Dartmoor on 28 Jun 1813.

Brutus, Mario - Seaman - Number: 2401 - Prize name: Rachel & Ann, prize of the Privateer Prince de Neufchatel - Ship type: P - How taken: HM Frigate Cydnus - When taken: 6 Jan 1814 - Where taken: at sea - Date received: 4 Feb 1814 - From what ship: HMS Cydnus - Born: New Orleans - Age: 29 - Discharged on 10 May 1814 and sent to Dartmoor.

Bryant, James - Seaman - Number: 999 - Prize name: Polly - Ship type: MV - How taken: HM Frigate Surveillante - When taken: 27 Mar 1813 - Where taken: Bay of Biscay - Date received: 16 Apr 1813 - From what ship: HMS Fairy - Born: Beverly - Age: 18 - Sent to Chatham on 8 Jul 1813 on HM Tender Neptune.

Buchanan, James - Master - Number: 232 - Prize name: Empress - Ship type: MV - How taken: HM Brig Rover - When taken: 29 Nov 1812 - Where taken: at Coruna, Spain - Date received: 11 Jan 1813 - From what ship: HMS Rover - Born: New York - Age: 28 - Sent on 12 Jan 1813 to Ashburton on parole.

Buchanan, John - Seaman - Number: 1987 - Prize name: Ned - Ship type: LM - How taken: HM Brig Royalist - When taken: 6 Sep 1812 - Where taken: coast of France - Date received: 22 Sep 1813 - From what ship: HMS Royalist - Born: Boston - Age: 24 - Discharged on 27 Sep 1813 and sent to Dartmoor.

Bucher, William Palmer - Seaman - Number: 910 - Prize name: Prompt - Ship type: MV - How taken: Chance, privateer - When taken: 22 Mar 1813 - Where taken: Bay of Biscay - Date received: 3 Apr 1813 - From what ship: Mary - Born: New Haven - Age: 16 - Sent on 21 Jun 1813 to Portsmouth on HMS Prometheus.

Buckhannon, Edwin - Master - Number: 1949 - Prize name: Joel Barlow - Ship type: LM - How taken: HM Frigate Briton - When taken: 3 Jul 1813 - Where taken: off Bordeaux - Date received: 30 Aug 1813 - From what ship: HMS Briton - Born: Westminster - Age: 27 - Discharged on 1 Sep 1813 and sent to Ashburton on parole.

Bunaberg, Nicholas - 2nd Mate - Number: 1062 - Prize name: Meteor - Ship type: MV - How taken: HM Frigate Briton - When taken: 12 Mar 1813 - Where taken: Bay of Biscay - Date received: 22 Apr 1813 - From what ship: HMS Superb - Born: Karlskrona, Sweden - Age: 25 - Liberated on 21 May 1813.

Bunker, Isaiah - Prize Master - Number: 2060 - Prize name: Betsy, prize to the Privateer True Blooded Yankee - Ship type: P - How taken: HM Frigate Eurotas - When taken: 26 Oct 1813 - Where taken: off Ushant - Date received: 1 Nov 1813 - From what ship: HMS Hannibal - Born: New Castle, DE - Age: 23 - Discharged on 3 Nov 1813 and sent to Dartmoor.

Bunker, James - Seaman - Number: 764 - Prize name: William Bayard - Ship type: MV - How taken: HM Ship-of-the-Line Warspite - When taken: 3 Mar 1813 - Where taken: Bay of Biscay - Date received: 19 Mar 1813 - From what ship: HMS Warspite - Born: Massachusetts - Age: 27 - Sent to Dartmoor on 28 Jun 1813.

Bunker, Peter - Seaman - Number: 321 - Prize name: Porcupine - Ship type: MV - How taken: HM Frigate Dryad - When taken: 8 Jan 1813 - Where taken: off Bordeaux - Date received: 21 Jan 1813 - From what ship: HMS Abercrombie - Born: Nantucket - Age: 32 - Sent to Chatham on 29 Mar 1813 on the HMS Braham.

Bunkerson, John - Seaman - Number: 2343 - Prize name: Apparencen, prize of the Privateer Bunker Hill - Ship type: P - How taken: Cartdian - When taken: 27 Jan 1814 - Where taken: off Ushant - Date received: 30 Jan 1814 - From what ship: Cardian - Born: New Orleans - Age: 18 - Discharged on 31 Jan 1814 and sent to Dartmoor.

Bunoth, Mansfield - Boy - Number: 208 - Prize name: Hunter - Ship type: P - How taken: HM Frigate Phoebe - When taken: 24 Dec 1812 - Where taken: off Western Islands - Date received: 9 Jan 1813 - From what ship: HMS Phoebe - Born: Salem - Age: 17 - Sent to Portsmouth on 8 Feb 1813 on the HMS Colossus.

Burbank, G. W. - 1st Lieutenant - Number: 1496 - Prize name: Paul Jones - Ship type: P - How taken: HM Frigate Leonidas - When taken: 23 May 1813 - Where taken: off Cape Clear - Date received: 27 May 1813 - From what ship: Paul Jones - Born: Suffield - Age: 35 - Discharged on 30 May 1813 and sent to Ashburton on parole.

Burch, James - Seaman - Number: 434 - Prize name: Resolution - Ship type: MV - How taken: Hibernia, letter of marque - When taken: 21 Sep 1812 - Where taken: off Bermuda - Date received: 5 Feb 1813 - From what ship: HMS Neptune - Born: New Jersey - Age: 26 - Sent to Chatham on 29 Mar 1813 on the HMS Braham.

Burdge, Samuel - Cabin Boy - Number: 826 - Prize name: Decornau - Ship type: MV - How taken: HM Sloop Pheasant - When taken: 15 Mar 1813 - Where taken: Bay of Biscay - Date received: 21 Mar 1813 - From what ship: HMS Pheasant - Born: New York - Age: 20 - Sent to Dartmoor on 28 Jun 1813.

Burgman, John - Seaman - Number: 1560 - Prize name: Courier - Ship type: LM - How taken: HM Brig Rover - When taken: 14 Mar 1813 - Where taken: Bay of Biscay - Date received: 29 May 1813 - From what ship: HMS Hannibal - Born: Maryland - Age: 25 - Race: Colored - Discharged on 30 Jun 1813 and sent to Stapleton.

Burn, Reuben - Seaman - Number: 1071 - Prize name: Meteor - Ship type: MV - How taken: HM Frigate Briton - When taken: 12 Mar 1813 - Where taken: Bay of Biscay - Date received: 22 Apr 1813 - From what ship: HMS Superb - Born: New York - Age: 16 - Sent to Dartmoor on 8 Sep 1813.

Burnham, Enoch - Seaman - Number: 1242 - Prize name: Essex - Ship type: MV - How taken: HM Frigate Pyramus - When taken: 2 Apr 1813 - Where taken: Bay of Biscay - Date received: 9 May 1813 - From what ship: HMS

Andromache - Born: Boston - Age: 22 - Discharged on 3 Jul 1813 and sent to Stapleton Prison.

Burnham, Francis A. - Master - Number: 560 - Prize name: Spitfire - Ship type: MV - How taken: HM Brig Achates - When taken: 14 Feb 1814 - Where taken: off Ushant - Date received: 21 Feb 1813 - From what ship: HMS Achates - Born: Marblehead - Age: 25 - Sent on 22 Feb 1813 to Ashburton on parole.

Burns, Charles - Seaman - Number: 1263 - Prize name: Caroline - Ship type: MV - How taken: HM Frigate Medusa - When taken: 12 Apr 1813 - Where taken: Bay of Biscay - Date received: 10 May 1813 - From what ship: HMS Medusa - Born: Salem - Age: 20 - Race: Negro - Discharged on 3 Jul 1813 and sent to Stapleton Prison.

Burr, Francis L. - Seaman - Number: 1091 - Prize name: Young Holkar - Ship type: MV - How taken: HM Ship-of-the-Line Superb - When taken: 10 Apr 1813 - Where taken: off Belle Isle - Date received: 22 Apr 1813 - From what ship: HMS Superb - Born: Connecticut - Age: 19 - Sent to Dartmoor on 1 Jul 1813.

Burr, Isaac - Seaman - Number: 2338 - Prize name: Young Holkar - Ship type: MV - How taken: HM Ship-of-the-Line Superb - When taken: 10 Apr 1813 - Where taken: off Belle Isle - Date received: 29 Jan 1814 - From what ship: Dartmoor prison - Born: Connecticut - Age: 19 - Discharged on 3 Feb 1814 and sent to Dartmoor.

Burrett, Benjamin - Seaman - Number: 1686 - Prize name: Revenge - Ship type: LM - How taken: HM Frigate Belle Poule - When taken: 11 May 1813 - Where taken: Bay of Biscay - Date received: 26 Jun 1813 - From what ship: HMS Duncan - Born: New Haven - Age: 34 - Discharged on 30 Jun 1813 and sent to Stapleton.

Bursted, John - Marine - Number: 1477 - Prize name: Paul Jones - Ship type: P - How taken: HM Frigate Leonidas - When taken: 23 May 1813 - Where taken: off Cape Clear - Date received: 26 May 1813 - From what ship: HMS Leonidas - Born: New York - Age: 20 - Discharged on 30 Jun 1813 and sent to Stapleton.

Burton, John - Seaman - Number: 839 - Prize name: Pallas - Ship type: MV - How taken: HM Brig Rebuff - When taken: 23 Dec 1812 - Where taken: off Cadiz - Date received: 22 Mar 1813 - From what ship: HMS Dauntless - Born: Trieste - Age: 29 - Sent to Dartmoor on 28 Jun 1813.

Butler, George - Seaman - Number: 1584 - Prize name: Orders in Council - Ship type: LM - How taken: HM Frigate Surveillante - When taken: 1 Jun 1813 - Where taken: Bay of Biscay - Date received: 13 Jun 1813 - From what ship: Forvey - Born: Massachusetts - Age: 28 - Discharged on 30 Jun 1813 and sent to Stapleton.

Butman, Charles - Seaman - Number: 1649 - Prize name: Tickler - Ship type: LM - How taken: HM Frigate Magiciene - When taken: 5 Jun 1813 - Where taken: Lat 47 Long 13 - Date received: 14 Jun 1813 - From what ship: HMS Orestes - Born: Massachusetts - Age: 24 - Discharged on 30 Jun 1813 and sent to Stapleton.

Butnet, William - Seaman - Number: 1922 - Prize name: US Brig Argus - Ship type: War - How taken: HM Brig Pelican - When taken: 14 Aug 1813 - Where taken: Irish Channel - Date received: 23 Aug 1813 - From what ship: HMS Pelican - Born: New Orleans - Age: 33 - Race: Black - Discharged on 8 Sep 1813 and sent to Dartmoor.

Butts, Joseph - Seaman - Number: 2457 - Prize name: Fair American - Ship type: MV - How taken: HM Frigate Andromache - When taken: 19 Jan 1814 - Where taken: Bay of Biscay - Date received: 22 Feb 1814 - From what ship: HMS York - Born: New York - Age: 22 - Discharged on 10 May 1814 and sent to Dartmoor.

Bymer, George - Seaman - Number: 2091 - Prize name: Hepa - Ship type: MV - How taken: HM Brig Zenobia - When taken: 28 Jun 1813 - Where taken: off Lisbon - Date received: 2 Nov 1813 - From what ship: HMS Bacchus - Born: Philadelphia - Age: 27 - Discharged on 3 Nov 1813 and sent to Dartmoor.

Cadwell, James - Seaman - Number: 973 - Prize name: Ferox - Ship type: MV - How taken: HM Frigate Medusa & HM Brig Lyra - When taken: 28 Mar 1813 - Where taken: off Cape Ortagle - Date received: 9 Apr 1813 - From what ship: HMS Lyra - Born: Hartford - Age: 25 - Sent to Chatham on 8 Jul 1813 on HM Tender Neptune.

Cadwell, James - Seaman - Number: 2683 - How taken: Sent into custody from HM Ship-of-the-Line Minden - Date received: 5 Jun 1814 - From what ship: HMS Gronville - Born: Newport - Age: 28 - Discharged on 14 Jun 1814 and sent to Dartmoor.

Cainton, Joseph - Seaman - Number: 1345 - Prize name: Fox - Ship type: LM - How taken: HM Sloop Pheasant - When taken: 23 Apr 1813 - Where taken: Bay of Biscay - Date received: 14 May 1813 - From what ship: HMS Pleasant - Born: Spain - Age: 31 - Discharged on 3 Jul 1813 and sent to Stapleton Prison.

Calder, John H. - Seaman - Number: 881 - Prize name: Tiger - Ship type: MV - How taken: HM Brig Scylla - When taken: 22 Mar 1813 - Where taken: Bay of Biscay - Date received: 1 Apr 1813 - From what ship: HMS Scylla -

Born: New York - Age: 22 - Sent on 21 Jun 1813 to Portsmouth.

Caldwell, James - Seaman - Number: 727 - Prize name: Mars - Ship type: MV - How taken: HM Ship-of-the-Line Warspite - When taken: 26 Feb 1813 - Where taken: Bay of Biscay - Date received: 19 Mar 1813 - From what ship: HMS Warspite - Born: West Nottingham - Age: 28 - Sent to Dartmoor on 2 Apr 1813.

Caldwell, Samuel - Seaman - Number: 891 - Prize name: Tiger - Ship type: MV - How taken: HM Brig Scylla - When taken: 22 Mar 1813 - Where taken: Bay of Biscay - Date received: 1 Apr 1813 - From what ship: HMS Scylla - Born: New York - Age: 19 - Sent on 21 Jun 1813 to Portsmouth on HMS Prometheus.

Calfax, William - Seaman - Number: 836 - Prize name: Pallas - Ship type: MV - How taken: HM Brig Rebuff - When taken: 23 Dec 1812 - Where taken: off Cadiz - Date received: 22 Mar 1813 - From what ship: HMS Dauntless - Born: New Jersey - Age: 21 - Sent to Dartmoor on 28 Jun 1813.

Calhoun, Richard - Seaman - Number: 1346 - Prize name: Fox - Ship type: LM - How taken: HM Sloop Pheasant - When taken: 23 Apr 1813 - Where taken: Bay of Biscay - Date received: 14 May 1813 - From what ship: HMS Pleasant - Born: New York - Age: 19 - Discharged on 3 Jul 1813 and sent to Stapleton Prison.

Calkings, Zera - Seaman - Number: 273 - Prize name: Brutus - Ship type: MV - How taken: HM Frigate Briton - When taken: Jan 1813 - Where taken: Bay of Biscay - Date received: 21 Jan 1813 - From what ship: HMS Briton - Born: Connecticut - Age: 26 - Sent to Portsmouth on 8 Feb 1813 on the HMS Colossus.

Cammon, Robert - Seaman - Number: 1083 - Prize name: John & Frances - Ship type: MV - How taken: HM Frigate Belle Poule - When taken: 19 Mar 1813 - Where taken: Bay of Biscay - Date received: 22 Apr 1813 - From what ship: HMS Superb - Born: Westchester, NY - Age: 27 - Sent to Dartmoor on 1 Jul 1813.

Campbell, Abraham - Seaman - Number: 2080 - Prize name: captured French frigate - How taken: HM Ship-of-the-Line Rippon, HM Brig Scylla & HM Brig Royalist - When taken: 21 Oct 1813 - Where taken: off Ushant - Date received: 31 Oct 1813 - From what ship: HMS Rippon - Discharged on 29 Nov 1813 and sent to Dartmoor.

Campbell, Alexander - Master - Number: 2348 - Prize name: Minerva - Ship type: MV - How taken: HM Ship-of-the-Line Conquestador - When taken: 20 Jan 1814 - Where taken: Lat 47.12 Long 7.19 - Date received: 31 Jan 1814 - From what ship: HMS Surveillante - Born: Charlestown - Age: 39 - Discharged on 31 Jan 1814 and sent to Ashburton on parole.

Campbell, James - Chief Mate - Number: 2258 - Prize name: Squirrel - Ship type: MV - How taken: HM Frigate Belle Poule - When taken: 14 Dec 1813 - Where taken: Bay of Biscay - Date received: 18 Jan 1814 - From what ship: HMS Teazer - Born: Philadelphia - Age: 24 - Discharged on 29 Jan 1814 and sent to Ashburton on parole.

Campbell, John - Seaman - Number: 1490 - How taken: Impressed by HM Transport Deptford - When taken: 3 May 1813 - Where taken: Dublin - Date received: 26 May 1813 - From what ship: HMS Neptune - Discharged on 30 Jun 1813 and sent to Stapleton.

Campbell, John - Seaman - Number: 1129 - Prize name: Viper - Ship type: MV - How taken: HM Ship-of-the-Line Superb - When taken: 15 Apr 1813 - Where taken: Bay of Biscay - Date received: 22 Apr 1813 - From what ship: HMS Superb - Born: Norfolk - Age: 46 - Sent to Dartmoor on 1 Jul 1813.

Campbell, John - Seaman - Number: 2139 - Prize name: Amiable - Ship type: LM - How taken: HM Ship-of-the-Line Magnificant - When taken: 30 Oct 1813 - Where taken: off Lorient - Date received: 29 Nov 1813 - From what ship: HMS Dublin - Born: Delaware - Age: 18 - Discharged on 4 Dec 1813 and sent to Dartmoor.

Campbell, Reynold - Seaman - Number: 1353 - Prize name: Shadow - Ship type: LM - How taken: HM Brig Reindeer & HM Schooner Helicon - When taken: 6 Apr 1813 - Where taken: Bay of Biscay - Date received: 14 May 1813 - From what ship: HMS Reindeer - Born: Philadelphia - Age: 35 - Discharged on 3 Jul 1813 and sent to Stapleton Prison.

Campbell, William - Cook - Number: 2372 - Prize name: Minerva - Ship type: MV - How taken: HM Ship-of-the-Line Conquestador - When taken: 19 Dec 1814 - Where taken: Bay of Biscay - Date received: 31 Jan 1814 - From what ship: HMS Surveillante - Born: Charlestown - Age: 27 - Race: Black - Discharged on 27 Feb 1814 and sent to Chatham on HMS Haleyon.

Can, Moses - 1st Mate - Number: 315 - Prize name: Porcupine - Ship type: MV - How taken: HM Frigate Dryad - When taken: 8 Jan 1813 - Where taken: off Bordeaux - Date received: 21 Jan 1813 - From what ship: HMS

Abercrombie - Born: Nantucket - Age: 34 - Sent on 22 Jan 1813 to Ashburton on parole.

Canada, Prince - Seaman - Number: 291 - Prize name: Columbia - Ship type: MV - How taken: HM Frigate Briton - When taken: 17 Dec 1812 - Where taken: off Bordeaux - Date received: 21 Jan 1813 - From what ship: HMS Abercrombie - Born: Rhode Island - Age: 39 - Race: Black - Sent to Portsmouth on 8 Feb 1813 on the HMS Colossus.

Cannoway, Charles - Seaman - Number: 1174 - Prize name: Hebe - Ship type: MV - How taken: HM Frigate Stag - When taken: 18 Apr 1813 - Where taken: Bay of Biscay - Date received: 6 May 1813 - From what ship: HMS Stag - Born: Baltimore - Age: 43 - Sent on 8 Sep 1813 to Dartmoor.

Cantrell, Norville - Armorer - Number: 1366 - Prize name: Tom - Ship type: LM - How taken: HM Frigate Surveillante - When taken: 27 Apr 1813 - Where taken: Bay of Biscay - Date received: 15 May 1813 - From what ship: HMS Foxhound - Born: New Orleans - Age: 18 - Discharged on 3 Jul 1813 and sent to Stapleton Prison.

Capewell, Bartholomew - Seaman - Number: 615 - Prize name: Governor McKean - Ship type: LM - How taken: HM Brig Rover - When taken: 26 Jun 1813 - Where taken: off Bordeaux - Date received: 2 Mar 1813 - From what ship: HMS Insolent - Born: New Orleans - Age: 25 - Sent to Dartmoor on 2 Apr 1813.

Capron, William - Seaman - Number: 272 - Prize name: Brutus - Ship type: MV - How taken: HM Frigate Briton - When taken: Jan 1813 - Where taken: Bay of Biscay - Date received: 21 Jan 1813 - From what ship: HMS Briton - Born: Nantucket - Age: 20 - Sent to Portsmouth on 8 Feb 1813 on the HMS Colossus.

Carbenett, John - Seaman - Number: 1841 - Prize name: US Brig Argus - Ship type: War - How taken: HM Brig Pelican - When taken: 14 Aug 1813 - Where taken: Irish Channel - Date received: 17 Aug 1813 - From what ship: USS Argus - Born: New York - Age: 16 - Discharged on 8 Sep 1813 and sent to Dartmoor.

Card, Nathaniel - Boatswain's Mate - Number: 163 - Prize name: Hunter - Ship type: P - How taken: HM Frigate Phoebe - When taken: 24 Dec 1812 - Where taken: off Western Islands - Date received: 9 Jan 1813 - From what ship: HMS Phoebe - Born: Marblehead - Age: 55 - Sent to Portsmouth on 8 Feb 1813 on the HMS Colossus.

Carland, Lewis - Seaman - Number: 2073 - Prize name: Avon, prize of the Privateer True Blooded Yankee - Ship type: P - How taken: HM Frigate Eurotas - When taken: 27 Oct 1813 - Where taken: off Ushant - Date received: 1 Nov 1813 - From what ship: HMS Hannibal - Born: Louisiana - Age: 28 - Discharged on 3 Nov 1813 and sent to Dartmoor.

Carles, Paul - Passenger - Number: 975 - Prize name: Tiger - Ship type: MV - How taken: HM Brig Scylla - When taken: 22 Mar 1813 - Where taken: Bay of Biscay - Date received: 11 Apr 1813 - From what ship: HMS Plymouth - Born: San Domingo (Haiti) - Age: 26 - Sent on 12 Apr 1813 to Ashburton on parole.

Carman, Francis - Seaman - Number: 456 - Prize name: Dolphin - Ship type: LM - How taken: HM Ship-of-the-Line Colossus, HM Frigate Rhin & HM Brig Goldfinch - When taken: 5 Jan 1813 - Where taken: off Western Islands - Date received: 6 Feb 1813 - From what ship: HMS Rhin - Born: New Orleans - Age: 23 - Sent to Chatham on 29 Mar 1813 on the HMS Braham.

Carnes, Joseph - Seaman - Number: 1077 - Prize name: John & Frances - Ship type: MV - How taken: HM Frigate Belle Poule - When taken: 19 Mar 1813 - Where taken: Bay of Biscay - Date received: 22 Apr 1813 - From what ship: HMS Superb - Born: Lisbon - Age: 18 - Sent to Dartmoor on 1 Jul 1813.

Carney, Edward - Boatswain's Mate - Number: 453 - Prize name: Dolphin - Ship type: LM - How taken: HM Ship-of-the-Line Colossus, HM Frigate Rhin & HM Brig Goldfinch - When taken: 5 Jan 1813 - Where taken: off Western Islands - Date received: 6 Feb 1813 - From what ship: HMS Rhin - Born: Philadelphia - Age: 39 - Sent to Chatham on 29 Mar 1813 on the HMS Braham.

Carns, John - Seaman - Number: 2178 - Prize name: Wolf, prize of the Privateer Grand Turk - Ship type: P - How taken: HM Frigate Briton - When taken: 1 Dec 1813 - Where taken: Bay of Biscay - Date received: 17 Dec 1813 - From what ship: HMS Briton - Born: Kennebunkport - Age: 20 - Discharged on 17 Dec 1813 and released to HMS Britton.

Carns, Richard - Seaman - Number: 413 - Prize name: Union - Ship type: MV - How taken: HM Frigate Iris - When taken: 17 Jan 1813 - Where taken: Lat 44 N Long 2.3 W - Date received: 5 Feb 1813 - From what ship: HMS San Josef - Born: Philadelphia - Age: 17 - Sent to Chatham on 29 Mar 1813 on the HMS Braham.

Carpenter, Henry - Seaman - Number: 1703 - How taken: Impressed at Dublin - When taken: 10 Jun 1813 - Date received: 3 Jul 1813 - From what ship: HMS Prince Frederick - Born: New York - Age: 31 - Discharged on 8 Sep 1813 and sent to Dartmoor.

Carr, James - Seaman - Number: 2358 - Prize name: Zephyr - Ship type: MV - How taken: HM Frigate Surveillante - When taken: 6 Jan 1814 - Where taken: Bay of Biscay - Date received: 31 Jan 1814 - From what ship: HMS Surveillante - Born: Cumberland - Age: 19 - Discharged on 10 May 1814 and sent to Dartmoor.

Carr, Laurence - Cook - Number: 118 - Prize name: Otter - Ship type: MV - How taken: HM Ship-Sloop Jalousie - When taken: 1 Dec 1812 - Where taken: off Cape Vincent - Date received: 30 Dec 1812 - From what ship: HMS Leonidas - Born: New York - Age: 30 - Sent to Portsmouth on 4 Jan 1813 on the HMS Revolutionnaire.

Carr, Richard - Seaman - Number: 69 - How taken: HM Battery Princess - When taken: 5 Aug 1812 - Where taken: Liverpool - Date received: 29 Nov 1812 - From what ship: HMS Salvador del Mundo - Born: Charlestown - Age: 29 - Sent to Portsmouth on 29 Dec 1812 on the HMS Northumberland.

Carrol, Robert - Seaman - Number: 194 - Prize name: Hunter - Ship type: P - How taken: HM Frigate Phoebe - When taken: 24 Dec 1812 - Where taken: off Western Islands - Date received: 9 Jan 1813 - From what ship: HMS Phoebe - Born: Boston - Age: 16 - Sent to Portsmouth on 8 Feb 1813 on the HMS Colossus.

Carrot, Charles - Seaman - Number: 2484 - How taken: Impressed at Liverpool - When taken: 8 Feb 1814 - Date received: 13 Mar 1814 - From what ship: HMS Bittern - Born: New York - Age: 25 - Race: Negro - Discharged on 10 May 1814 and sent to Dartmoor.

Carson, Robert - Seaman - Number: 236 - Prize name: John Barnes - Ship type: MV - How taken: Impressed at Liverpool - When taken: 17 Nov 1812 - Date received: 12 Jan 1813 - From what ship: HMS Bittern - Born: Cherry Hill, MD - Age: 23 - Sent to Portsmouth on 8 Feb 1813 on the HMS Colossus.

Carter, Daniel - Seaman - Number: 1542 - Prize name: Zebra - Ship type: LM - How taken: HM Frigate Pyramus - When taken: 20 Apr 1813 - Where taken: Bay of Biscay - Date received: 29 May 1813 - From what ship: HMS Hannibal - Born: Virginia - Age: 26 - Discharged on 30 Jun 1813 and sent to Stapleton.

Carter, Edward - Seaman - Number: 1541 - Prize name: Zebra - Ship type: LM - How taken: HM Frigate Pyramus - When taken: 20 Apr 1813 - Where taken: Bay of Biscay - Date received: 29 May 1813 - From what ship: HMS Hannibal - Born: Norfolk - Age: 23 - Discharged on 30 Jun 1813 and sent to Stapleton.

Carter, Enoch - Seaman - Number: 2380 - Prize name: Devon, prize to the Privateer Bunker Hill - Ship type: P - How taken: HM Brig Fly - When taken: 21 Jan 1814 - Where taken: at sea - Date received: 31 Jan 1814 - From what ship: HMS Fly - Born: Middletown - Age: 37 - Discharged on 27 Feb 1814 and sent to Chatham on HMS Haleyon.

Carter, Henry - Seaman - Number: 2395 - Prize name: Harvest, prize of the Privateer Bunker Hill - Ship type: P - How taken: HM Brig Orestes - When taken: 21 Jan 1814 - Date received: 3 Feb 1814 - From what ship: HMS Orestes - Born: New York - Age: 26 - Race: Mulatto - Discharged on 27 Feb 1814 and sent to Chatham on HMS Haleyon.

Carter, James W. - Seaman - Number: 2042 - How taken: Impressed at Liverpool - When taken: 16 Oct 1813 - Date received: 1 Nov 1813 - From what ship: HMS Bittern - Born: Long Island - Age: 20 - Discharged on 3 Nov 1813 and sent to Dartmoor.

Carter, Moses - Surgeon - Number: 145 - Prize name: Hunter - Ship type: P - How taken: HM Frigate Phoebe - When taken: 24 Dec 1812 - Where taken: off Western Islands - Date received: 9 Jan 1813 - From what ship: HMS Phoebe - Born: Concord - Age: 29 - Sent to Portsmouth on 8 Feb 1813 on the HMS Colossus.

Carton, Thomas - Carpenter - Number: 542 - Prize name: Spitfire - Ship type: MV - How taken: HM Brig Achates - When taken: 14 Feb 1813 - Where taken: off Ushant - Date received: 16 Feb 1813 - From what ship: HMS Achates - Born: Pembroke - Age: 26 - Sent to Dartmoor on 2 Apr 1813.

Cartwright, Alexander J. - Master - Number: 651 - Prize name: Amphitrite - Ship type: MV - How taken: HM Ketch Gleaner - When taken: 27 Feb 1813 - Where taken: Bay of Biscay - Date received: 5 Mar 1813 - From what ship: HMS Amphitrite - Born: Nantucket - Age: 28 - Sent on 7 Mar 1813 to Ashburton on parole.

Cartwright, George - Seaman - Number: 2161 - Prize name: Princess - Ship type: MV - How taken: Impressed at Liverpool - When taken: 1 Nov 1813 - Date received: 13 Dec 1813 - From what ship: HMS Bittern - Born: Rhode Island - Age: 34 - Discharged on 31 Jan 1814 and sent to Dartmoor.

Carver, Abraham - Seaman - Number: 329 - Prize name: Porcupine - Ship type: MV - How taken: HM Frigate Dryad - When taken: 8 Jan 1813 - Where taken: off Bordeaux - Date received: 21 Jan 1813 - From what ship: HMS Abercrombie - Born: New Jersey - Age: 26 - Sent to Chatham on 29 Mar 1813 on the HMS Braham.

Carves, John - Seaman - Number: 184 - Prize name: Hunter - Ship type: P - How taken: HM Frigate Phoebe - When taken: 24 Dec 1812 - Where taken: off Western Islands - Date received: 9 Jan 1813 - From what ship: HMS Phoebe - Born: Kennebunk - Age: 29 - Sent to Portsmouth on 8 Feb 1813 on the HMS Colossus.

Cary, John - Seaman - Number: 1634 - Prize name: Leo - Ship type: LM - How taken: HM Frigate Magiciene - When taken: 4 Jun 1813 - Where taken: Lat 45 Long 14 - Date received: 14 Jun 1813 - From what ship: HMS Orestes - Born: Brunswick - Age: 24 - Discharged on 30 Jun 1813 and sent to Stapleton.

Casey, Henry - Seaman - Number: 1004 - Prize name: Eliza - Ship type: MV - How taken: HM Frigate Surveillante - When taken: 27 Mar 1813 - Where taken: Bay of Biscay - Date received: 16 Apr 1813 - From what ship: HMS Fairy - Born: Washington - Age: 24 - Sent to Dartmoor on 8 Sep 1813.

Cassam, Michael - Seaman - Number: 2485 - How taken: Impressed at Liverpool - When taken: 8 Feb 1814 - Date received: 13 Mar 1814 - From what ship: HMS Bittern - Born: New York - Age: 28 - Race: Negro - Discharged on 10 May 1814 and sent to Dartmoor.

Castania, John Baptiste - Seaman - Number: 2347 - Prize name: Hall, prize of the Privateer Prince de Neufchatel - Ship type: P - How taken: Unknown - When taken: 10 Jan 1814 - Where taken: off Ushant - Date received: 30 Jan 1814 - From what ship: Hall - Born: Bordeaux - Age: 58 - Discharged on 31 Jan 1814 and sent to Dartmoor.

Castera, Jean S. - Passenger - Number: 2188 - Prize name: Charlotte - Ship type: MV - How taken: Cutter Dwarf - When taken: 4 Nov 1812 - Where taken: off Bordeaux - Date received: 19 Dec 1813 - From what ship: HMS Conquistador - Born: Valence, France - Age: 33 - Discharged on 25 Dec 1813 and sent to Ashburton on parole.

Castor, Charles - Boy - Number: 125 - Prize name: Columbia - Ship type: MV - How taken: HM Frigate Briton - When taken: 17 Dec 1812 - Where taken: off Bordeaux - Date received: 1 Jan 1813 - From what ship: HMS Briton - Born: Batavia - Age: 18 - Sent to Portsmouth on 4 Jan 1813 on the HMS Revolutionnaire.

Castor, Thomas - Seaman - Number: 2442 - Prize name: Mary, prize of the Privateer Prince de Neufchatel - Ship type: P - How taken: Retaken by the crew of the Mary - Date received: 15 Feb 1814 - From what ship: HMS Kangaroo - Born: Elsinore, Denmark - Age: 38 - Liberated on 23 Feb 1814.

Cato, John - Cook's Mate - Number: 1453 - Prize name: Paul Jones - Ship type: P - How taken: HM Frigate Leonidas - When taken: 23 May 1813 - Where taken: off Cape Clear - Date received: 26 May 1813 - From what ship: HMS Leonidas - Born: New London - Age: 35 - Race: Black - Discharged on 30 Jun 1813 and sent to Stapleton.

Catwood, Zenas - Seaman - Number: 2564 - How taken: Impressed at Cove - When taken: 28 Mar 1814 - Date received: 25 Apr 1814 - From what ship: HMS Helena - Born: Wellfleet - Age: 19 - Discharged on 10 May 1814 and sent to Dartmoor.

Caesar, Joseph - Seaman - Number: 832 - Prize name: Paulina - Ship type: MV - How taken: HMS Lavinia - When taken: 9 Aug 1812 - Where taken: Gibraltar - Date received: 22 Mar 1813 - From what ship: HMS Dauntless - Born: Massachusetts - Age: 17 - Sent to Dartmoor on 28 Jun 1813.

Cerk, Frederick - Boy - Number: 753 - Prize name: Charlotte - Ship type: MV - How taken: HM Ship-of-the-Line Warspite - When taken: 3 Mar 1813 - Where taken: Bay of Biscay - Date received: 19 Mar 1813 - From what ship: HMS Warspite - Born: San Domingo (Haiti) - Age: 15 - Race: Black - Sent on 4 Apr 1813 to Ashburton on parole.

Chauvet, Frederic George - Passenger - Number: 821 - Prize name: Decornau - Ship type: MV - How taken: HM Sloop Pheasant - When taken: 15 Mar 1813 - Where taken: Bay of Biscay - Date received: 20 Mar 1813 - From what ship: HMS Pheasant - Entered on 9 Apr 1813 into the French General Entry as prisoner number 39425.

Chadwick, John - Passenger - Number: 2594 - Prize name: Valentine - Ship type: MV - How taken: Detained at the Cape of Good Hope - Date received: 10 May 1814 - From what ship: HMS Lion - Born: Nantucket - Age: 32 - Discharged on 12 May 1814 and sent to Ashburton on parole.

Chadwick, M. T. - Surgeon's Mate - Number: 1416 - Prize name: Paul Jones - Ship type: P - How taken: HM Frigate Leonidas - When taken: 23 May 1813 - Where taken: off Cape Clear - Date received: 26 May 1813 - From what ship: HMS Leonidas - Born: Morris County - Age: 24 - Discharged on 30 May 1813 and sent to Ashburton on parole.

Chain, John - Seaman - Number: 2210 - Prize name: Squirrel - Ship type: MV - How taken: HM Frigate Belle Poule - When taken: 14 Dec 1813 - Where taken: Bay of Biscay - Date received: 27 Dec 1813 - From what ship: HMS Protector - Born: Maryland - Age: 23 - Discharged on 31 Jan 1814 and sent to Dartmoor.

Chambers, Elias - Boy - Number: 2417 - Prize name: Joseph - Ship type: MV - How taken: HM Brig Royalist - When taken: 18 Jan 1814 - Where taken: Bay of Biscay - Date received: 9 Feb 1814 - From what ship: HMS Sparrow - Born: Marblehead - Age: 13 - Discharged on 10 May 1814 and sent to Dartmoor.

Chambers, Henry - Seaman - Number: 1655 - Prize name: Criterion - Ship type: MV - How taken: HM Frigate Belle Poule - When taken: 14 Feb 1813 - Where taken: Bay of Biscay - Date received: 15 Jun 1813 - From what ship: Dartmoor Prison - Born: Oldenburg - Age: 24 - Discharged on 16 Jun 1813 and released to HMS Salvador del Mundo.

Champy, Peter Felix - Passenger - Number: 1577 - Prize name: Orders in Council - Ship type: LM - How taken: HM Frigate Surveillante - When taken: 1 Jun 1813 - Where taken: Bay of Biscay - Date received: 12 Jun 1813 - From what ship: Cutter Earl Wellington - Born: Charlestown - Age: 23 - Discharged on 19 Jun 1813 and sent on the American schooner Hope.

Chandler, Simon - Seaman - Number: 1233 - Prize name: Essex - Ship type: MV - How taken: HM Frigate Pyramus - When taken: 2 Apr 1813 - Where taken: Bay of Biscay - Date received: 9 May 1813 - From what ship: HMS Andromache - Born: Massachusetts - Age: 19 - Discharged on 3 Jul 1813 and sent to Stapleton Prison.

Chane, Daniel - Seaman - Number: 2657 - How taken: Sent into custody from HM Frigate Nereus - Date received: 5 Jun 1814 - From what ship: HMS Gronville - Born: Boston - Age: 25 - Discharged on 14 Jun 1814 and sent to Dartmoor.

Channing, John - Seaman - Number: 2515 - Prize name: Bunker Hill - Ship type: P - How taken: HM Frigate Pomone & HM Frigate Cydnus - When taken: 4 Mar 1814 - Where taken: Bay of Biscay - Date received: 4 Apr 1814 - From what ship: HMS Virago - Born: Petersburg - Age: 33 - Discharged on 10 May 1814 and sent to Dartmoor.

Chappell, William - Seaman - Number: 435 - Prize name: Resolution - Ship type: MV - How taken: Hibernia, letter of marque - When taken: 21 Sep 1812 - Where taken: off Bermuda - Date received: 5 Feb 1813 - From what ship: HMS Neptune - Born: Flourtown, PA - Age: 20 - Sent to Chatham on 29 Mar 1813 on the HMS Braham.

Chappell, John - Seaman - Number: 930 - Prize name: Weasel - Ship type: MV - How taken: HM Brig Foxhound - When taken: 25 Mar 1813 - Where taken: Bay of Biscay - Date received: 6 Apr 1813 - From what ship: HMS Foxhound - Born: Newport - Age: 18 - Sent on 21 Jun 1813 to Portsmouth on HMS Prometheus.

Charles, John - Seaman - Number: 2331 - How taken: Impressed at Dublin - When taken: 1 Oct 1813 - Date received: 28 Jan 1814 - From what ship: HMS Neptune - Born: Boston - Age: 22 - Race: Negro - Discharged on 31 Jan 1814 and sent to Dartmoor.

Charlies, James - Passenger - Number: 1943 - Prize name: Marmion - Ship type: MV - How taken: HM Frigate President - When taken: 14 Aug 1813 - Where taken: off Nantes - Date received: 27 Aug 1813 - From what ship: HMS Urgent - Born: North Carolina - Age: 21 - Discharged on 8 Sep 1813 and sent to Dartmoor.

Charter, Samuel - Passenger - Number: 1599 - Prize name: Governor Gerry - Ship type: MV - How taken: HM Brig Royalist - When taken: 31 May 1813 - Where taken: Bay of Biscay - Date received: 14 Jun 1813 - From what ship: HMS Royalist - Born: Maryland - Age: 30 - Discharged on 3 Jun 1813 and sent to Dartmoor.

Chase, Joseph - Seaman - Number: 2621 - How taken: HM Ship-of-the-Line Edinburgh - When taken: 28 Oct 1812 - Date received: 16 May 1814 - From what ship: HMS Repulse - Born: Rhode Island - Age: 26 - Discharged on 14 Jun 1814 and sent to Dartmoor.

Chase, Nathaniel - Seaman - Number: 908 - Prize name: Prompt - Ship type: MV - How taken: Chance, privateer - When taken: 22 Mar 1813 - Where taken: Bay of Biscay - Date received: 3 Apr 1813 - From what ship: Mary - Born: Cape Cod - Age: 24 - Sent on 21 Jun 1813 to Portsmouth on HMS Prometheus.

Chauvel, Thomas - Boy - Number: 2376 - Prize name: Devon, prize to the Privateer Bunker Hill - Ship type: P - How taken: HM Brig Fly - When taken: 21 Jan 1814 - Where taken: at sea - Date received: 31 Jan 1814 - From what ship: HMS Fly - Born: Lanion, France - Age: 11 - Discharged on 27 Feb 1814 and sent to Chatham on HMS

Haleyon.

Check, Stephen - Seaman - Number: 672 - How taken: Impressed at Liverpool - When taken: 10 Jan 1813 - Date received: 15 Mar 1813 - From what ship: HMS Bittern - Born: Kent County - Age: 31 - Sent to Dartmoor on 1 Jul 1813.

Chesbrough, Benjamin F. - Seaman - Number: 2531 - Prize name: Young Dixon - Ship type: MV - How taken: HM Transport Hydra - When taken: 3 Apr 1814 - Where taken: at sea - Date received: 7 Apr 1814 - From what ship: HMS Fly - Born: New York - Age: 32 - Discharged on 10 May 1814 and sent to Dartmoor.

Chekes, William - Surgeon - Number: 386 - Prize name: Union - Ship type: MV - How taken: HM Frigate Iris - When taken: 17 Jan 1813 - Where taken: Lat 44 N Long 2.3 W - Date received: 5 Feb 1813 - From what ship: HMS San Josef - Born: Philadelphia - Age: 27 - Sent on 9 Feb 1813 to Ashburton on parole.

Cheney, Daniel - Seaman - Number: 2268 - Prize name: Fanny - Ship type: MV - How taken: HM Frigate Eurotas - When taken: 25 Dec 1813 - Where taken: at sea - Date received: 20 Jan 1814 - From what ship: HMS Eurotas - Born: New York - Age: 20 - Discharged on 31 Jan 1814 and sent to Dartmoor.

Chestly, Amos - Seaman - Number: 359 - Prize name: Louisa, prize of the Privateer Decatur - Ship type: P - How taken: HM Frigate Andromache - When taken: 11 Jan 1813 - Where taken: off Bordeaux - Date received: 4 Feb 1813 - From what ship: HMS Cornwall - Born: Dover, DE - Age: 19 - Sent to Chatham on 29 Mar 1813 on the HMS Braham.

Chevers, Joseph - Seaman - Number: 2177 - Prize name: Wolf, prize of the Privateer Grand Turk - Ship type: P - How taken: HM Frigate Briton - When taken: 1 Dec 1813 - Where taken: Bay of Biscay - Date received: 17 Dec 1813 - From what ship: HMS Briton - Born: Marblehead - Age: 28 - Discharged on 17 Dec 1813 and released to HMS Britton.

Chew, Joseph - Seaman - Number: 464 - Prize name: Dolphin - Ship type: LM - How taken: HM Ship-of-the-Line Colossus, HM Frigate Rhin & HM Brig Goldfinch - When taken: 5 Jan 1813 - Where taken: off Western Islands - Date received: 6 Feb 1813 - From what ship: HMS Rhin - Born: New Castle, DE - Age: 32 - Race: Negro - Sent to Chatham on 29 Mar 1813 on the HMS Braham.

Child, Samuel - Seaman - Number: 200 - Prize name: Hunter - Ship type: P - How taken: HM Frigate Phoebe - When taken: 24 Dec 1812 - Where taken: off Western Islands - Date received: 9 Jan 1813 - From what ship: HMS Phoebe - Born: Roxborough - Age: 28 - Sent to Portsmouth on 8 Feb 1813 on the HMS Colossus.

Chiney, Amos - Seaman - Number: 1196 - Prize name: Zebra - Ship type: LM - How taken: HM Frigate Pyramus - When taken: 20 Apr 1813 - Where taken: Bay of Biscay - Date received: 9 May 1813 - From what ship: HMS Andromache - Born: New York - Age: 24 - Sent on 8 Sep 1813 to Dartmoor.

Chipman, Christopher - Seaman - Number: 1962 - Prize name: Joel Barlow - Ship type: LM - How taken: HM Frigate Briton - When taken: 3 Jul 1813 - Where taken: off Bordeaux - Date received: 31 Aug 1813 - From what ship: HMS Clarence - Born: New London - Age: 29 - Discharged on 8 Sep 1813 and sent to Dartmoor.

Chisselsine, John - Seaman - Number: 1175 - Prize name: Hebe - Ship type: MV - How taken: HM Frigate Stag - When taken: 18 Apr 1813 - Where taken: Bay of Biscay - Date received: 6 May 1813 - From what ship: HMS Stag - Born: Newbury - Age: 20 - Sent on 3 Jul 1813 to Stapleton prison.

Chambers, Henry - Seaman - Number: 639 - Prize name: Criterion - Ship type: MV - How taken: HM Frigate Belle Poule - When taken: 14 Feb 1813 - Where taken: Bay of Biscay - Date received: 4 Mar 1813 - From what ship: HMS Strenuous - Born: Oldenburg - Age: 24 - Sent to Dartmoor on 2 Apr 1813.

Christian, John - Seaman - Number: 2552 - Prize name: Bunker Hill - Ship type: P - How taken: HM Frigate Pomone & HM Frigate Cydnus - When taken: 4 Mar 1814 - Where taken: Bay of Biscay - Date received: 17 Apr 1814 - From what ship: HMS Teazer - Born: New Orleans - Age: 22 - Discharged on 10 May 1814 and sent to Dartmoor.

Christie, James - Seaman - Number: 1639 - Prize name: Tickler - Ship type: LM - How taken: HM Frigate Magiciene - When taken: 5 Jun 1813 - Where taken: Lat 47 Long 13 - Date received: 14 Jun 1813 - From what ship: HMS Orestes - Born: New York - Age: 23 - Discharged on 30 Jun 1813 and sent to Stapleton.

Christy, Alexander - Seaman - Number: 745 - Prize name: Charlotte - Ship type: MV - How taken: HM Ship-of-the-Line Warspite - When taken: 3 Mar 1813 - Where taken: Bay of Biscay - Date received: 19 Mar 1813 - From what ship: HMS Warspite - Born: Charleston - Age: 25 - Sent to Dartmoor on 2 Apr 1813.

Chuglar, John - Marine - Number: 1868 - Prize name: US Brig Argus - Ship type: War - How taken: HM Brig Pelican - When taken: 14 Aug 1813 - Where taken: Irish Channel - Date received: 23 Aug 1813 - From what ship: HMS Pelican - Born: New York - Age: 23 - Discharged on 8 Sep 1813 and sent to Dartmoor.

Church, William - Prize Master - Number: 2665 - Prize name: Traveler, prize of the Privateer Surprize - Ship type: P - How taken: HMS Cawser - When taken: 7 May 1814 - Where taken: off Cape Clear - Date received: 5 Jun 1814 - From what ship: HMS Gronville - Born: Rhode Island - Age: 34 - Discharged on 14 Jun 1814 and sent to Dartmoor.

Churchill, Manuel - Seaman - Number: 2008 - Prize name: Ned - Ship type: LM - How taken: HM Brig Royalist - When taken: 6 Sep 1812 - Where taken: coast of France - Date received: 24 Sep 181 - From what ship: HMS Rippon - Born: Plymouth - Age: 24 - Discharged on 27 Sep 1813 and sent to Dartmoor.

Churchill, Stephen - Seaman - Number: 1112 - Prize name: Viper - Ship type: MV - How taken: HM Ship-of-the-Line Superb - When taken: 15 Apr 1813 - Where taken: Bay of Biscay - Date received: 22 Apr 1813 - From what ship: HMS Superb - Born: Richmond - Age: 34 - Discharged on 17 Jul 1813 and sent on the HMS Salvador del Mundo.

Churchill, Timothy - Seaman - Number: 882 - Prize name: Tiger - Ship type: MV - How taken: HM Brig Scylla - When taken: 22 Mar 1813 - Where taken: Bay of Biscay - Date received: 1 Apr 1813 - From what ship: HMS Scylla - Born: Plymouth, MA - Age: 27 - Sent on 21 Jun 1813 to Portsmouth.

Clapp, Abraham - Seaman - Number: 1226 - Prize name: Grand Napoleon - Ship type: MV - How taken: HM Frigate Belle Poule - When taken: 3 Apr 1813 - Where taken: off Bordeaux - Date received: 9 May 1813 - From what ship: HMS Andromache - Born: Massachusetts - Age: 22 - Discharged on 3 Jul 1813 and sent to Stapleton Prison.

Clapp, George - Seaman - Number: 2459 - Prize name: Fair American - Ship type: MV - How taken: HM Frigate Andromache - When taken: 19 Jan 1814 - Where taken: Bay of Biscay - Date received: 22 Feb 1814 - From what ship: HMS York - Born: Boston - Age: 19 - Discharged on 10 May 1814 and sent to Dartmoor.

Clark, Elisha - Yeoman - Number: 350 - How taken: Taken off the HM Frigate Andromache - Date received: 29 Jan 1813 - From what ship: HMS Royal Sovereign - Born: New Bedford - Age: 35 - Sent to Chatham on 29 Mar 1813 on the HMS Braham.

Clark, Jacob - Seaman - Number: 2259 - How taken: Delivered himself up from HM Brig Basilisk - Date received: 18 Jan 1814 - From what ship: HMS Basilisk - Born: Philadelphia - Age: 38 - Race: Negro - Discharged on 31 Jan 1814 and sent to Dartmoor.

Clark, John - Seaman - Number: 1887 - Prize name: US Brig Argus - Ship type: War - How taken: HM Brig Pelican - When taken: 14 Aug 1813 - Where taken: Irish Channel - Date received: 23 Aug 1813 - From what ship: HMS Pelican - Born: Gloucester - Age: 24 - Discharged on 8 Sep 1813 and sent to Dartmoor.

Clark, Joseph Wanton - Master - Number: 875 - Prize name: Charlotte - Ship type: MV - How taken: HM Ship-of-the-Line Warspite - When taken: 3 Mar 1813 - Where taken: Bay of Biscay - Date received: 30 Mar 1813 - From what ship: HMS Plymouth - Born: Newport - Age: 44 - Sent on 31 Mar 1813 to Ashburton on parole.

Clawe, Maurice - Seaman - Number: 1701 - How taken: Delivered himself up from HM Ship-of-the-Line Ville de Paris - Date received: 26 Jun 1813 - From what ship: HMS Duncan - Born: Long Island - Age: 31 - Race: Black - Discharged on 8 Jul 1813 and sent to Chatham on tender Neptune.

Cleaveland, Davis - Seaman - Number: 369 - Prize name: Orbit - Ship type: MV - How taken: HM Brig Achates - When taken: 29 Jan 1813 - Where taken: Lat 49 N Long 13 W - Date received: 4 Feb 1813 - From what ship: HMS Achates - Born: Nantucket - Age: 19 - Sent to Chatham on 29 Mar 1813 on the HMS Braham.

Clements, John C. - Seaman - Number: 706 - Prize name: Star - Ship type: MV - How taken: HM Ship-of-the-Line Superb - When taken: 9 Feb 1813 - Where taken: Bay of Biscay - Date received: 19 Mar 1813 - From what ship: HMS Warspite - Born: New Jersey - Age: 23 - Sent to Dartmoor on 2 Apr 1813.

Clerk, George - Master at Arms - Number: 1816 - Prize name: US Brig Argus - Ship type: War - How taken: HM Brig Pelican - When taken: 14 Aug 1813 - Where taken: Irish Channel - Date received: 17 Aug 1813 - From what ship: USS Argus - Born: New York - Age: 22 - Discharged on 8 Sep 1813 and sent to Dartmoor.

Clerk, William - Seaman - Number: 701 - Prize name: Star - Ship type: MV - How taken: HM Ship-of-the-Line Superb

- When taken: 9 Feb 1813 - Where taken: Bay of Biscay - Date received: 19 Mar 1813 - From what ship: HMS Warspite - Born: Newport - Age: 19 - Sent to Dartmoor on 2 Apr 1813.

Cleveland, Lawrence - Chief Mate - Number: 853 - Prize name: Mars - Ship type: MV - How taken: HM Ship-of-the-Line Warspite - When taken: 26 Feb 1813 - Where taken: Bay of Biscay - Date received: 25 May 1813 - From what ship: HMS Warspite - Born: Providence, RI - Age: 32 - Sent on 26 Mar 1813 to Ashburton on parole.

Clothey, Thomas - Seaman - Number: 1159 - Prize name: Essex - Ship type: MV - How taken: HM Frigate Pyramus - When taken: 2 Apr 1813 - Where taken: Bay of Biscay - Date received: 2 Mar 1813 - From what ship: HMS Rota - Born: Marblehead - Age: 19 - Sent on 3 Jul 1813 to Stapleton prison.

Clough, Isaac - Seaman - Number: 2167 - Prize name: Wolf, prize of the Privateer Grand Turk - Ship type: P - How taken: HM Frigate Briton - When taken: 1 Dec 1813 - Where taken: Bay of Biscay - Date received: 17 Dec 1813 - From what ship: HMS Briton - Born: Marblehead - Age: 20 - Discharged on 17 Dec 1813 and released to HMS Britton.

Cloutman, Robert - Seaman - Number: 2165 - Prize name: Wolf, prize of the Privateer Grand Turk - Ship type: P - How taken: HM Frigate Briton - When taken: 1 Dec 1813 - Where taken: Bay of Biscay - Date received: 17 Dec 1813 - From what ship: HMS Briton - Born: Marblehead - Age: 50 - Discharged on 17 Dec 1813 and released to HMS Britton.

Cloutman, Samuel - Carpenter's Mate - Number: 165 - Prize name: Hunter - Ship type: P - How taken: HM Frigate Phoebe - When taken: 24 Dec 1812 - Where taken: off Western Islands - Date received: 9 Jan 1813 - From what ship: HMS Phoebe - Born: Salem - Age: 25 - Sent to Portsmouth on 8 Feb 1813 on the HMS Colossus.

Coader, Antonio - Steward - Number: 1950 - Prize name: Marmion - Ship type: MV - How taken: HM Frigate President - When taken: 14 Aug 1813 - Where taken: off Nantes - Date received: 30 Aug 1813 - From what ship: Marmion - Born: New Orleans - Age: 28 - Race: Mulatto - Discharged on 8 Sep 1813 and sent to Dartmoor.

Cobb, Samuel - 2nd Mate - Number: 126 - Prize name: Experiment - Ship type: MV - How taken: HM Brig Rover - When taken: 21 Dec 1812 - Where taken: off Bordeaux - Date received: 1 Jan 1813 - From what ship: HMS Magnicficent - Born: Massachusetts - Age: 28 - Sent to Portsmouth on 4 Jan 1813 on the HMS Revolutionnaire.

Cobb, Samuel - 2nd Mate - Number: 90 - Prize name: Experiment - Ship type: MV - How taken: HM Brig Rover - When taken: 21 Oct 1812 - Where taken: off Bordeaux - Date received: 25 Dec 1812 - From what ship: HMS Northumberland - Born: Massachusetts - Age: 20 - Sent to Portsmouth on 29 Dec 1812 on the HMS Northumberland.

Cochran, Peter - Sailing Master - Number: 317 - Prize name: Porcupine - Ship type: MV - How taken: HM Frigate Dryad - When taken: 8 Jan 1813 - Where taken: off Bordeaux - Date received: 21 Jan 1813 - From what ship: HMS Abercrombie - Born: Boston - Age: 49 - Sent to Chatham on 29 Mar 1813 on the HMS Braham.

Cochran, Stephen - Seaman - Number: 225 - Prize name: Vengeance - Ship type: LM - How taken: HM Frigate Phoebe - When taken: 1 Jan 1813 - Where taken: Lat 44.4 Long 23 - Date received: 9 Jan 1813 - From what ship: HMS Phoebe - Born: Wiscasset - Age: 20 - Sent to Portsmouth on 8 Feb 1813 on the HMS Colossus.

Codman, Richard - Seaman - Number: 1624 - Prize name: Leo - Ship type: LM - How taken: HM Frigate Magiciene - When taken: 4 Jun 1813 - Where taken: Lat 45 Long 14 - Date received: 14 Jun 1813 - From what ship: HMS Orestes - Born: Portland - Age: 20 - Discharged on 30 Jun 1813 and sent to Stapleton.

Codsifershall, Charles - Master - Number: 6 - Prize name: Science - Ship type: MV - How taken: HM Schooner Alphea - When taken: 14 Aug 1812 - Where taken: Lat 44 Long 43.3 - Date received: 17 Nov 1812 - From what ship: HMS Plymouth - Born: Philadelphia - Age: 24 - Returned to America on 17 Nov 1812 via London.

Coffer, William - Seaman - Number: 2305 - Prize name: Siro - Ship type: LM - How taken: HM Brig Pelican - When taken: 13 Jan 1814 - Where taken: at sea - Date received: 23 Jan 1814 - From what ship: HMS Pelican - Born: Georgetown - Age: 24 - Discharged on 31 Jan 1814 and sent to Dartmoor.

Coffin, Alexander - Master - Number: 338 - Prize name: Porcupine - Ship type: MV - How taken: HM Frigate Dryad - When taken: 8 Jan 1813 - Where taken: off Bordeaux - Date received: 22 Jan 1813 - From what ship: Porcupine - Born: Nantucket - Age: 48 - Sent on 23 Jan 1813 to Ashburton on parole.

Coffin, Alexander - Seaman - Number: 2224 - Prize name: Zephyr - Ship type: MV - How taken: HM Frigate Pyramus

- When taken: 30 Nov 1813 - Where taken: off Lorient - Date received: 3 Jan 1814 - From what ship: HMS Warspite - Born: Nantucket - Age: 20 - Discharged on 31 Jan 1814 and sent to Dartmoor.

Coffin, Frederick Henry - Seaman - Number: 243 - How taken: Impressed at Liverpool - When taken: 17 Nov 1812 - Date received: 12 Jan 1813 - From what ship: HMS Bittern - Born: Nantucket - Age: 39 - Sent to Portsmouth on 8 Feb 1813 on the HMS Colossus.

Coffin, George - Seaman - Number: 460 - Prize name: Dolphin - Ship type: LM - How taken: HM Ship-of-the-Line Colossus, HM Frigate Rhin & HM Brig Goldfinch - When taken: 5 Jan 1813 - Where taken: off Western Islands - Date received: 6 Feb 1813 - From what ship: HMS Rhin - Born: New York - Age: 23 - Sent to Chatham on 29 Mar 1813 on the HMS Braham.

Coffin, George P. - Boy - Number: 918 - Prize name: Tiger - Ship type: MV - How taken: HM Brig Scylla - When taken: 22 Mar 1813 - Where taken: Bay of Biscay - Date received: 5 Apr 1813 - From what ship: HMS Plymouth - Born: Nantucket - Age: 18 - Sent on 6 Apr 1813 to Ashburton on parole.

Coffin, John - Seaman - Number: 2003 - Prize name: Francis & Ann - Ship type: MV - How taken: HM Sloop Lightning - When taken: 20 Aug 1813 - Where taken: at sea - Date received: 22 Sep 1813 - From what ship: HMS Lightning - Born: New York - Age: 25 - Discharged on 27 Sep 1813 and sent to Dartmoor.

Coffin, Samuel - Cook - Number: 782 - Prize name: Cannoneer - Ship type: P - How taken: HM Ship-of-the-Line Warspite - When taken: 14 Mar 1813 - Where taken: Bay of Biscay - Date received: 19 Mar 1813 - From what ship: HMS Warspite - Born: Boston - Age: 26 - Sent to Dartmoor on 28 Jun 1813.

Coffin, William P. - Master - Number: 917 - Prize name: Tiger - Ship type: MV - How taken: HM Brig Scylla - When taken: 22 Mar 1813 - Where taken: Bay of Biscay - Date received: 5 Apr 1813 - From what ship: HMS Plymouth - Born: Nantucket - Age: 25 - Sent on 6 Apr 1813 to Ashburton on parole.

Cogan, John - Seaman - Number: 1711 - Prize name: Joseph - Ship type: MV - How taken: HM Frigate Iris - When taken: 8 Jun 1813 - Where taken: Bay of Biscay - Date received: 4 Jul 1813 - From what ship: HMS Iris - Born: Sandwich - Age: 22 - Discharged on 22 Aug 1813 and released to HM Brig Redpole.

Cogswell, Edward - Armorer's Mate - Number: 158 - Prize name: Hunter - Ship type: P - How taken: HM Frigate Phoebe - When taken: 24 Dec 1812 - Where taken: off Western Islands - Date received: 9 Jan 1813 - From what ship: HMS Phoebe - Born: Concord - Age: 21 - Sent to Portsmouth on 8 Feb 1813 on the HMS Colossus.

Colcocha, Anthony - Seaman - Number: 576 - Prize name: Rolla - Ship type: MV - How taken: HM Frigate Surveillante - When taken: 11 Feb 1813 - Where taken: Bay of Biscay - Date received: 23 Feb 1813 - From what ship: HMS Surveillante - Born: New Orleans - Age: 26 - Sent to Dartmoor on 2 Apr 1813.

Cole, John - Seaman - Number: 2627 - Prize name: Ateline - Ship type: MV - How taken: HM Frigate Magiciene - When taken: 14 Mar 1814 - Where taken: off Cape Finisterre, Spain - Date received: 17 May 1814 - From what ship: HMS Tortois - Born: Baltimore - Age: 38 - Race: Black - Discharged on 14 Jun 1814 and sent to Dartmoor.

Cole, Richard - Chief Mate - Number: 559 - Prize name: Terrible - Ship type: MV - How taken: HM Brig Foxhound - When taken: 8 Feb 1813 - Where taken: Channel - Date received: 17 Feb 1813 - From what ship: HMS Foxhound - Born: Monmouth, NJ - Age: 23 - Sent on 18 Feb 1813 to Ashburton on parole.

Cole, William - Seaman - Number: 904 - Prize name: Prompt - Ship type: MV - How taken: Chance, privateer - When taken: 22 Mar 1813 - Where taken: Bay of Biscay - Date received: 3 Apr 1813 - From what ship: Mary - Born: Cape Cod - Age: 23 - Sent on 21 Jun 1813 to Portsmouth on HMS Prometheus.

Coleman, Charles - Chief Mate - Number: 660 - Prize name: Manilla - Ship type: MV - How taken: Tiger, letter of marque - When taken: 16 Jan 1813 - Where taken: Lat 20N Long 61W - Date received: 8 Mar 1813 - From what ship: Manilla - Born: Nantucket - Age: 24 - Sent on 10 Feb 1813 to Ashburton on parole.

Coleman, David - Boy - Number: 1488 - Prize name: Paul Jones - Ship type: P - How taken: HM Frigate Leonidas - When taken: 23 May 1813 - Where taken: off Cape Clear - Date received: 26 May 1813 - From what ship: HMS Leonidas - Born: New York - Age: 14 - Discharged on 30 Jun 1813 and sent to Stapleton.

Coleman, John - Seaman - Number: 121 - How taken: Impressed at Dublin - When taken: 14 Nov 1812 - Date received: 30 Dec 1812 - From what ship: HMS Frederick - Born: Delaware - Age: 27 - Sent to Portsmouth on 4 Jan 1813 on the HMS Revolutionnaire.

Coleman, William - Seaman - Number: 2535 - Prize name: Lyon - Ship type: MV - How taken: Brilliant, privateer - When taken: 1 Jan 1814 - Date received: 13 Apr 1814 - From what ship: HMS Bittern - Born: Salem - Age: 23 - Race: Mulatto - Discharged on 10 May 1814 and sent to Dartmoor.

Collin, William - Seaman - Number: 1896 - Prize name: US Brig Argus - Ship type: War - How taken: HM Brig Pelican - When taken: 14 Aug 1813 - Where taken: Irish Channel - Date received: 23 Aug 1813 - From what ship: HMS Pelican - Born: Annapolis - Age: 25 - Race: Black - Discharged on 8 Sep 1813 and sent to Dartmoor.

Collins, Andrew - Seaman - Number: 1919 - Prize name: US Brig Argus - Ship type: War - How taken: HM Brig Pelican - When taken: 14 Aug 1813 - Where taken: Irish Channel - Date received: 23 Aug 1813 - From what ship: HMS Pelican - Born: Annapolis - Age: 29 - Discharged on 8 Sep 1813 and sent to Dartmoor.

Collison, Joseph - Chief Mate - Number: 984 - Prize name: Lightning - Ship type: MV - How taken: HM Frigate Medusa - When taken: 2 Apr 1813 - Where taken: Bay of Biscay - Date received: 16 Apr 1813 - From what ship: HMS Fairy - Born: Philadelphia - Age: 24 - Sent on 22 Apr 1813 to Ashburton on parole.

Colton, Walter - Marine Lieutenant - Number: 1500 - Prize name: Paul Jones - Ship type: P - How taken: HM Frigate Leonidas - When taken: 23 May 1813 - Where taken: off Cape Clear - Date received: 27 May 1813 - From what ship: HMS Leonidas - Born: Springfield, MA - Age: 30 - Discharged on 30 Jun 1813 and sent to Stapleton.

Colville, John - Seaman - Number: 794 - Prize name: Cannoniere - Ship type: P - How taken: HM Ship-of-the-Line Warspite - When taken: 14 Mar 1813 - Where taken: Bay of Biscay - Date received: 19 Mar 1813 - From what ship: HMS Warspite - Born: New York - Age: 23 - Sent to Dartmoor on 28 Jun 1813.

Combs, William - Seaman - Number: 1058 - Prize name: Virginia Planter - Ship type: MV - How taken: HM Frigate Pyramus - When taken: 17 Mar 1813 - Where taken: off Nantes - Date received: 22 Apr 1813 - From what ship: HMS Superb - Born: New York - Age: 19 - Sent to Dartmoor on 1 Jul 1813.

Conklin, Edward - Seaman - Number: 2630 - Prize name: Ateline - Ship type: MV - How taken: HM Frigate Magiciene - When taken: 14 Mar 1814 - Where taken: off Cape Finisterre, Spain - Date received: 17 May 1814 - From what ship: HMS Tortois - Born: Newark - Age: 31 - Discharged on 14 Jun 1814 and sent to Dartmoor.

Conklin, Robert - Gunner - Number: 2420 - Prize name: US Brig Argus - Ship type: War - How taken: Sent from HM Ship-of-the-Line Salvador del Mundo - When taken: 14 Aug 1813 - Where taken: Irish Channel - Date received: 12 Feb 1814 - From what ship: HMS Salvador del Mundo - Born: Kingston - Age: 26 - Discharged on 10 May 1814 and sent to Dartmoor.

Conklin, Robert - Gunner - Number: 1815 - Prize name: US Brig Argus - Ship type: War - How taken: HM Brig Pelican - When taken: 14 Aug 1813 - Where taken: Irish Channel - Date received: 17 Aug 1813 - From what ship: USS Argus - Born: Kingston - Age: 23 - Discharged on 24 Aug 1813 and released to HMS Salvador del Mundo.

Conklin, Smith - Seaman - Number: 2386 - How taken: Gave himself up - Date received: 1 Feb 1814 - From what ship: HMS Bittern - Born: New York - Age: 21 - Discharged on 27 Feb 1814 and sent to Chatham on HMS Haleyon.

Conklin, William - Boatswain - Number: 877 - Prize name: Tiger - Ship type: MV - How taken: HM Brig Scylla - When taken: 22 Mar 1813 - Where taken: Bay of Biscay - Date received: 1 Apr 1813 - From what ship: HMS Scylla - Born: New York - Age: 32 - Sent to Dartmoor on 28 Jun 1813.

Conley, Samuel - Seaman - Number: 1905 - Prize name: US Brig Argus - Ship type: War - How taken: HM Brig Pelican - When taken: 14 Aug 1813 - Where taken: Irish Channel - Date received: 23 Aug 1813 - From what ship: HMS Pelican - Born: Tolbert, Holland - Age: 23 - Discharged on 8 Sep 1813 and sent to Dartmoor.

Conner, Edward - Prize Master - Number: 2559 - Prize name: Bunker Hill - Ship type: P - How taken: HM Frigate Pomone & HM Frigate Cydnus - When taken: 4 Mar 1814 - Where taken: Bay of Biscay - Date received: 17 Apr 1814 - From what ship: HMS Teazer - Born: Philadelphia - Age: 25 - Discharged on 10 May 1814 and sent to Dartmoor.

Conner, Michael - Seaman - Number: 2370 - Prize name: Minerva - Ship type: MV - How taken: HM Ship-of-the-Line Conquestador - When taken: 19 Dec 1814 - Where taken: Bay of Biscay - Date received: 31 Jan 1814 - From what ship: HMS Surveillante - Born: Charlestown - Age: 26 - Discharged on 27 Feb 1814 and sent to Chatham on HMS Haleyon.

Conrad, Godfred - Seaman - Number: 2255 - How taken: Impressed at Cove of Cork - When taken: 26 Dec 1813 - Date received: 15 Jan 1814 - From what ship: HMS Teazer - Born: Altona, Germany - Age: 47 - Liberated on 7 Feb 1814.

Conroy, William M. - 2nd Mate - Number: 2190 - Prize name: Agnes, prize of the Privateer Rambler - Ship type: LM - How taken: Cutter Jane - When taken: 28 Nov 1813 - Where taken: Bay of Biscay - Date received: 21 Dec 1813 - From what ship: Agnes - Born: Boston - Age: 54 - Discharged on 31 Jan 1814 and sent to Dartmoor.

Conway, Samuel - Seaman - Number: 51 - How taken: Elizabeth, tender - When taken: 29 Oct 1812 - Where taken: Greenock - Date received: 26 Nov 1813 - From what ship: HMS Frederick - Born: Salem - Age: 28 - Sent to Portsmouth on 29 Dec 1812 on the HMS Northumberland.

Conway, William - Seaman - Number: 519 - Prize name: Allegany - Ship type: MV - How taken: Detained at Gibraltar - When taken: 8 Aug 1812 - Date received: 15 Feb 1813 - From what ship: HMS Andromeda - Born: Marblehead - Age: 22 - Sent to Dartmoor on 2 Apr 1813.

Cook, Benjamin - Seaman - Number: 2119 - Prize name: Chesapeake - Ship type: LM - How taken: HM Frigate Hotspur & HM Frigate Pyramus - When taken: 26 Oct 1813 - Where taken: Bay of Biscay - Date received: 22 Nov 1813 - From what ship: HMS Pyramus - Born: Baltimore - Age: 26 - Race: Black - Discharged on 29 Nov 1813 and sent to Dartmoor.

Cook, Charles H. - Seaman - Number: 1469 - Prize name: Paul Jones - Ship type: P - How taken: HM Frigate Leonidas - When taken: 23 May 1813 - Where taken: off Cape Clear - Date received: 26 May 1813 - From what ship: HMS Leonidas - Born: South Carolina - Age: 27 - Race: Black - Discharged on 30 Jun 1813 and sent to Stapleton.

Cook, James - Seaman - Number: 538 - Prize name: Terrible - Ship type: MV - How taken: HM Brig Foxhound - When taken: 8 Feb 1813 - Where taken: Channel - Date received: 15 Feb 1813 - From what ship: HMS Foxhound - Born: Norfolk - Age: 25 - Sent to Dartmoor on 2 Apr 1813.

Cook, Joseph B. - Master - Number: 37 - Prize name: Warren - Ship type: MV - How taken: HM Frigate Sybille & HM Frigate Fortunee- When taken: 5 Sep 1812 - Where taken: Lat 41.4 Long 33 - Date received: 23 Nov 1812 - From what ship: HMS Stork - Born: Rhode Island - Age: 47 - Sent on 8 Dec 1812 to Ashburton on parole.

Cooke, Samuel - Seaman - Number: 1291 - Prize name: Price - Ship type: LM - How taken: HM Frigate Medusa - When taken: 13 Apr 1813 - Where taken: Bay of Biscay - Date received: 10 May 1813 - From what ship: HMS Medusa - Born: Tiverton - Age: 22 - Discharged on 8 Sep 1813 and sent to Dartmoor.

Cooke, William - Seaman - Number: 1464 - Prize name: Paul Jones - Ship type: P - How taken: HM Frigate Leonidas - When taken: 23 May 1813 - Where taken: off Cape Clear - Date received: 26 May 1813 - From what ship: HMS Leonidas - Born: New York - Age: 18 - Discharged on 30 Jun 1813 and sent to Stapleton.

Coombes, James - Seaman - Number: 1913 - Prize name: US Brig Argus - Ship type: War - How taken: HM Brig Pelican - When taken: 14 Aug 1813 - Where taken: Irish Channel - Date received: 23 Aug 1813 - From what ship: HMS Pelican - Born: Wiscasset - Age: 22 - Discharged on 8 Sep 1813 and sent to Dartmoor.

Cooper, Andrew A. - Gunner's Mate - Number: 1455 - Prize name: Paul Jones - Ship type: P - How taken: HM Frigate Leonidas - When taken: 23 May 1813 - Where taken: off Cape Clear - Date received: 26 May 1813 - From what ship: HMS Leonidas - Born: Albany - Age: 28 - Discharged on 30 Jun 1813 and sent to Stapleton.

Cooper, Charles - Seaman - Number: 1519 - Prize name: Grand Napoleon - Ship type: MV - How taken: HM Brig Goldfinch - When taken: 17 Apr 1813 - Where taken: Bay of Biscay - Date received: 29 May 1813 - From what ship: HMS Hannibal - Born: New York - Age: 23 - Discharged on 30 Jun 1813 and sent to Stapleton.

Cooper, Daniel - Seaman - Number: 583 - Prize name: Cashiere - Ship type: LM - How taken: HM Brig Reindeer - When taken: 3 Feb 1813 - Where taken: Bay of Biscay - Date received: 23 Feb 1813 - From what ship: HMS Surveillante - Born: Baltimore - Age: 30 - Sent to Dartmoor on 2 Apr 1813.

Cooper, James - Seaman - Number: 824 - Prize name: King George - Ship type: MV - How taken: HM Brig Piercer - When taken: 9 Mar 1813 - Where taken: off Isle Ross - Date received: 21 Mar 1813 - From what ship: HMS Piercer - Born: Long Island - Age: 25 - Race: Black - Sent to Dartmoor on 28 Jun 1813.

Cooper, James - 2nd Gunner - Number: 2423 - Prize name: US Brig Argus - Ship type: War - How taken: Sent from HM Ship-of-the-Line Salvador del Mundo - When taken: 14 Aug 1813 - Where taken: Irish Channel - Date received: 12 Feb 1814 - From what ship: HMS Salvador del Mundo - Born: New York - Age: 25 - Discharged on 10 May 1814 and sent to Dartmoor.

Cooper, James - Quarter Gunner - Number: 1819 - Prize name: US Brig Argus - Ship type: War - How taken: HM Brig Pelican - When taken: 14 Aug 1813 - Where taken: Irish Channel - Date received: 17 Aug 1813 - From what ship: USS Argus - Born: Jamaica, NY - Age: 25 - Discharged on 24 Aug 1813 and released to HMS Salvador del Mundo.

Cooper, John - Seaman - Number: 2312 - Prize name: Siro - Ship type: LM - How taken: HM Brig Pelican - When taken: 13 Jan 1814 - Where taken: at sea - Date received: 23 Jan 1814 - From what ship: HMS Pelican - Born: Charlestown - Age: 27 - Discharged on 31 Jan 1814 and sent to Dartmoor.

Cooper, Tannock - Seaman - Number: 2487 - How taken: Sent into custody by own request - Date received: 15 Mar 1814 - From what ship: HMS Salvador del Mundo - Born: Baltimore - Age: 34 - Discharged on 10 May 1814 and sent to Dartmoor.

Coperris, Nicholas - Seaman - Number: 270 - Prize name: Brutus - Ship type: MV - How taken: HM Frigate Briton - When taken: Jan 1813 - Where taken: Bay of Biscay - Date received: 21 Jan 1813 - From what ship: HMS Briton - Born: Bosno, Turkey - Age: 44 - Sent to Portsmouth on 8 Feb 1813 on the HMS Colossus.

Copland, James - Chief Mate - Number: 911 - Prize name: Tiger - Ship type: MV - How taken: HM Brig Scylla - When taken: 22 Mar 1813 - Where taken: Bay of Biscay - Date received: 3 Apr 1813 - From what ship: HMS Scylla - Born: New York - Age: 23 - Sent on 6 Apr 1813 to Ashburton on parole.

Cornish, Charles - Seaman - Number: 2117 - Prize name: Chesapeake - Ship type: LM - How taken: HM Frigate Hotspur & HM Frigate Pyramus - When taken: 26 Oct 1813 - Where taken: Bay of Biscay - Date received: 22 Nov 1813 - From what ship: HMS Pyramus - Born: Maryland - Age: 40 - Race: Black - Discharged on 29 Nov 1813 and sent to Dartmoor.

Cornwall, Arthur - Seaman - Number: 1518 - Prize name: Grand Napoleon - Ship type: MV - How taken: HM Brig Goldfinch - When taken: 17 Apr 1813 - Where taken: Bay of Biscay - Date received: 29 May 1813 - From what ship: HMS Hannibal - Born: Philadelphia - Age: 26 - Discharged on 30 Jun 1813 and sent to Stapleton.

Cortu, John - Cook - Number: 941 - Prize name: Gleamer - Ship type: MV - How taken: Brothers, privateer - When taken: 26 Mar 1813 - Where taken: Bay of Biscay - Date received: 8 Apr 1813 - From what ship: HMS Wasp - Born: New York - Age: 25 - Sent on 21 Jun 1813 to Portsmouth on HMS Prometheus.

Cosevin, Pierre Mathrie - Passenger - Number: 1186 - Prize name: Zebra - Ship type: LM - How taken: HM Frigate Pyramus - When taken: 20 Apr 1813 - Where taken: Bay of Biscay - Date received: 9 May 1813 - From what ship: HMS Andromache - Born: Rochelle - Age: 38 - Sent on 11 Jun 1813 to the American ship Hope.

Cotterill, Henry - Master - Number: 2530 - Prize name: Young Dixon - Ship type: MV - How taken: HM Transport Hydra - When taken: 3 Apr 1814 - Where taken: at sea - Date received: 7 Apr 1814 - From what ship: HMS Fly - Born: Baltimore - Age: 28 - Discharged on 10 May 1814 and sent to Dartmoor.

Cotterill, James - Seaman - Number: 1403 - How taken: Delivered himself up from HM Hospital Ship Trent - When taken: 25 May 1813 - Date received: 18 May 1813 - From what ship: HMS Treazer - Born: Nobleboro - Age: 36 - Discharged on 8 Jul 1813 and sent to Chatham on tender Neptune.

Cottle, William - Captain - Number: 2505 - Prize name: Bunker Hill - Ship type: P - How taken: HM Frigate Pomone & HM Frigate Cydnus - When taken: 4 Mar 1814 - Where taken: Bay of Biscay - Date received: 4 Apr 1814 - From what ship: Prison - Born: Asbury - Age: 38 - Discharged on 6 Apr 1814 and sent to Ashburton on parole.

Couet, John - Seaman - Number: 218 - Prize name: Vengeance - Ship type: LM - How taken: HM Frigate Phoebe - When taken: 1 Jan 1813 - Where taken: Lat 44.4 Long 23 - Date received: 9 Jan 1813 - From what ship: HMS Phoebe - Born: Charles County - Age: 32 - Sent to Portsmouth on 8 Feb 1813 on the HMS Colossus.

Couret, Francis - Seaman - Number: 1879 - Prize name: Betsey, prize to the US Brig Argus - Ship type: War - How taken: HM Frigate Leonidas - When taken: 12 Aug 1813 - Where taken: Channel - Date received: 23 Aug 1813 - From what ship: HMS Pelican - Born: Pennsylvania - Age: 19 - Discharged on 8 Sep 1813 and sent to Dartmoor.

Courris, Francis - Seaman - Number: 1756 - Prize name: Friendship, prize to the Privateer America - Ship type: P - How taken: HM Schooner Whiting - When taken: 15 Jul 1815 - Where taken: Lat 47N Long 8W - Date received:

20 Jul 1813 - From what ship: HMS Whiting - Born: Massachusetts - Age: 21 - Discharged on 8 Sep 1813 and sent to Dartmoor.

Court, Robert - Seaman - Number: 736 - Prize name: Pert - Ship type: MV - How taken: HM Ship-of-the-Line Warspite - When taken: 1 Mar 1813 - Where taken: Bay of Biscay - Date received: 19 Mar 1813 - From what ship: HMS Warspite - Born: Philadelphia - Age: 19 - Sent to Dartmoor on 2 Apr 1813.

Courtis, Harry - Seaman - Number: 1757 - Prize name: Friendship, prize to the Privateer America - Ship type: P - How taken: HM Schooner Whiting - When taken: 15 Jul 1815 - Where taken: Lat 47N Long 8W - Date received: 20 Jul 1813 - From what ship: HMS Whiting - Born: New York - Age: 21 - Discharged on 8 Sep 1813 and sent to Dartmoor.

Courtney, James - Seaman - Number: 964 - Prize name: Ferox - Ship type: MV - How taken: HM Frigate Medusa & HM Brig Lyra - When taken: 28 Mar 1813 - Where taken: off Cape Ortagle - Date received: 9 Apr 1813 - From what ship: HMS Lyra - Born: New York - Age: 34 - Sent on 21 Jun 1813 to the American ship Hope.

Cousor, Adam - Seaman - Number: 282 - Prize name: Columbia - Ship type: MV - How taken: HM Frigate Briton - When taken: 17 Dec 1812 - Where taken: off Bordeaux - Date received: 21 Jan 1813 - From what ship: HMS Abercrombie - Born: Philadelphia - Age: 30 - Sent to Portsmouth on 8 Feb 1813 on the HMS Colossus.

Coutie, Thomas - 3rd Lieutenant - Number: 2351 - Prize name: Harvest, prize of the Privateer Bunker Hill - Ship type: P - How taken: HM Brig Orestes - When taken: 21 Jan 1814 - Date received: 31 Jan 1814 - From what ship: Harvest - Born: Marblehead - Age: 35 - Died on 7 Apr 1814 in the Mill Prison Hospital.

Covell, Isaac - Seaman - Number: 590 - Prize name: St. Martin's Plantation, prize of the Privateer Paul Jones - Ship type: P - How taken: HM Ship-of-the-Line Dublin - When taken: 9 Feb 1815 - Where taken: Lat 43 N Long 33.5 W - Date received: 25 Feb 1813 - From what ship: HMS Dublin - Born: Ellington - Age: 29 - Sent to Dartmoor on 2 Apr 1813.

Cowell, Slater - 2nd Mate - Number: 472 - Prize name: Print - Ship type: MV - How taken: HM Frigate Rhin - When taken: 15 Jan 1813 - Where taken: Lat 44 N Long 17 W - Date received: 9 Feb 1813 - From what ship: HMS Rhin - Born: Marblehead - Age: 22 - Sent on 15 Feb 1813 to Ashburton on parole.

Cowen, Robert - Seaman - Number: 1098 - Prize name: Young Holkar - Ship type: MV - How taken: HM Ship-of-the-Line Superb - When taken: 10 Apr 1813 - Where taken: off Belle Isle - Date received: 22 Apr 1813 - From what ship: HMS Superb - Born: Philadelphia - Age: 39 - Sent to Dartmoor on 1 Jul 1813.

Cox Miles - 2nd Mate - Number: 360 - Prize name: Orbit - Ship type: MV - How taken: HM Brig Achates - When taken: 29 Jan 1813 - Where taken: Lat 49 N Long 13 W - Date received: 4 Feb 1813 - From what ship: HMS Achates - Born: Philadelphia - Age: 25 - Sent to Dartmoor on 2 Apr 1813.

Cox, John - Seaman - Number: 713 - Prize name: Star - Ship type: MV - How taken: HM Ship-of-the-Line Superb - When taken: 9 Feb 1813 - Where taken: Bay of Biscay - Date received: 19 Mar 1813 - From what ship: HMS Warspite - Born: Chester - Age: 20 - Sent to Dartmoor on 2 Apr 1813.

Cox, John - Seaman - Number: 1741 - Prize name: Star - Ship type: MV - How taken: HM Ship-of-the-Line Superb - When taken: 9 Feb 1813 - Where taken: Bay of Biscay - Date received: 10 Jul 1813 - From what ship: Dartmoor Prison - Born: Chester - Age: 20 - Discharged on 10 Jul 1813 and released to HMS Salvador del Mundo.

Cox, John - Seaman - Number: 2618 - How taken: HM Ship-of-the-Line Elizabeth - Date received: 16 May 1814 - From what ship: HMS Repulse - Born: Portsmouth - Age: 44 - Discharged on 14 Jun 1814 and sent to Dartmoor.

Crafford, Robert - Soldier - Number: 2419 - Prize name: US Brig Argus - Ship type: War - How taken: Sent from HM Ship-of-the-Line Salvador del Mundo - When taken: 14 Aug 1813 - Where taken: Irish Channel - Date received: 12 Feb 1814 - From what ship: HMS Salvador del Mundo - Born: New York - Age: 41 - Discharged on 10 May 1814 and sent to Dartmoor.

Crafford, Robert - Marine - Number: 1870 - Prize name: US Brig Argus - Ship type: War - How taken: HM Brig Pelican - When taken: 14 Aug 1813 - Where taken: Irish Channel - Date received: 23 Aug 1813 - From what ship: HMS Pelican - Born: New York - Age: 41 - Discharged on 24 Aug 1813 and released to HMS Salvador del Mundo.

Craig, William - Seaman - Number: 2612 - How taken: MV Alemena - When taken: 25 Jan 1813 - Date received: 16 May 1814 - From what ship: HMS Repulse - Born: Dorset - Age: 29 - Race: Black - Discharged on 14 Jun 1814 and sent to Dartmoor.

Cramstead, James - Seaman - Number: 1456 - Prize name: Paul Jones - Ship type: P - How taken: HM Frigate Leonidas - When taken: 23 May 1813 - Where taken: off Cape Clear - Date received: 26 May 1813 - From what ship: HMS Leonidas - Born: New York - Age: 32 - Discharged on 30 Jun 1813 and sent to Stapleton.

Crandall, John - Seaman - Number: 970 - Prize name: Ferox - Ship type: MV - How taken: HM Frigate Medusa & HM Brig Lyra - When taken: 28 Mar 1813 - Where taken: off Cape Ortagle - Date received: 9 Apr 1813 - From what ship: HMS Lyra - Born: Duchess County, NY - Age: 29 - Sent to Chatham on 8 Jul 1813 on HM Tender Neptune.

Crandall, John - Passenger - Number: 865 - Prize name: Coxerien - Ship type: MV - How taken: HM Frigate Andromache - When taken: 14 Mar 1813 - Where taken: Bay of Biscay - Date received: 28 Mar 1813 - From what ship: HMS Andromache - Born: New Haven - Age: 35 - Sent on 30 Mar 1813 to Ashburton on parole.

Crawford, Nelson - Seaman - Number: 990 - Prize name: Lightning - Ship type: MV - How taken: HM Frigate Medusa - When taken: 2 Apr 1813 - Where taken: Bay of Biscay - Date received: 16 Apr 1813 - From what ship: HMS Fairy - Born: Richmond - Age: 26 - Sent to Chatham on 8 Jul 1813 on HM Tender Neptune.

Crawford, William - Seaman - Number: 2124 - Prize name: Chesapeake - Ship type: LM - How taken: HM Frigate Hotspur & HM Frigate Pyramus - When taken: 26 Oct 1813 - Where taken: Bay of Biscay - Date received: 22 Nov 1813 - From what ship: HMS Pyramus - Born: Delaware - Age: 27 - Discharged on 29 Nov 1813 and sent to Dartmoor.

Creighton, Manuel - Seaman - Number: 2403 - Prize name: Rachel & Ann, prize of the Privateer Prince de Neufchatel - Ship type: P - How taken: HM Frigate Cydnus - When taken: 6 Jan 1814 - Where taken: at sea - Date received: 4 Feb 1814 - From what ship: HMS Cydnus - Born: Philadelphia - Age: 27 - Discharged on 10 May 1814 and sent to Dartmoor.

Creping, Thomas - Seaman - Number: 2184 - Prize name: Charlotte - Ship type: MV - How taken: Cutter Dwarf - When taken: 4 Nov 1812 - Where taken: off Bordeaux - Date received: 18 Dec 1813 - From what ship: HMS Conquistador - Born: Wilmington, NC - Age: 38 - Discharged on 28 Feb 1814 and released to HMS Salvador del Mundo.

Creps, Warren - Seaman - Number: 2131 - Prize name: Amiable - Ship type: LM - How taken: HM Ship-of-the-Line Magnificant - When taken: 30 Oct 1813 - Where taken: off Lorient - Date received: 29 Nov 1813 - From what ship: HMS Dublin - Born: Massachusetts - Age: 26 - Discharged on 4 Dec 1813 and sent to Dartmoor.

Crete, John Nicholas - Seaman - Number: 684 - Prize name: Nope - Ship type: MV - How taken: Chance, privateer - When taken: 15 Feb 1813 - Where taken: off Bordeaux - Date received: 15 Mar 1813 - From what ship: Growler - Born: New Orleans - Age: 24 - Sent to Dartmoor on 2 Apr 1813.

Criger, Namon - Seaman - Number: 800 - Prize name: Cannoniere - Ship type: P - How taken: HM Ship-of-the-Line Warspite - When taken: 14 Mar 1813 - Where taken: Bay of Biscay - Date received: 19 Mar 1813 - From what ship: HMS Warspite - Born: New York - Age: 22 - Discharged on 21 Jun 1813 and released to the American ship Mount Hope.

Croft, George - Chief Mate - Number: 855 - Prize name: Charlotte - Ship type: MV - How taken: HM Ship-of-the-Line Warspite - When taken: 3 Mar 1813 - Where taken: Bay of Biscay - Date received: 25 May 1813 - From what ship: HMS Warspite - Born: Boston - Age: 25 - Sent on 26 Mar 1813 to Ashburton on parole.

Crofts, William - Sailmaker - Number: 396 - Prize name: Union - Ship type: MV - How taken: HM Frigate Iris - When taken: 17 Jan 1813 - Where taken: Lat 44 N Long 2.3 W - Date received: 5 Feb 1813 - From what ship: HMS San Josef - Born: Newport - Age: 28 - Sent to Chatham on 29 Mar 1813 on the HMS Braham.

Croker, Nathaniel - Seaman - Number: 1920 - Prize name: Betsey, prize to the US Brig Argus - Ship type: War - How taken: HM Frigate Leonidas - When taken: 12 Aug 1813 - Where taken: Channel - Date received: 23 Aug 1813 - From what ship: HMS Pelican - Born: Boston - Age: 32 - Discharged on 8 Sep 1813 and sent to Dartmoor.

Crosby, George - Seaman - Number: 2134 - Prize name: Amiable - Ship type: LM - How taken: HM Ship-of-the-Line Magnificant - When taken: 30 Oct 1813 - Where taken: off Lorient - Date received: 29 Nov 1813 - From what ship: HMS Dublin - Born: Fairfield - Age: 23 - Discharged on 4 Dec 1813 and sent to Dartmoor.

Crosby, John - Seaman - Number: 2571 - How taken: Sent into custody by HM Frigate Nisus - Date received: 6 May 1814 - From what ship: HMS Salvador del Mundo - Born: Titting - Age: 25 - Discharged on 10 May 1814 and sent to Dartmoor.

Cross, Oliver - Seaman - Number: 1604 - Prize name: Governor Gerry - Ship type: MV - How taken: HM Brig Royalist - When taken: 31 May 1813 - Where taken: Bay of Biscay - Date received: 14 Jun 1813 - From what ship: HMS Royalist - Born: New York - Age: 44 - Discharged on 30 Jun 1813 and sent to Stapleton.

Cross, Peter - Seaman - Number: 2253 - Prize name: Porcupine - Ship type: LM - How taken: HM Frigate Acasta - When taken: 17 Jun 1813 - Where taken: at sea - Date received: 15 Jan 1814 - From what ship: HMS Teazer - Born: Point Sicily - Age: 23 - Discharged on 31 Jan 1814 and sent to Dartmoor.

Crouch, Richard - Seaman - Number: 1128 - Prize name: Viper - Ship type: MV - How taken: HM Ship-of-the-Line Superb - When taken: 15 Apr 1813 - Where taken: Bay of Biscay - Date received: 22 Apr 1813 - From what ship: HMS Superb - Born: New York - Age: 31 - Sent to Dartmoor on 1 Jul 1813.

Crowder, James - Seaman - Number: 2502 - Prize name: Diamond, prize to the Privateer True Blooded Yankee - Ship type: P - How taken: HM Ship-of-the-Line Vengeur - When taken: 6 Mar 1814 - Where taken: Lat 47.4N Long 5.4W - Date received: 3 Apr 1814 - From what ship: HMS Rippon - Born: Colry - Age: 40 - Discharged on 10 May 1814 and sent to Dartmoor.

Crowell, Uriel - Ordinary Seaman - Number: 480 - How taken: Impressed at Greenock - When taken: 27 Dec 1812 - Date received: 10 Feb 1813 - From what ship: HMS Frederick - Born: Georgetown - Age: 33 - Sent to Chatham on 29 Mar 1813 on the HMS Braham.

Cudworth, Henry - Seaman - Number: 1601 - Prize name: Governor Gerry - Ship type: MV - How taken: HM Brig Royalist - When taken: 31 May 1813 - Where taken: Bay of Biscay - Date received: 14 Jun 1813 - From what ship: HMS Royalist - Born: Charlestown - Age: 19 - Discharged on 30 Jun 1813 and sent to Stapleton.

Cummings, Samuel - Master - Number: 2233 - Prize name: Squirrel - Ship type: MV - How taken: HM Frigate Belle Poule - When taken: 14 Dec 1813 - Where taken: Bay of Biscay - Date received: 15 Jan 1814 - From what ship: HMS Bellona - Born: New Hampshire - Age: 38 - Discharged on 16 Jan 1814 and sent to Ashburton on parole.

Cummings, James - Seaman - Number: 1391 - Prize name: Tom - Ship type: LM - How taken: HM Frigate Surveillante - When taken: 27 Apr 1813 - Where taken: Bay of Biscay - Date received: 15 May 1813 - From what ship: HMS Foxhound - Born: Connecticut - Age: 22 - Discharged on 3 Jul 1813 and sent to Stapleton Prison.

Cunningham, John - Seaman - Number: 47 - How taken: HM Battery Princess - When taken: 27 Oct 1812 - Where taken: Liverpool - Date received: 26 Nov 1813 - From what ship: HMS Frederick - Born: Charlestown - Age: 23 - Sent to Portsmouth on 29 Dec 1812 on the HMS Northumberland.

Currien, Stephen - Seaman - Number: 1754 - Prize name: Friendship, prize to the Privateer America - Ship type: P - How taken: HM Schooner Whiting - When taken: 15 Jul 1815 - Where taken: Lat 47N Long 8W - Date received: 20 Jul 1813 - From what ship: HMS Whiting - Born: Massachusetts - Age: 28 - Discharged on 8 Sep 1813 and sent to Dartmoor.

Curtis, John - Seaman - Number: 1770 - Prize name: Union, prize of the Privateer Brutus - Ship type: LM - How taken: HM Brig Goldfinch - When taken: 17 Jul 1813 - Where taken: Bay of Biscay - Date received: 25 Jul 1813 - From what ship: HMS Pyramus - Born: Woolwich, MA - Age: 20 - Discharged on 8 Sep 1813 and sent to Dartmoor.

Curtis, Joseph - Seaman - Number: 1826 - Prize name: US Brig Argus - Ship type: War - How taken: HM Brig Pelican - When taken: 14 Aug 1813 - Where taken: Irish Channel - Date received: 17 Aug 1813 - From what ship: USS Argus - Born: Arron in America - Age: 42 - Discharged on 8 Sep 1813 and sent to Dartmoor.

Custis, George - Seaman - Number: 431 - Prize name: Union - Ship type: MV - How taken: HM Frigate Iris - When taken: 17 Jan 1813 - Where taken: Lat 44 N Long 2.3 W - Date received: 5 Feb 1813 - From what ship: HMS San Josef - Born: Philadelphia - Age: 24 - Race: Mulatto - Sent to Chatham on 29 Mar 1813 on the HMS Braham.

Dagger, Robert - Prize Master - Number: 2560 - Prize name: Bunker Hill - Ship type: P - How taken: HM Frigate Pomone & HM Frigate Cydnus - When taken: 4 Mar 1814 - Where taken: Bay of Biscay - Date received: 17 Apr 1814 - From what ship: HMS Teazer - Born: Providence - Age: 21 - Discharged on 10 May 1814 and sent to Dartmoor.

Dalloway, John - Seaman - Number: 2007 - Prize name: Ned - Ship type: LM - How taken: HM Brig Royalist - When taken: 6 Sep 1812 - Where taken: coast of France - Date received: 24 Sep 181 - From what ship: HMS Rippon - Born: Boston - Age: 26 - Race: Black - Discharged on 27 Sep 1813 and sent to Dartmoor.

Daly, John - Master - Number: 445 - Prize name: Governor McKean - Ship type: LM - How taken: HM Brig Rover - When taken: 26 Jan 1813 - Where taken: off Bordeaux - Date received: 6 Feb 1813 - From what ship: Governor McKean - Born: Philadelphia - Age: 35 - Sent to 8 Feb 1813 to Ashburton on parole.

Damiere, Etienne - Seaman - Number: 2377 - Prize name: Devon, prize to the Privateer Bunker Hill - Ship type: P - How taken: HM Brig Fly - When taken: 21 Jan 1814 - Where taken: at sea - Date received: 31 Jan 1814 - From what ship: HMS Fly - Born: Paris - Age: 11 - Discharged on 27 Feb 1814 and sent to Chatham on HMS Haleyon.

Daniel, Robert - Seaman - Number: 2616 - How taken: Sent into custody from a merchant vessel - When taken: 17 Mar 1813 - Date received: 16 May 1814 - From what ship: HMS Repulse - Born: New York - Age: 27 - Discharged on 14 Jun 1814 and sent to Dartmoor.

Daniels, J. D. - Master - Number: 334 - Prize name: Rossie - Ship type: MV - How taken: Rochefort, France - When taken: 1 Jan 1813 - Where taken: Basque Roads, France - Date received: 22 Jan 1813 - From what ship: Rossie - Born: Massachusetts - Age: 29 - Sent on 23 Jan 1813 to Ashburton on parole.

Daniels, John - Seaman - Number: 1924 - Prize name: Betsey, prize to the US Brig Argus - Ship type: War - How taken: HM Frigate Leonidas - When taken: 12 Aug 1813 - Where taken: Channel - Date received: 23 Aug 1813 - From what ship: HMS Pelican - Born: Baltimore - Age: 45 - Race: Mulatto - Discharged on 8 Sep 1813 and sent to Dartmoor.

Dasing, Caesar - Seaman - Number: 8 - Prize name: Purse - Ship type: MV - How taken: HM Frigate Amide - When taken: 29 May 1812 - Where taken: off Bordeaux - Date received: 20 Nov 1812 - From what ship: HMS Salvador del Mundo - Born: Isle du France - Age: 19 - Sent to Portsmouth on 29 Dec 1812 on the HMS Northumberland.

Davey, Charles - Seaman - Number: 2486 - How taken: Impressed at Liverpool - When taken: 8 Feb 1814 - Date received: 13 Mar 1814 - From what ship: HMS Bittern - Born: New Orleans - Age: 19 - Discharged on 10 May 1814 and sent to Dartmoor.

David, William - Seaman - Number: 2322 - Prize name: Amity, prize of the Privateer Prince de Neufchatel - Ship type: P - How taken: HM Brig Achates - When taken: 22 Dec 1814 - Where taken: Bay of Biscay - Date received: 25 Jan 1814 - From what ship: HMS Conflict - Born: New Orleans - Age: 24 - Race: Black - Discharged on 31 Jan 1814 and sent to Dartmoor.

David, Thomas - Mate - Number: 341 - Prize name: Portsea, prize to the Privateer Thrasher - Ship type: P - How taken: HM Sloop Helena - When taken: 31 Dec 1813 - Where taken: off the Western Islands - Date received: 22 Jan 1813 - From what ship: HMS Helena - Born: Baltimore - Age: 29 - Sent to Chatham on 29 Mar 1813 on the HMS Braham.

Davis, Benjamin S. - Prize Master - Number: 624 - Prize name: Good Intent, prize of the Privateer Thrasher - Ship type: P - How taken: HM Frigate Pyramus - When taken: 26 Jun 1813 - Where taken: off Bordeaux - Date received: 2 Mar 1813 - From what ship: HMS Insolent - Born: Gloucester - Age: 25 - Sent to Dartmoor on 2 Apr 1813.

Davis, Charles - Seaman - Number: 746 - Prize name: Charlotte - Ship type: MV - How taken: HM Ship-of-the-Line Warspite - When taken: 3 Mar 1813 - Where taken: Bay of Biscay - Date received: 19 Mar 1813 - From what ship: HMS Warspite - Born: Norfolk - Age: 27 - Sent to Dartmoor on 2 Apr 1813.

Davis, Charles S. - Surgeon - Number: 2328 - Prize name: Siro - Ship type: LM - How taken: HM Brig Pelican - When taken: 13 Jan 1814 - Where taken: at sea - Date received: 27 Jan 1814 - From what ship: HMS Pelican - Born: Baltimore - Age: 19 - Discharged on 27 Jan 1814 and sent to Ashburton on parole.

Davis, Ezra - Passenger - Number: 852 - Prize name: Enterprise - Ship type: MV - How taken: HM Sloop Lyra - When taken: 12 Mar 1813 - Where taken: Bay of Biscay - Date received: 24 May 1813 - From what ship: Gold Coiner - Born: Charleston - Age: 38 - Sent on 11 Apr 1813 to Ashburton on parole.

Davis, Ezra - Passenger - Number: 1316 - Prize name: Enterprise - Ship type: MV - How taken: HM Brig Lyra - When taken: 12 Mar 1813 - Where taken: Bay of Biscay - Date received: 12 May 1813 - From what ship: HMS Exeter - Born: Charlton - Age: 38 - Discharged on 21 Jun 1813 and released to American ship Mount Hope.

Davis, George - Seaman - Number: 2038 - How taken: Impressed at Liverpool - When taken: 16 Oct 1813 - Date received: 1 Nov 1813 - From what ship: HMS Bittern - Born: Dayfield - Age: 23 - Discharged on 3 Nov 1813 and sent to Dartmoor.

Davis, Henry - Master - Number: 863 - Prize name: Decornau - Ship type: MV - How taken: HM Sloop Pheasant - When taken: 15 Mar 1813 - Where taken: Bay of Biscay - Date received: 27 Mar 1813 - From what ship: HMS Pheasant - Born: Boston - Age: 48 - Sent on 4 Apr 1813 to Ashburton on parole.

Davis, James - Seaman - Number: 1238 - Prize name: Essex - Ship type: MV - How taken: HM Frigate Pyramus - When taken: 2 Apr 1813 - Where taken: Bay of Biscay - Date received: 9 May 1813 - From what ship: HMS Andromache - Born: New Brunswick - Age: 26 - Discharged on 3 Jul 1813 and sent to Stapleton Prison.

Davis, James - Seaman - Number: 775 - Prize name: William Bayard - Ship type: MV - How taken: HM Ship-of-the-Line Warspite - When taken: 3 Mar 1813 - Where taken: Bay of Biscay - Date received: 19 Mar 1813 - From what ship: HMS Warspite - Born: Bristol - Age: 34 - Sent to Dartmoor on 28 Jun 1813.

Davis, John - Seaman - Number: 1368 - Prize name: Tom - Ship type: LM - How taken: HM Frigate Surveillante - When taken: 27 Apr 1813 - Where taken: Bay of Biscay - Date received: 15 May 1813 - From what ship: HMS Foxhound - Born: New Orleans - Age: 18 - Discharged on 3 Jul 1813 and sent to Stapleton Prison.

Davis, John - Seaman - Number: 66 - Prize name: Independence - Ship type: MV - How taken: HM Frigate Medusa - When taken: 9 Nov 1812 - Where taken: San Sebastian - Date received: 27 Nov 1812 - From what ship: HMS Wasp - Born: New York - Age: 24 - Sent to Portsmouth on 29 Dec 1812 on the HMS Northumberland.

Davis, John - Seaman - Number: 1387 - Prize name: Tom - Ship type: LM - How taken: HM Frigate Surveillante - When taken: 27 Apr 1813 - Where taken: Bay of Biscay - Date received: 15 May 1813 - From what ship: HMS Foxhound - Born: Massachusetts - Age: 24 - Discharged on 3 Jul 1813 and sent to Stapleton Prison.

Davis, John - Seaman - Number: 1626 - Prize name: Leo - Ship type: LM - How taken: HM Frigate Magiciene - When taken: 4 Jun 1813 - Where taken: Lat 45 Long 14 - Date received: 14 Jun 1813 - From what ship: HMS Orestes - Born: Biddeford - Age: 29 - Discharged on 30 Jun 1813 and sent to Stapleton.

Davis, John - Seaman - Number: 244 - Prize name: Janice & Lydia, prize of the Privateer General Armstrong - Ship type: P - How taken: Barton, letter of marque - When taken: 29 Nov 1812 - Where taken: off Bermuda - Date received: 12 Jan 1813 - From what ship: HMS Bittern - Born: New York - Age: 33 - Sent to Portsmouth on 8 Feb 1813 on the HMS Colossus.

Davis, John - Seaman - Number: 2550 - Prize name: Bunker Hill - Ship type: P - How taken: HM Frigate Pomone & HM Frigate Cydnus - When taken: 4 Mar 1814 - Where taken: Bay of Biscay - Date received: 17 Apr 1814 - From what ship: HMS Teazer - Born: Camden - Age: 22 - Discharged on 10 May 1814 and sent to Dartmoor.

Davis, Morris - 3rd Mate - Number: 1705 - Prize name: Joseph - Ship type: MV - How taken: HM Frigate Iris - When taken: 8 Jun 1813 - Where taken: Bay of Biscay - Date received: 4 Jul 1813 - From what ship: HMS Iris - Born: Manchester - Age: 21 - Discharged on 22 Aug 1813 and released to HM Brig Redpole.

Davis, Robert - Master - Number: 898 - Prize name: Courier - Ship type: MV - How taken: HM Frigate Andromache - When taken: 14 Mar 1813 - Where taken: Bay of Biscay - Date received: 2 Apr 1813 - From what ship: HMS Plymouth - Born: Eastham - Age: 35 - Sent on 4 Apr 1813 to Ashburton on parole.

Davis, William - Seaman - Number: 1212 - Prize name: Zebra - Ship type: LM - How taken: HM Frigate Pyramus - When taken: 20 Apr 1813 - Where taken: Bay of Biscay - Date received: 9 May 1813 - From what ship: HMS Andromache - Born: Charlestown - Age: 20 - Discharged on 3 Jul 1813 and sent to Stapleton Prison.

Davis, William - Seaman - Number: 1900 - Prize name: US Brig Argus - Ship type: War - How taken: HM Brig Pelican - When taken: 14 Aug 1813 - Where taken: Irish Channel - Date received: 23 Aug 1813 - From what ship: HMS Pelican - Born: Virginia - Age: 26 - Discharged on 8 Sep 1813 and sent to Dartmoor.

Davison, Henry - Carpenter - Number: 1013 - Prize name: Thrasher - Ship type: P - How taken: HM Frigate Magiciene - When taken: 17 Jan 1813 - Where taken: St. Mary, Western Island - Date received: 20 Apr 1813 - From what ship: HMS Libria - Born: Maryland - Age: 50 - Sent to Dartmoor on 1 Jul 1813.

Dawson, John - Seaman - Number: 105 - How taken: Impressed at Cove of Cork - Date received: 30 Dec 1812 - From what ship: HMS Leonidas - Born: Philadelphia - Age: 20 - Sent to Portsmouth on 4 Jan 1813 on the HMS

Revolutionnaire.

Day, Benjamin - Seaman - Number: 2555 - Prize name: Bunker Hill - Ship type: P - How taken: HM Frigate Pomone & HM Frigate Cydnus - When taken: 4 Mar 1814 - Where taken: Bay of Biscay - Date received: 17 Apr 1814 - From what ship: HMS Teazer - Born: Bath - Age: 34 - Discharged on 10 May 1814 and sent to Dartmoor.

Day, James - Seaman - Number: 2680 - How taken: Sent into custody from HM Frigate Phoenix - Date received: 5 Jun 1814 - From what ship: HMS Gronville - Born: Connecticut - Age: 20 - Discharged on 14 Jun 1814 and sent to Dartmoor.

Day, John - Seaman - Number: 2614 - How taken: MV Guadeloupe - When taken: 19 Dec 1812 - Date received: 16 May 1814 - From what ship: HMS Repulse - Born: Springfield - Age: 32 - Discharged on 14 Jun 1814 and sent to Dartmoor.

Day, Syles C. - Sailmaker's Mate - Number: 1818 - Prize name: US Brig Argus - Ship type: War - How taken: HM Brig Pelican - When taken: 14 Aug 1813 - Where taken: Irish Channel - Date received: 17 Aug 1813 - From what ship: USS Argus - Born: New York - Age: 31 - Discharged on 8 Sep 1813 and sent to Dartmoor.

De Colville, Laur - Seaman - Number: 437 - How taken: Impressed at Belfast - Date received: 5 Feb 1813 - From what ship: HMS Neptune - Born: Newbury - Age: 20 - Sent to Chatham on 29 Mar 1813 on the HMS Braham.

De Forest, John H. - Passenger - Number: 923 - Prize name: Prompt - Ship type: MV - How taken: Chance, privateer - When taken: 22 Mar 1813 - Where taken: Bay of Biscay - Date received: 6 Apr 1813 - From what ship: Fonvey - Born: Fairfield County, CT - Age: 36 - Sent on 9 Apr 1813 to Ashburton on parole.

Deal, John - Seaman - Number: 969 - Prize name: Ferox - Ship type: MV - How taken: HM Frigate Medusa & HM Brig Lyra - When taken: 28 Mar 1813 - Where taken: off Cape Ortagle - Date received: 9 Apr 1813 - From what ship: HMS Lyra - Born: Philadelphia - Age: 40 - Sent to Chatham on 8 Jul 1813 on HM Tender Neptune.

Deal, William - Seaman - Number: 487 - Prize name: Cashiere - Ship type: LM - How taken: HM Brig Reindeer - When taken: 3 Feb 1813 - Where taken: Bay of Biscay - Date received: 12 Feb 1813 - From what ship: HMS Reindeer - Born: Bell Haven - Age: 17 - Sent to Dartmoor on 2 Apr 1813.

Dean, Daniel - Gunner's Mate - Number: 155 - Prize name: Hunter - Ship type: P - How taken: HM Frigate Phoebe - When taken: 24 Dec 1812 - Where taken: off Western Islands - Date received: 9 Jan 1813 - From what ship: HMS Phoebe - Born: New York - Age: 32 - Sent to Portsmouth on 8 Feb 1813 on the HMS Colossus.

Dean, Jonas - Seaman - Number: 1299 - Prize name: Price - Ship type: LM - How taken: HM Frigate Medusa - When taken: 13 Apr 1813 - Where taken: Bay of Biscay - Date received: 10 May 1813 - From what ship: HMS Medusa - Born: Massachusetts - Age: 29 - Discharged on 3 Jul 1813 and sent to Stapleton Prison.

Dean, Moses - Seaman - Number: 2544 - Prize name: Mary, prize to the Privateer Blockhead - Ship type: P - How taken: HM Post Ship Crocodile - When taken: 6 Aug 1813 - Where taken: Coruna, Spain - Date received: 16 Apr 1814 - From what ship: Transport Fanny - Born: Boston - Age: 23 - Discharged on 10 May 1814 and sent to Dartmoor.

Dean, Nathaniel B. - Seaman - Number: 1394 - Prize name: Tom - Ship type: LM - How taken: HM Frigate Surveillante - When taken: 27 Apr 1813 - Where taken: Bay of Biscay - Date received: 15 May 1813 - From what ship: HMS Foxhound - Born: New Hampshire - Age: 21 - Discharged on 3 Jul 1813 and sent to Stapleton Prison.

Debaize, Francois Jean - Seaman - Number: 2373 - Prize name: Devon, prize to the Privateer Bunker Hill - Ship type: P - How taken: HM Brig Fly - When taken: 21 Jan 1814 - Where taken: at sea - Date received: 31 Jan 1814 - From what ship: HMS Fly - Born: Isle du France - Age: 21 - Discharged on 27 Feb 1814 and sent to Chatham on HMS Haleyon.

Deer, Andrew - Seaman - Number: 1028 - Prize name: Two Brothers - Ship type: MV - How taken: Bootle of Liverpool, letter of marque - When taken: 18 Mar 1813 - Where taken: Western Islands - Date received: 21 Apr 1813 - From what ship: HMS Bittern - Born: New Orleans - Age: 20 - Race: Black - Sent to Dartmoor on 1 Jul 1813.

Deham, Charles - Seaman - Number: 967 - Prize name: Ferox - Ship type: MV - How taken: HM Frigate Medusa & HM Brig Lyra - When taken: 28 Mar 1813 - Where taken: off Cape Ortagle - Date received: 9 Apr 1813 - From what ship: HMS Lyra - Born: New York - Age: 37 - Sent to Chatham on 8 Jul 1813 on HM Tender Neptune.

Dela Batiet, Francois Gabriel - Passenger - Number: 1578 - Prize name: Orders in Council - Ship type: LM - How taken: HM Frigate Surveillante - When taken: 1 Jun 1813 - Where taken: Bay of Biscay - Date received: 12 Jun 1813 - From what ship: Cutter Earl Wellington - Born: Martinique - Age: 23 - Discharged on 7 Jul 1813 and sent to Ashburton on parole.

Delaney, Matthew - Seaman - Number: 1016 - How taken: Delivered himself up from HM Ship-of-the-Line Malta - Where taken: Gibraltar - Date received: 20 Apr 1813 - From what ship: HMS Libria - Born: Philadelphia - Age: 22 - Sent to Dartmoor on 1 Jul 1813.

Delbos, Felix - Passenger - Number: 1191 - Prize name: Zebra - Ship type: LM - How taken: HM Frigate Pyramus - When taken: 20 Apr 1813 - Where taken: Bay of Biscay - Date received: 9 May 1813 - From what ship: HMS Andromache - Born: Guadeloupe - Age: 18 - Sent on 11 Jun 1813 to the American ship Hope.

Delphy, Richard - Midshipman - Number: 1814 - Prize name: US Brig Argus - Ship type: War - How taken: HM Brig Pelican - When taken: 14 Aug 1813 - Where taken: Irish Channel - Date received: 17 Aug 1813 - From what ship: Received dead - Born: Washington - Age: 18 - Died on 17 Aug 1813.

Dematra, George - Seaman - Number: 2271 - Prize name: Fanny - Ship type: MV - How taken: HM Frigate Eurotas - When taken: 25 Dec 1813 - Where taken: at sea - Date received: 20 Jan 1814 - From what ship: HMS Eurotas - Born: Virginia - Age: 30 - Discharged on 31 Jan 1814 and sent to Dartmoor.

Dempsey, John - Master - Number: 354 - Prize name: Charles - Ship type: MV - How taken: Detained at Belfast - When taken: 27 Oct 1812 - Date received: 1 Feb 1813 - From what ship: HMS Royal Sovereign - Age: 26 - Sent on 8 Feb 1813 to Ashburton on parole.

Denckle, Christian - Passenger - Number: 387 - Prize name: Union - Ship type: MV - How taken: HM Frigate Iris - When taken: 17 Jan 1813 - Where taken: Lat 44 N Long 2.3 W - Date received: 5 Feb 1813 - From what ship: HMS San Josef - Born: Philadelphia - Age: 18 - Sent on 18 Jan 1813 to Ashburton on parole.

Denham, Cornelius - Seaman - Number: 1764 - How taken: Impressed at Greenock - When taken: 13 Jun 1813 - Date received: 23 Jun 1813 - From what ship: HMS Prince Frederick - Born: Massachusetts - Age: 26 - Discharged on 8 Sep 1813 and sent to Dartmoor.

Denham, John - Seaman - Number: 1176 - Prize name: Hebe - Ship type: MV - How taken: HM Frigate Stag - When taken: 18 Apr 1813 - Where taken: Bay of Biscay - Date received: 6 May 1813 - From what ship: HMS Stag - Born: Philadelphia - Age: 25 - Sent on 8 Sep 1813 to Dartmoor.

Denison, George - Prize Master - Number: 2276 - Prize name: Siro - Ship type: LM - How taken: HM Brig Pelican - When taken: 13 Jan 1814 - Where taken: at sea - Date received: 23 Jan 1814 - From what ship: HMS Pelican - Born: Freeport - Age: 26 - Discharged on 31 Jan 1814 and sent to Dartmoor.

Dennison, Henry - Purser - Number: 1849 - Prize name: US Brig Argus - Ship type: War - How taken: HM Brig Pelican - When taken: 14 Aug 1813 - Where taken: Irish Channel - Date received: 18 Aug 1813 - From what ship: USS Argus - Born: Connecticut - Age: 32 - Discharged on 25 Aug 1813 and sent to Ashburton on parole.

Dennison, Jedidiah - Seaman - Number: 692 - Prize name: Star - Ship type: MV - How taken: HM Ship-of-the-Line Superb - When taken: 9 Feb 1813 - Where taken: Bay of Biscay - Date received: 19 Mar 1813 - From what ship: HMS Warspite - Born: Saybrook - Age: 25 - Sent to Dartmoor on 2 Apr 1813.

Dennison, Judah - Seaman - Number: 1653 - Prize name: Star - Ship type: MV - How taken: HM Ship-of-the-Line Superb - When taken: 9 Feb 1813 - Where taken: Bay of Biscay - Date received: 15 Jun 1813 - From what ship: Dartmoor Prison - Born: Saybrook - Age: 25 - Discharged on 16 Jun 1813 and released to HMS Salvador del Mundo.

Deparvier, John - Seaman - Number: 2402 - Prize name: Rachel & Ann, prize of the Privateer Prince de Neufchatel - Ship type: P - How taken: HM Frigate Cydnus - When taken: 6 Jan 1814 - Where taken: at sea - Date received: 4 Feb 1814 - From what ship: HMS Cydnus - Born: Erebough, Sweden - Age: 39 - Discharged on 10 May 1814 and sent to Dartmoor.

Deshays, Jean Francois Rine - Passenger - Number: 1594 - Prize name: Governor Gerry - Ship type: MV - How taken: HM Brig Royalist - When taken: 31 May 1813 - Where taken: Bay of Biscay - Date received: 14 Jun 1813 - From what ship: HMS Royalist - Born: Agenton, France - Age: 40 - Race: Black - Discharged on 7 Jul 1813 and sent to Ashburton on parole.

DeWitt, John - Boy - Number: 646 - Prize name: Criterion - Ship type: MV - How taken: HM Frigate Belle Poule - When taken: 14 Feb 1813 - Where taken: Bay of Biscay - Date received: 4 Mar 1813 - From what ship: HMS Strenuous - Born: New York - Age: 18 - Sent to Dartmoor on 2 Apr 1813.

Diamon, William - Seaman - Number: 2547 - Prize name: Mary, prize to the Privateer Blockhead - Ship type: P - How taken: HM Post Ship Crocodile - When taken: 6 Aug 1813 - Where taken: Coruna, Spain - Date received: 16 Apr 1814 - From what ship: Transport Fanny - Born: Rhode Island - Age: 29 - Discharged on 10 May 1814 and sent to Dartmoor.

Dibble, Reuben - Seaman - Number: 1471 - Prize name: Paul Jones - Ship type: P - How taken: HM Frigate Leonidas - When taken: 23 May 1813 - Where taken: off Cape Clear - Date received: 26 May 1813 - From what ship: HMS Leonidas - Born: Hartford, CT - Age: 23 - Discharged on 30 Jun 1813 and sent to Stapleton.

Dickenson, Francis - Ordinary Seaman - Number: 286 - Prize name: Columbia - Ship type: MV - How taken: HM Frigate Briton - When taken: 17 Dec 1812 - Where taken: off Bordeaux - Date received: 21 Jan 1813 - From what ship: HMS Abercrombie - Born: Philadelphia - Age: 17 - Sent to Portsmouth on 8 Feb 1813 on the HMS Colossus.

Dickinson, Chester - Seaman - Number: 1534 - Prize name: Courier - Ship type: LM - How taken: HM Brig Rover - When taken: 14 Mar 1813 - Where taken: Bay of Biscay - Date received: 29 May 1813 - From what ship: HMS Hannibal - Born: Massachusetts - Age: 24 - Discharged on 30 Jun 1813 and sent to Stapleton.

Dickson, Charles - Seaman - Number: 638 - Prize name: Criterion - Ship type: MV - How taken: HM Frigate Belle Poule - When taken: 14 Feb 1813 - Where taken: Bay of Biscay - Date received: 4 Mar 1813 - From what ship: HMS Strenuous - Born: New York - Age: 24 - Sent to Dartmoor on 2 Apr 1813.

Dickson, Richard - Seaman - Number: 304 - Prize name: Stephen - Ship type: MV - How taken: HM Frigate Briton & HM Frigate Andromache - When taken: 17 Dec 1812 - Where taken: off Bordeaux - Date received: 21 Jan 1813 - From what ship: HMS Abercrombie - Born: Long Island - Age: 24 - Sent to Chatham on 29 Mar 1813 on the HMS Braham.

Dieman, John - Seaman - Number: 2201 - Prize name: General Kempt, prize of the Privateer Grand Turk - Ship type: P - How taken: HM Brig Foxhound - When taken: 18 Dec 1813 - Where taken: Lat 48.4 Long 6 - Date received: 21 Dec 1813 - From what ship: HMS Foxhound - Born: Marblehead - Age: 20 - Discharged on 27 Feb 1814 and sent to Chatham on HMS Haleyon.

Dillin, Pierre - Seaman - Number: 1288 - Prize name: Price - Ship type: LM - How taken: HM Frigate Medusa - When taken: 13 Apr 1813 - Where taken: Bay of Biscay - Date received: 10 May 1813 - From what ship: HMS Medusa - Born: New York - Age: 18 - Discharged on 8 Sep 1813 and sent to Dartmoor.

Dillon, William - Marine - Number: 1869 - Prize name: US Brig Argus - Ship type: War - How taken: HM Brig Pelican - When taken: 14 Aug 1813 - Where taken: Irish Channel - Date received: 23 Aug 1813 - From what ship: HMS Pelican - Born: New Jersey - Age: 48 - Discharged on 8 Sep 1813 and sent to Dartmoor.

Dilus, Benjamin - Seaman - Number: 1232 - Prize name: Essex - Ship type: MV - How taken: HM Frigate Pyramus - When taken: 2 Apr 1813 - Where taken: Bay of Biscay - Date received: 9 May 1813 - From what ship: HMS Andromache - Born: Massachusetts - Age: 19 - Discharged on 3 Jul 1813 and sent to Stapleton Prison.

Dempsey, William - Mate - Number: 439 - Prize name: Charles - Ship type: MV - How taken: Stopped at Belfast - When taken: 8 Aug 1812 - Date received: 5 Feb 1813 - From what ship: HMS Neptune - Born: Beverly - Age: 21 - Sent on 8 Feb 1813 to Ashburton on parole.

Dingell, George - Seaman - Number: 673 - How taken: Taken out of Flag Truce Pennsylvania - Date received: 15 Mar 1813 - From what ship: HMS Bittern - Born: Maryland - Age: 29 - Sent to Dartmoor on 1 Jul 1813.

Dennison, Thomas - Seaman - Number: 2028 - Prize name: Friendship West, prize of Privateer True Blooded Yankee - Ship type: P - How taken: HM Schooner Helicon & HM Schooner Whiting - When taken: 25 Oct 1813 - Where taken: Bay of Biscay - Date received: 31 Oct 1813 - From what ship: HMS Whiting - Born: Christiansted, Denmark - Age: 27 - Discharged on 3 Nov 1813 and sent to Dartmoor.

Dinsmore, John - Seaman - Number: 1641 - Prize name: Tickler - Ship type: LM - How taken: HM Frigate Magiciene - When taken: 5 Jun 1813 - Where taken: Lat 47 Long 13 - Date received: 14 Jun 1813 - From what ship: HMS Orestes - Born: New Castle - Age: 23 - Discharged on 8 Sep 1813 and sent to Dartmoor.

Dishele, Alexander - Seaman - Number: 311 - Prize name: Blue Bird - Ship type: MV - How taken: HM Frigate Briton - When taken: 1 Jan 1813 - Where taken: off Bordeaux - Date received: 21 Jan 1813 - From what ship: HMS Abercrombie - Born: Little York - Age: 29 - Sent to Chatham on 29 Mar 1813 on the HMS Braham.

Divers, Charles - Seaman - Number: 797 - Prize name: Cannoniere - Ship type: P - How taken: HM Ship-of-the-Line Warspite - When taken: 14 Mar 1813 - Where taken: Bay of Biscay - Date received: 19 Mar 1813 - From what ship: HMS Warspite - Born: Lancaster - Age: 18 - Sent to Dartmoor on 28 Jun 1813.

Dixey, John - Master - Number: 482 - Prize name: Print - Ship type: MV - How taken: HM Frigate Rhin - When taken: 15 Jan 1813 - Where taken: Lat 44 N Long 17 W - Date received: 11 Feb 1813 - From what ship: HMS Plymouth - Born: Marblehead - Age: 36 - Sent on 12 Jan 1813 to Ashburton on parole.

Dixon, Benjamin - Seaman - Number: 2231 - How taken: Taken off an English merchant vessel - Date received: 10 Jan 1814 - From what ship: HMS Scylla - Born: Baltimore - Age: 30 - Discharged on 31 Jan 1814 and sent to Dartmoor.

Dixon, Michael - Master - Number: 123 - Prize name: Columbia - Ship type: MV - How taken: HM Frigate Briton - When taken: 17 Dec 1812 - Where taken: off Bordeaux - Date received: 1 Jan 1813 - From what ship: HMS Briton - Born: Wexford, Ireland - Age: 36 - Sent on 8 Dec 1812 to Ashburton on parole.

Dobbins, John - Seaman - Number: 628 - How taken: Taken off the HM Transport William No. 69 - Date received: 4 Mar 1813 - From what ship: HMS Plymouth - Born: Harford County - Age: 27 - Sent to Dartmoor on 2 Apr 1813.

Dodd, Samuel - Seaman - Number: 550 - Prize name: Spitfire - Ship type: MV - How taken: HM Brig Achates - When taken: 14 Feb 1813 - Where taken: off Ushant - Date received: 16 Feb 1813 - From what ship: HMS Achates - Born: Marblehead - Age: 17 - Sent to Dartmoor on 2 Apr 1813.

Dolinson, Andrew - Chief Mate - Number: 831 - Prize name: Allegany - Ship type: MV - How taken: Detained at Gibraltar - When taken: 8 Aug 1812 - Date received: 22 Mar 1813 - From what ship: HMS Dauntless - Born: New Berne - Age: 27 - Sent on 24 Mar 1813 to Ashburton on parole.

D'Olivera, Manuel (alias Jerry Jarvis) - Seaman - Number: 1762 - How taken: Impressed at Liverpool - When taken: 6 Jul 1813 - Date received: 23 Jun 1813 - From what ship: HMS Prince Frederick - Born: Connecticut - Age: 31 - Race: Negro - Discharged on 8 Sep 1813 and sent to Dartmoor.

Dolliver, Francis - Seaman - Number: 554 - Prize name: Spitfire - Ship type: MV - How taken: HM Brig Achates - When taken: 14 Feb 1813 - Where taken: off Ushant - Date received: 16 Feb 1813 - From what ship: HMS Achates - Born: Marblehead - Age: 28 - Sent to Dartmoor on 2 Apr 1813.

Dolliver, John - Seaman - Number: 1712 - Prize name: Joseph - Ship type: MV - How taken: HM Frigate Iris - When taken: 8 Jun 1813 - Where taken: Bay of Biscay - Date received: 4 Jul 1813 - From what ship: HMS Iris - Born: Gloucester - Age: 19 - Discharged on 22 Aug 1813 and released to HM Brig Redpole.

Dolliver, Richard - Seaman - Number: 2196 - Prize name: General Kempt, prize of the Privateer Grand Turk - Ship type: P - How taken: HM Brig Foxhound - When taken: 18 Dec 1813 - Where taken: Lat 48.4 Long 6 - Date received: 21 Dec 1813 - From what ship: HMS Foxhound - Born: Marblehead - Age: 19 - Discharged on 31 Jan 1814 and sent to Dartmoor.

Dolliver, William - Seaman - Number: 1157 - Prize name: Essex - Ship type: MV - How taken: HM Frigate Pyramus - When taken: 2 Apr 1813 - Where taken: Bay of Biscay - Date received: 2 Mar 1813 - From what ship: HMS Rota - Born: Cape Ann - Age: 16 - Sent on 3 Jul 1813 to Stapleton prison.

Dolorer, John - Seaman - Number: 183 - Prize name: Hunter - Ship type: P - How taken: HM Frigate Phoebe - When taken: 24 Dec 1812 - Where taken: off Western Islands - Date received: 9 Jan 1813 - From what ship: HMS Phoebe - Born: Marblehead - Age: 18 - Sent to Portsmouth on 8 Feb 1813 on the HMS Colossus.

Dominic, John - Seaman - Number: 134 - Prize name: Nope - Ship type: MV - How taken: HM Schooner Bramble - When taken: 3 Dec 1812 - Where taken: Coruna, Spain - Date received: 7 Jan 1813 - From what ship: Nope - Born: New Orleans - Age: 25 - Sent to Portsmouth on 8 Feb 1813 on the HMS Colossus.

Donaldson, Joseph - Seaman - Number: 901 - Prize name: Dick - Ship type: MV - How taken: HM Brig Dispatch - When taken: 17 Mar 1813 - Where taken: Bay of Biscay - Date received: 2 Apr 1813 - From what ship: HMS Warspite - Born: New York - Age: 21 - Sent on 21 Jun 1813 to Portsmouth on HMS Prometheus.

Donelson, Joseph - Seaman - Number: 2354 - Prize name: Zephyr - Ship type: MV - How taken: HM Frigate Surveillante - When taken: 6 Jan 1814 - Where taken: Bay of Biscay - Date received: 31 Jan 1814 - From what ship: HMS Surveillante - Born: Grodno, Poland - Age: 29 - Discharged on 10 May 1814 and sent to Dartmoor.

Donelson, Joseph - Seaman - Number: 2701 - Prize name: Zephyr - Ship type: MV - How taken: HM Frigate Surveillante - When taken: 6 Jan 1814 - Where taken: Bay of Biscay - Date received: 17 Jun 1814 - From what ship: Dartmouth - Discharged on 20 Jun 1814 and released to the Speedwell Cartel.

Donham, Ebenezer - Marine - Number: 1873 - Prize name: US Brig Argus - Ship type: War - How taken: HM Brig Pelican - When taken: 14 Aug 1813 - Where taken: Irish Channel - Date received: 23 Aug 1813 - From what ship: HMS Pelican - Born: Connecticut - Age: 18 - Discharged on 8 Sep 1813 and sent to Dartmoor.

Doolittle, Henry - Seaman - Number: 1390 - Prize name: Tom - Ship type: LM - How taken: HM Frigate Surveillante - When taken: 27 Apr 1813 - Where taken: Bay of Biscay - Date received: 15 May 1813 - From what ship: HMS Foxhound - Born: Connecticut - Age: 21 - Discharged on 3 Jul 1813 and sent to Stapleton Prison.

Doolittle, Isaac - Passenger - Number: 1315 - Prize name: Ducornau - Ship type: MV - How taken: HM Sloop Pheasant - When taken: 15 Mar 1813 - Where taken: Bay of Biscay - Date received: 12 May 1813 - From what ship: HMS Exeter - Born: New Haven - Age: 28 - Discharged on 21 Jun 1813 and released to American ship Mount Hope.

Doolittle, Isaac - Passenger - Number: 863 - Prize name: Decornau - Ship type: MV - How taken: HM Sloop Pheasant - When taken: 15 Mar 1813 - Where taken: Bay of Biscay - Date received: 27 Mar 1813 - From what ship: HMS Pheasant - Born: New Haven - Age: 28 - Sent on 12 Apr 1813 to Ashburton on parole.

Dorrell, John - Seaman - Number: 776 - Prize name: William Bayard - Ship type: MV - How taken: HM Ship-of-the-Line Warspite - When taken: 3 Mar 1813 - Where taken: Bay of Biscay - Date received: 19 Mar 1813 - From what ship: HMS Warspite - Born: Wethersfield - Age: 27 - Sent to Dartmoor on 28 Jun 1813.

Dorsey, John - Seaman - Number: 2382 - Prize name: Devon, prize to the Privateer Bunker Hill - Ship type: P - How taken: HM Brig Fly - When taken: 21 Jan 1814 - Where taken: at sea - Date received: 31 Jan 1814 - From what ship: HMS Fly - Born: Albany - Age: 30 - Race: Mulatto - Discharged on 21 Jun 1814 and sent to Royal Hospital.

Douarty, Angelo - Seaman - Number: 842 - Prize name: Pallas - Ship type: MV - How taken: HM Brig Rebuff - When taken: 23 Dec 1812 - Where taken: off Cadiz - Date received: 23 May 1813 - From what ship: HMS Dauntless - Born: New Orleans - Age: 42 - Sent to Dartmoor on 28 Jun 1813.

Douchney, Hiram - Ordinary Seaman - Number: 287 - Prize name: Columbia - Ship type: MV - How taken: HM Frigate Briton - When taken: 17 Dec 1812 - Where taken: off Bordeaux - Date received: 21 Jan 1813 - From what ship: HMS Abercrombie - Born: Bridgetown - Age: 24 - Sent to Portsmouth on 8 Feb 1813 on the HMS Colossus.

Dougal, Thomas - Seaman - Number: 1643 - Prize name: Tickler - Ship type: LM - How taken: HM Frigate Magiciene - When taken: 5 Jun 1813 - Where taken: Lat 47 Long 13 - Date received: 14 Jun 1813 - From what ship: HMS Orestes - Born: Waterford, Ireland - Age: 50 - Discharged on 30 Jun 1813 and sent to Stapleton.

Dougherty, Michael - Mate - Number: 1061 - Prize name: Meteor - Ship type: MV - How taken: HM Frigate Briton - When taken: 12 Mar 1813 - Where taken: Bay of Biscay - Date received: 22 Apr 1813 - From what ship: HMS Superb - Born: New York - Age: 24 - Sent on 25 Feb 1813 to Ashburton on parole.

Doughty, Jesse - Seaman - Number: 1629 - Prize name: Leo - Ship type: LM - How taken: HM Frigate Magiciene - When taken: 4 Jun 1813 - Where taken: Lat 45 Long 14 - Date received: 14 Jun 1813 - From what ship: HMS Orestes - Born: Massachusetts - Age: 21 - Discharged on 30 Jun 1813 and sent to Stapleton.

Doughty, Levi - Seaman - Number: 1585 - Prize name: Orders in Council - Ship type: LM - How taken: HM Frigate Surveillante - When taken: 1 Jun 1813 - Where taken: Bay of Biscay - Date received: 13 Jun 1813 - From what ship: Forvey - Born: Brunswick - Age: 22 - Discharged on 30 Jun 1813 and sent to Stapleton.

Douglas, Thomas - Master - Number: 124 - Prize name: Columbia - Ship type: MV - How taken: HM Frigate Briton - When taken: 17 Dec 1812 - Where taken: off Bordeaux - Date received: 1 Jan 1813 - From what ship: HMS Briton - Born: Rhode Island - Age: 25 - Sent to Portsmouth on 4 Jan 1813 on the HMS Revolutionnaire.

Douglas, Thomas - Boy - Number: 430 - Prize name: Union - Ship type: MV - How taken: HM Frigate Iris - When taken: 17 Jan 1813 - Where taken: Lat 44 N Long 2.3 W - Date received: 5 Feb 1813 - From what ship: HMS San Josef - Born: Alexandria - Age: 15 - Race: Black - Sent to Chatham on 29 Mar 1813 on the HMS Braham.

Dover, Charles - Seaman - Number: 1007 - Prize name: Courier - Ship type: LM - How taken: HM Frigate Andromache - When taken: 14 Mar 1813 - Where taken: Bay of Biscay - Date received: 16 Apr 1813 - From what ship: HMS Fairy - Born: Chapely - Age: 23 - Sent to Dartmoor on 8 Sep 1813.

Dow, John - Seaman - Number: 370 - Prize name: Orbit - Ship type: MV - How taken: HM Brig Achates - When taken: 29 Jan 1813 - Where taken: Lat 49 N Long 13 W - Date received: 4 Feb 1813 - From what ship: HMS Achates - Born: Massachusetts - Age: 22 - Sent to Chatham on 29 Mar 1813 on the HMS Braham.

Dowdell, George R. - Master - Number: 214 - Prize name: Vengeance - Ship type: LM - How taken: HM Frigate Phoebe - When taken: 1 Jan 1813 - Where taken: Lat 44.4 Long 23 - Date received: 9 Jan 1813 - From what ship: HMS Phoebe - Born: New York - Age: 30 - Sent on 13 Jan 1813 to Ashburton on parole.

Dowell, Isaac - 2nd Mate - Number: 953 - Prize name: Courier - Ship type: LM - How taken: HM Frigate Andromache - When taken: 14 Mar 1813 - Where taken: Bay of Biscay - Date received: 7 Apr 1813 - From what ship: HMS Sea Lark - Born: Virginia - Age: 25 - Sent to Dartmoor on 28 Jun 1813.

Dowling, Anthony - Seaman - Number: 249 - Prize name: Rising Sun - Ship type: MV - How taken: HM Ship-Sloop Jalousie - When taken: 12 Dec 1812 - Where taken: Lat 43 Long 20 - Date received: 16 Jan 1813 - From what ship: HMS Stork - Born: New York - Age: 28 - Race: Black - Sent to Portsmouth on 8 Feb 1813 on the HMS Colossus.

Downe, William - Seaman - Number: 2467 - Prize name: Fair American - Ship type: MV - How taken: HM Frigate Andromache - When taken: 19 Jan 1814 - Where taken: Bay of Biscay - Date received: 22 Feb 1814 - From what ship: HMS York - Born: Boston - Age: 25 - Discharged on 10 May 1814 and sent to Dartmoor.

Drake, Henry - Seaman - Number: 498 - Prize name: Cashiere - Ship type: LM - How taken: HM Brig Reindeer - When taken: 3 Feb 1813 - Where taken: Bay of Biscay - Date received: 12 Feb 1813 - From what ship: HMS Reindeer - Born: New York - Age: 23 - Race: Black - Sent to Dartmoor on 2 Apr 1813.

Drake, John - Seaman - Number: 2524 - Prize name: Hope, prize to the Privateer True Blooded Yankee - Ship type: P - How taken: HM Frigate Seahorse - When taken: 22 Mar 1814 - Where taken: at sea - Date received: 6 Apr 1814 - From what ship: HMS Queen Charlotte - Born: Philadelphia - Age: 33 - Discharged on 10 May 1814 and sent to Dartmoor.

Drew, Charles - Seaman - Number: 2557 - Prize name: Bunker Hill - Ship type: P - How taken: HM Frigate Pomone & HM Frigate Cydnus - When taken: 4 Mar 1814 - Where taken: Bay of Biscay - Date received: 17 Apr 1814 - From what ship: HMS Teazer - Born: Boston - Age: 31 - Discharged on 10 May 1814 and sent to Dartmoor.

Drew, William - Seaman - Number: 1769 - Prize name: Union, prize of the Privateer Brutus - Ship type: LM - How taken: HM Brig Goldfinch - When taken: 17 Jul 1813 - Where taken: Bay of Biscay - Date received: 25 Jul 1813 - From what ship: HMS Pyramus - Born: Massachusetts - Age: 17 - Discharged on 8 Sep 1813 and sent to Dartmoor.

Drinkwater, Peter - 3rd Lieutenant - Number: 2273 - Prize name: Siro - Ship type: LM - How taken: HM Brig Pelican - When taken: 13 Jan 1814 - Where taken: at sea - Date received: 23 Jan 1814 - From what ship: HMS Pelican - Born: Boston - Age: 25 - Discharged on 31 Jan 1814 and sent to Dartmoor.

Driver, John - Seaman - Number: 2620 - How taken: HM Ship-of-the-Line Edinburgh - When taken: 28 Oct 1812 - Date received: 16 May 1814 - From what ship: HMS Repulse - Born: New York - Age: 29 - Discharged on 14 Jun 1814 and sent to Dartmoor.

Drybourgh, James - 2nd Mate - Number: 102 - Prize name: Argus - Ship type: MV - How taken: Fancy, Cutter - When taken: 19 Dec 1812 - Where taken: Bay of Biscay - Date received: 27 Dec 1812 - From what ship: Fancy, Cutter - Born: Philadelphia - Age: 31 - Sent to Portsmouth on 29 Dec 1812 on the HMS Northumberland.

Dubois, Alexander - Seaman - Number: 2086 - Prize name: Sybella - Ship type: MV - How taken: HM Brig Zenobia - When taken: 17 Jun 1813 - Where taken: off Cape Henry - Date received: 2 Nov 1813 - From what ship: HMS Bacchus - Born: New York - Age: 25 - Race: Black - Discharged on 3 Nov 1813 and sent to Dartmoor.

Ducassau, Charles Arnold - Passenger & Merchant - Number: 77 - Prize name: Independence - Ship type: MV - How taken: HM Frigate Medusa - When taken: 9 Nov 1812 - Where taken: San Sebastian - Date received: 29 Nov 1812 - From what ship: HMS Wasp - Born: Bordeaux - Age: 22 - Sent to Portsmouth on 29 Dec 1812 on the HMS Northumberland.

Duff, James - Seaman - Number: 2323 - Prize name: Amity, prize of the Privateer Prince de Neufchatel - Ship type: P - How taken: HM Brig Achates - When taken: 22 Dec 1814 - Where taken: Bay of Biscay - Date received: 25 Jan 1814 - From what ship: HMS Conflict - Born: Philadelphia - Age: 26 - Discharged on 31 Jan 1814 and sent to Dartmoor.

Duffy, Nathaniel - Seaman - Number: 325 - Prize name: Porcupine - Ship type: MV - How taken: HM Frigate Dryad - When taken: 8 Jan 1813 - Where taken: off Bordeaux - Date received: 21 Jan 1813 - From what ship: HMS Abercrombie - Born: Pawtucket - Age: 21 - Sent to Chatham on 29 Mar 1813 on the HMS Braham.

Dumas, Raymond - Merchant & Passenger - Number: 2238 - Prize name: Squirrel - Ship type: MV - How taken: HM Frigate Belle Poule - When taken: 14 Dec 1813 - Where taken: Bay of Biscay - Date received: 15 Jan 1814 - From what ship: HMS Bellona - Born: France - Age: 54 - Discharged on 29 Jan 1814 and sent to Ashburton on parole.

Dunlap, James Alexander - Passenger - Number: 1187 - Prize name: Zebra - Ship type: LM - How taken: HM Frigate Pyramus - When taken: 20 Apr 1813 - Where taken: Bay of Biscay - Date received: 9 May 1813 - From what ship: HMS Andromache - Born: New York - Age: 22 - Sent on 11 May 1813 to Ashburton on parole.

Duncan, George - Seaman - Number: 1988 - Prize name: Ned - Ship type: LM - How taken: HM Brig Royalist - When taken: 6 Sep 1812 - Where taken: coast of France - Date received: 22 Sep 1813 - From what ship: HMS Royalist - Born: New York - Age: 18 - Discharged on 27 Sep 1813 and sent to Dartmoor.

Duncan, Peter - Seaman - Number: 1121 - Prize name: Viper - Ship type: MV - How taken: HM Ship-of-the-Line Superb - When taken: 15 Apr 1813 - Where taken: Bay of Biscay - Date received: 22 Apr 1813 - From what ship: HMS Superb - Born: Boston - Age: 30 - Discharged on 17 Jul 1813 and sent on the HMS Salvador del Mundo.

Dunchller, Isaac - Seaman - Number: 1003 - Prize name: Eliza - Ship type: MV - How taken: HM Frigate Surveillante - When taken: 27 Mar 1813 - Where taken: Bay of Biscay - Date received: 16 Apr 1813 - From what ship: HMS Fairy - Born: Boston - Age: 21 - Sent to Chatham on 8 Jul 1813 on HM Tender Neptune.

Dunn, David - Seaman - Number: 933 - Prize name: Weasel - Ship type: MV - How taken: HM Brig Foxhound - When taken: 25 Mar 1813 - Where taken: Bay of Biscay - Date received: 6 Apr 1813 - From what ship: HMS Foxhound - Born: New York - Age: 27 - Sent on 21 Jun 1813 to Portsmouth on HMS Prometheus.

Dunn, Hezekiah - Seaman - Number: 100 - Prize name: Experiment - Ship type: MV - How taken: HM Brig Rover - When taken: 21 Oct 1812 - Where taken: off Bordeaux - Date received: 25 Dec 1812 - From what ship: HMS Northumberland - Born: Maryland - Age: 24 - Sent to Portsmouth on 29 Dec 1812 on the HMS Northumberland.

Dunn, John - Seaman - Number: 1333 - Prize name: Fox - Ship type: LM - How taken: HM Sloop Pheasant - When taken: 23 Apr 1813 - Where taken: Bay of Biscay - Date received: 14 May 1813 - From what ship: HMS Pleasant - Born: Savannah - Age: 34 - Discharged on 3 Jul 1813 and sent to Stapleton Prison.

Dunn, William - Seaman - Number: 2263 - Prize name: Growler - Ship type: MV - How taken: HM Brig Wolf - When taken: 11 Aug 1813 - Where taken: at sea - Date received: 20 Jan 1814 - From what ship: MV Nero - Born: Alexandria - Age: 23 - Discharged on 31 Jan 1814 and sent to Dartmoor.

Duran, Pierre - Boy - Number: 1351 - Prize name: Fox - Ship type: LM - How taken: HM Sloop Pheasant - When taken: 23 Apr 1813 - Where taken: Bay of Biscay - Date received: 15 May 1813 - From what ship: HMS Pleasant - Born: Bordeaux, France - Age: 8 - Discharged on 8 Jun 1813 and entered into the French general entry book.

Durand, John - Seaman - Number: 1540 - Prize name: Zebra - Ship type: LM - How taken: HM Frigate Pyramus - When taken: 20 Apr 1813 - Where taken: Bay of Biscay - Date received: 29 May 1813 - From what ship: HMS Hannibal - Born: New Orleans - Age: 22 - Race: Colored - Discharged on 30 Jun 1813 and sent to Stapleton.

Durvolf, Stephen - Seaman - Number: 1965 - Prize name: Joel Barlow - Ship type: LM - How taken: HM Frigate Briton - When taken: 3 Jul 1813 - Where taken: off Bordeaux - Date received: 31 Aug 1813 - From what ship: HMS Clarence - Born: Connecticut - Age: 24 - Discharged on 8 Sep 1813 and sent to Dartmoor.

Dusheets, Arthur - Seaman - Number: 2113 - Prize name: Chesapeake - Ship type: LM - How taken: HM Frigate Hotspur & HM Frigate Pyramus - When taken: 26 Oct 1813 - Where taken: Bay of Biscay - Date received: 22 Nov 1813 - From what ship: HMS Pyramus - Born: Maryland - Age: 20 - Discharged on 29 Nov 1813 and sent to Dartmoor.

Duston, Peter - Seaman - Number: 693 - Prize name: Star - Ship type: MV - How taken: HM Ship-of-the-Line Superb

- When taken: 9 Feb 1813 - Where taken: Bay of Biscay - Date received: 19 Mar 1813 - From what ship: HMS Warspite - Born: New York - Age: 19 - Sent to Dartmoor on 2 Apr 1813.

Dwight, Samuel - Seaman - Number: 570 - Prize name: Rolla - Ship type: MV - How taken: HM Frigate Surveillante - When taken: 11 Feb 1813 - Where taken: Bay of Biscay - Date received: 23 Feb 1813 - From what ship: HMS Surveillante - Born: Evesham, England - Age: 26 - Sent to Dartmoor on 2 Apr 1813.

Dye, William - Master - Number: 629 - Prize name: Mars - Ship type: MV - How taken: HM Ship-of-the-Line Warspite - When taken: 26 Feb 1813 - Where taken: Bay of Biscay - Date received: 4 Mar 1813 - From what ship: Mars - Born: Charles County - Age: 37 - Sent on 6 Mar 1813 to Ashburton on parole.

Dyer, Johnathan - Seaman - Number: 2074 - Prize name: Avon, prize of the Privateer True Blooded Yankee - Ship type: P - How taken: HM Frigate Eurotas - When taken: 27 Oct 1813 - Where taken: off Ushant - Date received: 1 Nov 1813 - From what ship: HMS Hannibal - Born: Cape Cod - Age: 40 - Discharged on 3 Nov 1813 and sent to Dartmoor.

Dymoss, Peter - Prize Master - Number: 1750 - Prize name: Fox Packet, prize of the Privateer Fox - Ship type: P - How taken: Superior, letter of marque - When taken: 25 Jun 1813 - Where taken: Lat 50N Long 21W - Date received: 19 Jul 1813 - From what ship: HMS Bittern - Born: Bath - Age: 22 - Discharged on 8 Sep 1813 and sent to Dartmoor.

Easterly, George - Master - Number: 873 - Prize name: John & Frances - Ship type: MV - How taken: HM Frigate Belle Poule - When taken: 19 Mar 1813 - Where taken: Bay of Biscay - Date received: 30 Mar 1813 - From what ship: HMS Plymouth - Born: Baltimore - Age: 36 - Sent on 31 Mar 1813 to Ashburton on parole.

Eastland, James - Seaman - Number: 1154 - Prize name: Essex - Ship type: MV - How taken: HM Frigate Pyramus - When taken: 2 Apr 1813 - Where taken: Bay of Biscay - Date received: 2 Mar 1813 - From what ship: HMS Rota - Born: Marblehead - Age: 35 - Sent on 24 Apr 1813 to Ashburton on parole.

Easton Joseph - Seaman - Number: 2173 - Prize name: Wolf, prize of the Privateer Grand Turk - Ship type: P - How taken: HM Frigate Briton - When taken: 1 Dec 1813 - Where taken: Bay of Biscay - Date received: 17 Dec 1813 - From what ship: HMS Briton - Born: Beverly - Age: 20 - Discharged on 17 Dec 1813 and released to HMS Britton.

Eaton, George - Seaman - Number: 787 - Prize name: Cannoniere - Ship type: P - How taken: HM Ship-of-the-Line Warspite - When taken: 14 Mar 1813 - Where taken: Bay of Biscay - Date received: 19 Mar 1813 - From what ship: HMS Warspite - Born: Milton - Age: 27 - Sent to Dartmoor on 28 Jun 1813.

Eddey, John - Seaman - Number: 675 - How taken: Taken out of Flag Truce Pennsylvania - Date received: 15 Mar 1813 - From what ship: HMS Bittern - Born: Hampton - Age: 18 - Sent to Dartmoor on 28 Jun 1813.

Edgar, William - Seaman - Number: 2092 - Prize name: Hepa - Ship type: MV - How taken: HM Brig Zenobia - When taken: 28 Jun 1813 - Where taken: off Lisbon - Date received: 2 Nov 1813 - From what ship: HMS Bacchus - Born: New Jersey - Age: 36 - Discharged on 3 Nov 1813 and sent to Dartmoor.

Edgerly, George - Seaman - Number: 837 - Prize name: Pallas - Ship type: MV - How taken: HM Brig Rebuff - When taken: 23 Dec 1812 - Where taken: off Cadiz - Date received: 22 Mar 1813 - From what ship: HMS Dauntless - Born: Essex - Age: 19 - Sent to Dartmoor on 28 Jun 1813.

Edgerly, William - Seaman - Number: 1558 - How taken: Delivered himself up from HM Ship-of-the-Line Royal Sovereign - Date received: 29 May 1813 - From what ship: HMS Hannibal - Born: Portsmouth - Age: 22 - Discharged on 8 Jul 1813 and sent to Chatham on HM Tender Neptune.

Edsom, John - Seaman - Number: 1550 - Prize name: Zebra - Ship type: LM - How taken: HM Frigate Pyramus - When taken: 20 Apr 1813 - Where taken: Bay of Biscay - Date received: 29 May 1813 - From what ship: HMS Hannibal - Born: New Hampshire - Age: 28 - Discharged on 30 Jun 1813 and sent to Stapleton.

Edward, Prince - Cook - Number: 845 - Prize name: Pallas - Ship type: MV - How taken: HM Brig Rebuff - When taken: 23 Dec 1812 - Where taken: off Cadiz - Date received: 23 May 1813 - From what ship: HMS Dauntless - Born: Staten Island - Age: 40 - Race: Black - Sent to Dartmoor on 28 Jun 1813.

Edwards, David - Seaman - Number: 1254 - Prize name: Caroline - Ship type: MV - How taken: HM Frigate Medusa - When taken: 12 Apr 1813 - Where taken: Bay of Biscay - Date received: 10 May 1813 - From what ship: HMS

Medusa - Born: New York - Age: 28 - Race: Colored - Discharged on 3 Jul 1813 and sent to Stapleton Prison.

Edwards, John - Seaman - Number: 2390 - How taken: Delivered himself up from HM Frigate Crescent - When taken: 27 Dec 1813 - Where taken: St. Johns, New Brunswick - Date received: 2 Feb 1814 - From what ship: HMS Pheasant - Born: New York - Age: 28 - Race: Black - Discharged on 10 May 1814 and sent to Dartmoor.

Edwards, John - Carpenter - Number: 1420 - Prize name: Paul Jones - Ship type: P - How taken: HM Frigate Leonidas - When taken: 23 May 1813 - Where taken: off Cape Clear - Date received: 26 May 1813 - From what ship: HMS Leonidas - Born: New York - Age: 42 - Discharged on 3 Jul 1813 and sent to Stapleton Prison.

Edwards, John (alias Rouse) - Carpenter - Number: 1454 - Prize name: Paul Jones - Ship type: P - How taken: HM Frigate Leonidas - When taken: 23 May 1813 - Where taken: off Cape Clear - Date received: 26 May 1813 -

From what ship: HMS Leonidas - Born: Louisiana - Age: 26 - Discharged on 30 Jun 1813 and sent to Stapleton.

Eggert, Francis - Seaman - Number: 1804 - Prize name: US Brig Argus - Ship type: War - How taken: HM Brig Pelican - When taken: 14 Aug 1813 - Where taken: Irish Channel - Date received: 17 Aug 1813 - From what ship: USS Argus - Died on 31 Aug 1813 in Mill Prison Hospital.

Eldridge, Nathaniel - Seaman - Number: 1015 - How taken: Delivered himself up from HM Ship-of-the-Line Malta - Where taken: Gibraltar - Date received: 20 Apr 1813 - From what ship: HMS Libria - Born: Massachusetts - Age: 32 - Sent to Dartmoor on 1 Jul 1813.

Elles, William - Seaman - Number: 589 - Prize name: St. Martin's Plantation, prize of the Privateer Paul Jones - Ship type: P - How taken: HM Ship-of-the-Line Dublin - When taken: 9 Feb 1815 - Where taken: Lat 43 N Long 33.5 W - Date received: 25 Feb 1813 - From what ship: HMS Dublin - Born: Rhode Island - Age: 17 - Sent to Dartmoor on 2 Apr 1813.

Elliot, Francis - Seaman - Number: 403 - Prize name: Union - Ship type: MV - How taken: HM Frigate Iris - When taken: 17 Jan 1813 - Where taken: Lat 44 N Long 2.3 W - Date received: 5 Feb 1813 - From what ship: HMS San Josef - Born: Philadelphia - Age: 25 - Sent to Chatham on 29 Mar 1813 on the HMS Braham.

Elliot, Robert - Mate - Number: 2481 - Prize name: Three Brothers - Ship type: MV - How taken: Ringdoff - When taken: 11 Nov 1813 - Where taken: at sea - Date received: 13 Mar 1814 - From what ship: HMS Bittern - Born: New York - Age: 35 - Discharged on 10 May 1814 and sent to Dartmoor.

Elliott, Benjamin - Seaman - Number: 2058 - How taken: Taken off a Russian merchant vessel - Date received: 1 Nov 1813 - From what ship: HMS Rippon - Born: Beverly - Age: 23 - Discharged on 3 Nov 1813 and sent to Dartmoor.

Ellis, John - Seaman - Number: 96 - Prize name: Experiment - Ship type: MV - How taken: HM Brig Rover - When taken: 21 Oct 1812 - Where taken: off Bordeaux - Date received: 25 Dec 1812 - From what ship: HMS Northumberland - Born: New York - Age: 21 - Sent to Portsmouth on 29 Dec 1812 on the HMS Northumberland.

Elm, John - Seaman - Number: 780 - Prize name: William Bayard - Ship type: MV - How taken: HM Ship-of-the-Line Warspite - When taken: 3 Mar 1813 - Where taken: Bay of Biscay - Date received: 19 Mar 1813 - From what ship: HMS Warspite - Born: New York - Age: 14 - Sent to Dartmoor on 28 Jun 1813.

Elves, Joseph - Seaman - Number: 2522 - Prize name: Hope, prize to the Privateer True Blooded Yankee - Ship type: P - How taken: HM Frigate Seahorse - When taken: 22 Mar 1814 - Where taken: at sea - Date received: 6 Apr 1814 - From what ship: HMS Queen Charlotte - Born: Lisbon - Age: 20 - Discharged on 10 May 1814 and sent to Dartmoor.

Elves, Manuel - Seaman - Number: 2521 - Prize name: Hope, prize to the Privateer True Blooded Yankee - Ship type: P - How taken: HM Frigate Seahorse - When taken: 22 Mar 1814 - Where taken: at sea - Date received: 6 Apr 1814 - From what ship: HMS Queen Charlotte - Born: Lisbon - Age: 23 - Discharged on 10 May 1814 and sent to Dartmoor.

Elwell, Thomas - Seaman - Number: 1164 - Prize name: Good Intent, prize of the Privateer Thrasher - Ship type: P - How taken: HM Frigate Pyramus - When taken: 26 Jun 1813 - Where taken: off Bordeaux - Date received: 2 Mar 1813 - From what ship: HMS Rota - Born: Saco - Age: 19 - Sent on 3 Jul 1813 to Stapleton prison.

Emmerton, James - Chief Mate - Number: 2572 - Prize name: Alligator - Ship type: MV - How taken: Detained by the East India Company - When taken: 10 May 1813 - Where taken: Calcutta - Date received: 7 May 1814 - From what ship: Northam, East Indianman - Born: Massachusetts - Age: 25 - Discharged on 7 May 1814 and sent to

Ashburton on parole.

Enderson, James - Seaman - Number: 981 - How taken: Delivered himself up from Greenock - When taken: 22 Jan 1813 - Date received: 12 Apr 1813 - From what ship: HMS Elizabeth - Born: New York - Age: 29 - Sent to Chatham on 8 Jul 1813 on HM Tender Neptune.

English, Edward - Seaman - Number: 1608 - Prize name: Revenge - Ship type: LM - How taken: HM Frigate Belle Poule - When taken: 11 May 1813 - Where taken: Bay of Biscay - Date received: 14 Jun 1813 - From what ship: HMS Royalist - Born: New York - Age: 24 - Died on 2 Nov 1813 on the prison ship Caton.

Erlstroom, John - Seaman - Number: 792 - Prize name: Cannoniere - Ship type: P - How taken: HM Ship-of-the-Line Warspite - When taken: 14 Mar 1813 - Where taken: Bay of Biscay - Date received: 19 Mar 1813 - From what ship: HMS Warspite - Born: Sweden - Age: 36 - Sent to Dartmoor on 28 Jun 1813.

Ervin, Ely - Carpenter - Number: 1035 - How taken: Impressed at Liverpool - When taken: 29 Mar 1813 - Date received: 21 Apr 1813 - From what ship: HMS Bittern - Born: Norwich - Age: 21 - Sent to Dartmoor on 1 Jul 1813.

Ervin, William - Seaman - Number: 691 - Prize name: Star - Ship type: MV - How taken: HM Ship-of-the-Line Superb - When taken: 9 Feb 1813 - Where taken: Bay of Biscay - Date received: 19 Mar 1813 - From what ship: HMS Warspite - Born: Cumberland, NJ - Age: 33 - Sent to Dartmoor on 2 Apr 1813.

Esdale, Thomas - Seaman - Number: 1355 - Prize name: Shadow - Ship type: LM - How taken: HM Brig Reindeer & HM Schooner Helicon - When taken: 6 Apr 1813 - Where taken: Bay of Biscay - Date received: 14 May 1813 - From what ship: HMS Reindeer - Born: Boston - Age: 47 - Discharged on 8 Sep 1813 and sent to Dartmoor.

Eskindson, George - Seaman - Number: 1775 - Prize name: Matilda - Ship type: P - How taken: HM Frigate Revolutionnaire - When taken: 27 Jul 1813 - Where taken: off Larvine - Date received: 31 Jul 1813 - From what ship: Matilda, prize - Born: Cape Ann - Age: 19 - Discharged on 8 Sep 1813 and sent to Dartmoor.

Evans, Hale - Seaman - Number: 330 - Prize name: Porcupine - Ship type: MV - How taken: HM Frigate Dryad - When taken: 8 Jan 1813 - Where taken: off Bordeaux - Date received: 21 Jan 1813 - From what ship: HMS Abercrombie - Born: Nottingham - Age: 22 - Sent to Chatham on 29 Mar 1813 on the HMS Braham.

Evans, Henry - Seaman - Number: 500 - Prize name: Cashiere - Ship type: LM - How taken: HM Brig Reindeer - When taken: 3 Feb 1813 - Where taken: Bay of Biscay - Date received: 12 Feb 1813 - From what ship: HMS Reindeer - Born: Philadelphia - Age: 29 - Sent to Dartmoor on 2 Apr 1813.

Evans, Jacob - Prize Master - Number: 823 - Prize name: King George - Ship type: MV - How taken: HM Brig Piercer - When taken: 9 Mar 1813 - Where taken: off Isle Ross - Date received: 21 Mar 1813 - From what ship: HMS Piercer - Born: Baltimore - Age: 22 - Sent to Dartmoor on 28 Jun 1813.

Evans, James - Seaman - Number: 1493 - How taken: Delivered himself up from HM Brig Foxhound - Date received: 27 May 1813 - From what ship: Dartmoor prison - Born: Charlestown - Age: 30 - Discharged on 8 Jul 1813 and sent to Chatham on HM Tender Neptune.

Evans, James - Seaman - Number: 607 - How taken: Delivered himself up from HM Brig Foxhound - Date received: 27 Feb 1813 - From what ship: HMS Salvador del Mundo - Born: Charlestown - Age: 30 - Sent to Dartmoor on 2 Apr 1813.

Evans, John - Seaman - Number: 12 - Prize name: Thomas Wilson - Ship type: MV - How taken: Taken at Liverpool - When taken: 18 Oct 1812 - Date received: 21 Nov 1812 - From what ship: HMS Salvador del Mundo - Born: Norfolk - Age: 20 - Sent to Portsmouth on 29 Dec 1812 on the HMS Northumberland.

Evans, Moses - Seaman - Number: 1260 - Prize name: Caroline - Ship type: MV - How taken: HM Frigate Medusa - When taken: 12 Apr 1813 - Where taken: Bay of Biscay - Date received: 10 May 1813 - From what ship: HMS Medusa - Born: Madison County - Age: 25 - Race: Negro - Discharged on 3 Jul 1813 and sent to Stapleton Prison.

Evans, Thomas - Seaman - Number: 116 - Prize name: Otter - Ship type: MV - How taken: HM Ship-Sloop Jalousie - When taken: 1 Dec 1812 - Where taken: off Cape Vincent - Date received: 30 Dec 1812 - From what ship: HMS Leonidas - Born: Rhode Island - Age: 32 - Sent to Portsmouth on 4 Jan 1813 on the HMS Revolutionnaire.

Evans, Thomas L. - Master - Number: 132 - Prize name: Expectation - Ship type: MV - How taken: HM Frigate Briton & HM Frigate Andromache - When taken: 20 Dec 1812 - Where taken: off Bordeaux - Date received: 7 Jan 1813

- From what ship: Expectation - Born: Chester County - Age: 26 - Sent on 8 Dec 1812 to Ashburton on parole.

Evelett, Francois - 1st Mate - Number: 89 - Prize name: Experiment - Ship type: MV - How taken: HM Brig Rover - When taken: 21 Oct 1812 - Where taken: off Bordeaux - Date received: 25 Dec 1812 - From what ship: HMS Northumberland - Born: Massachusetts - Age: 26 - Sent on 8 Dec 1812 to Ashburton on parole.

Eveleth, William - Seaman - Number: 1611 - Prize name: Revenge - Ship type: LM - How taken: HM Frigate Belle Poule - When taken: 11 May 1813 - Where taken: Bay of Biscay - Date received: 14 Jun 1813 - From what ship: HMS Royalist - Born: Providence - Age: 17 - Discharged on 30 Jun 1813 and sent to Stapleton.

Everill, Daniel - Seaman - Number: 1220 - Prize name: Grand Napoleon - Ship type: MV - How taken: HM Frigate Belle Poule - When taken: 3 Apr 1813 - Where taken: off Bordeaux - Date received: 9 May 1813 - From what ship: HMS Andromache - Born: Connecticut - Age: 20 - Discharged on 3 Jul 1813 and sent to Stapleton Prison.

Everly, John - Seaman - Number: 864 - Prize name: Pert - Ship type: MV - How taken: HM Ship-of-the-Line Warspite - When taken: 1 Mar 1813 - Where taken: Bay of Biscay - Date received: 27 Mar 1813 - From what ship: HMS Warspite - Born: Pennsylvania - Age: 19 - Sent to Dartmoor on 28 Jun 1813.

Ewell, Edward - Seaman - Number: 2384 - Prize name: Devon, prize to the Privateer Bunker Hill - Ship type: P - How taken: HM Brig Fly - When taken: 21 Jan 1814 - Where taken: at sea - Date received: 31 Jan 1814 - From what ship: HMS Fly - Born: Norfolk - Age: 21 - Discharged on 27 Feb 1814 and sent to Chatham on HMS Haleyon.

Ewing, James - Master - Number: 2128 - Prize name: Amiable - Ship type: LM - How taken: HM Ship-of-the-Line Magnificant - When taken: 30 Oct 1813 - Where taken: off Lorient - Date received: 29 Nov 1813 - From what ship: HMS Dublin - Born: Princeton, NJ - Age: 38 - Discharged on 3 Dec 1813 and sent to Ashburton on parole.

Fadden, John - Seaman - Number: 1914 - Prize name: US Brig Argus - Ship type: War - How taken: HM Brig Pelican - When taken: 14 Aug 1813 - Where taken: Irish Channel - Date received: 23 Aug 1813 - From what ship: HMS Pelican - Born: Hampton - Age: 32 - Discharged on 8 Sep 1813 and sent to Dartmoor.

Fairchild, Hamlet - Master - Number: 1149 - Prize name: Eliza - Ship type: MV - How taken: HM Frigate Surveillante - When taken: 27 Mar 1813 - Where taken: Bay of Biscay - Date received: 23 Apr 1813 - From what ship: HMS Plymouth - Born: New Haven - Age: 40 - Sent on 24 Apr 1813 to Ashburton on parole.

Fairchild, Robert - Mate - Number: 1001 - Prize name: Eliza - Ship type: MV - How taken: HM Frigate Surveillante - When taken: 27 Mar 1813 - Where taken: Bay of Biscay - Date received: 16 Apr 1813 - From what ship: HMS Fairy - Born: Middletown - Age: 35 - Sent on 22 Apr 1813 to Ashburton on parole.

Fairweather, Robert - Gunner - Number: 393 - Prize name: Union - Ship type: MV - How taken: HM Frigate Iris - When taken: 17 Jan 1813 - Where taken: Lat 44 N Long 2.3 W - Date received: 5 Feb 1813 - From what ship: HMS San Josef - Born: Philadelphia - Age: 28 - Sent to Chatham on 29 Mar 1813 on the HMS Braham.

Famer, John - Seaman - Number: 1811 - Prize name: US Brig Argus - Ship type: War - How taken: HM Brig Pelican - When taken: 14 Aug 1813 - Where taken: Irish Channel - Date received: 17 Aug 1813 - From what ship: USS Argus - Born: Groningen, Holland - Age: 32 - Discharged on 3 Nov 1813 and sent to Dartmoor.

Fannol, Augustus - Seaman - Number: 193 - Prize name: Hunter - Ship type: P - How taken: HM Frigate Phoebe - When taken: 24 Dec 1812 - Where taken: off Western Islands - Date received: 9 Jan 1813 - From what ship: HMS Phoebe - Born: New York - Age: 19 - Race: Black - Sent to Portsmouth on 8 Feb 1813 on the HMS Colossus.

Fardell, Thomas - Seaman - Number: 1773 - Prize name: Minerva - Ship type: MV - How taken: HM Brig Goldfinch - When taken: 27 Jun 1813 - Where taken: off Nantes - Date received: 25 Jul 1813 - From what ship: HMS Pyramus - Born: Virginia - Age: 30 - Discharged on 3 Nov 1813 and sent to Dartmoor.

Fardy, Anthony - Seaman - Number: 1444 - Prize name: Paul Jones - Ship type: P - How taken: HM Frigate Leonidas - When taken: 23 May 1813 - Where taken: off Cape Clear - Date received: 26 May 1813 - From what ship: HMS Leonidas - Born: Palermo, Italy - Age: 32 - Discharged on 30 Jun 1813 and sent to Stapleton.

Faritte, Francis - Seaman - Number: 1308 - Prize name: Eliza - Ship type: MV - How taken: HM Frigate Surveillante - When taken: 22 Apr 1813 - Where taken: at sea - Date received: 10 May 1813 - From what ship: HMS Medusa - Born: Marblehead - Age: 22 - Discharged on 3 Jul 1813 and sent to Stapleton Prison.

Farmer, George - Cook - Number: 2676 - Prize name: John, prize to Amelia - Ship type: P - How taken: HM Ship-of-the-Line Sterling Castle - When taken: 10 May 1814 - Where taken: Lat 36 Long 37 - Date received: 5 Jun 1814 -

From what ship: HMS Gronville - Born: New Jersey - Age: 30 - Race: Black - Discharged on 14 Jun 1814 and sent to Dartmoor.

Farrell, Andrew - Seaman - Number: 1185 - How taken: Delivered himself up from HM Ship-of-the-Line Clarence - Date received: 8 May 1813 - From what ship: HMS Dublin - Born: Nordea, NY - Age: 25 - Sent to Chatham on 8 Jul 1813 on HM Tender Neptune.

Farrell, Jeffery - Master - Number: 899 - Prize name: Messenger - Ship type: MV - How taken: HM Schooner Helicon - When taken: 10 Mar 1813 - Where taken: Bay of Biscay - Date received: 2 Apr 1813 - From what ship: HMS Plymouth - Born: Virginia - Age: 31 - Sent on 4 Apr 1813 to Ashburton on parole.

Farrell, John - Seaman - Number: 1114 - Prize name: Viper - Ship type: MV - How taken: HM Ship-of-the-Line Superb - When taken: 15 Apr 1813 - Where taken: Bay of Biscay - Date received: 22 Apr 1813 - From what ship: HMS Superb - Born: New Orleans - Age: 29 - Sent to Dartmoor on 1 Jul 1813.

Farrell, John - Seaman - Number: 60 - Prize name: Independence - Ship type: MV - How taken: HM Frigate Medusa - When taken: 9 Nov 1812 - Where taken: San Sebastian - Date received: 27 Nov 1812 - From what ship: HMS Wasp - Born: New Jersey - Age: 24 - Sent to Portsmouth on 29 Dec 1812 on the HMS Northumberland.

Farret, George - Seaman - Number: 825 - Prize name: King George - Ship type: MV - How taken: HM Brig Piercer - When taken: 9 Mar 1813 - Where taken: off Isle Ross - Date received: 21 Mar 1813 - From what ship: HMS Piercer - Born: Georgetown - Age: 34 - Race: Black - Sent to Dartmoor on 28 Jun 1813.

Farrell, Francis - Seaman - Number: 618 - Prize name: Governor McKean - Ship type: LM - How taken: HM Brig Rover - When taken: 26 Jun 1813 - Where taken: off Bordeaux - Date received: 2 Mar 1813 - From what ship: HMS Insolent - Born: Philadelphia - Age: 23 - Sent to Dartmoor on 2 Apr 1813.

Fate, Thomas - Seaman - Number: 91 - Prize name: Experiment - Ship type: MV - How taken: HM Brig Rover - When taken: 21 Oct 1812 - Where taken: off Bordeaux - Date received: 25 Dec 1812 - From what ship: HMS Northumberland - Born: Maryland - Age: 34 - Sent to Portsmouth on 29 Dec 1812 on the HMS Northumberland.

Fawcett, William - Seaman - Number: 653 - How taken: Delivered himself up from HMS Lavinia - When taken: 8 Feb 1813 - Date received: 5 Mar 1813 - From what ship: HMS Salvador del Monde - Born: Peterborough - Age: 29 - Sent to Dartmoor on 2 Apr 1813.

Fay, Salmon - Seaman - Number: 1208 - Prize name: Zebra - Ship type: LM - How taken: HM Frigate Pyramus - When taken: 20 Apr 1813 - Where taken: Bay of Biscay - Date received: 9 May 1813 - From what ship: HMS Andromache - Born: Massachusetts - Age: 23 - Discharged on 3 Jul 1813 and sent to Stapleton Prison.

Fayolle, Eugene - Passenger - Number: 1074 - Prize name: John & Frances - Ship type: MV - How taken: HM Frigate Belle Poule - When taken: 19 Mar 1813 - Where taken: Bay of Biscay - Date received: 22 Apr 1813 - From what ship: HMS Superb - Born: Charlestown - Age: 21 - Sent on 25 Apr 1813 to Ashburton on parole.

Fedie, John - Seaman - Number: 1937 - Prize name: Marmion - Ship type: MV - How taken: HM Frigate President - When taken: 14 Aug 1813 - Where taken: off Nantes - Date received: 27 Aug 1813 - From what ship: HMS Urgent - Born: Charlestown - Age: 27 - Discharged on 8 Sep 1813 and sent to Dartmoor.

Farley, John - Seaman - Number: 1312 - How taken: Delivered himself up from HM Ship-of-the-Line Clarence - Date received: 10 May 1813 - From what ship: HMS Clarence - Born: Baltimore - Age: 23 - Discharged on 8 Jul 1813 and sent to Chatham on tender Neptune.

Ferrald, John - Seaman - Number: 850 - How taken: Delivered himself up from HM Ship-of-the-Line Leyden - Date received: 23 May 1813 - From what ship: HMS Dauntless - Born: Massachusetts - Age: 27 - Sent to Chatham on 8 Jul 1813 on HM Tender Neptune.

Ferris, Joseph - Seaman - Number: 2408 - Prize name: Rachel & Ann, prize of the Privateer Prince de Neufchatel - Ship type: P - How taken: HM Frigate Cydnus - When taken: 6 Jan 1814 - Where taken: at sea - Date received: 4 Feb 1814 - From what ship: HMS Cydnus - Born: New York - Age: 25 - Discharged on 10 May 1814 and sent to Dartmoor.

Finch, William - Seaman - Number: 424 - Prize name: Union - Ship type: MV - How taken: HM Frigate Iris - When taken: 17 Jan 1813 - Where taken: Lat 44 N Long 2.3 W - Date received: 5 Feb 1813 - From what ship: HMS San Josef - Born: Orange County, NY - Age: 25 - Sent to Dartmoor on 2 Apr 1813.

Fink, Johannes - Seaman - Number: 1424 - Prize name: Paul Jones - Ship type: P - How taken: HM Frigate Leonidas - When taken: 23 May 1813 - Where taken: off Cape Clear - Date received: 26 May 1813 - From what ship: HMS Leonidas - Born: Germany - Age: 23 - Discharged on 3 Jul 1813 and sent to Stapleton Prison.

Fish, James - Seaman - Number: 1173 - Prize name: Hebe - Ship type: MV - How taken: HM Frigate Stag - When taken: 18 Apr 1813 - Where taken: Bay of Biscay - Date received: 6 May 1813 - From what ship: HMS Stag - Born: Boston - Age: 22 - Sent on 3 Jul 1813 to Stapleton prison.

Fisher, James - Seaman - Number: 2541 - How taken: Taken at Liverpool off a Portuguese merchant vessel - When taken: 2 Apr 1814 - Date received: 13 Apr 1814 - From what ship: HMS Bittern - Born: Philadelphia - Age: 25 - Discharged on 10 May 1814 and sent to Dartmoor.

Fisher, John - Seaman - Number: 2193 - Prize name: General Kempt, prize of the Privateer Grand Turk - Ship type: P - How taken: HM Brig Foxhound - When taken: 18 Dec 1813 - Where taken: Lat 48.4 Long 6 - Date received: 21 Dec 1813 - From what ship: HMS Foxhound - Born: New York - Age: 22 - Race: Negro - Discharged on 31 Jan 1814 and sent to Dartmoor.

Fisher, John - Seaman - Number: 2554 - Prize name: Bunker Hill - Ship type: P - How taken: HM Frigate Pomone & HM Frigate Cydnus - When taken: 4 Mar 1814 - Where taken: Bay of Biscay - Date received: 17 Apr 1814 - From what ship: HMS Teazer - Born: Salem - Age: 16 - Discharged on 10 May 1814 and sent to Dartmoor.

Fisher, Lewis - Seaman - Number: 1162 - Prize name: Essex - Ship type: MV - How taken: HM Frigate Pyramus - When taken: 2 Apr 1813 - Where taken: Bay of Biscay - Date received: 2 Mar 1813 - From what ship: HMS Rota - Born: Milford - Age: 25 - Sent on 3 Jul 1813 to Stapleton prison.

Fisher, Richard D. - Carpenter - Number: 450 - Prize name: Dolphin - Ship type: LM - How taken: HM Ship-of-the-Line Colossus, HM Frigate Rhin & HM Brig Goldfinch - When taken: 5 Jan 1813 - Where taken: off Western Islands - Date received: 6 Feb 1813 - From what ship: HMS Rhin - Born: Philadelphia - Age: 25 - Sent to Chatham on 29 Mar 1813 on the HMS Braham.

Fitzpatrick, William - Master - Number: 1151 - Prize name: Essex - Ship type: MV - How taken: HM Frigate Pyramus - When taken: 2 Apr 1813 - Where taken: Bay of Biscay - Date received: 23 Apr 1813 - From what ship: HMS Plymouth - Born: Boston - Age: 27 - Sent on 24 Apr 1813 to Ashburton on parole.

Flatt, Robert - Seaman - Number: 2633 - Prize name: Ateline - Ship type: MV - How taken: HM Frigate Magiciene - When taken: 14 Mar 1814 - Where taken: off Cape Finisterre, Spain - Date received: 17 May 1814 - From what ship: HMS Tortois - Born: North Carolina - Age: 37 - Discharged on 14 Jun 1814 and sent to Dartmoor.

Fleming, John Joseph - Sailmaker - Number: 1817 - Prize name: US Brig Argus - Ship type: War - How taken: HM Brig Pelican - When taken: 14 Aug 1813 - Where taken: Irish Channel - Date received: 17 Aug 1813 - From what ship: USS Argus - Born: New York - Age: 20 - Discharged on 8 Sep 1813 and sent to Dartmoor.

Fletcher, Henry - Quartermaster - Number: 161 - Prize name: Hunter - Ship type: P - How taken: HM Frigate Phoebe - When taken: 24 Dec 1812 - Where taken: off Western Islands - Date received: 9 Jan 1813 - From what ship: HMS Phoebe - Born: Boston - Age: 29 - Sent to Portsmouth on 8 Feb 1813 on the HMS Colossus.

Fletcher, James - Seaman - Number: 1359 - Prize name: Fox - Ship type: LM - How taken: HM Sloop Pheasant - When taken: 23 Apr 1813 - Where taken: Bay of Biscay - Date received: 15 May 1813 - From what ship: HMS Foxhound - Born: Massachusetts - Age: 21 - Discharged on 3 Jul 1813 and sent to Stapleton Prison.

Fletcher, John - Seaman - Number: 2126 - Prize name: Chesapeake - Ship type: LM - How taken: HM Frigate Hotspur & HM Frigate Pyramus - When taken: 26 Oct 1813 - Where taken: Bay of Biscay - Date received: 22 Nov 1813 - From what ship: HMS Pyramus - Born: New York - Age: 18 - Discharged on 29 Nov 1813 and sent to Dartmoor.

Fletcher, John William - Seaman - Number: 2498 - Prize name: Mary, prize of the Privateer Rattle Snake - Ship type: P - How taken: Retaken by the crew of the Mary - When taken: 18 Dec 1813 - Date received: 26 Mar 1814 - From what ship: Orr - Born: Alexandria - Age: 26 - Discharged on 10 May 1814 and sent to Dartmoor.

Fletcher, William B. - Seaman - Number: 557 - Prize name: Spitfire - Ship type: MV - How taken: HM Brig Achates - When taken: 14 Feb 1813 - Where taken: off Ushant - Date received: 16 Feb 1813 - From what ship: HMS Achates - Born: Marblehead - Age: 45 - Sent to Dartmoor on 2 Apr 1813.

Fleur, Francois Xavier - Merchant & Passenger - Number: 2217 - Prize name: Zephyr - Ship type: MV - How taken:

HM Frigate Pyramus - When taken: 30 Nov 1813 - Where taken: off Lorient - Date received: 30 Dec 1813 - From what ship: Martial - Born: Martinique - Age: 26 - Discharged on 30 Dec 1813 and sent to Ashburton on parole.

Flynn, Abraham - Seaman - Number: 1612 - Prize name: Revenge - Ship type: LM - How taken: HM Frigate Belle Poule - When taken: 11 May 1813 - Where taken: Bay of Biscay - Date received: 14 Jun 1813 - From what ship: HMS Royalist - Born: Boston - Age: 23 - Discharged on 30 Jun 1813 and sent to Stapleton.

Flynn, Pierre - Seaman - Number: 2667 - Prize name: Traveler, prize of the Privateer Surprize - Ship type: P - How taken: HMS Cawser - When taken: 7 May 1814 - Where taken: off Cape Clear - Date received: 5 Jun 1814 - From what ship: HMS Gronville - Born: Charleston - Age: 23 - Discharged on 14 Jun 1814 and sent to Dartmoor.

Flood, David - Seaman - Number: 677 - How taken: Impressed at Liverpool - When taken: 10 Feb 1813 - Date received: 15 Mar 1813 - From what ship: HMS Bittern - Born: Portland - Age: 25 - Sent to Dartmoor on 28 Jun 1813.

Floyd, James - Seaman - Number: 48 - How taken: HM Battery Princess - When taken: 29 Oct 1812 - Where taken: Liverpool - Date received: 26 Nov 1813 - From what ship: HMS Frederick - Born: New York - Age: 28 - Sent to Portsmouth on 29 Dec 1812 on the HMS Northumberland.

Folger, Shubert - 2nd Mate - Number: 870 - Prize name: William Bayard - Ship type: MV - How taken: HM Ship-of-the-Line Warspite - When taken: 3 Mar 1813 - Where taken: Bay of Biscay - Date received: 28 Mar 1813 - From what ship: HMS Warspite - Born: Nantucket - Age: 21 - Sent to Dartmoor on 28 Jun 1813.

Follet, William - Seaman - Number: 2171 - Prize name: Wolf, prize of the Privateer Grand Turk - Ship type: P - How taken: HM Frigate Briton - When taken: 1 Dec 1813 - Where taken: Bay of Biscay - Date received: 17 Dec 1813 - From what ship: HMS Briton - Born: Marblehead - Age: 30 - Discharged on 17 Dec 1813 and released to HMS Britton.

Forbes, James - Mate - Number: 474 - Prize name: Charles - Ship type: MV - How taken: Detained at Belfast - When taken: 8 Aug 1812 - Date received: 10 Feb 1813 - From what ship: HMS Neptune - Born: Philadelphia - Age: 28 - Sent to Chatham on 29 Mar 1813 on the HMS Braham.

Forbes, William - Seaman - Number: 744 - Prize name: Charlotte - Ship type: MV - How taken: HM Ship-of-the-Line Warspite - When taken: 3 Mar 1813 - Where taken: Bay of Biscay - Date received: 19 Mar 1813 - From what ship: HMS Warspite - Born: Boston - Age: 20 - Sent to Dartmoor on 2 Apr 1813.

Ford, Charles - Seaman - Number: 2623 - How taken: HM Ship-of-the-Line Edinburgh - When taken: 28 Oct 1812 - Date received: 16 May 1814 - From what ship: HMS Repulse - Born: North Carolina - Age: 23 - Discharged on 14 Jun 1814 and sent to Dartmoor.

Ford, George L. - 2nd Mate - Number: 1362 - Prize name: Tom - Ship type: LM - How taken: HM Frigate Surveillante - When taken: 27 Apr 1813 - Where taken: Bay of Biscay - Date received: 15 May 1813 - From what ship: HMS Foxhound - Born: Connecticut - Age: 26 - Discharged on 3 Jul 1813 and sent to Stapleton Prison.

Ford, Philip - Seaman - Number: 2609 - How taken: MV Sultan - When taken: 16 Oct 1812 - Date received: 16 May 1814 - From what ship: HMS Repulse - Born: Wilmington - Age: 55 - Discharged on 14 Jun 1814 and sent to Dartmoor.

Forester, Francis - Seaman - Number: 1064 - Prize name: Meteor - Ship type: MV - How taken: HM Frigate Briton - When taken: 12 Mar 1813 - Where taken: Bay of Biscay - Date received: 22 Apr 1813 - From what ship: HMS Superb - Born: Charlestown - Age: 19 - Sent to Dartmoor on 8 Sep 1813.

Foss, Edmond - Seaman - Number: 1632 - Prize name: Leo - Ship type: LM - How taken: HM Frigate Magiciene - When taken: 4 Jun 1813 - Where taken: Lat 45 Long 14 - Date received: 14 Jun 1813 - From what ship: HMS Orestes - Born: Limington - Age: 21 - Discharged on 30 Jun 1813 and sent to Stapleton.

Foss, Joseph - Seaman - Number: 1635 - Prize name: Leo - Ship type: LM - How taken: HM Frigate Magiciene - When taken: 4 Jun 1813 - Where taken: Lat 45 Long 14 - Date received: 14 Jun 1813 - From what ship: HMS Orestes - Born: Scarborough - Age: 22 - Discharged on 30 Jun 1813 and sent to Stapleton.

Fossender, William - Seaman - Number: 1652 - How taken: Delivered himself up from HMS Lavinia - Date received: 15 Jun 1813 - From what ship: Dartmoor Prison - Born: Hamburg - Age: 24 - Discharged on 16 Jun 1813 and released to HMS Salvador del Mundo.

Fossender, William - Seaman - Number: 654 - How taken: Delivered himself up from HMS Lavinia - When taken: 8 Feb 1813 - Date received: 5 Mar 1813 - From what ship: HMS Salvador del Monde - Born: Plymouth - Age: 24 - Sent to Dartmoor on 2 Apr 1813.

Foster, Asa - Seaman - Number: 188 - Prize name: Hunter - Ship type: P - How taken: HM Frigate Phoebe - When taken: 24 Dec 1812 - Where taken: off Western Islands - Date received: 9 Jan 1813 - From what ship: HMS Phoebe - Born: New Salisbury - Age: 26 - Sent to Portsmouth on 8 Feb 1813 on the HMS Colossus.

Foster, Cato - Seaman - Number: 658 - How taken: Delivered himself up from HMS Lavinia - Date received: 5 Mar 1813 - From what ship: HMS Salvador del Monde - Born: Marblehead - Age: 33 - Race: Negro - Sent to Dartmoor on 2 Apr 1813.

Foster, David - Seaman - Number: 1105 - Prize name: Viper - Ship type: MV - How taken: HM Ship-of-the-Line Superb - When taken: 15 Apr 1813 - Where taken: Bay of Biscay - Date received: 22 Apr 1813 - From what ship: HMS Superb - Born: Maryland - Age: 38 - Sent to Dartmoor on 1 Jul 1813.

Foster, John Thomas - 2nd Mate - Number: 1636 - Prize name: Tickler - Ship type: LM - How taken: HM Frigate Magiciene - When taken: 5 Jun 1813 - Where taken: Lat 47 Long 13 - Date received: 14 Jun 1813 - From what ship: HMS Orestes - Born: Gloucester - Age: 23 - Discharged on 30 Jun 1813 and sent to Stapleton.

Foster, Joseph - Seaman - Number: 524 - Prize name: Phoenix - Ship type: MV - How taken: Detained at Gibraltar - When taken: 8 Aug 1812 - Date received: 15 Feb 1813 - From what ship: HMS Andromeda - Born: Beverly - Age: 23 - Sent to Dartmoor on 2 Apr 1813.

Foster, Joseph - Seaman - Number: 488 - Prize name: Cashiere - Ship type: LM - How taken: HM Brig Reindeer - When taken: 3 Feb 1813 - Where taken: Bay of Biscay - Date received: 12 Feb 1813 - From what ship: HMS Reindeer - Born: Talbot County - Age: 18 - Sent to Dartmoor on 28 Jun 1813.

Foster, Thomas - Seaman - Number: 1134 - Prize name: Magdalen - Ship type: MV - How taken: HM Ship-of-the-Line Superb - When taken: 15 Apr 1813 - Where taken: off Belle Isle - Date received: 22 Apr 1813 - From what ship: HMS Superb - Born: Portsmouth - Age: 24 - Sent to Dartmoor on 1 Jul 1813.

Fountain, Isaac - Seaman - Number: 58 - Prize name: Independence - Ship type: MV - How taken: HM Frigate Medusa - When taken: 9 Nov 1812 - Where taken: San Sebastian - Date received: 27 Nov 1812 - From what ship: HMS Wasp - Born: New York - Age: 25 - Sent to Portsmouth on 29 Dec 1812 on the HMS Northumberland.

Fox, Richard - Chief Mate - Number: 1720 - Prize name: Hannah & Eliza - Ship type: MV - How taken: HM Brig Lyra - When taken: 29 May 1813 - Where taken: off north coast of Spain - Date received: 4 Jul 1813 - From what ship: HMS Iris - Born: New York - Age: 34 - Discharged on 9 Jul 1813 and sent to Ashburton on parole.

Fox, Washington - Seaman - Number: 669 - Prize name: Margaret, prize of the Privateer True Blooded Yankee - Ship type: P - How taken: HM Brig Nimrod - When taken: 9 Mar 1813 - Where taken: off Morant Bay, Jamaica - Date received: 14 Mar 1813 - From what ship: HMS Salvador del Mundo - Born: Alexandria - Age: 38 - Sent to Dartmoor on 1 Jul 1813.

Fox, Washington - Seaman - Number: 1565 - Prize name: Margaret, prize of the Privateer True Blooded Yankee - Ship type: P - How taken: HM Brig Nimrod - When taken: 9 Mar 1813 - Where taken: off Morant Bay, Jamaica - Date received: 6 Jun 1813 - From what ship: Dartmoor prison - Born: Alexandria - Age: 38 - Discharged on 8 Sep 1813 and sent to Dartmoor.

Francis, Benjamin - Seaman - Number: 770 - Prize name: William Bayard - Ship type: MV - How taken: HM Ship-of-the-Line Warspite - When taken: 3 Mar 1813 - Where taken: Bay of Biscay - Date received: 19 Mar 1813 - From what ship: HMS Warspite - Born: Philadelphia - Age: 29 - Sent to Dartmoor on 28 Jun 1813.

Francis, James - Seaman - Number: 1296 - Prize name: Price - Ship type: LM - How taken: HM Frigate Medusa - When taken: 13 Apr 1813 - Where taken: Bay of Biscay - Date received: 10 May 1813 - From what ship: HMS Medusa - Born: New York - Age: 25 - Discharged on 3 Jul 1813 and sent to Stapleton Prison.

Francis, John - Seaman - Number: 1294 - Prize name: Price - Ship type: LM - How taken: HM Frigate Medusa - When taken: 13 Apr 1813 - Where taken: Bay of Biscay - Date received: 10 May 1813 - From what ship: HMS Medusa - Born: Rhode Island - Age: 23 - Discharged on 3 Jul 1813 and sent to Stapleton Prison.

Francis, Joseph Butler - 2nd Mate - Number: 1166 - Prize name: Hebe - Ship type: MV - How taken: HM Frigate Stag

- When taken: 18 Apr 1813 - Where taken: Bay of Biscay - Date received: 6 May 1813 - From what ship: HMS Stag - Born: Providence, RI - Age: 26 - Sent on 3 Jul 1813 to Stapleton prison.

Francois, James - Seaman - Number: 65 - Prize name: Independence - Ship type: MV - How taken: HM Frigate Medusa - When taken: 9 Nov 1812 - Where taken: San Sebastian - Date received: 27 Nov 1812 - From what ship: HMS Wasp - Born: New Orleans - Age: 16 - Sent to Portsmouth on 29 Dec 1812 on the HMS Northumberland.

Franks, Francis - Steward - Number: 2097 - Prize name: Amiable - Ship type: LM - How taken: HM Ship-of-the-Line Magnificant - When taken: 30 Oct 1813 - Where taken: off Lorient - Date received: 9 Nov 1813 - From what ship: Amiable - Born: New York - Age: 37 - Race: Black - Discharged on 29 Nov 1813 and sent to Dartmoor.

Frash, James - Seaman - Number: 705 - Prize name: Star - Ship type: MV - How taken: HM Ship-of-the-Line Superb - When taken: 9 Feb 1813 - Where taken: Bay of Biscay - Date received: 19 Mar 1813 - From what ship: HMS Warspite - Born: New Castle - Age: 24 - Sent to Dartmoor on 2 Apr 1813.

Frazer, John - Seaman - Number: 1370 - Prize name: Tom - Ship type: LM - How taken: HM Frigate Surveillante - When taken: 27 Apr 1813 - Where taken: Bay of Biscay - Date received: 15 May 1813 - From what ship: HMS Foxhound - Born: Maryland - Age: 25 - Discharged on 3 Jul 1813 and sent to Stapleton Prison.

Freddle, John - Seaman - Number: 733 - Prize name: Pert - Ship type: MV - How taken: HM Ship-of-the-Line Warspite - When taken: 1 Mar 1813 - Where taken: Bay of Biscay - Date received: 19 Mar 1813 - From what ship: HMS Warspite - Born: Long Island - Age: 28 - Race: Black - Sent to Dartmoor on 2 Apr 1813.

Frederick, Charles - Seaman - Number: 2443 - Prize name: Mary, prize of the Privateer Prince de Neufchatel - Ship type: P - How taken: Retaken by the crew of the Mary - Date received: 15 Feb 1814 - From what ship: HMS Kangaroo - Born: New York - Age: 25 - Discharged on 10 May 1814 and sent to Dartmoor.

Freely, Henry - Seaman - Number: 1796 - How taken: Impressed at Liverpool - When taken: 12 Jul 1813 - Date received: 16 Aug 1813 - From what ship: HMS Bittern - Born: Pennsylvania - Age: 29 - Discharged on 8 Sep 1813 and sent to Dartmoor.

Freeman, Charles - Seaman - Number: 972 - Prize name: Ferox - Ship type: MV - How taken: HM Frigate Medusa & HM Brig Lyra - When taken: 28 Mar 1813 - Where taken: off Cape Ortagle - Date received: 9 Apr 1813 - From what ship: HMS Lyra - Born: Delaware - Age: 23 - Race: Black - Sent to Chatham on 8 Jul 1813 on HM Tender Neptune.

Freeman, David - Seaman - Number: 2480 - How taken: Impressed at Liverpool - When taken: 2 Feb 1814 - Date received: 13 Mar 1814 - From what ship: HMS Bittern - Born: New Bedford - Age: 24 - Race: Mulatto - Discharged on 10 May 1814 and sent to Dartmoor.

Freeman, John - Seaman - Number: 1427 - Prize name: Paul Jones - Ship type: P - How taken: HM Frigate Leonidas - When taken: 23 May 1813 - Where taken: off Cape Clear - Date received: 26 May 1813 - From what ship: HMS Leonidas - Born: Boston - Age: 30 - Race: Black - Discharged on 3 Jul 1813 and sent to Stapleton Prison.

Freeman, John - Quarter Gunner - Number: 1812 - Prize name: US Brig Argus - Ship type: War - How taken: HM Brig Pelican - When taken: 14 Aug 1813 - Where taken: Irish Channel - Date received: 17 Aug 1813 - From what ship: USS Argus - Born: Maryland - Age: 35 - Discharged on 27 Sep 1813 and sent to Dartmoor.

Freeman, Prince - Seaman - Number: 1491 - How taken: Delivered himself up from HM Ship-of-the-Line Boyne - Date received: 27 May 1813 - From what ship: Dartmoor prison - Born: Gloucester - Age: 34 - Discharged on 8 Jul 1813 and sent to Chatham on HM Tender Neptune.

Frees, James - Seaman - Number: 365 - Prize name: Orbit - Ship type: MV - How taken: HM Brig Achates - When taken: 29 Jan 1813 - Where taken: Lat 49 N Long 13 W - Date received: 4 Feb 1813 - From what ship: HMS Achates - Born: Lancaster - Age: 40 - Sent to Chatham on 29 Mar 1813 on the HMS Braham.

Ferguson, Henry - Seaman - Number: 1877 - Prize name: US Brig Argus - Ship type: War - How taken: HM Brig Pelican - When taken: 14 Aug 1813 - Where taken: Irish Channel - Date received: 23 Aug 1813 - From what ship: HMS Pelican - Born: Philadelphia - Age: 26 - Discharged on 8 Sep 1813 and sent to Dartmoor.

Freeman, Prince - Seaman - Number: 540 - How taken: Delivered himself up from HM Ship-of-the-Line Boyne - Date received: 15 Feb 1813 - From what ship: HMS Salvador del Mundo - Born: Brewster - Age: 34 - Sent to Dartmoor on 2 Apr 1813.

Frenarye, Pierre - Merchant & Passenger - Number: 2236 - Prize name: Squirrel - Ship type: MV - How taken: HM Frigate Belle Poule - When taken: 14 Dec 1813 - Where taken: Bay of Biscay - Date received: 15 Jan 1814 - From what ship: HMS Bellona - Born: Gannet, France - Age: 24 - Discharged on 16 Jan 1814 and sent to Ashburton on parole.

Friday, John - Steward - Number: 1421 - Prize name: Paul Jones - Ship type: P - How taken: HM Frigate Leonidas - When taken: 23 May 1813 - Where taken: off Cape Clear - Date received: 26 May 1813 - From what ship: HMS Leonidas - Born: New Orleans - Age: 21 - Discharged on 3 Jul 1813 and sent to Stapleton Prison.

Fritts, Joseph - Seaman - Number: 1257 - Prize name: Caroline - Ship type: MV - How taken: HM Frigate Medusa - When taken: 12 Apr 1813 - Where taken: Bay of Biscay - Date received: 10 May 1813 - From what ship: HMS Medusa - Born: Philadelphia - Age: 41 - Discharged on 3 Jul 1813 and sent to Stapleton Prison.

Fry, John - Carpenter - Number: 1724 - Prize name: Hannah & Eliza - Ship type: MV - How taken: HM Brig Lyra - When taken: 29 May 1813 - Where taken: off north coast of Spain - Date received: 4 Jul 1813 - From what ship: HMS Iris - Born: Andover - Age: 27 - Discharged on 8 Sep 1813 and sent to Dartmoor.

Fryer, John - Gunner - Number: 449 - Prize name: Dolphin - Ship type: LM - How taken: HM Ship-of-the-Line Colossus, HM Frigate Rhin & HM Brig Goldfinch - When taken: 5 Jan 1813 - Where taken: off Western Islands - Date received: 6 Feb 1813 - From what ship: HMS Rhin - Born: Pennsylvania - Age: 29 - Sent to Chatham on 29 Mar 1813 on the HMS Braham.

Fuller, Enoch - Seaman - Number: 2568 - How taken: Impressed at Cork - When taken: 28 Mar 1814 - Date received: 25 Apr 1814 - From what ship: HMS Helena - Born: Norfolk - Age: 22 - Race: Black - Discharged on 10 May 1814 and sent to Dartmoor.

Fuller, Zachariah - Seaman - Number: 313 - Prize name: Blue Bird - Ship type: MV - How taken: HM Frigate Briton - When taken: 1 Jan 1813 - Where taken: off Bordeaux - Date received: 21 Jan 1813 - From what ship: HMS Abercrombie - Born: Pennsylvania - Age: 30 - Sent to Chatham on 29 Mar 1813 on the HMS Braham.

Furlong, William - 2nd Mate - Number: 2319 - Prize name: Siro - Ship type: LM - How taken: HM Brig Pelican - When taken: 13 Jan 1814 - Where taken: at sea - Date received: 25 Jan 1814 - From what ship: HMS Pelican - Born: Baltimore - Age: 19 - Discharged on 31 Jan 1814 and sent to Dartmoor.

Gable, John - Seaman - Number: 2216 - Prize name: Squirrel - Ship type: MV - How taken: HM Frigate Belle Poule - When taken: 14 Dec 1813 - Where taken: Bay of Biscay - Date received: 27 Dec 1813 - From what ship: HMS Protector - Born: New York - Age: 29 - Race: Mulatto - Discharged on 31 Jan 1814 and sent to Dartmoor.

Gabriel, Joshua - Steward - Number: 1689 - Prize name: Governor Gerry - Ship type: MV - How taken: HM Brig Royalist - When taken: 31 May 1813 - Where taken: Bay of Biscay - Date received: 26 Jun 1813 - From what ship: HMS Duncan - Born: New Orleans - Age: 30 - Race: Black - Discharged on 30 Jun 1813 and sent to Stapleton.

Gadet, Martin - Passenger & Merchant - Number: 131 - Prize name: Empress - Ship type: MV - How taken: HM Brig Rover - When taken: 29 Nov 1812 - Where taken: off Bordeaux - Date received: 5 Jan 1813 - From what ship: HMS Rover - Born: Lyon, France - Age: 24 - Entered on 12 Jan 1813 into the French General Entry Book as prisoner number 39251.

Gage, Isaac - Passenger - Number: 1598 - Prize name: Governor Gerry - Ship type: MV - How taken: HM Brig Royalist - When taken: 31 May 1813 - Where taken: Bay of Biscay - Date received: 14 Jun 1813 - From what ship: HMS Royalist - Born: Barnstable - Age: 24 - Discharged on 30 Jun 1813 and sent to Stapleton.

Gall, Michael - Boatswain's Mate - Number: 2280 - Prize name: Siro - Ship type: LM - How taken: HM Brig Pelican - When taken: 13 Jan 1814 - Where taken: at sea - Date received: 23 Jan 1814 - From what ship: HMS Pelican - Born: Palermo - Age: 25 - Discharged on 31 Jan 1814 and sent to Dartmoor.

Galloway, Joseph - Seaman - Number: 492 - Prize name: Cashiere - Ship type: LM - How taken: HM Brig Reindeer - When taken: 3 Feb 1813 - Where taken: Bay of Biscay - Date received: 12 Feb 1813 - From what ship: HMS Reindeer - Born: Cabot County - Age: 21 - Sent to Dartmoor on 2 Apr 1813.

Gallibrandt, Bernard - Seaman - Number: 1397 - Prize name: Henry Clements - Ship type: MV - How taken: HM Brig Orestes - When taken: 13 Apr 1813 - Where taken: Bay of Biscay - Date received: 15 May 1813 - From what ship: HMS Orestes - Born: Leghorn, Italy - Age: 45 - Discharged on 8 Sep 1813 and sent to Dartmoor.

Gallin, John - Seaman - Number: 2187 - Prize name: Charlotte - Ship type: MV - How taken: Cutter Dwarf - When taken: 4 Nov 1812 - Where taken: off Bordeaux - Date received: 18 Dec 1813 - From what ship: HMS Conquistador - Born: Africa - Age: 21 - Race: Black - Discharged on 31 Jan 1814 and sent to Dartmoor.

Gallo, Joseph - Ordinary Seaman - Number: 479 - How taken: Impressed at Greenock - When taken: 19 Dec 1812 - Date received: 10 Feb 1813 - From what ship: HMS Frederick - Born: Norwich - Age: 23 - Died on 14 Mar 1813 on HM Prison Ship Caton.

Gandell, Epine - Seaman - Number: 1504 - How taken: Impressed at Belfast - When taken: 16 Jan 1813 - Date received: 27 May 1813 - From what ship: HMS Prince Frederick - Born: Boston - Age: 24 - Discharged on 8 Sep 1813 and sent to Dartmoor.

Ganson, Richard - Seaman - Number: 2186 - How taken: Impressed at Falmouth - When taken: 10 Dec 1813 - Date received: 18 Dec 1813 - From what ship: HMS Salvador del Mundo - Born: Virginia - Age: 18 - Discharged on 10 May 1814 and sent to Dartmoor.

Ganzler, George - Seaman - Number: 2650 - How taken: Delivered himself up from HM Ship-Sloop Cygnet - Date received: 23 May 1814 - From what ship: HMS Cygnet - Born: Boston - Age: 29 - Discharged on 14 Jun 1814 and sent to Dartmoor.

Garcia, Anthony - Seaman - Number: 2527 - Prize name: Maria Christiana, prize to the Privateer True Blooded Yankee - Ship type: P - How taken: Pactolus - When taken: 25 Mar 1814 - Where taken: coast of France - Date received: 6 Apr 1814 - From what ship: HMS Queen Charlotte - Born: Malaga - Age: 31 - Discharged on 10 May 1814 and sent to Dartmoor.

Garcia, Francois - Seaman - Number: 1199 - Prize name: Zebra - Ship type: LM - How taken: HM Frigate Pyramus - When taken: 20 Apr 1813 - Where taken: Bay of Biscay - Date received: 9 May 1813 - From what ship: HMS Andromache - Born: New Orleans - Age: 39 - Sent on 8 Sep 1813 to Dartmoor.

Gardiner, Charles - Mate - Number: 1 - Prize name: Asia - Ship type: MV - How taken: HM Ship-of-the-Line Royal Sovereign - When taken: 13 Aug 1812 - Where taken: off the Edystone - Date received: 22 Oct 1812 - From what ship: HMS Frederick - Born: Nantucket - Age: 29 - Sent on 8 Dec 1812 to Ashburton on parole.

Gardner, Anthony - Seaman - Number: 300 - Prize name: Stephen - Ship type: MV - How taken: HM Frigate Briton & HM Frigate Andromache - When taken: 17 Dec 1812 - Where taken: off Bordeaux - Date received: 21 Jan 1813 - From what ship: HMS Abercrombie - Born: Londonderry, NH - Age: 27 - Race: Black - Sent to Portsmouth on 8 Feb 1813 on the HMS Colossus.

Gardner, Edward - Seaman - Number: 2414 - Prize name: Joseph - Ship type: MV - How taken: HM Brig Royalist - When taken: 18 Jan 1814 - Where taken: Bay of Biscay - Date received: 9 Feb 1814 - From what ship: HMS Sparrow - Born: South Kingston - Age: 23 - Discharged on 10 May 1814 and sent to Dartmoor.

Gardner, Edward - 2nd Mate - Number: 2456 - Prize name: Fair American - Ship type: MV - How taken: HM Frigate Andromache - When taken: 19 Jan 1814 - Where taken: Bay of Biscay - Date received: 22 Feb 1814 - From what ship: HMS York - Born: Nantucket - Age: 23 - Discharged on 10 May 1814 and sent to Dartmoor.

Gardner, James - Seaman - Number: 279 - Prize name: Leader - Ship type: MV - How taken: HM Frigate Briton - When taken: 10 Dec 1812 - Where taken: off Bordeaux - Date received: 21 Jan 1813 - From what ship: HMS Abercrombie - Born: Newburyport - Age: 33 - Sent to Portsmouth on 8 Feb 1813 on the HMS Colossus.

Gardner, Jerry - Seaman - Number: 2385 - How taken: Gave himself up - Date received: 1 Feb 1814 - From what ship: HMS Bittern - Born: Rhode Island - Age: 27 - Race: Black - Discharged on 27 Feb 1814 and sent to Chatham on HMS Haleyon.

Gardner, Joseph - Seaman - Number: 1688 - Prize name: Revenge - Ship type: LM - How taken: HM Frigate Belle Poule - When taken: 11 May 1813 - Where taken: Bay of Biscay - Date received: 26 Jun 1813 - From what ship: HMS Duncan - Born: Boston - Age: 21 - Race: Black - Discharged on 30 Jun 1813 and sent to Stapleton.

Gardner, Joseph - Seaman - Number: 2510 - Prize name: Bunker Hill - Ship type: P - How taken: HM Frigate Pomone & HM Frigate Cydnus - When taken: 4 Mar 1814 - Where taken: Bay of Biscay - Date received: 4 Apr 1814 - From what ship: HMS Virago - Born: Beverly - Age: 21 - Discharged on 10 May 1814 and sent to Dartmoor.

Gardner, Peter - Seaman - Number: 1559 - How taken: Delivered himself up from HM Ship-of-the-Line Royal

Sovereign - Date received: 29 May 1813 - From what ship: HMS Hannibal - Born: New York - Age: 24 - Discharged on 8 Jul 1813 and sent to Chatham on HM Tender Neptune.

Gardner, Samuel - Gunner's Mate - Number: 164 - Prize name: Hunter - Ship type: P - How taken: HM Frigate Phoebe - When taken: 24 Dec 1812 - Where taken: off Western Islands - Date received: 9 Jan 1813 - From what ship: HMS Phoebe - Born: Salem - Age: 37 - Sent to Portsmouth on 8 Feb 1813 on the HMS Colossus.

Gardner, William - Seaman - Number: 1716 - Prize name: Orders in Council - Ship type: LM - How taken: HM Frigate Surveillante - When taken: 1 Jun 1813 - Where taken: Bay of Biscay - Date received: 4 Jul 1813 - From what ship: HMS Iris - Born: Boston - Age: 20 - Discharged on 22 Aug 1813 and released to HM Brig Redpole.

Garner, James - Seaman - Number: 1846 - How taken: Impressed at Dublin - When taken: 25 Jul 1813 - Date received: 17 Aug 1813 - From what ship: HMS Prince Frederick - Born: Boston - Age: 23 - Discharged on 8 Sep 1813 and sent to Dartmoor.

Garret, James - Seaman - Number: 817 - Prize name: Decornau - Ship type: MV - How taken: HM Sloop Pheasant - When taken: 15 Mar 1813 - Where taken: Bay of Biscay - Date received: 19 Mar 1813 - From what ship: HMS Pheasant - Born: Charlestown - Age: 20 - Sent to Dartmoor on 28 Jun 1813.

Garrison, John - Seaman - Number: 62 - Prize name: Independence - Ship type: MV - How taken: HM Frigate Medusa - When taken: 9 Nov 1812 - Where taken: San Sebastian - Date received: 27 Nov 1812 - From what ship: HMS Wasp - Born: New York - Age: 22 - Sent to Portsmouth on 29 Dec 1812 on the HMS Northumberland.

Garrot, William - Seaman - Number: 2029 - Prize name: Friendship West, prize of Privateer True Blooded Yankee - Ship type: P - How taken: HM Schooner Helicon & HM Schooner Whiting - When taken: 25 Oct 1813 - Where taken: Bay of Biscay - Date received: 31 Oct 1813 - From what ship: HMS Whiting - Born: New Jersey - Age: 23 - Discharged on 3 Nov 1813 and sent to Dartmoor.

Garthon, Willey - Seaman - Number: 289 - Prize name: Columbia - Ship type: MV - How taken: HM Frigate Briton - When taken: 17 Dec 1812 - Where taken: off Bordeaux - Date received: 21 Jan 1813 - From what ship: HMS Abercrombie - Born: Philadelphia - Age: 34 - Sent to Portsmouth on 8 Feb 1813 on the HMS Colossus.

Garthy, James - Seaman - Number: 988 - Prize name: Lightning - Ship type: MV - How taken: HM Frigate Medusa - When taken: 2 Apr 1813 - Where taken: Bay of Biscay - Date received: 16 Apr 1813 - From what ship: HMS Fairy - Born: Philadelphia - Age: 21 - Sent to Chatham on 8 Jul 1813 on HM Tender Neptune.

Gassy, Raymond - Seaman - Number: 2444 - Prize name: Mary, prize of the Privateer Prince de Neufchatel - Ship type: P - How taken: Retaken by the crew of the Mary - Date received: 15 Feb 1814 - From what ship: HMS Kangaroo - Born: New Orleans - Age: 57 - Discharged on 10 May 1814 and sent to Dartmoor.

Gatewood, James - Seaman - Number: 2514 - Prize name: Bunker Hill - Ship type: P - How taken: HM Frigate Pomone & HM Frigate Cydnus - When taken: 4 Mar 1814 - Where taken: Bay of Biscay - Date received: 4 Apr 1814 - From what ship: HMS Virago - Born: Portsmouth - Age: 33 - Race: Mulatto - Discharged on 10 May 1814 and sent to Dartmoor.

Gault, William - Seaman - Number: 795 - Prize name: Cannoniere - Ship type: P - How taken: HM Ship-of-the-Line Warspite - When taken: 14 Mar 1813 - Where taken: Bay of Biscay - Date received: 19 Mar 1813 - From what ship: HMS Warspite - Born: New York - Age: 21 - Sent to Chatham on 8 Jul 1813 on HM Tender Neptune.

Gadson, John - Seaman - Number: 245 - Prize name: Janice & Lydia, prize of the Privateer General Armstrong - Ship type: P - How taken: Barton, letter of marque - When taken: 29 Nov 1812 - Where taken: off Bermuda - Date received: 12 Jan 1813 - From what ship: HMS Bittern - Born: Fredericktown - Age: 39 - Sent to Portsmouth on 8 Feb 1813 on the HMS Colossus.

Gee, Thomas - Seaman - Number: 1371 - Prize name: Tom - Ship type: LM - How taken: HM Frigate Surveillante - When taken: 27 Apr 1813 - Where taken: Bay of Biscay - Date received: 15 May 1813 - From what ship: HMS Foxhound - Born: Virginia - Age: 25 - Discharged on 3 Jul 1813 and sent to Stapleton Prison.

Gellens, William - Seaman - Number: 735 - Prize name: Pert - Ship type: MV - How taken: HM Ship-of-the-Line Warspite - When taken: 1 Mar 1813 - Where taken: Bay of Biscay - Date received: 19 Mar 1813 - From what ship: HMS Warspite - Born: Philadelphia - Age: 20 - Sent to Dartmoor on 2 Apr 1813.

Gellie, Thomas - Seaman - Number: 1543 - Prize name: Zebra - Ship type: LM - How taken: HM Frigate Pyramus -

When taken: 20 Apr 1813 - Where taken: Bay of Biscay - Date received: 29 May 1813 - From what ship: HMS Hannibal - Born: New York - Age: 23 - Race: Black - Discharged on 30 Jun 1813 and sent to Stapleton.

Gerard, William - Seaman - Number: 2469 - Prize name: Fair American - Ship type: MV - How taken: HM Frigate Andromache - When taken: 19 Jan 1814 - Where taken: Bay of Biscay - Date received: 22 Feb 1814 - From what ship: HMS York - Born: Boston - Age: 44 - Discharged on 10 May 1814 and sent to Dartmoor.

Geyer, Joshua - Carpenter - Number: 1677 - Prize name: Revenge - Ship type: LM - How taken: HM Frigate Belle Poule - When taken: 11 May 1813 - Where taken: Bay of Biscay - Date received: 26 Jun 1813 - From what ship: HMS Duncan - Born: Boston - Age: 28 - Discharged on 30 Jun 1813 and sent to Stapleton.

Gibb, James - Seaman - Number: 438 - How taken: Impressed at Belfast - Date received: 5 Feb 1813 - From what ship: HMS Neptune - Born: Philadelphia - Age: 32 - Race: Black - Sent to Chatham on 29 Mar 1813 on the HMS Braham.

Gibbons, Andrew - Seaman - Number: 85 - Prize name: Argus - Ship type: MV - How taken: Fancy, Cutter - When taken: 19 Dec 1812 - Where taken: Bay of Biscay - Date received: 24 Dec 1812 - From what ship: Fancy, Cutter - Born: New York - Age: 22 - Sent to Portsmouth on 29 Dec 1812 on the HMS Northumberland.

Gibbs, Daniel - Seaman - Number: 926 - Prize name: Weasel - Ship type: MV - How taken: HM Brig Foxhound - When taken: 25 Mar 1813 - Where taken: Bay of Biscay - Date received: 6 Apr 1813 - From what ship: HMS Foxhound - Born: Newport - Age: 19 - Sent on 21 Jun 1813 to Portsmouth on HMS Prometheus.

Gibbs, Henry - Seaman - Number: 1443 - Prize name: Paul Jones - Ship type: P - How taken: HM Frigate Leonidas - When taken: 23 May 1813 - Where taken: off Cape Clear - Date received: 26 May 1813 - From what ship: HMS Leonidas - Born: Massachusetts - Age: 18 - Discharged on 30 Jun 1813 and sent to Stapleton.

Gibbs, William - Seaman - Number: 1774 - How taken: Impressed at Cork - When taken: 21 Jul 1813 - Date received: 29 Jul 1813 - From what ship: Cutter Earl Spence - Born: Springfield - Age: 48 - Discharged on 8 Sep 1813 and sent to Dartmoor.

Gibby, John - Seaman - Number: 240 - How taken: Impressed at Liverpool - When taken: 17 Nov 1812 - Date received: 12 Jan 1813 - From what ship: HMS Bittern - Born: Baltimore - Age: 21 - Sent to Portsmouth on 8 Feb 1813 on the HMS Colossus.

Gifford, Barry - Seaman - Number: 26 - Prize name: Catharine - Ship type: MV - How taken: HM Frigate Leonidas - When taken: 31 Jul 1812 - Where taken: off Ireland - Date received: 23 Nov 1812 - From what ship: HMS Stork - Born: Westport - Age: 22 - Sent to Portsmouth on 29 Dec 1812 on the HMS Northumberland.

Gifford, Thomas - Chief Mate - Number: 943 - Prize name: Good Friends - Ship type: MV - How taken: HM Frigate Andromache - When taken: 2 Apr 1813 - Where taken: Bay of Biscay - Date received: 7 Apr 1813 - From what ship: HMS Sea Lark - Born: New York - Age: 32 - Sent on 10 Apr 1813 to Ashburton on parole.

Giggord, Robert - Seaman - Number: 1032 - How taken: Impressed at Liverpool - When taken: 4 Apr 1813 - Date received: 21 Apr 1813 - From what ship: HMS Bittern - Born: Philadelphia - Age: 29 - Sent to Dartmoor on 1 Jul 1813.

Gilbert, George - Boy - Number: 974 - Prize name: Ferox - Ship type: MV - How taken: HM Frigate Medusa & HM Brig Lyra - When taken: 28 Mar 1813 - Where taken: off Cape Ortagle - Date received: 9 Apr 1813 - From what ship: HMS Lyra - Born: Gloucester - Age: 16 - Sent to Chatham on 8 Jul 1813 on HM Tender Neptune.

Gilbert, John - Seaman - Number: 599 - Prize name: St. Martin's Plantation, prize of the Privateer Paul Jones - Ship type: P - How taken: HM Ship-of-the-Line Dublin - When taken: 9 Feb 1815 - Where taken: Lat 43 N Long 33.5 W - Date received: 25 Feb 1813 - From what ship: HMS Dublin - Born: New York - Age: 20 - Sent to Dartmoor on 2 Apr 1813.

Gilbert, Jurdonet - Seaman - Number: 2427 - Prize name: US Brig Argus - Ship type: War - How taken: HM Brig Pelican - When taken: 14 Aug 1813 - Where taken: Irish Channel - Date received: 12 Feb 1814 - From what ship: HMS Salvador del Mundo - Born: Baltimore - Age: 26 - Discharged on 10 May 1814 and sent to Dartmoor.

Gilbert, Thomas - Seaman - Number: 676 - How taken: Impressed at Liverpool - When taken: 15 Jan 1813 - Date received: 15 Mar 1813 - From what ship: HMS Bittern - Born: Norfolk - Age: 50 - Sent to Dartmoor on 28 Jun 1813.

Gilbert, Thomas - Seaman - Number: 497 - Prize name: Cashiere - Ship type: LM - How taken: HM Brig Reindeer - When taken: 3 Feb 1813 - Where taken: Bay of Biscay - Date received: 12 Feb 1813 - From what ship: HMS Reindeer - Born: Philadelphia - Age: 52 - Sent to Dartmoor on 2 Apr 1813.

Gilchrist, John - Seaman - Number: 2209 - Prize name: Squirrel - Ship type: MV - How taken: HM Frigate Belle Poule - When taken: 14 Dec 1813 - Where taken: Bay of Biscay - Date received: 27 Dec 1813 - From what ship: HMS Protector - Born: New York - Age: 24 - Discharged on 31 Jan 1814 and sent to Dartmoor.

Giles, Edward - Seaman - Number: 1758 - Prize name: Friendship, prize to the Privateer America - Ship type: P - How taken: HM Schooner Whiting - When taken: 15 Jul 1815 - Where taken: Lat 47N Long 8W - Date received: 20 Jul 1813 - From what ship: HMS Whiting - Born: Marblehead - Age: 26 - Discharged on 8 Sep 1813 and sent to Dartmoor.

Gillet, John Francis - Boy - Number: 2520 - Prize name: Hope, prize to the Privateer True Blooded Yankee - Ship type: P - How taken: HM Frigate Seahorse - When taken: 22 Mar 1814 - Where taken: at sea - Date received: 6 Apr 1814 - From what ship: HMS Queen Charlotte - Born: Brest - Age: 17 - Discharged on 10 May 1814 and sent to Dartmoor.

Gilligin, William - Seaman - Number: 406 - Prize name: Union - Ship type: MV - How taken: HM Frigate Iris - When taken: 17 Jan 1813 - Where taken: Lat 44 N Long 2.3 W - Date received: 5 Feb 1813 - From what ship: HMS San Josef - Born: Limerick - Age: 31 - Sent to Chatham on 29 Mar 1813 on the HMS Braham.

Gilmore, William H. - Seaman - Number: 1385 - Prize name: Tom - Ship type: LM - How taken: HM Frigate Surveillante - When taken: 27 Apr 1813 - Where taken: Bay of Biscay - Date received: 15 May 1813 - From what ship: HMS Foxhound - Born: Pennsylvania - Age: 23 - Discharged on 8 Sep 1813 and sent to Dartmoor.

Gilpin, William - Seaman - Number: 2651 - How taken: Delivered himself up a Spanish MV - Where taken: Liverpool - Date received: 23 May 1814 - From what ship: HMS Bittern - Born: Amboy - Age: 24 - Discharged on 14 Jun 1814 and sent to Dartmoor.

Ginkins, William - Seaman - Number: 1904 - Prize name: US Brig Argus - Ship type: War - How taken: HM Brig Pelican - When taken: 14 Aug 1813 - Where taken: Irish Channel - Date received: 23 Aug 1813 - From what ship: HMS Pelican - Born: Pennsylvania - Age: 26 - Race: Black - Discharged on 8 Sep 1813 and sent to Dartmoor.

Gisby, John - 1st Mate - Number: 107 - Prize name: Otter - Ship type: MV - How taken: HM Ship-Sloop Jalousie - When taken: 1 Dec 1812 - Where taken: off Cape Vincent - Date received: 30 Dec 1812 - From what ship: HMS Leonidas - Born: Middleborough - Age: 33 - Sent on 8 Dec 1812 to Ashburton on parole.

Gisenado, Samuel H. - Prize Master - Number: 2436 - Prize name: Prince of Wales, prize of the Privateer Prince de Neufchatel - Ship type: P - How taken: Transport Nelson - When taken: 4 Feb 1814 - Where taken: at sea - Date received: 14 Feb 1814 - From what ship: HMS Halcyon - Born: Rhode Island - Age: 32 - Discharged on 10 May 1814 and sent to Dartmoor.

Givel, Oliver - Seaman - Number: 1832 - Prize name: US Brig Argus - Ship type: War - How taken: HM Brig Pelican - When taken: 14 Aug 1813 - Where taken: Irish Channel - Date received: 17 Aug 1813 - From what ship: USS Argus - Born: New Orleans - Age: 28 - Discharged on 8 Sep 1813 and sent to Dartmoor.

Gland, Benjamin - Seaman - Number: 1609 - Prize name: Revenge - Ship type: LM - How taken: HM Frigate Belle Poule - When taken: 11 May 1813 - Where taken: Bay of Biscay - Date received: 14 Jun 1813 - From what ship: HMS Royalist - Born: Philadelphia - Age: 23 - Discharged on 30 Jun 1813 and sent to Stapleton.

Glasgow, John - Seaman - Number: 191 - Prize name: Hunter - Ship type: P - How taken: HM Frigate Phoebe - When taken: 24 Dec 1812 - Where taken: off Western Islands - Date received: 9 Jan 1813 - From what ship: HMS Phoebe - Born: Boston - Age: 23 - Race: Black - Sent to Portsmouth on 8 Feb 1813 on the HMS Colossus.

Glass, Reuben - 2nd Mate - Number: 866 - Prize name: Mars - Ship type: MV - How taken: HM Ship-of-the-Line Warspite - When taken: 26 Feb 1813 - Where taken: Bay of Biscay - Date received: 28 Mar 1813 - From what ship: HMS Warspite - Born: Duxbury - Age: 30 - Died on 21 May 1813 on HM Prison Ship Caton.

Glenn, Robert - Seaman - Number: 137 - Prize name: Nope - Ship type: MV - How taken: HM Sloop Pheasant - When taken: 13 Dec 1812 - Where taken: off Western Islands - Date received: 8 Jan 1813 - From what ship: HMS Pheasant - Born: York County, PA - Age: 53 - Sent to Portsmouth on 8 Feb 1813 on the HMS Colossus.

Glover, John - Chief Mate - Number: 2412 - Prize name: Joseph - Ship type: MV - How taken: HM Brig Royalist - When taken: 18 Jan 1814 - Where taken: Bay of Biscay - Date received: 9 Feb 1814 - From what ship: HMS Sparrow - Born: Marblehead - Age: 24 - Discharged on 15 Apr 1814 and sent to Ashburton on parole.

Glover, John - Seaman - Number: 522 - Prize name: Phoenix - Ship type: MV - How taken: Detained at Gibraltar - When taken: 8 Aug 1812 - Date received: 15 Feb 1813 - From what ship: HMS Andromeda - Born: Massachusetts - Age: 35 - Sent to Dartmoor on 2 Apr 1813.

Godfry, Edward - Seaman - Number: 1915 - Prize name: US Brig Argus - Ship type: War - How taken: HM Brig Pelican - When taken: 14 Aug 1813 - Where taken: Irish Channel - Date received: 23 Aug 1813 - From what ship: HMS Pelican - Born: Pennsylvania - Age: 21 - Race: Black - Discharged on 8 Sep 1813 and sent to Dartmoor.

Godfry, William - Seaman - Number: 1437 - Prize name: Paul Jones - Ship type: P - How taken: HM Frigate Leonidas - When taken: 23 May 1813 - Where taken: off Cape Clear - Date received: 26 May 1813 - From what ship: HMS Leonidas - Born: Providence, RI - Age: 18 - Race: Black - Discharged on 30 Jun 1813 and sent to Stapleton.

Godshall, John - Seaman - Number: 421 - Prize name: Union - Ship type: MV - How taken: HM Frigate Iris - When taken: 17 Jan 1813 - Where taken: Lat 44 N Long 2.3 W - Date received: 5 Feb 1813 - From what ship: HMS San Josef - Born: Philadelphia - Age: 16 - Sent to Chatham on 29 Mar 1813 on the HMS Braham.

Godwin, William - Seaman - Number: 1583 - Prize name: Orders in Council - Ship type: LM - How taken: HM Frigate Surveillante - When taken: 1 Jun 1813 - Where taken: Bay of Biscay - Date received: 13 Jun 1813 - From what ship: Forvey - Born: New York - Age: 29 - Discharged on 30 Jun 1813 and sent to Stapleton.

Goetz, Lewis - Passenger - Number: 391 - Prize name: Union - Ship type: MV - How taken: HM Frigate Iris - When taken: 17 Jan 1813 - Where taken: Lat 44 N Long 2.3 W - Date received: 5 Feb 1813 - From what ship: HMS San Josef - Born: Lichtenan, Germany - Age: 31 - Entered on 14 Feb 1813 into the French General Entry as prisoner number 39379.

Goff, James - Seaman - Number: 747 - Prize name: Charlotte - Ship type: MV - How taken: HM Ship-of-the-Line Warspite - When taken: 3 Mar 1813 - Where taken: Bay of Biscay - Date received: 19 Mar 1813 - From what ship: HMS Warspite - Born: New York - Age: 20 - Sent to Dartmoor on 2 Apr 1813.

Golandre, John - Seaman - Number: 2360 - Prize name: Zephyr - Ship type: MV - How taken: HM Frigate Surveillante - When taken: 6 Jan 1814 - Where taken: Bay of Biscay - Date received: 31 Jan 1814 - From what ship: HMS Surveillante - Born: Stockholm, Sweden - Age: 18 - Discharged on 10 May 1814 and sent to Dartmoor.

Goldsbury, William - Seaman - Number: 2111 - Prize name: Chesapeake - Ship type: LM - How taken: HM Frigate Hotspur & HM Frigate Pyramus - When taken: 26 Oct 1813 - Where taken: Bay of Biscay - Date received: 22 Nov 1813 - From what ship: HMS Pyramus - Born: Maryland - Age: 32 - Discharged on 29 Nov 1813 and sent to Dartmoor.

Golfin, John - Seaman - Number: 250 - How taken: Delivered himself up from HM Hospital Ship Trent - Date received: 16 Jan 1813 - From what ship: HMS Stork - Born: Washington - Age: 35 - Race: Black - Sent to Portsmouth on 8 Feb 1813 on the HMS Colossus.

Gomez, Manuel - Seaman - Number: 263 - Prize name: Brutus - Ship type: MV - How taken: HM Frigate Briton - When taken: Jan 1813 - Where taken: Bay of Biscay - Date received: 21 Jan 1813 - From what ship: HMS Briton - Born: Oporto, Portugal - Age: 24 - Sent to Portsmouth on 8 Feb 1813 on the HMS Colossus.

Gonnett, Felix - Passenger - Number: 1593 - Prize name: Governor Gerry - Ship type: MV - How taken: HM Brig Royalist - When taken: 31 May 1813 - Where taken: Bay of Biscay - Date received: 14 Jun 1813 - From what ship: HMS Royalist - Born: France - Age: 31 - Discharged on 7 Jul 1813 and sent to Ashburton on parole.

Goodday, James - Master - Number: 874 - Prize name: Virginia Planter - Ship type: MV - How taken: HM Frigate Pyramus - When taken: 17 Mar 1813 - Where taken: off Nantes - Date received: 30 Mar 1813 - From what ship: HMS Plymouth - Born: Sandwich - Age: 36 - Sent on 31 Mar 1813 to Ashburton on parole.

Goodnow, Saal B. - Gunner's Mate - Number: 2279 - Prize name: Siro - Ship type: LM - How taken: HM Brig Pelican - When taken: 13 Jan 1814 - Where taken: at sea - Date received: 23 Jan 1814 - From what ship: HMS Pelican - Born: Boston - Age: 25 - Discharged on 31 Jan 1814 and sent to Dartmoor.

Gordon, Thomas - Seaman - Number: 698 - Prize name: Star - Ship type: MV - How taken: HM Ship-of-the-Line

Superb - When taken: 9 Feb 1813 - Where taken: Bay of Biscay - Date received: 19 Mar 1813 - From what ship: HMS Warspite - Born: New Orleans - Age: 22 - Sent to Dartmoor on 2 Apr 1813.

Gorham, John - Chief Mate - Number: 1247 - Prize name: Messenger - Ship type: MV - How taken: HM Frigate Iris & HM Frigate Medusa - When taken: 10 Mar 1813 - Where taken: Cape Ortegal, Spain - Date received: 10 May 1813 - From what ship: HMS Medusa - Born: Easton - Age: 28 - Discharged on 14 May 1813 and sent to Ashburton on parole.

Gorling, George C. - Seaman - Number: 762 - Prize name: William Bayard - Ship type: MV - How taken: HM Ship-of-the-Line Warspite - When taken: 3 Mar 1813 - Where taken: Bay of Biscay - Date received: 19 Mar 1813 - From what ship: HMS Warspite - Born: Swinemunde, Prussia - Age: 36 - Sent to Dartmoor on 2 Apr 1813.

Gorling, George C. - Seaman - Number: 2697 - Prize name: William Bayard - Ship type: MV - How taken: HM Ship-of-the-Line Warspite - When taken: 3 Mar 1813 - Where taken: Bay of Biscay - Date received: 17 Jun 1814 - From what ship: Dartmouth - Born: Prussia - Age: 36 - Discharged on 20 Jun 1814 and released to the Speedwell Cartel.

Goss, Joshua - Boy - Number: 1000 - Prize name: Polly - Ship type: MV - How taken: HM Frigate Surveillante - When taken: 27 Mar 1813 - Where taken: Bay of Biscay - Date received: 16 Apr 1813 - From what ship: HMS Fairy - Born: Marblehead - Age: 14 - Sent to Chatham on 8 Jul 1813 on HM Tender Neptune.

Gould, Thomas - Boy - Number: 336 - Prize name: Rossie - Ship type: MV - How taken: Rochefort, France - When taken: 1 Jan 1813 - Where taken: Basque Roads, France - Date received: 22 Jan 1813 - From what ship: Rossie - Born: Baltimore - Age: 14 - Sent on 1 Feb 1813 to Ashburton on parole.

Gove, William - Seaman - Number: 1631 - Prize name: Leo - Ship type: LM - How taken: HM Frigate Magiciene - When taken: 4 Jun 1813 - Where taken: Lat 45 Long 14 - Date received: 14 Jun 1813 - From what ship: HMS Orestes - Born: Waterford - Age: 20 - Discharged on 30 Jun 1813 and sent to Stapleton.

Grace, Allen - Seaman - Number: 1763 - How taken: Impressed at Liverpool - When taken: 6 Jul 1813 - Date received: 23 Jun 1813 - From what ship: HMS Prince Frederick - Born: Maryland - Age: 22 - Race: Negro - Discharged on 8 Sep 1813 and sent to Dartmoor.

Graham, David - Seaman - Number: 1838 - Prize name: US Brig Argus - Ship type: War - How taken: HM Brig Pelican - When taken: 14 Aug 1813 - Where taken: Irish Channel - Date received: 17 Aug 1813 - From what ship: USS Argus - Born: Baltimore - Age: 21 - Discharged on 8 Sep 1813 and sent to Dartmoor.

Graham, George - Seaman - Number: 303 - Prize name: Stephen - Ship type: MV - How taken: HM Frigate Briton & HM Frigate Andromache - When taken: 17 Dec 1812 - Where taken: off Bordeaux - Date received: 21 Jan 1813 - From what ship: HMS Abercrombie - Born: New York - Age: 19 - Sent to Chatham on 29 Mar 1813 on the HMS Braham.

Graham, James - Seaman - Number: 2223 - Prize name: Zephyr - Ship type: MV - How taken: HM Frigate Pyramus - When taken: 30 Nov 1813 - Where taken: off Lorient - Date received: 3 Jan 1814 - From what ship: HMS Warspite - Born: New York - Age: 18 - Discharged on 31 Jan 1814 and sent to Dartmoor.

Graham, John - Seaman - Number: 2446 - How taken: Serving on board the Transport Mary - Date received: 15 Feb 1814 - From what ship: HMS Kangaroo - Born: Cape Ann - Age: 59 - Discharged on 10 May 1814 and sent to Dartmoor.

Graham, William - Seaman - Number: 423 - Prize name: Union - Ship type: MV - How taken: HM Frigate Iris - When taken: 17 Jan 1813 - Where taken: Lat 44 N Long 2.3 W - Date received: 5 Feb 1813 - From what ship: HMS San Josef - Born: Delaware - Age: 18 - Sent to Chatham on 29 Mar 1813 on the HMS Braham.

Grimsby, George - Mate - Number: 1165 - Prize name: Hebe - Ship type: MV - How taken: HM Frigate Stag - When taken: 18 Apr 1813 - Where taken: Bay of Biscay - Date received: 6 May 1813 - From what ship: HMS Stag - Born: New York - Age: 38 - Sent on 14 Apr 1813 to Ashburton on parole.

Grand, William - Seaman - Number: 2168 - Prize name: Wolf, prize of the Privateer Grand Turk - Ship type: P - How taken: HM Frigate Briton - When taken: 1 Dec 1813 - Where taken: Bay of Biscay - Date received: 17 Dec 1813 - From what ship: HMS Briton - Born: Marblehead - Age: 17 - Discharged on 17 Dec 1813 and released to HMS Britton.

Grant, James - Seaman - Number: 2542 - How taken: Sent into custody from HM Sloop Volentaire - When taken: 2

Feb 1814 - Date received: 16 Apr 1814 - From what ship: Transport Fanny - Born: Salem - Age: 27 - Race: Mulatto - Discharged on 10 May 1814 and sent to Dartmoor.

Grant, Peter - Seaman - Number: 785 - Prize name: Cannoniere - Ship type: P - How taken: HM Ship-of-the-Line Warspite - When taken: 14 Mar 1813 - Where taken: Bay of Biscay - Date received: 19 Mar 1813 - From what ship: HMS Warspite - Born: Maryland - Age: 36 - Sent to Dartmoor on 28 Jun 1813.

Gravely, Joseph - Seaman - Number: 2375 - Prize name: Devon, prize to the Privateer Bunker Hill - Ship type: P - How taken: HM Brig Fly - When taken: 21 Jan 1814 - Where taken: at sea - Date received: 31 Jan 1814 - From what ship: HMS Fly - Born: Lorient, France - Age: 71 - Discharged on 27 Feb 1814 and sent to Chatham on HMS Haleyon.

Graves, Ebenezer - Seaman - Number: 2512 - Prize name: Bunker Hill - Ship type: P - How taken: HM Frigate Pomone & HM Frigate Cydnus - When taken: 4 Mar 1814 - Where taken: Bay of Biscay - Date received: 4 Apr 1814 - From what ship: HMS Virago - Born: Marblehead - Age: 24 - Discharged on 10 May 1814 and sent to Dartmoor.

Gray, Isaac - Seaman - Number: 1767 - Prize name: Union, prize of the Privateer Brutus - Ship type: LM - How taken: HM Brig Goldfinch - When taken: 17 Jul 1813 - Where taken: Bay of Biscay - Date received: 25 Jul 1813 - From what ship: HMS Pyramus - Born: New Hampshire - Age: 18 - Discharged on 8 Sep 1813 and sent to Dartmoor.

Gray, Morehouse - Seaman - Number: 1222 - Prize name: Grand Napoleon - Ship type: MV - How taken: HM Frigate Belle Poule - When taken: 3 Apr 1813 - Where taken: off Bordeaux - Date received: 9 May 1813 - From what ship: HMS Andromache - Born: Connecticut - Age: 25 - Discharged on 3 Jul 1813 and sent to Stapleton Prison.

Gray, William - Seaman - Number: 1637 - Prize name: Tickler - Ship type: LM - How taken: HM Frigate Magiciene - When taken: 5 Jun 1813 - Where taken: Lat 47 Long 13 - Date received: 14 Jun 1813 - From what ship: HMS Orestes - Born: Rhode Island - Age: 27 - Discharged on 28 Jun 1813 and released to HMS Salvador del Mundo.

Green, Horace - Seaman - Number: 223 - Prize name: Vengeance - Ship type: LM - How taken: HM Frigate Phoebe - When taken: 1 Jan 1813 - Where taken: Lat 44.4 Long 23 - Date received: 9 Jan 1813 - From what ship: HMS Phoebe - Born: Hampshire, MA - Age: 19 - Sent to Portsmouth on 8 Feb 1813 on the HMS Colossus.

Green, John - Seaman - Number: 35 - Prize name: Wasp - Ship type: MV - How taken: Earl Spencer, cutter - When taken: 4 Aug 1812 - Where taken: off Cape Clear - Date received: 23 Nov 1812 - From what ship: HMS Stork - Born: Albany - Age: 23 - Race: Black - Sent to Portsmouth on 29 Dec 1812 on the HMS Northumberland.

Green, Moses - Seaman - Number: 1753 - Prize name: Friendship, prize to the Privateer America - Ship type: P - How taken: HM Schooner Whiting - When taken: 15 Jul 1815 - Where taken: Lat 47N Long 8W - Date received: 20 Jul 1813 - From what ship: HMS Whiting - Born: Massachusetts - Age: 19 - Discharged on 8 Sep 1813 and sent to Dartmoor.

Green, Peter - Seaman - Number: 2130 - Prize name: Amiable - Ship type: LM - How taken: HM Ship-of-the-Line Magnificant - When taken: 30 Oct 1813 - Where taken: off Lorient - Date received: 29 Nov 1813 - From what ship: HMS Dublin - Born: Philadelphia - Age: 47 - Discharged on 4 Dec 1813 and sent to Dartmoor.

Green, Reuben - Seaman - Number: 2065 - Prize name: Betsy, prize to the Privateer True Blooded Yankee - Ship type: P - How taken: HM Frigate Eurotas - When taken: 26 Oct 1813 - Where taken: off Ushant - Date received: 1 Nov 1813 - From what ship: HMS Hannibal - Born: Connecticut - Age: 22 - Discharged on 3 Nov 1813 and sent to Dartmoor.

Green, Samuel - Seaman - Number: 84 - Prize name: Argus - Ship type: MV - How taken: Fancy, Cutter - When taken: 19 Dec 1812 - Where taken: Bay of Biscay - Date received: 24 Dec 1812 - From what ship: Fancy, Cutter - Born: Massachusetts - Age: 29 - Sent to Portsmouth on 29 Dec 1812 on the HMS Northumberland.

Green, Solomon - Seaman - Number: 2607 - How taken: MV Thomas - When taken: 15 Nov 1812 - Date received: 16 May 1814 - From what ship: HMS Repulse - Born: Baltimore - Age: 38 - Race: Black - Discharged on 14 Jun 1814 and sent to Dartmoor.

Green, William - Seaman - Number: 1468 - Prize name: Paul Jones - Ship type: P - How taken: HM Frigate Leonidas - When taken: 23 May 1813 - Where taken: off Cape Clear - Date received: 26 May 1813 - From what ship: HMS Leonidas - Born: New Jersey - Age: 38 - Race: Black - Discharged on 30 Jun 1813 and sent to Stapleton.

Greenland, Stephen - 2nd Mate - Number: 2590 - Prize name: Monticello - Ship type: MV - How taken: HM Brig

Racehorse - When taken: 12 Nov 1812 - Where taken: Cape of Good Hope - Date received: 10 May 1814 - From what ship: HMS Lion - Born: Billerica - Age: 39 - Discharged on 14 Jun 1814 and sent to Dartmoor.

Greenleaf, Thomas - Seaman - Number: 375 - Prize name: Orbit - Ship type: MV - How taken: HM Brig Achates - When taken: 29 Jan 1813 - Where taken: Lat 49 N Long 13 W - Date received: 4 Feb 1813 - From what ship: HMS Achates - Born: America - Age: 17 - Sent to Chatham on 29 Mar 1813 on the HMS Braham.

Greenwood, Thales - Seaman - Number: 786 - Prize name: Cannoniere - Ship type: P - How taken: HM Ship-of-the-Line Warspite - When taken: 14 Mar 1813 - Where taken: Bay of Biscay - Date received: 19 Mar 1813 - From what ship: HMS Warspite - Born: Providence - Age: 23 - Sent to Dartmoor on 28 Jun 1813.

Gregory, Elizah - Seaman - Number: 2689 - How taken: Sent into custody from HM Ship-of-the-Line Cornwallis - Date received: 7 Jun 1814 - From what ship: Inap - Born: New York - Age: 28 - Discharged on 14 Jun 1814 and sent to Dartmoor.

Gregory, George - Seaman - Number: 2207 - Prize name: Squirrel - Ship type: MV - How taken: HM Frigate Belle Poule - When taken: 14 Dec 1813 - Where taken: Bay of Biscay - Date received: 27 Dec 1813 - From what ship: HMS Protector - Born: Maryland - Age: 23 - Discharged on 27 Feb 1814 and sent to Chatham on HMS Haleyon.

Grenaux, Jean Yves - Volunteer - Number: 2399 - Prize name: Harvest, prize of the Privateer Bunker Hill - Ship type: P - How taken: HM Brig Orestes - When taken: 21 Jan 1814 - Date received: 3 Feb 1814 - From what ship: HMS Orestes - Born: Brest - Age: 16 - Discharged on 10 May 1814 and sent to Dartmoor.

Grey, John - Seaman - Number: 587 - Prize name: St. Martin's Plantation, prize of the Privateer Paul Jones - Ship type: P - How taken: HM Ship-of-the-Line Dublin - When taken: 9 Feb 1815 - Where taken: Lat 43 N Long 33.5 W - Date received: 25 Feb 1813 - From what ship: HMS Dublin - Born: Richmond - Age: 20 - Sent to Dartmoor on 2 Apr 1813.

Grey, Thomas - Cook - Number: 609 - Prize name: Rolla - Ship type: MV - How taken: HM Frigate Surveillante - When taken: 11 Feb 1813 - Where taken: Bay of Biscay - Date received: 28 Feb 1813 - From what ship: HMS Surveillante - Born: Baltimore - Age: 28 - Sent to Dartmoor on 2 Apr 1813.

Griffin, John - Seaman - Number: 1080 - Prize name: John & Frances - Ship type: MV - How taken: HM Frigate Belle Poule - When taken: 19 Mar 1813 - Where taken: Bay of Biscay - Date received: 22 Apr 1813 - From what ship: HMS Superb - Born: Louisiana - Age: 23 - Sent to Dartmoor on 1 Jul 1813.

Griffin, William - Seaman - Number: 1047 - Prize name: Independence - Ship type: MV - How taken: HM Ship-of-the-Line Superb - When taken: 16 Mar 1813 - Where taken: Bay of Biscay - Date received: 22 Apr 1813 - From what ship: HMS Superb - Born: Marblehead - Age: 20 - Sent to Dartmoor on 1 Jul 1813.

Griffin, William - Seaman - Number: 1957 - Prize name: Joel Barlow - Ship type: LM - How taken: HM Frigate Briton - When taken: 3 Jul 1813 - Where taken: off Bordeaux - Date received: 31 Aug 1813 - From what ship: HMS Clarence - Born: Freeport - Age: 21 - Discharged on 8 Sep 1813 and sent to Dartmoor.

Griffith, Benjamin - Seaman - Number: 1910 - Prize name: US Brig Argus - Ship type: War - How taken: HM Brig Pelican - When taken: 14 Aug 1813 - Where taken: Irish Channel - Date received: 23 Aug 1813 - From what ship: HMS Pelican - Born: Chestertown - Age: 23 - Discharged on 8 Sep 1813 and sent to Dartmoor.

Griffiths, Thomas - Seaman - Number: 637 - Prize name: Criterion - Ship type: MV - How taken: HM Frigate Belle Poule - When taken: 14 Feb 1813 - Where taken: Bay of Biscay - Date received: 4 Mar 1813 - From what ship: HMS Strenuous - Born: New York - Age: 31 - Sent to Dartmoor on 2 Apr 1813.

Grisse, William - Supercargo - Number: 469 - Prize name: Union - Ship type: MV - How taken: HM Frigate Iris - When taken: 17 Jan 1813 - Where taken: Lat 44 N Long 2.3 W - Date received: 8 Feb 1813 - From what ship: Union - Born: Germany - Age: 30 - Sent on 9 Feb 1813 to Ashburton on parole.

Groger, Henry - Seaman - Number: 1874 - Prize name: US Brig Argus - Ship type: War - How taken: HM Brig Pelican - When taken: 14 Aug 1813 - Where taken: Irish Channel - Date received: 23 Aug 1813 - From what ship: HMS Pelican - Born: Lubeck, Germany - Age: 20 - Discharged on 8 Sep 1813 and sent to Dartmoor.

Gronard, Peter - Seaman - Number: 441 - How taken: Impressed at Belfast - Date received: 5 Feb 1813 - From what ship: HMS Neptune - Born: Portsmouth - Age: 46 - Sent to Dartmoor on 28 Jun 1813.

Grosette, Jean Maurice - Seaman - Number: 2378 - Prize name: Devon, prize to the Privateer Bunker Hill - Ship type:

P - How taken: HM Brig Fly - When taken: 21 Jan 1814 - Where taken: at sea - Date received: 31 Jan 1814 - From what ship: HMS Fly - Born: France - Age: 15 - Discharged on 27 Feb 1814 and sent to Chatham on HMS Haleyon.

Grosse, William - Boy - Number: 1487 - Prize name: Paul Jones - Ship type: P - How taken: HM Frigate Leonidas - When taken: 23 May 1813 - Where taken: off Cape Clear - Date received: 26 May 1813 - From what ship: HMS Leonidas - Born: New York - Age: 14 - Discharged on 30 Jun 1813 and sent to Stapleton.

Grover, Edmond - Prize Master - Number: 1413 - Prize name: Miranda, prize of the Privateer Paul Jones - Ship type: P - How taken: HM Frigate Unicorn - When taken: 21 May 1813 - Where taken: off Ushant - Date received: 25 May 1813 - From what ship: Miranda - Born: Cape Ann - Age: 40 - Discharged on 8 Sep 1813 and sent to Dartmoor.

Grubb, Andrews - Seaman - Number: 2704 - Prize name: Margaret, prize to the Privateer Surprize - Ship type: P - How taken: HM Brig Foxhound - When taken: 27 May 1814 - Where taken: off Scylly - Date received: 16 Jun 1814 - From what ship: HMS Foxhound - Born: Baltimore - Age: 29 - Discharged on 20 Jun 1814 and sent to Dartmoor.

Grubb, James Julian - Merchant's Clerk & Passenger - Number: 483 - Prize name: Orbit - Ship type: MV - How taken: HM Brig Achates - When taken: 29 Jan 1813 - Where taken: Lat 49 N Long 13 W - Date received: 13 Feb 1813 - From what ship: HMS Plymouth - Born: Lorient, France - Age: 25 - Liberated on 24 Jul 1813.

Grunlief, Timothy - Steward's Mate - Number: 169 - Prize name: Hunter - Ship type: P - How taken: HM Frigate Phoebe - When taken: 24 Dec 1812 - Where taken: off Western Islands - Date received: 9 Jan 1813 - From what ship: HMS Phoebe - Born: Newburyport - Age: 35 - Sent to Portsmouth on 8 Feb 1813 on the HMS Colossus.

Guillard, Joseph - Steward - Number: 1932 - Prize name: Ellen & Emeline - Ship type: MV - How taken: HM Schooner Telegraph - When taken: 13 Aug 1813 - Where taken: off St. Andrews - Date received: 25 Aug 1813 - From what ship: Ellen & Emeline - Born: Natchez - Age: 24 - Race: Black - Discharged on 8 Sep 1813 and sent to Dartmoor.

Guillard, Louis - Marine - Number: 1482 - Prize name: Paul Jones - Ship type: P - How taken: HM Frigate Leonidas - When taken: 23 May 1813 - Where taken: off Cape Clear - Date received: 26 May 1813 - From what ship: HMS Leonidas - Born: New Orleans - Age: 19 - Discharged on 30 Jun 1813 and sent to Stapleton.

Guillard, Peter - Seaman - Number: 1476 - Prize name: Paul Jones - Ship type: P - How taken: HM Frigate Leonidas - When taken: 23 May 1813 - Where taken: off Cape Clear - Date received: 26 May 1813 - From what ship: HMS Leonidas - Born: Nantes, France - Age: 20 - Discharged on 30 Jun 1813 and sent to Stapleton.

Guilmot, Richard - Seaman - Number: 2313 - Prize name: Siro - Ship type: LM - How taken: HM Brig Pelican - When taken: 13 Jan 1814 - Where taken: at sea - Date received: 23 Jan 1814 - From what ship: HMS Pelican - Born: Portsmouth - Age: 25 - Race: Mulatto - Discharged on 31 Jan 1814 and sent to Dartmoor.

Gundy, James - Boy - Number: 2018 - Prize name: Ned - Ship type: LM - How taken: HM Brig Royalist - When taken: 6 Sep 1812 - Where taken: coast of France - Date received: 30 Sep 1813 - From what ship: HMS Royalist - Born: Maryland - Age: 12 - Discharged on 3 Nov 1813 and sent to Dartmoor.

Gustave, Peter - Seaman - Number: 2105 - Prize name: Chesapeake - Ship type: LM - How taken: HM Frigate Hotspur & HM Frigate Pyramus - When taken: 26 Oct 1813 - Where taken: Bay of Biscay - Date received: 22 Nov 1813 - From what ship: HMS Pyramus - Born: Sweden - Age: 18 - Discharged on 29 Nov 1813 and sent to Dartmoor.

Hacket, William - Master - Number: 2016 - Prize name: Ned - Ship type: LM - How taken: HM Brig Royalist - When taken: 6 Sep 1812 - Where taken: coast of France - Date received: 30 Sep 1813 - From what ship: HMS Royalist - Born: Virginia - Age: 26 - Discharged on 30 Sep 1813 and sent to Ashburton on parole.

Hacking, Robert - Seaman - Number: 1648 - Prize name: Tickler - Ship type: LM - How taken: HM Frigate Magiciene - When taken: 5 Jun 1813 - Where taken: Lat 47 Long 13 - Date received: 14 Jun 1813 - From what ship: HMS Orestes - Born: Hudson, NY - Age: 33 - Discharged on 30 Jun 1813 and sent to Stapleton.

Hale, Horace - Chief Mate - Number: 861 - Prize name: Decornau - Ship type: MV - How taken: HM Sloop Pheasant - When taken: 15 Mar 1813 - Where taken: Bay of Biscay - Date received: 27 Mar 1813 - From what ship: HMS Pheasant - Age: 26 - Sent on 28 Mar 1813 to Ashburton on parole.

Halfpenny, Robert - Seaman - Number: 489 - Prize name: Cashiere - Ship type: LM - How taken: HM Brig Reindeer - When taken: 3 Feb 1813 - Where taken: Bay of Biscay - Date received: 12 Feb 1813 - From what ship: HMS

Reindeer - Born: Baltimore - Age: 19 - Sent to Dartmoor on 2 Apr 1813.

Hall, David - 2nd Mate - Number: 954 - Prize name: Courier - Ship type: LM - How taken: HM Frigate Andromache - When taken: 14 Mar 1813 - Where taken: Bay of Biscay - Date received: 7 Apr 1813 - From what ship: HMS Sea Lark - Born: Maryland - Age: 22 - Sent to Dartmoor on 28 Jun 1813.

Hall, James - Seaman - Number: 769 - Prize name: William Bayard - Ship type: MV - How taken: HM Ship-of-the-Line Warspite - When taken: 3 Mar 1813 - Where taken: Bay of Biscay - Date received: 19 Mar 1813 - From what ship: HMS Warspite - Born: New York - Age: 22 - Sent to Chatham on 8 Jul 1813 on HM Tender Neptune.

Hall, James - Seaman - Number: 1809 - Prize name: US Brig Argus - Ship type: War - How taken: HM Brig Pelican - When taken: 14 Aug 1813 - Where taken: Irish Channel - Date received: 17 Aug 1813 - From what ship: USS Argus - Born: Cambridge, MA - Age: 26 - Discharged on 8 Sep 1813 and sent to Dartmoor.

Hall, James - Seaman - Number: 1779 - Prize name: Orders in Council - Ship type: LM - How taken: HM Frigate Surveillante - When taken: 1 Jun 1813 - Where taken: Bay of Biscay - Date received: 7 Aug 1813 - From what ship: HMS Gleaner - Born: Virginia - Age: 29 - Discharged on 8 Sep 1813 and sent to Dartmoor.

Hall, John - Seaman - Number: 1771 - Prize name: Minerva - Ship type: MV - How taken: HM Brig Goldfinch - When taken: 27 Jun 1813 - Where taken: off Nantes - Date received: 25 Jul 1813 - From what ship: HMS Pyramus - Born: Maryland - Age: 28 - Discharged on 10 Aug 1813 and released to HMS Salvador del Mundo.

Hall, Richard - 2nd Mate - Number: 955 - Prize name: Courier - Ship type: LM - How taken: HM Frigate Andromache - When taken: 14 Mar 1813 - Where taken: Bay of Biscay - Date received: 7 Apr 1813 - From what ship: HMS Sea Lark - Born: Snowhill - Age: 25 - Sent to Dartmoor on 28 Jun 1813.

Hall, William - Prize Master - Number: 147 - Prize name: Hunter - Ship type: P - How taken: HM Frigate Phoebe - When taken: 24 Dec 1812 - Where taken: off Western Islands - Date received: 9 Jan 1813 - From what ship: HMS Phoebe - Born: New Jersey - Age: 29 - Sent to Portsmouth on 8 Feb 1813 on the HMS Colossus.

Ham, John - Seaman - Number: 2437 - Prize name: Prince of Wales, prize of the Privateer Prince de Neufchatel - Ship type: P - How taken: Transport Nelson - When taken: 4 Feb 1814 - Where taken: at sea - Date received: 14 Feb 1814 - From what ship: HMS Halcyon - Born: Portsmouth - Age: 24 - Discharged on 10 May 1814 and sent to Dartmoor.

Hamill, William - Seaman - Number: 1702 - How taken: Delivered himself up from HM Ship-of-the-Line Clarence - Date received: 28 Jun 1813 - From what ship: HMS Royal Sovereign - Born: New York - Age: 22 - Discharged on 8 Jul 1813 and sent to Chatham on tender Neptune.

Hamilton, Anthony M. - Boy - Number: 1485 - Prize name: Paul Jones - Ship type: P - How taken: HM Frigate Leonidas - When taken: 23 May 1813 - Where taken: off Cape Clear - Date received: 26 May 1813 - From what ship: HMS Leonidas - Born: Boston - Age: 14 - Discharged on 30 Jun 1813 and sent to Stapleton.

Hamilton, John - Seaman - Number: 458 - Prize name: Dolphin - Ship type: LM - How taken: HM Ship-of-the-Line Colossus, HM Frigate Rhin & HM Brig Goldfinch - When taken: 5 Jan 1813 - Where taken: off Western Islands - Date received: 6 Feb 1813 - From what ship: HMS Rhin - Born: New London - Age: 23 - Race: Negro - Sent to Chatham on 29 Mar 1813 on the HMS Braham.

Hammet, John - Seaman - Number: 173 - Prize name: Hunter - Ship type: P - How taken: HM Frigate Phoebe - When taken: 24 Dec 1812 - Where taken: off Western Islands - Date received: 9 Jan 1813 - From what ship: HMS Phoebe - Born: Plymouth - Age: 25 - Sent to Portsmouth on 8 Feb 1813 on the HMS Colossus.

Hammond, Joseph - Seaman - Number: 1309 - Prize name: Eliza - Ship type: MV - How taken: HM Frigate Surveillante - When taken: 22 Apr 1813 - Where taken: at sea - Date received: 10 May 1813 - From what ship: HMS Medusa - Born: Marblehead - Age: 21 - Discharged on 3 Jul 1813 and sent to Stapleton Prison.

Hammond, William - Seaman - Number: 2413 - Prize name: Joseph - Ship type: MV - How taken: HM Brig Royalist - When taken: 18 Jan 1814 - Where taken: Bay of Biscay - Date received: 9 Feb 1814 - From what ship: HMS Sparrow - Born: Marblehead - Age: 22 - Discharged on 10 May 1814 and sent to Dartmoor.

Hancock, William - Seaman - Number: 2222 - Prize name: Zephyr - Ship type: MV - How taken: HM Frigate Pyramus - When taken: 30 Nov 1813 - Where taken: off Lorient - Date received: 3 Jan 1814 - From what ship: HMS Warspite - Born: New York - Age: 40 - Discharged on 31 Jan 1814 and sent to Dartmoor.

Handel, Henry - Seaman - Number: 1002 - Prize name: Eliza - Ship type: MV - How taken: HM Frigate Surveillante - When taken: 27 Mar 1813 - Where taken: Bay of Biscay - Date received: 16 Apr 1813 - From what ship: HMS Fairy - Born: Philadelphia - Age: 23 - Sent on 21 Jun 1813 to the American ship Hope.

Handfield, Robert - Seaman - Number: 2493 - How taken: Impressed out a merchant vessel - When taken: 2 Mar 1814 - Date received: 26 Mar 1814 - From what ship: Earl Spencer, cutter - Born: Savannah - Age: 30 - Discharged on 10 May 1814 and sent to Dartmoor.

Hanford, William - Seaman - Number: 1216 - Prize name: Zebra - Ship type: LM - How taken: HM Frigate Pyramus - When taken: 20 Apr 1813 - Where taken: Bay of Biscay - Date received: 9 May 1813 - From what ship: HMS Andromache - Born: Connecticut - Age: 20 - Discharged on 3 Jul 1813 and sent to Stapleton Prison.

Hanscom, Moses - Seaman - Number: 1908 - Prize name: US Brig Argus - Ship type: War - How taken: HM Brig Pelican - When taken: 14 Aug 1813 - Where taken: Irish Channel - Date received: 23 Aug 1813 - From what ship: HMS Pelican - Born: Kittery - Age: 27 - Discharged on 8 Sep 1813 and sent to Dartmoor.

Hansell, John - Seaman - Number: 1320 - Prize name: Shadow - Ship type: LM - How taken: HM Brig Reindeer & HM Schooner Helicon - When taken: 6 Apr 1813 - Where taken: Bay of Biscay - Date received: 12 May 1813 - From what ship: HMS Helicon - Born: Philadelphia - Age: 26 - Discharged on 3 Jul 1813 and sent to Stapleton Prison.

Hanson, Christian - Seaman - Number: 1170 - Prize name: Hebe - Ship type: MV - How taken: HM Frigate Stag - When taken: 18 Apr 1813 - Where taken: Bay of Biscay - Date received: 6 May 1813 - From what ship: HMS Stag - Born: Copenhagen - Age: 28 - Sent on 3 Jul 1813 to Stapleton prison.

Hanson, Henry - Seaman - Number: 2297 - Prize name: Siro - Ship type: LM - How taken: HM Brig Pelican - When taken: 13 Jan 1814 - Where taken: at sea - Date received: 23 Jan 1814 - From what ship: HMS Pelican - Born: Boston - Age: 42 - Discharged on 31 Jan 1814 and sent to Dartmoor.

Hanson, Peter - Seaman - Number: 2298 - Prize name: Siro - Ship type: LM - How taken: HM Brig Pelican - When taken: 13 Jan 1814 - Where taken: at sea - Date received: 23 Jan 1814 - From what ship: HMS Pelican - Born: Gothenburg, Sweden - Age: 27 - Discharged on 31 Jan 1814 and sent to Dartmoor.

Harding, John - Seaman - Number: 1059 - Prize name: Virginia Planter - Ship type: MV - How taken: HM Frigate Pyramus - When taken: 17 Mar 1813 - Where taken: off Nantes - Date received: 22 Apr 1813 - From what ship: HMS Superb - Born: Newport - Age: 19 - Sent to Dartmoor on 1 Jul 1813.

Harens, William - Seaman - Number: 774 - Prize name: William Bayard - Ship type: MV - How taken: HM Ship-of-the-Line Warspite - When taken: 3 Mar 1813 - Where taken: Bay of Biscay - Date received: 19 Mar 1813 - From what ship: HMS Warspite - Born: New Jersey - Age: 27 - Sent to Chatham on 8 Jul 1813 on HM Tender Neptune.

Harker, John - Boy - Number: 2670 - Prize name: Traveler, prize of the Privateer Surprize - Ship type: P - How taken: HMS Cawser - When taken: 7 May 1814 - Where taken: off Cape Clear - Date received: 5 Jun 1814 - From what ship: HMS Gronville - Born: Maryland - Age: 17 - Discharged on 14 Jun 1814 and sent to Dartmoor.

Harper, Nicholas - Supercargo - Number: 1406 - Prize name: Henry Clements - Ship type: MV - How taken: HM Brig Orestes - When taken: 15 Apr 1813 - Where taken: Bay of Biscay - Date received: 20 May 1813 - From what ship: HMS Orestes - Born: Dublin - Age: 39 - Discharged on 23 May 1813 and sent to Ashburton on parole.

Harper, Robert - Master - Number: 230 - Prize name: Hope - Ship type: MV - How taken: HM Schooner Bramble - When taken: 3 Dec 1812 - Where taken: at Coruna, Spain - Date received: 11 Jan 1813 - From what ship: HMS Pheasant - Born: Virginia - Age: 28 - Sent on 13 Jan 1813 to Ashburton on parole.

Harre, John - Seaman - Number: 1135 - Prize name: Magdalen - Ship type: MV - How taken: HM Ship-of-the-Line Superb - When taken: 15 Apr 1813 - Where taken: off Belle Isle - Date received: 22 Apr 1813 - From what ship: HMS Superb - Born: Bluefield, VA - Age: 23 - Sent to Dartmoor on 1 Jul 1813.

Harrington, John - Seaman - Number: 2066 - Prize name: Betsy, prize to the Privateer True Blooded Yankee - Ship type: P - How taken: HM Frigate Eurotas - When taken: 26 Oct 1813 - Where taken: off Ushant - Date received: 1 Nov 1813 - From what ship: HMS Hannibal - Born: New York - Age: 23 - Discharged on 3 Nov 1813 and sent to Dartmoor.

Harrington, Simon - Seaman - Number: 1194 - Prize name: Zebra - Ship type: LM - How taken: HM Frigate Pyramus

- When taken: 20 Apr 1813 - Where taken: Bay of Biscay - Date received: 9 May 1813 - From what ship: HMS Andromache - Born: Eastport - Age: 23 - Sent on 3 Jul 1813 to Stapleton prison.

Harris, Bradley - Boy - Number: 2286 - Prize name: Siro - Ship type: LM - How taken: HM Brig Pelican - When taken: 13 Jan 1814 - Where taken: at sea - Date received: 23 Jan 1814 - From what ship: HMS Pelican - Born: Boston - Age: 18 - Discharged on 31 Jan 1814 and sent to Dartmoor.

Harris, Ebenezer - Seaman - Number: 2367 - Prize name: Minerva - Ship type: MV - How taken: HM Ship-of-the-Line Conquestador - When taken: 19 Dec 1814 - Where taken: Bay of Biscay - Date received: 31 Jan 1814 - From what ship: HMS Surveillante - Born: North Yarmouth - Age: 25 - Discharged on 27 Feb 1814 and sent to Chatham on HMS Haleyon.

Harris, George - Seaman - Number: 67 - Prize name: Independence - Ship type: MV - How taken: HM Frigate Medusa - When taken: 9 Nov 1812 - Where taken: San Sebastian - Date received: 27 Nov 1812 - From what ship: HMS Wasp - Born: Baltimore - Age: 32 - Sent to Portsmouth on 29 Dec 1812 on the HMS Northumberland.

Harris, James - Seaman - Number: 1691 - Prize name: Governor Gerry - Ship type: MV - How taken: HM Brig Royalist - When taken: 31 May 1813 - Where taken: Bay of Biscay - Date received: 26 Jun 1813 - From what ship: HMS Duncan - Born: New Orleans - Age: 19 - Race: Black - Discharged on 30 Jun 1813 and sent to Stapleton.

Harris, James - Seaman - Number: 99 - Prize name: Experiment - Ship type: MV - How taken: HM Brig Rover - When taken: 21 Oct 1812 - Where taken: off Bordeaux - Date received: 25 Dec 1812 - From what ship: HMS Northumberland - Born: Maryland - Sent to Portsmouth on 29 Dec 1812 on the HMS Northumberland.

Harris, John - Seaman - Number: 1339 - Prize name: Fox - Ship type: LM - How taken: HM Sloop Pheasant - When taken: 23 Apr 1813 - Where taken: Bay of Biscay - Date received: 14 May 1813 - From what ship: HMS Pleasant - Born: Philadelphia - Age: 26 - Race: Black - Discharged on 3 Jul 1813 and sent to Stapleton Prison.

Harris, John - Seaman - Number: 580 - Prize name: Cashiere - Ship type: LM - How taken: HM Brig Reindeer - When taken: 3 Feb 1813 - Where taken: Bay of Biscay - Date received: 23 Feb 1813 - From what ship: HMS Surveillante - Born: Maryland - Age: 25 - Sent to Dartmoor on 2 Apr 1813.

Harris, John - Soldier - Number: 2387 - How taken: Gave himself up at Belfast as a militiaman - Date received: 1 Feb 1814 - From what ship: HMS Salvador del Mundo - Born: Virginia - Age: 26 - Race: Mulatto - Discharged on 27 Feb 1814 and sent to Chatham on HMS Haleyon.

Harris, Simon - Seaman - Number: 2359 - Prize name: Zephyr - Ship type: MV - How taken: HM Frigate Surveillante - When taken: 6 Jan 1814 - Where taken: Bay of Biscay - Date received: 31 Jan 1814 - From what ship: HMS Surveillante - Born: Virginia - Age: 40 - Race: Black - Discharged on 27 Feb 1814 and sent to Chatham on HMS Haleyon.

Harris, William - Seaman - Number: 299 - Prize name: Stephen - Ship type: MV - How taken: HM Frigate Briton & HM Frigate Andromache - When taken: 17 Dec 1812 - Where taken: off Bordeaux - Date received: 21 Jan 1813 - From what ship: HMS Abercrombie - Born: Pennsylvania - Age: 32 - Race: Black - Sent to Portsmouth on 8 Feb 1813 on the HMS Colossus.

Harris, William - 2nd Mate - Number: 1329 - Prize name: Fox - Ship type: LM - How taken: HM Sloop Pheasant - When taken: 23 Apr 1813 - Where taken: Bay of Biscay - Date received: 14 May 1813 - From what ship: HMS Pleasant - Born: New Jersey - Age: 28 - Discharged on 3 Jul 1813 and sent to Stapleton Prison.

Harris, William - Seaman - Number: 996 - Prize name: Polly - Ship type: MV - How taken: HM Frigate Surveillante - When taken: 27 Mar 1813 - Where taken: Bay of Biscay - Date received: 16 Apr 1813 - From what ship: HMS Fairy - Born: Marblehead - Age: 43 - Sent to Chatham on 8 Jul 1813 on HM Tender Neptune.

Harris, William - Seaman - Number: 331 - Prize name: Porcupine - Ship type: MV - How taken: HM Frigate Dryad - When taken: 8 Jan 1813 - Where taken: off Bordeaux - Date received: 21 Jan 1813 - From what ship: HMS Abercrombie - Born: Philadelphia - Age: 33 - Sent to Chatham on 29 Mar 1813 on the HMS Braham.

Harrison, John - Seaman - Number: 402 - Prize name: Union - Ship type: MV - How taken: HM Frigate Iris - When taken: 17 Jan 1813 - Where taken: Lat 44 N Long 2.3 W - Date received: 5 Feb 1813 - From what ship: HMS San Josef - Born: Portsmouth, VA - Age: 25 - Sent to Chatham on 29 Mar 1813 on the HMS Braham.

Harry, John (alias Avery) - Seaman - Number: 652 - How taken: Delivered himself up from HM Ship of the Line

Abercrombie - When taken: 1 Feb 1813 - Date received: 5 Mar 1813 - From what ship: HMS Salvador del Monde - Born: New Orleans - Age: 45 - Sent to Dartmoor on 2 Apr 1813.

Hart, Bartholomew - Seaman - Number: 1954 - Prize name: Joel Barlow - Ship type: LM - How taken: HM Frigate Briton - When taken: 3 Jul 1813 - Where taken: off Bordeaux - Date received: 31 Aug 1813 - From what ship: HMS Clarence - Born: New York - Age: 42 - Discharged on 8 Sep 1813 and sent to Dartmoor.

Hart, Joshua - Seaman - Number: 1939 - Prize name: Marmion - Ship type: MV - How taken: HM Frigate President - When taken: 14 Aug 1813 - Where taken: off Nantes - Date received: 27 Aug 1813 - From what ship: HMS Urgent - Born: Baltimore - Age: 22 - Race: Negro - Discharged on 8 Sep 1813 and sent to Dartmoor.

Harter, Henry - Seaman - Number: 1060 - Prize name: Virginia Planter - Ship type: MV - How taken: HM Frigate Pyramus - When taken: 17 Mar 1813 - Where taken: off Nantes - Date received: 22 Apr 1813 - From what ship: HMS Superb - Born: Charlestown - Age: 20 - Sent to Dartmoor on 1 Jul 1813.

Hartfield, James - Seaman - Number: 1790 - How taken: Delivered himself up from HM Ship-of-the-Line Scipion - When taken: 1 Dec 1812 - Date received: 13 Aug 1813 - From what ship: HMS Protection - Born: New York - Age: 36 - Discharged on 8 Sep 1813 and sent to Dartmoor.

Hartford, James - Seaman - Number: 104 - How taken: Impressed at Cove of Cork - Date received: 30 Dec 1812 - From what ship: HMS Leonidas - Born: New Market - Age: 29 - Sent to Portsmouth on 4 Jan 1813 on the HMS Revolutionnaire.

Hartford, John - Ordinary Seaman - Number: 288 - Prize name: Columbia - Ship type: MV - How taken: HM Frigate Briton - When taken: 17 Dec 1812 - Where taken: off Bordeaux - Date received: 21 Jan 1813 - From what ship: HMS Abercrombie - Born: New Hampshire - Age: 22 - Sent to Portsmouth on 8 Feb 1813 on the HMS Colossus.

Horton, Henry - Seaman - Number: 1778 - Prize name: Virginia Planter - Ship type: MV - How taken: HM Frigate Pyramus - When taken: 17 Mar 1813 - Where taken: off Nantes - Date received: 3 Aug 1813 - From what ship: Dartmoor Prison - Born: Charlestown - Age: 20 - Discharged on 8 Sep 1813 and sent to Dartmoor.

Harts, William - Seaman - Number: 1144 - Prize name: Magdalen - Ship type: MV - How taken: HM Ship-of-the-Line Superb - When taken: 15 Apr 1813 - Where taken: off Belle Isle - Date received: 22 Apr 1813 - From what ship: HMS Superb - Born: New York - Age: 24 - Sent to Dartmoor on 1 Jul 1813.

Hartwell, Barry - Seaman - Number: 2608 - How taken: Thomas - When taken: 28 Dec 1812 - Date received: 16 May 1814 - From what ship: HMS Repulse - Born: Philadelphia - Age: 37 - Discharged on 14 Jun 1814 and sent to Dartmoor.

Harvey, John - Seaman - Number: 1495 - How taken: Delivered himself up from HM Brig Foxhound - Date received: 27 May 1813 - From what ship: Dartmoor prison - Born: New Orleans - Age: 45 - Discharged on 8 Jul 1813 and sent to Chatham on HM Tender Neptune.

Harvey, Joseph - Seaman - Number: 1051 - Prize name: Independence - Ship type: MV - How taken: HM Ship-of-the-Line Superb - When taken: 16 Mar 1813 - Where taken: Bay of Biscay - Date received: 22 Apr 1813 - From what ship: HMS Superb - Born: Salem - Age: 19 - Race: Black - Sent to Dartmoor on 1 Jul 1813.

Harvey, Joseph - Seaman - Number: 2364 - Prize name: Hannah - Ship type: MV - How taken: HM Ship-of-the-Line Conquestador - When taken: 14 Jan 1814 - Where taken: Lat 47.3 Long 7 - Date received: 31 Jan 1814 - From what ship: HMS Surveillante - Born: Beverly - Age: 33 - Discharged on 27 Feb 1814 and sent to Chatham on HMS Haleyon.

Hasem, John - Seaman - Number: 398 - Prize name: Union - Ship type: MV - How taken: HM Frigate Iris - When taken: 17 Jan 1813 - Where taken: Lat 44 N Long 2.3 W - Date received: 5 Feb 1813 - From what ship: HMS San Josef - Born: Boston - Age: 30 - Sent to Chatham on 29 Mar 1813 on the HMS Braham.

Haskell, Thomas - Seaman - Number: 2095 - Prize name: Fire Fly - Ship type: MV - How taken: HM Frigate Revolutionnaire - When taken: 20 Oct 1813 - Where taken: off Cape Ortagle - Date received: 6 Nov 1813 - From what ship: Fire Fly - Born: Cape Ann - Age: 38 - Discharged on 29 Nov 1813 and sent to Dartmoor.

Hataway, William N. - Boy - Number: 43 - Prize name: Perseverance - Ship type: MV - How taken: HM Frigate Sybille - When taken: 12 Aug 1812 - Where taken: off Cape Clear - Date received: 23 Nov 1812 - From what ship: HMS Stork - Born: New York - Age: 14 - Sent to Portsmouth on 29 Dec 1812 on the HMS Northumberland.

Hatch, Abraham - Seaman - Number: 275 - Prize name: Leader - Ship type: MV - How taken: HM Frigate Briton - When taken: 10 Dec 1812 - Where taken: off Bordeaux - Date received: 21 Jan 1813 - From what ship: HMS Abercrombie - Born: Bristol - Age: 29 - Sent to Portsmouth on 8 Feb 1813 on the HMS Colossus.

Hatch, Samuel - Seaman - Number: 2539 - Prize name: Caroline - Ship type: MV - How taken: Brilliant, privateer - When taken: 12 Aug 1813 - Where taken: off Charleston - Date received: 13 Apr 1814 - From what ship: HMS Bittern - Born: South Carolina - Age: 27 - Discharged on 10 May 1814 and sent to Dartmoor.

Hatchess, Levi - 2nd Lieutenant - Number: 142 - Prize name: Hunter - Ship type: P - How taken: HM Frigate Phoebe - When taken: 24 Dec 1812 - Where taken: off Western Islands - Date received: 9 Jan 1813 - From what ship: HMS Phoebe - Born: Boston - Age: 25 - Sent to Portsmouth on 8 Feb 1813 on the HMS Colossus.

Hatterson, Nathan - Seaman - Number: 1297 - Prize name: Price - Ship type: LM - How taken: HM Frigate Medusa - When taken: 13 Apr 1813 - Where taken: Bay of Biscay - Date received: 10 May 1813 - From what ship: HMS Medusa - Born: New York - Age: 27 - Discharged on 21 Jun 1813 and released to American ship Mount Hope.

Haught, John - Seaman - Number: 1516 - Prize name: Grand Napoleon - Ship type: MV - How taken: HM Brig Goldfinch - When taken: 17 Apr 1813 - Where taken: Bay of Biscay - Date received: 29 May 1813 - From what ship: HMS Hannibal - Born: New York - Age: 39 - Discharged on 30 Jun 1813 and sent to Stapleton.

Hawking, John - Seaman - Number: 848 - How taken: Delivered himself up from HM Ship-of-the-Line Leyden - Date received: 23 May 1813 - From what ship: HMS Dauntless - Born: Philadelphia - Age: 22 - Sent to Chatham on 8 Jul 1813 on HM Tender Neptune.

Hay, Cornelius - Boatswain - Number: 1126 - Prize name: Viper - Ship type: MV - How taken: HM Ship-of-the-Line Superb - When taken: 15 Apr 1813 - Where taken: Bay of Biscay - Date received: 22 Apr 1813 - From what ship: HMS Superb - Born: Chester - Age: 33 - Sent to Dartmoor on 1 Jul 1813.

Hayes, Benjamin - Seaman - Number: 2267 - Prize name: Fanny - Ship type: MV - How taken: HM Frigate Eurotas - When taken: 25 Dec 1813 - Where taken: at sea - Date received: 20 Jan 1814 - From what ship: HMS Eurotas - Born: Maryland - Age: 24 - Discharged on 31 Jan 1814 and sent to Dartmoor.

Hayes, Elias Warner - Chief Mate - Number: 1045 - Prize name: Independence - Ship type: MV - How taken: HM Ship-of-the-Line Superb - When taken: 16 Mar 1813 - Where taken: Bay of Biscay - Date received: 22 Apr 1813 - From what ship: HMS Superb - Born: Gloucester - Age: 31 - Sent on 27 Feb 1813 to Ashburton on parole.

Hayes, Simon - Seaman - Number: 619 - Prize name: Governor McKean - Ship type: LM - How taken: HM Brig Rover - When taken: 26 Jun 1813 - Where taken: off Bordeaux - Date received: 2 Mar 1813 - From what ship: HMS Insolent - Born: Virginia - Age: 19 - Sent to Dartmoor on 2 Apr 1813.

Hays, Moses - Boy - Number: 1304 - Prize name: Price - Ship type: LM - How taken: HM Frigate Medusa - When taken: 13 Apr 1813 - Where taken: Bay of Biscay - Date received: 10 May 1813 - From what ship: HMS Medusa - Born: Savannah - Age: 15 - Discharged on 3 Jul 1813 and sent to Stapleton Prison.

Haywood, Simon - Seaman - Number: 246 - Prize name: Janice & Lydia, prize of the Privateer General Armstrong - Ship type: P - How taken: Barton, letter of marque - When taken: 29 Nov 1812 - Where taken: off Bermuda - Date received: 12 Jan 1813 - From what ship: HMS Bittern - Born: Lenox - Age: 21 - Sent to Portsmouth on 8 Feb 1813 on the HMS Colossus.

Hazard, Thomas - Seaman - Number: 41 - Prize name: Warren - Ship type: MV - How taken: HM Frigate Sybille & HM Frigate Fortunee- When taken: 5 Sep 1812 - Where taken: Lat 41.4 Long 33 - Date received: 23 Nov 1812 - From what ship: HMS Stork - Born: South Kingston, RI - Age: 24 - Race: Black - Sent to Portsmouth on 29 Dec 1812 on the HMS Northumberland.

Hazard, William - Master - Number: 859 - Prize name: Cannoniere - Ship type: P - How taken: HM Ship-of-the-Line Warspite - When taken: 14 Mar 1813 - Where taken: Bay of Biscay - Date received: 26 Mar 1813 - From what ship: HMS Warspite - Born: New Fairfield - Age: 34 - Sent on 28 Mar 1813 to Ashburton on parole.

Hazel, George - Chief Mate - Number: 1402 - Prize name: Revenge - Ship type: LM - How taken: HM Frigate Belle Poule - When taken: 11 May 1813 - Where taken: Bay of Biscay - Date received: 18 May 1813 - From what ship: HMS Revenge - Born: South Eustat - Age: 33 - Discharged on 23 May 1813 and sent to Ashburton on parole.

Healy, John - Seaman - Number: 1682 - Prize name: Revenge - Ship type: LM - How taken: HM Frigate Belle Poule

- When taken: 11 May 1813 - Where taken: Bay of Biscay - Date received: 26 Jun 1813 - From what ship: HMS Duncan - Born: Boston - Age: 20 - Discharged on 30 Jun 1813 and sent to Stapleton.

Heard, Thomas - Seaman - Number: 1392 - Prize name: Tom - Ship type: LM - How taken: HM Frigate Surveillante - When taken: 27 Apr 1813 - Where taken: Bay of Biscay - Date received: 15 May 1813 - From what ship: HMS Foxhound - Born: New Jersey - Age: 28 - Discharged on 3 Jul 1813 and sent to Stapleton Prison.

Hebron, John - Seaman - Number: 2005 - Prize name: Francis & Ann - Ship type: MV - How taken: HM Sloop Lightning - When taken: 20 Aug 1813 - Where taken: at sea - Date received: 22 Sep 1813 - From what ship: HMS Lightning - Born: Philadelphia - Age: 19 - Discharged on 27 Sep 1813 and sent to Dartmoor.

Hedenburg, Jacob - Seaman - Number: 2509 - Prize name: Bunker Hill - Ship type: P - How taken: HM Frigate Pomone & HM Frigate Cydnus - When taken: 4 Mar 1814 - Where taken: Bay of Biscay - Date received: 4 Apr 1814 - From what ship: HMS Virago - Born: Philadelphia - Age: 35 - Discharged on 10 May 1814 and sent to Dartmoor.

Helenburg, James - Seaman - Number: 2699 - Prize name: Bunker Hill - Ship type: P - How taken: HM Frigate Pomone & HM Frigate Cydnus - When taken: 4 Mar 1814 - Where taken: Bay of Biscay - Date received: 17 Jun 1814 - From what ship: Dartmouth - Discharged on 20 Jun 1814 and released to the Speedwell Cartel.

Helm, Charles - Seaman - Number: 296 - Prize name: Columbia - Ship type: MV - How taken: HM Frigate Briton - When taken: 17 Dec 1812 - Where taken: off Bordeaux - Date received: 21 Jan 1813 - From what ship: HMS Abercrombie - Born: Philadelphia - Age: 35 - Sent to Portsmouth on 8 Feb 1813 on the HMS Colossus.

Henderson, Alexander - Seaman - Number: 721 - Prize name: Criterion - Ship type: MV - How taken: HM Frigate Belle Poule - When taken: 14 Feb 1813 - Where taken: Bay of Biscay - Date received: 19 Mar 1813 - From what ship: HMS Warspite - Born: Connecticut - Age: 26 - Sent to Chatham on 8 Jul 1813 on HM Tender Neptune.

Henderson, Benjamin - Captain - Number: 2581 - Prize name: James - Ship type: MV - How taken: HM Brig Harpy - When taken: 19 Dec 1812 - Where taken: off Isle du France - Date received: 10 May 1814 - From what ship: HMS Lion - Born: Philadelphia - Age: 33 - Discharged on 12 May 1814 and sent to Ashburton on parole.

Henderson, David - Ordinary Seaman - Number: 285 - Prize name: Columbia - Ship type: MV - How taken: HM Frigate Briton - When taken: 17 Dec 1812 - Where taken: off Bordeaux - Date received: 21 Jan 1813 - From what ship: HMS Abercrombie - Born: Philadelphia - Age: 21 - Sent to Portsmouth on 8 Feb 1813 on the HMS Colossus.

Hennery, James - Seaman - Number: 1831 - Prize name: US Brig Argus - Ship type: War - How taken: HM Brig Pelican - When taken: 14 Aug 1813 - Where taken: Irish Channel - Date received: 17 Aug 1813 - From what ship: USS Argus - Born: New York - Age: 18 - Discharged on 8 Sep 1813 and sent to Dartmoor.

Hennay, Jacob - Seaman - Number: 2032 - Prize name: Friendship West, prize of Privateer True Blooded Yankee - Ship type: P - How taken: HM Schooner Helicon & HM Schooner Whiting - When taken: 25 Oct 1813 - Where taken: Bay of Biscay - Date received: 31 Oct 1813 - From what ship: HMS Whiting - Born: Pillau, Prussia - Age: 28 - Discharged on 3 Nov 1813 and sent to Dartmoor.

Henry, Edward - Boy - Number: 211 - Prize name: Hunter - Ship type: P - How taken: HM Frigate Phoebe - When taken: 24 Dec 1812 - Where taken: off Western Islands - Date received: 9 Jan 1813 - From what ship: HMS Phoebe - Born: Newbury - Age: 15 - Sent to Portsmouth on 8 Feb 1813 on the HMS Colossus.

Henry, John - Seaman - Number: 318 - Prize name: Porcupine - Ship type: MV - How taken: HM Frigate Dryad - When taken: 8 Jan 1813 - Where taken: off Bordeaux - Date received: 21 Jan 1813 - From what ship: HMS Abercrombie - Born: New Orleans - Age: 36 - Sent to Chatham on 29 Mar 1813 on the HMS Braham.

Henry, John - Seaman - Number: 1911 - Prize name: US Brig Argus - Ship type: War - How taken: HM Brig Pelican - When taken: 14 Aug 1813 - Where taken: Irish Channel - Date received: 23 Aug 1813 - From what ship: HMS Pelican - Born: New York - Age: 24 - Discharged on 8 Sep 1813 and sent to Dartmoor.

Henry, William - Boy - Number: 1921 - Prize name: US Brig Argus - Ship type: War - How taken: HM Brig Pelican - When taken: 14 Aug 1813 - Where taken: Irish Channel - Date received: 23 Aug 1813 - From what ship: HMS Pelican - Born: New Orleans - Age: 16 - Race: Black - Discharged on 8 Sep 1813 and sent to Dartmoor.

Henshaw, Jacob S. - Chief Mate - Number: 829 - Prize name: Criterion - Ship type: MV - How taken: HM Frigate Belle Poule - When taken: 14 Feb 1813 - Where taken: Bay of Biscay - Date received: 22 Mar 1813 - From what ship: HMS Warspite - Born: New Bedford - Age: 29 - Sent on 23 Mar 1813 to Ashburton on parole.

Haydon, Eli - Seaman - Number: 815 - Prize name: Decornau - Ship type: MV - How taken: HM Sloop Pheasant - When taken: 15 Mar 1813 - Where taken: Bay of Biscay - Date received: 19 Mar 1813 - From what ship: HMS Pheasant - Born: Hartford - Age: 35 - Sent to Dartmoor on 28 Jun 1813.

Hickman, Joseph - Seaman - Number: 1292 - Prize name: Price - Ship type: LM - How taken: HM Frigate Medusa - When taken: 13 Apr 1813 - Where taken: Bay of Biscay - Date received: 10 May 1813 - From what ship: HMS Medusa - Born: New Jersey - Age: 22 - Discharged on 3 Jul 1813 and sent to Stapleton Prison.

Higgins, George - Seaman - Number: 520 - Prize name: Allegany - Ship type: MV - How taken: Detained at Gibraltar - When taken: 8 Aug 1812 - Date received: 15 Feb 1813 - From what ship: HMS Andromeda - Born: Boston - Age: 24 - Race: Black - Sent to Dartmoor on 2 Apr 1813.

Higgins, James - Seaman - Number: 1084 - Prize name: John & Frances - Ship type: MV - How taken: HM Frigate Belle Poule - When taken: 19 Mar 1813 - Where taken: Bay of Biscay - Date received: 22 Apr 1813 - From what ship: HMS Superb - Born: Philadelphia - Age: 27 - Sent to Dartmoor on 1 Jul 1813.

Higgins, William - Seaman - Number: 20 - Prize name: Charles - Ship type: MV - How taken: Taken at Liverpool - When taken: 18 Oct 1812 - Date received: 21 Nov 1812 - From what ship: HMS Salvador del Mundo - Born: Virginia - Age: 24 - Sent to Portsmouth on 29 Dec 1812 on the HMS Northumberland.

Hilaire, Gasper - Seaman - Number: 1563 - Prize name: Courier - Ship type: LM - How taken: HM Brig Rover - When taken: 14 Mar 1813 - Where taken: Bay of Biscay - Date received: 29 May 1813 - From what ship: HMS Hannibal - Born: Philadelphia - Age: 52 - Discharged on 8 Sep 1813 and sent to Dartmoor.

Hile, Shadrick - Seaman - Number: 1277 - Prize name: Caroline - Ship type: MV - How taken: HM Frigate Medusa - When taken: 12 Apr 1813 - Where taken: Bay of Biscay - Date received: 10 May 1813 - From what ship: HMS Medusa - Born: Baltimore - Age: 40 - Race: Black - Discharged on 2 Sep 1813 and sent to Dartmoor.

Hiler, George - Carpenter - Number: 2281 - Prize name: Siro - Ship type: LM - How taken: HM Brig Pelican - When taken: 13 Jan 1814 - Where taken: at sea - Date received: 23 Jan 1814 - From what ship: HMS Pelican - Born: Boston - Age: 27 - Discharged on 31 Jan 1814 and sent to Dartmoor.

Hill, Daniel - Seaman - Number: 2371 - Prize name: Minerva - Ship type: MV - How taken: HM Ship-of-the-Line Conquestador - When taken: 19 Dec 1814 - Where taken: Bay of Biscay - Date received: 31 Jan 1814 - From what ship: HMS Surveillante - Born: Saco - Age: 22 - Discharged on 27 Feb 1814 and sent to Chatham on HMS Haleyon.

Hill, Ephraim - Seaman - Number: 896 - Prize name: Tiger - Ship type: MV - How taken: HM Brig Scylla - When taken: 22 Mar 1813 - Where taken: Bay of Biscay - Date received: 1 Apr 1813 - From what ship: HMS Scylla - Born: Hartford - Age: 24 - Race: Black - Sent on 21 Jun 1813 to Portsmouth on HMS Prometheus.

Hill, John - Seaman - Number: 1696 - Prize name: Governor Gerry - Ship type: MV - How taken: HM Brig Royalist - When taken: 31 May 1813 - Where taken: Bay of Biscay - Date received: 26 Jun 1813 - From what ship: HMS Duncan - Born: Philadelphia - Age: 21 - Race: Black - Discharged on 30 Jun 1813 and sent to Stapleton.

Hill, Joseph - Seaman - Number: 888 - Prize name: Tiger - Ship type: MV - How taken: HM Brig Scylla - When taken: 22 Mar 1813 - Where taken: Bay of Biscay - Date received: 1 Apr 1813 - From what ship: HMS Scylla - Born: New Orleans - Age: 33 - Race: Black - Sent on 21 Jun 1813 to Portsmouth on HMS Prometheus.

Hill, Leonard - Seaman - Number: 1069 - Prize name: Meteor - Ship type: MV - How taken: HM Frigate Briton - When taken: 12 Mar 1813 - Where taken: Bay of Biscay - Date received: 22 Apr 1813 - From what ship: HMS Superb - Born: Alexandria - Age: 19 - Sent to Dartmoor on 8 Sep 1813.

Hill, Thomas - Seaman - Number: 2431 - Prize name: US Brig Argus - Ship type: War - How taken: HM Brig Pelican - When taken: 14 Aug 1813 - Where taken: Irish Channel - Date received: 12 Feb 1814 - From what ship: HMS Salvador del Mundo - Born: Brooklyn - Age: 27 - Discharged on 10 May 1814 and sent to Dartmoor.

Hill, William - Seaman - Number: 416 - Prize name: Union - Ship type: MV - How taken: HM Frigate Iris - When taken: 17 Jan 1813 - Where taken: Lat 44 N Long 2.3 W - Date received: 5 Feb 1813 - From what ship: HMS San Josef - Born: Spring Mills, NJ - Age: 34 - Sent to Chatham on 29 Mar 1813 on the HMS Braham.

Hilliard, Robert B. - Master - Number: 101 - Prize name: Argus - Ship type: MV - How taken: Fancy, Cutter - When taken: 19 Dec 1812 - Where taken: Bay of Biscay - Date received: 27 Dec 1812 - From what ship: Fancy, Cutter

- Born: Maryland - Age: 32 - Sent on 8 Dec 1812 to Ashburton on parole.

Hillman, John - Seaman - Number: 2155 - Prize name: Agnes, prize of the Privateer Rambler - Ship type: LM - How taken: Cutter Jane - When taken: 28 Nov 1813 - Where taken: Bay of Biscay - Date received: 10 Dec 1813 - From what ship: Cutter Jane - Born: Boston - Age: 17 - Discharged on 27 Feb 1814 and sent to Chatham on HMS Haleyon.

Himes, Walter - Marine - Number: 1863 - Prize name: US Brig Argus - Ship type: War - How taken: HM Brig Pelican - When taken: 14 Aug 1813 - Where taken: Irish Channel - Date received: 23 Aug 1813 - From what ship: HMS Pelican - Born: New Hampshire - Age: 21 - Discharged on 8 Sep 1813 and sent to Dartmoor.

Hinkley, Aaron - Seaman - Number: 1111 - Prize name: Viper - Ship type: MV - How taken: HM Ship-of-the-Line Superb - When taken: 15 Apr 1813 - Where taken: Bay of Biscay - Date received: 22 Apr 1813 - From what ship: HMS Superb - Born: Massachusetts - Age: 25 - Sent to Dartmoor on 1 Jul 1813.

Hinkley, Joseph - Chief Mate - Number: 2454 - Prize name: Fair American - Ship type: MV - How taken: HM Frigate Andromache - When taken: 19 Jan 1814 - Where taken: Bay of Biscay - Date received: 22 Feb 1814 - From what ship: HMS York - Born: Hallowell, MA - Age: 28 - Discharged on 24 Feb 1814 and sent to Ashburton on parole.

Hinkley, Thomas - Master - Number: 2472 - Prize name: Fair American - Ship type: MV - How taken: HM Frigate Andromache - When taken: 19 Jan 1814 - Where taken: Bay of Biscay - Date received: 24 Feb 1814 - From what ship: Plymouth - Discharged on 24 Feb 1814 and sent to Ashburton on parole.

Hitch, Ebenezer - Master - Number: 33 - Prize name: Wasp - Ship type: MV - How taken: Earl Spencer, cutter - When taken: 4 Aug 1812 - Where taken: off Cape Clear - Date received: 23 Nov 1812 - From what ship: HMS Stork - Born: Massachusetts - Age: 32 - Sent on 8 Dec 1812 to Ashburton on parole.

Hitch, Joshua - Passenger - Number: 1591 - Prize name: Governor Gerry - Ship type: MV - How taken: HM Brig Royalist - When taken: 31 May 1813 - Where taken: Bay of Biscay - Date received: 14 Jun 1813 - From what ship: HMS Royalist - Born: Massachusetts - Age: 35 - Discharged on 30 Jun 1813 and sent to Ashburton on parole.

Hitchi, John - Boy - Number: 119 - Prize name: Otter - Ship type: MV - How taken: HM Ship-Sloop Jalousie - When taken: 1 Dec 1812 - Where taken: off Cape Vincent - Date received: 30 Dec 1812 - From what ship: HMS Leonidas - Born: Africa - Age: 15 - Race: Mulatto - Sent to Portsmouth on 4 Jan 1813 on the HMS Revolutionnaire.

Hitre, Dempsey - Seaman - Number: 1462 - Prize name: Paul Jones - Ship type: P - How taken: HM Frigate Leonidas - When taken: 23 May 1813 - Where taken: off Cape Clear - Date received: 26 May 1813 - From what ship: HMS Leonidas - Born: North Carolina - Age: 25 - Race: Black - Discharged on 8 Sep 1813 and sent to Dartmoor.

Hoage, Rufus - Seaman - Number: 499 - Prize name: Cashiere - Ship type: LM - How taken: HM Brig Reindeer - When taken: 3 Feb 1813 - Where taken: Bay of Biscay - Date received: 12 Feb 1813 - From what ship: HMS Reindeer - Born: Wiscasset, MA - Age: 22 - Sent to Dartmoor on 2 Apr 1813.

Hobert, George - Seaman - Number: 178 - Prize name: Hunter - Ship type: P - How taken: HM Frigate Phoebe - When taken: 24 Dec 1812 - Where taken: off Western Islands - Date received: 9 Jan 1813 - From what ship: HMS Phoebe - Born: Monmouth - Age: 19 - Sent to Portsmouth on 8 Feb 1813 on the HMS Colossus.

Hobert, William - Prize Master - Number: 2320 - Prize name: Amity, prize of the Privateer Prince de Neufchatel - Ship type: P - How taken: HM Brig Achates - When taken: 22 Dec 1814 - Where taken: Bay of Biscay - Date received: 25 Jan 1814 - From what ship: HMS Conflict - Born: Connecticut - Age: 26 - Discharged on 31 Jan 1814 and sent to Dartmoor.

Hock, J. Nicholas - Seaman - Number: 788 - Prize name: Cannoniere - Ship type: P - How taken: HM Ship-of-the-Line Warspite - When taken: 14 Mar 1813 - Where taken: Bay of Biscay - Date received: 19 Mar 1813 - From what ship: HMS Warspite - Born: Denmark - Age: 23 - Sent to Dartmoor on 28 Jun 1813.

Hockman, William - Seaman - Number: 844 - Prize name: Pallas - Ship type: MV - How taken: HM Brig Rebuff - When taken: 23 Dec 1812 - Where taken: off Cadiz - Date received: 23 May 1813 - From what ship: HMS Dauntless - Born: Prussia - Age: 28 - Sent to Dartmoor on 28 Jun 1813.

Hoden, George - Seaman - Number: 2031 - Prize name: Friendship West, prize of Privateer True Blooded Yankee - Ship type: P - How taken: HM Schooner Helicon & HM Schooner Whiting - When taken: 25 Oct 1813 - Where taken: Bay of Biscay - Date received: 31 Oct 1813 - From what ship: HMS Whiting - Born: Connecticut - Age: 20

- Discharged on 3 Nov 1813 and sent to Dartmoor.

Hodgins, Daniel - Seaman - Number: 1050 - Prize name: Independence - Ship type: MV - How taken: HM Ship-of-the-Line Superb - When taken: 16 Mar 1813 - Where taken: Bay of Biscay - Date received: 22 Apr 1813 - From what ship: HMS Superb - Born: Norwich - Age: 24 - Sent to Dartmoor on 1 Jul 1813.

Hogabets, John - 2nd Mate - Number: 945 - Prize name: Good Friends - Ship type: MV - How taken: HM Frigate Andromache - When taken: 2 Apr 1813 - Where taken: Bay of Biscay - Date received: 7 Apr 1813 - From what ship: HMS Sea Lark - Born: Philadelphia - Age: 25 - Sent to Dartmoor on 28 Jun 1813.

Hokinson, Nathaniel - Seaman - Number: 1642 - Prize name: Tickler - Ship type: LM - How taken: HM Frigate Magiciene - When taken: 5 Jun 1813 - Where taken: Lat 47 Long 13 - Date received: 14 Jun 1813 - From what ship: HMS Orestes - Born: New York - Age: 28 - Discharged on 30 Jun 1813 and sent to Stapleton.

Holberg, Emanuel - Seaman - Number: 2619 - How taken: HM Ship-of-the-Line Edinburgh - When taken: 28 Oct 1812 - Date received: 16 May 1814 - From what ship: HMS Repulse - Born: New York - Age: 28 - Discharged on 14 Jun 1814 and sent to Dartmoor.

Holbrook, Elias - Seaman - Number: 2635 - Prize name: Ateline - Ship type: MV - How taken: HM Frigate Magiciene - When taken: 14 Mar 1814 - Where taken: off Cape Finisterre, Spain - Date received: 17 May 1814 - From what ship: HMS Tortois - Born: Freeport - Age: 30 - Discharged on 14 Jun 1814 and sent to Dartmoor.

Holden, Andrew - Seaman - Number: 2392 - Prize name: Harvest, prize of the Privateer Bunker Hill - Ship type: P - How taken: HM Brig Orestes - When taken: 21 Jan 1814 - Date received: 3 Feb 1814 - From what ship: HMS Orestes - Born: New York - Age: 18 - Discharged on 10 May 1814 and sent to Dartmoor.

Holden, Charles - Boatswain's Mate - Number: 156 - Prize name: Hunter - Ship type: P - How taken: HM Frigate Phoebe - When taken: 24 Dec 1812 - Where taken: off Western Islands - Date received: 9 Jan 1813 - From what ship: HMS Phoebe - Born: Salem - Age: 37 - Sent to Portsmouth on 8 Feb 1813 on the HMS Colossus.

Holland, James - Seaman - Number: 2661 - Prize name: Indian Lass, prize of the Privateer Grand Turk - Ship type: P - How taken: HM Transport Akbar - When taken: 29 Apr 1814 - Where taken: at sea - Date received: 5 Jun 1814 - From what ship: HMS Gronville - Born: Cape Ann - Age: 27 - Discharged on 14 Jun 1814 and sent to Dartmoor.

Holland, Richard - Seaman - Number: 1616 - Prize name: Revenge - Ship type: LM - How taken: HM Frigate Belle Poule - When taken: 11 May 1813 - Where taken: Bay of Biscay - Date received: 14 Jun 1813 - From what ship: HMS Royalist - Born: Maryland - Age: 25 - Discharged on 30 Jun 1813 and sent to Stapleton.

Hollinger, William - Boatswain - Number: 1364 - Prize name: Tom - Ship type: LM - How taken: HM Frigate Surveillante - When taken: 27 Apr 1813 - Where taken: Bay of Biscay - Date received: 15 May 1813 - From what ship: HMS Foxhound - Born: Virginia - Age: 28 - Discharged on 3 Jul 1813 and sent to Stapleton Prison.

Holmes, Caleb K. - Seaman - Number: 791 - Prize name: Cannoniere - Ship type: P - How taken: HM Ship-of-the-Line Warspite - When taken: 14 Mar 1813 - Where taken: Bay of Biscay - Date received: 19 Mar 1813 - From what ship: HMS Warspite - Born: Connecticut - Age: 25 - Sent to Dartmoor on 28 Jun 1813.

Holmes, Charles - Seaman - Number: 2055 - How taken: Pallas - Where taken: St. Johns, New Brunswick - Date received: 1 Nov 1813 - From what ship: HMS Bittern - Born: Thomastown - Age: 20 - Discharged on 3 Nov 1813 and sent to Dartmoor.

Holmes, James - Seaman - Number: 1137 - Prize name: Magdalen - Ship type: MV - How taken: HM Ship-of-the-Line Superb - When taken: 15 Apr 1813 - Where taken: off Belle Isle - Date received: 22 Apr 1813 - From what ship: HMS Superb - Born: Portsmouth - Age: 23 - Sent to Dartmoor on 1 Jul 1813.

Holms, Andres - Seaman - Number: 742 - Prize name: Charlotte - Ship type: MV - How taken: HM Ship-of-the-Line Warspite - When taken: 3 Mar 1813 - Where taken: Bay of Biscay - Date received: 19 Mar 1813 - From what ship: HMS Warspite - Born: New York - Age: 21 - Sent to Dartmoor on 2 Apr 1813.

Holmstrom, Alexander - Boy - Number: 1618 - Prize name: Revenge - Ship type: LM - How taken: HM Frigate Belle Poule - When taken: 11 May 1813 - Where taken: Bay of Biscay - Date received: 14 Jun 1813 - From what ship: HMS Royalist - Born: New York - Age: 10 - Discharged on 9 Sep 1813.

Holmstrom, Lorentz - Seaman - Number: 1617 - Prize name: Revenge - Ship type: LM - How taken: HM Frigate Belle Poule - When taken: 11 May 1813 - Where taken: Bay of Biscay - Date received: 14 Jun 1813 - From what ship:

HMS Royalist - Born: Stockholm - Age: 40 - Discharged on 9 Sep 1813.

Holsten, Peter - Seaman - Number: 2595 - How taken: Impressed at Aliant - When taken: 28 Aug 1813 - Date received: 16 May 1814 - From what ship: HMS Repulse - Born: Norway - Age: 30 - Discharged on 14 Jun 1814 and sent to Dartmoor.

Holstade, Joseph - Seaman - Number: 2597 - How taken: HM Frigate Havannah - Date received: 16 May 1814 - From what ship: HMS Repulse - Born: New York - Age: 48 - Discharged on 14 Jun 1814 and sent to Dartmoor.

Holts, Daniel - Boatswain - Number: 687 - Prize name: Star - Ship type: MV - How taken: HM Ship-of-the-Line Superb - When taken: 9 Feb 1813 - Where taken: Bay of Biscay - Date received: 19 Mar 1813 - From what ship: HMS Warspite - Born: New London - Age: 24 - Sent to Dartmoor on 2 Apr 1813.

Holts, Peter - Seaman - Number: 963 - Prize name: Ferox - Ship type: MV - How taken: HM Frigate Medusa & HM Brig Lyra - When taken: 28 Mar 1813 - Where taken: off Cape Ortagle - Date received: 9 Apr 1813 - From what ship: HMS Lyra - Born: Sweden - Age: 31 - Liberated on 21 May 1813.

Holtz, Daniel - Boatswain - Number: 1661 - Prize name: Star - Ship type: MV - How taken: HM Ship-of-the-Line Superb - When taken: 9 Feb 1813 - Where taken: Bay of Biscay - Date received: 15 Jun 1813 - From what ship: Dartmoor Prison - Born: New London - Age: 26 - Discharged on 16 Jun 1813 and released to HMS Salvador del Mundo.

Homell, Christopher - Seaman - Number: 2121 - Prize name: Chesapeake - Ship type: LM - How taken: HM Frigate Hotspur & HM Frigate Pyramus - When taken: 26 Oct 1813 - Where taken: Bay of Biscay - Date received: 22 Nov 1813 - From what ship: HMS Pyramus - Born: North Carolina - Age: 43 - Discharged on 29 Nov 1813 and sent to Dartmoor.

Homes, Michael - Seaman - Number: 2466 - Prize name: Fair American - Ship type: MV - How taken: HM Frigate Andromache - When taken: 19 Jan 1814 - Where taken: Bay of Biscay - Date received: 22 Feb 1814 - From what ship: HMS York - Born: Boston - Age: 19 - Discharged on 10 May 1814 and sent to Dartmoor.

Honson, J. A. - Seaman - Number: 2538 - Prize name: Caroline - Ship type: MV - How taken: Brilliant, privateer - When taken: 12 Aug 1813 - Where taken: off Charleston - Date received: 13 Apr 1814 - From what ship: HMS Bittern - Born: New York - Age: 42 - Discharged on 10 May 1814 and sent to Dartmoor.

Hook, Aaron - Seaman - Number: 114 - Prize name: Otter - Ship type: MV - How taken: HM Ship-Sloop Jalousie - When taken: 1 Dec 1812 - Where taken: off Cape Vincent - Date received: 30 Dec 1812 - From what ship: HMS Leonidas - Born: Maryland - Age: 20 - Sent to Portsmouth on 4 Jan 1813 on the HMS Revolutionnaire.

Hook, John - Seaman - Number: 617 - Prize name: Governor McKean - Ship type: LM - How taken: HM Brig Rover - When taken: 26 Jun 1813 - Where taken: off Bordeaux - Date received: 2 Mar 1813 - From what ship: HMS Insolent - Born: Pennsylvania - Age: 24 - Sent to Dartmoor on 2 Apr 1813.

Hooper, Benjamin E. - Seaman - Number: 1697 - Prize name: Governor Gerry - Ship type: MV - How taken: HM Brig Royalist - When taken: 31 May 1813 - Where taken: Bay of Biscay - Date received: 26 Jun 1813 - From what ship: HMS Duncan - Born: New York - Age: 23 - Discharged on 30 Jun 1813 and sent to Stapleton.

Hooper, William - Seaman - Number: 2617 - How taken: HM Ship-of-the-Line Elizabeth - When taken: 4 Aug 1812 - Date received: 16 May 1814 - From what ship: HMS Repulse - Born: Marblehead - Age: 33 - Discharged on 14 Jun 1814 and sent to Dartmoor.

Hopkins, Daniel - Seaman - Number: 1537 - Prize name: Courier - Ship type: LM - How taken: HM Brig Rover - When taken: 14 Mar 1813 - Where taken: Bay of Biscay - Date received: 29 May 1813 - From what ship: HMS Hannibal - Born: Maryland - Age: 28 - Discharged on 30 Jun 1813 and sent to Stapleton.

Hopkins, Elisha - Seaman - Number: 1671 - Prize name: Omer - Ship type: MV - How taken: New Briton, letter of marque - Where taken: off the Mercy Lights - Date received: 17 Jun 1813 - From what ship: HMS Bittern - Born: Massachusetts - Age: 18 - Discharged on 30 Jun 1813 and sent to Stapleton.

Hopkins, William - Seaman - Number: 1828 - Prize name: US Brig Argus - Ship type: War - How taken: HM Brig Pelican - When taken: 14 Aug 1813 - Where taken: Irish Channel - Date received: 17 Aug 1813 - From what ship: USS Argus - Born: Boston - Age: 49 - Discharged on 8 Sep 1813 and sent to Dartmoor.

Hoppin, William - Seaman - Number: 2154 - Prize name: Agnes, prize of the Privateer Rambler - Ship type: LM -

How taken: Cutter Jane - When taken: 28 Nov 1813 - Where taken: Bay of Biscay - Date received: 10 Dec 1813 - From what ship: Cutter Jane - Born: Newburyport - Age: 23 - Discharged on 31 Jan 1814 and sent to Dartmoor.

Hopson, Ebenezer - Seaman - Number: 1225 - Prize name: Grand Napoleon - Ship type: MV - How taken: HM Frigate Belle Poule - When taken: 3 Apr 1813 - Where taken: off Bordeaux - Date received: 9 May 1813 - From what ship: HMS Andromache - Born: Connecticut - Age: 21 - Discharged on 3 Jul 1813 and sent to Stapleton Prison.

Horn, Abner - Seaman - Number: 2089 - How taken: Delivered himself up from HM Frigate Fortunee - Date received: 2 Nov 1813 - From what ship: HMS Bacchus - Born: Massachusetts - Age: 33 - Discharged on 3 Nov 1813 and sent to Dartmoor.

Horsydise, William - Chief Mate - Number: 1408 - Prize name: Henry Clements - Ship type: MV - How taken: HM Brig Orestes - When taken: 15 Apr 1813 - Where taken: Bay of Biscay - Date received: 20 May 1813 - From what ship: HMS Orestes - Born: Maryland - Age: 31 - Discharged on 23 May 1813 and sent to Ashburton on parole.

Hoselquist, John - Seaman - Number: 2266 - Prize name: Fanny - Ship type: MV - How taken: HM Frigate Eurotas - When taken: 25 Dec 1813 - Where taken: at sea - Date received: 20 Jan 1814 - From what ship: HMS Eurotas - Born: Wexchy, Sweden - Age: 38 - Discharged on 31 Jan 1814 and sent to Dartmoor.

Hoskins, James - Seaman - Number: 768 - Prize name: William Bayard - Ship type: MV - How taken: HM Ship-of-the-Line Warspite - When taken: 3 Mar 1813 - Where taken: Bay of Biscay - Date received: 19 Mar 1813 - From what ship: HMS Warspite - Born: Massachusetts - Age: 20 - Sent to Dartmoor on 28 Jun 1813.

Hosmer, Joseph - Quartermaster - Number: 160 - Prize name: Hunter - Ship type: P - How taken: HM Frigate Phoebe - When taken: 24 Dec 1812 - Where taken: off Western Islands - Date received: 9 Jan 1813 - From what ship: HMS Phoebe - Born: Salem - Age: 21 - Sent to Portsmouth on 8 Feb 1813 on the HMS Colossus.

Hough, Ebenezer - Seaman - Number: 515 - Prize name: Allegany - Ship type: MV - How taken: Detained at Gibraltar - When taken: 8 Aug 1812 - Date received: 15 Feb 1813 - From what ship: HMS Andromeda - Born: Avondale - Age: 22 - Sent to Dartmoor on 2 Apr 1813.

Hovey, Joseph - Seaman - Number: 523 - Prize name: Phoenix - Ship type: MV - How taken: Detained at Gibraltar - When taken: 8 Aug 1812 - Date received: 15 Feb 1813 - From what ship: HMS Andromeda - Born: Massachusetts - Age: 22 - Sent to Dartmoor on 2 Apr 1813.

Hovrington, William - Seaman - Number: 1808 - Prize name: US Brig Argus - Ship type: War - How taken: HM Brig Pelican - When taken: 14 Aug 1813 - Where taken: Irish Channel - Date received: 17 Aug 1813 - From what ship: USS Argus - Born: Delaware - Age: 25 - Race: Blackman - Discharged on 27 Sep 1813 and sent to Dartmoor.

Howard, John - Master - Number: 1150 - Prize name: Grand Napoleon - Ship type: MV - How taken: HM Frigate Belle Poule - When taken: 3 Apr 1813 - Where taken: off Bordeaux - Date received: 23 Apr 1813 - From what ship: HMS Plymouth - Born: Connecticut - Age: 33 - Sent on 24 Apr 1813 to Ashburton on parole.

Howard, Samuel - Seaman - Number: 1085 - Prize name: John & Frances - Ship type: MV - How taken: HM Frigate Belle Poule - When taken: 19 Mar 1813 - Where taken: Bay of Biscay - Date received: 22 Apr 1813 - From what ship: HMS Superb - Born: New York - Age: 31 - Sent to Dartmoor on 1 Jul 1813.

Howard, William - Boatswain - Number: 1330 - Prize name: Fox - Ship type: LM - How taken: HM Sloop Pheasant - When taken: 23 Apr 1813 - Where taken: Bay of Biscay - Date received: 14 May 1813 - From what ship: HMS Pleasant - Born: Baltimore - Age: 24 - Discharged on 3 Jul 1813 and sent to Stapleton Prison.

Howe, Phineas - Seaman - Number: 2707 - How taken: Sent into custody from HM Ship-of-the-Line Dublin - Date received: 18 Jun 1814 - From what ship: HMS Foxhound - Born: New Bedford - Age: 26 - Discharged on 20 Jun 1814 and sent to Dartmoor.

Howell, Thomas - Seaman - Number: 1048 - Prize name: Independence - Ship type: MV - How taken: HM Ship-of-the-Line Superb - When taken: 16 Mar 1813 - Where taken: Bay of Biscay - Date received: 22 Apr 1813 - From what ship: HMS Superb - Born: Beverly - Age: 24 - Sent to Dartmoor on 1 Jul 1813.

Howey, Artemas - Seaman - Number: 83 - Prize name: Argus - Ship type: MV - How taken: Fancy, Cutter - When taken: 19 Dec 1812 - Where taken: Bay of Biscay - Date received: 24 Dec 1812 - From what ship: Fancy, Cutter - Sent to Portsmouth on 29 Dec 1812 on the HMS Northumberland.

Howland, Samuel - Seaman - Number: 112 - Prize name: Otter - Ship type: MV - How taken: HM Ship-Sloop Jalousie

- When taken: 1 Dec 1812 - Where taken: off Cape Vincent - Date received: 30 Dec 1812 - From what ship: HMS Leonidas - Born: Plymouth - Age: 25 - Sent to Portsmouth on 4 Jan 1813 on the HMS Revolutionnaire.

Hoyt, James - Seaman - Number: 2656 - How taken: Sent into custody from HM Frigate Nereus - Date received: 5 Jun 1814 - From what ship: HMS Gronville - Born: Norwalk - Age: 46 - Discharged on 14 Jun 1814 and sent to Dartmoor.

Hubbard, Alfred - 2nd Mate - Number: 903 - Prize name: Prompt - Ship type: MV - How taken: Chance, privateer - When taken: 22 Mar 1813 - Where taken: Bay of Biscay - Date received: 3 Apr 1813 - From what ship: Mary - Born: New Haven - Age: 20 - Sent on 21 Jun 1813 to Portsmouth on HMS Prometheus.

Hubbard, John - Seaman - Number: 1651 - Prize name: Tyger - Ship type: MV - How taken: Detained at Gibraltar - When taken: 8 Aug 1813 - Date received: 15 Jun 1813 - From what ship: Dartmoor Prison - Born: New York - Discharged on 16 Jun 1813 and released to HMS Salvador del Mundo.

Hubbard, John - Seaman - Number: 512 - Prize name: Tyger - Ship type: MV - How taken: Detained at Gibraltar - When taken: 8 Aug 1812 - Date received: 15 Feb 1813 - From what ship: HMS Andromeda - Born: Philadelphia - Age: 24 - Sent to Dartmoor on 2 Apr 1813.

Hubbard, John - Seaman - Number: 1792 - How taken: Delivered himself up from HM Frigate Bombay - When taken: 30 Nov 1812 - Date received: 13 Aug 1813 - From what ship: HMS Protection - Born: Norfolk - Age: 27 - Discharged on 8 Sep 1813 and sent to Dartmoor.

Hubbell, James - Seaman - Number: 2383 - Prize name: Devon, prize to the Privateer Bunker Hill - Ship type: P - How taken: HM Brig Fly - When taken: 21 Jan 1814 - Where taken: at sea - Date received: 31 Jan 1814 - From what ship: HMS Fly - Born: Fairfield - Age: 29 - Discharged on 27 Feb 1814 and sent to Chatham on HMS Halcyon.

Hudleback, George - Supercargo - Number: 830 - Prize name: Mars - Ship type: MV - How taken: HM Ship-of-the-Line Warspite - When taken: 26 Feb 1813 - Where taken: Bay of Biscay - Date received: 22 Mar 1813 - From what ship: HMS Warspite - Released on 27 Mar 1813.

Hudson, James - Seaman - Number: 2441 - Prize name: Mary, prize of the Privateer Prince de Neufchatel - Ship type: P - How taken: Retaken by the crew of the Mary - Date received: 14 Feb 1814 - From what ship: HMS Halcyon - Born: Baltimore - Age: 24 - Discharged on 10 May 1814 and sent to Dartmoor.

Hudson, John - Sailing Master - Number: 1852 - Prize name: US Brig Argus - Ship type: War - How taken: HM Brig Pelican - When taken: 14 Aug 1813 - Where taken: Irish Channel - Date received: 19 Aug 1813 - From what ship: USS Argus - Born: Pennsylvania - Age: 26 - Discharged on 25 Aug 1813 and sent to Ashburton on parole.

Hudson, Thomas - Boy - Number: 1605 - Prize name: Revenge - Ship type: LM - How taken: HM Frigate Belle Poule - When taken: 11 May 1813 - Where taken: Bay of Biscay - Date received: 14 Jun 1813 - From what ship: HMS Royalist - Born: Richmond, VA - Age: 19 - Discharged on 30 Jun 1813 and sent to Stapleton.

Hudson, William - 2nd Lieutenant - Number: 506 - Prize name: Cashiere - Ship type: LM - How taken: HM Brig Reindeer - When taken: 3 Feb 1813 - Where taken: Bay of Biscay - Date received: 15 Feb 1813 - From what ship: Cashier - Born: North Carolina - Age: 38 - Sent to Chatham on 29 Mar 1813 on the HMS Braham.

Huggins, Daniel - Carpenter - Number: 1324 - Prize name: Shadow - Ship type: LM - How taken: HM Brig Reindeer & HM Schooner Helicon - When taken: 6 Apr 1813 - Where taken: Bay of Biscay - Date received: 12 May 1813 - From what ship: HMS Helicon - Born: Hamburg - Age: 39 - Discharged on 8 Sep 1813 and sent to Dartmoor.

Hughes, John - Seaman - Number: 689 - Prize name: Star - Ship type: MV - How taken: HM Ship-of-the-Line Superb - When taken: 9 Feb 1813 - Where taken: Bay of Biscay - Date received: 19 Mar 1813 - From what ship: HMS Warspite - Born: Philadelphia - Age: 24 - Sent to Dartmoor on 2 Apr 1813.

Hughes, John - Seaman - Number: 1743 - Prize name: Star - Ship type: MV - How taken: HM Ship-of-the-Line Superb - When taken: 9 Feb 1813 - Where taken: Bay of Biscay - Date received: 10 Jul 1813 - From what ship: Dartmoor Prison - Born: Philadelphia - Age: 26 - Discharged on 10 Jul 1813 and released to HMS Salvador del Mundo.

Hughes, John - Seaman - Number: 2206 - Prize name: Squirrel - Ship type: MV - How taken: HM Frigate Belle Poule - When taken: 14 Dec 1813 - Where taken: Bay of Biscay - Date received: 27 Dec 1813 - From what ship: HMS Protector - Born: New Jersey - Age: 29 - Discharged on 31 Jan 1814 and sent to Dartmoor.

Hulbert, Henry - Seaman - Number: 2513 - Prize name: Bunker Hill - Ship type: P - How taken: HM Frigate Pomone & HM Frigate Cydnus - When taken: 4 Mar 1814 - Where taken: Bay of Biscay - Date received: 4 Apr 1814 - From what ship: HMS Virago - Born: Philadelphia - Age: 17 - Discharged on 10 May 1814 and sent to Dartmoor.

Hulet, Michael - Seaman - Number: 594 - Prize name: St. Martin's Plantation, prize of the Privateer Paul Jones - Ship type: P - How taken: HM Ship-of-the-Line Dublin - When taken: 9 Feb 1815 - Where taken: Lat 43 N Long 33.5 W - Date received: 25 Feb 1813 - From what ship: HMS Dublin - Born: Shrewsbury - Age: 21 - Sent to Dartmoor on 2 Apr 1813.

Hull, Edward - Seaman - Number: 927 - Prize name: Weasel - Ship type: MV - How taken: HM Brig Foxhound - When taken: 25 Mar 1813 - Where taken: Bay of Biscay - Date received: 6 Apr 1813 - From what ship: HMS Foxhound - Born: Newport - Age: 19 - Sent on 21 Jun 1813 to Portsmouth on HMS Prometheus.

Huma, John Lewis - Seaman - Number: 174 - Prize name: Hunter - Ship type: P - How taken: HM Frigate Phoebe - When taken: 24 Dec 1812 - Where taken: off Western Islands - Date received: 9 Jan 1813 - From what ship: HMS Phoebe - Born: New Orleans - Age: 32 - Sent to Portsmouth on 8 Feb 1813 on the HMS Colossus.

Hume, George - Seaman - Number: 1623 - Prize name: Leo - Ship type: LM - How taken: HM Frigate Magiciene - When taken: 4 Jun 1813 - Where taken: Lat 45 Long 14 - Date received: 14 Jun 1813 - From what ship: HMS Orestes - Born: Winslow - Age: 35 - Discharged on 8 Sep 1813 and sent to Dartmoor.

Humphries, James - Seaman - Number: 622 - Prize name: Governor McKean - Ship type: LM - How taken: HM Brig Rover - When taken: 26 Jun 1813 - Where taken: off Bordeaux - Date received: 2 Mar 1813 - From what ship: HMS Insolent - Born: Philadelphia - Age: 28 - Sent to Dartmoor on 2 Apr 1813.

Humphries, Thomas - Master - Number: 122 - Prize name: Stephen - Ship type: MV - How taken: HM Frigate Briton & HM Frigate Andromache - When taken: 17 Dec 1812 - Where taken: off Bordeaux - Date received: 1 Jan 1813 - From what ship: HMS Andromache - Born: New York - Age: 28 - Sent on 8 Dec 1812 to Ashburton on parole.

Humphries, Warren - Seaman - Number: 2030 - Prize name: Friendship West, prize of Privateer True Blooded Yankee - Ship type: P - How taken: HM Schooner Helicon & HM Schooner Whiting - When taken: 25 Oct 1813 - Where taken: Bay of Biscay - Date received: 31 Oct 1813 - From what ship: HMS Whiting - Born: Connecticut - Age: 25 - Discharged on 3 Nov 1813 and sent to Dartmoor.

Hunt, Charles - Marine - Number: 1860 - Prize name: US Brig Argus - Ship type: War - How taken: HM Brig Pelican - When taken: 14 Aug 1813 - Where taken: Irish Channel - Date received: 23 Aug 1813 - From what ship: HMS Pelican - Born: New York - Age: 22 - Discharged on 8 Sep 1813 and sent to Dartmoor.

Hunt, David - Seaman - Number: 1009 - How taken: Delivered himself up from HM Sloop Buster - Date received: 18 Apr 1813 - From what ship: HMS Buster - Born: Brunswick - Age: 23 - Sent to Chatham on 8 Jul 1813 on HM Tender Neptune.

Hunter, Charles - Seaman - Number: 1918 - Prize name: US Brig Argus - Ship type: War - How taken: HM Brig Pelican - When taken: 14 Aug 1813 - Where taken: Irish Channel - Date received: 23 Aug 1813 - From what ship: HMS Pelican - Born: New Jersey - Age: 22 - Discharged on 8 Sep 1813 and sent to Dartmoor.

Hunter, George - Seaman - Number: 1120 - Prize name: Viper - Ship type: MV - How taken: HM Ship-of-the-Line Superb - When taken: 15 Apr 1813 - Where taken: Bay of Biscay - Date received: 22 Apr 1813 - From what ship: HMS Superb - Born: Fredericktown - Age: 27 - Sent to Dartmoor on 1 Jul 1813.

Hunter, William - Seaman - Number: 1295 - Prize name: Price - Ship type: LM - How taken: HM Frigate Medusa - When taken: 13 Apr 1813 - Where taken: Bay of Biscay - Date received: 10 May 1813 - From what ship: HMS Medusa - Born: New York - Age: 18 - Discharged on 3 Jul 1813 and sent to Stapleton Prison.

Huntress, Robert - Boatswain - Number: 361 - Prize name: Orbit - Ship type: MV - How taken: HM Brig Achates - When taken: 29 Jan 1813 - Where taken: Lat 49 N Long 13 W - Date received: 4 Feb 1813 - From what ship: HMS Achates - Born: Portsmouth - Age: 27 - Sent to Chatham on 29 Mar 1813 on the HMS Braham.

Hurd, William - Seaman - Number: 813 - Prize name: Decornau - Ship type: MV - How taken: HM Sloop Pheasant - When taken: 15 Mar 1813 - Where taken: Bay of Biscay - Date received: 19 Mar 1813 - From what ship: HMS Pheasant - Born: New York - Age: 24 - Sent to Dartmoor on 28 Jun 1813.

Hurtall, Peter - Merchant & Passenger - Number: 337 - Prize name: Brutus - Ship type: MV - How taken: HM Frigate

Briton - When taken: Jan 1813 - Where taken: Bay of Biscay - Date received: 22 Jan 1813 - From what ship: HMS Briton - Born: Port Au Prince, Haiti - Age: 40 - Sent on 23 Jan 1813 to Ashburton on parole.

Haskell, Robert - 2nd Mate - Number: 1586 - Prize name: Orders in Council - Ship type: LM - How taken: HM Frigate Surveillante - When taken: 1 Jun 1813 - Where taken: Bay of Biscay - Date received: 13 Jun 1813 - From what ship: Forvey - Born: Manchester - Age: 29 - Discharged on 30 Jun 1813 and sent to Stapleton.

Hutchins, Edward - Seaman - Number: 1525 - Prize name: Courier - Ship type: LM - How taken: HM Brig Rover - When taken: 14 Mar 1813 - Where taken: Bay of Biscay - Date received: 29 May 1813 - From what ship: HMS Hannibal - Born: Maryland - Age: 22 - Race: Negro - Discharged on 30 Jun 1813 and sent to Stapleton.

Hutchins, Henry - Seaman - Number: 1509 - Prize name: Grand Napoleon - Ship type: MV - How taken: HM Brig Goldfinch - When taken: 17 Apr 1813 - Where taken: Bay of Biscay - Date received: 29 May 1813 - From what ship: HMS Hannibal - Born: New York - Age: 21 - Discharged on 30 Jun 1813 and sent to Stapleton.

Hutchins, William - Armorer - Number: 2278 - Prize name: Siro - Ship type: LM - How taken: HM Brig Pelican - When taken: 13 Jan 1814 - Where taken: at sea - Date received: 23 Jan 1814 - From what ship: HMS Pelican - Born: Philadelphia - Age: 32 - Discharged on 31 Jan 1814 and sent to Dartmoor.

Hutchinson, Henry P. - Clerk - Number: 146 - Prize name: Hunter - Ship type: P - How taken: HM Frigate Phoebe - When taken: 24 Dec 1812 - Where taken: off Western Islands - Date received: 9 Jan 1813 - From what ship: HMS Phoebe - Born: New Jersey - Age: 21 - Sent to Portsmouth on 8 Feb 1813 on the HMS Colossus.

Ide, John Henry - Supercargo - Number: 2583 - Prize name: James - Ship type: MV - How taken: HM Brig Harpy - When taken: 19 Dec 1812 - Where taken: off Isle du France - Date received: 10 May 1814 - From what ship: HMS Lion - Born: Stanbury - Age: 42 - Discharged on 12 May 1814 and sent to Ashburton on parole.

Ingbriton, Nicholas - Seaman - Number: 581 - Prize name: Cashiere - Ship type: LM - How taken: HM Brig Reindeer - When taken: 3 Feb 1813 - Where taken: Bay of Biscay - Date received: 23 Feb 1813 - From what ship: HMS Surveillante - Born: Sweden - Age: 22 - Sent to Dartmoor on 2 Apr 1813.

Ingalls, Edmond - Seaman - Number: 187 - Prize name: Hunter - Ship type: P - How taken: HM Frigate Phoebe - When taken: 24 Dec 1812 - Where taken: off Western Islands - Date received: 9 Jan 1813 - From what ship: HMS Phoebe - Born: Lynn - Age: 21 - Sent to Portsmouth on 8 Feb 1813 on the HMS Colossus.

Ingerson, Michael - Seaman - Number: 1293 - Prize name: Price - Ship type: LM - How taken: HM Frigate Medusa - When taken: 13 Apr 1813 - Where taken: Bay of Biscay - Date received: 10 May 1813 - From what ship: HMS Medusa - Born: New Jersey - Age: 21 - Discharged on 3 Jul 1813 and sent to Stapleton Prison.

Ingle, John - Seaman - Number: 1271 - Prize name: Caroline - Ship type: MV - How taken: HM Frigate Medusa - When taken: 12 Apr 1813 - Where taken: Bay of Biscay - Date received: 10 May 1813 - From what ship: HMS Medusa - Born: Baltimore - Age: 26 - Discharged on 3 Jul 1813 and sent to Stapleton Prison.

Ingles, Daniel - Steward - Number: 2156 - Prize name: Zephyr - Ship type: MV - How taken: HM Frigate Pyramus - When taken: 30 Nov 1813 - Where taken: off Lorient - Date received: 10 Dec 1813 - From what ship: Zephyr - Born: Dover - Age: 39 - Race: Black - Discharged on 31 Jan 1814 and sent to Dartmoor.

Ingles, David - Seaman - Number: 1666 - Ship type: MV - How taken: Impressed from the Nile English - When taken: 15 May 1813 - Date received: 17 Jun 1813 - From what ship: HMS Bittern - Born: Chester - Age: 30 - Discharged on 30 Jun 1813 and sent to Stapleton.

Innes, John - Seaman - Number: 1356 - Prize name: Shadow - Ship type: LM - How taken: HM Brig Reindeer & HM Schooner Helicon - When taken: 6 Apr 1813 - Where taken: Bay of Biscay - Date received: 14 May 1813 - From what ship: HMS Reindeer - Born: Guadeloupe - Age: 16 - Race: Negro - Discharged on 3 Jul 1813 and sent to Stapleton Prison.

Ireland, James - Seaman - Number: 568 - Prize name: Rolla - Ship type: MV - How taken: HM Frigate Surveillante - When taken: 11 Feb 1813 - Where taken: Bay of Biscay - Date received: 23 Feb 1813 - From what ship: HMS Surveillante - Born: Egg Harbor - Age: 28 - Sent to Dartmoor on 2 Apr 1813.

Irvin, Anthony - Boy - Number: 1486 - Prize name: Paul Jones - Ship type: P - How taken: HM Frigate Leonidas - When taken: 23 May 1813 - Where taken: off Cape Clear - Date received: 26 May 1813 - From what ship: HMS Leonidas - Born: Philadelphia - Age: 13 - Discharged on 30 Jun 1813 and sent to Stapleton.

Irvin, Mathew - Boy - Number: 1350 - Prize name: Fox - Ship type: LM - How taken: HM Sloop Pheasant - When taken: 23 Apr 1813 - Where taken: Bay of Biscay - Date received: 15 May 183 - From what ship: HMS Pleasant - Born: Philadelphia - Age: 14 - Discharged on 8 Sep 1813 and sent to Dartmoor.

Irving, William - Seaman - Number: 1055 - Prize name: Virginia Planter - Ship type: MV - How taken: HM Frigate Pyramus - When taken: 17 Mar 1813 - Where taken: off Nantes - Date received: 22 Apr 1813 - From what ship: HMS Superb - Born: Eddington - Age: 22 - Discharged on 17 Jul 1813 and sent on the HMS Salvador del Mundo.

Irwin, Magnus - Boy - Number: 429 - Prize name: Union - Ship type: MV - How taken: HM Frigate Iris - When taken: 17 Jan 1813 - Where taken: Lat 44 N Long 2.3 W - Date received: 5 Feb 1813 - From what ship: HMS San Josef - Born: Philadelphia - Age: 16 - Sent to Chatham on 29 Mar 1813 on the HMS Braham.

Isaac, Moses - Seaman - Number: 700 - Prize name: Star - Ship type: MV - How taken: HM Ship-of-the-Line Superb - When taken: 9 Feb 1813 - Where taken: Bay of Biscay - Date received: 19 Mar 1813 - From what ship: HMS Warspite - Born: New York - Age: 21 - Sent to Dartmoor on 2 Apr 1813.

Jackson, Curtis - Seaman - Number: 2152 - Prize name: Agnes, prize of the Privateer Rambler - Ship type: LM - How taken: Cutter Jane - When taken: 28 Nov 1813 - Where taken: Bay of Biscay - Date received: 10 Dec 1813 - From what ship: Cutter Jane - Born: Boston - Age: 17 - Discharged on 31 Jan 1814 and sent to Dartmoor.

Jackson, Daniel - Boy - Number: 1030 - Prize name: Two Brothers - Ship type: MV - How taken: Bootle of Liverpool, letter of marque - When taken: 18 Mar 1813 - Where taken: Western Islands - Date received: 21 Apr 1813 - From what ship: HMS Bittern - Born: Boston - Age: 15 - Sent to Dartmoor on 1 Jul 1813.

Jackson, Daniel - Captain - Number: 1039 - How taken: Delivered himself up from HM Ship-of-the-Line Ajax - Date received: 22 Apr 1813 - From what ship: HMS Ajax - Born: Connecticut - Age: 37 - Sent to Chatham on 8 Jul 1813 on HM Tender Neptune.

Jackson, David - Cook - Number: 1977 - Prize name: Ann - Ship type: MV - How taken: Tenedos - When taken: 5 May 1813 - Where taken: Boston Bay - Date received: 19 Sep 1813 - From what ship: HMS Bittern - Born: Albany - Age: 20 - Race: Black - Discharged on 27 Sep 1813 and sent to Dartmoor.

Jackson, Henry - 3rd Lieutenant - Number: 1414 - Prize name: Paul Jones - Ship type: P - How taken: HM Frigate Leonidas - When taken: 23 May 1813 - Where taken: off Cape Clear - Date received: 26 May 1813 - From what ship: HMS Leonidas - Born: Maryland - Age: 25 - Discharged on 30 Jun 1813 and sent to Stapleton.

Jackson, James - Seaman - Number: 1432 - Prize name: Paul Jones - Ship type: P - How taken: HM Frigate Leonidas - When taken: 23 May 1813 - Where taken: off Cape Clear - Date received: 26 May 1813 - From what ship: HMS Leonidas - Born: New York - Age: 26 - Discharged on 30 Jun 1813 and sent to Stapleton.

Jackson, James - Seaman - Number: 2536 - Prize name: Lyon - Ship type: MV - How taken: Brilliant, privateer - When taken: 1 Jan 1814 - Date received: 13 Apr 1814 - From what ship: HMS Bittern - Born: Salem - Age: 40 - Race: Black - Discharged on 10 May 1814 and sent to Dartmoor.

Jackson, James - Seaman - Number: 1968 - Prize name: Paul Jones - Ship type: P - How taken: HM Frigate Leonidas - When taken: 23 May 1813 - Where taken: off Cape Clear - Date received: 13 Sep 1813 - From what ship: Dartmoor prison - Born: New York - Age: 27 - Discharged on 14 Sep 1813 and sent to Dartmoor.

Jackson, John - Seaman - Number: 1195 - Prize name: Zebra - Ship type: LM - How taken: HM Frigate Pyramus - When taken: 20 Apr 1813 - Where taken: Bay of Biscay - Date received: 9 May 1813 - From what ship: HMS Andromache - Born: New Brunswick - Age: 25 - Sent on 3 Jul 1813 to Stapleton prison.

Jackson, John - Seaman - Number: 297 - Prize name: Columbia - Ship type: MV - How taken: HM Frigate Briton - When taken: 17 Dec 1812 - Where taken: off Bordeaux - Date received: 21 Jan 1813 - From what ship: HMS Abercrombie - Born: New York - Age: 27 - Sent to Portsmouth on 8 Feb 1813 on the HMS Colossus.

Jackson, John - Seaman - Number: 765 - Prize name: William Bayard - Ship type: MV - How taken: HM Ship-of-the-Line Warspite - When taken: 3 Mar 1813 - Where taken: Bay of Biscay - Date received: 19 Mar 1813 - From what ship: HMS Warspite - Born: Dover, DE - Age: 38 - Sent to Dartmoor on 28 Jun 1813.

Jackson, John W. - Boy - Number: 805 - Prize name: Cannoniere - Ship type: P - How taken: HM Ship-of-the-Line Warspite - When taken: 14 Mar 1813 - Where taken: Bay of Biscay - Date received: 19 Mar 1813 - From what ship: HMS Warspite - Born: Philadelphia - Age: 15 - Sent on 8 Apr 1813 to Ashburton on parole.

Jackson, Lambert - Seaman - Number: 1447 - Prize name: Paul Jones - Ship type: P - How taken: HM Frigate Leonidas - When taken: 23 May 1813 - Where taken: off Cape Clear - Date received: 26 May 1813 - From what ship: HMS Leonidas - Born: Middleton, NY - Age: 23 - Discharged on 30 Jun 1813 and sent to Stapleton.

Jackson, Samuel - Seaman - Number: 2315 - Prize name: Siro - Ship type: LM - How taken: HM Brig Pelican - When taken: 13 Jan 1814 - Where taken: at sea - Date received: 23 Jan 1814 - From what ship: HMS Pelican - Born: New York - Age: 29 - Discharged on 31 Jan 1814 and sent to Dartmoor.

Jackson, Thomas - Seaman - Number: 1139 - Prize name: Magdalen - Ship type: MV - How taken: HM Ship-of-the-Line Superb - When taken: 15 Apr 1813 - Where taken: off Belle Isle - Date received: 22 Apr 1813 - From what ship: HMS Superb - Born: Charlestown - Age: 19 - Sent to Dartmoor on 1 Jul 1813.

Jackson, Thomas - Ordinary Seaman - Number: 476 - How taken: Impressed at Greenock - When taken: 27 Nov 1812 - Date received: 10 Feb 1813 - From what ship: HMS Frederick - Born: New Jersey - Age: 27 - Race: Black - Sent on 20 Feb 1813 to Ashburton on parole.

Jackson, Thomas - Boy - Number: 381 - Prize name: Orbit - Ship type: MV - How taken: HM Brig Achates - When taken: 29 Jan 1813 - Where taken: Lat 49 N Long 13 W - Date received: 4 Feb 1813 - From what ship: HMS Achates - Born: Jersey - Age: 14 - Race: Black - Sent to Chatham on 29 Mar 1813 on the HMS Braham.

Jackson, Thomas - Cook - Number: 2573 - How taken: Delivered himself up from the MV Hebrus - Date received: 8 May 1814 - From what ship: MV Hebrus - Born: New York - Age: 24 - Race: Black - Discharged on 7 May 1814 and sent to Ashburton on parole.

Jackson, William - Prize Master - Number: 1749 - Prize name: Fox Packet, prize of the Privateer Fox - Ship type: P - How taken: Superior, letter of marque - When taken: 25 Jun 1813 - Where taken: Lat 50N Long 21W - Date received: 19 Jul 1813 - From what ship: HMS Bittern - Born: Levering - Age: 25 - Discharged on 8 Sep 1813 and sent to Dartmoor.

Jackson, William - Seaman - Number: 2146 - Prize name: Amiable - Ship type: LM - How taken: HM Ship-of-the-Line Magnificant - When taken: 30 Oct 1813 - Where taken: off Lorient - Date received: 29 Nov 1813 - From what ship: HMS Dublin - Born: Philadelphia - Age: 21 - Discharged on 4 Dec 1813 and sent to Dartmoor.

Jackson, William - Sailmaker - Number: 2644 - How taken: Delivered himself up from HM Frigate Pyramus - Date received: 19 May 1814 - From what ship: HMS Salvador del Mundo - Born: Cambridge - Age: 32 - Race: Black - Discharged on 14 Jun 1814 and sent to Dartmoor.

James, John - Seaman - Number: 463 - Prize name: Dolphin - Ship type: LM - How taken: HM Ship-of-the-Line Colossus, HM Frigate Rhin & HM Brig Goldfinch - When taken: 5 Jan 1813 - Where taken: off Western Islands - Date received: 6 Feb 1813 - From what ship: HMS Rhin - Born: New Orleans - Age: 30 - Sent to Chatham on 29 Mar 1813 on the HMS Braham.

James, Daniel - Seaman - Number: 1341 - Prize name: Fox - Ship type: LM - How taken: HM Sloop Pheasant - When taken: 23 Apr 1813 - Where taken: Bay of Biscay - Date received: 14 May 1813 - From what ship: HMS Pleasant - Born: Boston - Age: 25 - Discharged on 3 Jul 1813 and sent to Stapleton Prison.

James, Peter - Seaman - Number: 56 - Prize name: Independence - Ship type: MV - How taken: HM Frigate Medusa - When taken: 9 Nov 1812 - Where taken: San Sebastian - Date received: 27 Nov 1812 - From what ship: HMS Wasp - Born: Boston - Age: 20 - Sent to Portsmouth on 29 Dec 1812 on the HMS Northumberland.

James, Sacket - Seaman - Number: 1037 - How taken: Delivered himself up from HM Ship-of-the-Line Dublin - Date received: 21 Apr 1813 - From what ship: HMS Dublin - Born: Virginia - Age: 50 - Sent to Chatham on 8 Jul 1813 on HM Tender Neptune.

James, Thomas - Seaman - Number: 1092 - Prize name: Young Holkar - Ship type: MV - How taken: HM Ship-of-the-Line Superb - When taken: 10 Apr 1813 - Where taken: off Belle Isle - Date received: 22 Apr 1813 - From what ship: HMS Superb - Born: New Orleans - Age: 29 - Sent to Dartmoor on 1 Jul 1813.

Jamison, Peter - Seaman - Number: 1503 - How taken: Impressed at Belfast - When taken: 16 Jan 1813 - Date received: 27 May 1813 - From what ship: HMS Prince Frederick - Born: Northumberland - Age: 53 - Race: Black - Discharged on 8 Sep 1813 and sent to Dartmoor.

Jamison, Robert - Seaman - Number: 1810 - Prize name: US Brig Argus - Ship type: War - How taken: HM Brig

Pelican - When taken: 14 Aug 1813 - Where taken: Irish Channel - Date received: 17 Aug 1813 - From what ship: USS Argus - Born: New Castle, DE - Age: 54 - Discharged on 29 Nov 1813 and sent to Dartmoor.

Jamison, William - Midshipman - Number: 1929 - Prize name: US Brig Argus - Ship type: War - How taken: HM Brig Pelican - When taken: 14 Aug 1813 - Where taken: Irish Channel - Date received: 23 Aug 1813 - From what ship: HMS Pelican - Born: Virginia - Age: 19 - Discharged on 26 Aug 1813 and sent to Ashburton on parole.

Jane, Joseph - Seaman - Number: 39 - Prize name: Warren - Ship type: MV - How taken: HM Frigate Sybille & HM Frigate Fortunee- When taken: 5 Sep 1812 - Where taken: Lat 41.4 Long 33 - Date received: 23 Nov 1812 - From what ship: HMS Stork - Born: Providence - Age: 20 - Sent to Portsmouth on 29 Dec 1812 on the HMS Northumberland.

Jansen, Knut - Seaman - Number: 2046 - How taken: Impressed at Liverpool - When taken: 16 Oct 1813 - Date received: 1 Nov 1813 - From what ship: HMS Bittern - Born: New York - Age: 21 - Discharged on 3 Nov 1813 and sent to Dartmoor.

Jarvis, James - Seaman - Number: 2599 - How taken: MV Hauphy - Date received: 16 May 1814 - From what ship: HMS Repulse - Born: Long Island - Age: 27 - Race: Black - Discharged on 14 Jun 1814 and sent to Dartmoor.

Jasseciu, Louis - Seaman - Number: 1451 - Prize name: Paul Jones - Ship type: P - How taken: HM Frigate Leonidas - When taken: 23 May 1813 - Where taken: off Cape Clear - Date received: 26 May 1813 - From what ship: HMS Leonidas - Born: Bordeaux - Age: 18 - Discharged on 30 Jun 1813 and sent to Stapleton.

Jeffery, Job - Boy - Number: 1967 - Prize name: Joel Barlow - Ship type: LM - How taken: HM Frigate Briton - When taken: 3 Jul 1813 - Where taken: off Bordeaux - Date received: 8 Sep 1813 - From what ship: Plymouth - Born: New London - Age: 12 - Discharged on 27 Sep 1813 and sent to Dartmoor.

Jenkins, Joseph - Prize Master - Number: 1747 - Prize name: Fox Packet, prize of the Privateer Fox - Ship type: P - How taken: Superior, letter of marque - When taken: 25 Jun 1813 - Where taken: Lat 50N Long 21W - Date received: 19 Jul 1813 - From what ship: HMS Bittern - Born: Massachusetts - Age: 39 - Discharged on 8 Sep 1813 and sent to Dartmoor.

Jenkins, Nathaniel - Seaman - Number: 1393 - Prize name: Tom - Ship type: LM - How taken: HM Frigate Surveillante - When taken: 27 Apr 1813 - Where taken: Bay of Biscay - Date received: 15 May 1813 - From what ship: HMS Foxhound - Born: Virginia - Age: 19 - Race: Black - Discharged on 3 Jul 1813 and sent to Stapleton Prison.

Jenkins, Richard - Seaman - Number: 171 - Prize name: Hunter - Ship type: P - How taken: HM Frigate Phoebe - When taken: 24 Dec 1812 - Where taken: off Western Islands - Date received: 9 Jan 1813 - From what ship: HMS Phoebe - Born: Boston - Age: 26 - Sent to Portsmouth on 8 Feb 1813 on the HMS Colossus.

Jennin, Samuel - Seaman - Number: 358 - Prize name: Louisa, prize of the Privateer Decatur - Ship type: P - How taken: HM Frigate Andromache - When taken: 11 Jan 1813 - Where taken: off Bordeaux - Date received: 4 Feb 1813 - From what ship: HMS Cornwall - Born: Baltimore - Age: 40 - Race: Negro - Sent to Chatham on 29 Mar 1813 on the HMS Braham.

Jennings, Nathaniel - Seaman - Number: 781 - Prize name: Cannoniere - Ship type: P - How taken: HM Ship-of-the-Line Warspite - When taken: 14 Mar 1813 - Where taken: Bay of Biscay - Date received: 19 Mar 1813 - From what ship: HMS Warspite - Born: Connecticut - Age: 26 - Sent to Dartmoor on 28 Jun 1813.

Jentle, Joseph - Seaman - Number: 494 - Prize name: Cashiere - Ship type: LM - How taken: HM Brig Reindeer - When taken: 3 Feb 1813 - Where taken: Bay of Biscay - Date received: 12 Feb 1813 - From what ship: HMS Reindeer - Born: New Orleans - Age: 28 - Sent to Dartmoor on 2 Apr 1813.

Jervis, John - Seaman - Number: 548 - Prize name: Spitfire - Ship type: MV - How taken: HM Brig Achates - When taken: 14 Feb 1813 - Where taken: off Ushant - Date received: 16 Feb 1813 - From what ship: HMS Achates - Born: Marblehead - Age: 21 - Sent to Dartmoor on 2 Apr 1813.

Jewell, Samuel - Seaman - Number: 98 - Prize name: Experiment - Ship type: MV - How taken: HM Brig Rover - When taken: 21 Oct 1812 - Where taken: off Bordeaux - Date received: 25 Dec 1812 - From what ship: HMS Northumberland - Age: 20 - Sent to Portsmouth on 29 Dec 1812 on the HMS Northumberland.

Jiffs, Joseph - Seaman - Number: 936 - Prize name: Weasel - Ship type: MV - How taken: HM Brig Foxhound - When taken: 25 Mar 1813 - Where taken: Bay of Biscay - Date received: 6 Apr 1813 - From what ship: HMS Foxhound

- Born: Gloucester - Age: 22 - Sent on 21 Jun 1813 to Portsmouth on HMS Prometheus.

Johns, Bellona - Seaman - Number: 269 - Prize name: Brutus - Ship type: MV - How taken: HM Frigate Briton - When taken: Jan 1813 - Where taken: Bay of Biscay - Date received: 21 Jan 1813 - From what ship: HMS Briton - Born: New Orleans - Age: 24 - Sent to Portsmouth on 8 Feb 1813 on the HMS Colossus.

Johns, Thomas - Seaman - Number: 1899 - Prize name: US Brig Argus - Ship type: War - How taken: HM Brig Pelican - When taken: 14 Aug 1813 - Where taken: Irish Channel - Date received: 23 Aug 1813 - From what ship: HMS Pelican - Born: Norfolk - Age: 23 - Race: Black - Discharged on 8 Sep 1813 and sent to Dartmoor.

Johnson, Andrew - Seaman - Number: 2610 - How taken: MV Ajax - When taken: 22 Jan 1813 - Date received: 16 May 1814 - From what ship: HMS Repulse - Born: Portland - Age: 24 - Discharged on 14 Jun 1814 and sent to Dartmoor.

Johnson, Edward - Seaman - Number: 574 - Prize name: Rolla - Ship type: MV - How taken: HM Frigate Surveillante - When taken: 11 Feb 1813 - Where taken: Bay of Biscay - Date received: 23 Feb 1813 - From what ship: HMS Surveillante - Born: Darlington, England - Age: 68 - Sent to Dartmoor on 2 Apr 1813.

Johnson, Edward - Seaman - Number: 2626 - Prize name: Ateline - Ship type: MV - How taken: HM Frigate Magiciene - When taken: 14 Mar 1814 - Where taken: off Cape Finisterre, Spain - Date received: 17 May 1814 - From what ship: HMS Tortois - Born: Virginia - Age: 33 - Discharged on 14 Jun 1814 and sent to Dartmoor.

Johnson, Edward - Seaman - Number: 2575 - How taken: Delivered himself up from the MV Hebrus - Date received: 8 May 1814 - From what ship: MV Hebrus - Born: Philadelphia - Age: 20 - Race: Black - Discharged on 7 May 1814 and sent to Ashburton on parole.

Johnson, Elias - Seaman - Number: 2078 - Prize name: Avon, prize of the Privateer True Blooded Yankee - Ship type: P - How taken: HM Frigate Eurotas - When taken: 27 Oct 1813 - Where taken: off Ushant - Date received: 1 Nov 1813 - From what ship: HMS Hannibal - Born: Sweden - Age: 59 - Discharged on 3 Nov 1813 and sent to Dartmoor.

Johnson, Henry - Boy - Number: 2677 - Prize name: John, prize to Amelia - Ship type: P - How taken: HM Ship-of-the-Line Sterling Castle - When taken: 10 May 1814 - Where taken: Lat 36 Long 37 - Date received: 5 Jun 1814 - From what ship: HMS Gronville - Born: Baltimore - Age: 15 - Race: Black - Discharged on 14 Jun 1814 and sent to Dartmoor.

Johnson, Jacob - Seaman - Number: 378 - Prize name: Orbit - Ship type: MV - How taken: HM Brig Achates - When taken: 29 Jan 1813 - Where taken: Lat 49 N Long 13 W - Date received: 4 Feb 1813 - From what ship: HMS Achates - Born: Long Island - Age: 23 - Sent to Chatham on 29 Mar 1813 on the HMS Braham.

Johnson, James - Seaman - Number: 1267 - Prize name: Caroline - Ship type: MV - How taken: HM Frigate Medusa - When taken: 12 Apr 1813 - Where taken: Bay of Biscay - Date received: 10 May 1813 - From what ship: HMS Medusa - Born: Talbot County - Age: 19 - Discharged on 3 Jul 1813 and sent to Stapleton Prison.

Johnson, John - Boatswain - Number: 631 - Prize name: Criterion - Ship type: MV - How taken: HM Frigate Belle Poule - When taken: 14 Feb 1813 - Where taken: Bay of Biscay - Date received: 4 Mar 1813 - From what ship: HMS Strenuous - Born: Rhode Island - Age: 23 - Sent to Dartmoor on 2 Apr 1813.

Johnson, John - Seaman - Number: 565 - Prize name: Rolla - Ship type: MV - How taken: HM Frigate Surveillante - When taken: 11 Feb 1813 - Where taken: Bay of Biscay - Date received: 23 Feb 1813 - From what ship: HMS Surveillante - Born: Newcastle - Age: 28 - Sent to Dartmoor on 2 Apr 1813.

Johnson, John - Sergeant - Number: 2019 - Prize name: 14th U.S. Infantry - Ship type: LF - How taken: British army - When taken: 24 Jun 1813 - Where taken: Forty Mile Creek - Date received: 13 Oct 1813 - From what ship: Transport Lord Cathcart - Born: Amsterdam - Age: 27 - Discharged on 3 Nov 1813 and sent to Dartmoor.

Johnson, John - Seaman - Number: 2112 - Prize name: Chesapeake - Ship type: LM - How taken: HM Frigate Hotspur & HM Frigate Pyramus - When taken: 26 Oct 1813 - Where taken: Bay of Biscay - Date received: 22 Nov 1813 - From what ship: HMS Pyramus - Born: New York - Age: 29 - Discharged on 29 Nov 1813 and sent to Dartmoor.

Johnson, John (alias Daniel) - Seaman - Number: 1412 - Prize name: Rolla - Ship type: MV - How taken: HM Frigate Surveillante - When taken: 11 Feb 1813 - Where taken: Bay of Biscay - Date received: 22 May 1813 - From what ship: Dartmoor prison - Born: New Castle - Age: 27 - Discharged on 24 May 1813 and released to HMS Salvador del Mundo.

Johnson, Peter - Seaman - Number: 2647 - How taken: Delivered himself up from HM Brig Scylla - Date received: 22 May 1814 - From what ship: HMS Scylla - Born: Bienne, France - Age: 25 - Discharged on 22 Jun 1814.

Johnson, Robert - Seaman - Number: 72 - Prize name: Ceres - Ship type: MV - How taken: HM Battery Princess - When taken: 30 Aug 1812 - Where taken: Liverpool - Date received: 29 Nov 1812 - From what ship: HMS Salvador del Mundo - Born: New York - Age: 23 - Sent to Portsmouth on 29 Dec 1812 on the HMS Northumberland.

Johnson, Robert - Chief Mate - Number: 1948 - Prize name: Marmion - Ship type: MV - How taken: HM Frigate President - When taken: 14 Aug 1813 - Where taken: off Nantes - Date received: 30 Aug 1813 - From what ship: Marmion - Born: Salem - Age: 33 - Discharged on 31 Aug 1813 and sent to Ashburton on parole.

Johnson, Samuel - Seaman - Number: 950 - Prize name: Good Friends - Ship type: MV - How taken: HM Frigate Andromache - When taken: 2 Apr 1813 - Where taken: Bay of Biscay - Date received: 7 Apr 1813 - From what ship: HMS Sea Lark - Born: Suffolk - Age: 26 - Race: Black - Sent to Dartmoor on 28 Jun 1813.

Johnson, Samuel - Seaman - Number: 1791 - How taken: Delivered himself up from HM Ship-of-the-Line Scipion - When taken: 1 Dec 1812 - Date received: 13 Aug 1813 - From what ship: HMS Protection - Born: Staten Island - Age: 28 - Race: Negro - Discharged on 8 Sep 1813 and sent to Dartmoor.

Johnson, Samuel B. - Seaman - Number: 2391 - How taken: Gave himself up from HMS Eridanus - Date received: 3 Feb 1814 - From what ship: Eridanus - Born: Salem - Age: 39 - Discharged on 27 Feb 1814 and sent to Chatham on HMS Haleyon.

Johnson, William - Seaman - Number: 661 - How taken: Delivered himself up from HM Schooner Arrow - Date received: 8 Mar 1813 - From what ship: HMS Arrow - Born: New York - Age: 26 - Sent to Dartmoor on 2 Apr 1813.Johnson, William - Seaman - Number: 1963 - Prize name: Joel Barlow - Ship type: LM - How taken: HM Frigate Briton - When taken: 3 Jul 1813 - Where taken: off Bordeaux - Date received: 31 Aug 1813 - From what ship: HMS Clarence - Born: Connecticut - Age: 23 - Discharged on 8 Sep 1813 and sent to Dartmoor.

Johnston, Edward - Seaman - Number: 444 - Prize name: Resolution - Ship type: MV - How taken: Hibernia, letter of marque - When taken: 21 Sep 1812 - Where taken: off Bermuda - Date received: 5 Feb 1813 - From what ship: HMS Neptune - Born: Kent County - Age: 25 - Sent to Chatham on 29 Mar 1813 on the HMS Braham.

Johnston, Gershom - Seaman - Number: 696 - Prize name: Star - Ship type: MV - How taken: HM Ship-of-the-Line Superb - When taken: 9 Feb 1813 - Where taken: Bay of Biscay - Date received: 19 Mar 1813 - From what ship: HMS Warspite - Born: New York - Age: 26 - Sent to Dartmoor on 2 Apr 1813.

Johnston, James - Seaman - Number: 376 - Prize name: Orbit - Ship type: MV - How taken: HM Brig Achates - When taken: 29 Jan 1813 - Where taken: Lat 49 N Long 13 W - Date received: 4 Feb 1813 - From what ship: HMS Achates - Born: Northumberland - Age: 34 - Race: Black - Sent to Chatham on 29 Mar 1813 on the HMS Braham.

Johnston, John - Seaman - Number: 1146 - Prize name: Magdalen - Ship type: MV - How taken: HM Ship-of-the-Line Superb - When taken: 15 Apr 1813 - Where taken: off Belle Isle - Date received: 22 Apr 1813 - From what ship: HMS Superb - Born: Norfolk - Age: 36 - Sent to Dartmoor on 1 Jul 1813.

Johnston, Joseph - Seaman - Number: 1463 - Prize name: Paul Jones - Ship type: P - How taken: HM Frigate Leonidas - When taken: 23 May 1813 - Where taken: off Cape Clear - Date received: 26 May 1813 - From what ship: HMS Leonidas - Born: Connecticut - Age: 19 - Discharged on 30 Jun 1813 and sent to Stapleton.

Johnston, Peter - Seaman - Number: 1147 - Prize name: Magdalen - Ship type: MV - How taken: HM Ship-of-the-Line Superb - When taken: 15 Apr 1813 - Where taken: off Belle Isle - Date received: 22 Apr 1813 - From what ship: HMS Superb - Born: New Orleans - Age: 32 - Sent to Dartmoor on 1 Jul 1813.

Johnston, Samuel - Boatswain's Mate - Number: 395 - Prize name: Union - Ship type: MV - How taken: HM Frigate Iris - When taken: 17 Jan 1813 - Where taken: Lat 44 N Long 2.3 W - Date received: 5 Feb 1813 - From what ship: HMS San Josef - Born: New Jersey - Age: 23 - Sent to Chatham on 29 Mar 1813 on the HMS Braham.

Johnston, Simon - Seaman - Number: 1354 - Prize name: Shadow - Ship type: LM - How taken: HM Brig Reindeer & HM Schooner Helicon - When taken: 6 Apr 1813 - Where taken: Bay of Biscay - Date received: 14 May 1813 - From what ship: HMS Reindeer - Born: Hamburg - Age: 33 - Discharged on 3 Jul 1813 and sent to Stapleton

Prison.

Johnston, Thomas - Seaman - Number: 716 - Prize name: Criterion - Ship type: MV - How taken: HM Frigate Belle Poule - When taken: 14 Feb 1813 - Where taken: Bay of Biscay - Date received: 19 Mar 1813 - From what ship: HMS Warspite - Born: Albany - Age: 22 - Sent to Chatham on 8 Jul 1813 on HM Tender Neptune.

Johnston, William - Seaman - Number: 401 - Prize name: Union - Ship type: MV - How taken: HM Frigate Iris - When taken: 17 Jan 1813 - Where taken: Lat 44 N Long 2.3 W - Date received: 5 Feb 1813 - From what ship: HMS San Josef - Born: Philadelphia - Age: 23 - Sent to Chatham on 29 Mar 1813 on the HMS Braham.

Johnstone, Henry - Seaman - Number: 16 - Prize name: Hanna - Ship type: MV - How taken: Taken at Liverpool - When taken: 18 Oct 1812 - Date received: 21 Nov 1812 - From what ship: HMS Salvador del Mundo - Born: Philadelphia - Age: 34 - Sent to Portsmouth on 29 Dec 1812 on the HMS Northumberland.

Johnstone, Samuel - Seaman - Number: 212 - Prize name: Hunter - Ship type: P - How taken: HM Frigate Phoebe - When taken: 24 Dec 1812 - Where taken: off Western Islands - Date received: 9 Jan 1813 - From what ship: HMS Phoebe - Born: Boston - Age: 60 - Sent to Chatham on 29 Mar 1813 on the HMS Braham.

Joles, Robert - Seaman - Number: 1645 - Prize name: Tickler - Ship type: LM - How taken: HM Frigate Magiciene - When taken: 5 Jun 1813 - Where taken: Lat 47 Long 13 - Date received: 14 Jun 1813 - From what ship: HMS Orestes - Born: Rhode Island - Age: 37 - Discharged on 30 Jun 1813 and sent to Stapleton.

Jones, Benjamin - Cook - Number: 723 - Prize name: Mars - Ship type: MV - How taken: HM Ship-of-the-Line Warspite - When taken: 26 Feb 1813 - Where taken: Bay of Biscay - Date received: 19 Mar 1813 - From what ship: HMS Warspite - Born: Washington - Age: 25 - Sent to Dartmoor on 2 Apr 1813.

Jones, Benjamin - Seaman - Number: 761 - Prize name: William Bayard - Ship type: MV - How taken: HM Ship-of-the-Line Warspite - When taken: 3 Mar 1813 - Where taken: Bay of Biscay - Date received: 19 Mar 1813 - From what ship: HMS Warspite - Born: Milford, CT - Age: 27 - Sent to Dartmoor on 2 Apr 1813.

Jones, Charles - Seaman - Number: 1200 - Prize name: Zebra - Ship type: LM - How taken: HM Frigate Pyramus - When taken: 20 Apr 1813 - Where taken: Bay of Biscay - Date received: 9 May 1813 - From what ship: HMS Andromache - Born: New York - Age: 32 - Sent on 8 Sep 1813 to Dartmoor.

Jones, Charles - Seaman - Number: 1845 - How taken: Impressed at Dublin - When taken: 13 Jun 1813 - Date received: 17 Aug 1813 - From what ship: HMS Prince Frederick - Born: Connecticut - Age: 19 - Discharged on 8 Sep 1813 and sent to Dartmoor.

Jones, Francis - Seaman - Number: 546 - Prize name: Spitfire - Ship type: MV - How taken: HM Brig Achates - When taken: 14 Feb 1813 - Where taken: off Ushant - Date received: 16 Feb 1813 - From what ship: HMS Achates - Born: Marblehead - Age: 38 - Sent to Dartmoor on 2 Apr 1813.

Jones, Francis Vickery - Seaman - Number: 547 - Prize name: Spitfire - Ship type: MV - How taken: HM Brig Achates - When taken: 14 Feb 1813 - Where taken: off Ushant - Date received: 16 Feb 1813 - From what ship: HMS Achates - Born: Marblehead - Age: 21 - Sent to Dartmoor on 2 Apr 1813.

Jones, Gardner - Seaman - Number: 1282 - Prize name: Price - Ship type: LM - How taken: HM Frigate Medusa - When taken: 13 Apr 1813 - Where taken: Bay of Biscay - Date received: 10 May 1813 - From what ship: HMS Medusa - Born: Swansea - Age: 23 - Discharged on 21 Jun 1813 and released to American ship Mount Hope.

Jones, George - Seaman - Number: 1127 - Prize name: Viper - Ship type: MV - How taken: HM Ship-of-the-Line Superb - When taken: 15 Apr 1813 - Where taken: Bay of Biscay - Date received: 22 Apr 1813 - From what ship: HMS Superb - Born: New Orleans - Age: 24 - Sent to Dartmoor on 1 Jul 1813.

Jones, Hensley - Seaman - Number: 2706 - Prize name: Margaret, prize to the Privateer Surprize - Ship type: P - How taken: HM Brig Foxhound - When taken: 27 May 1814 - Where taken: off Scylly - Date received: 16 Jun 1814 - From what ship: HMS Foxhound - Born: Not readable - Age: 20 - Race: Black - Discharged on 20 Jun 1814 and sent to Dartmoor.

Jones, James - Passenger - Number: 897 - Prize name: Tiger - Ship type: MV - How taken: HM Brig Scylla - When taken: 22 Mar 1813 - Where taken: Bay of Biscay - Date received: 1 Apr 1813 - From what ship: HMS Scylla - Born: Connecticut - Age: 35 - Sent on 21 Jun 1813 to Portsmouth on HMS Prometheus.

Jones, John - Seaman - Number: 356 - Prize name: Louisa, prize of the Privateer Decatur - Ship type: P - How taken:

HM Frigate Andromache - When taken: 11 Jan 1813 - Where taken: off Bordeaux - Date received: 4 Feb 1813 - From what ship: HMS Cornwall - Born: Norfolk - Age: 17 - Sent to Chatham on 29 Mar 1813 on the HMS Braham.

Jones, John - Seaman - Number: 778 - Prize name: William Bayard - Ship type: MV - How taken: HM Ship-of-the-Line Warspite - When taken: 3 Mar 1813 - Where taken: Bay of Biscay - Date received: 19 Mar 1813 - From what ship: HMS Warspite - Born: New Castle - Age: 54 - Sent to Dartmoor on 28 Jun 1813.

Jones, John - Seaman - Number: 415 - Prize name: Union - Ship type: MV - How taken: HM Frigate Iris - When taken: 17 Jan 1813 - Where taken: Lat 44 N Long 2.3 W - Date received: 5 Feb 1813 - From what ship: HMS San Josef - Born: Philadelphia - Age: 17 - Sent to Chatham on 29 Mar 1813 on the HMS Braham.

Jones, John - Seaman - Number: 312 - Prize name: Blue Bird - Ship type: MV - How taken: HM Frigate Briton - When taken: 1 Jan 1813 - Where taken: off Bordeaux - Date received: 21 Jan 1813 - From what ship: HMS Abercrombie - Born: New Jersey - Age: 28 - Sent to Chatham on 29 Mar 1813 on the HMS Braham.

Jones, Peter (alias Benjamin) - Seaman - Number: 1178 - How taken: Delivered himself up from HM Ship-of-the-Line Magnificent - Date received: 6 May 1813 - From what ship: HMS Stag - Born: Maryland - Age: 24 - Race: Black - Sent to Chatham on 8 Jul 1813 on HM Tender Neptune.

Jones, Richard - Seaman - Number: 991 - Prize name: Lightning - Ship type: MV - How taken: HM Frigate Medusa - When taken: 2 Apr 1813 - Where taken: Bay of Biscay - Date received: 16 Apr 1813 - From what ship: HMS Fairy - Born: Baltimore - Age: 25 - Sent to Dartmoor on 27 Jul 1813.

Jones, Thomas - Seaman - Number: 1670 - Prize name: Good Intent - Ship type: MV - How taken: Impressed at Liverpool - When taken: 4 May 1813 - Date received: 17 Jun 1813 - From what ship: HMS Bittern - Born: Baltimore - Age: 22 - Discharged on 30 Jun 1813 and sent to Stapleton.

Jones, Thomas - Chief Mate - Number: 1251 - Prize name: Caroline - Ship type: MV - How taken: HM Frigate Medusa - When taken: 12 Apr 1813 - Where taken: Bay of Biscay - Date received: 10 May 1813 - From what ship: HMS Medusa - Born: Maryland - Age: 27 - Discharged on 9 Jul 1813 and sent to Ashburton on parole.

Jones, Thomas - Seaman - Number: 1699 - Prize name: Governor Gerry - Ship type: MV - How taken: HM Brig Royalist - When taken: 31 May 1813 - Where taken: Bay of Biscay - Date received: 26 Jun 1813 - From what ship: HMS Duncan - Born: New York - Age: 28 - Race: Black - Discharged on 30 Jun 1813 and sent to Stapleton.

Jones, Thomas - Seaman - Number: 2228 - Prize name: Zephyr - Ship type: MV - How taken: HM Frigate Pyramus - When taken: 30 Nov 1813 - Where taken: off Lorient - Date received: 3 Jan 1814 - From what ship: HMS Warspite - Born: Philadelphia - Age: 48 - Discharged on 31 Jan 1814 and sent to Dartmoor.

Jones, Walter - Seaman - Number: 2026 - Prize name: Friendship West, prize of Privateer True Blooded Yankee - Ship type: P - How taken: HM Schooner Helicon & HM Schooner Whiting - When taken: 25 Oct 1813 - Where taken: Bay of Biscay - Date received: 31 Oct 1813 - From what ship: HMS Whiting - Born: Litchfield - Age: 28 - Discharged on 3 Nov 1813 and sent to Dartmoor.

Jones, William - Seaman - Number: 968 - Prize name: Ferox - Ship type: MV - How taken: HM Frigate Medusa & HM Brig Lyra - When taken: 28 Mar 1813 - Where taken: off Cape Ortagle - Date received: 9 Apr 1813 - From what ship: HMS Lyra - Born: Baltimore - Age: 25 - Sent on 21 Jun 1813 to the American ship Hope.

Jones, William - Seaman - Number: 1182 - How taken: Delivered himself up from HM Ship-of-the-Line Clarence - Date received: 6 May 1813 - From what ship: HMS Stag - Born: Baltimore - Age: 28 - Race: Black - Sent to Chatham on 8 Jul 1813 on HM Tender Neptune.

Jones, William - Seaman - Number: 789 - Prize name: Cannoniere - Ship type: P - How taken: HM Ship-of-the-Line Warspite - When taken: 14 Mar 1813 - Where taken: Bay of Biscay - Date received: 19 Mar 1813 - From what ship: HMS Warspite - Born: Philadelphia - Age: 18 - Sent to Dartmoor on 28 Jun 1813.

Jones, William P. - Prize Master - Number: 586 - Prize name: St. Martin's Plantation, prize of the Privateer Paul Jones - Ship type: P - How taken: HM Ship-of-the-Line Dublin - When taken: 9 Feb 1815 - Where taken: Lat 43 N Long 33.5W - Date received: 25 Feb 1813 - From what ship: HMS Dublin - Born: Haverhill - Age: 35 - Sent to Dartmoor on 2 Apr 1813.

Jordan, Joseph - Seaman - Number: 1813 - Prize name: US Brig Argus - Ship type: War - How taken: HM Brig Pelican - When taken: 14 Aug 1813 - Where taken: Irish Channel - Date received: 17 Aug 1813 - From what ship: Received dead - Born: Maryland - Age: 30 - Died on 17 Aug 1813.

Jordan, Peter - Seaman - Number: 987 - Prize name: Lightning - Ship type: MV - How taken: HM Frigate Medusa - When taken: 2 Apr 1813 - Where taken: Bay of Biscay - Date received: 16 Apr 1813 - From what ship: HMS Fairy - Born: Messina, Sicily - Age: 31 - Sent to Chatham on 8 Jul 1813 on HM Tender Neptune.

Jordan, Richard - Seaman - Number: 1783 - Prize name: Orders in Council - Ship type: LM - How taken: HM Frigate Surveillante - When taken: 1 Jun 1813 - Where taken: Bay of Biscay - Date received: 7 Aug 1813 - From what ship: HMS Gleaner - Born: New York - Age: 17 - Discharged on 8 Sep 1813 and sent to Dartmoor.

Jordon, David - Seaman - Number: 436 - Prize name: Resolution - Ship type: MV - How taken: Hibernia, letter of marque - When taken: 21 Sep 1812 - Where taken: off Bermuda - Date received: 5 Feb 1813 - From what ship: HMS Neptune - Born: Portland - Age: 38 - Sent to Chatham on 29 Mar 1813 on the HMS Braham.

Joseph, Francis - Seaman - Number: 1430 - Prize name: Paul Jones - Ship type: P - How taken: HM Frigate Leonidas - When taken: 23 May 1813 - Where taken: off Cape Clear - Date received: 26 May 1813 - From what ship: HMS Leonidas - Born: Lisbon, Portugal - Age: 23 - Discharged on 30 Jun 1813 and sent to Stapleton.

Joseph, Francis - Seaman - Number: 707 - Prize name: Star - Ship type: MV - How taken: HM Ship-of-the-Line Superb - When taken: 9 Feb 1813 - Where taken: Bay of Biscay - Date received: 19 Mar 1813 - From what ship: HMS Warspite - Born: New Orleans - Age: 29 - Sent to Dartmoor on 2 Apr 1813.

Joseph, Francis - Seaman - Number: 1735 - Prize name: Star - Ship type: MV - How taken: HM Ship-of-the-Line Superb - When taken: 9 Feb 1813 - Where taken: Bay of Biscay - Date received: 10 Jul 1813 - From what ship: Dartmoor Prison - Born: New Orleans - Age: 20 - Discharged on 10 Jul 1813 and released to HMS Salvador del Mundo.

Joseph, Francis - 2nd Lieutenant - Number: 2558 - Prize name: Bunker Hill - Ship type: P - How taken: HM Frigate Pomone & HM Frigate Cydnus - When taken: 4 Mar 1814 - Where taken: Bay of Biscay - Date received: 17 Apr 1814 - From what ship: HMS Teazer - Born: Salem - Age: 38 - Discharged on 10 May 1814 and sent to Dartmoor.

Joseph, John - Seaman - Number: 40 - Prize name: Warren - Ship type: MV - How taken: HM Frigate Sybille & HM Frigate Fortunee- When taken: 5 Sep 1812 - Where taken: Lat 41.4 Long 33 - Date received: 23 Nov 1812 - From what ship: HMS Stork - Born: Lisbon - Age: 25 - Sent to Portsmouth on 29 Dec 1812 on the HMS Northumberland.

Joseph, Michael - Cook - Number: 87 - Prize name: Argus - Ship type: MV - How taken: Fancy, Cutter - When taken: 19 Dec 1812 - Where taken: Bay of Biscay - Date received: 24 Dec 1812 - From what ship: Fancy, Cutter - Born: New York - Age: 24 - Race: Negro - Sent to Portsmouth on 29 Dec 1812 on the HMS Northumberland.

Joseph, Thomas - Seaman - Number: 1078 - Prize name: John & Frances - Ship type: MV - How taken: HM Frigate Belle Poule - When taken: 19 Mar 1813 - Where taken: Bay of Biscay - Date received: 22 Apr 1813 - From what ship: HMS Superb - Born: Charlestown - Age: 27 - Sent to Dartmoor on 1 Jul 1813.

Jolyon, Robert - Seaman - Number: 695 - Prize name: Star - Ship type: MV - How taken: HM Ship-of-the-Line Superb - When taken: 9 Feb 1813 - Where taken: Bay of Biscay - Date received: 19 Mar 1813 - From what ship: HMS Warspite - Born: Connecticut - Age: 27 - Sent to Dartmoor on 28 Jun 1813.

Juderwicks, James - Surgeon - Number: 1848 - Prize name: US Brig Argus - Ship type: War - How taken: HM Brig Pelican - When taken: 14 Aug 1813 - Where taken: Irish Channel - Date received: 18 Aug 1813 - From what ship: USS Argus - Born: New York - Age: 23 - Discharged on 25 Aug 1813 and sent to Ashburton on parole.

Jutt, Benjamin - Seaman - Number: 553 - Prize name: Spitfire - Ship type: MV - How taken: HM Brig Achates - When taken: 14 Feb 1813 - Where taken: off Ushant - Date received: 16 Feb 1813 - From what ship: HMS Achates - Born: Marblehead - Age: 22 - Sent to Dartmoor on 2 Apr 1813.

Kain, Peter - Seaman - Number: 290 - Prize name: Columbia - Ship type: MV - How taken: HM Frigate Briton - When taken: 17 Dec 1812 - Where taken: off Bordeaux - Date received: 21 Jan 1813 - From what ship: HMS Abercrombie - Born: France - Age: 29 - Sent to Portsmouth on 8 Feb 1813 on the HMS Colossus.

Kalm, John - Chief Mate - Number: 959 - Prize name: Ferox - Ship type: MV - How taken: HM Frigate Medusa & HM Brig Lyra - When taken: 28 Mar 1813 - Where taken: off Cape Ortagle - Date received: 9 Apr 1813 - From what ship: HMS Lyra - Born: Connecticut - Age: 21 - Sent on 11 Apr 1813 to Ashburton on parole.

Keen, Joseph - Seaman - Number: 2357 - Prize name: Zephyr - Ship type: MV - How taken: HM Frigate Surveillante

- When taken: 6 Jan 1814 - Where taken: Bay of Biscay - Date received: 31 Jan 1814 - From what ship: HMS Surveillante - Born: Massachusetts - Age: 26 - Discharged on 27 Feb 1814 and sent to Chatham on HMS Haleyon.

Keen, Nathaniel - Boatswain - Number: 1006 - Prize name: Courier - Ship type: LM - How taken: HM Frigate Andromache - When taken: 14 Mar 1813 - Where taken: Bay of Biscay - Date received: 16 Apr 1813 - From what ship: HMS Fairy - Born: Kittery - Age: 40 - Sent to Dartmoor on 8 Sep 1813.

Keith, James - Seaman - Number: 277 - Prize name: Leader - Ship type: MV - How taken: HM Frigate Briton - When taken: 10 Dec 1812 - Where taken: off Bordeaux - Date received: 21 Jan 1813 - From what ship: HMS Abercrombie - Born: Warren - Age: 20 - Sent to Portsmouth on 8 Feb 1813 on the HMS Colossus.

Kellam, James - Seaman - Number: 1807 - Prize name: US Brig Argus - Ship type: War - How taken: HM Brig Pelican - When taken: 14 Aug 1813 - Where taken: Irish Channel - Date received: 17 Aug 1813 - From what ship: USS Argus - Born: Northampton, VA - Age: 39 - Discharged on 3 Nov 1813 and sent to Dartmoor.

Keller, John - Seaman - Number: 1690 - Prize name: Governor Gerry - Ship type: MV - How taken: HM Brig Royalist - When taken: 31 May 1813 - Where taken: Bay of Biscay - Date received: 26 Jun 1813 - From what ship: HMS Duncan - Born: Boston - Age: 40 - Discharged on 30 Jun 1813 and sent to Stapleton.

Kellinger, John - Seaman - Number: 1680 - Prize name: Revenge - Ship type: LM - How taken: HM Frigate Belle Poule - When taken: 11 May 1813 - Where taken: Bay of Biscay - Date received: 26 Jun 1813 - From what ship: HMS Duncan - Born: New York - Age: 22 - Discharged on 30 Jun 1813 and sent to Stapleton.

Kellum, John C. - Seaman - Number: 767 - Prize name: William Bayard - Ship type: MV - How taken: HM Ship-of-the-Line Warspite - When taken: 3 Mar 1813 - Where taken: Bay of Biscay - Date received: 19 Mar 1813 - From what ship: HMS Warspite - Born: Virginia - Age: 23 - Sent to Dartmoor on 28 Jun 1813.

Kennedy, John - Seaman - Number: 1136 - Prize name: Magdalen - Ship type: MV - How taken: HM Ship-of-the-Line Superb - When taken: 15 Apr 1813 - Where taken: off Belle Isle - Date received: 22 Apr 1813 - From what ship: HMS Superb - Born: New Jersey - Age: 25 - Sent to Dartmoor on 1 Jul 1813.

Kennedy, John - Seaman - Number: 1068 - Prize name: Meteor - Ship type: MV - How taken: HM Frigate Briton - When taken: 12 Mar 1813 - Where taken: Bay of Biscay - Date received: 22 Apr 1813 - From what ship: HMS Superb - Born: Hudson, NY - Age: 24 - Sent to Dartmoor on 8 Sep 1813.

Kennedy, John - Seaman - Number: 138 - Prize name: Nope - Ship type: MV - How taken: HM Sloop Pheasant - When taken: 13 Dec 1812 - Where taken: off Western Islands - Date received: 8 Jan 1813 - From what ship: HMS Pheasant - Born: Newburgh, NY - Age: 24 - Sent to Portsmouth on 8 Feb 1813 on the HMS Colossus.

Kennedy, Peter - Seaman - Number: 367 - Prize name: Orbit - Ship type: MV - How taken: HM Brig Achates - When taken: 29 Jan 1813 - Where taken: Lat 49 N Long 13 W - Date received: 4 Feb 1813 - From what ship: HMS Achates - Born: New Jersey - Age: 20 - Sent to Chatham on 29 Mar 1813 on the HMS Braham.

Kennedy, Richard - Seaman - Number: 2432 - Prize name: US Brig Argus - Ship type: War - How taken: HM Brig Pelican - When taken: 14 Aug 1813 - Where taken: Irish Channel - Date received: 12 Feb 1814 - From what ship: HMS Salvador del Mundo - Born: Brooklyn - Age: 21 - Discharged on 10 May 1814 and sent to Dartmoor.

Kennedy, William (alias Freeman) - Seaman - Number: 1041 - How taken: Delivered himself up from HM Ship-of-the-Line Ajax - Date received: 22 Apr 1813 - From what ship: HMS Ajax - Born: Boston - Age: 29 - Sent to Chatham on 8 Jul 1813 on HM Tender Neptune.

Kennisen, William - Seaman - Number: 2158 - Prize name: Fortune - Ship type: MV - How taken: Impressed at Plymouth - When taken: 10 Nov 1813 - Date received: 12 Dec 1813 - From what ship: HMS Halcyon - Born: Deerfield - Age: 23 - Discharged on 31 Jan 1814 and sent to Dartmoor.

Kenny, Harris - Pilot - Number: 1576 - Prize name: Orders in Council - Ship type: LM - How taken: HM Frigate Surveillante - When taken: 1 Jun 1813 - Where taken: Bay of Biscay - Date received: 12 Jun 1813 - From what ship: Cutter Earl Wellington - Born: New London - Age: 30 - Discharged on 8 Sep 1813 and sent to Dartmoor.

Kerban, Thomas - Seaman - Number: 70 - How taken: Impressed at Belfast - When taken: 15 Sep 1812 - Date received: 29 Nov 1812 - From what ship: HMS Salvador del Mundo - Born: New Orleans - Age: 29 - Sent to Portsmouth on 29 Dec 1812 on the HMS Northumberland.

Kershon, Abraham - Seaman - Number: 886 - Prize name: Tiger - Ship type: MV - How taken: HM Brig Scylla -

When taken: 22 Mar 1813 - Where taken: Bay of Biscay - Date received: 1 Apr 1813 - From what ship: HMS Scylla - Born: Long Island - Age: 32 - Sent on 21 Jun 1813 to Portsmouth.

Keven, Edward - Gunner - Number: 1418 - Prize name: Paul Jones - Ship type: P - How taken: HM Frigate Leonidas - When taken: 23 May 1813 - Where taken: off Cape Clear - Date received: 26 May 1813 - From what ship: HMS Leonidas - Born: Petersburg - Age: 25 - Discharged on 8 Sep 1813 and sent to Dartmoor.

Kimball, Nathaniel - Seaman - Number: 2679 - How taken: Sent into custody from HM Frigate Phoenix - Date received: 5 Jun 1814 - From what ship: HMS Gronville - Born: East Harford - Age: 19 - Discharged on 14 Jun 1814 and sent to Dartmoor.

Kimmins, John - Seaman - Number: 502 - Prize name: Cashiere - Ship type: LM - How taken: HM Brig Reindeer - When taken: 3 Feb 1813 - Where taken: Bay of Biscay - Date received: 12 Feb 1813 - From what ship: HMS Reindeer - Born: New York - Age: 25 - Sent to Dartmoor on 2 Apr 1813.

King, John - Seaman - Number: 931 - Prize name: Weasel - Ship type: MV - How taken: HM Brig Foxhound - When taken: 25 Mar 1813 - Where taken: Bay of Biscay - Date received: 6 Apr 1813 - From what ship: HMS Foxhound - Born: New York - Age: 20 - Sent on 21 Jun 1813 to Portsmouth on HMS Prometheus.

King, Joseph - Seaman - Number: 1422 - Prize name: Paul Jones - Ship type: P - How taken: HM Frigate Leonidas - When taken: 23 May 1813 - Where taken: off Cape Clear - Date received: 26 May 1813 - From what ship: HMS Leonidas - Born: Lisbon, Portugal - Age: 48 - Discharged on 3 Jul 1813 and sent to Stapleton Prison.

Kingley, Benjamin - Able Seaman - Number: 351 - How taken: Taken off the HM Frigate Andromache - Date received: 29 Jan 1813 - From what ship: HMS Royal Sovereign - Born: Nobleboro - Age: 35 - Sent to Chatham on 29 Mar 1813 on the HMS Braham.

Kinlay, Joseph - Seaman - Number: 327 - Prize name: Porcupine - Ship type: MV - How taken: HM Frigate Dryad - When taken: 8 Jan 1813 - Where taken: off Bordeaux - Date received: 21 Jan 1813 - From what ship: HMS Abercrombie - Born: New Haven - Age: 19 - Sent to Chatham on 29 Mar 1813 on the HMS Braham.

Kirby, Benjamin - 2nd Mate - Number: 384 - Prize name: Union - Ship type: MV - How taken: HM Frigate Iris - When taken: 17 Jan 1813 - Where taken: Lat 44 N Long 2.3 W - Date received: 5 Feb 1813 - From what ship: HMS San Josef - Born: Philadelphia - Age: 24 - Sent to Chatham on 29 Mar 1813 on the HMS Braham.

Kitchen, Richard - Seaman - Number: 2142 - Prize name: Amiable - Ship type: LM - How taken: HM Ship-of-the-Line Magnificant - When taken: 30 Oct 1813 - Where taken: off Lorient - Date received: 29 Nov 1813 - From what ship: HMS Dublin - Born: New York - Age: 22 - Discharged on 4 Dec 1813 and sent to Dartmoor.

Knapp, Samuel - Seaman - Number: 2175 - Prize name: Wolf, prize of the Privateer Grand Turk - Ship type: P - How taken: HM Frigate Briton - When taken: 1 Dec 1813 - Where taken: Bay of Biscay - Date received: 17 Dec 1813 - From what ship: HMS Briton - Born: Marblehead - Age: 24 - Discharged on 17 Dec 1813 and released to HMS Britton.

Knapp, Walker - Seaman - Number: 4 - Prize name: Rhode & Betsey - Ship type: MV - How taken: HM Sloop Talbot - When taken: Aug 1812 - Where taken: off Cape Clear - Date received: 24 Oct 1812 - From what ship: HMS Frederick - Born: Stamford - Age: 21 - Sent to Portsmouth on 29 Dec 1812 on the HMS Northumberland.

Kneeland, Ebenezer - Merchant & Passenger - Number: 339 - Prize name: Porcupine - Ship type: MV - How taken: HM Frigate Dryad - When taken: 8 Jan 1813 - Where taken: off Bordeaux - Date received: 22 Jan 1813 - From what ship: Porcupine - Born: New York - Age: 22 - Sent on 23 Jan 1813 to Ashburton on parole.

Knight, Isaac D. - Boy - Number: 433 - Prize name: Union - Ship type: MV - How taken: HM Frigate Iris - When taken: 17 Jan 1813 - Where taken: Lat 44 N Long 2.3 W - Date received: 5 Feb 1813 - From what ship: HMS San Josef - Born: Philadelphia - Age: 17 - Sent to Chatham on 29 Mar 1813 on the HMS Braham.

Knight, John - Chief Mate - Number: 2330 - Prize name: Hannah - Ship type: MV - How taken: HM Ship-of-the-Line Conquestador - When taken: 14 Jan 1814 - Where taken: Lat 47.3 Long 7 - Date received: 28 Jan 1814 - From what ship: Hanna - Born: Marblehead - Age: 21 - Discharged on 3 Feb 1814 and sent to Ashburton on parole.

Knox John - Seaman - Number: 2654 - How taken: Sent into custody from HM Frigate Nereus - Date received: 5 Jun 1814 - From what ship: HMS Gronville - Born: New York - Age: 47 - Discharged on 14 Jun 1814 and sent to Dartmoor.

Knox, Thomas - Seaman - Number: 176 - Prize name: Hunter - Ship type: P - How taken: HM Frigate Phoebe - When taken: 24 Dec 1812 - Where taken: off Western Islands - Date received: 9 Jan 1813 - From what ship: HMS Phoebe - Born: Boston - Age: 52 - Sent to Portsmouth on 8 Feb 1813 on the HMS Colossus.

Kromhout, Barney - Seaman - Number: 532 - Prize name: Terrible - Ship type: MV - How taken: HM Brig Foxhound - When taken: 8 Feb 1813 - Where taken: Channel - Date received: 15 Feb 1813 - From what ship: HMS Foxhound - Born: Baltimore - Age: 23 - Sent to Dartmoor on 2 Apr 1813.

Labery, Pierre Joseph - Passenger - Number: 130 - Prize name: Empress - Ship type: MV - How taken: HM Brig Rover - When taken: 29 Nov 1812 - Where taken: off Bordeaux - Date received: 5 Jan 1813 - From what ship: HMS Rover - Born: Bern - Age: 49 - Entered on 12 Jan 1813 into the French General Entry Book as prisoner number 39250.

Laborde, Peter - Seaman - Number: 2252 - Prize name: Volante - Ship type: MV - How taken: Callaloo - When taken: 10 Mar 1814 - Where taken: at sea - Date received: 15 Jan 1814 - From what ship: HMS Teazer - Born: Genoa - Age: 20 - Discharged on 31 Jan 1814 and sent to Dartmoor.

Lacoen, John B. - Seaman - Number: 1428 - Prize name: Paul Jones - Ship type: P - How taken: HM Frigate Leonidas - When taken: 23 May 1813 - Where taken: off Cape Clear - Date received: 26 May 1813 - From what ship: HMS Leonidas - Born: Naples - Age: 51 - Discharged on 3 Jul 1813 and sent to Stapleton Prison.

Lactortier, Charles Felid - Passenger - Number: 2189 - Prize name: Charlotte - Ship type: MV - How taken: Cutter Dwarf - When taken: 4 Nov 1812 - Where taken: off Bordeaux - Date received: 19 Dec 1813 - From what ship: HMS Conquistador - Born: Guadeloupe - Age: 21 - Discharged on 25 Dec 1813 and sent to Ashburton on parole.

Lambert, Joseph - Seaman - Number: 1264 - Prize name: Caroline - Ship type: MV - How taken: HM Frigate Medusa - When taken: 12 Apr 1813 - Where taken: Bay of Biscay - Date received: 10 May 1813 - From what ship: HMS Medusa - Born: New York - Age: 24 - Race: Negro - Discharged on 3 Jul 1813 and sent to Stapleton Prison.

Lambert, William - Prize Master - Number: 2264 - Prize name: Fanny - Ship type: MV - How taken: HM Frigate Eurotas - When taken: 25 Dec 1813 - Where taken: at sea - Date received: 20 Jan 1814 - From what ship: HMS Eurotas - Born: New York - Age: 42 - Discharged on 31 Jan 1814 and sent to Dartmoor.

Lammers, Henry - Seaman - Number: 743 - Prize name: Charlotte - Ship type: MV - How taken: HM Ship-of-the-Line Warspite - When taken: 3 Mar 1813 - Where taken: Bay of Biscay - Date received: 19 Mar 1813 - From what ship: HMS Warspite - Born: New York - Age: 23 - Sent to Dartmoor on 2 Apr 1813.

Lamoure, Peter - Chief Mate - Number: 1410 - Prize name: Fox - Ship type: LM - How taken: HM Sloop Pheasant - When taken: 23 Apr 1813 - Where taken: Bay of Biscay - Date received: 20 May 1813 - From what ship: HMS Pleasant - Born: New York - Age: 22 - Discharged on 23 May 1813 and sent to Ashburton on parole.

Lamson, Charles - 2nd Mate - Number: 541 - Prize name: Spitfire - Ship type: MV - How taken: HM Brig Achates - When taken: 14 Feb 1813 - Where taken: off Ushant - Date received: 16 Feb 1813 - From what ship: HMS Achates - Born: Beverly - Age: 23 - Sent to Dartmoor on 2 Apr 1813.

Landerman, Joseph - Steward - Number: 783 - Prize name: Cannoniere - Ship type: P - How taken: HM Ship-of-the-Line Warspite - When taken: 14 Mar 1813 - Where taken: Bay of Biscay - Date received: 19 Mar 1813 - From what ship: HMS Warspite - Born: Havana - Age: 23 - Race: Black - Sent to Dartmoor on 28 Jun 1813.

Landford, John - Prize Master - Number: 2059 - Prize name: Betsy, prize to the Privateer True Blooded Yankee - Ship type: P - How taken: HM Frigate Eurotas - When taken: 26 Oct 1813 - Where taken: off Ushant - Date received: 1 Nov 1813 - From what ship: HMS Hannibal - Born: Somerset County - Age: 25 - Discharged on 3 Nov 1813 and sent to Dartmoor.

Lane, James - Seaman - Number: 1510 - Prize name: Grand Napoleon - Ship type: MV - How taken: HM Brig Goldfinch - When taken: 17 Apr 1813 - Where taken: Bay of Biscay - Date received: 29 May 1813 - From what ship: HMS Hannibal - Born: New York - Age: 24 - Discharged on 30 Jun 1813 and sent to Stapleton.

Lane, James - Chief Mate - Number: 2096 - Prize name: Amiable - Ship type: LM - How taken: HM Ship-of-the-Line Magnificent - When taken: 30 Oct 1813 - Where taken: off Lorient - Date received: 9 Nov 1813 - From what ship: Amiable - Born: Maryland - Age: 22 - Discharged on 11 Nov 1813 and sent to Ashburton on parole.

Langoth, Francis - Seaman - Number: 432 - How taken: Delivered himself up from HMS Martral - Date received: 5

Feb 1813 - From what ship: HMS San Josef - Born: Hungary - Age: 33 - Sent to Chatham on 29 Mar 1813 on the HMS Braham.

Lapavous, Philip - Mate - Number: 448 - Prize name: Point - Ship type: MV - How taken: HM Frigate Rhin - When taken: 15 Jan 1813 - Where taken: Lat 44 N Long 15 W - Date received: 6 Feb 1813 - From what ship: HMS Rhin - Born: Marblehead - Age: 36 - Sent on 11 Feb 1813 to Ashburton on parole.

Lapierre, Etienne - Boy - Number: 2646 - Prize name: Squirrel - Ship type: MV - How taken: HM Frigate Belle Poule - When taken: 14 Dec 1813 - Where taken: Bay of Biscay - Date received: 23 May 1814 - From what ship: Ashburton from parole - Born: Bayonne - Age: 16 - Discharged on 8 Jun 1814 and sent to France on HMS Rapid.

Lapierre, Etienne - Merchant & Passenger - Number: 2235 - Prize name: Squirrel - Ship type: MV - How taken: HM Frigate Belle Poule - When taken: 14 Dec 1813 - Where taken: Bay of Biscay - Date received: 15 Jan 1814 - From what ship: HMS Bellona - Born: Bayonne - Age: 16 - Discharged on 16 Jan 1814 and sent to Ashburton on parole.

Laquerene, Peter Louis - Passenger - Number: 1075 - Prize name: John & Frances - Ship type: MV - How taken: HM Frigate Belle Poule - When taken: 19 Mar 1813 - Where taken: Bay of Biscay - Date received: 22 Apr 1813 - From what ship: HMS Superb - Born: New York - Age: 18 - Sent on 25 Apr 1813 to Ashburton on parole.

Larrabee, William - Seaman - Number: 379 - Prize name: Orbit - Ship type: MV - How taken: HM Brig Achates - When taken: 29 Jan 1813 - Where taken: Lat 49 N Long 13 W - Date received: 4 Feb 1813 - From what ship: HMS Achates - Born: Portland - Age: 44 - Sent to Chatham on 29 Mar 1813 on the HMS Braham.

Latimore, Matthew - Seaman - Number: 1063 - Prize name: Meteor - Ship type: MV - How taken: HM Frigate Briton - When taken: 12 Mar 1813 - Where taken: Bay of Biscay - Date received: 22 Apr 1813 - From what ship: HMS Superb - Born: New York - Age: 19 - Sent to Dartmoor on 8 Sep 1813.

Latish, Joseph - Boy - Number: 9 - Prize name: Purse - Ship type: MV - How taken: HM Frigate Amide - When taken: 29 May 1812 - Where taken: off Bordeaux - Date received: 20 Nov 1812 - From what ship: HMS Salvador del Mundo - Born: Philadelphia - Age: 10 - Sent to Portsmouth on 29 Dec 1812 on the HMS Northumberland.

Laurencau, Jean - Boy - Number: 1574 - Prize name: Hannah & Eliza - Ship type: MV - How taken: HM Brig Lyra - When taken: 29 May 1813 - Where taken: off north coast of Spain - Date received: 10 May 1813 - From what ship: Hannah & Eliza - Born: Bordeaux - Age: 9 - Discharged on 8 Sep 1813 and sent to Dartmoor.

Laurnsberry, Benjamin - Servant - Number: 1854 - Prize name: US Brig Argus - Ship type: War - How taken: HM Brig Pelican - When taken: 14 Aug 1813 - Where taken: Irish Channel - Date received: 19 Aug 1813 - From what ship: USS Argus - Born: West Chester, NY - Age: 16 - Discharged on 8 Sep 1813 and sent to Dartmoor.

Law, John - Seaman - Number: 228 - Prize name: Vengeance - Ship type: LM - How taken: HM Frigate Phoebe - When taken: 1 Jan 1813 - Where taken: Lat 44.4 Long 23 - Date received: 9 Jan 1813 - From what ship: HMS Phoebe - Age: 23 - Sent to Portsmouth on 8 Feb 1813 on the HMS Colossus.

Law, William - Master - Number: 958 - Prize name: Independence - Ship type: MV - How taken: HM Ship-of-the-Line Superb - When taken: 16 Mar 1813 - Where taken: Bay of Biscay - Date received: 9 Apr 1813 - From what ship: HMS Plymouth - Born: Roxford - Age: 35 - Sent on 10 Apr 1813 to Ashburton on parole.

Lawrence, David - Seaman - Number: 447 - Prize name: Industry, prize to Privateer Decatur - Ship type: P - How taken: Channel Fleet - When taken: 24 Dec 1812 - Date received: 6 Feb 1813 - From what ship: Labrador - Born: Newburyport - Age: 18 - Sent to Chatham on 29 Mar 1813 on the HMS Braham.

Lawrence, John - Seaman - Number: 1207 - Prize name: Zebra - Ship type: LM - How taken: HM Frigate Pyramus - When taken: 20 Apr 1813 - Where taken: Bay of Biscay - Date received: 9 May 1813 - From what ship: HMS Andromache - Born: Isle du France - Age: 26 - Sent on 3 Jul 1813 to Stapleton prison.

Lawson, James - Steward - Number: 722 - Prize name: Mars - Ship type: MV - How taken: HM Ship-of-the-Line Warspite - When taken: 26 Feb 1813 - Where taken: Bay of Biscay - Date received: 19 Mar 1813 - From what ship: HMS Warspite - Born: Africa - Age: 27 - Race: Black - Sent to Dartmoor on 2 Apr 1813.

Lawson, Thomas - Seaman - Number: 734 - Prize name: Pert - Ship type: MV - How taken: HM Ship-of-the-Line Warspite - When taken: 1 Mar 1813 - Where taken: Bay of Biscay - Date received: 19 Mar 1813 - From what ship: HMS Warspite - Born: Stamford - Age: 36 - Sent to Dartmoor on 2 Apr 1813.

Layfield, Littleton - Seaman - Number: 1373 - Prize name: Tom - Ship type: LM - How taken: HM Frigate Surveillante

- When taken: 27 Apr 1813 - Where taken: Bay of Biscay - Date received: 15 May 1813 - From what ship: HMS Foxhound - Born: Maryland - Age: 20 - Discharged on 3 Jul 1813 and sent to Stapleton Prison.

Layton, William - Boy - Number: 942 - Prize name: Gleamer - Ship type: MV - How taken: Brothers, privateer - When taken: 26 Mar 1813 - Where taken: Bay of Biscay - Date received: 8 Apr 1813 - From what ship: Wasp, sloop from Guernsey - Born: East Town - Age: 18 - Sent to Chatham on 8 Jul 1813 on HM Tender Neptune.

Le Baith, John J. - Seaman - Number: 1090 - Prize name: Young Holkar - Ship type: MV - How taken: HM Ship-of-the-Line Superb - When taken: 10 Apr 1813 - Where taken: off Belle Isle - Date received: 22 Apr 1813 - From what ship: HMS Superb - Born: New Orleans - Age: 57 - Sent to Dartmoor on 1 Jul 1813.

Le Core, Peter Murry - Pilot - Number: 2333 - Prize name: Narius, prize of the Privateer Bunker Hill - Ship type: P - How taken: HM Brig Orestes - When taken: 20 Jan 1814 - Where taken: Bay of Biscay - Date received: 29 Jan 1814 - From what ship: HMS Orestes - Born: Roscoff, France - Age: 63 - Discharged on 31 Jan 1814 and sent to Dartmoor.

Le Graet, Peter - Passenger - Number: 1190 - Prize name: Zebra - Ship type: LM - How taken: HM Frigate Pyramus - When taken: 20 Apr 1813 - Where taken: Bay of Biscay - Date received: 9 May 1813 - From what ship: HMS Andromache - Born: Guadeloupe - Age: 28 - Sent on 11 May 1813 to Ashburton on parole.

Leach, Charles - Sailing Master - Number: 143 - Prize name: Hunter - Ship type: P - How taken: HM Frigate Phoebe - When taken: 24 Dec 1812 - Where taken: off Western Islands - Date received: 9 Jan 1813 - From what ship: HMS Phoebe - Born: Salem - Age: 26 - Sent to Portsmouth on 8 Feb 1813 on the HMS Colossus.

Leach, Daniel - Seaman - Number: 1844 - How taken: Impressed at Cork - When taken: 29 Jul 1813 - Date received: 17 Aug 1813 - From what ship: HMS Prince Frederick - Born: Massachusetts - Age: 24 - Discharged on 2 Sep 1813 and released to HMS Salvador del Mundo.

Leach, Josem - Seaman - Number: 1894 - Prize name: US Brig Argus - Ship type: War - How taken: HM Brig Pelican - When taken: 14 Aug 1813 - Where taken: Irish Channel - Date received: 23 Aug 1813 - From what ship: HMS Pelican - Born: Onondaga - Age: 34 - Discharged on 8 Sep 1813 and sent to Dartmoor.

Leach, William - Seaman - Number: 772 - Prize name: William Bayard - Ship type: MV - How taken: HM Ship-of-the-Line Warspite - When taken: 3 Mar 1813 - Where taken: Bay of Biscay - Date received: 19 Mar 1813 - From what ship: HMS Warspite - Born: Newburgh - Age: 19 - Sent to Dartmoor on 28 Jun 1813.

Leach, William - Seaman - Number: 536 - Prize name: Terrible - Ship type: MV - How taken: HM Brig Foxhound - When taken: 8 Feb 1813 - Where taken: Channel - Date received: 15 Feb 1813 - From what ship: HMS Foxhound - Born: Derby - Age: 20 - Sent to Dartmoor on 2 Apr 1813.

Leary, Daniel - Marine - Number: 1866 - Prize name: US Brig Argus - Ship type: War - How taken: HM Brig Pelican - When taken: 14 Aug 1813 - Where taken: Irish Channel - Date received: 23 Aug 1813 - From what ship: HMS Pelican - Born: New York - Age: 23 - Discharged on 24 Aug 1813 and released to HMS Salvador del Mundo.

Leary, Daniel - Seaman - Number: 2426 - Prize name: US Brig Argus - Ship type: War - How taken: Sent from HM Ship-of-the-Line Salvador del Mundo - When taken: 14 Aug 1813 - Where taken: Irish Channel - Date received: 12 Feb 1814 - From what ship: HMS Salvador del Mundo - Born: New York - Age: 24 - Discharged on 10 May 1814 and sent to Dartmoor.

Leas, Anthony - Seaman - Number: 1022 - Prize name: Two Brothers - Ship type: MV - How taken: Bootle of Liverpool, letter of marque - When taken: 18 Mar 1813 - Where taken: Western Islands - Date received: 21 Apr 1813 - From what ship: HMS Bittern - Born: New Orleans - Age: 30 - Sent to Dartmoor on 1 Jul 1813.

Lebarth, John - Seaman - Number: 2336 - Prize name: Young Holkar - Ship type: MV - How taken: HM Ship-of-the-Line Superb - When taken: 10 Apr 1813 - Where taken: off Belle Isle - Date received: 29 Jan 1814 - From what ship: Dartmoor prison - Born: New Orleans - Age: 20 - Discharged on 3 Feb 1814 and sent to Dartmoor.

Lechler, Anthony - Seaman - Number: 2687 - How taken: Sent into custody from HM Ship-of-the-Line Cornwallis - Date received: 7 Jun 1814 - From what ship: Inap - Born: Lancaster - Age: 26 - Discharged on 14 Jun 1814 and sent to Dartmoor.

Lee, Ely - Seaman - Number: 1830 - Prize name: US Brig Argus - Ship type: War - How taken: HM Brig Pelican - When taken: 14 Aug 1813 - Where taken: Irish Channel - Date received: 17 Aug 1813 - From what ship: USS

Argus - Born: Connecticut - Age: 34 - Discharged on 8 Sep 1813 and sent to Dartmoor.

Lee, George - Seaman - Number: 965 - Prize name: Ferox - Ship type: MV - How taken: HM Frigate Medusa & HM Brig Lyra - When taken: 28 Mar 1813 - Where taken: off Cape Ortagle - Date received: 9 Apr 1813 - From what ship: HMS Lyra - Born: New York - Age: 33 - Sent to Chatham on 8 Jul 1813 on HM Tender Neptune.

Lee, Nathaniel - Seaman - Number: 185 - Prize name: Hunter - Ship type: P - How taken: HM Frigate Phoebe - When taken: 24 Dec 1812 - Where taken: off Western Islands - Date received: 9 Jan 1813 - From what ship: HMS Phoebe - Born: Massachusetts - Age: 48 - Sent to Portsmouth on 8 Feb 1813 on the HMS Colossus.

Leeds, Leonard - Seaman - Number: 513 - Prize name: Tyger - Ship type: MV - How taken: Detained at Gibraltar - When taken: 8 Aug 1812 - Date received: 15 Feb 1813 - From what ship: HMS Andromeda - Born: Boston - Age: 18 - Sent to Dartmoor on 2 Apr 1813.

Leger, Andrew - Seaman - Number: 2326 - Prize name: Amity, prize of the Privateer Prince de Neufchatel - Ship type: P - How taken: HM Brig Achates - When taken: 22 Dec 1814 - Where taken: Bay of Biscay - Date received: 25 Jan 1814 - From what ship: HMS Conflict - Born: New Orleans - Age: 23 - Discharged on 31 Jan 1814 and sent to Dartmoor.

Leggett, William - Seaman - Number: 2507 - Prize name: Bunker Hill - Ship type: P - How taken: HM Frigate Pomone & HM Frigate Cydnus - When taken: 4 Mar 1814 - Where taken: Bay of Biscay - Date received: 4 Apr 1814 - From what ship: HMS Virago - Born: Boston - Age: 21 - Discharged on 10 May 1814 and sent to Dartmoor.

Legrand, Claudius Francis - Passenger - Number: 389 - Prize name: Union - Ship type: MV - How taken: HM Frigate Iris - When taken: 17 Jan 1813 - Where taken: Lat 44 N Long 2.3 W - Date received: 5 Feb 1813 - From what ship: HMS San Josef - Born: Philadelphia - Age: 27 - Sent to Chatham on 29 Mar 1813 on the HMS Braham.

Lehart, Jacob - Seaman - Number: 2151 - Prize name: Agnes, prize of the Privateer Rambler - Ship type: LM - How taken: Cutter Jane - When taken: 28 Nov 1813 - Where taken: Bay of Biscay - Date received: 10 Dec 1813 - From what ship: Cutter Jane - Born: Cape Cod - Age: 21 - Discharged on 31 Jan 1814 and sent to Dartmoor.

Lellifer, Jacob - Cook - Number: 860 - Prize name: Dick - Ship type: MV - How taken: HM Brig Dispatch - When taken: 17 Mar 1813 - Where taken: Bay of Biscay - Date received: 26 Mar 1813 - From what ship: HMS Dispatch - Born: Pennsylvania - Age: 35 - Race: Black - Sent to Dartmoor on 28 Jun 1813.

Lemercier, Peter - Seaman - Number: 810 - Prize name: Decornau - Ship type: MV - How taken: HM Sloop Pheasant - When taken: 15 Mar 1813 - Where taken: Bay of Biscay - Date received: 19 Mar 1813 - From what ship: HMS Pheasant - Born: Boston - Age: 25 - Sent to Dartmoor on 28 Jun 1813.

Lemmon, Henry - Seaman - Number: 925 - Prize name: Weasel - Ship type: MV - How taken: HM Brig Foxhound - When taken: 25 Mar 1813 - Where taken: Bay of Biscay - Date received: 6 Apr 1813 - From what ship: HMS Foxhound -

Born: Savannah - Age: 22 - Sent on 21 Jun 1813 to Portsmouth on HMS Prometheus.

Lemond, John - Seaman - Number: 1606 - Prize name: Revenge - Ship type: LM - How taken: HM Frigate Belle Poule - When taken: 11 May 1813 - Where taken: Bay of Biscay - Date received: 14 Jun 1813 - From what ship: HMS Royalist - Born: Philadelphia - Age: 37 - Discharged on 30 Jun 1813 and sent to Stapleton.

Lendison, Henry - Carpenter - Number: 50 - How taken: HM Battery Princess - When taken: 26 Oct 1812 - Where taken: Liverpool - Date received: 26 Nov 1813 - From what ship: HMS Frederick - Born: New York - Age: 30 - Sent to Portsmouth on 29 Dec 1812 on the HMS Northumberland.

Lepberg, John P. - Seaman - Number: 2369 - Prize name: Minerva - Ship type: MV - How taken: HM Ship-of-the-Line Conquestador - When taken: 19 Dec 1814 - Where taken: Bay of Biscay - Date received: 31 Jan 1814 - From what ship: HMS Surveillante - Born: Gothenburg, Sweden - Age: 20 - Discharged on 10 May 1814 and sent to Dartmoor.

Lerna, John - Seaman - Number: 1553 - Prize name: Good Friends - Ship type: MV - How taken: HM Frigate Andromache - When taken: 2 Apr 1813 - Where taken: Bay of Biscay - Date received: 29 May 1813 - From what ship: HMS Hannibal - Born: Leghorn, Italy - Age: 28 - Discharged on 30 Jun 1813 and sent to Stapleton.

Leron, Alexander - Seaman - Number: 1314 - How taken: Delivered himself up from HM Ship-of-the-Line Clarence - Date received: 11 May 1813 - From what ship: HMS Clarence - Born: Boston - Age: 20 - Discharged on 8 Jul

1813 and sent to Chatham on tender Neptune.

LeRoy, Alexander - Seaman - Number: 801 - Prize name: Cannoniere - Ship type: P - How taken: HM Ship-of-the-Line Warspite - When taken: 14 Mar 1813 - Where taken: Bay of Biscay - Date received: 19 Mar 1813 - From what ship: HMS Warspite - Born: Rhode Island - Age: 18 - Sent to Dartmoor on 28 Jun 1813.

Lerves, Robert - Seaman - Number: 1372 - Prize name: Tom - Ship type: LM - How taken: HM Frigate Surveillante - When taken: 27 Apr 1813 - Where taken: Bay of Biscay - Date received: 15 May 1813 - From what ship: HMS Foxhound - Born: Rhode Island - Age: 23 - Discharged on 3 Jul 1813 and sent to Stapleton Prison.

Lerwick, John - Seaman - Number: 2543 - Prize name: Indostan - Ship type: MV - How taken: HM Brig Zenobia - When taken: 25 Jun 1813 - Where taken: off Lisbon - Date received: 16 Apr 1814 - From what ship: Transport Fanny - Born: Virginia - Age: 23 - Discharged on 10 May 1814 and sent to Dartmoor.

Lessalle, Francisco - Seaman - Number: 575 - Prize name: Rolla - Ship type: MV - How taken: HM Frigate Surveillante - When taken: 11 Feb 1813 - Where taken: Bay of Biscay - Date received: 23 Feb 1813 - From what ship: HMS Surveillante - Born: San Sebastian - Age: 22 - Sent to Dartmoor on 2 Apr 1813.

Lessalle, Francisco - Seaman - Number: 1739 - Prize name: Rolla - Ship type: MV - How taken: HM Frigate Surveillante - When taken: 11 Feb 1813 - Where taken: Bay of Biscay - Date received: 10 Jul 1813 - From what ship: Dartmoor Prison - Born: San Sebastian, Spain - Age: 22 - Discharged on 10 Jul 1813 and released to HMS Salvador del Mundo.

Lettington, James - Seaman - Number: 2004 - Prize name: Francis & Ann - Ship type: MV - How taken: HM Sloop Lightning - When taken: 20 Aug 1813 - Where taken: at sea - Date received: 22 Sep 1813 - From what ship: HMS Lightning - Born: Jersey - Age: 25 - Discharged on 27 Sep 1813 and sent to Dartmoor.

Leveatt, William - Surgeon's Mate - Number: 154 - Prize name: Hunter - Ship type: P - How taken: HM Frigate Phoebe - When taken: 24 Dec 1812 - Where taken: off Western Islands - Date received: 9 Jan 1813 - From what ship: HMS Phoebe - Born: Salisbury - Age: 21 - Sent to Portsmouth on 8 Feb 1813 on the HMS Colossus.

Leverage, William - Seaman - Number: 1133 - Prize name: Magdalen - Ship type: MV - How taken: HM Ship-of-the-Line Superb - When taken: 15 Apr 1813 - Where taken: off Belle Isle - Date received: 22 Apr 1813 - From what ship: HMS Superb - Born: Pennsylvania - Age: 19 - Sent to Dartmoor on 1 Jul 1813.

Levi, Uriah Phillips - Supernumery Master - Number: 1859 - Prize name: US Brig Argus - Ship type: War - How taken: HM Brig Pelican - When taken: 14 Aug 1813 - Where taken: Irish Channel - Date received: 23 Aug 1813 - From what ship: HMS Pelican - Born: Philadelphia - Age: 23 - Discharged on 25 Aug 1813 and sent to Ashburton on parole.

Levis, Raymond - Seaman - Number: 182 - Prize name: Hunter - Ship type: P - How taken: HM Frigate Phoebe - When taken: 24 Dec 1812 - Where taken: off Western Islands - Date received: 9 Jan 1813 - From what ship: HMS Phoebe - Born: Eastham - Age: 20 - Sent to Portsmouth on 8 Feb 1813 on the HMS Colossus.

Lewis, Angus T. - Master - Number: 558 - Prize name: Terrible - Ship type: MV - How taken: HM Brig Foxhound - When taken: 8 Feb 1813 - Where taken: Channel - Date received: 17 Feb 1813 - From what ship: HMS Foxhound - Born: Stratford, CT - Age: 35 - Sent on 18 Feb 1813 to Ashburton on parole.

Lewis, Edward - Steward - Number: 847 - Prize name: Pallas - Ship type: MV - How taken: HM Brig Rebuff - When taken: 23 Dec 1812 - Where taken: off Cadiz - Date received: 23 May 1813 - From what ship: HMS Dauntless - Born: Philadelphia - Age: 30 - Race: Black - Sent to Dartmoor on 28 Jun 1813.

Lewis, Francis B. - Seaman - Number: 1138 - Prize name: Magdalen - Ship type: MV - How taken: HM Ship-of-the-Line Superb - When taken: 15 Apr 1813 - Where taken: off Belle Isle - Date received: 22 Apr 1813 - From what ship: HMS Superb - Born: New York - Age: 34 - Sent to Dartmoor on 1 Jul 1813.

Lewis, George - Seaman - Number: 377 - Prize name: Orbit - Ship type: MV - How taken: HM Brig Achates - When taken: 29 Jan 1813 - Where taken: Lat 49 N Long 13 W - Date received: 4 Feb 1813 - From what ship: HMS Achates - Born: Delaware - Age: 25 - Race: Black - Sent to Chatham on 29 Mar 1813 on the HMS Braham.

Lewis, James - 2nd Lieutenant & Prize Master - Number: 2526 - Prize name: Maria Christiana, prize to the Privateer True Blooded Yankee - Ship type: P - How taken: Pactolus - When taken: 25 Mar 1814 - Where taken: coast of France - Date received: 6 Apr 1814 - From what ship: HMS Queen Charlotte - Born: Southey Town - Age: 30 -

Discharged on 10 May 1814 and sent to Dartmoor.

Lewis, John - Seaman - Number: 759 - Prize name: William Bayard - Ship type: MV - How taken: HM Ship-of-the-Line Warspite - When taken: 3 Mar 1813 - Where taken: Bay of Biscay - Date received: 19 Mar 1813 - From what ship: HMS Warspite - Born: New York - Age: 20 - Race: Black - Sent to Dartmoor on 2 Apr 1813.

Lewis, John - Seaman - Number: 715 - Prize name: Star - Ship type: MV - How taken: HM Ship-of-the-Line Superb - When taken: 9 Feb 1813 - Where taken: Bay of Biscay - Date received: 19 Mar 1813 - From what ship: HMS Warspite - Born: Staten Island - Age: 18 - Sent to Dartmoor on 2 Apr 1813.

Lewis, John - Seaman - Number: 61 - Prize name: Independence - Ship type: MV - How taken: HM Frigate Medusa - When taken: 9 Nov 1812 - Where taken: San Sebastian - Date received: 27 Nov 1812 - From what ship: HMS Wasp - Born: Maryland - Age: 25 - Sent to Portsmouth on 29 Dec 1812 on the HMS Northumberland.

Lewis, John - Cook - Number: 266 - Prize name: Brutus - Ship type: MV - How taken: HM Frigate Briton - When taken: Jan 1813 - Where taken: Bay of Biscay - Date received: 21 Jan 1813 - From what ship: HMS Briton - Born: New Orleans - Age: 45 - Race: Black - Sent to Portsmouth on 8 Feb 1813 on the HMS Colossus.

Lewis, John - Seaman - Number: 2491 - Prize name: Sherbrook - Ship type: MV - How taken: Impressed at Falmouth - When taken: 20 Mar 1814 - Date received: 26 Mar 1814 - From what ship: HMS Prince Frederick - Born: Rhode Island - Age: 26 - Race: Mulatto - Discharged on 10 May 1814 and sent to Dartmoor.

Lewis, Thomas - Boatswain - Number: 1989 - Prize name: Ned - Ship type: LM - How taken: HM Brig Royalist - When taken: 6 Sep 1812 - Where taken: coast of France - Date received: 22 Sep 1813 - From what ship: HMS Royalist - Born: Norfolk - Age: 35 - Discharged on 27 Sep 1813 and sent to Dartmoor.

Ley, John - Seaman - Number: 2587 - Prize name: James - Ship type: MV - How taken: HM Brig Harpy - When taken: 19 Dec 1812 - Where taken: off Isle du France - Date received: 10 May 1814 - From what ship: HMS Lion - Born: Philadelphia - Age: 19 - Discharged on 14 Jun 1814 and sent to Dartmoor.

Libby, Moses - Seaman - Number: 1243 - Prize name: Essex - Ship type: MV - How taken: HM Frigate Pyramus - When taken: 2 Apr 1813 - Where taken: Bay of Biscay - Date received: 9 May 1813 - From what ship: HMS Andromache - Born: New Hampshire - Age: 28 - Discharged on 3 Jul 1813 and sent to Stapleton Prison.

Ligan, Pierre - Seaman - Number: 1475 - Prize name: Paul Jones - Ship type: P - How taken: HM Frigate Leonidas - When taken: 23 May 1813 - Where taken: off Cape Clear - Date received: 26 May 1813 - From what ship: HMS Leonidas - Born: Nantes - Age: 28 - Discharged on 30 Jun 1813 and sent to Stapleton.

Likens, Andrew - Seaman - Number: 614 - Prize name: Governor McKean - Ship type: LM - How taken: HM Brig Rover - When taken: 26 Jun 1813 - Where taken: off Bordeaux - Date received: 2 Mar 1813 - From what ship: HMS Insolent - Born: Medford - Age: 25 - Sent to Dartmoor on 2 Apr 1813.

Lilley, Simon - Seaman - Number: 1325 - Prize name: Shadow - Ship type: LM - How taken: HM Brig Reindeer & HM Schooner Helicon - When taken: 6 Apr 1813 - Where taken: Bay of Biscay - Date received: 12 May 1813 - From what ship: HMS Helicon - Born: Massachusetts - Age: 19 - Discharged on 3 Jul 1813 and sent to Stapleton Prison.

Limonden, William - Seaman - Number: 2079 - Prize name: captured French frigate - How taken: HM Ship-of-the-Line Rippon, HM Brig Scylla & HM Brig Royalist - When taken: 21 Oct 1813 - Where taken: off Ushant - Date received: 31 Oct 1813 - From what ship: HMS Rippon - Discharged on 29 Nov 1813 and sent to Dartmoor.

Lindsborough, Charles - Seaman - Number: 2024 - Prize name: Friendship West, prize of Privateer True Blooded Yankee - Ship type: P - How taken: HM Schooner Helicon & HM Schooner Whiting - When taken: 25 Oct 1813 - Where taken: Bay of Biscay - Date received: 31 Oct 1813 - From what ship: HMS Whiting - Born: Gotland, Sweden - Age: 36 - Discharged on 3 Nov 1813 and sent to Dartmoor.

Lingrin, Peter - Seaman - Number: 2257 - Prize name: US Frigate Chesapeake - Ship type: War - How taken: HM Frigate Shannon - When taken: 1 Jul 1813 - Where taken: off Boston - Date received: 15 Jan 1814 - From what ship: HMS Teazer - Born: Lisbon - Age: 33 - Discharged on 31 Jan 1814 and sent to Dartmoor.

Linsey, Alexander - Seaman - Number: 1695 - Prize name: Governor Gerry - Ship type: MV - How taken: HM Brig Royalist - When taken: 31 May 1813 - Where taken: Bay of Biscay - Date received: 26 Jun 1813 - From what ship: HMS Duncan - Born: Baltimore - Age: 16 - Discharged on 30 Jun 1813 and sent to Stapleton.

Lippart, Thomas D. - Prize Master - Number: 1415 - Prize name: Paul Jones - Ship type: P - How taken: HM Frigate Leonidas - When taken: 23 May 1813 - Where taken: off Cape Clear - Date received: 26 May 1813 - From what ship: HMS Leonidas - Born: Pennsylvania - Age: 49 - Discharged on 8 Sep 1813 and sent to Dartmoor.

Littlefield, Rufus - Seaman - Number: 1570 - Prize name: Miranda, prize of the Privateer Paul Jones - Ship type: P - How taken: HM Frigate Unicorn - When taken: 21 Mar 1813 - Where taken: off Ushant - Date received: 9 Jun 1813 - From what ship: HMS Conquestador - Born: Massachusetts - Age: 17 - Discharged on 30 Jun 1813 and sent to Stapleton.

Little, George - Sailing Master - Number: 1498 - Prize name: Paul Jones - Ship type: P - How taken: HM Frigate Leonidas - When taken: 23 May 1813 - Where taken: off Cape Clear - Date received: 27 May 1813 - From what ship: HMS Leonidas - Born: Roxbury, MA - Age: 25 - Discharged on 30 Jun 1813 and sent to Stapleton.

Little, John - Seaman - Number: 1772 - Prize name: Minerva - Ship type: MV - How taken: HM Brig Goldfinch - When taken: 27 Jun 1813 - Where taken: off Nantes - Date received: 25 Jul 1813 - From what ship: HMS Pyramus - Born: Philadelphia - Age: 29 - Discharged on 8 Sep 1813 and sent to Dartmoor.

Livermore, Arthur - Seaman - Number: 2025 - Prize name: Friendship West, prize of Privateer True Blooded Yankee - Ship type: P - How taken: HM Schooner Helicon & HM Schooner Whiting - When taken: 25 Oct 1813 - Where taken: Bay of Biscay - Date received: 31 Oct 1813 - From what ship: HMS Whiting - Born: Rockingham - Age: 22 - Discharged on 3 Nov 1813 and sent to Dartmoor.

Lockwood, Rufus - Clerk - Number: 1417 - Prize name: Paul Jones - Ship type: P - How taken: HM Frigate Leonidas - When taken: 23 May 1813 - Where taken: off Cape Clear - Date received: 26 May 1813 - From what ship: HMS Leonidas - Born: Connecticut - Age: 22 - Discharged on 3 Jul 1813 and sent to Stapleton Prison.

Logan, William - Seaman - Number: 1533 - Prize name: Courier - Ship type: LM - How taken: HM Brig Rover - When taken: 14 Mar 1813 - Where taken: Bay of Biscay - Date received: 29 May 1813 - From what ship: HMS Hannibal - Born: Baltimore - Age: 23 - Discharged on 30 Jun 1813 and sent to Stapleton.

Long, Charles - Seaman - Number: 2477 - Prize name: Squirrel - Ship type: MV - How taken: HM Frigate Belle Poule - When taken: 14 Dec 1813 - Where taken: Bay of Biscay - Date received: 28 Feb 1814 - From what ship: Dartmoor prison - Born: Swinehunde - Age: 21 - Discharged on 4 Mar 1814 and sent to Harwich for HMS Rosemond.

Long, Charles - Seaman - Number: 2214 - Prize name: Squirrel - Ship type: MV - How taken: HM Frigate Belle Poule - When taken: 14 Dec 1813 - Where taken: Bay of Biscay - Date received: 27 Dec 1813 - From what ship: HMS Protector - Born: Swinehunde - Age: 21 - Discharged on 31 Jan 1814 and sent to Dartmoor.

Long, Joseph - Seaman - Number: 2669 - Prize name: Traveler, prize of the Privateer Surprize - Ship type: P - How taken: HMS Cawser - When taken: 7 May 1814 - Where taken: off Cape Clear - Date received: 5 Jun 1814 - From what ship: HMS Gronville - Born: Baltimore - Age: 36 - Discharged on 14 Jun 1814 and sent to Dartmoor.

Long, R. S. - Prize Master - Number: 1497 - Prize name: Paul Jones - Ship type: P - How taken: HM Frigate Leonidas - When taken: 23 May 1813 - Where taken: off Cape Clear - Date received: 27 May 1813 - From what ship: HMS Leonidas - Born: Charlestown - Age: 30 - Discharged on 8 Sep 1813 and sent to Dartmoor.

Longfield, Thomas - Chief Mate - Number: 939 - Prize name: Weasel - Ship type: MV - How taken: HM Brig Foxhound - When taken: 25 Mar 1813 - Where taken: Bay of Biscay - Date received: 7 Apr 1813 - From what ship: HMS Foxhound - Born: New York - Age: 23 - Sent on 8 Apr 1813 to Ashburton on parole.

Longford, Samuel - Seaman - Number: 1307 - Prize name: Eliza - Ship type: MV - How taken: HM Frigate Surveillante - When taken: 22 Apr 1813 - Where taken: at sea - Date received: 10 May 1813 - From what ship: HMS Medusa - Born: Maryland - Age: 23 - Discharged on 3 Jul 1813 and sent to Stapleton Prison.

Longworthy, John - Seaman - Number: 579 - Prize name: Cashiere - Ship type: LM - How taken: HM Brig Reindeer - When taken: 3 Feb 1813 - Where taken: Bay of Biscay - Date received: 23 Feb 1813 - From what ship: HMS Surveillante - Born: Newburn - Age: 41 - Sent to Dartmoor on 2 Apr 1813.

Lopaus, William - Seaman - Number: 906 - Prize name: Prompt - Ship type: MV - How taken: Chance, privateer - When taken: 22 Mar 1813 - Where taken: Bay of Biscay - Date received: 3 Apr 1813 - From what ship: Mary - Born: Boston - Age: 21 - Sent on 21 Jun 1813 to Portsmouth on HMS Prometheus.

Lopez, Joseph - Seaman - Number: 2528 - Prize name: Maria Christiana, prize to the Privateer True Blooded Yankee

- Ship type: P - How taken: Pactolus - When taken: 25 Mar 1814 - Where taken: coast of France - Date received: 6 Apr 1814 - From what ship: HMS Queen Charlotte - Born: Malaga - Age: 20 - Discharged on 10 May 1814 and sent to Dartmoor.

Lorenze, Thomas - Seaman - Number: 2631 - Prize name: Ateline - Ship type: MV - How taken: HM Frigate Magiciene - When taken: 14 Mar 1814 - Where taken: off Cape Finisterre, Spain - Date received: 17 May 1814 - From what ship: HMS Tortois - Born: Not readable - Age: 17 - Race: Black - Discharged on 14 Jun 1814 and sent to Dartmoor.

Lothrop, James - Seaman - Number: 1615 - Prize name: Revenge - Ship type: LM - How taken: HM Frigate Belle Poule - When taken: 11 May 1813 - Where taken: Bay of Biscay - Date received: 14 Jun 1813 - From what ship: HMS Royalist - Born: Boston - Age: 26 - Discharged on 30 Jun 1813 and sent to Stapleton.

Louis, Nicholas - Seaman - Number: 1438 - Prize name: Paul Jones - Ship type: P - How taken: HM Frigate Leonidas - When taken: 23 May 1813 - Where taken: off Cape Clear - Date received: 26 May 1813 - From what ship: HMS Leonidas - Born: Portsmouth, NH - Age: 43 - Discharged on 30 Jun 1813 and sent to Stapleton.

Louring, Henry - Seaman - Number: 709 - Prize name: Star - Ship type: MV - How taken: HM Ship-of-the-Line Superb - When taken: 9 Feb 1813 - Where taken: Bay of Biscay - Date received: 19 Mar 1813 - From what ship: HMS Warspite - Born: Norfolk - Age: 35 - Sent to Dartmoor on 2 Apr 1813.

Lovely, Henry - Master - Number: 2327 - Prize name: Siro - Ship type: LM - How taken: HM Brig Pelican - When taken: 13 Jan 1814 - Where taken: at sea - Date received: 27 Jan 1814 - From what ship: HMS Pelican - Born: Baltimore - Age: 29 - Discharged on 27 Jan 1814 and sent to Ashburton on parole.

Lovering, William - Seaman - Number: 465 - Prize name: Dolphin - Ship type: LM - How taken: HM Ship-of-the-Line Colossus, HM Frigate Rhin & HM Brig Goldfinch - When taken: 5 Jan 1813 - Where taken: off Western Islands - Date received: 6 Feb 1813 - From what ship: HMS Rhin - Born: Pennsylvania - Age: 21 - Sent to Chatham on 29 Mar 1813 on the HMS Braham.

Lovet, John - 2nd Mate - Number: 529 - Prize name: Terrible - Ship type: MV - How taken: HM Brig Foxhound - When taken: 8 Feb 1813 - Where taken: Channel - Date received: 15 Feb 1813 - From what ship: HMS Foxhound - Born: Newbury - Age: 40 - Sent to Dartmoor on 2 Apr 1813.

Lovett, Henry - Servant - Number: 1855 - Prize name: US Brig Argus - Ship type: War - How taken: HM Brig Pelican - When taken: 14 Aug 1813 - Where taken: Irish Channel - Date received: 19 Aug 1813 - From what ship: USS Argus - Born: Providence - Age: 20 - Race: Negro - Discharged on 8 Sep 1813 and sent to Dartmoor.

Lovett, Thomas L. - Master - Number: 1663 - Prize name: Flash - Ship type: MV - How taken: HM Ship-of-the-Line Warspite - When taken: 23 May 1813 - Where taken: coast of France - Date received: 16 Jun 1813 - From what ship: Flash - Born: Massachusetts - Age: 30 - Discharged on 19 Jun 1813 and sent to Ashburton on parole.

Lovett, William - Seaman - Number: 555 - Prize name: Spitfire - Ship type: MV - How taken: HM Brig Achates - When taken: 14 Feb 1813 - Where taken: off Ushant - Date received: 16 Feb 1813 - From what ship: HMS Achates - Born: Marblehead - Age: 21 - Sent to Dartmoor on 2 Apr 1813.

Low, Thomas - Seaman - Number: 454 - Prize name: Dolphin - Ship type: LM - How taken: HM Ship-of-the-Line Colossus, HM Frigate Rhin & HM Brig Goldfinch - When taken: 5 Jan 1813 - Where taken: off Western Islands - Date received: 6 Feb 1813 - From what ship: HMS Rhin - Born: Delaware - Age: 29 - Sent to Chatham on 29 Mar 1813 on the HMS Braham.

Lowe, John - Seaman - Number: 1245 - Prize name: Courier - Ship type: LM - How taken: HM Frigate Andromache - When taken: 14 Mar 1813 - Where taken: Bay of Biscay - Date received: 9 May 1813 - From what ship: HMS Andromache - Born: Baltimore - Age: 18 - Discharged on 3 Jul 1813 and sent to Stapleton Prison.

Lownsburg, Carpenter - Carpenter - Number: 399 - Prize name: Union - Ship type: MV - How taken: HM Frigate Iris - When taken: 17 Jan 1813 - Where taken: Lat 44 N Long 2.3 W - Date received: 5 Feb 1813 - From what ship: HMS San Josef - Born: Delaware - Age: 45 - Sent to Chatham on 29 Mar 1813 on the HMS Braham.

Luce, Prispery - Seaman - Number: 2027 - Prize name: Friendship West, prize of Privateer True Blooded Yankee - Ship type: P - How taken: HM Schooner Helicon & HM Schooner Whiting - When taken: 25 Oct 1813 - Where taken: Bay of Biscay - Date received: 31 Oct 1813 - From what ship: HMS Whiting - Born: Massachusetts - Age: 21 - Discharged on 3 Nov 1813 and sent to Dartmoor.

Ludlow, Charles - Boy - Number: 649 - Prize name: Criterion - Ship type: MV - How taken: HM Frigate Belle Poule - When taken: 14 Feb 1813 - Where taken: Bay of Biscay - Date received: 4 Mar 1813 - From what ship: HMS Strenuous - Born: New Jersey - Age: 17 - Sent to Dartmoor on 2 Apr 1813.

Ludlow, Reuben - Seaman - Number: 889 - Prize name: Tiger - Ship type: MV - How taken: HM Brig Scylla - When taken: 22 Mar 1813 - Where taken: Bay of Biscay - Date received: 1 Apr 1813 - From what ship: HMS Scylla - Born: Philadelphia - Age: 33 - Sent on 21 Jun 1813 to Portsmouth on HMS Prometheus.

Ludson, Daniel - Seaman - Number: 1555 - Prize name: Elvia - Ship type: MV - How taken: HM Frigate Surveillante - When taken: 21 Mar 1813 - Where taken: Bay of Biscay - Date received: 29 May 1813 - From what ship: HMS Hannibal - Born: Rhode Island - Age: 28 - Discharged on 30 Jun 1813 and sent to Stapleton.

Lunberge, Berg - Seaman - Number: 2445 - Prize name: Mary, prize of the Privateer Prince de Neufchatel - Ship type: P - How taken: Retaken by the crew of the Mary - Date received: 15 Feb 1814 - From what ship: HMS Kangaroo - Born: Gothenburg, Sweden - Age: 22 - Discharged on 10 May 1814 and sent to Dartmoor.

Lund, Peter M. - Seaman - Number: 1727 - Prize name: Hannah & Eliza - Ship type: MV - How taken: HM Brig Lyra - When taken: 29 May 1813 - Where taken: off north coast of Spain - Date received: 4 Jul 1813 - From what ship: HMS Iris - Born: Gelfia, Sweden - Age: 26 - Liberated on 3 Sep 1813.

Lymones, Procter - Seaman - Number: 2182 - Prize name: Wolf, prize of the Privateer Grand Turk - Ship type: P - How taken: HM Frigate Briton - When taken: 1 Dec 1813 - Where taken: Bay of Biscay - Date received: 17 Dec 1813 - From what ship: HMS Briton - Born: Not readable - Age: 16 - Discharged on 17 Dec 1813 and released to HMS Britton.

Lynch, Joseph - Seaman - Number: 217 - Prize name: Vengeance - Ship type: LM - How taken: HM Frigate Phoebe - When taken: 1 Jan 1813 - Where taken: Lat 44.4 Long 23 - Date received: 9 Jan 1813 - From what ship: HMS Phoebe - Born: Philadelphia - Age: 35 - Sent to Portsmouth on 8 Feb 1813 on the HMS Colossus.

Lynch, William - Seaman - Number: 97 - Prize name: Experiment - Ship type: MV - How taken: HM Brig Rover - When taken: 21 Oct 1812 - Where taken: off Bordeaux - Date received: 25 Dec 1812 - From what ship: HMS Northumberland - Born: Maryland - Age: 21 - Sent to Portsmouth on 29 Dec 1812 on the HMS Northumberland.

Lyon, Charles - Seaman - Number: 1206 - Prize name: Zebra - Ship type: LM - How taken: HM Frigate Pyramus - When taken: 20 Apr 1813 - Where taken: Bay of Biscay - Date received: 9 May 1813 - From what ship: HMS Andromache - Born: New Jersey - Age: 21 - Sent on 3 Jul 1813 to Stapleton prison.

Lyon, Charles - Seaman - Number: 2181 - Prize name: Wolf, prize of the Privateer Grand Turk - Ship type: P - How taken: HM Frigate Briton - When taken: 1 Dec 1813 - Where taken: Bay of Biscay - Date received: 17 Dec 1813 - From what ship: HMS Briton - Born: Marblehead - Age: 19 - Discharged on 17 Dec 1813 and released to HMS Britton.

Lyon, John - Supercargo - Number: 2455 - Prize name: Fair American - Ship type: MV - How taken: HM Frigate Andromache - When taken: 19 Jan 1814 - Where taken: Bay of Biscay - Date received: 22 Feb 1814 - From what ship: HMS York - Born: Fredericksburg, VA - Age: 33 - Discharged on 24 Feb 1814 and sent to Ashburton on parole.

Lyons, Henry - Seaman - Number: 256 - Prize name: Ocean, prized to the Privateer Diligent - Ship type: P - How taken: HM Frigate Surveillante - When taken: 20 Dec 1812 - Where taken: Lat 44 N, Long 6 W - Date received: 21 Jan 1813 - From what ship: HMS Ocean - Born: New Jersey - Age: 18 - Sent to Portsmouth on 8 Feb 1813 on the HMS Colossus.

Lyons, Peter - Seaman - Number: 320 - Prize name: Porcupine - Ship type: MV - How taken: HM Frigate Dryad - When taken: 8 Jan 1813 - Where taken: off Bordeaux - Date received: 21 Jan 1813 - From what ship: HMS Abercrombie - Born: Baltimore - Age: 37 - Sent to Chatham on 29 Mar 1813 on the HMS Braham.

Lyons, Samuel - Seaman - Number: 332 - Prize name: Porcupine - Ship type: MV - How taken: HM Frigate Dryad - When taken: 8 Jan 1813 - Where taken: off Bordeaux - Date received: 21 Jan 1813 - From what ship: HMS Abercrombie - Born: Connecticut - Age: 32 - Sent to Chatham on 29 Mar 1813 on the HMS Braham.

Mack, Theron - Seaman - Number: 1349 - Prize name: Fox - Ship type: LM - How taken: HM Sloop Pheasant - When taken: 23 Apr 1813 - Where taken: Bay of Biscay - Date received: 14 May 1813 - From what ship: HMS Pleasant - Born: New York - Age: 24 - Discharged on 3 Jul 1813 and sent to Stapleton Prison.

Macomb, James - Master - Number: 938 - Prize name: Weasel - Ship type: MV - How taken: HM Brig Foxhound - When taken: 25 Mar 1813 - Where taken: Bay of Biscay - Date received: 7 Apr 1813 - From what ship: HMS Foxhound - Born: Charlestown - Age: 24 - Sent on 8 Apr 1813 to Ashburton on parole.

Madden, Fredrick - Seaman - Number: 503 - Prize name: Cashiere - Ship type: LM - How taken: HM Brig Reindeer - When taken: 3 Feb 1813 - Where taken: Bay of Biscay - Date received: 12 Feb 1813 - From what ship: HMS Reindeer - Born: Alexandria - Age: 19 - Sent to Dartmoor on 2 Apr 1813.

Madey, Gustave Charles D'Eocouthant - Passenger - Number: 1579 - Prize name: Orders in Council - Ship type: LM - How taken: HM Frigate Surveillante - When taken: 1 Jun 1813 - Where taken: Bay of Biscay - Date received: 12 Jun 1813 - From what ship: Cutter Earl Wellington - Born: Martinique - Age: 20 - Discharged on 23 Jun 1813 and sent to London.

Main, Henry - Boy - Number: 209 - Prize name: Hunter - Ship type: P - How taken: HM Frigate Phoebe - When taken: 24 Dec 1812 - Where taken: off Western Islands - Date received: 9 Jan 1813 - From what ship: HMS Phoebe - Born: Boston - Age: 18 - Race: Black - Sent to Portsmouth on 8 Feb 1813 on the HMS Colossus.

Main, William - Seaman - Number: 1163 - Prize name: Essex - Ship type: MV - How taken: HM Frigate Pyramus - When taken: 2 Apr 1813 - Where taken: Bay of Biscay - Date received: 2 Mar 1813 - From what ship: HMS Rota - Born: Marblehead - Age: 20 - Sent on 3 Jul 1813 to Stapleton prison.

Mains, Darius - Seaman - Number: 592 - Prize name: St. Martin's Plantation, prize of the Privateer Paul Jones - Ship type: P - How taken: HM Ship-of-the-Line Dublin - When taken: 9 Feb 1815 - Where taken: Lat 43 N Long 33.5 W - Date received: 25 Feb 1813 - From what ship: HMS Dublin - Born: Georgetown - Age: 24 - Sent to Dartmoor on 2 Apr 1813.

Mains, John - Seaman - Number: 343 - How taken: Impressed at Falmouth - When taken: 17 Jan 1813 - Date received: 23 Jan 1813 - From what ship: HMS Frederick - Born: Rhinebeck - Age: 26 - Sent to Chatham on 29 Mar 1813 on the HMS Braham.

Mains, John - Seaman - Number: 2388 - How taken: Gave himself up from HM Schooner Whiting - Date received: 2 Feb 1814 - From what ship: HMS Whiting - Born: Shrewsbury - Age: 39 - Discharged on 27 Feb 1814 and sent to Chatham on HMS Haleyon.

Malbone, Evan - Chief Mate - Number: 1785 - Prize name: Jane Barns - Ship type: MV - How taken: HM Frigate Comus - When taken: 14 May 1813 - Where taken: off Lisbon - Date received: 13 Aug 1813 - From what ship: HMS Protection - Born: Pomfret, CT - Age: 31 - Discharged on 15 Aug 1813 and sent to Ashburton on parole.

Malbony, William T. - Chief Mate - Number: 857 - Prize name: Cannoniere - Ship type: P - How taken: HM Ship-of-the-Line Warspite - When taken: 14 Mar 1813 - Where taken: Bay of Biscay - Date received: 25 May 1813 - From what ship: HMS Warspite - Born: Newport - Age: 30 - Sent on 26 Mar 1813 to Ashburton on parole.

Mallery, Michael - Gunner - Number: 688 - Prize name: Star - Ship type: MV - How taken: HM Ship-of-the-Line Superb - When taken: 9 Feb 1813 - Where taken: Bay of Biscay - Date received: 19 Mar 1813 - From what ship: HMS Warspite - Born: New Orleans - Age: 32 - Sent to Dartmoor on 2 Apr 1813.

Man, Tabez - Seaman - Number: 2307 - Prize name: Siro - Ship type: LM - How taken: HM Brig Pelican - When taken: 13 Jan 1814 - Where taken: at sea - Date received: 23 Jan 1814 - From what ship: HMS Pelican - Born: Boston - Age: 30 - Discharged on 31 Jan 1814 and sent to Dartmoor.

Manett, Richard - Seaman - Number: 642 - Prize name: Criterion - Ship type: MV - How taken: HM Frigate Belle Poule - When taken: 14 Feb 1813 - Where taken: Bay of Biscay - Date received: 4 Mar 1813 - From what ship: HMS Strenuous - Born: New York - Age: 26 - Sent to Dartmoor on 2 Apr 1813.

Maniers, Benjamin - Seaman - Number: 428 - Prize name: Union - Ship type: MV - How taken: HM Frigate Iris - When taken: 17 Jan 1813 - Where taken: Lat 44 N Long 2.3 W - Date received: 5 Feb 1813 - From what ship: HMS San Josef - Born: New London - Age: 22 - Sent to Chatham on 29 Mar 1813 on the HMS Braham.

Mann, John - 2nd Mate - Number: 869 - Prize name: Cannoniere - Ship type: P - How taken: HM Ship-of-the-Line Warspite - When taken: 14 Mar 1813 - Where taken: Bay of Biscay - Date received: 28 Mar 1813 - From what ship: HMS Warspite - Born: Philadelphia - Age: 27 - Sent to Chatham on 8 Jul 1813 on HM Tender Neptune.

Mansfield, James - Seaman - Number: 1383 - Prize name: Tom - Ship type: LM - How taken: HM Frigate Surveillante

- When taken: 27 Apr 1813 - Where taken: Bay of Biscay - Date received: 15 May 1813 - From what ship: HMS Foxhound - Born: Boston - Age: 20 - Discharged on 3 Jul 1813 and sent to Stapleton Prison.

Manson, Jeremiah - 2nd Mate - Number: 1193 - Prize name: Zebra - Ship type: LM - How taken: HM Frigate Pyramus - When taken: 20 Apr 1813 - Where taken: Bay of Biscay - Date received: 9 May 1813 - From what ship: HMS Andromache - Born: Philadelphia - Age: 25 - Sent on 3 Jul 1813 to Stapleton prison.

Manson, Nathaniel - Seaman - Number: 1625 - Prize name: Leo - Ship type: LM - How taken: HM Frigate Magiciene - When taken: 4 Jun 1813 - Where taken: Lat 45 Long 14 - Date received: 14 Jun 1813 - From what ship: HMS Orestes - Born: Limington - Age: 22 - Discharged on 30 Jun 1813 and sent to Stapleton.

Manson, William - Seaman - Number: 1524 - Prize name: Courier - Ship type: LM - How taken: HM Brig Rover - When taken: 14 Mar 1813 - Where taken: Bay of Biscay - Date received: 29 May 1813 - From what ship: HMS Hannibal - Born: New London - Age: 25 - Discharged on 30 Jun 1813 and sent to Stapleton.

Manson, William - Seaman - Number: 1529 - Prize name: Courier - Ship type: LM - How taken: HM Brig Rover - When taken: 14 Mar 1813 - Where taken: Bay of Biscay - Date received: 29 May 1813 - From what ship: HMS Hannibal - Born: Massachusetts - Age: 16 - Discharged on 30 Jun 1813 and sent to Stapleton.

Manuel, Diego - Seaman - Number: 1787 - Prize name: Jane Barns - Ship type: MV - How taken: HM Frigate Comus - When taken: 14 May 1813 - Where taken: off Lisbon - Date received: 13 Aug 1813 - From what ship: HMS Protection - Born: New Orleans - Age: 23 - Liberated on 3 Sep 1813.

Manuel, John - Seaman - Number: 64 - Prize name: Independence - Ship type: MV - How taken: HM Frigate Medusa - When taken: 9 Nov 1812 - Where taken: San Sebastian - Date received: 27 Nov 1812 - From what ship: HMS Wasp - Born: New Orleans - Age: 57 - Sent to Portsmouth on 29 Dec 1812 on the HMS Northumberland.

Manuel, Joseph - Seaman - Number: 42 - Prize name: Warren - Ship type: MV - How taken: HM Frigate Sybille & HM Frigate Fortunee- When taken: 5 Sep 1812 - Where taken: Lat 41.4 Long 33 - Date received: 23 Nov 1812 - From what ship: HMS Stork - Born: Oporto, Portugal - Age: 24 - Died on 29 Dec 1812 on HM Prison Ship Caton.

Marble, Jabez - Seaman - Number: 636 - Prize name: Criterion - Ship type: MV - How taken: HM Frigate Belle Poule - When taken: 14 Feb 1813 - Where taken: Bay of Biscay - Date received: 4 Mar 1813 - From what ship: HMS Strenuous - Born: Massachusetts - Age: 23 - Sent to Dartmoor on 2 Apr 1813.

Marchant, William - Seaman - Number: 2615 - How taken: MV Guadeloupe - Date received: 16 May 1814 - From what ship: HMS Repulse - Born: Martha's Vineyard - Age: 33 - Discharged on 14 Jun 1814 and sent to Dartmoor.

Marchley, William - Master - Number: 247 - Prize name: Rising Sun - Ship type: MV - How taken: HM Ship-Sloop Jalousie - When taken: 12 Dec 1812 - Where taken: Lat 43 Long 20 - Date received: 16 Jan 1813 - From what ship: HMS Stork - Born: Concord - Age: 36 - Sent on 18 Jan 1813 to Ashburton on parole.

Mars, George - Seaman - Number: 887 - Prize name: Tiger - Ship type: MV - How taken: HM Brig Scylla - When taken: 22 Mar 1813 - Where taken: Bay of Biscay - Date received: 1 Apr 1813 - From what ship: HMS Scylla - Born: Massachusetts - Age: 39 - Sent on 21 Jun 1813 to Portsmouth on HMS Prometheus.

Marsh, Beverly - Seaman - Number: 2037 - How taken: Impressed at Liverpool - When taken: 16 Oct 1813 - Date received: 1 Nov 1813 - From what ship: HMS Bittern - Born: New Jersey - Age: 20 - Discharged on 3 Nov 1813 and sent to Dartmoor.

Marsh, Jesse - Boy - Number: 1052 - Prize name: Independence - Ship type: MV - How taken: HM Ship-of-the-Line Superb - When taken: 16 Mar 1813 - Where taken: Bay of Biscay - Date received: 22 Apr 1813 - From what ship: HMS Superb - Born: Boston - Age: 14 - Sent to Dartmoor on 1 Jul 1813.

Marshall, Alexander - Seaman - Number: 1319 - Prize name: Shadow - Ship type: LM - How taken: HM Brig Reindeer & HM Schooner Helicon - When taken: 6 Apr 1813 - Where taken: Bay of Biscay - Date received: 12 May 1813 - From what ship: HMS Helicon - Born: Philadelphia - Age: 23 - Discharged on 3 Jul 1813 and sent to Stapleton Prison.

Marshall, David - Chief Mate - Number: 603 - Prize name: Rolla - Ship type: MV - How taken: HM Frigate Surveillante - When taken: 11 Feb 1813 - Where taken: Bay of Biscay - Date received: 25 Feb 1813 - From what ship: HMS Plymouth - Born: Massachusetts - Age: 23 - Sent on 27 Feb 1813 to Ashburton on parole.

Marshall, Emanuel - Seaman - Number: 2123 - Prize name: Chesapeake - Ship type: LM - How taken: HM Frigate

Hotspur & HM Frigate Pyramus - When taken: 26 Oct 1813 - Where taken: Bay of Biscay - Date received: 22 Nov 1813 - From what ship: HMS Pyramus - Born: Providence - Age: 40 - Discharged on 29 Nov 1813 and sent to Dartmoor.

Marshall, John - Mate - Number: 308 - Prize name: Blue Bird - Ship type: MV - How taken: HM Frigate Briton - When taken: 1 Jan 1813 - Where taken: off Bordeaux - Date received: 21 Jan 1813 - From what ship: HMS Abercrombie - Born: Nantucket - Age: 31 - Sent on 22 Jan 1813 to Ashburton on parole.

Marshall, John - Mate - Number: 1980 - How taken: Impressed at Liverpool - When taken: 10 Sep 1812 - Date received: 19 Sep 1813 - From what ship: HMS Bittern - Born: Nantucket - Age: 31 - Discharged on 29 Nov 1813 and sent to Dartmoor.

Marshall, Thomas - Seaman - Number: 1794 - How taken: Delivered himself up from HM Ship-of-the-Line Pompee - When taken: 31 Oct 1812 - Date received: 13 Aug 1813 - From what ship: HMS Protection - Born: Drummond Town - Age: 26 - Discharged on 8 Sep 1813 and sent to Dartmoor.

Mart, James - Seaman - Number: 1523 - Prize name: Courier - Ship type: LM - How taken: HM Brig Rover - When taken: 14 Mar 1813 - Where taken: Bay of Biscay - Date received: 29 May 1813 - From what ship: HMS Hannibal - Born: New London - Age: 21 - Discharged on 30 Jun 1813 and sent to Stapleton.

Martain, Ephraim - Seaman - Number: 242 - How taken: Impressed at Liverpool - When taken: 17 Nov 1812 - Date received: 12 Jan 1813 - From what ship: HMS Bittern - Born: Wiscasset - Age: 40 - Sent to Portsmouth on 8 Feb 1813 on the HMS Colossus.

Martain, John B. - Seaman - Number: 1374 - Prize name: Tom - Ship type: LM - How taken: HM Frigate Surveillante - When taken: 27 Apr 1813 - Where taken: Bay of Biscay - Date received: 15 May 1813 - From what ship: HMS Foxhound - Born: Boston - Age: 18 - Discharged on 3 Jul 1813 and sent to Stapleton Prison.

Martain, Peter - Steward - Number: 531 - Prize name: Terrible - Ship type: MV - How taken: HM Brig Foxhound - When taken: 8 Feb 1813 - Where taken: Channel - Date received: 15 Feb 1813 - From what ship: HMS Foxhound - Born: Arlington - Age: 24 - Sent to Dartmoor on 2 Apr 1813.

Martin, Anthony - Seaman - Number: 1255 - Prize name: Caroline - Ship type: MV - How taken: HM Frigate Medusa - When taken: 12 Apr 1813 - Where taken: Bay of Biscay - Date received: 10 May 1813 - From what ship: HMS Medusa - Born: New Orleans - Age: 32 - Discharged on 3 Jul 1813 and sent to Stapleton Prison.

Martin, Henry - Seaman - Number: 1551 - Prize name: Good Friends - Ship type: MV - How taken: HM Frigate Andromache - When taken: 2 Apr 1813 - Where taken: Bay of Biscay - Date received: 29 May 1813 - From what ship: HMS Hannibal - Born: Prussia - Age: 25 - Discharged on 30 Jun 1813 and sent to Stapleton.

Martin, Isaac - Boy - Number: 1483 - Prize name: Paul Jones - Ship type: P - How taken: HM Frigate Leonidas - When taken: 23 May 1813 - Where taken: off Cape Clear - Date received: 26 May 1813 - From what ship: HMS Leonidas - Born: Baltimore - Age: 16 - Discharged on 30 Jun 1813 and sent to Stapleton.

Martin, John - Seaman - Number: 1431 - Prize name: Paul Jones - Ship type: P - How taken: HM Frigate Leonidas - When taken: 23 May 1813 - Where taken: off Cape Clear - Date received: 26 May 1813 - From what ship: HMS Leonidas - Born: Salem - Age: 27 - Discharged on 30 Jun 1813 and sent to Stapleton.

Martin, John - Passenger - Number: 390 - Prize name: Union - Ship type: MV - How taken: HM Frigate Iris - When taken: 17 Jan 1813 - Where taken: Lat 44 N Long 2.3 W - Date received: 5 Feb 1813 - From what ship: HMS San Josef - Born: Bordeaux - Age: 21 - Entered on 14 Feb 1813 into the French General Entry Book as prisoner number 39378.

Martin, John - Seaman - Number: 2212 - Prize name: Squirrel - Ship type: MV - How taken: HM Frigate Belle Poule - When taken: 14 Dec 1813 - Where taken: Bay of Biscay - Date received: 27 Dec 1813 - From what ship: HMS Protector - Born: Dolliver - Age: 27 - Discharged on 31 Jan 1814 and sent to Dartmoor.

Martin, Manuel - Seaman - Number: 1460 - Prize name: Paul Jones - Ship type: P - How taken: HM Frigate Leonidas - When taken: 23 May 1813 - Where taken: off Cape Clear - Date received: 26 May 1813 - From what ship: HMS Leonidas - Born: New Orleans - Age: 17 - Discharged on 30 Jun 1813 and sent to Stapleton.

Martin, Stephen - Boy - Number: 2125 - Prize name: Chesapeake - Ship type: LM - How taken: HM Frigate Hotspur & HM Frigate Pyramus - When taken: 26 Oct 1813 - Where taken: Bay of Biscay - Date received: 22 Nov 1813 -

From what ship: HMS Pyramus - Born: New York - Age: 13 - Discharged on 29 Nov 1813 and sent to Dartmoor.

Martin, William - Seaman - Number: 1719 - Prize name: Orders in Council - Ship type: LM - How taken: HM Frigate Surveillante - When taken: 1 Jun 1813 - Where taken: Bay of Biscay - Date received: 4 Jul 1813 - From what ship: HMS Iris - Born: Norfolk - Age: 25 - Race: Mulatto - Discharged on 22 Aug 1813 and released to HM Brig Redpole.

Martin, William - Seaman - Number: 1760 - How taken: Impressed at Belfast - When taken: 29 Jun 1813 - Date received: 22 Jul 1813 - From what ship: HMS Elizabeth - Born: New York - Age: 22 - Race: Negro - Discharged on 8 Sep 1813 and sent to Dartmoor.

Mason, Francis - Seaman - Number: 421 - Prize name: Union - Ship type: MV - How taken: HM Frigate Iris - When taken: 17 Jan 1813 - Where taken: Lat 44 N Long 2.3 W - Date received: 5 Feb 1813 - From what ship: HMS San Josef - Born: Philadelphia - Age: 16 - Sent to Chatham on 29 Mar 1813 on the HMS Braham.

Mason, John B. - Seaman - Number: 2054 - How taken: Pallas - Where taken: St. Johns, New Brunswick - Date received: 1 Nov 1813 - From what ship: HMS Bittern - Born: Marblehead - Age: 22 - Discharged on 3 Nov 1813 and sent to Dartmoor.

Masson, William - Seaman - Number: 2303 - Prize name: Siro - Ship type: LM - How taken: HM Brig Pelican - When taken: 13 Jan 1814 - Where taken: at sea - Date received: 23 Jan 1814 - From what ship: HMS Pelican - Born: New York - Age: 40 - Discharged on 31 Jan 1814 and sent to Dartmoor.

Mather, Allyn - Master - Number: 858 - Prize name: William Bayard - Ship type: MV - How taken: HM Ship-of-the-Line Warspite - When taken: 3 Mar 1813 - Where taken: Bay of Biscay - Date received: 26 Mar 1813 - From what ship: HMS Warspite - Born: New Haven - Age: 35 - Sent on 28 Mar 1813 to Ashburton on parole.

Mathews, Henry - Seaman - Number: 2232 - How taken: Taken off an English merchant vessel - Date received: 10 Jan 1814 - From what ship: HMS Scylla - Born: Virginia - Age: 32 - Discharged on 31 Jan 1814 and sent to Dartmoor.

Mathews, Joseph - Seaman - Number: 657 - How taken: Delivered himself up from HMS Lavinia - When taken: 15 Feb 1813 - Date received: 5 Mar 1813 - From what ship: HMS Salvador del Monde - Born: Rhode Island - Age: 30 - Sent to Dartmoor on 2 Apr 1813.

Mathews, Joseph - Seaman - Number: 2225 - Prize name: Zephyr - Ship type: MV - How taken: HM Frigate Pyramus - When taken: 30 Nov 1813 - Where taken: off Lorient - Date received: 3 Jan 1814 - From what ship: HMS Warspite - Born: Providence - Age: 32 - Discharged on 31 Jan 1814 and sent to Dartmoor.

Mathews, Richard - Seaman - Number: 1246 - How taken: Impressed at Bristol - When taken: 1 May 1813 - Date received: 9 May 1813 - From what ship: HMS Prince Frederick - Born: New Jersey - Age: 34 - Discharged on 3 Jul 1813 and sent to Stapleton Prison.

Mathews, William - Seaman - Number: 2465 - Prize name: Fair American - Ship type: MV - How taken: HM Frigate Andromache - When taken: 19 Jan 1814 - Where taken: Bay of Biscay - Date received: 22 Feb 1814 - From what ship: HMS York - Born: Essex - Age: 25 - Discharged on 10 May 1814 and sent to Dartmoor.

Mathis, Hendrick - Seaman - Number: 1990 - Prize name: Ned - Ship type: LM - How taken: HM Brig Royalist - When taken: 6 Sep 1812 - Where taken: coast of France - Date received: 22 Sep 1813 - From what ship: HMS Royalist - Born: New York - Age: 18 - Discharged on 27 Sep 1813 and sent to Dartmoor.

Matthews, Leonard - Supercargo - Number: 2316 - Prize name: Siro - Ship type: LM - How taken: HM Brig Pelican - When taken: 13 Jan 1814 - Where taken: at sea - Date received: 25 Jan 1814 - From what ship: HMS Pelican - Born: Maryland - Age: 28 - Discharged on 25 Jan 1814 and sent to Ashburton on parole.

May, Henry - Seaman - Number: 1975 - Prize name: Montgomery - Ship type: P - How taken: HM Frigate Nymphe - When taken: 5 May 1813 - Where taken: Boston Bay - Date received: 19 Sep 1813 - From what ship: HMS Bittern - Born: Staten Island - Age: 27 - Discharged on 27 Sep 1813 and sent to Dartmoor.

May, Henry - Seaman - Number: 2700 - Prize name: Montgomery - Ship type: P - How taken: HM Frigate Nymphe - When taken: 5 May 1813 - Where taken: Boston Bay - Date received: 17 Jun 1814 - From what ship: Dartmouth - Born: Staten Island - Age: 27 - Discharged on 20 Jun 1814 and released to the Speedwell Cartel.

May, Walter - Seaman - Number: 1286 - Prize name: Price - Ship type: LM - How taken: HM Frigate Medusa - When taken: 13 Apr 1813 - Where taken: Bay of Biscay - Date received: 10 May 1813 - From what ship: HMS Medusa - Born: Norfolk - Age: 28 - Discharged on 3 Jul 1813 and sent to Stapleton Prison.

Mayor, James - Seaman - Number: 2460 - Prize name: Fair American - Ship type: MV - How taken: HM Frigate Andromache - When taken: 19 Jan 1814 - Where taken: Bay of Biscay - Date received: 22 Feb 1814 - From what ship: HMS York - Born: Salem - Age: 32 - Discharged on 10 May 1814 and sent to Dartmoor.

Mazeb, James - Seaman - Number: 1018 - How taken: Delivered himself up from HM Ship-of-the-Line Malta - Where taken: Gibraltar - Date received: 20 Apr 1813 - From what ship: HMS Libria - Born: Norristown - Age: 20 - Sent to Chatham on 8 Jul 1813 on HM Tender Neptune.

McBride, James - Seaman - Number: 1313 - How taken: Delivered himself up from HM Ship-of-the-Line Clarence - Date received: 11 May 1813 - From what ship: HMS Clarence - Born: Baltimore - Age: 27 - Discharged on 8 Jul 1813 and sent to Chatham on tender Neptune.

McCarthy, John - Seaman - Number: 2606 - How taken: MV Nautilus - When taken: 20 Dec 1812 - Date received: 16 May 1814 - From what ship: HMS Repulse - Born: Philadelphia - Age: 35 - Discharged on 14 Jun 1814 and sent to Dartmoor.

McCaulle, George M. - Boy - Number: 577 - Prize name: Rolla - Ship type: MV - How taken: HM Frigate Surveillante - When taken: 11 Feb 1813 - Where taken: Bay of Biscay - Date received: 23 Feb 1813 - From what ship: HMS Surveillante - Born: Philadelphia - Age: 14 - Sent to Dartmoor on 2 Apr 1813.

McConnell, John - Seaman - Number: 2494 - How taken: Impressed out of MV Friend - When taken: 5 Mar 1814 - Date received: 26 Mar 1814 - From what ship: Earl Spencer, cutter - Born: Wilmington - Age: 25 - Discharged on 10 May 1814 and sent to Dartmoor.

McCormick, Simon - Passenger - Number: 827 - Prize name: Decornau - Ship type: MV - How taken: HM Sloop Pheasant - When taken: 15 Mar 1813 - Where taken: Bay of Biscay - Date received: 21 Mar 1813 - From what ship: HMS Pheasant - Born: Boston - Age: 54 - Race: Black - Sent to Dartmoor on 28 Jun 1813.

McCoy, James Abercrombie - Seaman - Number: 1344 - Prize name: Fox - Ship type: LM - How taken: HM Sloop Pheasant - When taken: 23 Apr 1813 - Where taken: Bay of Biscay - Date received: 14 May 1813 - From what ship: HMS Pleasant - Born: Philadelphia - Age: 21 - Discharged on 3 Jul 1813 and sent to Stapleton Prison.

McDonald, John - Seaman - Number: 407 - Prize name: Union - Ship type: MV - How taken: HM Frigate Iris - When taken: 17 Jan 1813 - Where taken: Lat 44 N Long 2.3 W - Date received: 5 Feb 1813 - From what ship: HMS San Josef - Born: Philadelphia - Age: 24 - Sent to Chatham on 29 Mar 1813 on the HMS Braham.

McDonald, John - Seaman - Number: 2114 - Prize name: Chesapeake - Ship type: LM - How taken: HM Frigate Hotspur & HM Frigate Pyramus - When taken: 26 Oct 1813 - Where taken: Bay of Biscay - Date received: 22 Nov 1813 - From what ship: HMS Pyramus - Born: Norfolk - Age: 27 - Discharged on 29 Nov 1813 and sent to Dartmoor.

McDonald, John - Seaman - Number: 1897 - Prize name: US Brig Argus - Ship type: War - How taken: HM Brig Pelican - When taken: 14 Aug 1813 - Where taken: Irish Channel - Date received: 23 Aug 1813 - From what ship: HMS Pelican - Born: Baltimore - Age: 25 - Discharged on 8 Sep 1813 and sent to Dartmoor.

McDowall, John - Seaman - Number: 2076 - Prize name: Avon, prize of the Privateer True Blooded Yankee - Ship type: P - How taken: HM Frigate Eurotas - When taken: 27 Oct 1813 - Where taken: off Ushant - Date received: 1 Nov 1813 - From what ship: HMS Hannibal - Born: Pennsylvania - Age: 26 - Discharged on 3 Nov 1813 and sent to Dartmoor.

McElroy, William - Cooper - Number: 397 - Prize name: Union - Ship type: MV - How taken: HM Frigate Iris - When taken: 17 Jan 1813 - Where taken: Lat 44 N Long 2.3 W - Date received: 5 Feb 1813 - From what ship: HMS San Josef - Born: Philadelphia - Age: 24 - Sent to Chatham on 29 Mar 1813 on the HMS Braham.

McFadden, James - 2nd Prize Master - Number: 2265 - Prize name: Fanny - Ship type: MV - How taken: HM Frigate Eurotas - When taken: 25 Dec 1813 - Where taken: at sea - Date received: 20 Jan 1814 - From what ship: HMS Eurotas - Born: Baltimore - Age: 26 - Discharged on 31 Jan 1814 and sent to Dartmoor.

McFall, William - Seaman - Number: 1971 - How taken: Out of a Spanish merchant vessel - When taken: 10 Sep 1813 - Date received: 19 Sep 1813 - From what ship: HMS Bittern - Born: New York - Age: 33 - Discharged on 27 Sep

1813 and sent to Dartmoor.

McFarland, John - Seaman - Number: 2213 - Prize name: Squirrel - Ship type: MV - How taken: HM Frigate Belle Poule - When taken: 14 Dec 1813 - Where taken: Bay of Biscay - Date received: 27 Dec 1813 - From what ship: HMS Protector - Born: Massachusetts - Age: 21 - Discharged on 31 Jan 1814 and sent to Dartmoor.

McFuel, James - Mate - Number: 54 - Prize name: Independence - Ship type: MV - How taken: HM Frigate Medusa - When taken: 9 Nov 1812 - Where taken: San Sebastian - Date received: 27 Nov 1812 - From what ship: HMS Wasp - Born: New York - Age: 32 - Sent on 8 Dec 1812 to Ashburton on parole.

McIntire, Samuel - Seaman - Number: 1377 - Prize name: Tom - Ship type: LM - How taken: HM Frigate Surveillante - When taken: 27 Apr 1813 - Where taken: Bay of Biscay - Date received: 15 May 1813 - From what ship: HMS Foxhound - Born: Massachusetts - Age: 23 - Discharged on 3 Jul 1813 and sent to Stapleton Prison.

McKenny, John - Seaman - Number: 1610 - Prize name: Revenge - Ship type: LM - How taken: HM Frigate Belle Poule - When taken: 11 May 1813 - Where taken: Bay of Biscay - Date received: 14 Jun 1813 - From what ship: HMS Royalist - Born: Georgetown - Age: 35 - Discharged on 30 Jun 1813 and sent to Stapleton.

McKenzie, Alexander - Seaman - Number: 1008 - Prize name: Courier - Ship type: LM - How taken: HM Frigate Andromache - When taken: 14 Mar 1813 - Where taken: Bay of Biscay - Date received: 16 Apr 1813 - From what ship: HMS Fairy - Born: New York - Age: 30 - Sent to Dartmoor on 8 Sep 1813.

McKerbie, Alexander - Prize Master - Number: 2671 - Prize name: John, prize to Amelia - Ship type: P - How taken: HM Ship-of-the-Line Sterling Castle - When taken: 10 May 1814 - Where taken: Lat 36 Long 37 - Date received: 5 Jun 1814 - From what ship: HMS Gronville - Born: Baltimore - Age: 28 - Discharged on 14 Jun 1814 and sent to Dartmoor.

McKinzey, John - Seaman - Number: 36 - Prize name: Wasp - Ship type: MV - How taken: Earl Spencer, cutter - When taken: 4 Aug 1812 - Where taken: off Cape Clear - Date received: 23 Nov 1812 - From what ship: HMS Stork - Born: Massachusetts - Age: 25 - Race: Black - Sent to Portsmouth on 29 Dec 1812 on the HMS Northumberland.

McLellan, William - Seaman - Number: 655 - How taken: Delivered himself up from HMS Lavinia - When taken: 15 Feb 1813 - Date received: 5 Mar 1813 - From what ship: HMS Salvador del Monde - Born: Schenectady - Age: 31 - Sent to Dartmoor on 2 Apr 1813.

McLeod, Collin - Boatswain - Number: 1801 - Prize name: US Brig Argus - Ship type: War - How taken: HM Brig Pelican - When taken: 14 Aug 1813 - Where taken: Irish Channel - Date received: 17 Aug 1813 - From what ship: USS Argus - Born: Philadelphia - Age: 30 - Discharged on 29 Nov 1813 and sent to Dartmoor.

McMaken, John - Seaman - Number: 1358 - Prize name: Fox - Ship type: LM - How taken: HM Sloop Pheasant - When taken: 23 Apr 1813 - Where taken: Bay of Biscay - Date received: 15 May 1813 - From what ship: HMS Foxhound - Born: Pennsylvania - Age: 22 - Discharged on 3 Jul 1813 and sent to Stapleton Prison.

McNab, James - Seaman - Number: 2012 - Prize name: Ned - Ship type: LM - How taken: HM Brig Royalist - When taken: 6 Sep 1812 - Where taken: coast of France - Date received: 24 Sep 181 - From what ship: HMS Rippon - Born: New York - Age: 19 - Discharged on 27 Sep 1813 and sent to Dartmoor.

McNeil, John - Seaman - Number: 2205 - How taken: Impressed at Cork - When taken: 15 Dec 1813 - Date received: 27 Dec 1813 - From what ship: HMS Protector - Born: Salem - Age: 20 - Discharged on 10 May 1814 and sent to Dartmoor.

McNelly, Thomas - Carpenter - Number: 1116 - Prize name: Viper - Ship type: MV - How taken: HM Ship-of-the-Line Superb - When taken: 15 Apr 1813 - Where taken: Bay of Biscay - Date received: 22 Apr 1813 - From what ship: HMS Superb - Born: Maryland - Age: 57 - Sent to Dartmoor on 1 Jul 1813.

McQuillan, James - 2nd Lieutenant - Number: 2685 - Prize name: Margaret, prize to the Privateer Surprize - Ship type: P - How taken: HM Brig Foxhound - When taken: 27 May 1814 - Where taken: off Scylly - Date received: 7 Jun 1814 - From what ship: Maynard - Born: South Carolina - Age: 31 - Discharged on 14 Jun 1814 and sent to Dartmoor.

Mecoh, David - Seaman - Number: 2499 - Prize name: Mary, prize of the Privateer Rattle Snake - Ship type: P - How taken: Retaken by the crew of the Mary - When taken: 18 Dec 1813 - Date received: 26 Mar 1814 - From what ship: Orr - Born: New London - Age: 24 - Discharged on 10 May 1814 and sent to Dartmoor.

Meden, William - Seaman - Number: 2069 - Prize name: Betsy, prize to the Privateer True Blooded Yankee - Ship type: P - How taken: HM Frigate Eurotas - When taken: 26 Oct 1813 - Where taken: off Ushant - Date received: 1 Nov 1813 - From what ship: HMS Hannibal - Born: Philadelphia - Age: 24 - Discharged on 3 Nov 1813 and sent to Dartmoor.

Meigs, John - 2nd Mate - Number: 851 - Prize name: Gold Coiner - Ship type: MV - How taken: HM Sloop Lyra - When taken: 20 Mar 1813 - Where taken: Lat 44.2 N Long 7.10 W - Date received: 24 May 1813 - From what ship: Gold Coiner - Born: Thomas Town - Age: 27 - Sent to Dartmoor on 28 Jun 1813.

Meilleur, Simon - Passenger - Number: 1580 - Prize name: Orders in Council - Ship type: LM - How taken: HM Frigate Surveillante - When taken: 1 Jun 1813 - Where taken: Bay of Biscay - Date received: 12 Jun 1813 - From what ship: Cutter Earl Wellington - Born: New Orleans - Age: 19 - Discharged on 31 Aug 1813 and sent to Ashburton on parole.

Meiniffe, Charles - Seaman - Number: 2002 - Prize name: Ned - Ship type: LM - How taken: HM Brig Royalist - When taken: 6 Sep 1812 - Where taken: coast of France - Date received: 22 Sep 1813 - From what ship: HMS Royalist - Born: Baltimore - Age: 21 - Discharged on 27 Sep 1813 and sent to Dartmoor.

Melville, John - Seaman - Number: 326 - Prize name: Porcupine - Ship type: MV - How taken: HM Frigate Dryad - When taken: 8 Jan 1813 - Where taken: off Bordeaux - Date received: 21 Jan 1813 - From what ship: HMS Abercrombie - Born: Baltimore - Age: 19 - Sent to Chatham on 29 Mar 1813 on the HMS Braham.

Membelly, William - Seaman - Number: 1209 - Prize name: Zebra - Ship type: LM - How taken: HM Frigate Pyramus - When taken: 20 Apr 1813 - Where taken: Bay of Biscay - Date received: 9 May 1813 - From what ship: HMS Andromache - Born: Savannah - Age: 24 - Discharged on 3 Jul 1813 and sent to Stapleton Prison.

Menard, Augustus - Passenger - Number: 1945 - Prize name: Marmion - Ship type: MV - How taken: HM Frigate President - When taken: 14 Aug 1813 - Where taken: off Nantes - Date received: 27 Aug 1813 - From what ship: HMS Urgent - Born: Niort, France - Age: 12 - Discharged on 8 Sep 1813 and sent to Dartmoor.

Menard, Nicholas - Master - Number: 231 - Prize name: Empress - Ship type: MV - How taken: HM Brig Rover - When taken: 29 Nov 1812 - Where taken: at Coruna, Spain - Date received: 11 Jan 1813 - From what ship: HMS Rover - Born: Brackland - Age: 38 - Sent on 12 Jan 1813 to Ashburton on parole.

Mercer, Chaumont - Steward - Number: 400 - Prize name: Union - Ship type: MV - How taken: HM Frigate Iris - When taken: 17 Jan 1813 - Where taken: Lat 44 N Long 2.3 W - Date received: 5 Feb 1813 - From what ship: HMS San Josef - Born: Prince Georges County - Age: 25 - Sent to Chatham on 29 Mar 1813 on the HMS Braham.

Merrill, John - Seaman - Number: 2663 - Prize name: Columbia - Ship type: MV - How taken: Sir John Sherbrook, privateer - Where taken: near Boston - Date received: 5 Jun 1814 - From what ship: HMS Gronville - Born: North Yarmouth - Age: 33 - Discharged on 14 Jun 1814 and sent to Dartmoor.

Morrel, Enoch - Seaman - Number: 2366 - Prize name: Minerva - Ship type: MV - How taken: HM Ship-of-the-Line Conquestador - When taken: 19 Dec 1814 - Where taken: Bay of Biscay - Date received: 31 Jan 1814 - From what ship: HMS Surveillante - Born: Falmouth - Age: 19 - Discharged on 27 Feb 1814 and sent to Chatham on HMS Haleyon.

Merrett, Robert - Passenger - Number: 1597 - Prize name: Governor Gerry - Ship type: MV - How taken: HM Brig Royalist - When taken: 31 May 1813 - Where taken: Bay of Biscay - Date received: 14 Jun 1813 - From what ship: HMS Royalist - Born: New York - Age: 35 - Discharged on 30 Jun 1813 and sent to Stapleton.

Merrett, Thomas - Seaman - Number: 1160 - Prize name: Essex - Ship type: MV - How taken: HM Frigate Pyramus - When taken: 2 Apr 1813 - Where taken: Bay of Biscay - Date received: 2 Mar 1813 - From what ship: HMS Rota - Born: Marblehead - Age: 19 - Sent on 3 Jul 1813 to Stapleton prison.

Merit, Almon - Seaman - Number: 596 - Prize name: St. Martin's Plantation, prize of the Privateer Paul Jones - Ship type: P - How taken: HM Ship-of-the-Line Dublin - When taken: 9 Feb 1815 - Where taken: Lat 43 N Long 33.5 W - Date received: 25 Feb 1813 - From what ship: HMS Dublin - Born: Blandford - Age: 21 - Sent to Dartmoor on 2 Apr 1813.

Merritt, Jonathan - Seaman - Number: 1685 - Prize name: Revenge - Ship type: LM - How taken: HM Frigate Belle

Poule - When taken: 11 May 1813 - Where taken: Bay of Biscay - Date received: 26 Jun 1813 - From what ship: HMS Duncan - Born: New York - Age: 19 - Discharged on 30 Jun 1813 and sent to Stapleton.

Mercer, Benjamin - Seaman - Number: 1202 - Prize name: Zebra - Ship type: LM - How taken: HM Frigate Pyramus - When taken: 20 Apr 1813 - Where taken: Bay of Biscay - Date received: 9 May 1813 - From what ship: HMS Andromache - Born: New York - Age: 21 - Sent on 8 Sep 1813 to Dartmoor.

Mesey, William - Mate - Number: 76 - Prize name: Independence - Ship type: MV - How taken: HM Frigate Medusa - When taken: 9 Nov 1812 - Where taken: San Sebastian - Date received: 29 Nov 1812 - From what ship: HMS Wasp - Born: Nantucket - Age: 52 - Sent on 8 Dec 1812 to Ashburton on parole.

Metley, Thomas - Seaman - Number: 1268 - Prize name: Caroline - Ship type: MV - How taken: HM Frigate Medusa - When taken: 12 Apr 1813 - Where taken: Bay of Biscay - Date received: 10 May 1813 - From what ship: HMS Medusa - Born: Philadelphia - Age: 25 - Discharged on 3 Jul 1813 and sent to Stapleton Prison.

Mezick, Elihu - Seaman - Number: 1681 - Prize name: Revenge - Ship type: LM - How taken: HM Frigate Belle Poule - When taken: 11 May 1813 - Where taken: Bay of Biscay - Date received: 26 Jun 1813 - From what ship: HMS Duncan - Born: Maryland - Age: 38 - Race: Mulatto - Discharged on 30 Jun 1813 and sent to Stapleton.

Michael, Car. - Seaman - Number: 1713 - Prize name: Joseph - Ship type: MV - How taken: HM Frigate Iris - When taken: 8 Jun 1813 - Where taken: Bay of Biscay - Date received: 4 Jul 1813 - From what ship: HMS Iris - Born: Maryland - Age: 17 - Race: Mulatto - Discharged on 22 Aug 1813 and released to HM Brig Redpole.

Michael, Peter - Seaman - Number: 129 - Prize name: Nope - Ship type: MV - How taken: HM Sloop Pheasant - When taken: 13 Dec 1812 - Where taken: off Oporto - Date received: 2 Jan 1813 - From what ship: Nope - Born: Scio - Age: 25 - Sent to Portsmouth on 4 Jan 1813 on the HMS Revolutionnaire.

Michell, Fairly - Seaman - Number: 2087 - Prize name: Sybella - Ship type: MV - How taken: HM Brig Zenobia - When taken: 17 Jun 1813 - Where taken: off Cape Henry - Date received: 2 Nov 1813 - From what ship: HMS Bacchus - Born: New York - Age: 19 - Discharged on 3 Nov 1813 and sent to Dartmoor.

Michell, James - Seaman - Number: 202 - Prize name: Hunter - Ship type: P - How taken: HM Frigate Phoebe - When taken: 24 Dec 1812 - Where taken: off Western Islands - Date received: 9 Jan 1813 - From what ship: HMS Phoebe - Born: Warrington - Age: 20 - Sent to Portsmouth on 8 Feb 1813 on the HMS Colossus.

Miles, William - Seaman - Number: 2508 - Prize name: Bunker Hill - Ship type: P - How taken: HM Frigate Pomone & HM Frigate Cydnus - When taken: 4 Mar 1814 - Where taken: Bay of Biscay - Date received: 4 Apr 1814 - From what ship: HMS Virago - Born: Pennsylvania - Age: 24 - Discharged on 10 May 1814 and sent to Dartmoor.

Miller, Charles - Seaman - Number: 1547 - Prize name: Zebra - Ship type: LM - How taken: HM Frigate Pyramus - When taken: 20 Apr 1813 - Where taken: Bay of Biscay - Date received: 29 May 1813 - From what ship: HMS Hannibal - Born: Gothenburg, Sweden - Age: 28 - Discharged on 30 Jun 1813 and sent to Stapleton.

Miller, Henry - Seaman - Number: 563 - Prize name: Rolla - Ship type: MV - How taken: HM Frigate Surveillante - When taken: 11 Feb 1813 - Where taken: Bay of Biscay - Date received: 23 Feb 1813 - From what ship: HMS Surveillante - Born: Germantown - Age: 29 - Sent to Dartmoor on 2 Apr 1813.

Miller, John - Cook - Number: 633 - Prize name: Criterion - Ship type: MV - How taken: HM Frigate Belle Poule - When taken: 14 Feb 1813 - Where taken: Bay of Biscay - Date received: 4 Mar 1813 - From what ship: HMS Strenuous - Born: Queen Anne's county - Age: 25 - Race: Black - Sent to Dartmoor on 2 Apr 1813.

Miller, John - Passenger - Number: 1592 - Prize name: Governor Gerry - Ship type: MV - How taken: HM Brig Royalist - When taken: 31 May 1813 - Where taken: Bay of Biscay - Date received: 14 Jun 1813 - From what ship: HMS Royalist - Born: Portland - Age: 25 - Discharged on 30 Jun 1813 and sent to Ashburton on parole.

Miller, Stephen - Seaman - Number: 1079 - Prize name: John & Frances - Ship type: MV - How taken: HM Frigate Belle Poule - When taken: 19 Mar 1813 - Where taken: Bay of Biscay - Date received: 22 Apr 1813 - From what ship: HMS Superb - Born: Hudson, NY - Age: 28 - Sent to Dartmoor on 1 Jul 1813.

Miller, Thomas - Seaman - Number: 1429 - Prize name: Paul Jones - Ship type: P - How taken: HM Frigate Leonidas - When taken: 23 May 1813 - Where taken: off Cape Clear - Date received: 26 May 1813 - From what ship: HMS Leonidas - Born: New York - Age: 19 - Discharged on 3 Jul 1813 and sent to Stapleton Prison.

Miller, William - Chief Mate - Number: 1620 - Prize name: Leo - Ship type: LM - How taken: HM Frigate Magiciene

- When taken: 4 Jun 1813 - Where taken: Lat 45 Long 14 - Date received: 14 Jun 1813 - From what ship: HMS Orestes - Born: Kennebunkport - Age: 25 - Discharged on 19 Jun 1813 and sent to Ashburton on parole.

Miller, William - Seaman - Number: 913 - How taken: Delivered himself up as a prisoner of war - Date received: 4 Apr 1813 - From what ship: HMS Warspite - Born: Boston - Age: 22 - Sent on 21 Jun 1813 to Portsmouth on HMS Prometheus.

Miller, William - Seaman - Number: 2309 - Prize name: Siro - Ship type: LM - How taken: HM Brig Pelican - When taken: 13 Jan 1814 - Where taken: at sea - Date received: 23 Jan 1814 - From what ship: HMS Pelican - Born: Portsmouth, NH - Age: 29 - Discharged on 31 Jan 1814 and sent to Dartmoor.

Millet, James - Seaman - Number: 626 - Prize name: Good Intent, prize of the Privateer Thrasher - Ship type: P - How taken: HM Frigate Pyramus - When taken: 26 Jun 1813 - Where taken: off Bordeaux - Date received: 2 Mar 1813 - From what ship: HMS Insolent - Born: Cape Ann - Age: 21 - Sent to Dartmoor on 2 Apr 1813.

Mills, John - Seaman - Number: 78 - How taken: Delivered himself up from HM Frigate Belle Poule - When taken: 10 Dec 1812 - Date received: 10 Dec 1812 - From what ship: HMS Belle Poule - Born: Portsmouth - Age: 21 - Sent to Portsmouth on 29 Dec 1812 on the HMS Northumberland.

Mills, William - Seaman - Number: 1536 - Prize name: Courier - Ship type: LM - How taken: HM Brig Rover - When taken: 14 Mar 1813 - Where taken: Bay of Biscay - Date received: 29 May 1813 - From what ship: HMS Hannibal - Born: Maryland - Age: 30 - Discharged on 30 Jun 1813 and sent to Stapleton.

Mills, William - Seaman - Number: 1198 - Prize name: Zebra - Ship type: LM - How taken: HM Frigate Pyramus - When taken: 20 Apr 1813 - Where taken: Bay of Biscay - Date received: 9 May 1813 - From what ship: HMS Andromache - Born: Westfield, NJ - Age: 21 - Sent on 3 Jul 1813 to Stapleton prison.

Mines, Artemas - Steward - Number: 2000 - Prize name: Ned - Ship type: LM - How taken: HM Brig Royalist - When taken: 6 Sep 1812 - Where taken: coast of France - Date received: 22 Sep 1813 - From what ship: HMS Royalist - Born: Maryland - Age: 22 - Race: Black - Discharged on 27 Sep 1813 and sent to Dartmoor.

Mingle, Thomas - Seaman - Number: 1335 - Prize name: Fox - Ship type: LM - How taken: HM Sloop Pheasant - When taken: 23 Apr 1813 - Where taken: Bay of Biscay - Date received: 14 May 1813 - From what ship: HMS Pleasant - Born: Africa - Age: 54 - Race: Negro - Discharged on 3 Jul 1813 and sent to Stapleton Prison.

Mingle, William - Seaman - Number: 1323 - Prize name: Shadow - Ship type: LM - How taken: HM Brig Reindeer & HM Schooner Helicon - When taken: 6 Apr 1813 - Where taken: Bay of Biscay - Date received: 12 May 1813 - From what ship: HMS Helicon - Born: Philadelphia - Age: 19 - Discharged on 3 Jul 1813 and sent to Stapleton Prison.

Minois, Pierre - Marine - Number: 2567 - Prize name: Carantine, prize of the Privateer True Blooded Yankee - Ship type: P - How taken: HM Ship-of-the-Line Vengeur - When taken: 24 Feb 1814 - Where taken: off Belle Isle - Date received: 25 Apr 1814 - From what ship: HMS Helena - Born: Belle Isle, France - Age: 16 - Discharged on 10 May 1814 and sent to Dartmoor.

Minugh, John - Seaman - Number: 1201 - Prize name: Zebra - Ship type: LM - How taken: HM Frigate Pyramus - When taken: 20 Apr 1813 - Where taken: Bay of Biscay - Date received: 9 May 1813 - From what ship: HMS Andromache - Born: New York - Age: 20 - Sent on 21 Jun 1813 to the American ship Hope.

Mires, John - Seaman - Number: 521 - Prize name: Allegany - Ship type: MV - How taken: Detained at Gibraltar - When taken: 8 Aug 1812 - Date received: 15 Feb 1813 - From what ship: HMS Andromeda - Born: Hralsound, Sweden - Age: 24 - Sent to Dartmoor on 2 Apr 1813.

Mitchel, Jacob - Seaman - Number: 1265 - Prize name: Caroline - Ship type: MV - How taken: HM Frigate Medusa - When taken: 12 Apr 1813 - Where taken: Bay of Biscay - Date received: 10 May 1813 - From what ship: HMS Medusa - Born: Baltimore - Age: 24 - Race: Black - Discharged on 3 Jul 1813 and sent to Stapleton Prison.

Mitchell, Ezekiel - Seaman - Number: 750 - Prize name: Charlotte - Ship type: MV - How taken: HM Ship-of-the-Line Warspite - When taken: 3 Mar 1813 - Where taken: Bay of Biscay - Date received: 19 Mar 1813 - From what ship: HMS Warspite - Born: Massachusetts - Age: 23 - Sent to Dartmoor on 2 Apr 1813.

Mitchell, John - Seaman - Number: 192 - Prize name: Hunter - Ship type: P - How taken: HM Frigate Phoebe - When taken: 24 Dec 1812 - Where taken: off Western Islands - Date received: 9 Jan 1813 - From what ship: HMS Phoebe

- Born: Baltimore - Age: 18 - Sent to Portsmouth on 8 Feb 1813 on the HMS Colossus.

Mitchell, John - Seaman - Number: 2523 - Prize name: Hope, prize to the Privateer True Blooded Yankee - Ship type: P - How taken: HM Frigate Seahorse - When taken: 22 Mar 1814 - Where taken: at sea - Date received: 6 Apr 1814 - From what ship: HMS Queen Charlotte - Born: Maryland - Age: 22 - Race: Negro - Discharged on 10 May 1814 and sent to Dartmoor.

Mix, William A. - Seaman - Number: 711 - Prize name: Star - Ship type: MV - How taken: HM Ship-of-the-Line Superb - When taken: 9 Feb 1813 - Where taken: Bay of Biscay - Date received: 19 Mar 1813 - From what ship: HMS Warspite - Born: New Haven - Age: 18 - Sent to Dartmoor on 2 Apr 1813.

Mode, David - Cook - Number: 1167 - Prize name: Hebe - Ship type: MV - How taken: HM Frigate Stag - When taken: 18 Apr 1813 - Where taken: Bay of Biscay - Date received: 6 May 1813 - From what ship: HMS Stag - Born: Delaware - Age: 21 - Sent on 3 Jul 1813 to Stapleton prison.

Modre, Joseph - Seaman - Number: 2355 - Prize name: Zephyr - Ship type: MV - How taken: HM Frigate Surveillante - When taken: 6 Jan 1814 - Where taken: Bay of Biscay - Date received: 31 Jan 1814 - From what ship: HMS Surveillante - Born: Portugal - Age: 26 - Discharged on 27 Feb 1814 and sent to Chatham on HMS Haleyon.

Moffit, John - Steward - Number: 978 - Prize name: Grand Napoleon - Ship type: MV - How taken: HM Frigate Belle Poule - When taken: 3 Apr 1813 - Where taken: off Bordeaux - Date received: 11 Apr 1813 - From what ship: Napoleon - Born: New York - Age: 38 - Sent to Chatham on 8 Jul 1813 on HM Tender Neptune.

Molineux, William - Seaman - Number: 1788 - Prize name: Jane Barns - Ship type: MV - How taken: HM Frigate Comus - When taken: 14 May 1813 - Where taken: off Lisbon - Date received: 13 Aug 1813 - From what ship: HMS Protection - Born: Frankford, PA - Age: 23 - Discharged on 8 Sep 1813 and sent to Dartmoor.

Mone, Denis - Seaman - Number: 2194 - Prize name: General Kempt, prize of the Privateer Grand Turk - Ship type: P - How taken: HM Brig Foxhound - When taken: 18 Dec 1813 - Where taken: Lat 48.4 Long 6 - Date received: 21 Dec 1813 - From what ship: HMS Foxhound - Born: Salem - Age: 28 - Discharged on 31 Jan 1814 and sent to Dartmoor.

Monett, Samuel - Seaman - Number: 204 - Prize name: Hunter - Ship type: P - How taken: HM Frigate Phoebe - When taken: 24 Dec 1812 - Where taken: off Western Islands - Date received: 9 Jan 1813 - From what ship: HMS Phoebe - Born: Salem - Age: 19 - Sent to Portsmouth on 8 Feb 1813 on the HMS Colossus.

Moneys, Robert - Seaman - Number: 345 - Prize name: Phoebe & Jane - Ship type: MV - How taken: HMS Rhodian - When taken: Aug 1812 - Where taken: off Charleston - Date received: 27 Jan 1813 - From what ship: Malvina - Born: Stockholm - Age: 28 - Died on 12 Mar 1813 at Mill Prison.

Monks, John - Seaman - Number: 718 - Prize name: Criterion - Ship type: MV - How taken: HM Frigate Belle Poule - When taken: 14 Feb 1813 - Where taken: Bay of Biscay - Date received: 19 Mar 1813 - From what ship: HMS Warspite - Born: Massachusetts - Age: 29 - Sent to Dartmoor on 2 Apr 1813.

Monroe, John - Seaman - Number: 1647 - Prize name: Tickler - Ship type: LM - How taken: HM Frigate Magiciene - When taken: 5 Jun 1813 - Where taken: Lat 47 Long 13 - Date received: 14 Jun 1813 - From what ship: HMS Orestes - Born: New York - Age: 30 - Discharged on 30 Jun 1813 and sent to Stapleton.

Montaindon, Julie - Daughter - Number: 2245 - Prize name: Squirrel - Ship type: MV - How taken: HM Frigate Belle Poule - When taken: 14 Dec 1813 - Where taken: Bay of Biscay - Date received: 15 Jan 1814 - From what ship: HMS Bellona - Born: Switzerland - Liberated on 5 Feb 1814.

Montaindon, Louise - Daughter - Number: 2246 - Prize name: Squirrel - Ship type: MV - How taken: HM Frigate Belle Poule - When taken: 14 Dec 1813 - Where taken: Bay of Biscay - Date received: 15 Jan 1814 - From what ship: HMS Bellona - Born: Switzerland - Liberated on 5 Feb 1814.

Montaindon, Aimee - Passenger - Number: 2242 - Prize name: Squirrel - Ship type: MV - How taken: HM Frigate Belle Poule - When taken: 14 Dec 1813 - Where taken: Bay of Biscay - Date received: 15 Jan 1814 - From what ship: HMS Bellona - Born: Switzerland - Age: 17 - Liberated on 5 Feb 1814.

Montaindon, Henry - Passenger - Number: 2243 - Prize name: Squirrel - Ship type: MV - How taken: HM Frigate Belle Poule - When taken: 14 Dec 1813 - Where taken: Bay of Biscay - Date received: 15 Jan 1814 - From what ship: HMS Bellona - Born: Switzerland - Age: 14 - Liberated on 5 Feb 1814.

Montaindon, Julien - Goldsmith & Passenger - Number: 2241 - Prize name: Squirrel - Ship type: MV - How taken: HM Frigate Belle Poule - When taken: 14 Dec 1813 - Where taken: Bay of Biscay - Date received: 15 Jan 1814 - From what ship: HMS Bellona - Born: Switzerland - Age: 26 - Liberated on 5 Feb 1814.

Montaindon, Marianne - Wife - Number: 2244 - Prize name: Squirrel - Ship type: MV - How taken: HM Frigate Belle Poule - When taken: 14 Dec 1813 - Where taken: Bay of Biscay - Date received: 15 Jan 1814 - From what ship: HMS Bellona - Born: Switzerland - Liberated on 5 Feb 1814.

Montaindon, Pierre Hans - Artist & Passenger - Number: 2240 - Prize name: Squirrel - Ship type: MV - How taken: HM Frigate Belle Poule - When taken: 14 Dec 1813 - Where taken: Bay of Biscay - Date received: 15 Jan 1814 - From what ship: HMS Bellona - Born: Switzerland - Age: 26 - Liberated on 5 Feb 1814.

Montel, Pierre - Passenger - Number: 1941 - Prize name: Marmion - Ship type: MV - How taken: HM Frigate President - When taken: 14 Aug 1813 - Where taken: off Nantes - Date received: 27 Aug 1813 - From what ship: HMS Urgent - Born: Guadeloupe - Age: 40 - Race: Negro - Died on 13 Apr 1813 in the Mill Prison Hospital.

Montgomery, John - Seaman - Number: 2039 - How taken: Impressed at Liverpool - When taken: 16 Oct 1813 - Date received: 1 Nov 1813 - From what ship: HMS Bittern - Born: New York - Age: 21 - Race: Black - Discharged on 3 Nov 1813 and sent to Dartmoor.

Montierne, Francois - Seaman - Number: 1086 - Prize name: John & Frances - Ship type: MV - How taken: HM Frigate Belle Poule - When taken: 19 Mar 1813 - Where taken: Bay of Biscay - Date received: 22 Apr 1813 - From what ship: HMS Superb - Born: Plymouth - Age: 23 - Sent to Dartmoor on 1 Jul 1813.

Montierne, Francois - Seaman - Number: 2570 - Prize name: John & Francis - Ship type: MV - How taken: HM Frigate Belle Poule - When taken: 13 Mar 1813 - Where taken: Bay of Biscay - Date received: 28 Apr 1814 - From what ship: Transport Mercury - Born: Sardinia - Age: 23 - Discharged on 10 May 1814 and sent to Dartmoor.

Mooney, Peter - Marine - Number: 1481 - Prize name: Paul Jones - Ship type: P - How taken: HM Frigate Leonidas - When taken: 23 May 1813 - Where taken: off Cape Clear - Date received: 26 May 1813 - From what ship: HMS Leonidas - Born: New Orleans - Age: 18 - Race: Black - Discharged on 30 Jun 1813 and sent to Stapleton.

Moor, Lawrence - Seaman - Number: 846 - Prize name: Pallas - Ship type: MV - How taken: HM Brig Rebuff - When taken: 23 Dec 1812 - Where taken: off Cadiz - Date received: 23 May 1813 - From what ship: HMS Dauntless - Born: New York - Age: 47 - Sent to Dartmoor on 28 Jun 1813.

Moor, Thomas - Cook's Mate - Number: 168 - Prize name: Hunter - Ship type: P - How taken: HM Frigate Phoebe - When taken: 24 Dec 1812 - Where taken: off Western Islands - Date received: 9 Jan 1813 - From what ship: HMS Phoebe - Born: Long Island - Age: 25 - Race: Black - Sent to Portsmouth on 8 Feb 1813 on the HMS Colossus.

Moore, Alexander - Seaman - Number: 2310 - Prize name: Siro - Ship type: LM - How taken: HM Brig Pelican - When taken: 13 Jan 1814 - Where taken: at sea - Date received: 23 Jan 1814 - From what ship: HMS Pelican - Born: New York - Age: 21 - Race: Black - Discharged on 31 Jan 1814 and sent to Dartmoor.

Moore, Benjamin - Seaman - Number: 373 - Prize name: Orbit - Ship type: MV - How taken: HM Brig Achates - When taken: 29 Jan 1813 - Where taken: Lat 49 N Long 13 W - Date received: 4 Feb 1813 - From what ship: HMS Achates - Born: Staten Island - Age: 21 - Sent to Chatham on 29 Mar 1813 on the HMS Braham.

Moore, George - Seaman - Number: 2034 - Prize name: Friendship West, prize of Privateer True Blooded Yankee - Ship type: P - How taken: HM Schooner Helicon & HM Schooner Whiting - When taken: 25 Oct 1813 - Where taken: Bay of Biscay - Date received: 31 Oct 1813 - From what ship: HMS Whiting - Born: New York - Age: 22 - Race: Black - Discharged on 3 Nov 1813 and sent to Dartmoor.

Moore, James - Seaman - Number: 2496 - How taken: Impressed out of MV Francis - When taken: 26 Jan 1814 - Date received: 26 Mar 1814 - From what ship: Earl Spencer, cutter - Born: North Carolina - Age: 40 - Discharged on 10 May 1814 and sent to Dartmoor.

Moore, John - Seaman - Number: 2116 - Prize name: Chesapeake - Ship type: LM - How taken: HM Frigate Hotspur & HM Frigate Pyramus - When taken: 26 Oct 1813 - Where taken: Bay of Biscay - Date received: 22 Nov 1813 - From what ship: HMS Pyramus - Born: Maryland - Age: 29 - Race: Black - Discharged on 29 Nov 1813 and sent to Dartmoor.

Moore, John - Seaman - Number: 2325 - Prize name: Amity, prize of the Privateer Prince de Neufchatel - Ship type:

P - How taken: HM Brig Achates - When taken: 22 Dec 1814 - Where taken: Bay of Biscay - Date received: 25 Jan 1814 - From what ship: HMS Conflict - Born: Messina - Age: 56 - Discharged on 31 Jan 1814 and sent to Dartmoor.

Moore, Richard - Seaman - Number: 1337 - Prize name: Fox - Ship type: LM - How taken: HM Sloop Pheasant - When taken: 23 Apr 1813 - Where taken: Bay of Biscay - Date received: 14 May 1813 - From what ship: HMS Pleasant - Born: Pennsylvania - Age: 35 - Race: Black - Discharged on 3 Jul 1813 and sent to Stapleton Prison.

Moore, William - Seaman - Number: 1985 - Prize name: Ned - Ship type: LM - How taken: HM Brig Royalist - When taken: 6 Sep 1812 - Where taken: coast of France - Date received: 22 Sep 1813 - From what ship: HMS Royalist - Born: Portsmouth - Age: 40 - Discharged on 27 Sep 1813 and sent to Dartmoor.

More, Francis - Steward - Number: 1573 - Prize name: Hannah & Eliza - Ship type: MV - How taken: HM Brig Lyra - When taken: 29 May 1813 - Where taken: off north coast of Spain - Date received: 10 May 1813 - From what ship: Hannah & Eliza - Born: Philadelphia - Age: 22 - Discharged on 30 Jun 1813 and sent to Stapleton.

More, Henry - Seaman - Number: 1938 - Prize name: Marmion - Ship type: MV - How taken: HM Frigate President - When taken: 14 Aug 1813 - Where taken: off Nantes - Date received: 27 Aug 1813 - From what ship: HMS Urgent - Born: New York - Age: 30 - Race: Negro - Discharged on 8 Sep 1813 and sent to Dartmoor.

Morell, Benjamin - Seaman - Number: 1958 - Prize name: Joel Barlow - Ship type: LM - How taken: HM Frigate Briton - When taken: 3 Jul 1813 - Where taken: off Bordeaux - Date received: 31 Aug 1813 - From what ship: HMS Clarence - Born: Connecticut - Age: 18 - Discharged on 8 Sep 1813 and sent to Dartmoor.

Morell, Thomas - Seaman - Number: 501 - Prize name: Cashiere - Ship type: LM - How taken: HM Brig Reindeer - When taken: 3 Feb 1813 - Where taken: Bay of Biscay - Date received: 12 Feb 1813 - From what ship: HMS Reindeer - Born: New York - Age: 32 - Sent to Dartmoor on 2 Apr 1813.

Morgan, Henry - 2nd Mate - Number: 1218 - Prize name: Grand Napoleon - Ship type: MV - How taken: HM Frigate Belle Poule - When taken: 3 Apr 1813 - Where taken: off Bordeaux - Date received: 9 May 1813 - From what ship: HMS Andromache - Born: Albany - Age: 20 - Discharged on 3 Jul 1813 and sent to Stapleton Prison.

Morgan, John - Seaman - Number: 2406 - Prize name: Rachel & Ann, prize of the Privateer Prince de Neufchatel - Ship type: P - How taken: HM Frigate Cydnus - When taken: 6 Jan 1814 - Where taken: at sea - Date received: 4 Feb 1814 - From what ship: HMS Cydnus - Born: Boston - Age: 32 - Discharged on 10 May 1814 and sent to Dartmoor.

Morgan, John - Seaman - Number: 1273 - Prize name: Caroline - Ship type: MV - How taken: HM Frigate Medusa - When taken: 12 Apr 1813 - Where taken: Bay of Biscay - Date received: 10 May 1813 - From what ship: HMS Medusa - Born: Wilmot - Age: 25 - Discharged on 3 Jul 1813 and sent to Stapleton Prison.

Morgan, Jonathan - Seaman - Number: 2471 - Prize name: Fair American - Ship type: MV - How taken: HM Frigate Andromache - When taken: 19 Jan 1814 - Where taken: Bay of Biscay - Date received: 22 Feb 1814 - From what ship: HMS York - Born: Manchester - Age: 21 - Discharged on 10 May 1814 and sent to Dartmoor.

Morgan, Joseph - Seaman - Number: 1100 - Prize name: Viper - Ship type: MV - How taken: HM Ship-of-the-Line Superb - When taken: 15 Apr 1813 - Where taken: Bay of Biscay - Date received: 22 Apr 1813 - From what ship: HMS Superb - Born: New Jersey - Age: 26 - Sent to Dartmoor on 1 Jul 1813.

Morrie, Joseph - Seaman - Number: 1694 - Prize name: Governor Gerry - Ship type: MV - How taken: HM Brig Royalist - When taken: 31 May 1813 - Where taken: Bay of Biscay - Date received: 26 Jun 1813 - From what ship: HMS Duncan - Born: Lisbon, Portugal - Age: 24 - Discharged on 30 Jun 1813 and sent to Stapleton.

Morris, James - Steward - Number: 920 - Prize name: Tiger - Ship type: MV - How taken: HM Brig Scylla - When taken: 22 Mar 1813 - Where taken: Bay of Biscay - Date received: 5 Apr 1813 - From what ship: HMS Scylla - Born: New York - Age: 29 - Sent on 21 Jun 1813 to Portsmouth on HMS Prometheus.

Morisau, Jean Baptiste - Seaman - Number: 2397 - Prize name: Harvest, prize of the Privateer Bunker Hill - Ship type: P - How taken: HM Brig Orestes - When taken: 21 Jan 1814 - Date received: 3 Feb 1814 - From what ship: HMS Orestes - Born: Cape Francois, Haiti - Age: 18 - Discharged on 10 May 1814 and sent to Dartmoor.

Morrey, Henry - Seaman - Number: 1132 - Prize name: Magdalen - Ship type: MV - How taken: HM Ship-of-the-Line Superb - When taken: 15 Apr 1813 - Where taken: off Belle Isle - Date received: 22 Apr 1813 - From what ship: HMS Superb - Born: New Orleans - Age: 18 - Sent to Dartmoor on 1 Jul 1813.

Morris, George - Seaman - Number: 951 - Prize name: Good Friends - Ship type: MV - How taken: HM Frigate Andromache - When taken: 2 Apr 1813 - Where taken: Bay of Biscay - Date received: 7 Apr 1813 - From what ship: HMS Sea Lark - Born: Portsmouth - Age: 19 - Sent to Dartmoor on 28 Jun 1813.

Morris, Isaac - Seaman - Number: 2625 - Prize name: Ateline - Ship type: MV - How taken: HM Frigate Magiciene - When taken: 14 Mar 1814 - Where taken: off Cape Finisterre, Spain - Date received: 17 May 1814 - From what ship: HMS Tortois - Born: Virginia - Age: 21 - Race: Black - Discharged on 14 Jun 1814 and sent to Dartmoor.

Morris, John - Seaman - Number: 409 - Prize name: Union - Ship type: MV - How taken: HM Frigate Iris - When taken: 17 Jan 1813 - Where taken: Lat 44 N Long 2.3 W - Date received: 5 Feb 1813 - From what ship: HMS San Josef - Born: Delaware - Age: 24 - Sent to Chatham on 29 Mar 1813 on the HMS Braham.

Morris, John - Seaman - Number: 566 - Prize name: Rolla - Ship type: MV - How taken: HM Frigate Surveillante - When taken: 11 Feb 1813 - Where taken: Bay of Biscay - Date received: 23 Feb 1813 - From what ship: HMS Surveillante - Born: Natchez, NY - Age: 28 - Sent to Dartmoor on 2 Apr 1813.

Morris, John - Seaman - Number: 948 - Prize name: Good Friends - Ship type: MV - How taken: HM Frigate Andromache - When taken: 2 Apr 1813 - Where taken: Bay of Biscay - Date received: 7 Apr 1813 - From what ship: HMS Sea Lark - Born: New Orleans - Age: 48 - Sent to Dartmoor on 28 Jun 1813.

Morris, John - Seaman - Number: 2072 - Prize name: Avon, prize of the Privateer True Blooded Yankee - Ship type: P - How taken: HM Frigate Eurotas - When taken: 27 Oct 1813 - Where taken: off Ushant - Date received: 1 Nov 1813 - From what ship: HMS Hannibal - Born: Virginia - Age: 23 - Discharged on 3 Nov 1813 and sent to Dartmoor.

Morris, Patterson - Master - Number: 663 - Prize name: Nope - Ship type: MV - How taken: Chance, privateer - When taken: 15 Feb 1813 - Where taken: off Gorduan - Date received: 10 Mar 1813 - From what ship: Sloop Ann - Born: New Jersey - Age: 34 - Sent on 12 Feb 1813 to Ashburton on parole.

Morris, Robert - Master - Number: 1248 - Prize name: Lightning - Ship type: MV - How taken: HM Frigate Medusa - When taken: 2 Apr 1813 - Where taken: Bay of Biscay - Date received: 10 May 1813 - From what ship: HMS Medusa - Born: Gosport, VA - Age: 36 - Discharged on 14 May 1813 and sent to Ashburton on parole.

Morris, Thomas - Seaman - Number: 422 - Prize name: Union - Ship type: MV - How taken: HM Frigate Iris - When taken: 17 Jan 1813 - Where taken: Lat 44 N Long 2.3 W - Date received: 5 Feb 1813 - From what ship: HMS San Josef - Born: Maryland - Age: 22 - Sent to Chatham on 29 Mar 1813 on the HMS Braham.

Morris, William - Seaman - Number: 237 - How taken: Delivered himself up as a prisoner of war - Date received: 12 Jan 1813 - From what ship: HMS Bittern - Born: Charlestown - Age: 30 - Sent to Portsmouth on 8 Feb 1813 on the HMS Colossus.

Morrison, David - Seaman - Number: 600 - Prize name: St. Martin's Plantation, prize of the Privateer Paul Jones - Ship type: P - How taken: HM Ship-of-the-Line Dublin - When taken: 9 Feb 1815 - Where taken: Lat 43 N Long 33.5 W - Date received: 25 Feb 1813 - From what ship: HMS Dublin - Born: Northumberland - Age: 21 - Sent to Dartmoor on 2 Apr 1813.

Morse, Henry - Seaman - Number: 952 - Prize name: Good Friends - Ship type: MV - How taken: HM Frigate Andromache - When taken: 2 Apr 1813 - Where taken: Bay of Biscay - Date received: 7 Apr 1813 - From what ship: HMS Sea Lark - Born: Vermont - Age: 19 - Sent to Dartmoor on 28 Jun 1813.

Morslender, Reuben - Seaman - Number: 884 - Prize name: Tiger - Ship type: MV - How taken: HM Brig Scylla - When taken: 22 Mar 1813 - Where taken: Bay of Biscay - Date received: 1 Apr 1813 - From what ship: HMS Scylla - Born: Nantucket - Age: 22 - Sent on 21 Jun 1813 to Portsmouth.

Morweek, Andre - Seaman - Number: 2001 - Prize name: Ned - Ship type: LM - How taken: HM Brig Royalist - When taken: 6 Sep 1812 - Where taken: coast of France - Date received: 22 Sep 1813 - From what ship: HMS Royalist - Born: Portland - Age: 21 - Discharged on 27 Sep 1813 and sent to Dartmoor.

Moss, John - Seaman - Number: 949 - Prize name: Good Friends - Ship type: MV - How taken: HM Frigate Andromache - When taken: 2 Apr 1813 - Where taken: Bay of Biscay - Date received: 7 Apr 1813 - From what ship: HMS Sea Lark - Born: Stralsund, Prussia - Age: 34 - Sent to Dartmoor on 28 Jun 1813.

Moss, Thomas - Seaman - Number: 1240 - Prize name: Essex - Ship type: MV - How taken: HM Frigate Pyramus - When taken: 2 Apr 1813 - Where taken: Bay of Biscay - Date received: 9 May 1813 - From what ship: HMS Andromache - Born: Marblehead - Age: 33 - Discharged on 3 Jul 1813 and sent to Stapleton Prison.

Moss, William - Seaman - Number: 241 - How taken: Impressed at Liverpool - When taken: 17 Nov 1812 - Date received: 12 Jan 1813 - From what ship: HMS Bittern - Born: Norfolk - Age: 29 - Sent to Portsmouth on 8 Feb 1813 on the HMS Colossus.

Mouls, Benjamin - Seaman - Number: 1589 - Prize name: Orders in Council - Ship type: LM - How taken: HM Frigate Surveillante - When taken: 1 Jun 1813 - Where taken: Bay of Biscay - Date received: 13 Jun 1813 - From what ship: Forvey - Born: Virginia - Age: 38 - Discharged on 3 Jun 1813 and sent to Dartmoor.

Muggin, John - Seaman - Number: 2052 - How taken: Pallas - Where taken: St. Johns, New Brunswick - Date received: 1 Nov 1813 - From what ship: HMS Bittern - Born: Warren - Age: 39 - Discharged on 3 Nov 1813 and sent to Dartmoor.

Mullan, John - Seaman - Number: 486 - Prize name: Cashiere - Ship type: LM - How taken: HM Brig Reindeer - When taken: 3 Feb 1813 - Where taken: Bay of Biscay - Date received: 12 Feb 1813 - From what ship: HMS Reindeer - Born: New York - Age: 23 - Sent to Dartmoor on 2 Apr 1813.

Mullen, Edward - Seaman - Number: 1357 - Prize name: Fox - Ship type: LM - How taken: HM Sloop Pheasant - When taken: 23 Apr 1813 - Where taken: Bay of Biscay - Date received: 15 May 1813 - From what ship: HMS Foxhound - Born: Pennsylvania - Age: 39 - Discharged on 3 Jul 1813 and sent to Stapleton Prison.

Mullins, James - Seaman - Number: 1530 - Prize name: Courier - Ship type: LM - How taken: HM Brig Rover - When taken: 14 Mar 1813 - Where taken: Bay of Biscay - Date received: 29 May 1813 - From what ship: HMS Hannibal - Born: North Carolina - Age: 40 - Discharged on 30 Jun 1813 and sent to Stapleton.

Mully, Etienne - Seaman - Number: 1436 - Prize name: Paul Jones - Ship type: P - How taken: HM Frigate Leonidas - When taken: 23 May 1813 - Where taken: off Cape Clear - Date received: 26 May 1813 - From what ship: HMS Leonidas - Born: Lorient, France - Age: 23 - Discharged on 30 Jun 1813 and sent to Stapleton.

Mumford, Thomas - 2nd Mate - Number: 1082 - Prize name: John & Frances - Ship type: MV - How taken: HM Frigate Belle Poule - When taken: 19 Mar 1813 - Where taken: Bay of Biscay - Date received: 22 Apr 1813 - From what ship: HMS Superb - Born: Newport - Age: 41 - Sent to Dartmoor on 1 Jul 1813.

Murphy, William - Seaman - Number: 2393 - Prize name: Harvest, prize of the Privateer Bunker Hill - Ship type: P - How taken: HM Brig Orestes - When taken: 21 Jan 1814 - Date received: 3 Feb 1814 - From what ship: HMS Orestes - Born: Philadelphia - Age: 26 - Discharged on 10 May 1814 and sent to Dartmoor.

Murrall, Mark - Seaman - Number: 1326 - Prize name: Shadow - Ship type: LM - How taken: HM Brig Reindeer & HM Schooner Helicon - When taken: 6 Apr 1813 - Where taken: Bay of Biscay - Date received: 12 May 1813 - From what ship: HMS Helicon - Born: Marblehead - Age: 28 - Discharged on 3 Jul 1813 and sent to Stapleton Prison.

Murray, Jacob - Seaman - Number: 1306 - Prize name: Eliza - Ship type: MV - How taken: HM Frigate Surveillante - When taken: 22 Apr 1813 - Where taken: at sea - Date received: 10 May 1813 - From what ship: HMS Medusa - Born: South Carolina - Age: 23 - Race: Black - Discharged on 3 Jul 1813 and sent to Stapleton Prison.

Murray, John - Seaman - Number: 1278 - Prize name: Caroline - Ship type: MV - How taken: HM Frigate Medusa - When taken: 12 Apr 1813 - Where taken: Bay of Biscay - Date received: 10 May 1813 - From what ship: HMS Medusa - Born: Long Island - Age: 35 - Race: Black - Discharged on 2 Sep 1813 and sent to Dartmoor.

Murray, John - Seaman - Number: 73 - Prize name: Experiment - Ship type: MV - How taken: HM Transport Deptford - When taken: 2 Nov 1812 - Where taken: Dublin - Date received: 29 Nov 1812 - From what ship: HMS Salvador del Mundo - Born: New Castle, DE - Age: 22 - Sent to Portsmouth on 29 Dec 1812 on the HMS Northumberland.

Murray, Nathaniel - Seaman - Number: 2585 - Prize name: James - Ship type: MV - How taken: HM Brig Harpy - When taken: 19 Dec 1812 - Where taken: off Isle du France - Date received: 10 May 1814 - From what ship: HMS Lion - Born: Lancaster - Age: 33 - Race: Black - Discharged on 14 Jun 1814 and sent to Dartmoor.

Murry, James - Seaman - Number: 1276 - Prize name: Caroline - Ship type: MV - How taken: HM Frigate Medusa - When taken: 12 Apr 1813 - Where taken: Bay of Biscay - Date received: 10 May 1813 - From what ship: HMS

Medusa - Born: Kent County - Age: 24 - Race: Black - Discharged on 2 Sep 1813 and sent to Dartmoor.

Musick, Joseph - Seaman - Number: 2601 - How taken: MV Furuise - When taken: 23 Sep 1812 - Date received: 16 May 1814 - From what ship: HMS Repulse - Born: Charlestown - Age: 30 - Race: Mulatto - Discharged on 14 Jun 1814 and sent to Dartmoor.

Myatt, Frederick - Seaman - Number: 1835 - Prize name: US Brig Argus - Ship type: War - How taken: HM Brig Pelican - When taken: 14 Aug 1813 - Where taken: Irish Channel - Date received: 17 Aug 1813 - From what ship: USS Argus - Born: Baltimore - Age: 25 - Discharged on 8 Sep 1813 and sent to Dartmoor.

Myer, John - Cooper - Number: 452 - Prize name: Dolphin - Ship type: LM - How taken: HM Ship-of-the-Line Colossus, HM Frigate Rhin & HM Brig Goldfinch - When taken: 5 Jan 1813 - Where taken: off Western Islands - Date received: 6 Feb 1813 - From what ship: HMS Rhin - Born: Philadelphia - Age: 24 - Sent to Chatham on 29 Mar 1813 on the HMS Braham.

Myers, Daniel - Seaman - Number: 2461 - Prize name: Fair American - Ship type: MV - How taken: HM Frigate Andromache - When taken: 19 Jan 1814 - Where taken: Bay of Biscay - Date received: 22 Feb 1814 - From what ship: HMS York - Born: Philadelphia - Age: 24 - Discharged on 10 May 1814 and sent to Dartmoor.

Myers, David - Seaman - Number: 966 - Prize name: Ferox - Ship type: MV - How taken: HM Frigate Medusa & HM Brig Lyra - When taken: 28 Mar 1813 - Where taken: off Cape Ortagle - Date received: 9 Apr 1813 - From what ship: HMS Lyra - Born: Thomastown, MS - Age: 28 - Sent to Chatham on 8 Jul 1813 on HM Tender Neptune.

Myers, Frederick - Seaman - Number: 324 - Prize name: Porcupine - Ship type: MV - How taken: HM Frigate Dryad - When taken: 8 Jan 1813 - Where taken: off Bordeaux - Date received: 21 Jan 1813 - From what ship: HMS Abercrombie - Born: Baltimore - Age: 37 - Sent to Chatham on 29 Mar 1813 on the HMS Braham.

Nagel, George - Seaman - Number: 1836 - Prize name: US Brig Argus - Ship type: War - How taken: HM Brig Pelican - When taken: 14 Aug 1813 - Where taken: Irish Channel - Date received: 17 Aug 1813 - From what ship: USS Argus - Born: Maryland - Age: 32 - Discharged on 8 Sep 1813 and sent to Dartmoor.

Nargney, James - Seaman - Number: 461 - Prize name: Dolphin - Ship type: LM - How taken: HM Ship-of-the-Line Colossus, HM Frigate Rhin & HM Brig Goldfinch - When taken: 5 Jan 1813 - Where taken: off Western Islands - Date received: 6 Feb 1813 - From what ship: HMS Rhin - Born: Pennsylvania - Age: 28 - Sent to Chatham on 29 Mar 1813 on the HMS Braham.

Norris, Augusta George - Seaman - Number: 1669 - Prize name: Commerce - Ship type: MV - How taken: Impressed at Liverpool - When taken: 4 May 1813 - Date received: 17 Jun 1813 - From what ship: HMS Bittern - Born: Rhode Island - Age: 25 - Discharged on 30 Jun 1813 and sent to Stapleton.

Nartigue, John - 2nd Mate - Number: 260 - Prize name: Brutus - Ship type: MV - How taken: HM Frigate Briton - When taken: Jan 1813 - Where taken: Bay of Biscay - Date received: 21 Jan 1813 - From what ship: HMS Briton - Born: Georgia - Age: 24 - Sent to Portsmouth on 8 Feb 1813 on the HMS Colossus.

Nash, Manuel - Seaman - Number: 1880 - Prize name: US Brig Argus - Ship type: War - How taken: HM Brig Pelican - When taken: 14 Aug 1813 - Where taken: Irish Channel - Date received: 23 Aug 1813 - From what ship: HMS Pelican - Born: New Orleans - Age: 18 - Discharged on 8 Sep 1813 and sent to Dartmoor.

Neal, Dennis - Seaman - Number: 1562 - Prize name: Courier - Ship type: LM - How taken: HM Brig Rover - When taken: 14 Mar 1813 - Where taken: Bay of Biscay - Date received: 29 May 1813 - From what ship: HMS Hannibal - Born: Maryland - Age: 25 - Discharged on 30 Jun 1813 and sent to Stapleton.

Neal, Henry - Seaman - Number: 197 - Prize name: Hunter - Ship type: P - How taken: HM Frigate Phoebe - When taken: 24 Dec 1812 - Where taken: off Western Islands - Date received: 9 Jan 1813 - From what ship: HMS Phoebe - Born: Boston - Age: 18 - Sent to Portsmouth on 8 Feb 1813 on the HMS Colossus.

Neal, John - Gunner - Number: 2275 - Prize name: Siro - Ship type: LM - How taken: HM Brig Pelican - When taken: 13 Jan 1814 - Where taken: at sea - Date received: 23 Jan 1814 - From what ship: HMS Pelican - Born: Baltimore - Age: 36 - Discharged on 31 Jan 1814 and sent to Dartmoor.

Neall, Daniel - Seaman - Number: 2341 - Prize name: Apparencen, prize of the Privateer Bunker Hill - Ship type: P - How taken: Cartdian - When taken: 27 Jan 1814 - Where taken: off Ushant - Date received: 30 Jan 1814 - From what ship: Cardian - Born: Portsmouth - Age: 21 - Discharged on 31 Jan 1814 and sent to Dartmoor.

Neath, Henry - Seaman - Number: 799 - Prize name: Cannoniere - Ship type: P - How taken: HM Ship-of-the-Line Warspite - When taken: 14 Mar 1813 - Where taken: Bay of Biscay - Date received: 19 Mar 1813 - From what ship: HMS Warspite - Born: New York - Age: 18 - Sent to Dartmoor on 28 Jun 1813.

Neaver, Thomas - Seaman - Number: 283 - Prize name: Columbia - Ship type: MV - How taken: HM Frigate Briton - When taken: 17 Dec 1812 - Where taken: off Bordeaux - Date received: 21 Jan 1813 - From what ship: HMS Abercrombie - Born: Philadelphia - Age: 20 - Sent to Portsmouth on 8 Feb 1813 on the HMS Colossus.

Neel, David - Seaman - Number: 640 - Prize name: Criterion - Ship type: MV - How taken: HM Frigate Belle Poule - When taken: 14 Feb 1813 - Where taken: Bay of Biscay - Date received: 4 Mar 1813 - From what ship: HMS Strenuous - Born: Philadelphia - Age: 27 - Sent to Dartmoor on 2 Apr 1813.

Neel, William - Seaman - Number: 2545 - Prize name: Mary, prize to the Privateer Blockhead - Ship type: P - How taken: HM Post Ship Crocodile - When taken: 6 Aug 1813 - Where taken: Coruna, Spain - Date received: 16 Apr 1814 - From what ship: Transport Fanny - Born: Fredericktown - Age: 23 - Discharged on 10 May 1814 and sent to Dartmoor.

Neeve, Thomas - Seaman - Number: 2580 - How taken: Delivered himself up from the HMS Seminomas - When taken: 20 Jan 1814 - Where taken: Cape of Good Hope - Date received: 10 May 1814 - From what ship: HMS Lion - Age: 29 - Liberated on 13 Jun 1814.

Neil, Henry - Seaman - Number: 1986 - Prize name: Ned - Ship type: LM - How taken: HM Brig Royalist - When taken: 6 Sep 1812 - Where taken: coast of France - Date received: 22 Sep 1813 - From what ship: HMS Royalist - Born: New York - Age: 25 - Discharged on 27 Sep 1813 and sent to Dartmoor.

Neil, James - Mater - Number: 34 - Prize name: Wasp - Ship type: MV - How taken: Earl Spencer, cutter - When taken: 4 Aug 1812 - Where taken: off Cape Clear - Date received: 23 Nov 1812 - From what ship: HMS Stork - Born: New Bedford - Age: 25 - Sent on 8 Dec 1812 to Ashburton on parole.

Nelley, Richard John - Seaman - Number: 1379 - Prize name: Tom - Ship type: LM - How taken: HM Frigate Surveillante - When taken: 27 Apr 1813 - Where taken: Bay of Biscay - Date received: 15 May 1813 - From what ship: HMS Foxhound - Born: Pennsylvania - Age: 33 - Discharged on 3 Jul 1813 and sent to Stapleton Prison.

Nelson, Richard - Seaman - Number: 1044 - How taken: Delivered himself up from HM Ship-of-the-Line Ajax - Date received: 22 Apr 1813 - From what ship: HMS Ajax - Born: New York - Age: 25 - Sent to Chatham on 8 Jul 1813 on HM Tender Neptune.

Nelson, Thomas - Seaman - Number: 2136 - Prize name: Amiable - Ship type: LM - How taken: HM Ship-of-the-Line Magnificant - When taken: 30 Oct 1813 - Where taken: off Lorient - Date received: 29 Nov 1813 - From what ship: HMS Dublin - Born: Delaware - Age: 18 - Discharged on 4 Dec 1813 and sent to Dartmoor.

Nesbitt, Richard - Seaman - Number: 2565 - Prize name: Dorotha - Ship type: MV - How taken: HM Brig Teazer - When taken: 26 Mar 1814 - Where taken: off Cork - Date received: 25 Apr 1814 - From what ship: HMS Helena - Born: Philadelphia - Age: 27 - Discharged on 10 May 1814 and sent to Dartmoor.

Newby, Lamb - Seaman - Number: 1973 - How taken: Out of a British merchant vessel - When taken: 28 Aug 1813 - Date received: 19 Sep 1813 - From what ship: HMS Bittern - Born: New York - Age: 40 - Race: Black - Discharged on 27 Sep 1813 and sent to Dartmoor.

Newell, Isaac - Seaman - Number: 1449 - Prize name: Paul Jones - Ship type: P - How taken: HM Frigate Leonidas - When taken: 23 May 1813 - Where taken: off Cape Clear - Date received: 26 May 1813 - From what ship: HMS Leonidas - Born: York, MA - Age: 20 - Discharged on 30 Jun 1813 and sent to Stapleton.

Newell, John - Cook - Number: 510 - Prize name: Tyger - Ship type: MV - How taken: Detained at Gibralter - When taken: 8 Aug 1812 - Date received: 15 Feb 1813 - From what ship: HMS Andromeda - Born: Africa - Age: 32 - Race: Negro - Sent to Dartmoor on 2 Apr 1813.

Newell, Nathaniel - Carpenter - Number: 2145 - Prize name: Amiable - Ship type: LM - How taken: HM Ship-of-the-Line Magnificant - When taken: 30 Oct 1813 - Where taken: off Lorient - Date received: 29 Nov 1813 - From what ship: HMS Dublin - Born: Providence - Age: 30 - Discharged on 4 Dec 1813 and sent to Dartmoor.

Newman, Daniel - Seaman - Number: 2628 - Prize name: Ateline - Ship type: MV - How taken: HM Frigate Magiciene - When taken: 14 Mar 1814 - Where taken: off Cape Finisterre, Spain - Date received: 17 May 1814 - From what ship: HMS Tortois - Born: Newburyport - Age: 26 - Discharged on 14 Jun 1814 and sent to Dartmoor.

Nickerson, Josiah - Seaman - Number: 1588 - Prize name: Orders in Council - Ship type: LM - How taken: HM Frigate Surveillante - When taken: 1 Jun 1813 - Where taken: Bay of Biscay - Date received: 13 Jun 1813 - From what ship: Forvey - Born: Boston - Age: 19 - Discharged on 30 Jun 1813 and sent to Stapleton.

Nicholas, John - Seaman - Number: 894 - Prize name: Tiger - Ship type: MV - How taken: HM Brig Scylla - When taken: 22 Mar 1813 - Where taken: Bay of Biscay - Date received: 1 Apr 1813 - From what ship: HMS Scylla - Born: New York - Age: 18 - Sent on 21 Jun 1813 to Portsmouth on HMS Prometheus.

Nicholas, John Bruss - Seaman - Number: 2098 - Prize name: Dart - Ship type: MV - How taken: HM Frigate Niger - When taken: 12 Nov 1813 - Where taken: off Cape Finisterre, Spain - Date received: 19 Nov 1813 - From what ship: Dart - Born: Massachusetts - Age: 24 - Discharged on 31 Jan 1814 and sent to Dartmoor.

Nicholas, Stephen - Seaman - Number: 1520 - Prize name: Grand Napoleon - Ship type: MV - How taken: HM Brig Goldfinch - When taken: 17 Apr 1813 - Where taken: Bay of Biscay - Date received: 29 May 1813 - From what ship: HMS Hannibal - Born: Prussia - Age: 29 - Discharged on 30 Jun 1813 and sent to Stapleton.

Nicholls, John - Seaman - Number: 2479 - How taken: Impressed at Liverpool - When taken: 2 Feb 1814 - Date received: 13 Mar 1814 - From what ship: HMS Bittern - Born: New Orleans - Age: 27 - Discharged on 10 May 1814 and sent to Dartmoor.

Nicholson, James - Seaman - Number: 2047 - How taken: Impressed at Liverpool - When taken: 16 Oct 1813 - Date received: 1 Nov 1813 - From what ship: HMS Bittern - Born: Jersey - Age: 24 - Discharged on 3 Nov 1813 and sent to Dartmoor.

Nicholson, James - 2nd Mate - Number: 2056 - How taken: Taken off a Russian merchant vessel - Date received: 1 Nov 1813 - From what ship: HMS Rippon - Born: Chatham - Age: 22 - Discharged on 3 Nov 1813 and sent to Dartmoor.

Nones, A. B. - Supercargo - Number: 610 - Prize name: Governor McKean - Ship type: LM - How taken: HM Brig Rover - When taken: 26 Jun 1813 - Where taken: off Bordeaux - Date received: 1 Mar 1813 - From what ship: HMS Insolent - Born: Philadelphia - Age: 19 - Sent on 3 Mar 1813 to Ashburton on parole.

Nooney, Peter - Boy - Number: 1916 - Prize name: US Brig Argus - Ship type: War - How taken: HM Brig Pelican - When taken: 14 Aug 1813 - Where taken: Irish Channel - Date received: 23 Aug 1813 - From what ship: HMS Pelican - Born: Bordeaux - Age: 16 - Discharged on 8 Sep 1813 and sent to Dartmoor.

Nooney, William - Seaman - Number: 2529 - Prize name: Bunker Hill - Ship type: P - How taken: HM Frigate Pomone & HM Frigate Cydnus - When taken: 4 Mar 1814 - Where taken: Bay of Biscay - Date received: 6 Apr 1814 - From what ship: HMS Primrose - Born: Philadelphia - Age: 25 - Discharged on 10 May 1814 and sent to Dartmoor.

Norcross, Nehemiah - Seaman - Number: 2450 - How taken: Taken out of a Russian ship - When taken: 28 Jan 1814 - Where taken: Cove of Cork - Date received: 15 Feb 1814 - From what ship: HMS Zealous - Born: Boston - Age: 27 - Discharged on 10 May 1814 and sent to Dartmoor.

Norgrave, Jeremiah - 1st Mate - Number: 2582 - Prize name: James - Ship type: MV - How taken: HM Brig Harpy - When taken: 19 Dec 1812 - Where taken: off Isle du France - Date received: 10 May 1814 - From what ship: HMS Lion - Age: 34 - Discharged on 12 May 1814 and sent to Ashburton on parole.

Norman, Michael - Seaman - Number: 2686 - How taken: Sent into custody from HM Ship-of-the-Line Cornwallis - Date received: 7 Jun 1814 - From what ship: Inap - Born: Baltimore - Age: 25 - Discharged on 14 Jun 1814 and sent to Dartmoor.

Norman, Peter - Seaman - Number: 740 - Prize name: Charlotte - Ship type: MV - How taken: HM Ship-of-the-Line Warspite - When taken: 3 Mar 1813 - Where taken: Bay of Biscay - Date received: 19 Mar 1813 - From what ship: HMS Warspite - Born: Falmouth - Age: 36 - Sent to Dartmoor on 2 Apr 1813.

Norris, Lament - Seaman - Number: 1141 - Prize name: Magdalen - Ship type: MV - How taken: HM Ship-of-the-Line Superb - When taken: 15 Apr 1813 - Where taken: off Belle Isle - Date received: 22 Apr 1813 - From what ship: HMS Superb - Born: Massachusetts - Age: 20 - Sent to Dartmoor on 1 Jul 1813.

Norris, Robert - Seaman - Number: 1995 - Prize name: Ned - Ship type: LM - How taken: HM Brig Royalist - When taken: 6 Sep 1812 - Where taken: coast of France - Date received: 22 Sep 1813 - From what ship: HMS Royalist

- Born: New York - Age: 17 - Discharged on 27 Sep 1813 and sent to Dartmoor.

Norton, George - Seaman - Number: 809 - Prize name: Decornau - Ship type: MV - How taken: HM Sloop Pheasant - When taken: 15 Mar 1813 - Where taken: Bay of Biscay - Date received: 19 Mar 1813 - From what ship: HMS Pheasant - Born: New Bedford - Age: 44 - Sent to Dartmoor on 28 Jun 1813.

Noyes, Charles - Seaman - Number: 1889 - Prize name: Betsey, prize to the US Brig Argus - Ship type: War - How taken: HM Frigate Leonidas - When taken: 12 Aug 1813 - Where taken: Channel - Date received: 23 Aug 1813 - From what ship: HMS Pelican - Born: Andover - Age: 28 - Discharged on 8 Sep 1813 and sent to Dartmoor.

Nuganes, Daniel - Seaman - Number: 2698 - Prize name: Shadow - Ship type: LM - How taken: HM Brig Reindeer & HM Schooner Helicon - When taken: 6 Apr 1813 - Where taken: Bay of Biscay - Date received: 17 Jun 1814 - From what ship: Dartmouth - Discharged on 24 Jun 1814 and sent to the Royal Hospital.

Nugent, John - Seaman - Number: 2062 - Prize name: Betsy, prize to the Privateer True Blooded Yankee - Ship type: P - How taken: HM Frigate Eurotas - When taken: 26 Oct 1813 - Where taken: off Ushant - Date received: 1 Nov 1813 - From what ship: HMS Hannibal - Born: New York - Age: 18 - Discharged on 3 Nov 1813 and sent to Dartmoor.

Nugent, John - Seaman - Number: 1805 - Prize name: US Brig Argus - Ship type: War - How taken: HM Brig Pelican - When taken: 14 Aug 1813 - Where taken: Irish Channel - Date received: 17 Aug 1813 - From what ship: USS Argus - Born: Philadelphia - Age: 27 - Discharged on 8 Sep 1813 and sent to Dartmoor.

Null, William - Seaman - Number: 737 - Prize name: Pert - Ship type: MV - How taken: HM Ship-of-the-Line Warspite - When taken: 1 Mar 1813 - Where taken: Bay of Biscay - Date received: 19 Mar 1813 - From what ship: HMS Warspite - Born: New York - Age: 31 - Sent to Dartmoor on 2 Apr 1813.

Nye, Charles N. - Boatswain - Number: 166 - Prize name: Hunter - Ship type: P - How taken: HM Frigate Phoebe - When taken: 24 Dec 1812 - Where taken: off Western Islands - Date received: 9 Jan 1813 - From what ship: HMS Phoebe - Born: Norwalk - Age: 24 - Sent to Portsmouth on 8 Feb 1813 on the HMS Colossus.

Nye, Cornelius - Seaman - Number: 2296 - Prize name: Siro - Ship type: LM - How taken: HM Brig Pelican - When taken: 13 Jan 1814 - Where taken: at sea - Date received: 23 Jan 1814 - From what ship: HMS Pelican - Born: Boston - Age: 21 - Discharged on 31 Jan 1814 and sent to Dartmoor.

Oakley, Jacob - Seaman - Number: 2021 - How taken: Delivered himself up from HM Frigate Armide - When taken: 10 Mar 1813 - Date received: 15 Oct 1813 - From what ship: Transport Lord Cathcart - Born: New York - Age: 25 - Discharged on 3 Nov 1813 and sent to Dartmoor.

Oakley, Joseph - Seaman - Number: 2118 - Prize name: Chesapeake - Ship type: LM - How taken: HM Frigate Hotspur & HM Frigate Pyramus - When taken: 26 Oct 1813 - Where taken: Bay of Biscay - Date received: 22 Nov 1813 - From what ship: HMS Pyramus - Born: Delaware - Age: 27 - Discharged on 29 Nov 1813 and sent to Dartmoor.

Ober, Ezra - Passenger - Number: 2418 - Prize name: Deborah - Ship type: MV - How taken: Cutter Nimble - When taken: 19 Jan 1814 - Where taken: off Saint Andrea - Date received: 10 Feb 1814 - From what ship: HMS Racer - Born: Beverly - Age: 21 - Discharged on 13 Feb 1814 and sent to Ashburton on parole.

Obrion, Jeuth. - Marine Officer - Number: 144 - Prize name: Hunter - Ship type: P - How taken: HM Frigate Phoebe - When taken: 24 Dec 1812 - Where taken: off Western Islands - Date received: 9 Jan 1813 - From what ship: HMS Phoebe - Born: Newbury - Age: 22 - Sent to Portsmouth on 8 Feb 1813 on the HMS Colossus.

Odeen, John - Seaman - Number: 516 - Prize name: Allegany - Ship type: MV - How taken: Detained at Gibraltar - When taken: 8 Aug 1812 - Date received: 15 Feb 1813 - From what ship: HMS Andromeda - Born: Alexandria - Age: 21 - Sent to Dartmoor on 2 Apr 1813.

Ogden, James - Seaman - Number: 2035 - How taken: Impressed at Liverpool - When taken: 16 Sep 1813 - Date received: 1 Nov 1813 - From what ship: HMS Bittern - Born: Eastport - Age: 27 - Discharged on 3 Nov 1813 and sent to Dartmoor.

Olcott, Jedediah - Master - Number: 2435 - Prize name: Zephyr - Ship type: MV - How taken: HM Frigate Pyramus - When taken: 30 Nov 1813 - Where taken: off Lorient - Date received: 14 Feb 1814 - From what ship: Plymouth - Born: Hartford - Age: 43 - Discharged on 14 Feb 1814 and sent to London.

Oleily, Thomas - Mate - Number: 127 - Prize name: Nope - Ship type: MV - How taken: HM Sloop Pheasant - When taken: 13 Dec 1812 - Where taken: off Oporto - Date received: 2 Jan 1813 - From what ship: Nope - Born: Virginia - Age: 28 - Sent to Portsmouth on 4 Jan 1813 on the HMS Revolutionnaire.

Olmstead, Moss - Master - Number: 468 - Prize name: Union - Ship type: MV - How taken: HM Frigate Iris - When taken: 17 Jan 1813 - Where taken: Lat 44 N Long 2.3 W - Date received: 8 Feb 1813 - From what ship: Union - Born: Connecticut - Age: 34 - Sent on 9 Feb 1813 to Ashburton on parole.

Olsen, Elias - Seaman - Number: 2077 - Prize name: Avon, prize of the Privateer True Blooded Yankee - Ship type: P - How taken: HM Frigate Eurotas - When taken: 27 Oct 1813 - Where taken: off Ushant - Date received: 1 Nov 1813 - From what ship: HMS Hannibal - Born: Sweden - Age: 23 - Discharged on 3 Nov 1813 and sent to Dartmoor.

Orne, Oliver P. - Boy - Number: 2365 - Prize name: Hannah - Ship type: MV - How taken: HM Ship-of-the-Line Conquestador - When taken: 14 Jan 1814 - Where taken: Lat 47.3 Long 7 - Date received: 31 Jan 1814 - From what ship: HMS Surveillante - Born: Marblehead - Age: 13 - Discharged on 1 Mar 1814 and sent to Ashburton on parole.

Orne, William B. - 3rd Lieutenant - Number: 2506 - Prize name: Bunker Hill - Ship type: P - How taken: HM Frigate Pomone & HM Frigate Cydnus - When taken: 4 Mar 1814 - Where taken: Bay of Biscay - Date received: 4 Apr 1814 - From what ship: HMS Virago - Born: Marblehead - Age: 29 - Discharged on 10 May 1814 and sent to Dartmoor.

Osborne, Peter - Seaman - Number: 15 - Prize name: Rising Sun - Ship type: MV - How taken: Taken at Liverpool - When taken: 18 Oct 1812 - Date received: 21 Nov 1812 - From what ship: HMS Salvador del Mundo - Born: New London - Age: 23 - Sent to Portsmouth on 29 Dec 1812 on the HMS Northumberland.

Ostend, Michael - Seaman - Number: 1521 - Prize name: Grand Napoleon - Ship type: MV - How taken: HM Brig Goldfinch - When taken: 17 Apr 1813 - Where taken: Bay of Biscay - Date received: 29 May 1813 - From what ship: HMS Hannibal - Born: Paris - Age: 22 - Discharged on 30 Jun 1813 and sent to Stapleton.

Ostin, John - Seaman - Number: 605 - How taken: Delivered himself up from HM Brig Foxhound - Date received: 27 Feb 1813 - From what ship: HMS Salvador del Mundo - Born: Boston - Age: 23 - Sent to Dartmoor on 2 Apr 1813.

Otis, Charles - Mate - Number: 1361 - Prize name: Tom - Ship type: LM - How taken: HM Frigate Surveillante - When taken: 27 Apr 1813 - Where taken: Bay of Biscay - Date received: 15 May 1813 - From what ship: HMS Foxhound - Born: Massachusetts - Age: 34 - Discharged on 23 May 1813 and sent to Ashburton on parole.

Oudin, Pierre Francois - Match Maker & Passenger - Number: 2103 - Prize name: Chesapeake - Ship type: LM - How taken: HM Frigate Hotspur & HM Frigate Pyramus - When taken: 26 Oct 1813 - Where taken: Bay of Biscay - Date received: 22 Nov 1813 - From what ship: HMS Pyramus - Born: Sedan, France - Age: 37 - Discharged on 25 Aug 1813 and sent to Ashburton on parole.

Pack, Abraham - Seaman - Number: 584 - Prize name: Cashiere - Ship type: LM - How taken: HM Brig Reindeer - When taken: 3 Feb 1813 - Where taken: Bay of Biscay - Date received: 23 Feb 1813 - From what ship: HMS Surveillante - Born: Harford - Age: 26 - Race: Black - Sent to Dartmoor on 2 Apr 1813.

Packard, William - Seaman - Number: 79 - How taken: Delivered himself up from HM Frigate Belle Poule - When taken: 10 Dec 1812 - Date received: 10 Dec 1812 - From what ship: HMS Belle Poule - Born: Bridgewater - Age: 23 - Sent to Portsmouth on 29 Dec 1812 on the HMS Northumberland.

Paddock, Benjamin Mead - Seaman - Number: 346 - Prize name: Argus - Ship type: MV - How taken: Fancy, Cutter - When taken: 19 Dec 1812 - Where taken: Bay of Biscay - Date received: 28 Jan 1813 - From what ship: HMS Plymouth - Born: Rhinebeck - Age: 24 - Sent to Chatham on 29 Mar 1813 on the HMS Braham.

Page, John - Seaman - Number: 1614 - Prize name: Revenge - Ship type: LM - How taken: HM Frigate Belle Poule - When taken: 11 May 1813 - Where taken: Bay of Biscay - Date received: 14 Jun 1813 - From what ship: HMS Royalist - Born: Massachusetts - Age: 25 - Discharged on 30 Jun 1813 and sent to Stapleton.

Page, John - Seaman - Number: 1745 - Prize name: Revenge - Ship type: LM - How taken: HM Frigate Belle Poule - When taken: 11 May 1813 - Where taken: Bay of Biscay - Date received: 13 Jun 1813 - From what ship: Retaken after escape - Born: Massachusetts - Age: 25 - Discharged on 8 Sep 1813 and sent to Dartmoor.

Page, Josiah Clark - Master - Number: 2533 - Prize name: Sally - Ship type: MV - How taken: HM Brig Derwent - When taken: 2 Feb 1814 - Where taken: Bay of Biscay - Date received: 11 Apr 1814 - From what ship: Transport Hydra - Born: Salem - Age: 36 - Discharged on 11 Apr 1814 and sent to Ashburton on parole.

Page, Thomas - Seaman - Number: 1172 - Prize name: Hebe - Ship type: MV - How taken: HM Frigate Stag - When taken: 18 Apr 1813 - Where taken: Bay of Biscay - Date received: 6 May 1813 - From what ship: HMS Stag - Born: Penobscot - Age: 20 - Sent on 3 Jul 1813 to Stapleton prison.

Pagechin, Henry - Seaman - Number: 2215 - Prize name: Squirrel - Ship type: MV - How taken: HM Frigate Belle Poule - When taken: 14 Dec 1813 - Where taken: Bay of Biscay - Date received: 27 Dec 1813 - From what ship: HMS Protector - Born: Bremen - Age: 21 - Discharged on 31 Jan 1814 and sent to Dartmoor.

Pain, George - Seaman - Number: 2294 - Prize name: Siro - Ship type: LM - How taken: HM Brig Pelican - When taken: 13 Jan 1814 - Where taken: at sea - Date received: 23 Jan 1814 - From what ship: HMS Pelican - Born: Boston - Age: 24 - Discharged on 31 Jan 1814 and sent to Dartmoor.

Pain, James - Unknown - Number: 912 - How taken: Delivered himself up as a prisoner of war - Date received: 4 Apr 1813 - From what ship: HMS Warspite - Born: Block Island - Age: 22 - Sent on 21 Jun 1813 to Portsmouth on HMS Prometheus.

Paine, Joseph - Seaman - Number: 1125 - Prize name: Viper - Ship type: MV - How taken: HM Ship-of-the-Line Superb - When taken: 15 Apr 1813 - Where taken: Bay of Biscay - Date received: 22 Apr 1813 - From what ship: HMS Superb - Born: New Orleans - Age: 33 - Sent to Dartmoor on 1 Jul 1813.

Palmer, John - Seaman - Number: 793 - Prize name: Cannoniere - Ship type: P - How taken: HM Ship-of-the-Line Warspite - When taken: 14 Mar 1813 - Where taken: Bay of Biscay - Date received: 19 Mar 1813 - From what ship: HMS Warspite - Born: Wilmington - Age: 27 - Sent to Dartmoor on 28 Jun 1813.

Palmer, John - Seaman - Number: 2115 - Prize name: Chesapeake - Ship type: LM - How taken: HM Frigate Hotspur & HM Frigate Pyramus - When taken: 26 Oct 1813 - Where taken: Bay of Biscay - Date received: 22 Nov 1813 - From what ship: HMS Pyramus - Born: Maryland - Age: 27 - Discharged on 29 Nov 1813 and sent to Dartmoor.

Palmer, Pero - Seaman - Number: 1955 - Prize name: Joel Barlow - Ship type: LM - How taken: HM Frigate Briton - When taken: 3 Jul 1813 - Where taken: off Bordeaux - Date received: 31 Aug 1813 - From what ship: HMS Clarence - Born: Connecticut - Age: 22 - Race: Black - Discharged on 8 Sep 1813 and sent to Dartmoor.

Palmer, Robert - 2nd Mate - Number: 1960 - Prize name: Joel Barlow - Ship type: LM - How taken: HM Frigate Briton - When taken: 3 Jul 1813 - Where taken: off Bordeaux - Date received: 31 Aug 1813 - From what ship: HMS Clarence - Born: Stonington - Age: 28 - Discharged on 14 Sep 1814 and sent to Ashburton on parole.

Pannell, Hugh - Seaman - Number: 455 - Prize name: Dolphin - Ship type: LM - How taken: HM Ship-of-the-Line Colossus, HM Frigate Rhin & HM Brig Goldfinch - When taken: 5 Jan 1813 - Where taken: off Western Islands - Date received: 6 Feb 1813 - From what ship: HMS Rhin - Born: Baltimore - Age: 28 - Sent to Chatham on 29 Mar 1813 on the HMS Braham.

Pardoe, John - Seaman - Number: 49 - How taken: HM Battery Princess - When taken: 5 Nov 1812 - Where taken: Liverpool - Date received: 26 Nov 1813 - From what ship: HMS Frederick - Born: Bridgeport - Age: 23 - Sent to Portsmouth on 29 Dec 1812 on the HMS Northumberland.

Paris, William - Seaman - Number: 1898 - Prize name: Betsey, prize to the US Brig Argus - Ship type: War - How taken: HM Frigate Leonidas - When taken: 12 Aug 1813 - Where taken: Channel - Date received: 23 Aug 1813 - From what ship: HMS Pelican - Born: South Hanistead - Age: 23 - Race: Black - Discharged on 8 Sep 1813 and sent to Dartmoor.

Parish, Samuel - Seaman - Number: 1515 - Prize name: Grand Napoleon - Ship type: MV - How taken: HM Brig Goldfinch - When taken: 17 Apr 1813 - Where taken: Bay of Biscay - Date received: 29 May 1813 - From what ship: HMS Hannibal - Born: Norfolk - Age: 31 - Discharged on 30 Jun 1813 and sent to Stapleton.

Parker, David - Seaman - Number: 426 - Prize name: Union - Ship type: MV - How taken: HM Frigate Iris - When taken: 17 Jan 1813 - Where taken: Lat 44 N Long 2.3 W - Date received: 5 Feb 1813 - From what ship: HMS San Josef - Born: Massachusetts - Age: 18 - Sent to Chatham on 29 Mar 1813 on the HMS Braham.

Parker, Edward - Seaman - Number: 2688 - How taken: Sent into custody from HM Ship-of-the-Line Cornwallis -

Date received: 7 Jun 1814 - From what ship: Inap - Born: Elizabethtown - Age: 23 - Discharged on 14 Jun 1814 and sent to Dartmoor.

Parker, George - Seaman - Number: 724 - Prize name: Mars - Ship type: MV - How taken: HM Ship-of-the-Line Warspite - When taken: 26 Feb 1813 - Where taken: Bay of Biscay - Date received: 19 Mar 1813 - From what ship: HMS Warspite - Born: New York - Age: 25 - Sent to Dartmoor on 2 Apr 1813.

Parker, John A. - Seaman - Number: 1234 - Prize name: Essex - Ship type: MV - How taken: HM Frigate Pyramus - When taken: 2 Apr 1813 - Where taken: Bay of Biscay - Date received: 9 May 1813 - From what ship: HMS Andromache - Born: Boston - Age: 16 - Discharged on 3 Jul 1813 and sent to Stapleton Prison.

Parker, Peter - Seaman - Number: 2540 - Prize name: Vengeance (French) - Ship type: MV - How taken: HM Frigate Herald - When taken: 13 Jun 1813 - Date received: 13 Apr 1814 - From what ship: HMS Bittern - Born: Boston - Age: 24 - Discharged on 10 May 1814 and sent to Dartmoor.

Parker, Samuel - Seaman - Number: 267 - Prize name: Brutus - Ship type: MV - How taken: HM Frigate Briton - When taken: Jan 1813 - Where taken: Bay of Biscay - Date received: 21 Jan 1813 - From what ship: HMS Briton - Born: New Orleans - Age: 23 - Race: Black - Sent to Portsmouth on 8 Feb 1813 on the HMS Colossus.

Parley, William - Seaman - Number: 2195 - Prize name: General Kempt, prize of the Privateer Grand Turk - Ship type: P - How taken: HM Brig Foxhound - When taken: 18 Dec 1813 - Where taken: Lat 48.4 Long 6 - Date received: 21 Dec 1813 - From what ship: HMS Foxhound - Born: Exeter - Age: 20 - Discharged on 31 Jan 1814 and sent to Dartmoor.

Parmell, Benjamin - Seaman - Number: 1567 - Prize name: Miranda, prize of the Privateer Paul Jones - Ship type: P - How taken: HM Frigate Unicorn - When taken: 21 Mar 1813 - Where taken: off Ushant - Date received: 9 Jun 1813 - From what ship: HMS Conquestador - Born: Connecticut - Age: 29 - Discharged on 30 Jun 1813 and sent to Stapleton.

Parr, James - Seaman - Number: 328 - Prize name: Porcupine - Ship type: MV - How taken: HM Frigate Dryad - When taken: 8 Jan 1813 - Where taken: off Bordeaux - Date received: 21 Jan 1813 - From what ship: HMS Abercrombie - Born: Philadelphia - Age: 20 - Sent to Chatham on 29 Mar 1813 on the HMS Braham.

Parsons, Daniel - Seaman - Number: 1707 - Prize name: Joseph - Ship type: MV - How taken: HM Frigate Iris - When taken: 8 Jun 1813 - Where taken: Bay of Biscay - Date received: 4 Jul 1813 - From what ship: HMS Iris - Born: Cape Ann - Age: 29 - Discharged on 22 Aug 1813 and released to HM Brig Redpole.

Parsons, Ignatius - Seaman - Number: 1494 - How taken: Delivered himself up from HM Brig Foxhound - Date received: 27 May 1813 - From what ship: Dartmoor prison - Born: Gloucester - Age: 24 - Discharged on 8 Jul 1813 and sent to Chatham on HM Tender Neptune.

Parsons, Ignatius - Seaman - Number: 608 - How taken: Delivered himself up from HM Brig Foxhound - Date received: 27 Feb 1813 - From what ship: HMS Salvador del Mundo - Born: Gloucester - Age: 24 - Sent to Dartmoor on 2 Apr 1813.

Parsons, Samuel - Seaman - Number: 170 - Prize name: Hunter - Ship type: P - How taken: HM Frigate Phoebe - When taken: 24 Dec 1812 - Where taken: off Western Islands - Date received: 9 Jan 1813 - From what ship: HMS Phoebe - Born: Virginia - Age: 32 - Sent to Portsmouth on 8 Feb 1813 on the HMS Colossus.

Parsons, Samuel D. - Seaman - Number: 1474 - Prize name: Paul Jones - Ship type: P - How taken: HM Frigate Leonidas - When taken: 23 May 1813 - Where taken: off Cape Clear - Date received: 26 May 1813 - From what ship: HMS Leonidas - Born: Springfield - Age: 29 - Discharged on 8 Sep 1813 and sent to Dartmoor.

Pass, Samuel William - Seaman - Number: 2099 - Prize name: Dart - Ship type: MV - How taken: HM Frigate Niger - When taken: 12 Nov 1813 - Where taken: off Cape Finisterre, Spain - Date received: 19 Nov 1813 - From what ship: Dart - Born: Warwick, RI - Age: 24 - Discharged on 31 Jan 1814 and sent to Dartmoor.

Patten, James - Seaman - Number: 1784 - Prize name: Orders in Council - Ship type: LM - How taken: HM Frigate Surveillante - When taken: 1 Jun 1813 - Where taken: Bay of Biscay - Date received: 7 Aug 1813 - From what ship: HMS Gleaner - Born: New York - Age: 21 - Discharged on 8 Sep 1813 and sent to Dartmoor.

Patterson, Joshua - Seaman - Number: 2300 - Prize name: Siro - Ship type: LM - How taken: HM Brig Pelican - When taken: 13 Jan 1814 - Where taken: at sea - Date received: 23 Jan 1814 - From what ship: HMS Pelican - Born:

Stettin, Prussia - Age: 32 - Discharged on 31 Jan 1814 and sent to Dartmoor.

Patterson, Andrew - Seaman - Number: 2284 - Prize name: Siro - Ship type: LM - How taken: HM Brig Pelican - When taken: 13 Jan 1814 - Where taken: at sea - Date received: 23 Jan 1814 - From what ship: HMS Pelican - Born: Pillau, Prussia - Age: 42 - Discharged on 31 Jan 1814 and sent to Dartmoor.

Patterson, John - Seaman - Number: 2474 - Prize name: Siro - Ship type: LM - How taken: HM Brig Pelican - When taken: 13 Jan 1814 - Where taken: at sea - Date received: 28 Feb 1814 - From what ship: Dartmoor prison - Born: Stettin, Prussia - Age: 32 - Discharged on 4 Mar 1814 and sent to Harwich for HMS Rosemond.

Pattingale, Enoch - Seaman - Number: 24 - Prize name: Philips Burgh - Ship type: MV - How taken: Taken at Liverpool - When taken: 9 Nov 1812 - Date received: 21 Nov 1812 - From what ship: HMS Salvador del Mundo - Born: Boston - Age: 24 - Sent to Portsmouth on 29 Dec 1812 on the HMS Northumberland.

Patton, Robert - Seaman - Number: 1184 - How taken: Delivered himself up from HM Ship-of-the-Line Dublin - Date received: 8 May 1813 - From what ship: HMS Dublin - Born: Charleston - Age: 34 - Sent to Chatham on 8 Jul 1813 on HM Tender Neptune.

Paulin, Nicholas - Seaman - Number: 1024 - Prize name: Two Brothers - Ship type: MV - How taken: Bootle of Liverpool, letter of marque - When taken: 18 Mar 1813 - Where taken: Western Islands - Date received: 21 Apr 1813 - From what ship: HMS Bittern - Born: New Orleans - Age: 36 - Sent to Dartmoor on 1 Jul 1813.

Paulin, Nicholas - Seaman - Number: 2569 - Prize name: Two Brothers - Ship type: MV - How taken: Bootle of Liverpool, letter of marque - When taken: 18 Mar 1813 - Where taken: Western Islands - Date received: 28 Apr 1814 - From what ship: Transport Mercury - Born: New Orleans - Age: 36 - Discharged on 10 May 1814 and sent to Dartmoor.

Payechin, Henry - Seaman - Number: 2693 - Prize name: Squirrel - Ship type: MV - How taken: HM Frigate Belle Poule - When taken: 14 Dec 1813 - Where taken: Bay of Biscay - Date received: 17 Jun 1814 - From what ship: Dartmouth - Born: Bremen - Age: 21 - Discharged on 20 Jun 1814 and released to the Speedwell Cartel.

Payer, Walter - Seaman - Number: 1197 - Prize name: Zebra - Ship type: LM - How taken: HM Frigate Pyramus - When taken: 20 Apr 1813 - Where taken: Bay of Biscay - Date received: 9 May 1813 - From what ship: HMS Andromache - Born: New Orleans - Age: 34 - Sent on 3 Jul 1813 to Stapleton prison.

Payne, Joseph Smith - 2nd Mate - Number: 1401 - Prize name: Tom - Ship type: LM - How taken: HM Frigate Surveillante - When taken: 27 Apr 1813 - Where taken: Bay of Biscay - Date received: 18 May 1813 - From what ship: Tom - Born: Charlestown - Age: 16 - Discharged on 3 Jul 1813 and sent to Stapleton Prison.

Peane, Henry - Seaman - Number: 2140 - Prize name: Amiable - Ship type: LM - How taken: HM Ship-of-the-Line Magnificant - When taken: 30 Oct 1813 - Where taken: off Lorient - Date received: 29 Nov 1813 - From what ship: HMS Dublin - Born: Philadelphia - Age: 26 - Discharged on 4 Dec 1813 and sent to Dartmoor.

Pearce, Alexander - Seaman - Number: 2100 - Prize name: Dart - Ship type: MV - How taken: HM Frigate Niger - When taken: 12 Nov 1813 - Where taken: off Cape Finisterre, Spain - Date received: 20 Nov 1813 - From what ship: Dart - Born: Philadelphia - Age: 17 - Discharged on 29 Nov 1813 and sent to Dartmoor.

Pearce, Edward - Seaman - Number: 46 - How taken: Delivered himself up from HM Frigate Circe - Date received: 25 Nov 1813 - From what ship: HMS Stork - Born: Baltimore - Age: 26 - Sent to Portsmouth on 29 Dec 1812 on the HMS Northumberland.

Pearer, John - Boatswain - Number: 2110 - Prize name: Chesapeake - Ship type: LM - How taken: HM Frigate Hotspur & HM Frigate Pyramus - When taken: 26 Oct 1813 - Where taken: Bay of Biscay - Date received: 22 Nov 1813 - From what ship: HMS Pyramus - Born: Pennsylvania - Age: 36 - Discharged on 29 Nov 1813 and sent to Dartmoor.

Pearson, Samuel - Seaman - Number: 3 - Prize name: Charles - Ship type: MV - How taken: HM Brig Intelligent - When taken: 1 Aug 1812 - Where taken: Channel - Date received: 22 Oct 1812 - From what ship: HMS Frederick - Born: Gloucester - Age: 22 - Sent to Portsmouth on 29 Dec 1812 on the HMS Northumberland.

Peck, James - Seaman - Number: 1439 - Prize name: Paul Jones - Ship type: P - How taken: HM Frigate Leonidas - When taken: 23 May 1813 - Where taken: off Cape Clear - Date received: 26 May 1813 - From what ship: HMS Leonidas - Born: New London - Age: 39 - Race: Black - Discharged on 30 Jun 1813 and sent to Stapleton.

Packham, Henry - Chief Mate - Number: 1799 - Prize name: Godfrey & Mary - Ship type: MV - How taken: Robert

Todd, privateer - When taken: 23 Jun 1813 - Where taken: Lat 21.3N Long 53W - Date received: 16 Aug 1813 - From what ship: HMS Bittern - Born: Stonington - Age: 20 - Discharged on 8 Sep 1813 and sent to Dartmoor.

Peirce, Joseph - Chief Mate - Number: 2150 - Prize name: Agnes, prize of the Privateer Rambler - Ship type: LM - How taken: Cutter Jane - When taken: 28 Nov 1813 - Where taken: Bay of Biscay - Date received: 10 Dec 1813 - From what ship: Cutter Jane - Born: Boston - Age: 21 - Discharged on 31 Jan 1814 and sent to Dartmoor.

Pelham, Robert - Lieutenant & Prize Master - Number: 2339 - Prize name: Apparencen, prize of the Privateer Bunker Hill - Ship type: P - How taken: Cartdian - When taken: 27 Jan 1814 - Where taken: off Ushant - Date received: 30 Jan 1814 - From what ship: Cardian - Born: Boston - Age: 29 - Discharged on 31 Jan 1814 and sent to Dartmoor.

Penman, Richard - Seaman - Number: 1389 - Prize name: Tom - Ship type: LM - How taken: HM Frigate Surveillante - When taken: 27 Apr 1813 - Where taken: Bay of Biscay - Date received: 15 May 1813 - From what ship: HMS Foxhound - Born: New York - Age: 23 - Discharged on 3 Jul 1813 and sent to Stapleton Prison.

Peregrine, Taggert - Seaman - Number: 1266 - Prize name: Caroline - Ship type: MV - How taken: HM Frigate Medusa - When taken: 12 Apr 1813 - Where taken: Bay of Biscay - Date received: 10 May 1813 - From what ship: HMS Medusa - Born: Maryland - Age: 28 - Discharged on 3 Jul 1813 and sent to Stapleton Prison.

Perez, Joseph - Seaman - Number: 264 - Prize name: Brutus - Ship type: MV - How taken: HM Frigate Briton - When taken: Jan 1813 - Where taken: Bay of Biscay - Date received: 21 Jan 1813 - From what ship: HMS Briton - Born: Figueria, Portugal - Age: 22 - Sent to Portsmouth on 8 Feb 1813 on the HMS Colossus.

Peridis, Pierre - Pilot - Number: 2346 - Prize name: Apparencen, prize of the Privateer Bunker Hill - Ship type: P - How taken: Cartdian - When taken: 27 Jan 1814 - Where taken: off Ushant - Date received: 30 Jan 1814 - From what ship: Cardian - Born: Brest - Age: 57 - Discharged on 31 Jan 1814 and sent to Dartmoor.

Perkins, Elijah - Seaman - Number: 2200 - Prize name: General Kempt, prize of the Privateer Grand Turk - Ship type: P - How taken: HM Brig Foxhound - When taken: 18 Dec 1813 - Where taken: Lat 48.4 Long 6 - Date received: 21 Dec 1813 - From what ship: HMS Foxhound - Born: Salem - Age: 23 - Discharged on 31 Jan 1814 and sent to Dartmoor.

Perkins, John - Carpenter's Mate - Number: 2282 - Prize name: Siro - Ship type: LM - How taken: HM Brig Pelican - When taken: 13 Jan 1814 - Where taken: at sea - Date received: 23 Jan 1814 - From what ship: HMS Pelican - Born: New Hampton - Age: 25 - Discharged on 31 Jan 1814 and sent to Dartmoor.

Perkins, Thomas - Seaman - Number: 2135 - Prize name: Amiable - Ship type: LM - How taken: HM Ship-of-the-Line Magnificant - When taken: 30 Oct 1813 - Where taken: off Lorient - Date received: 29 Nov 1813 - From what ship: HMS Dublin - Born: Massachusetts - Age: 28 - Discharged on 4 Dec 1813 and sent to Dartmoor.

Perne, George - Seaman - Number: 1072 - Prize name: Meteor - Ship type: MV - How taken: HM Frigate Briton - When taken: 12 Mar 1813 - Where taken: Bay of Biscay - Date received: 22 Apr 1813 - From what ship: HMS Superb - Born: New Orleans - Age: 30 - Sent to Dartmoor on 8 Sep 1813.

Perney, William - Seaman - Number: 1065 - Prize name: Meteor - Ship type: MV - How taken: HM Frigate Briton - When taken: 12 Mar 1813 - Where taken: Bay of Biscay - Date received: 22 Apr 1813 - From what ship: HMS Superb - Born: New York - Age: 19 - Sent to Dartmoor on 8 Sep 1813.

Perrott, Jean Francois - Seaman - Number: 2374 - Prize name: Devon, prize to the Privateer Bunker Hill - Ship type: P - How taken: HM Brig Fly - When taken: 21 Jan 1814 - Where taken: at sea - Date received: 31 Jan 1814 - From what ship: HMS Fly - Born: France - Age: 69 - Discharged on 27 Feb 1814 and sent to Chatham on HMS Haleyon.

Perry, Charles - Seaman - Number: 1332 - Prize name: Fox - Ship type: LM - How taken: HM Sloop Pheasant - When taken: 23 Apr 1813 - Where taken: Bay of Biscay - Date received: 14 May 1813 - From what ship: HMS Pleasant - Born: Norfolk - Age: 29 - Discharged on 3 Jul 1813 and sent to Stapleton Prison.

Peter, John - Seaman - Number: 136 - Prize name: Nope - Ship type: MV - How taken: HM Schooner Bramble - When taken: 3 Dec 1812 - Where taken: Coruna, Spain - Date received: 7 Jan 1813 - From what ship: Nope - Born: Wilmington - Age: 32 - Race: Negro - Sent to Portsmouth on 8 Feb 1813 on the HMS Colossus.

Peters, Aaron - Seaman - Number: 1951 - Prize name: Joel Barlow - Ship type: LM - How taken: HM Frigate Briton - When taken: 3 Jul 1813 - Where taken: off Bordeaux - Date received: 31 Aug 1813 - From what ship: HMS Clarence - Born: Rhode Island - Age: 21 - Race: Black - Discharged on 8 Sep 1813 and sent to Dartmoor.

Peters, Jacob - Seaman - Number: 198 - Prize name: Hunter - Ship type: P - How taken: HM Frigate Phoebe - When taken: 24 Dec 1812 - Where taken: off Western Islands - Date received: 9 Jan 1813 - From what ship: HMS Phoebe - Born: Holland - Age: 25 - Sent to Portsmouth on 8 Feb 1813 on the HMS Colossus.

Peters, John - Seaman - Number: 1627 - Prize name: Leo - Ship type: LM - How taken: HM Frigate Magiciene - When taken: 4 Jun 1813 - Where taken: Lat 45 Long 14 - Date received: 14 Jun 1813 - From what ship: HMS Orestes - Born: Gothenburg, Sweden - Age: 24 - Discharged on 3 Sep 1813.

Peters, Peter - Seaman - Number: 2043 - How taken: Impressed at Liverpool - When taken: 16 Oct 1813 - Date received: 1 Nov 1813 - From what ship: HMS Bittern - Born: Exeter - Age: 20 - Discharged on 3 Nov 1813 and sent to Dartmoor.

Peterson, Andre - Seaman - Number: 2473 - Prize name: Siro - Ship type: LM - How taken: HM Brig Pelican - When taken: 13 Jan 1814 - Where taken: at sea - Date received: 28 Feb 1814 - From what ship: Dartmoor prison - Born: Pillau, Prussia - Age: 42 - Discharged on 4 Mar 1814 and sent to Harwich for HMS Rosemond.

Peterson, Hance - Seaman - Number: 526 - Prize name: Phoenix - Ship type: MV - How taken: Detained at Gibraltar - When taken: 8 Aug 1812 - Date received: 15 Feb 1813 - From what ship: HMS Andromeda - Born: Salem - Age: 17 - Sent to Dartmoor on 2 Apr 1813.

Peterson, James - Boatswain's Mate - Number: 162 - Prize name: Hunter - Ship type: P - How taken: HM Frigate Phoebe - When taken: 24 Dec 1812 - Where taken: off Western Islands - Date received: 9 Jan 1813 - From what ship: HMS Phoebe - Born: Foxborough - Age: 28 - Sent to Portsmouth on 8 Feb 1813 on the HMS Colossus.

Peterson, John - Seaman - Number: 234 - How taken: Delivered himself up from HM Hospital Ship Prince Frederick - When taken: 11 Jan 1813 - Date received: 11 Jan 1813 - From what ship: HMS Frederick - Born: New York - Age: 34 - Race: Black - Sent to Portsmouth on 8 Feb 1813 on the HMS Colossus.

Peterson, John - Mate - Number: 38 - Prize name: Warren - Ship type: MV - How taken: HM Frigate Sybille & HM Frigate Fortunee- When taken: 5 Sep 1812 - Where taken: Lat 41.4 Long 33 - Date received: 23 Nov 1812 - From what ship: HMS Stork - Born: Salem - Age: 29 - Sent on 8 Dec 1812 to Ashburton on parole.

Peterson, Laurence - Seaman - Number: 2138 - Prize name: Amiable - Ship type: LM - How taken: HM Ship-of-the-Line Magnificant - When taken: 30 Oct 1813 - Where taken: off Lorient - Date received: 29 Nov 1813 - From what ship: HMS Dublin - Born: Norrkoping, Sweden - Age: 20 - Discharged on 4 Dec 1813 and sent to Dartmoor.

Peterson, Nicholas - Seaman - Number: 177 - Prize name: Hunter - Ship type: P - How taken: HM Frigate Phoebe - When taken: 24 Dec 1812 - Where taken: off Western Islands - Date received: 9 Jan 1813 - From what ship: HMS Phoebe - Born: New Orleans - Age: 38 - Sent to Portsmouth on 8 Feb 1813 on the HMS Colossus.

Peterson, Peter - Seaman - Number: 1450 - Prize name: Paul Jones - Ship type: P - How taken: HM Frigate Leonidas - When taken: 23 May 1813 - Where taken: off Cape Clear - Date received: 26 May 1813 - From what ship: HMS Leonidas - Born: Gothenburg, Sweden - Age: 32 - Discharged on 30 Jun 1813 and sent to Stapleton.

Peterson, Peter - Seaman - Number: 986 - Prize name: Lightning - Ship type: MV - How taken: HM Frigate Medusa - When taken: 2 Apr 1813 - Where taken: Bay of Biscay - Date received: 16 Apr 1813 - From what ship: HMS Fairy - Born: Sweden - Age: 23 - Liberated on 21 May 1813.

Peterson, Alexander - Seaman - Number: 680 - How taken: Impressed at Liverpool - When taken: 23 Feb 1813 - Date received: 15 Mar 1813 - From what ship: HMS Bittern - Born: New York - Age: 29 - Sent to Dartmoor on 2 Apr 1813.

Philen, Richard - 2nd Mate - Number: 806 - Prize name: Decornau - Ship type: MV - How taken: HM Sloop Pheasant - When taken: 15 Mar 1813 - Where taken: Bay of Biscay - Date received: 19 Mar 1813 - From what ship: HMS Pheasant - Born: Wilmington - Age: 18 - Sent to Dartmoor on 28 Jun 1813.

Phillips, Benjamin - Seaman - Number: 1827 - Prize name: US Brig Argus - Ship type: War - How taken: HM Brig Pelican - When taken: 14 Aug 1813 - Where taken: Irish Channel - Date received: 17 Aug 1813 - From what ship: USS Argus - Born: Massachusetts - Age: 56 - Discharged on 8 Sep 1813 and sent to Dartmoor.

Phillips, James - Seaman - Number: 2497 - How taken: Impressed on shore - When taken: 23 Mar 1814 - Date received: 26 Mar 1814 - From what ship: Earl Spencer, cutter - Born: Harwich - Age: 33 - Discharged on 10 May 1814 and sent to Dartmoor.

Phillips, John - Seaman - Number: 1318 - Prize name: Shadow - Ship type: LM - How taken: HM Brig Reindeer & HM Schooner Helicon - When taken: 6 Apr 1813 - Where taken: Bay of Biscay - Date received: 12 May 1813 - From what ship: HMS Helicon - Born: Pennsylvania - Age: 25 - Discharged on 3 Jul 1813 and sent to Stapleton Prison.

Phillips, John - Chief Mate - Number: 1192 - Prize name: Zebra - Ship type: LM - How taken: HM Frigate Pyramus - When taken: 20 Apr 1813 - Where taken: Bay of Biscay - Date received: 9 May 1813 - From what ship: HMS Andromache - Born: New York - Age: 30 - Sent on 11 May 1813 to Ashburton on parole.

Phillips, John - Seaman - Number: 2673 - Prize name: John, prize to Amelia - Ship type: P - How taken: HM Ship-of-the-Line Sterling Castle - When taken: 10 May 1814 - Where taken: Lat 36 Long 37 - Date received: 5 Jun 1814 - From what ship: HMS Gronville - Born: Philadelphia - Age: 22 - Race: Black - Discharged on 14 Jun 1814 and sent to Dartmoor.

Phippen, John - Seaman - Number: 2166 - Prize name: Wolf, prize of the Privateer Grand Turk - Ship type: P - How taken: HM Frigate Briton - When taken: 1 Dec 1813 - Where taken: Bay of Biscay - Date received: 17 Dec 1813 - From what ship: HMS Briton - Born: Salem - Age: 26 - Discharged on 17 Dec 1813 and released to HMS Britton.

Phipping, Nathaniel - Seaman - Number: 2482 - Prize name: Three Brothers - Ship type: MV - How taken: Ringdoff - When taken: 11 Nov 1813 - Where taken: at sea - Date received: 13 Mar 1814 - From what ship: HMS Bittern - Born: Salem - Age: 23 - Discharged on 10 May 1814 and sent to Dartmoor.

Phipps, Stephen - Seaman - Number: 1211 - Prize name: Zebra - Ship type: LM - How taken: HM Frigate Pyramus - When taken: 20 Apr 1813 - Where taken: Bay of Biscay - Date received: 9 May 1813 - From what ship: HMS Andromache - Born: Massachusetts - Age: 20 - Discharged on 3 Jul 1813 and sent to Stapleton Prison.

Pickering, George - Cook - Number: 1029 - Prize name: Two Brothers - Ship type: MV - How taken: Bootle of Liverpool, letter of marque - When taken: 18 Mar 1813 - Where taken: Western Islands - Date received: 21 Apr 1813 - From what ship: HMS Bittern - Born: Boston - Age: 20 - Race: Black - Sent to Dartmoor on 1 Jul 1813.

Pickrage, George - Seaman - Number: 1646 - Prize name: Tickler - Ship type: LM - How taken: HM Frigate Magiciene - When taken: 5 Jun 1813 - Where taken: Lat 47 Long 13 - Date received: 14 Jun 1813 - From what ship: HMS Orestes - Born: Messina - Age: 24 - Discharged on 8 Sep 1813 and sent to Dartmoor.

Pierce, Amos - Seaman - Number: 659 - How taken: Delivered himself up from HMS Lavinia - Date received: 5 Mar 1813 - From what ship: HMS Salvador del Monde - Born: New Hampshire - Age: 29 - Sent to Dartmoor on 2 Apr 1813.

Pierson, Elijah - Merchant & Passenger - Number: 2218 - Prize name: Zephyr - Ship type: MV - How taken: HM Frigate Pyramus - When taken: 30 Nov 1813 - Where taken: off Lorient - Date received: 30 Dec 1813 - From what ship: Martial - Born: Morristown - Age: 27 - Discharged on 30 Dec 1813 and sent to Ashburton on parole.

Pigott, James - Seaman - Number: 258 - Prize name: Ocean, prized to the Privateer Diligent - Ship type: P - How taken: HM Frigate Surveillante - When taken: 20 Dec 1812 - Where taken: Lat 44 N, Long 6 W - Date received: 21 Jan 1813 - From what ship: HMS Ocean - Born: New Jersey - Age: 20 - Sent to Portsmouth on 8 Feb 1813 on the HMS Colossus.

Pike, John - Seaman - Number: 748 - Prize name: Charlotte - Ship type: MV - How taken: HM Ship-of-the-Line Warspite - When taken: 3 Mar 1813 - Where taken: Bay of Biscay - Date received: 19 Mar 1813 - From what ship: HMS Warspite - Born: Washington - Age: 18 - Sent to Dartmoor on 2 Apr 1813.

Pilton, Alexander - Seaman - Number: 1380 - Prize name: Tom - Ship type: LM - How taken: HM Frigate Surveillante - When taken: 27 Apr 1813 - Where taken: Bay of Biscay - Date received: 15 May 1813 - From what ship: HMS Foxhound - Born: Massachusetts - Age: 21 - Discharged on 3 Jul 1813 and sent to Stapleton Prison.

Pinkham, Amos - Seaman - Number: 2293 - Prize name: Siro - Ship type: LM - How taken: HM Brig Pelican - When taken: 13 Jan 1814 - Where taken: at sea - Date received: 23 Jan 1814 - From what ship: HMS Pelican - Born: Bristol - Age: 30 - Discharged on 31 Jan 1814 and sent to Dartmoor.

Pitts, William - Seaman - Number: 1546 - Prize name: Zebra - Ship type: LM - How taken: HM Frigate Pyramus - When taken: 20 Apr 1813 - Where taken: Bay of Biscay - Date received: 29 May 1813 - From what ship: HMS Hannibal - Born: Massachusetts - Age: 29 - Discharged on 30 Jun 1813 and sent to Stapleton.

Place, John - Armorer - Number: 1823 - Prize name: US Brig Argus - Ship type: War - How taken: HM Brig Pelican - When taken: 14 Aug 1813 - Where taken: Irish Channel - Date received: 17 Aug 1813 - From what ship: USS Argus - Born: New York - Age: 25 - Discharged on 8 Sep 1813 and sent to Dartmoor.

Plan, Madame G. - Woman - Number: 2250 - Prize name: Squirrel - Ship type: MV - How taken: HM Frigate Belle Poule - When taken: 14 Dec 1813 - Where taken: Bay of Biscay - Date received: 15 Jan 1814 - From what ship: HMS Bellona - Born: Switzerland - Discharged on 16 Jan 1814 and sent to Ashburton on parole.

Plumer, William Reed - Seaman - Number: 2353 - Prize name: Zephyr - Ship type: MV - How taken: HM Frigate Surveillante - When taken: 6 Jan 1814 - Where taken: Bay of Biscay - Date received: 31 Jan 1814 - From what ship: HMS Surveillante - Born: Connecticut - Age: 32 - Discharged on 27 Feb 1814 and sent to Chatham on HMS Haleyon.

Plumere, Joseph - Boatswain - Number: 1253 - Prize name: Caroline - Ship type: MV - How taken: HM Frigate Medusa - When taken: 12 Apr 1813 - Where taken: Bay of Biscay - Date received: 10 May 1813 - From what ship: HMS Medusa - Born: Virginia - Age: 23 - Discharged on 3 Jul 1813 and sent to Stapleton Prison.

Poland, David - Prize Master - Number: 149 - Prize name: Hunter - Ship type: P - How taken: HM Frigate Phoebe - When taken: 24 Dec 1812 - Where taken: off Western Islands - Date received: 9 Jan 1813 - From what ship: HMS Phoebe - Born: Ipswich - Age: 22 - Sent to Portsmouth on 8 Feb 1813 on the HMS Colossus.

Poland, Joshua - Seaman - Number: 1513 - Prize name: Grand Napoleon - Ship type: MV - How taken: HM Brig Goldfinch - When taken: 17 Apr 1813 - Where taken: Bay of Biscay - Date received: 29 May 1813 - From what ship: HMS Hannibal - Born: Beverly, MA - Age: 22 - Discharged on 30 Jun 1813 and sent to Stapleton.

Poland, Thomas - Seaman - Number: 2332 - How taken: Impressed at Bristol - When taken: 13 Oct 1813 - Date received: 28 Jan 1814 - From what ship: HMS Salvador del Mundo - Born: Salem - Age: 27 - Discharged on 31 Jan 1814 and sent to Dartmoor.

Polease, Jean M. - Seaman - Number: 2525 - Prize name: Hope, prize to the Privateer True Blooded Yankee - Ship type: P - How taken: HM Frigate Seahorse - When taken: 22 Mar 1814 - Where taken: at sea - Date received: 6 Apr 1814 - From what ship: HMS Queen Charlotte - Born: Brest - Age: 28 - Discharged on 10 May 1814 and sent to Dartmoor.

Pomeril, Robert - Seaman - Number: 1236 - Prize name: Essex - Ship type: MV - How taken: HM Frigate Pyramus - When taken: 2 Apr 1813 - Where taken: Bay of Biscay - Date received: 9 May 1813 - From what ship: HMS Andromache - Born: Massachusetts - Age: 19 - Discharged on 3 Jul 1813 and sent to Stapleton Prison.

Pomp, William - Seaman - Number: 1215 - Prize name: Zebra - Ship type: LM - How taken: HM Frigate Pyramus - When taken: 20 Apr 1813 - Where taken: Bay of Biscay - Date received: 9 May 1813 - From what ship: HMS Andromache - Born: Long Island - Age: 20 - Race: Black - Discharged on 3 Jul 1813 and sent to Stapleton Prison.

Popal, Richard - Seaman - Number: 2321 - Prize name: Amity, prize of the Privateer Prince de Neufchatel - Ship type: P - How taken: HM Brig Achates - When taken: 22 Dec 1814 - Where taken: Bay of Biscay - Date received: 25 Jan 1814 - From what ship: HMS Conflict - Born: Philadelphia - Age: 17 - Discharged on 31 Jan 1814 and sent to Dartmoor.

Porgan, Theodore - Seaman - Number: 307 - Prize name: Stephen - Ship type: MV - How taken: HM Frigate Briton & HM Frigate Andromache - When taken: 17 Dec 1812 - Where taken: off Bordeaux - Date received: 21 Jan 1813 - From what ship: HMS Abercrombie - Born: Stralsund, Prussia - Age: 16 - Sent to Chatham on 29 Mar 1813 on the HMS Braham.

Port, John - Seaman - Number: 2394 - Prize name: Harvest, prize of the Privateer Bunker Hill - Ship type: P - How taken: HM Brig Orestes - When taken: 21 Jan 1814 - Date received: 3 Feb 1814 - From what ship: HMS Orestes - Born: Rhode Island - Age: 21 - Race: Black - Discharged on 27 Feb 1814 and sent to Chatham on HMS Halcyon.

Porter, William - Seaman - Number: 719 - Prize name: Criterion - Ship type: MV - How taken: HM Frigate Belle Poule - When taken: 14 Feb 1813 - Where taken: Bay of Biscay - Date received: 19 Mar 1813 - From what ship: HMS Warspite - Born: Rhode Island - Age: 25 - Sent to Dartmoor on 2 Apr 1813.

Posnrill, John (alias Pavon) - Boy - Number: 2439 - Prize name: Prince of Wales, prize of the Privateer Prince de Neufchatel - Ship type: P - How taken: Transport Nelson - When taken: 4 Feb 1814 - Where taken: at sea - Date received: 14 Feb 1814 - From what ship: HMS Halcyon - Born: Long Island - Age: 16 - Discharged on 10 May

1814 and sent to Dartmoor.

Post, Anthony Oswald - Mate - Number: 81 - Prize name: Argus - Ship type: MV - How taken: Fancy, Cutter - When taken: 19 Dec 1812 - Where taken: Bay of Biscay - Date received: 24 Dec 1812 - From what ship: Fancy, Cutter - Born: New York - Age: 30 - Sent on 8 Dec 1812 to Ashburton on parole.

Postlock, William - 2nd Mate - Number: 2221 - Prize name: Zephyr - Ship type: MV - How taken: HM Frigate Pyramus - When taken: 30 Nov 1813 - Where taken: off Lorient - Date received: 3 Jan 1814 - From what ship: HMS Warspite - Born: Charlestown - Age: 32 - Discharged on 31 Jan 1814 and sent to Dartmoor.

Pote, Jeremiah - Seaman - Number: 1512 - Prize name: Grand Napoleon - Ship type: MV - How taken: HM Brig Goldfinch - When taken: 17 Apr 1813 - Where taken: Bay of Biscay - Date received: 29 May 1813 - From what ship: HMS Hannibal - Born: Massachusetts - Age: 18 - Discharged on 30 Jun 1813 and sent to Stapleton.

Pottenger, William - Midshipman - Number: 1928 - Prize name: US Brig Argus - Ship type: War - How taken: HM Brig Pelican - When taken: 14 Aug 1813 - Where taken: Irish Channel - Date received: 23 Aug 1813 - From what ship: HMS Pelican - Born: Maryland - Age: 20 - Discharged on 26 Aug 1813 and sent to Ashburton on parole.

Potter, John - Seaman - Number: 2500 - How taken: Impressed at Bristol - When taken: 25 Feb 1814 - Date received: 29 Mar 1814 - From what ship: HMS Prince Frederick - Born: Nantucket - Age: 26 - Discharged on 10 May 1814 and sent to Dartmoor.

Potter, Richard H. - 2nd Mate - Number: 2592 - Prize name: Valentine - Ship type: MV - How taken: HM Brig Racehorse - When taken: 16 Nov 1812 - Where taken: Cape of Good Hope - Date received: 10 May 1814 - From what ship: HMS Lion - Born: New London - Age: 30 - Discharged on 14 Jun 1814 and sent to Dartmoor.

Pottigal, John - Seaman - Number: 2629 - Prize name: Ateline - Ship type: MV - How taken: HM Frigate Magiciene - When taken: 14 Mar 1814 - Where taken: off Cape Finisterre, Spain - Date received: 17 May 1814 - From what ship: HMS Tortois - Born: Boston - Age: 46 - Discharged on 14 Jun 1814 and sent to Dartmoor.

Powers, James - 2nd Mate - Number: 956 - Prize name: Courier - Ship type: LM - How taken: HM Frigate Andromache - When taken: 14 Mar 1813 - Where taken: Bay of Biscay - Date received: 7 Apr 1813 - From what ship: HMS Sea Lark - Born: Philadelphia - Age: 26 - Race: Black - Sent to Dartmoor on 28 Jun 1813.

Pratt, Benjamin - Seaman - Number: 1224 - Prize name: Grand Napoleon - Ship type: MV - How taken: HM Frigate Belle Poule - When taken: 3 Apr 1813 - Where taken: off Bordeaux - Date received: 9 May 1813 - From what ship: HMS Andromache - Born: Connecticut - Age: 22 - Discharged on 3 Jul 1813 and sent to Stapleton Prison.

Pratt, Paris - Chief Mate - Number: 1800 - Prize name: Godfrey & Mary - Ship type: MV - How taken: Robert Todd, privateer - When taken: 23 Jun 1813 - Where taken: Lat 21.3N Long 53W - Date received: 16 Aug 1813 - From what ship: HMS Bittern - Born: New London - Age: 24 - Discharged on 8 Sep 1813 and sent to Dartmoor.

Preston, Jonas - Seaman - Number: 1876 - Prize name: Betsey, prize to the US Brig Argus - Ship type: War - How taken: HM Frigate Leonidas - When taken: 12 Aug 1813 - Where taken: Channel - Date received: 23 Aug 1813 - From what ship: HMS Pelican - Born: Wilmington - Age: 20 - Discharged on 8 Sep 1813 and sent to Dartmoor.

Price, David - Seaman - Number: 2180 - Prize name: Wolf, prize of the Privateer Grand Turk - Ship type: P - How taken: HM Frigate Briton - When taken: 1 Dec 1813 - Where taken: Bay of Biscay - Date received: 17 Dec 1813 - From what ship: HMS Briton - Born: Marblehead - Age: 19 - Discharged on 17 Dec 1813 and released to HMS Britton.

Price, Jacob - Seaman - Number: 1289 - Prize name: Price - Ship type: LM - How taken: HM Frigate Medusa - When taken: 13 Apr 1813 - Where taken: Bay of Biscay - Date received: 10 May 1813 - From what ship: HMS Medusa - Born: Connecticut - Age: 23 - Discharged on 3 Jul 1813 and sent to Stapleton Prison.

Price, John - 2nd Mate - Number: 1352 - Prize name: Shadow - Ship type: LM - How taken: HM Brig Reindeer & HM Schooner Helicon - When taken: 6 Apr 1813 - Where taken: Bay of Biscay - Date received: 14 May 1813 - From what ship: HMS Reindeer - Born: Delaware - Age: 22 - Discharged on 3 Jul 1813 and sent to Stapleton Prison.

Price, Samuel - Seaman - Number: 1613 - Prize name: Revenge - Ship type: LM - How taken: HM Frigate Belle Poule - When taken: 11 May 1813 - Where taken: Bay of Biscay - Date received: 14 Jun 1813 - From what ship: HMS Royalist - Born: Boston - Age: 21 - Discharged on 30 Jun 1813 and sent to Stapleton.

Price, William - Seaman - Number: 1891 - Prize name: US Brig Argus - Ship type: War - How taken: HM Brig Pelican - When taken: 14 Aug 1813 - Where taken: Irish Channel - Date received: 23 Aug 1813 - From what ship: HMS Pelican - Born: Philadelphia - Age: 38 - Discharged on 24 Aug 1813 and released to HMS Salvador del Mundo.

Price, William - Seaman - Number: 2424 - Prize name: US Brig Argus - Ship type: War - How taken: Sent from HM Ship-of-the-Line Salvador del Mundo - When taken: 14 Aug 1813 - Where taken: Irish Channel - Date received: 12 Feb 1814 - From what ship: HMS Salvador del Mundo - Born: Philadelphia - Age: 39 - Discharged on 10 May 1814 and sent to Dartmoor.

Priestly, James - Prize Master - Number: 2277 - Prize name: Siro - Ship type: LM - How taken: HM Brig Pelican - When taken: 13 Jan 1814 - Where taken: at sea - Date received: 23 Jan 1814 - From what ship: HMS Pelican - Born: Massachusetts - Age: 21 - Discharged on 31 Jan 1814 and sent to Dartmoor.

Prince, Benjamin - Seaman - Number: 1145 - Prize name: Magdalen - Ship type: MV - How taken: HM Ship-of-the-Line Superb - When taken: 15 Apr 1813 - Where taken: off Belle Isle - Date received: 22 Apr 1813 - From what ship: HMS Superb - Born: Massachusetts - Age: 27 - Sent to Dartmoor on 1 Jul 1813.

Prince, George - Seaman - Number: 1457 - Prize name: Paul Jones - Ship type: P - How taken: HM Frigate Leonidas - When taken: 23 May 1813 - Where taken: off Cape Clear - Date received: 26 May 1813 - From what ship: HMS Leonidas - Born: New Jersey - Age: 32 - Discharged on 30 Jun 1813 and sent to Stapleton.

Prior, Essa - 2nd Prize Master - Number: 2083 - Prize name: Colin West, prize of Privateer True Blooded Yankee - Ship type: P - How taken: HM Schooner Helicon & HM Schooner Whiting - When taken: 26 Oct 1813 - Where taken: Channel - Date received: 2 Nov 1813 - From what ship: HMS Whiting - Born: Connecticut - Age: 25 - Discharged on 3 Nov 1813 and sent to Dartmoor.

Pritchard, Israel - Seaman - Number: 2361 - Prize name: Hannah - Ship type: MV - How taken: HM Ship-of-the-Line Conquestador - When taken: 14 Jan 1814 - Where taken: Lat 47.3 Long 7 - Date received: 31 Jan 1814 - From what ship: HMS Surveillante - Born: Marblehead - Age: 22 - Discharged on 27 Feb 1814 and sent to Chatham on HMS Halcyon.

Prosper, John - Boy - Number: 2289 - Prize name: Siro - Ship type: LM - How taken: HM Brig Pelican - When taken: 13 Jan 1814 - Where taken: at sea - Date received: 23 Jan 1814 - From what ship: HMS Pelican - Born: Savannah - Age: 14 - Discharged on 6 May 1814 and entered on the French General Entry Book.

Prutty, Henry - Seaman - Number: 319 - Prize name: Porcupine - Ship type: MV - How taken: HM Frigate Dryad - When taken: 8 Jan 1813 - Where taken: off Bordeaux - Date received: 21 Jan 1813 - From what ship: HMS Abercrombie - Born: Plymouth, England - Age: 25 - Sent to Chatham on 29 Mar 1813 on the HMS Braham.

Prymroy, Richard - Seaman - Number: 2143 - Prize name: Amiable - Ship type: LM - How taken: HM Ship-of-the-Line Magnificant - When taken: 30 Oct 1813 - Where taken: off Lorient - Date received: 29 Nov 1813 - From what ship: HMS Dublin - Born: North Carolina - Age: 31 - Discharged on 4 Dec 1813 and sent to Dartmoor.

Putman, Charles - Seaman - Number: 1678 - Prize name: Revenge - Ship type: LM - How taken: HM Frigate Belle Poule - When taken: 11 May 1813 - Where taken: Bay of Biscay - Date received: 26 Jun 1813 - From what ship: HMS Duncan - Born: Massachusetts - Age: 30 - Discharged on 30 Jun 1813 and sent to Stapleton.

Pynne, James R. - Seaman - Number: 2049 - How taken: Impressed at Liverpool - When taken: 16 Oct 1813 - Date received: 1 Nov 1813 - From what ship: HMS Bittern - Born: Cheesnen - Age: 26 - Discharged on 3 Nov 1813 and sent to Dartmoor.

Quan, George - Seaman - Number: 306 - Prize name: Stephen - Ship type: MV - How taken: HM Frigate Briton & HM Frigate Andromache - When taken: 17 Dec 1812 - Where taken: off Bordeaux - Date received: 21 Jan 1813 - From what ship: HMS Abercrombie - Born: New Jersey - Age: 25 - Sent to Chatham on 29 Mar 1813 on the HMS Braham.

Quackenbush, William - Seaman - Number: 885 - Prize name: Tiger - Ship type: MV - How taken: HM Brig Scylla - When taken: 22 Mar 1813 - Where taken: Bay of Biscay - Date received: 1 Apr 1813 - From what ship: HMS Scylla - Born: New York - Age: 33 - Sent on 21 Jun 1813 to Portsmouth.

Quinan, Stephen - Seaman - Number: 2179 - Prize name: Wolf, prize of the Privateer Grand Turk - Ship type: P - How taken: HM Frigate Briton - When taken: 1 Dec 1813 - Where taken: Bay of Biscay - Date received: 17 Dec 1813 - From what ship: HMS Briton - Born: Marblehead - Age: 17 - Discharged on 17 Dec 1813 and released to HMS

Britton.

Ramblett, Charles - Mate - Number: 259 - Prize name: Brutus - Ship type: MV - How taken: HM Frigate Briton - When taken: Jan 1813 - Where taken: Bay of Biscay - Date received: 21 Jan 1813 - From what ship: HMS Briton - Born: Virginia - Age: 32 - Sent on 22 Jan 1813 to Ashburton on parole.

Ramsden, John - Boy - Number: 2329 - Prize name: Hannah - Ship type: MV - How taken: HM Ship-of-the-Line Conquestador - When taken: 14 Jan 1814 - Where taken: Lat 47.3 Long 7 - Date received: 28 Jan 1814 - From what ship: Hanna - Born: Marblehead - Age: 14 - Discharged on 31 Jan 1814 and sent to Dartmoor.

Ramsdell, John - Boy - Number: 473 - Prize name: Print - Ship type: MV - How taken: HM Frigate Rhin - When taken: 15 Jan 1813 - Where taken: Lat 44 N Long 17 W - Date received: 9 Feb 1813 - From what ship: HMS Rhin - Born: Marblehead - Age: 11 - Sent on 20 Feb 1813 to Ashburton on parole.

Randale, Frederick - Seaman - Number: 419 - Prize name: Union - Ship type: MV - How taken: HM Frigate Iris - When taken: 17 Jan 1813 - Where taken: Lat 44 N Long 2.3 W - Date received: 5 Feb 1813 - From what ship: HMS San Josef - Born: Maryland - Age: 30 - Sent to Chatham on 29 Mar 1813 on the HMS Braham.

Randol, Thomas - Seaman - Number: 1906 - Prize name: Betsey, prize to the US Brig Argus - Ship type: War - How taken: HM Frigate Leonidas - When taken: 12 Aug 1813 - Where taken: Channel - Date received: 23 Aug 1813 - From what ship: HMS Pelican - Born: New Bedford - Age: 24 - Died on 13 Sep 1813 on the Prison ship Caton.

Ranson, Joseph - Seaman - Number: 1981 - Prize name: Ned - Ship type: LM - How taken: HM Brig Royalist - When taken: 6 Sep 1812 - Where taken: coast of France - Date received: 22 Sep 1813 - From what ship: HMS Royalist - Born: Philadelphia - Age: 23 - Discharged on 27 Sep 1813 and sent to Dartmoor.

Ranter, Thomas - Mate - Number: 248 - Prize name: Rising Sun - Ship type: MV - How taken: HM Ship-Sloop Jalousie - When taken: 12 Dec 1812 - Where taken: Lat 43 Long 20 - Date received: 16 Jan 1813 - From what ship: HMS Stork - Born: Maryland - Age: 23 - Sent on 18 Jan 1813 to Ashburton on parole.

Rape, Nicholas - Seaman - Number: 366 - Prize name: Orbit - Ship type: MV - How taken: HM Brig Achates - When taken: 29 Jan 1813 - Where taken: Lat 49 N Long 13 W - Date received: 4 Feb 1813 - From what ship: HMS Achates - Born: Philadelphia - Age: 40 - Sent to Chatham on 29 Mar 1813 on the HMS Braham.

Raquinis, Francois - Boy - Number: 2345 - Prize name: Apparencen, prize of the Privateer Bunker Hill - Ship type: P - How taken: Cartdian - When taken: 27 Jan 1814 - Where taken: off Ushant - Date received: 30 Jan 1814 - From what ship: Cardian - Born: Brest - Age: 16 - Discharged on 31 Jan 1814 and sent to Dartmoor.

Ratoon, Thomas - Seaman - Number: 835 - Prize name: Pallas - Ship type: MV - How taken: HM Brig Rebuff - When taken: 23 Dec 1812 - Where taken: off Cadiz - Date received: 22 Mar 1813 - From what ship: HMS Dauntless - Born: New York - Age: 20 - Sent to Dartmoor on 28 Jun 1813.

Rawle, Benjamin - Carpenter - Number: 1331 - Prize name: Fox - Ship type: LM - How taken: HM Sloop Pheasant - When taken: 23 Apr 1813 - Where taken: Bay of Biscay - Date received: 14 May 1813 - From what ship: HMS Pleasant - Born: Philadelphia - Age: 48 - Discharged on 3 Jul 1813 and sent to Stapleton Prison.

Ray, Nathaniel - Master - Number: 662 - Prize name: Pert - Ship type: MV - How taken: HM Ship-of-the-Line Warspite - When taken: 1 Mar 1813 - Where taken: Bay of Biscay - Date received: 8 Mar 1813 - From what ship: Pert - Born: Nantucket - Age: 44 - Sent on 10 Feb 1813 to Ashburton on parole.

Reardon, Anthony - Seaman - Number: 840 - Prize name: Pallas - Ship type: MV - How taken: HM Brig Rebuff - When taken: 23 Dec 1812 - Where taken: off Cadiz - Date received: 22 Mar 1813 - From what ship: HMS Dauntless - Born: Trieste - Age: 27 - Sent to Dartmoor on 28 Jun 1813.

Redding, John - Seaman - Number: 1514 - Prize name: Grand Napoleon - Ship type: MV - How taken: HM Brig Goldfinch - When taken: 17 Apr 1813 - Where taken: Bay of Biscay - Date received: 29 May 1813 - From what ship: HMS Hannibal - Born: Falmouth, MA - Age: 23 - Discharged on 30 Jun 1813 and sent to Stapleton.

Redman, William - Seaman - Number: 2164 - Prize name: Princess - Ship type: MV - How taken: Impressed at Liverpool - When taken: 28 Nov 1813 - Date received: 13 Dec 1813 - From what ship: HMS Bittern - Born: Philadelphia - Age: 27 - Discharged on 31 Jan 1814 and sent to Dartmoor.

Reed, Hartwell - Master - Number: 686 - Prize name: Star - Ship type: MV - How taken: HM Ship-of-the-Line Superb - When taken: 9 Feb 1813 - Where taken: Bay of Biscay - Date received: 18 Mar 1813 - From what ship: Star -

Born: New Haven, CT - Age: 29 - Sent on 20 Mar 1813 to Ashburton on parole.

Reeves, James - Seaman - Number: 1676 - Prize name: Rolla - Ship type: MV - How taken: HM Frigate Surveillante - When taken: 11 Feb 1813 - Where taken: Bay of Biscay - Date received: 24 Jun 1813 - From what ship: Dartmoor Prison - Born: Salem - Age: 25 - Discharged on 27 Sep 1813 and sent to Dartmoor.

Reeves, James - Seaman - Number: 572 - Prize name: Rolla - Ship type: MV - How taken: HM Frigate Surveillante - When taken: 11 Feb 1813 - Where taken: Bay of Biscay - Date received: 23 Feb 1813 - From what ship: HMS Surveillante - Born: Salem - Age: 25 - Sent to Dartmoor on 2 Apr 1813.

Reeves, Joseph - Seaman - Number: 404 - Prize name: Union - Ship type: MV - How taken: HM Frigate Iris - When taken: 17 Jan 1813 - Where taken: Lat 44 N Long 2.3 W - Date received: 5 Feb 1813 - From what ship: HMS San Josef - Born: New Jersey - Age: 33 - Sent to Chatham on 29 Mar 1813 on the HMS Braham.

Reeves, William - Boy - Number: 647 - Prize name: Criterion - Ship type: MV - How taken: HM Frigate Belle Poule - When taken: 14 Feb 1813 - Where taken: Bay of Biscay - Date received: 4 Mar 1813 - From what ship: HMS Strenuous - Born: West Chester - Age: 16 - Sent to Dartmoor on 2 Apr 1813.

Regens, Jonathan - Seaman - Number: 616 - Prize name: Governor McKean - Ship type: LM - How taken: HM Brig Rover - When taken: 26 Jun 1813 - Where taken: off Bordeaux - Date received: 2 Mar 1813 - From what ship: HMS Insolent - Born: Salem - Age: 26 - Sent to Dartmoor on 2 Apr 1813.

Reid, Joseph - Ordinary Seaman - Number: 475 - How taken: Impressed at Greenock - When taken: 27 Nov 1812 - Date received: 10 Feb 1813 - From what ship: HMS Frederick - Born: Alexandria, VA - Age: 19 - Sent on 20 Feb 1813 to Ashburton on parole.

Relavoine, Louis - Seaman - Number: 2334 - Prize name: Narius, prize of the Privateer Bunker Hill - Ship type: P - How taken: HM Brig Orestes - When taken: 20 Jan 1814 - Where taken: Bay of Biscay - Date received: 29 Jan 1814 - From what ship: HMS Orestes - Born: Nantes - Age: 27 - Discharged on 31 Jan 1814 and sent to Dartmoor.

Rennabin, Benjamin - Seaman - Number: 1336 - Prize name: Fox - Ship type: LM - How taken: HM Sloop Pheasant - When taken: 23 Apr 1813 - Where taken: Bay of Biscay - Date received: 14 May 1813 - From what ship: HMS Pleasant - Born: Guadeloupe - Age: 28 - Race: Negro - Discharged on 8 Sep 1813 and sent to Dartmoor.

Rerette, John Francis - Passenger - Number: 977 - Prize name: Grand Napoleon - Ship type: MV - How taken: HM Frigate Belle Poule - When taken: 3 Apr 1813 - Where taken: off Bordeaux - Date received: 11 Apr 1813 - From what ship: Napoleon - Born: New Orleans - Age: 25 - Sent on 12 Apr 1813 to Ashburton on parole.

Rey, David - Seaman - Number: 2546 - Prize name: Mary, prize to the Privateer Blockhead - Ship type: P - How taken: HM Post Ship Crocodile - When taken: 6 Aug 1813 - Where taken: Coruna, Spain - Date received: 16 Apr 1814 - From what ship: Transport Fanny - Born: Adam - Age: 23 - Discharged on 10 May 1814 and sent to Dartmoor.

Reynolds, Frederick - Boy - Number: 979 - Prize name: Grand Napoleon - Ship type: MV - How taken: HM Frigate Belle Poule - When taken: 3 Apr 1813 - Where taken: off Bordeaux - Date received: 11 Apr 1813 - From what ship: Napoleon - Born: New York - Age: 18 - Sent to Chatham on 8 Jul 1813 on HM Tender Neptune.

Reynolds, James - Seaman - Number: 1549 - Prize name: Zebra - Ship type: LM - How taken: HM Frigate Pyramus - When taken: 20 Apr 1813 - Where taken: Bay of Biscay - Date received: 29 May 1813 - From what ship: HMS Hannibal - Born: Virginia - Age: 28 - Discharged on 30 Jun 1813 and sent to Stapleton.

Reynolds, Owen - Seaman - Number: 1143 - Prize name: Magdalen - Ship type: MV - How taken: HM Ship-of-the-Line Superb - When taken: 15 Apr 1813 - Where taken: off Belle Isle - Date received: 22 Apr 1813 - From what ship: HMS Superb - Born: Newburyport - Age: 32 - Sent to Dartmoor on 1 Jul 1813.

Rhodrick, Joseph - Seaman - Number: 681 - Prize name: Nope - Ship type: MV - How taken: Chance, privateer - When taken: 15 Feb 1813 - Where taken: off Bordeaux - Date received: 15 Mar 1813 - From what ship: Growler - Born: St. Marys - Age: 39 - Sent to Dartmoor on 2 Apr 1813.

Ricco, Joachim - Seaman - Number: 495 - Prize name: Cashiere - Ship type: LM - How taken: HM Brig Reindeer - When taken: 3 Feb 1813 - Where taken: Bay of Biscay - Date received: 12 Feb 1813 - From what ship: HMS Reindeer - Born: Lima, Peru - Age: 28 - Sent to Dartmoor on 2 Apr 1813.

Rice, Francis - Seaman - Number: 1054 - Prize name: Virginia Planter - Ship type: MV - How taken: HM Frigate

Pyramus - When taken: 17 Mar 1813 - Where taken: off Nantes - Date received: 22 Apr 1813 - From what ship: HMS Superb - Born: Brookfield - Age: 19 - Sent to Dartmoor on 1 Jul 1813.

Rice, John - Seaman - Number: 443 - Prize name: Caroline, prize to the Privateer Industry - Ship type: P - How taken: Recaptured by the crew - Date received: 5 Feb 1813 - From what ship: HMS Neptune - Born: Boston - Age: 23 - Sent to Chatham on 29 Mar 1813 on the HMS Braham.

Rice, John - 2nd Mate - Number: 2584 - Prize name: James - Ship type: MV - How taken: HM Brig Harpy - When taken: 19 Dec 1812 - Where taken: off Isle du France - Date received: 10 May 1814 - From what ship: HMS Lion - Born: Philadelphia - Age: 28 - Discharged on 14 Jun 1814 and sent to Dartmoor.

Rice, Thomas - Seaman - Number: 442 - Prize name: Caroline, prize to the Privateer Industry - Ship type: P - How taken: Recaptured by the crew - Date received: 5 Feb 1813 - From what ship: HMS Neptune - Born: Boston - Age: 25 - Sent to Chatham on 29 Mar 1813 on the HMS Braham.

Rich, Ebenezer - Chief Mate - Number: 1664 - Prize name: Flash - Ship type: MV - How taken: HM Ship-of-the-Line Warspite - When taken: 23 May 1813 - Where taken: coast of France - Date received: 16 Jun 1813 - From what ship: Flash - Born: Portland - Age: 37 - Discharged on 19 Jun 1813 and sent to Ashburton on parole.

Rich, Francis - Seaman - Number: 262 - Prize name: Brutus - Ship type: MV - How taken: HM Frigate Briton - When taken: Jan 1813 - Where taken: Bay of Biscay - Date received: 21 Jan 1813 - From what ship: HMS Briton - Born: New Orleans - Age: 27 - Sent to Portsmouth on 8 Feb 1813 on the HMS Colossus.

Rich, William - Seaman - Number: 1017 - How taken: Delivered himself up from HM Ship-of-the-Line Malta - Where taken: Gibraltar - Date received: 20 Apr 1813 - From what ship: HMS Libria - Born: Maryland - Age: 23 - Race: Black - Sent to Dartmoor on 1 Jul 1813.

Richard, John - Owner - Number: 75 - Prize name: Independence - Ship type: MV - How taken: HM Frigate Medusa - When taken: 9 Nov 1812 - Where taken: San Sebastian - Date received: 29 Nov 1812 - From what ship: HMS Wasp - Born: San Domingo (Haiti) - Age: 34 - Sent on 8 Dec 1812 to Ashburton on parole.

Richards, George - Marine - Number: 1867 - Prize name: US Brig Argus - Ship type: War - How taken: HM Brig Pelican - When taken: 14 Aug 1813 - Where taken: Irish Channel - Date received: 23 Aug 1813 - From what ship: HMS Pelican - Born: New York - Age: 26 - Discharged on 8 Sep 1813 and sent to Dartmoor.

Richards, James - Seaman - Number: 1404 - How taken: Delivered himself up from HM Hospital Ship Trent - When taken: 25 May 1813 - Date received: 18 May 1813 - From what ship: HMS Treazer - Born: Newburyport - Age: 44 - Discharged on 8 Jul 1813 and sent to Chatham on tender Neptune.

Richardson, Cheney - Seaman - Number: 2458 - Prize name: Fair American - Ship type: MV - How taken: HM Frigate Andromache - When taken: 19 Jan 1814 - Where taken: Bay of Biscay - Date received: 22 Feb 1814 - From what ship: HMS York - Born: Brookfield - Age: 24 - Discharged on 10 May 1814 and sent to Dartmoor.

Richardson, Daniel - Seaman - Number: 280 - Prize name: Leader - Ship type: MV - How taken: HM Frigate Briton - When taken: 10 Dec 1812 - Where taken: off Bordeaux - Date received: 21 Jan 1813 - From what ship: HMS Abercrombie - Born: Maryland - Age: 24 - Sent to Portsmouth on 8 Feb 1813 on the HMS Colossus.

Richardson, Francis - Seaman - Number: 779 - Prize name: William Bayard - Ship type: MV - How taken: HM Ship-of-the-Line Warspite - When taken: 3 Mar 1813 - Where taken: Bay of Biscay - Date received: 19 Mar 1813 - From what ship: HMS Warspite - Born: New Town - Age: 22 - Race: Black - Sent to Dartmoor on 28 Jun 1813.

Richardson, John - Able Seaman - Number: 349 - How taken: Taken off the HM Frigate Andromache - Date received: 29 Jan 1813 - From what ship: HMS Royal Sovereign - Born: Dresden - Age: 25 - Sent to Chatham on 29 Mar 1813 on the HMS Braham.

Richardson, Samuel - Sailmaker - Number: 451 - Prize name: Dolphin - Ship type: LM - How taken: HM Ship-of-the-Line Colossus, HM Frigate Rhin & HM Brig Goldfinch - When taken: 5 Jan 1813 - Where taken: off Western Islands - Date received: 6 Feb 1813 - From what ship: HMS Rhin - Born: Philadelphia - Age: 23 - Sent to Chatham on 29 Mar 1813 on the HMS Braham.

Richardson, William - Seaman - Number: 802 - Prize name: Cannoniere - Ship type: P - How taken: HM Ship-of-the-Line Warspite - When taken: 14 Mar 1813 - Where taken: Bay of Biscay - Date received: 19 Mar 1813 - From what ship: HMS Warspite - Born: Boston - Age: 18 - Sent to Dartmoor on 28 Jun 1813.

Richman, Joshua - Seaman - Number: 699 - Prize name: Star - Ship type: MV - How taken: HM Ship-of-the-Line Superb - When taken: 9 Feb 1813 - Where taken: Bay of Biscay - Date received: 19 Mar 1813 - From what ship: HMS Warspite - Born: Maryland - Age: 21 - Sent to Dartmoor on 2 Apr 1813.

Riddle, Thomas - Seaman - Number: 1619 - Prize name: Revenge - Ship type: LM - How taken: HM Frigate Belle Poule - When taken: 11 May 1813 - Where taken: Bay of Biscay - Date received: 14 Jun 1813 - From what ship: HMS Royalist - Born: Pennsylvania - Age: 18 - Discharged on 28 Jun 1813 and released to HMS Salvador del Mundo.

Riddlefield, James - Seaman - Number: 1461 - Prize name: Paul Jones - Ship type: P - How taken: HM Frigate Leonidas - When taken: 23 May 1813 - Where taken: off Cape Clear - Date received: 26 May 1813 - From what ship: HMS Leonidas - Born: New York - Age: 22 - Discharged on 30 Jun 1813 and sent to Stapleton.

Rideout, James - Seaman - Number: 2549 - How taken: Impressed in Scotland - When taken: 25 Jan 1814 - Date received: 16 Apr 1814 - From what ship: HMS Prince Frederick - Born: Annapolis - Age: 37 - Race: Negro - Discharged on 10 May 1814 and sent to Dartmoor.

Right, Joshua - Seaman - Number: 690 - Prize name: Star - Ship type: MV - How taken: HM Ship-of-the-Line Superb - When taken: 9 Feb 1813 - Where taken: Bay of Biscay - Date received: 19 Mar 1813 - From what ship: HMS Warspite - Born: Saybrook - Age: 24 - Sent to Dartmoor on 2 Apr 1813.

Riley, Michael - Seaman - Number: 773 - Prize name: William Bayard - Ship type: MV - How taken: HM Ship-of-the-Line Warspite - When taken: 3 Mar 1813 - Where taken: Bay of Biscay - Date received: 19 Mar 1813 - From what ship: HMS Warspite - Born: Virginia - Age: 25 - Race: Black - Sent to Dartmoor on 28 Jun 1813.

Riley, William - Seaman - Number: 45 - How taken: Delivered himself up from HM Frigate Circe - Date received: 25 Nov 1813 - From what ship: HMS Stork - Born: New Jersey - Age: 22 - Sent to Portsmouth on 29 Dec 1812 on the HMS Northumberland.

Ringgold, Peregrine - Seaman - Number: 2308 - Prize name: Siro - Ship type: LM - How taken: HM Brig Pelican - When taken: 13 Jan 1814 - Where taken: at sea - Date received: 23 Jan 1814 - From what ship: HMS Pelican - Born: Baltimore - Age: 36 - Discharged on 31 Jan 1814 and sent to Dartmoor.

Ripley, Eden - Seaman - Number: 1231 - Prize name: Essex - Ship type: MV - How taken: HM Frigate Pyramus - When taken: 2 Apr 1813 - Where taken: Bay of Biscay - Date received: 9 May 1813 - From what ship: HMS Andromache - Born: Massachusetts - Age: 19 - Discharged on 3 Jul 1813 and sent to Stapleton Prison.

Ripley, William - Seaman - Number: 674 - How taken: Taken out of Flag Truce Pennsylvania - Date received: 15 Mar 1813 - From what ship: HMS Bittern - Born: Pennsylvania - Age: 40 - Sent to Dartmoor on 28 Jun 1813.

Risings, John - Seaman - Number: 1473 - Prize name: Paul Jones - Ship type: P - How taken: HM Frigate Leonidas - When taken: 23 May 1813 - Where taken: off Cape Clear - Date received: 26 May 1813 - From what ship: HMS Leonidas - Born: Albany - Age: 29 - Race: Black - Discharged on 30 Jun 1813 and sent to Stapleton.

Riva, Antonio - Seaman - Number: 2396 - Prize name: Harvest, prize of the Privateer Bunker Hill - Ship type: P - How taken: HM Brig Orestes - When taken: 21 Jan 1814 - Date received: 3 Feb 1814 - From what ship: HMS Orestes - Born: Milan, Italy - Age: 28 - Discharged on 27 Apr 1814 and sent to Harwich for Transport Portland.

Roach, Nicholas - 2nd Mate - Number: 281 - Prize name: Columbia - Ship type: MV - How taken: HM Frigate Briton - When taken: 17 Dec 1812 - Where taken: off Bordeaux - Date received: 21 Jan 1813 - From what ship: HMS Abercrombie - Born: Philadelphia - Age: 34 - Sent to Portsmouth on 8 Feb 1813 on the HMS Colossus.

Roach, Reuben - Seaman - Number: 1183 - How taken: Delivered himself up from HM Ship-of-the-Line Dublin - Date received: 8 May 1813 - From what ship: HMS Dublin - Born: Maryland - Age: 29 - Sent to Chatham on 8 Jul 1813 on HM Tender Neptune.

Road, John - Seaman - Number: 2211 - Prize name: Squirrel - Ship type: MV - How taken: HM Frigate Belle Poule - When taken: 14 Dec 1813 - Where taken: Bay of Biscay - Date received: 27 Dec 1813 - From what ship: HMS Protector - Born: Hamburg - Age: 24 - Discharged on 31 Jan 1814 and sent to Dartmoor.

Road, John - Seaman - Number: 2694 - Date received: 17 Jun 1814 - From what ship: Dartmouth - Discharged on 20 Jun 1814 and released to the Speedwell Cartel.

Roberson, Henry - Seaman - Number: 1940 - Prize name: Marmion - Ship type: MV - How taken: HM Frigate

President - When taken: 14 Aug 1813 - Where taken: off Nantes - Date received: 27 Aug 1813 - From what ship: HMS Urgent - Born: Richmond - Age: 17 - Race: Negro - Discharged on 8 Sep 1813 and sent to Dartmoor.

Roberson, John - Seaman - Number: 1425 - Prize name: Paul Jones - Ship type: P - How taken: HM Frigate Leonidas - When taken: 23 May 1813 - Where taken: off Cape Clear - Date received: 26 May 1813 - From what ship: HMS Leonidas - Born: Connecticut - Age: 49 - Discharged on 3 Jul 1813 and sent to Stapleton Prison.

Roberson, John - Seaman - Number: 1834 - Prize name: US Brig Argus - Ship type: War - How taken: HM Brig Pelican - When taken: 14 Aug 1813 - Where taken: Irish Channel - Date received: 17 Aug 1813 - From what ship: USS Argus - Born: New York - Age: 21 - Discharged on 24 Aug 1813 and released to HMS Salvador del Mundo.

Roberts, Hugh - Seaman - Number: 1766 - Prize name: Union, prize of the Privateer Brutus - Ship type: LM - How taken: HM Brig Goldfinch - When taken: 17 Jul 1813 - Where taken: Bay of Biscay - Date received: 25 Jul 1813 - From what ship: HMS Pyramus - Born: Pennsylvania - Age: 39 - Discharged on 8 Sep 1813 and sent to Dartmoor.

Roberts, Joel - Seaman - Number: 2261 - Prize name: Growler - Ship type: MV - How taken: HM Brig Wolf - When taken: 11 Aug 1813 - Where taken: at sea - Date received: 20 Jan 1814 - From what ship: MV Nero - Born: Littleton - Age: 28 - Discharged on 31 Jan 1814 and sent to Dartmoor.

Roberts, John - Seaman - Number: 1411 - How taken: Delivered himself up at the Cove of Cork - When taken: 10 May 1813 - Date received: 22 May 1813 - From what ship: HMS Talbot - Born: Baltimore - Age: 23 - Race: Black - Discharged on 8 Sep 1813 and sent to Dartmoor.

Roberts, Moses - Seaman - Number: 2642 - Prize name: Ateline - Ship type: MV - How taken: HM Frigate Magiciene - When taken: 14 Mar 1814 - Where taken: off Cape Finisterre, Spain - Date received: 17 May 1814 - From what ship: HMS Tortois - Born: Portsmouth - Age: 30 - Discharged on 14 Jun 1814 and sent to Dartmoor.

Robertson, James - Seaman - Number: 1287 - Prize name: Price - Ship type: LM - How taken: HM Frigate Medusa - When taken: 13 Apr 1813 - Where taken: Bay of Biscay - Date received: 10 May 1813 - From what ship: HMS Medusa - Born: Massachusetts - Age: 21 - Discharged on 3 Jul 1813 and sent to Stapleton Prison.

Robertson, Robert - Seaman - Number: 1568 - Prize name: Miranda, prize of the Privateer Paul Jones - Ship type: P - How taken: HM Frigate Unicorn - When taken: 21 Mar 1813 - Where taken: off Ushant - Date received: 9 Jun 1813 - From what ship: HMS Conquestador - Born: North Carolina - Age: 23 - Discharged on 30 Jun 1813 and sent to Stapleton.

Robin, John - Seaman - Number: 2400 - Prize name: Rachel & Ann, prize of the Privateer Prince de Neufchatel - Ship type: P - How taken: HM Frigate Cydnus - When taken: 6 Jan 1814 - Where taken: at sea - Date received: 4 Feb 1814 - From what ship: HMS Cydnus - Born: New Orleans - Age: 30 - Discharged on 10 May 1814 and sent to Dartmoor.

Robins, Jeremiah - Seaman - Number: 195 - Prize name: Hunter - Ship type: P - How taken: HM Frigate Phoebe - When taken: 24 Dec 1812 - Where taken: off Western Islands - Date received: 9 Jan 1813 - From what ship: HMS Phoebe - Born: Sudbury - Age: 28 - Sent to Portsmouth on 8 Feb 1813 on the HMS Colossus.

Robinson, Benjamin - Seaman - Number: 293 - Prize name: Columbia - Ship type: MV - How taken: HM Frigate Briton - When taken: 17 Dec 1812 - Where taken: off Bordeaux - Date received: 21 Jan 1813 - From what ship: HMS Abercrombie - Born: Philadelphia - Age: 24 - Sent to Portsmouth on 8 Feb 1813 on the HMS Colossus.

Robinson, Charles - Seaman - Number: 1123 - Prize name: Viper - Ship type: MV - How taken: HM Ship-of-the-Line Superb - When taken: 15 Apr 1813 - Where taken: Bay of Biscay - Date received: 22 Apr 1813 - From what ship: HMS Superb - Born: Gothenburg, Sweden - Age: 22 - Sent to Dartmoor on 1 Jul 1813.

Robinson, David - Seaman - Number: 425 - Prize name: Union - Ship type: MV - How taken: HM Frigate Iris - When taken: 17 Jan 1813 - Where taken: Lat 44 N Long 2.3 W - Date received: 5 Feb 1813 - From what ship: HMS San Josef - Born: Massachusetts - Age: 28 - Sent to Chatham on 29 Mar 1813 on the HMS Braham.

Robinson, Ebenezer - Seaman - Number: 833 - Prize name: Pallas - Ship type: MV - How taken: HM Brig Rebuff - When taken: 23 Dec 1812 - Where taken: off Cadiz - Date received: 22 Mar 1813 - From what ship: HMS Dauntless - Born: Pennsylvania - Age: 36 - Discharged on 21 Jun 1813 and released to the American ship Mount Hope.

Robinson, Elias - Seaman - Number: 1587 - Prize name: Orders in Council - Ship type: LM - How taken: HM Frigate

Surveillante - When taken: 1 Jun 1813 - Where taken: Bay of Biscay - Date received: 13 Jun 1813 - From what ship: Forvey - Born: Norwich - Age: 20 - Discharged on 30 Jun 1813 and sent to Stapleton.

Robinson, James - Seaman - Number: 305 - Prize name: Stephen - Ship type: MV - How taken: HM Frigate Briton & HM Frigate Andromache - When taken: 17 Dec 1812 - Where taken: off Bordeaux - Date received: 21 Jan 1813 - From what ship: HMS Abercrombie - Born: Philadelphia - Age: 22 - Sent to Chatham on 29 Mar 1813 on the HMS Braham.

Robinson, James - Seaman - Number: 2425 - Prize name: US Brig Argus - Ship type: War - How taken: Sent from HM Ship-of-the-Line Salvador del Mundo - When taken: 14 Aug 1813 - Where taken: Irish Channel - Date received: 12 Feb 1814 - From what ship: HMS Salvador del Mundo - Born: New York - Age: 19 - Discharged on 10 May 1814 and sent to Dartmoor.

Robinson, John - Seaman - Number: 466 - Prize name: Dolphin - Ship type: LM - How taken: HM Ship-of-the-Line Colossus, HM Frigate Rhin & HM Brig Goldfinch - When taken: 5 Jan 1813 - Where taken: off Western Islands - Date received: 6 Feb 1813 - From what ship: HMS Rhin - Born: Delaware - Age: 21 - Sent to Chatham on 29 Mar 1813 on the HMS Braham.

Robinson, John - Chief Mate & Prize Master - Number: 1765 - Prize name: Union, prize of the Privateer Brutus - Ship type: LM - How taken: HM Brig Goldfinch - When taken: 17 Jul 1813 - Where taken: Bay of Biscay - Date received: 25 Jul 1813 - From what ship: HMS Pyramus - Born: New London - Age: 30 - Discharged on 8 Sep 1813 and sent to Dartmoor.

Robinson, John - Seaman - Number: 2563 - Prize name: US Brig Argus - Ship type: War - How taken: HM Brig Pelican - When taken: 14 Aug 1813 - Where taken: Irish Channel - Date received: 23 Apr 1814 - From what ship: HMS Salvador del Mundo - Born: Dundee - Age: 49 - Discharged on 10 May 1814 and sent to Dartmoor.

Robinson, Joseph - Seaman - Number: 1049 - Prize name: Independence - Ship type: MV - How taken: HM Ship-of-the-Line Superb - When taken: 16 Mar 1813 - Where taken: Bay of Biscay - Date received: 22 Apr 1813 - From what ship: HMS Superb - Born: Beverly - Age: 21 - Sent to Dartmoor on 1 Jul 1813.

Robinson, Michael - Seaman - Number: 227 - Prize name: Vengeance - Ship type: LM - How taken: HM Frigate Phoebe - When taken: 1 Jan 1813 - Where taken: Lat 44.4 Long 23 - Date received: 9 Jan 1813 - From what ship: HMS Phoebe - Born: Bristol - Age: 18 - Sent to Portsmouth on 8 Feb 1813 on the HMS Colossus.

Robinson, Nicholas - Gunner - Number: 1984 - Prize name: Ned - Ship type: LM - How taken: HM Brig Royalist - When taken: 6 Sep 1812 - Where taken: coast of France - Date received: 22 Sep 1813 - From what ship: HMS Royalist - Born: Stockholm - Age: 24 - Discharged on 27 Sep 1813 and sent to Dartmoor.

Robinson, Samuel - Seaman - Number: 811 - Prize name: Decornau - Ship type: MV - How taken: HM Sloop Pheasant - When taken: 15 Mar 1813 - Where taken: Bay of Biscay - Date received: 19 Mar 1813 - From what ship: HMS Pheasant - Born: Boston - Age: 33 - Sent to Dartmoor on 28 Jun 1813.

Robinson, William - Seaman - Number: 838 - Prize name: Pallas - Ship type: MV - How taken: HM Brig Rebuff - When taken: 23 Dec 1812 - Where taken: off Cadiz - Date received: 22 Mar 1813 - From what ship: HMS Dauntless - Born: New York - Age: 29 - Sent to Dartmoor on 28 Jun 1813.

Rock, Oliver - Seaman - Number: 670 - Prize name: Margaret, prize of the Privateer True Blooded Yankee - Ship type: P - How taken: HM Brig Nimrod - When taken: 9 Mar 1813 - Where taken: off Morant Bay, Jamaica - Date received: 14 Mar 1813 - From what ship: HMS Salvador del Mundo - Born: Morbaix, France - Age: 34 - Sent to Dartmoor on 1 Jul 1813.

Rodden, Charles - 1st Lieutenant - Number: 2488 - Prize name: Bunker Hill - Ship type: P - How taken: HM Frigate Pomone & HM Frigate Cydnus - When taken: 4 Mar 1814 - Where taken: Bay of Biscay - Date received: 24 Mar 1814 - From what ship: Bunker Hill - Born: Gothenburg, Sweden - Age: 37 - Discharged on 27 Mar 1814 and sent to Ashburton on parole.

Rodwin, Walter - Seaman - Number: 1934 - Prize name: Marmion - Ship type: MV - How taken: HM Frigate President - When taken: 14 Aug 1813 - Where taken: off Nantes - Date received: 27 Aug 1813 - From what ship: HMS Urgent - Born: Newport - Age: 27 - Discharged on 8 Sep 1813 and sent to Dartmoor.

Rodyman, John - Seaman - Number: 2141 - Prize name: Amiable - Ship type: LM - How taken: HM Ship-of-the-Line Magnificent - When taken: 30 Oct 1813 - Where taken: off Lorient - Date received: 29 Nov 1813 - From what ship: HMS Dublin - Born: New York - Age: 21 - Discharged on 4 Dec 1813 and sent to Dartmoor.

Roe, John - Seaman - Number: 755 - Prize name: William Bayard - Ship type: MV - How taken: HM Ship-of-the-Line Warspite - When taken: 3 Mar 1813 - Where taken: Bay of Biscay - Date received: 19 Mar 1813 - From what ship: HMS Warspite - Born: New Orleans - Age: 27 - Sent to Dartmoor on 2 Apr 1813.

Roff, Isaac - Seaman - Number: 1548 - Prize name: Zebra - Ship type: LM - How taken: HM Frigate Pyramus - When taken: 20 Apr 1813 - Where taken: Bay of Biscay - Date received: 29 May 1813 - From what ship: HMS Hannibal - Born: New Jersey - Age: 22 - Discharged on 30 Jun 1813 and sent to Stapleton.

Rogers, Gorham - Seaman - Number: 1709 - Prize name: Joseph - Ship type: MV - How taken: HM Frigate Iris - When taken: 8 Jun 1813 - Where taken: Bay of Biscay - Date received: 4 Jul 1813 - From what ship: HMS Iris - Born: Gloucester - Age: 22 - Discharged on 22 Aug 1813 and released to HM Brig Redpole.

Rogers, Henry - Seaman - Number: 2404 - Prize name: Rachel & Ann, prize of the Privateer Prince de Neufchatel - Ship type: P - How taken: HM Frigate Cydnus - When taken: 6 Jan 1814 - Where taken: at sea - Date received: 4 Feb 1814 - From what ship: HMS Cydnus - Born: New Orleans - Age: 25 - Discharged on 10 May 1814 and sent to Dartmoor.

Rogers, Henry - Chief Mate - Number: 2148 - Prize name: Charlotte - Ship type: MV - How taken: Cutter Dwarf - When taken: 4 Nov 1812 - Where taken: off Bordeaux - Date received: 7 Dec 1813 - From what ship: Cutter Dwarf - Born: Charleston - Age: 24 - Discharged on 8 Dec 1813 and sent to Ashburton on parole.

Rogers, James - Seaman - Number: 1081 - Prize name: John & Frances - Ship type: MV - How taken: HM Frigate Belle Poule - When taken: 19 Mar 1813 - Where taken: Bay of Biscay - Date received: 22 Apr 1813 - From what ship: HMS Superb - Born: Philadelphia - Age: 25 - Sent to Dartmoor on 1 Jul 1813.

Rogers, Nathaniel - Seaman - Number: 997 - Prize name: Polly - Ship type: MV - How taken: HM Frigate Surveillante - When taken: 27 Mar 1813 - Where taken: Bay of Biscay - Date received: 16 Apr 1813 - From what ship: HMS Fairy - Born: Marblehead - Age: 23 - Sent to Chatham on 8 Jul 1813 on HM Tender Neptune.

Rogers, Robert - Seaman - Number: 2199 - Prize name: General Kempt, prize of the Privateer Grand Turk - Ship type: P - How taken: HM Brig Foxhound - When taken: 18 Dec 1813 - Where taken: Lat 48.4 Long 6 - Date received: 21 Dec 1813 - From what ship: HMS Foxhound - Born: Virginia - Age: 23 - Discharged on 31 Jan 1814 and sent to Dartmoor.

Roles, John - Seaman - Number: 1532 - Prize name: Courier - Ship type: LM - How taken: HM Brig Rover - When taken: 14 Mar 1813 - Where taken: Bay of Biscay - Date received: 29 May 1813 - From what ship: HMS Hannibal - Born: Maryland - Age: 22 - Discharged on 30 Jun 1813 and sent to Stapleton.

Roley, John - Seaman - Number: 1472 - Prize name: Paul Jones - Ship type: P - How taken: HM Frigate Leonidas - When taken: 23 May 1813 - Where taken: off Cape Clear - Date received: 26 May 1813 - From what ship: HMS Leonidas - Born: France - Age: 38 - Race: Black - Discharged on 30 Jun 1813 and sent to Stapleton.

Rollands, Joseph - Seaman - Number: 777 - Prize name: William Bayard - Ship type: MV - How taken: HM Ship-of-the-Line Warspite - When taken: 3 Mar 1813 - Where taken: Bay of Biscay - Date received: 19 Mar 1813 - From what ship: HMS Warspite - Born: Pennsylvania - Age: 25 - Sent to Dartmoor on 28 Jun 1813.

Romain, Samuel - Seaman - Number: 694 - Prize name: Star - Ship type: MV - How taken: HM Ship-of-the-Line Superb - When taken: 9 Feb 1813 - Where taken: Bay of Biscay - Date received: 19 Mar 1813 - From what ship: HMS Warspite - Born: New York - Age: 30 - Sent to Dartmoor on 2 Apr 1813.

Roman, John - Seaman - Number: 1726 - Prize name: Hannah & Eliza - Ship type: MV - How taken: HM Brig Lyra - When taken: 29 May 1813 - Where taken: off north coast of Spain - Date received: 4 Jul 1813 - From what ship: HMS Iris - Born: New York - Age: 25 - Discharged on 8 Sep 1813 and sent to Dartmoor.

Rommel, Francis - Seaman - Number: 2108 - Prize name: Chesapeake - Ship type: LM - How taken: HM Frigate Hotspur & HM Frigate Pyramus - When taken: 26 Oct 1813 - Where taken: Bay of Biscay - Date received: 22 Nov 1813 - From what ship: HMS Pyramus - Born: San Sebastian, Spain - Age: 26 - Discharged on 29 Nov 1813 and sent to Dartmoor.

Root, Eliakim F. - Seaman - Number: 533 - Prize name: Terrible - Ship type: MV - How taken: HM Brig Foxhound - When taken: 8 Feb 1813 - Where taken: Channel - Date received: 15 Feb 1813 - From what ship: HMS Foxhound

- Born: Hartford - Age: 20 - Sent to Dartmoor on 2 Apr 1813.

Roper, Thomas William - Physician & Passenger - Number: 2219 - Prize name: Zephyr - Ship type: MV - How taken: HM Frigate Pyramus - When taken: 30 Nov 1813 - Where taken: off Lorient - Date received: 30 Dec 1813 - From what ship: Martial - Born: Charlestown - Age: 22 - Discharged on 30 Dec 1813 and sent to Ashburton on parole.

Rosah, Samuel - Seaman - Number: 1886 - Prize name: US Brig Argus - Ship type: War - How taken: HM Brig Pelican - When taken: 14 Aug 1813 - Where taken: Irish Channel - Date received: 23 Aug 1813 - From what ship: HMS Pelican - Born: New York - Age: 22 - Discharged on 8 Sep 1813 and sent to Dartmoor.

Rose, William - Seaman - Number: 2603 - How taken: MV Nautilus - When taken: 20 Dec 1812 - Date received: 16 May 1814 - From what ship: HMS Repulse - Born: Vienna - Age: 30 - Discharged on 14 Jun 1814 and sent to Dartmoor.

Rosette, Samuel - Seaman - Number: 679 - How taken: Impressed at Liverpool - When taken: 23 Feb 1813 - Date received: 15 Mar 1813 - From what ship: HMS Bittern - Born: New York - Age: 31 - Sent to Dartmoor on 2 Apr 1813.

Rosings, Cornelius - Seaman - Number: 1458 - Prize name: Paul Jones - Ship type: P - How taken: HM Frigate Leonidas - When taken: 23 May 1813 - Where taken: off Cape Clear - Date received: 26 May 1813 - From what ship: HMS Leonidas - Born: Gothenburg, Sweden - Age: 37 - Race: Black - Discharged on 30 Jun 1813 and sent to Stapleton.

Ross, George - Seaman - Number: 175 - Prize name: Hunter - Ship type: P - How taken: HM Frigate Phoebe - When taken: 24 Dec 1812 - Where taken: off Western Islands - Date received: 9 Jan 1813 - From what ship: HMS Phoebe - Born: Boston - Age: 55 - Sent to Portsmouth on 8 Feb 1813 on the HMS Colossus.

Ross, John - Seaman - Number: 440 - How taken: Impressed at Belfast - Date received: 5 Feb 1813 - From what ship: HMS Neptune - Born: Age: 23 - Sent to Chatham on 29 Mar 1813 on the HMS Braham.

Ross, Philip - Seaman - Number: 2085 - Prize name: Sybella - Ship type: MV - How taken: HM Brig Zenobia - When taken: 17 Jun 1813 - Where taken: off Cape Henry - Date received: 2 Nov 1813 - From what ship: HMS Bacchus - Born: New Orleans - Age: 24 - Discharged on 3 Nov 1813 and sent to Dartmoor.

Ross, Richard - Steward - Number: 347 - Prize name: Friendship - Ship type: MV - How taken: HM Frigate Rosamond - When taken: Aug 1812 - Where taken: off Halifax - Date received: 28 Jan 1813 - From what ship: HMS Plymouth - Born: Savanah, GA - Age: 19 - Sent to Chatham on 29 Mar 1813 on the HMS Braham.

Ratner, James - Seaman - Number: 571 - Prize name: Rolla - Ship type: MV - How taken: HM Frigate Surveillante - When taken: 11 Feb 1813 - Where taken: Bay of Biscay - Date received: 23 Feb 1813 - From what ship: HMS Surveillante - Born: Philadelphia - Age: 25 - Sent to Dartmoor on 2 Apr 1813.

Rouger, Desire - Volunteer - Number: 2398 - Prize name: Harvest, prize of the Privateer Bunker Hill - Ship type: P - How taken: HM Brig Orestes - When taken: 21 Jan 1814 - Date received: 3 Feb 1814 - From what ship: HMS Orestes - Born: Vangere, France - Age: 17 - Discharged on 10 May 1814 and sent to Dartmoor.

Roulemmon, Bernard - Passenger - Number: 2692 - Prize name: Fanny - Ship type: MV - How taken: HM Frigate Eurotas - When taken: 25 Dec 1813 - Where taken: at sea - Date received: 17 Jun 1814 - From what ship: Dartmouth - Born: Oldenburg, Germany - Age: 27 - Discharged on 20 Jun 1814 and released to the Speedwell Cartel.

Roulemmon, Bernard - Passenger - Number: 2272 - Prize name: Fanny - Ship type: MV - How taken: HM Frigate Eurotas - When taken: 25 Dec 1813 - Where taken: at sea - Date received: 20 Jan 1814 - From what ship: HMS Eurotas - Born: Oldenburg, Germany - Age: 27 - Discharged on 31 Jan 1814 and sent to Dartmoor.

Rowe, John - Seaman - Number: 1388 - Prize name: Tom - Ship type: LM - How taken: HM Frigate Surveillante - When taken: 27 Apr 1813 - Where taken: Bay of Biscay - Date received: 15 May 1813 - From what ship: HMS Foxhound - Born: Connecticut - Age: 26 - Discharged on 3 Jul 1813 and sent to Stapleton Prison.

Rowe, Stephen - Seaman - Number: 201 - Prize name: Hunter - Ship type: P - How taken: HM Frigate Phoebe - When taken: 24 Dec 1812 - Where taken: off Western Islands - Date received: 9 Jan 1813 - From what ship: HMS Phoebe - Born: Martha's Vineyard - Age: 30 - Sent to Portsmouth on 8 Feb 1813 on the HMS Colossus.

Rowell, James - Carpenter's Mate - Number: 157 - Prize name: Hunter - Ship type: P - How taken: HM Frigate Phoebe

- When taken: 24 Dec 1812 - Where taken: off Western Islands - Date received: 9 Jan 1813 - From what ship: HMS Phoebe - Born: Salem - Age: 22 - Sent to Portsmouth on 8 Feb 1813 on the HMS Colossus.

Rowley, Henry - Carpenter - Number: 1365 - Prize name: Tom - Ship type: LM - How taken: HM Frigate Surveillante - When taken: 27 Apr 1813 - Where taken: Bay of Biscay - Date received: 15 May 1813 - From what ship: HMS Foxhound - Born: Pennsylvania - Age: 36 - Race: Colored - Discharged on 3 Jul 1813 and sent to Stapleton Prison.

Roy, Charles - Seaman - Number: 271 - Prize name: Brutus - Ship type: MV - How taken: HM Frigate Briton - When taken: Jan 1813 - Where taken: Bay of Biscay - Date received: 21 Jan 1813 - From what ship: HMS Briton - Born: Nantucket - Age: 19 - Sent to Portsmouth on 8 Feb 1813 on the HMS Colossus.

Royal, John - Seaman - Number: 635 - Prize name: Criterion - Ship type: MV - How taken: HM Frigate Belle Poule - When taken: 14 Feb 1813 - Where taken: Bay of Biscay - Date received: 4 Mar 1813 - From what ship: HMS Strenuous - Born: Petersburg - Age: 28 - Sent to Dartmoor on 2 Apr 1813.

Rozier, William - Seaman - Number: 1506 - Prize name: Grand Napoleon - Ship type: MV - How taken: HM Brig Goldfinch - When taken: 17 Apr 1813 - Where taken: Bay of Biscay - Date received: 29 May 1813 - From what ship: HMS Hannibal - Born: New Orleans - Age: 24 - Discharged on 8 Sep 1813 and sent to Dartmoor.

Ruddick, William - 2nd Mate - Number: 985 - Prize name: Lightning - Ship type: MV - How taken: HM Frigate Medusa - When taken: 2 Apr 1813 - Where taken: Bay of Biscay - Date received: 16 Apr 1813 - From what ship: HMS Fairy - Born: Philadelphia - Age: 25 - Sent to Chatham on 8 Jul 1813 on HM Tender Neptune.

Ruffield, Samuel - Seaman - Number: 1465 - Prize name: Paul Jones - Ship type: P - How taken: HM Frigate Leonidas - When taken: 23 May 1813 - Where taken: off Cape Clear - Date received: 26 May 1813 - From what ship: HMS Leonidas - Born: Norwich, CT - Age: 20 - Discharged on 30 Jun 1813 and sent to Stapleton.

Ruliff, London - Cook - Number: 446 - Prize name: Hunter - Ship type: P - How taken: HM Frigate Phoebe - When taken: 24 Dec 1812 - Where taken: off Western Islands - Date received: 6 Feb 1813 - From what ship: HMS Phoebe - Born: Salem - Age: 25 - Sent to Chatham on 29 Mar 1813 on the HMS Braham.

Russell, Frederick - Seaman - Number: 1878 - Prize name: US Brig Argus - Ship type: War - How taken: HM Brig Pelican - When taken: 14 Aug 1813 - Where taken: Irish Channel - Date received: 23 Aug 1813 - From what ship: HMS Pelican - Born: Norway - Age: 30 - Discharged on 8 Sep 1813 and sent to Dartmoor.

Russell, George - Gunner - Number: 1363 - Prize name: Tom - Ship type: LM - How taken: HM Frigate Surveillante - When taken: 27 Apr 1813 - Where taken: Bay of Biscay - Date received: 15 May 1813 - From what ship: HMS Foxhound - Born: New Jersey - Age: 22 - Discharged on 3 Jul 1813 and sent to Stapleton Prison.

Russell, Isaac - Seaman - Number: 593 - Prize name: St. Martin's Plantation, prize of the Privateer Paul Jones - Ship type: P - How taken: HM Ship-of-the-Line Dublin - When taken: 9 Feb 1815 - Where taken: Lat 43 N Long 33.5 W - Date received: 25 Feb 1813 - From what ship: HMS Dublin - Born: New Bedford - Age: 24 - Sent to Dartmoor on 2 Apr 1813.

Russell, James - Seaman - Number: 1780 - Prize name: Orders in Council - Ship type: LM - How taken: HM Frigate Surveillante - When taken: 1 Jun 1813 - Where taken: Bay of Biscay - Date received: 7 Aug 1813 - From what ship: HMS Gleaner - Born: New York - Age: 25 - Discharged on 8 Sep 1813 and sent to Dartmoor.

Russell, John - Seaman - Number: 1087 - Prize name: John & Frances - Ship type: MV - How taken: HM Frigate Belle Poule - When taken: 19 Mar 1813 - Where taken: Bay of Biscay - Date received: 22 Apr 1813 - From what ship: HMS Superb - Born: New Hampshire - Age: 23 - Discharged on 7 May 1813 and sent on the HMS Salvador del Mundo.

Russell, Morris - Seaman - Number: 1665 - How taken: Delivered himself up from HM Brig Royalist - Date received: 16 Jun 1813 - From what ship: HMS Royalist - Born: Savannah - Age: 22 - Race: Black - Discharged on 8 Jul 1813 and sent to Chatham on tender Neptune.

Russell, Patten - Seaman - Number: 1161 - Prize name: Essex - Ship type: MV - How taken: HM Frigate Pyramus - When taken: 2 Apr 1813 - Where taken: Bay of Biscay - Date received: 2 Mar 1813 - From what ship: HMS Rota - Born: Cambridge - Age: 17 - Sent on 3 Jul 1813 to Stapleton prison.

Russell, William - Master - Number: 1407 - Prize name: Henry Clements - Ship type: MV - How taken: HM Brig Orestes - When taken: 15 Apr 1813 - Where taken: Bay of Biscay - Date received: 20 May 1813 - From what ship:

HMS Orestes - Born: Massachusetts - Age: 42 - Discharged on 23 May 1813 and sent to Ashburton on parole.

Ryan, Thomas - Seaman - Number: 1153 - How taken: Delivered himself up from HM Ship-of-the-Line Salvador del Mundo - Date received: 23 Apr 1813 - From what ship: HMS Salvador del Monde - Born: Philadelphia - Age: 34 - Sent to Chatham on 8 Jul 1813 on HM Tender Neptune.

Ryan, William - Seaman - Number: 567 - Prize name: Rolla - Ship type: MV - How taken: HM Frigate Surveillante - When taken: 11 Feb 1813 - Where taken: Bay of Biscay - Date received: 23 Feb 1813 - From what ship: HMS Surveillante - Born: Philadelphia - Age: 19 - Sent to Dartmoor on 2 Apr 1813.

Ryder, Philip - Master - Number: 88 - Prize name: Experiment - Ship type: MV - How taken: HM Brig Rover - When taken: 21 Oct 1812 - Where taken: off Bordeaux - Date received: 25 Dec 1812 - From what ship: HMS Northumberland - Born: Rhode Island - Age: 50 - Sent on 8 Dec 1812 to Ashburton on parole.

Ryer, Peter - Seaman - Number: 890 - Prize name: Tiger - Ship type: MV - How taken: HM Brig Scylla - When taken: 22 Mar 1813 - Where taken: Bay of Biscay - Date received: 1 Apr 1813 - From what ship: HMS Scylla - Born: Boston - Age: 27 - Sent on 21 Jun 1813 to Portsmouth on HMS Prometheus.

Saff, Francis - Seaman - Number: 1272 - Prize name: Caroline - Ship type: MV - How taken: HM Frigate Medusa - When taken: 12 Apr 1813 - Where taken: Bay of Biscay - Date received: 10 May 1813 - From what ship: HMS Medusa - Born: Cape Francois, Haiti - Age: 20 - Discharged on 3 Jul 1813 and sent to Stapleton Prison.

Sage, Moses - Seaman - Number: 645 - Prize name: Criterion - Ship type: MV - How taken: HM Frigate Belle Poule - When taken: 14 Feb 1813 - Where taken: Bay of Biscay - Date received: 4 Mar 1813 - From what ship: HMS Strenuous - Born: Middletown - Age: 29 - Sent to Dartmoor on 2 Apr 1813.

Sailes, James - Cook - Number: 1221 - Prize name: Grand Napoleon - Ship type: MV - How taken: HM Frigate Belle Poule - When taken: 3 Apr 1813 - Where taken: off Bordeaux - Date received: 9 May 1813 - From what ship: HMS Andromache - Born: New York - Age: 19 - Race: Black - Discharged on 3 Jul 1813 and sent to Stapleton Prison.

Salley, James - Quartermaster - Number: 1822 - Prize name: US Brig Argus - Ship type: War - How taken: HM Brig Pelican - When taken: 14 Aug 1813 - Where taken: Irish Channel - Date received: 17 Aug 1813 - From what ship: USS Argus - Born: Massachusetts - Age: 27 - Discharged on 8 Sep 1813 and sent to Dartmoor.

Sampson, Jacob - Seaman - Number: 895 - Prize name: Tiger - Ship type: MV - How taken: HM Brig Scylla - When taken: 22 Mar 1813 - Where taken: Bay of Biscay - Date received: 1 Apr 1813 - From what ship: HMS Scylla - Born: New York - Age: 24 - Race: Black - Sent on 21 Jun 1813 to Portsmouth on HMS Prometheus.

Samsuillen, John - Marine - Number: 1480 - Prize name: Paul Jones - Ship type: P - How taken: HM Frigate Leonidas - When taken: 23 May 1813 - Where taken: off Cape Clear - Date received: 26 May 1813 - From what ship: HMS Leonidas - Born: New Orleans - Age: 39 - Race: Black - Discharged on 30 Jun 1813 and sent to Stapleton.

Samuel, Nicholas - Seaman - Number: 2301 - Prize name: Siro - Ship type: LM - How taken: HM Brig Pelican - When taken: 13 Jan 1814 - Where taken: at sea - Date received: 23 Jan 1814 - From what ship: HMS Pelican - Born: Philadelphia - Age: 37 - Discharged on 31 Jan 1814 and sent to Dartmoor.

Sandburn, John - Seaman - Number: 2051 - How taken: Pallas - Where taken: St. Johns, New Brunswick - Date received: 1 Nov 1813 - From what ship: HMS Bittern - Born: Hampshire - Age: 23 - Discharged on 3 Nov 1813 and sent to Dartmoor.

Sanderhin, Daniel - Seaman - Number: 2014 - Prize name: Ned - Ship type: LM - How taken: HM Brig Royalist - When taken: 6 Sep 1812 - Where taken: coast of France - Date received: 24 Sep 181 - From what ship: HMS Rippon - Born: Not legible - Age: 23 - Discharged on 27 Sep 1813 and sent to Dartmoor.

Sanders, Joseph - Chief Mate - Number: 611 - Prize name: Governor McKean - Ship type: LM - How taken: HM Brig Rover - When taken: 26 Jun 1813 - Where taken: off Bordeaux - Date received: 2 Mar 1813 - From what ship: HMS Insolent - Born: Massachusetts - Age: 22 - Sent on 9 Mar 1813 to Ashburton on parole.

Sandford, John W. - Seaman - Number: 1441 - Prize name: Paul Jones - Ship type: P - How taken: HM Frigate Leonidas - When taken: 23 May 1813 - Where taken: off Cape Clear - Date received: 26 May 1813 - From what ship: HMS Leonidas - Born: New York - Age: 25 - Discharged on 8 Sep 1813 and sent to Dartmoor.

Sanford, William - Seaman - Number: 1423 - Prize name: Paul Jones - Ship type: P - How taken: HM Frigate Leonidas

- When taken: 23 May 1813 - Where taken: off Cape Clear - Date received: 26 May 1813 - From what ship:

HMS Leonidas - Born: New York - Age: 19 - Discharged on 3 Jul 1813 and sent to Stapleton Prison.

Sanis, Lewis - Seaman - Number: 1942 - Prize name: Marmion - Ship type: MV - How taken: HM Frigate President - When taken: 14 Aug 1813 - Where taken: off Nantes - Date received: 27 Aug 1813 - From what ship: HMS Urgent - Born: New Orleans - Age: 21 - Race: Mulatto - Discharged on 8 Sep 1813 and sent to Dartmoor.

Sargant, Samuel Allen - 2nd Prize Master - Number: 625 - Prize name: Good Intent, prize of the Privateer Thrasher - Ship type: P - How taken: HM Frigate Pyramus - When taken: 26 Jun 1813 - Where taken: off Bordeaux - Date received: 2 Mar 1813 - From what ship: HMS Insolent - Born: Cape Ann - Age: 20 - Sent to Dartmoor on 2 Apr 1813.

Saunders, William - Seaman - Number: 730 - Prize name: Mars - Ship type: MV - How taken: HM Ship-of-the-Line Warspite - When taken: 26 Feb 1813 - Where taken: Bay of Biscay - Date received: 19 Mar 1813 - From what ship: HMS Warspite - Born: Massachusetts - Age: 19 - Sent to Dartmoor on 2 Apr 1813.

Scannell, Cornell - Seaman - Number: 17 - Prize name: John - Ship type: MV - How taken: Taken at Liverpool - When taken: 18 Oct 1812 - Date received: 21 Nov 1812 - From what ship: HMS Salvador del Mundo - Born: Darby - Age: 27 - Sent to Portsmouth on 29 Dec 1812 on the HMS Northumberland.

Schoeman, Frederick - Seaman - Number: 490 - Prize name: Cashiere - Ship type: LM - How taken: HM Brig Reindeer - When taken: 3 Feb 1813 - Where taken: Bay of Biscay - Date received: 12 Feb 1813 - From what ship: HMS Reindeer - Born: Baltimore - Age: 19 - Sent to Dartmoor on 2 Apr 1813.

Schoeman, Frederick - Seaman - Number: 1742 - Prize name: Cashiere - Ship type: LM - How taken: HM Brig Reindeer - When taken: 3 Feb 1813 - Where taken: Bay of Biscay - Date received: 10 Jul 1813 - From what ship: Dartmoor Prison - Born: Baltimore - Age: 19 - Discharged on 10 Jul 1813 and released to HMS Salvador del Mundo.

Scheen, John - Master - Number: 2229 - Prize name: Antoinette, schooner - Ship type: MV - How taken: HM Brig Royalist - When taken: 17 Dec 1813 - Where taken: Basque Roads, France - Date received: 6 Jan 1814 - From what ship: Antoinette, schooner - Born: Prussia - Age: 40 - Discharged on 25 Jan 1814 and sent to Ashburton on parole.

Schew, Richard - Seaman - Number: 2129 - Prize name: Amiable - Ship type: LM - How taken: HM Ship-of-the-Line Magnificant - When taken: 30 Oct 1813 - Where taken: off Lorient - Date received: 29 Nov 1813 - From what ship: HMS Dublin - Born: New York - Age: 32 - Discharged on 4 Dec 1813 and sent to Dartmoor.

Scholle, John - Seaman - Number: 1966 - Prize name: Meteor - Ship type: MV - How taken: HM Frigate Briton - When taken: 12 Mar 1813 - Where taken: Bay of Biscay - Date received: 22 Apr 1813 - From what ship: HMS Superb - Born: Long Island - Age: 30 - Sent to Dartmoor on 8 Sep 1813.

Scholle, John (alias George Myers) - Seaman - Number: 2695 - Prize name: Meteor - Ship type: MV - How taken: HM Frigate Briton - When taken: 12 Mar 1813 - Where taken: Bay of Biscay - Date received: 17 Jun 1814 - From what ship:

Dartmouth - Born: Long Island - Age: 30 - Discharged on 20 Jun 1814 and released to the Speedwell Cartel.

Schrader, Martin - Seaman - Number: 2476 - Prize name: Fanny - Ship type: MV - How taken: HM Frigate Eurotas - When taken: 25 Dec 1813 - Where taken: at sea - Date received: 28 Feb 1814 - From what ship: Dartmoor prison - Born: Memel - Age: 22 - Discharged on 4 Mar 1814 and sent to Harwich for HMS Rosemond.

Schroder, Henrick - Cook - Number: 167 - Prize name: Hunter - Ship type: P - How taken: HM Frigate Phoebe - When taken: 24 Dec 1812 - Where taken: off Western Islands - Date received: 9 Jan 1813 - From what ship: HMS Phoebe - Born: Germantown - Age: 38 - Sent to Portsmouth on 8 Feb 1813 on the HMS Colossus.

Schultz, John - Seaman - Number: 1654 - Prize name: Criterion - Ship type: MV - How taken: HM Frigate Belle Poule - When taken: 14 Feb 1813 - Where taken: Bay of Biscay - Date received: 15 Jun 1813 - From what ship: Dartmoor Prison - Born: Stralsund - Age: 20 - Discharged on 16 Jun 1813 and released to HMS Salvador del Mundo.

Schultz, John - Seaman - Number: 641 - Prize name: Criterion - Ship type: MV - How taken: HM Frigate Belle Poule - When taken: 14 Feb 1813 - Where taken: Bay of Biscay - Date received: 4 Mar 1813 - From what ship: HMS Strenuous - Born: Stralsund, Prussia - Age: 20 - Sent to Dartmoor on 2 Apr 1813.

Schyder, Jacob Knapp - Seaman - Number: 410 - Prize name: Union - Ship type: MV - How taken: HM Frigate Iris - When taken: 17 Jan 1813 - Where taken: Lat 44 N Long 2.3 W - Date received: 5 Feb 1813 - From what ship: HMS San Josef - Born: Philadelphia - Age: 20 - Sent to Chatham on 29 Mar 1813 on the HMS Braham.

Scot, Benjamin - Seaman - Number: 238 - Prize name: John Barnes - Ship type: MV - How taken: Impressed at Liverpool - When taken: 17 Nov 1812 - Date received: 12 Jan 1813 - From what ship: HMS Bittern - Born: Virginia - Age: 24 - Sent to Portsmouth on 8 Feb 1813 on the HMS Colossus.

Scott, Andrew - Seaman - Number: 261 - Prize name: Brutus - Ship type: MV - How taken: HM Frigate Briton - When taken: Jan 1813 - Where taken: Bay of Biscay - Date received: 21 Jan 1813 - From what ship: HMS Briton - Born: New Orleans - Age: 28 - Sent to Portsmouth on 8 Feb 1813 on the HMS Colossus.

Scott, Ezekiel - Ordinary Seaman - Number: 478 - How taken: Impressed at Greenock - When taken: 2 Dec 1812 - Date received: 10 Feb 1813 - From what ship: HMS Frederick - Born: Virginia - Age: 45 - Race: Black - Sent to Chatham on 29 Mar 1813 on the HMS Braham.

Scott, James - Steward - Number: 11 - Prize name: Elizabeth - Ship type: MV - How taken: Taken at Liverpool - When taken: 18 Oct 1812 - Date received: 21 Nov 1812 - From what ship: HMS Salvador del Mundo - Born: Boston - Age: 27 - Sent to Portsmouth on 29 Dec 1812 on the HMS Northumberland.

Scott, John - Seaman - Number: 1040 - How taken: Delivered himself up from HM Ship-of-the-Line Ajax - Date received: 22 Apr 1813 - From what ship: HMS Ajax - Born: Essex - Age: 35 - Sent to Chatham on 8 Jul 1813 on HM Tender Neptune.

Scott, John - Seaman - Number: 1672 - How taken: Taken at Liverpool - When taken: 26 May 1813 - Date received: 17 Jun 1813 - From what ship: HMS Bittern - Born: Rhode Island - Age: 48 - Discharged on 30 Jun 1813 and sent to Stapleton.

Scott, John - Seaman - Number: 1927 - Prize name: US Brig Argus - Ship type: War - How taken: HM Brig Pelican - When taken: 14 Aug 1813 - Where taken: Irish Channel - Date received: 23 Aug 1813 - From what ship: HMS Pelican -

Born: Richmond - Discharged on 3 Nov 1813 and sent to Dartmoor.

Scott, John - Seaman - Number: 1925 - Prize name: US Brig Argus - Ship type: War - How taken: HM Brig Pelican - When taken: 14 Aug 1813 - Where taken: Irish Channel - Date received: 23 Aug 1813 - From what ship: HMS Pelican - Born: Philadelphia - Age: 25 - Race: Black - Discharged on 8 Sep 1813 and sent to Dartmoor.

Scott, Robert - Seaman - Number: 1837 - Prize name: US Brig Argus - Ship type: War - How taken: HM Brig Pelican - When taken: 14 Aug 1813 - Where taken: Irish Channel - Date received: 17 Aug 1813 - From what ship: USS Argus - Born: New York - Age: 20 - Discharged on 8 Sep 1813 and sent to Dartmoor.

Scribner, Elijah - Prize Master - Number: 148 - Prize name: Hunter - Ship type: P - How taken: HM Frigate Phoebe - When taken: 24 Dec 1812 - Where taken: off Western Islands - Date received: 9 Jan 1813 - From what ship: HMS Phoebe - Born: Westfield - Age: 29 - Sent to Portsmouth on 8 Feb 1813 on the HMS Colossus.

Seabold, John - Seaman - Number: 414 - Prize name: Union - Ship type: MV - How taken: HM Frigate Iris - When taken: 17 Jan 1813 - Where taken: Lat 44 N Long 2.3 W - Date received: 5 Feb 1813 - From what ship: HMS San Josef - Born: Philadelphia - Age: 24 - Sent to Chatham on 29 Mar 1813 on the HMS Braham.

Sears, Abram - Steward - Number: 1825 - Prize name: US Brig Argus - Ship type: War - How taken: HM Brig Pelican - When taken: 14 Aug 1813 - Where taken: Irish Channel - Date received: 17 Aug 1813 - From what ship: USS Argus - Born: New York - Age: 26 - Discharged on 8 Sep 1813 and sent to Dartmoor.

Sears, Charles - Passenger - Number: 2057 - How taken: Taken off a Russian merchant vessel - Date received: 1 Nov 1813 - From what ship: HMS Rippon - Born: Boston - Age: 17 - Discharged on 3 Nov 1813 and sent to Dartmoor.

Seimandy, Gasper - Passenger - Number: 392 - Prize name: Union - Ship type: MV - How taken: HM Frigate Iris - When taken: 17 Jan 1813 - Where taken: Lat 44 N Long 2.3 W - Date received: 5 Feb 1813 - From what ship: HMS San Josef - Age: 52 - Entered on 14 Feb 1813 into the French General Entry as prisoner number 39380.

Selman, John - Seaman - Number: 2362 - Prize name: Hannah - Ship type: MV - How taken: HM Ship-of-the-Line Conquestador - When taken: 14 Jan 1814 - Where taken: Lat 47.3 Long 7 - Date received: 31 Jan 1814 - From what ship: HMS Surveillante - Born: Marblehead - Age: 19 - Discharged on 27 Feb 1814 and sent to Chatham on

HMS Haleyon.

Selston, Isaac - Seaman - Number: 2063 - Prize name: Betsy, prize to the Privateer True Blooded Yankee - Ship type: P - How taken: HM Frigate Eurotas - When taken: 26 Oct 1813 - Where taken: off Ushant - Date received: 1 Nov 1813 - From what ship: HMS Hannibal - Born: Sweden - Age: 28 - Discharged on 3 Nov 1813 and sent to Dartmoor.

Sennet, John - Seaman - Number: 760 - Prize name: William Bayard - Ship type: MV - How taken: HM Ship-of-the-Line Warspite - When taken: 3 Mar 1813 - Where taken: Bay of Biscay - Date received: 19 Mar 1813 - From what ship: HMS Warspite - Born: Philadelphia - Age: 34 - Sent to Dartmoor on 2 Apr 1813.

Sentille, Francis - Seaman - Number: 1459 - Prize name: Paul Jones - Ship type: P - How taken: HM Frigate Leonidas - When taken: 23 May 1813 - Where taken: off Cape Clear - Date received: 26 May 1813 - From what ship: HMS Leonidas - Born: Charleston - Age: 17 - Discharged on 30 Jun 1813 and sent to Stapleton.

Sereder, Martin - Seaman - Number: 2270 - Prize name: Fanny - Ship type: MV - How taken: HM Frigate Eurotas - When taken: 25 Dec 1813 - Where taken: at sea - Date received: 20 Jan 1814 - From what ship: HMS Eurotas - Born: Menzel, Prussia - Age: 22 - Discharged on 31 Jan 1814 and sent to Dartmoor.

Shares, Samuel - 2nd Mate - Number: 1010 - Prize name: Amphitrite - Ship type: MV - How taken: HM Ketch Gleaner - When taken: 27 Feb 1813 - Where taken: Bay of Biscay - Date received: 20 Apr 1813 - From what ship: HMS Libria - Born: Stratford - Age: 25 - Sent to Dartmoor on 8 Sep 1813.

Shammton, Samuel - Seaman - Number: 585 - Prize name: Cashiere - Ship type: LM - How taken: HM Brig Reindeer - When taken: 3 Feb 1813 - Where taken: Bay of Biscay - Date received: 23 Feb 1813 - From what ship: HMS Surveillante - Born: Baltimore - Age: 35 - Sent to Dartmoor on 2 Apr 1813.

Shaw, Edward - Chief Mate - Number: 2101 - Prize name: Chesapeake - Ship type: LM - How taken: HM Frigate Hotspur & HM Frigate Pyramus - When taken: 26 Oct 1813 - Where taken: Bay of Biscay - Date received: 22 Nov 1813 - From what ship: HMS Pyramus - Born: Machias, MA - Age: 29 - Discharged on 25 Aug 1813 and sent to Ashburton on parole.

Shaw, Jacob - Passenger - Number: 961 - Prize name: Ferox - Ship type: MV - How taken: HM Frigate Medusa & HM Brig Lyra - When taken: 28 Mar 1813 - Where taken: off Cape Ortagle - Date received: 9 Apr 1813 - From what ship: HMS Lyra - Born: Wilmington - Age: 24 - Sent on 11 Apr 1813 to Ashburton on parole.

Shaw, Richard - Seaman - Number: 1527 - Prize name: Courier - Ship type: LM - How taken: HM Brig Rover - When taken: 14 Mar 1813 - Where taken: Bay of Biscay - Date received: 29 May 1813 - From what ship: HMS Hannibal - Born: Maryland - Age: 26 - Discharged on 30 Jun 1813 and sent to Stapleton.

Shaw, William - Seaman - Number: 2429 - Prize name: US Brig Argus - Ship type: War - How taken: HM Brig Pelican - When taken: 14 Aug 1813 - Where taken: Irish Channel - Date received: 12 Feb 1814 - From what ship: HMS Salvador del Mundo - Born: Philadelphia - Age: 23 - Discharged on 10 May 1814 and sent to Dartmoor.

Sheardon, Ellison - Seaman - Number: 2285 - Prize name: Siro - Ship type: LM - How taken: HM Brig Pelican - When taken: 13 Jan 1814 - Where taken: at sea - Date received: 23 Jan 1814 - From what ship: HMS Pelican - Born: Rhode Island - Age: 18 - Discharged on 31 Jan 1814 and sent to Dartmoor.

Sheffers, Steuben - Seaman - Number: 1386 - Prize name: Tom - Ship type: LM - How taken: HM Frigate Surveillante - When taken: 27 Apr 1813 - Where taken: Bay of Biscay - Date received: 15 May 1813 - From what ship: HMS Foxhound - Born: Connecticut - Age: 19 - Discharged on 3 Jul 1813 and sent to Stapleton Prison.

Shepherd, Daniel - Seaman - Number: 110 - Prize name: Otter - Ship type: MV - How taken: HM Ship-Sloop Jalousie - When taken: 1 Dec 1812 - Where taken: off Cape Vincent - Date received: 30 Dec 1812 - From what ship: HMS Leonidas - Born: New York - Age: 24 - Sent to Portsmouth on 4 Jan 1813 on the HMS Revolutionnaire.

Shepherd, William - Chief Mate - Number: 1797 - Prize name: Godfrey & Mary - Ship type: MV - How taken: Robert Todd, privateer - When taken: 23 Jun 1813 - Where taken: Lat 21.3N Long 53W - Date received: 16 Aug 1813 - From what ship: HMS Bittern - Born: New London - Age: 27 - Discharged on 26 Aug 1813 and sent to Ashburton on parole.

Shepherd, William - Seaman - Number: 2577 - How taken: Delivered himself up from the MV Hebrus - Date received: 8 May 1814 - From what ship: MV Hebrus - Born: Lancaster - Age: 25 - Discharged on 7 May 1814 and sent to

Ashburton on parole.

Sheppard, William - Seaman - Number: 1023 - Prize name: Two Brothers - Ship type: MV - How taken: Bootle of Liverpool, letter of marque - When taken: 18 Mar 1813 - Where taken: Western Islands - Date received: 21 Apr 1813 - From what ship: HMS Bittern - Born: New York - Age: 41 - Sent to Dartmoor on 1 Jul 1813.

Sherman, Reuben - 2nd Mate - Number: 871 - Prize name: Criterion - Ship type: MV - How taken: HM Frigate Belle Poule - When taken: 14 Feb 1813 - Where taken: off Bordeaux - Date received: 28 Mar 1813 - From what ship: HMS Warspite - Born: New Bedford - Age: 25 - Sent to Dartmoor on 28 Jun 1813.

Sherwood, John - Seaman - Number: 766 - Prize name: William Bayard - Ship type: MV - How taken: HM Ship-of-the-Line Warspite - When taken: 3 Mar 1813 - Where taken: Bay of Biscay - Date received: 19 Mar 1813 - From what ship: HMS Warspite - Born: Poughkeepsie - Age: 23 - Sent to Dartmoor on 28 Jun 1813.

Shipley, Daniel - Seaman - Number: 2040 - How taken: Impressed at Liverpool - When taken: 16 Oct 1813 - Date received: 1 Nov 1813 - From what ship: HMS Bittern - Born: New Castle - Age: 24 - Discharged on 3 Nov 1813 and sent to Dartmoor.

Short, Clement - Steward - Number: 2586 - Prize name: James - Ship type: MV - How taken: HM Brig Harpy - When taken: 19 Dec 1812 - Where taken: off Isle du France - Date received: 10 May 1814 - From what ship: HMS Lion - Born: New York - Age: 27 - Race: Black - Discharged on 14 Jun 1814 and sent to Dartmoor.

Shovel, John - Seaman - Number: 1237 - Prize name: Essex - Ship type: MV - How taken: HM Frigate Pyramus - When taken: 2 Apr 1813 - Where taken: Bay of Biscay - Date received: 9 May 1813 - From what ship: HMS Andromache - Born: Boston - Age: 19 - Discharged on 3 Jul 1813 and sent to Stapleton Prison.

Sickless, James M. - Seaman - Number: 1993 - Prize name: Ned - Ship type: LM - How taken: HM Brig Royalist - When taken: 6 Sep 1812 - Where taken: coast of France - Date received: 22 Sep 1813 - From what ship: HMS Royalist - Born: New York - Age: 22 - Discharged on 27 Sep 1813 and sent to Dartmoor.

Signard, Samuel Francis - Seaman - Number: 179 - Prize name: Hunter - Ship type: P - How taken: HM Frigate Phoebe - When taken: 24 Dec 1812 - Where taken: off Western Islands - Date received: 9 Jan 1813 - From what ship: HMS Phoebe - Born: New Orleans - Age: 29 - Race: Black - Sent to Portsmouth on 8 Feb 1813 on the HMS Colossus.

Sillock, Amos - Unknown - Number: 914 - How taken: Delivered himself up as a prisoner of war - Date received: 4 Apr 1813 - From what ship: HMS Salvador del Mundo - Born: Sheffield - Age: 30 - Sent on 21 Jun 1813 to Portsmouth on HMS Prometheus.

Silva, Francis - Seaman - Number: 1935 - Prize name: Marmion - Ship type: MV - How taken: HM Frigate President - When taken: 14 Aug 1813 - Where taken: off Nantes - Date received: 27 Aug 1813 - From what ship: HMS Urgent - Born: New York - Age: 22 - Discharged on 8 Sep 1813 and sent to Dartmoor.

Simons, William - Seaman - Number: 2489 - Prize name: Sherbrook - Ship type: MV - How taken: Impressed at Falmouth - When taken: 20 Mar 1814 - Date received: 26 Mar 1814 - From what ship: HMS Prince Frederick - Born: Nantucket - Age: 24 - Race: Negro - Discharged on 10 May 1814 and sent to Dartmoor.

Simpson, Thomas - Seaman - Number: 1274 - Prize name: Caroline - Ship type: MV - How taken: HM Frigate Medusa - When taken: 12 Apr 1813 - Where taken: Bay of Biscay - Date received: 10 May 1813 - From what ship: HMS Medusa - Born: Georgetown - Age: 35 - Discharged on 3 Jul 1813 and sent to Stapleton Prison.

Simpson, William - Seaman - Number: 219 - Prize name: Vengeance - Ship type: LM - How taken: HM Frigate Phoebe - When taken: 1 Jan 1813 - Where taken: Lat 44.4 Long 23 - Date received: 9 Jan 1813 - From what ship: HMS Phoebe - Born: New York - Age: 24 - Sent to Portsmouth on 8 Feb 1813 on the HMS Colossus.

Simpson, William - Seaman - Number: 2674 - Prize name: John, prize to Amelia - Ship type: P - How taken: HM Ship-of-the-Line Sterling Castle - When taken: 10 May 1814 - Where taken: Lat 36 Long 37 - Date received: 5 Jun 1814 - From what ship: HMS Gronville - Born: Philadelphia - Age: 27 - Discharged on 14 Jun 1814 and sent to Dartmoor.

Sims, Oliver - Seaman - Number: 982 - How taken: Impressed at Dublin - When taken: 12 Feb 1813 - Date received: 13 Apr 1813 - From what ship: HMS Frederick - Born: Philadelphia - Age: 19 - Sent to Chatham on 8 Jul 1813 on HM Tender Neptune.

Singleton, Samuel - Master - Number: 1409 - Prize name: Fox - Ship type: LM - How taken: HM Sloop Pheasant - When taken: 23 Apr 1813 - Where taken: Bay of Biscay - Date received: 20 May 1813 - From what ship: HMS Pleasant - Born: Wilmington - Age: 34 - Discharged on 23 May 1813 and sent to Ashburton on parole.

Sennett, John - Seaman - Number: 1660 - Prize name: William Bayard - Ship type: MV - How taken: HM Ship-of-the-Line Warspite - When taken: 3 Mar 1813 - Where taken: Bay of Biscay - Date received: 15 Jun 1813 - From what ship: Dartmoor Prison - Born: Philadelphia - Discharged on 16 Jun 1813 and released to HMS Salvador del Mundo.

Sketchly, William - Master - Number: 1328 - Prize name: Magdalen - Ship type: MV - How taken: HM Ship-of-the-Line Superb - When taken: 15 Apr 1813 - Where taken: off Belle Isle - Date received: 14 May 1813 - From what ship: HMS Superb - Born: New York - Age: 35 - Discharged on 15 May 1813 and sent to Liverpool on parole.

Skiddy, William - Mate - Number: 298 - Prize name: Stephen - Ship type: MV - How taken: HM Frigate Briton & HM Frigate Andromache - When taken: 17 Dec 1812 - Where taken: off Bordeaux - Date received: 21 Jan 1813 - From what ship: HMS Abercrombie - Born: New York - Age: 19 - Sent on 22 Jan 1813 to Ashburton on parole.

Skidder, Samuel - Seaman - Number: 2064 - Prize name: Betsy, prize to the Privateer True Blooded Yankee - Ship type: P - How taken: HM Frigate Eurotas - When taken: 26 Oct 1813 - Where taken: off Ushant - Date received: 1 Nov 1813 - From what ship: HMS Hannibal - Born: Connecticut - Age: 22 - Discharged on 3 Nov 1813 and sent to Dartmoor.

Skinner, Tilman - Seaman - Number: 2678 - Prize name: John, prize to Amelia - Ship type: P - How taken: HM Ship-of-the-Line Sterling Castle - When taken: 10 May 1814 - Where taken: Lat 36 Long 37 - Date received: 5 Jun 1814 - From what ship: HMS Gronville - Born: Philadelphia - Age: 22 - Race: Black - Discharged on 14 Jun 1814 and sent to Dartmoor.

Slate, Henry - Pilot - Number: 265 - Prize name: Brutus - Ship type: MV - How taken: HM Frigate Briton - When taken: Jan 1813 - Where taken: Bay of Biscay - Date received: 21 Jan 1813 - From what ship: HMS Briton - Born: New York - Age: 24 - Sent to Portsmouth on 8 Feb 1813 on the HMS Colossus.

Slayter, John - Master - Number: 44 - Prize name: Rising States - Ship type: MV - How taken: HM Frigate Fortunee - When taken: 12 Aug 1812 - Where taken: Western Islands - Date received: 23 Nov 1812 - From what ship: HMS Stork - Born: Delaware - Age: 30 - Sent on 8 Dec 1812 to Ashburton on parole.

Slocum, William - Seaman - Number: 1717 - Prize name: Orders in Council - Ship type: LM - How taken: HM Frigate Surveillante - When taken: 1 Jun 1813 - Where taken: Bay of Biscay - Date received: 4 Jul 1813 - From what ship: HMS Iris - Born: Newport - Age: 23 - Discharged on 22 Aug 1813 and released to HM Brig Redpole.

Slowin, Abraham - Seaman - Number: 1340 - Prize name: Fox - Ship type: LM - How taken: HM Sloop Pheasant - When taken: 23 Apr 1813 - Where taken: Bay of Biscay - Date received: 14 May 1813 - From what ship: HMS Pleasant - Born: New Orleans - Age: 31 - Discharged on 3 Jul 1813 and sent to Stapleton Prison.

Small, Enoch - 2nd Mate - Number: 2084 - Prize name: Sybella - Ship type: MV - How taken: HM Brig Zenobia - When taken: 17 Jun 1813 - Where taken: off Cape Henry - Date received: 2 Nov 1813 - From what ship: HMS Bacchus - Born: Massachusetts - Age: 21 - Discharged on 3 Nov 1813 and sent to Dartmoor.

Small, George D. - Seaman - Number: 893 - Prize name: Tiger - Ship type: MV - How taken: HM Brig Scylla - When taken: 22 Mar 1813 - Where taken: Bay of Biscay - Date received: 1 Apr 1813 - From what ship: HMS Scylla - Born: New York - Age: 24 - Sent on 21 Jun 1813 to Portsmouth on HMS Prometheus.

Small, Richard - Seaman - Number: 1124 - Prize name: Viper - Ship type: MV - How taken: HM Ship-of-the-Line Superb - When taken: 15 Apr 1813 - Where taken: Bay of Biscay - Date received: 22 Apr 1813 - From what ship: HMS Superb - Born: Baltimore - Age: 18 - Sent to Dartmoor on 1 Jul 1813.

Smally, Thomas - Seaman - Number: 1110 - Prize name: Viper - Ship type: MV - How taken: HM Ship-of-the-Line Superb - When taken: 15 Apr 1813 - Where taken: Bay of Biscay - Date received: 22 Apr 1813 - From what ship: HMS Superb - Born: Philadelphia - Age: 56 - Sent to Dartmoor on 1 Jul 1813.

Smasher, Allen - Seaman - Number: 459 - Prize name: Dolphin - Ship type: LM - How taken: HM Ship-of-the-Line Colossus, HM Frigate Rhin & HM Brig Goldfinch - When taken: 5 Jan 1813 - Where taken: off Western Islands - Date received: 6 Feb 1813 - From what ship: HMS Rhin - Born: Philadelphia - Age: 32 - Race: Negro - Sent to Chatham on 29 Mar 1813 on the HMS Braham.

Smiles, John - Seaman - Number: 2314 - Prize name: Siro - Ship type: LM - How taken: HM Brig Pelican - When taken: 13 Jan 1814 - Where taken: at sea - Date received: 23 Jan 1814 - From what ship: HMS Pelican - Born: Batavia - Age: 24 - Discharged on 31 Jan 1814 and sent to Dartmoor.

Smith, Adam - Seaman - Number: 1888 - Prize name: US Brig Argus - Ship type: War - How taken: HM Brig Pelican - When taken: 14 Aug 1813 - Where taken: Irish Channel - Date received: 23 Aug 1813 - From what ship: HMS Pelican - Born: Amsterdam - Age: 24 - Discharged on 8 Sep 1813 and sent to Dartmoor.

Smith, Andrew - Seaman - Number: 1378 - Prize name: Tom - Ship type: LM - How taken: HM Frigate Surveillante - When taken: 27 Apr 1813 - Where taken: Bay of Biscay - Date received: 15 May 1813 - From what ship: HMS Foxhound - Born: Maryland - Age: 23 - Discharged on 3 Jul 1813 and sent to Stapleton Prison.

Smith, Charles - Seaman - Number: 685 - How taken: Delivered himself up from HM Schooner Mackerel - Date received: 15 Mar 1813 - From what ship: Growler - Born: Albany - Age: 26 - Sent to Dartmoor on 2 Apr 1813.

Smith, Chester - Seaman - Number: 229 - Prize name: Vengeance - Ship type: LM - How taken: HM Frigate Phoebe - When taken: 1 Jan 1813 - Where taken: Lat 44.4 Long 23 - Date received: 9 Jan 1813 - From what ship: HMS Phoebe - Age: 24 - Sent to Portsmouth on 8 Feb 1813 on the HMS Colossus.

Smith, Chris - Seaman - Number: 1557 - Prize name: Elvia - Ship type: MV - How taken: HM Frigate Surveillante - When taken: 21 Mar 1813 - Where taken: Bay of Biscay - Date received: 29 May 1813 - From what ship: HMS Hannibal - Born: New Jersey - Age: 26 - Discharged on 30 Jun 1813 and sent to Stapleton.

Smith, Eldridge - Seaman - Number: 534 - Prize name: Terrible - Ship type: MV - How taken: HM Brig Foxhound - When taken: 8 Feb 1813 - Where taken: Channel - Date received: 15 Feb 1813 - From what ship: HMS Foxhound - Born: Plymouth - Age: 28 - Sent to Dartmoor on 2 Apr 1813.

Smith, Elisha - Seaman - Number: 1983 - Prize name: Ned - Ship type: LM - How taken: HM Brig Royalist - When taken: 6 Sep 1812 - Where taken: coast of France - Date received: 22 Sep 1813 - From what ship: HMS Royalist - Born: East Haven - Age: 25 - Discharged on 27 Sep 1813 and sent to Dartmoor.

Smith, George - Seaman - Number: 843 - Prize name: Pallas - Ship type: MV - How taken: HM Brig Rebuff - When taken: 23 Dec 1812 - Where taken: off Cadiz - Date received: 23 May 1813 - From what ship: HMS Dauntless - Born: Pennsylvania - Age: 30 - Sent to Dartmoor on 28 Jun 1813.

Smith, Henry - Chief Mate - Number: 922 - Prize name: Prompt - Ship type: MV - How taken: Chance, privateer - When taken: 22 Mar 1813 - Where taken: Bay of Biscay - Date received: 6 Apr 1813 - From what ship: Fonvey - Born: Hudson - Age: 25 - Sent on 6 Apr 1813 to Ashburton on parole.

Smith, Jacob - Seaman - Number: 80 - Prize name: Experiment - Ship type: MV - How taken: HM Brig Rover - When taken: 21 Oct 1812 - Where taken: off Bordeaux - Date received: 13 Dec 1812 - From what ship: HMS Martial - Born: Prussia - Age: 36 - Sent to Portsmouth on 29 Dec 1812 on the HMS Northumberland.

Smith, James - Seaman - Number: 1842 - Prize name: US Brig Argus - Ship type: War - How taken: HM Brig Pelican - When taken: 14 Aug 1813 - Where taken: Irish Channel - Date received: 17 Aug 1813 - From what ship: USS Argus - Born: Connecticut - Age: 21 - Discharged on 8 Sep 1813 and sent to Dartmoor.

Smith, James - Seaman - Number: 2162 - Prize name: Princess - Ship type: MV - How taken: Impressed at Liverpool - When taken: 1 Nov 1813 - Date received: 13 Dec 1813 - From what ship: HMS Bittern - Born: New York - Age: 22 - Discharged on 31 Jan 1814 and sent to Dartmoor.

Smith, John - Seaman - Number: 1700 - How taken: Delivered himself up from HM Ship-of-the-Line Ville de Paris - Date received: 26 Jun 1813 - From what ship: HMS Duncan - Born: Massachusetts - Age: 28 - Discharged on 8 Jul 1813 and sent to Chatham on tender Neptune.

Smith, John - Seaman - Number: 1442 - Prize name: Paul Jones - Ship type: P - How taken: HM Frigate Leonidas - When taken: 23 May 1813 - Where taken: off Cape Clear - Date received: 26 May 1813 - From what ship: HMS Leonidas - Born: Rhode Island - Age: 19 - Discharged on 30 Jun 1813 and sent to Stapleton.

Smith, John - Seaman - Number: 1602 - Prize name: Governor Gerry - Ship type: MV - How taken: HM Brig Royalist - When taken: 31 May 1813 - Where taken: Bay of Biscay - Date received: 14 Jun 1813 - From what ship: HMS Royalist - Born: Virginia - Age: 27 - Discharged on 30 Jun 1813 and sent to Stapleton.

Smith, John - 2nd Gunner - Number: 2422 - Prize name: US Brig Argus - Ship type: War - How taken: Sent from HM

Ship-of-the-Line Salvador del Mundo - When taken: 14 Aug 1813 - Where taken: Irish Channel - Date received: 12 Feb 1814 - From what ship: HMS Salvador del Mundo - Born: Middletown - Age: 38 - Discharged on 10 May 1814 and sent to Dartmoor.

Smith, John - Carpenter - Number: 530 - Prize name: Terrible - Ship type: MV - How taken: HM Brig Foxhound - When taken: 8 Feb 1813 - Where taken: Channel - Date received: 15 Feb 1813 - From what ship: HMS Foxhound - Born: Groton - Age: 23 - Sent to Dartmoor on 2 Apr 1813.

Smith, John - Seaman - Number: 1446 - Prize name: Paul Jones - Ship type: P - How taken: HM Frigate Leonidas - When taken: 23 May 1813 - Where taken: off Cape Clear - Date received: 26 May 1813 - From what ship: HMS Leonidas - Born: Massachusetts - Age: 23 - Race: Mulatto - Discharged on 30 Jun 1813 and sent to Stapleton.

Smith, John - Seaman - Number: 1692 - Prize name: Governor Gerry - Ship type: MV - How taken: HM Brig Royalist - When taken: 31 May 1813 - Where taken: Bay of Biscay - Date received: 26 Jun 1813 - From what ship: HMS Duncan - Born: Boston - Age: 25 - Race: Black - Discharged on 30 Jun 1813 and sent to Stapleton.

Smith, John - Seaman - Number: 2405 - Prize name: Rachel & Ann, prize of the Privateer Prince de Neufchatel - Ship type: P - How taken: HM Frigate Cydnus - When taken: 6 Jan 1814 - Where taken: at sea - Date received: 4 Feb 1814 - From what ship: HMS Cydnus - Born: New York - Age: 37 - Discharged on 10 May 1814 and sent to Dartmoor.

Smith, John - Seaman - Number: 667 - How taken: Impressed at Exeter - When taken: 12 Mar 1813 - Date received: 14 Mar 1813 - From what ship: HMS Salvador del Mundo - Born: Philadelphia - Age: 20 - Sent to Dartmoor on 1 Jul 1813.

Smith, John - Quarter Gunner - Number: 1820 - Prize name: US Brig Argus - Ship type: War - How taken: HM Brig Pelican - When taken: 14 Aug 1813 - Where taken: Irish Channel - Date received: 17 Aug 1813 - From what ship: USS Argus - Born: New Jersey - Age: 38 - Discharged on 24 Aug 1813 and released to HMS Salvador del Mundo.

Smith, John - Seaman - Number: 2534 - How taken: Impressed at Belfast - When taken: 18 Dec 1813 - Date received: 13 Apr 1814 - From what ship: HMS Bittern - Born: New York - Age: 36 - Race: Negro - Discharged on 10 May 1814 and sent to Dartmoor.

Smith, John - Seaman - Number: 2075 - Prize name: Avon, prize of the Privateer True Blooded Yankee - Ship type: P - How taken: HM Frigate Eurotas - When taken: 27 Oct 1813 - Where taken: off Ushant - Date received: 1 Nov 1813 - From what ship: HMS Hannibal - Born: Massachusetts - Age: 34 - Discharged on 3 Nov 1813 and sent to Dartmoor.

Smith, John - Seaman - Number: 2133 - Prize name: Amiable - Ship type: LM - How taken: HM Ship-of-the-Line Magnificant - When taken: 30 Oct 1813 - Where taken: off Lorient - Date received: 29 Nov 1813 - From what ship: HMS Dublin - Born: New York - Age: 28 - Discharged on 4 Dec 1813 and sent to Dartmoor.

Smith, John D. - Prize Master - Number: 2519 - Prize name: Hope, prize to the Privateer True Blooded Yankee - Ship type: P - How taken: HM Frigate Seahorse - When taken: 22 Mar 1814 - Where taken: at sea - Date received: 6 Apr 1814 - From what ship: HMS Queen Charlotte - Born: Stonington - Age: 27 - Discharged on 10 May 1814 and sent to Dartmoor.

Smith, Richard - Seaman - Number: 1603 - Prize name: Governor Gerry - Ship type: MV - How taken: HM Brig Royalist - When taken: 31 May 1813 - Where taken: Bay of Biscay - Date received: 14 Jun 1813 - From what ship: HMS Royalist - Born: Boston - Age: 29 - Discharged on 30 Jun 1813 and sent to Stapleton.

Smith, Richard - Gunner - Number: 484 - Prize name: Cashiere - Ship type: LM - How taken: HM Brig Reindeer - When taken: 3 Feb 1813 - Where taken: Bay of Biscay - Date received: 12 Feb 1813 - From what ship: HMS Reindeer - Born: New York - Age: 22 - Sent to Dartmoor on 2 Apr 1813.

Smith, Richard - Seaman - Number: 2203 - Prize name: General Kempt, prize of the Privateer Grand Turk - Ship type: P - How taken: HM Brig Foxhound - When taken: 18 Dec 1813 - Where taken: Lat 48.4 Long 6 - Date received: 21 Dec 1813 - From what ship: HMS Foxhound - Born: Salem - Age: 24 - Discharged on 27 Feb 1814 and sent to Chatham on HMS Haleyon.

Smith, Thomas - Seaman - Number: 117 - Prize name: Otter - Ship type: MV - How taken: HM Ship-Sloop Jalousie - When taken: 1 Dec 1812 - Where taken: off Cape Vincent - Date received: 30 Dec 1812 - From what ship: HMS Leonidas - Born: Maryland - Age: 22 - Sent to Portsmouth on 4 Jan 1813 on the HMS Revolutionnaire.

Smith, Thomas - Boatswain - Number: 1419 - Prize name: Paul Jones - Ship type: P - How taken: HM Frigate Leonidas - When taken: 23 May 1813 - Where taken: off Cape Clear - Date received: 26 May 1813 - From what ship: HMS Leonidas - Born: New York - Age: 29 - Discharged on 3 Jul 1813 and sent to Stapleton Prison.

Smith, Thomas - Seaman - Number: 1398 - Prize name: Henry Clements - Ship type: MV - How taken: HM Brig Orestes - When taken: 13 Apr 1813 - Where taken: Bay of Biscay - Date received: 15 May 1813 - From what ship: HMS Orestes - Born: Massachusetts - Age: 17 - Discharged on 3 Jul 1813 and sent to Stapleton Prison.

Smith, William - Seaman - Number: 1012 - How taken: Detained at Gibraltar - When taken: Oct 1812 - Date received: 20 Apr 1813 - From what ship: HMS Libria - Born: Baltimore - Age: 30 - Sent to Dartmoor on 8 Sep 1813.

Smith, William - Seaman - Number: 537 - Prize name: Terrible - Ship type: MV - How taken: HM Brig Foxhound - When taken: 8 Feb 1813 - Where taken: Channel - Date received: 15 Feb 1813 - From what ship: HMS Foxhound - Born: New York - Age: 18 - Sent to Dartmoor on 2 Apr 1813.

Smith, William - Seaman - Number: 1662 - Prize name: Terrible - Ship type: MV - How taken: HM Brig Foxhound - When taken: 8 Feb 1813 - Where taken: Channel - Date received: 15 Jun 1813 - From what ship: Dartmoor Prison - Born: New York - Age: 18 - Discharged on 16 Jun 1813 and released to HMS Salvador del Mundo.

Smith, William - Cook - Number: 731 - Prize name: Pert - Ship type: MV - How taken: HM Ship-of-the-Line Warspite - When taken: 1 Mar 1813 - Where taken: Bay of Biscay - Date received: 19 Mar 1813 - From what ship: HMS Warspite - Born: New York - Age: 26 - Race: Black - Sent to Dartmoor on 2 Apr 1813.

Smith, William - Seaman - Number: 1907 - Prize name: US Brig Argus - Ship type: War - How taken: HM Brig Pelican - When taken: 14 Aug 1813 - Where taken: Irish Channel - Date received: 23 Aug 1813 - From what ship: HMS Pelican - Born: Charlestown - Age: 25 - Discharged on 3 Nov 1813 and sent to Dartmoor.

Smith, William - Seaman - Number: 2605 - How taken: MV Nautilus - When taken: 20 Dec 1812 - Date received: 16 May 1814 - From what ship: HMS Repulse - Born: Rhode Island - Age: 33 - Discharged on 14 Jun 1814 and sent to Dartmoor.

Smith, William - Chief Mate - Number: 1959 - Prize name: Joel Barlow - Ship type: LM - How taken: HM Frigate Briton - When taken: 3 Jul 1813 - Where taken: off Bordeaux - Date received: 31 Aug 1813 - From what ship: HMS Clarence - Born: Middletown - Age: 38 - Discharged on 10 Sep 1813 and sent to Ashburton on parole.

Smith, William - Seaman - Number: 2430 - Prize name: US Brig Argus - Ship type: War - How taken: HM Brig Pelican - When taken: 14 Aug 1813 - Where taken: Irish Channel - Date received: 12 Feb 1814 - From what ship: HMS Salvador del Mundo - Born: Charlestown - Age: 38 - Discharged on 10 May 1814 and sent to Dartmoor.

Snate, George - Seaman - Number: 1070 - Prize name: Meteor - Ship type: MV - How taken: HM Frigate Briton - When taken: 12 Mar 1813 - Where taken: Bay of Biscay - Date received: 22 Apr 1813 - From what ship: HMS Superb - Born: New York - Age: 29 - Sent to Dartmoor on 8 Sep 1813.

Snell, David - 1st Mate - Number: 2589 - Prize name: Monticello - Ship type: MV - How taken: HM Brig Racehorse - When taken: 12 Nov 1812 - Where taken: Cape of Good Hope - Date received: 10 May 1814 - From what ship: HMS Lion - Born: Norfolk - Age: 29 - Discharged on 12 May 1814 and sent to Ashburton on parole.

Snelson, Robert Lewis - Midshipman - Number: 1851 - Prize name: US Brig Argus - Ship type: War - How taken: HM Brig Pelican - When taken: 14 Aug 1813 - Where taken: Irish Channel - Date received: 19 Aug 1813 - From what ship: USS Argus - Born: Petersburg - Age: 18 - Discharged on 25 Aug 1813 and sent to Ashburton on parole.

Sniffen, John - Carpenter's Mate - Number: 1926 - Prize name: US Brig Argus - Ship type: War - How taken: HM Brig Pelican - When taken: 14 Aug 1813 - Where taken: Irish Channel - Date received: 23 Aug 1813 - From what ship: HMS Pelican - Born: New York - Age: 30 - Discharged on 8 Sep 1813 and sent to Dartmoor.

Snow, Joseph - Mate - Number: 508 - Prize name: Tyger - Ship type: MV - How taken: Detained at Gibraltar - When taken: 8 Aug 1812 - Date received: 15 Feb 1813 - From what ship: HMS Andromeda - Born: Massachusetts - Age: 27 - Sent on 17 Feb 1813 to Ashburton on parole.

Snowdon, Jacob - Seaman - Number: 1994 - Prize name: Ned - Ship type: LM - How taken: HM Brig Royalist - When taken: 6 Sep 1812 - Where taken: coast of France - Date received: 22 Sep 1813 - From what ship: HMS Royalist - Born: Jersey - Age: 34 - Race: Black - Discharged on 27 Sep 1813 and sent to Dartmoor.

Snyder, William - Seaman - Number: 2691 - How taken: Sent into custody from HM Brig Barracouta - Date received:

7 Jun 1814 - From what ship: Inap - Born: Maryland - Age: 32 - Discharged on 14 Jun 1814 and sent to Dartmoor.

Soddart, Reuben - Seaman - Number: 739 - Prize name: Charlotte - Ship type: MV - How taken: HM Ship-of-the-Line Warspite - When taken: 3 Mar 1813 - Where taken: Bay of Biscay - Date received: 19 Mar 1813 - From what ship: HMS Warspite - Born: Massachusetts - Age: 17 - Sent to Dartmoor on 2 Apr 1813.

Soderback, Andrew - Seaman - Number: 2299 - Prize name: Siro - Ship type: LM - How taken: HM Brig Pelican - When taken: 13 Jan 1814 - Where taken: at sea - Date received: 23 Jan 1814 - From what ship: HMS Pelican - Born: Gothenburg, Sweden - Age: 24 - Discharged on 31 Jan 1814 and sent to Dartmoor.

Soderstrum, Ola - Seaman - Number: 1067 - Prize name: Meteor - Ship type: MV - How taken: HM Frigate Briton - When taken: 12 Mar 1813 - Where taken: Bay of Biscay - Date received: 22 Apr 1813 - From what ship: HMS Superb - Born: Sweden - Age: 28 - Liberated on 21 May 1813.

Sodoburgh, John - Boy - Number: 210 - Prize name: Hunter - Ship type: P - How taken: HM Frigate Phoebe - When taken: 24 Dec 1812 - Where taken: off Western Islands - Date received: 9 Jan 1813 - From what ship: HMS Phoebe - Born: Boston - Age: 13 - Sent to Portsmouth on 8 Feb 1813 on the HMS Colossus.

Sole, Edward - Seaman - Number: 2634 - Prize name: Ateline - Ship type: MV - How taken: HM Frigate Magiciene - When taken: 14 Mar 1814 - Where taken: off Cape Finisterre, Spain - Date received: 17 May 1814 - From what ship: HMS Tortois - Born: Freeport - Age: 30 - Discharged on 14 Jun 1814 and sent to Dartmoor.

Sole, Elias - Seaman - Number: 2636 - Prize name: Ateline - Ship type: MV - How taken: HM Frigate Magiciene - When taken: 14 Mar 1814 - Where taken: off Cape Finisterre, Spain - Date received: 17 May 1814 - From what ship: HMS Tortois - Born: Freeport - Age: 22 - Discharged on 14 Jun 1814 and sent to Dartmoor.

Sole, Thomas - Seaman - Number: 2639 - Prize name: Ateline - Ship type: MV - How taken: HM Frigate Magiciene - When taken: 14 Mar 1814 - Where taken: off Cape Finisterre, Spain - Date received: 17 May 1814 - From what ship: HMS Tortois - Born: Freeport - Age: 24 - Discharged on 14 Jun 1814 and sent to Dartmoor.

Sone, John - Seaman - Number: 2675 - Prize name: John, prize to Amelia - Ship type: P - How taken: HM Ship-of-the-Line Sterling Castle - When taken: 10 May 1814 - Where taken: Lat 36 Long 37 - Date received: 5 Jun 1814 - From what ship: HMS Gronville - Born: Trieste, Italy - Age: 27 - Discharged on 14 Jun 1814 and sent to Dartmoor.

Soreby, Robert - Seaman - Number: 2566 - Prize name: Dorotha - Ship type: MV - How taken: HM Brig Teazer - When taken: 26 Mar 1814 - Where taken: off Cork - Date received: 25 Apr 1814 - From what ship: HMS Helena - Born: Philadelphia - Age: 21 - Discharged on 10 May 1814 and sent to Dartmoor.

Southcombie, Plummer - Chief Mate - Number: 1249 - Prize name: Price - Ship type: LM - How taken: HM Frigate Medusa - When taken: 13 Apr 1813 - Where taken: Bay of Biscay - Date received: 10 May 1813 - From what ship: HMS Medusa - Born: Virginia - Age: 28 - Discharged on 14 May 1813 and sent to Ashburton on parole.

Spafford, Samuel - Master - Number: 1250 - Prize name: Caroline - Ship type: MV - How taken: HM Frigate Medusa - When taken: 12 Apr 1813 - Where taken: Bay of Biscay - Date received: 10 May 1813 - From what ship: HMS Medusa - Born: Massachusetts - Age: 32 - Discharged on 30 Jun 1813 and sent to Ashburton on parole.

Spangle, Frederick - Seaman - Number: 2015 - Prize name: Ned - Ship type: LM - How taken: HM Brig Royalist - When taken: 6 Sep 1812 - Where taken: coast of France - Date received: 24 Sep 181 - From what ship: HMS Rippon - Born: Carolina - Age: 19 - Discharged on 27 Sep 1813 and sent to Dartmoor.

Spark, Samuel - Seaman - Number: 2548 - Prize name: Mary, prize to the Privateer Blockhead - Ship type: P - How taken: HM Post Ship Crocodile - When taken: 6 Aug 1813 - Where taken: Coruna, Spain - Date received: 16 Apr 1814 - From what ship: Transport Fanny - Born: Philadelphia - Age: 19 - Discharged on 10 May 1814 and sent to Dartmoor.

Sparks, Henry - 2nd Mate - Number: 385 - Prize name: Union - Ship type: MV - How taken: HM Frigate Iris - When taken: 17 Jan 1813 - Where taken: Lat 44 N Long 2.3 W - Date received: 5 Feb 1813 - From what ship: HMS San Josef - Born: Philadelphia - Age: 24 - Sent to Chatham on 29 Mar 1813 on the HMS Braham.

Sparks, Robert - Seaman - Number: 2470 - Prize name: Fair American - Ship type: MV - How taken: HM Frigate Andromache - When taken: 19 Jan 1814 - Where taken: Bay of Biscay - Date received: 22 Feb 1814 - From what ship: HMS York - Born: New York - Age: 36 - Race: Black - Discharged on 10 May 1814 and sent to Dartmoor.

Sparrow, James - Seaman - Number: 1303 - Prize name: Price - Ship type: LM - How taken: HM Frigate Medusa -

When taken: 13 Apr 1813 - Where taken: Bay of Biscay - Date received: 10 May 1813 - From what ship: HMS Medusa - Born: Virginia - Age: 35 - Discharged on 3 Jul 1813 and sent to Stapleton Prison.

Spear, Samuel - Carpenter - Number: 485 - Prize name: Cashiere - Ship type: LM - How taken: HM Brig Reindeer - When taken: 3 Feb 1813 - Where taken: Bay of Biscay - Date received: 12 Feb 1813 - From what ship: HMS Reindeer - Born: Boston - Age: 26 - Sent to Dartmoor on 2 Apr 1813.

Spencer, Leonard - Seaman - Number: 880 - Prize name: Tiger - Ship type: MV - How taken: HM Brig Scylla - When taken: 22 Mar 1813 - Where taken: Bay of Biscay - Date received: 1 Apr 1813 - From what ship: HMS Scylla - Born: Connecticut - Age: 32 - Sent on 21 Jun 1813 to Portsmouth.

Spencer, Samuel - Seaman - Number: 457 - Prize name: Dolphin - Ship type: LM - How taken: HM Ship-of-the-Line Colossus, HM Frigate Rhin & HM Brig Goldfinch - When taken: 5 Jan 1813 - Where taken: off Western Islands - Date received: 6 Feb 1813 - From what ship: HMS Rhin - Born: Delaware - Age: 22 - Sent to Chatham on 29 Mar 1813 on the HMS Braham.

Spires, Thomas - 2nd Mate - Number: 2006 - Prize name: Ned - Ship type: LM - How taken: HM Brig Royalist - When taken: 6 Sep 1812 - Where taken: coast of France - Date received: 24 Sep 181 - From what ship: HMS Rippon - Born: Virginia - Age: 39 - Discharged on 27 Sep 1813 and sent to Dartmoor.

Spofford, Jacob - Master - Number: 602 - Prize name: Rolla - Ship type: MV - How taken: HM Frigate Surveillante - When taken: 11 Feb 1813 - Where taken: Bay of Biscay - Date received: 25 Feb 1813 - From what ship: HMS Plymouth - Born: Biddeford, MA - Age: 26 - Sent on 27 Feb 1813 to Ashburton on parole.

Sproson, James - Carpenter - Number: 878 - Prize name: Tiger - Ship type: MV - How taken: HM Brig Scylla - When taken: 22 Mar 1813 - Where taken: Bay of Biscay - Date received: 1 Apr 1813 - From what ship: HMS Scylla - Born: New York - Age: 26 - Sent to Dartmoor on 28 Jun 1813.

Sprunwell, Peter - Seaman - Number: 2389 - How taken: Delivered himself up from HM Frigate Crescent - When taken: 27 Dec 1813 - Where taken: St. Johns, New Brunswick - Date received: 2 Feb 1814 - From what ship: HMS Pheasant - Born: Bergen, Norway - Age: 23 - Liberated on 8 Feb 1814.

Spryon, John - Seaman - Number: 314 - Prize name: Blue Bird - Ship type: MV - How taken: HM Frigate Briton - When taken: 1 Jan 1813 - Where taken: off Bordeaux - Date received: 21 Jan 1813 - From what ship: HMS Abercrombie - Born: New Orleans - Age: 30 - Sent to Chatham on 29 Mar 1813 on the HMS Braham.

Spurr, Elijah - Seaman - Number: 2198 - Prize name: General Kempt, prize of the Privateer Grand Turk - Ship type: P - How taken: HM Brig Foxhound - When taken: 18 Dec 1813 - Where taken: Lat 48.4 Long 6 - Date received: 21 Dec 1813 - From what ship: HMS Foxhound - Born: Boston - Age: 18 - Discharged on 27 Feb 1814 and sent to Chatham on HMS Haleyon.

Stacey, Benjamin - Seaman - Number: 2169 - Prize name: Wolf, prize of the Privateer Grand Turk - Ship type: P - How taken: HM Frigate Briton - When taken: 1 Dec 1813 - Where taken: Bay of Biscay - Date received: 17 Dec 1813 - From what ship: HMS Briton - Born: Marblehead - Age: 19 - Discharged on 17 Dec 1813 and released to HMS Britton.

Stacey, Samuel - Seaman - Number: 2176 - Prize name: Wolf, prize of the Privateer Grand Turk - Ship type: P - How taken: HM Frigate Briton - When taken: 1 Dec 1813 - Where taken: Bay of Biscay - Date received: 17 Dec 1813 - From what ship: HMS Briton - Born: Marblehead - Age: 23 - Discharged on 17 Dec 1813 and released to HMS Britton.

Stacy, William - Seaman - Number: 2415 - Prize name: Joseph - Ship type: MV - How taken: HM Brig Royalist - When taken: 18 Jan 1814 - Where taken: Bay of Biscay - Date received: 9 Feb 1814 - From what ship: HMS Sparrow - Born: Marblehead - Age: 21 - Discharged on 10 May 1814 and sent to Dartmoor.

Stafford, John - Seaman - Number: 2452 - How taken: Taken out of a Russian ship - When taken: 28 Jan 1814 - Where taken: Cove of Cork - Date received: 15 Feb 1814 - From what ship: HMS Zealous - Born: Boston - Age: 29 - Discharged on 10 May 1814 and sent to Dartmoor.

Stag, John B. - Seaman - Number: 1213 - Prize name: Zebra - Ship type: LM - How taken: HM Frigate Pyramus - When taken: 20 Apr 1813 - Where taken: Bay of Biscay - Date received: 9 May 1813 - From what ship: HMS Andromache - Born: New York - Age: 26 - Discharged on 3 Jul 1813 and sent to Stapleton Prison.

Staitman, James - Seaman - Number: 656 - How taken: Delivered himself up from HMS Lavinia - When taken: 15 Feb 1813 - Date received: 5 Mar 1813 - From what ship: HMS Salvador del Monde - Born: Salisbury - Age: 26 - Sent to Dartmoor on 2 Apr 1813.

Stanley, Joseph - Steward - Number: 363 - Prize name: Orbit - Ship type: MV - How taken: HM Brig Achates - When taken: 29 Jan 1813 - Where taken: Lat 49 N Long 13 W - Date received: 4 Feb 1813 - From what ship: HMS Achates - Born: New Orleans - Age: 25 - Sent to Chatham on 29 Mar 1813 on the HMS Braham.

Stanberry, James - Chief Mate - Number: 2017 - Prize name: Ned - Ship type: LM - How taken: HM Brig Royalist - When taken: 6 Sep 1812 - Where taken: coast of France - Date received: 30 Sep 1813 - From what ship: HMS Royalist - Born: Maryland - Age: 27 - Discharged on 30 Sep 1813 and sent to Ashburton on parole.

Stanton, Henry - 2nd Mate - Number: 1019 - Prize name: Two Brothers - Ship type: MV - How taken: Bootle of Liverpool, letter of marque - When taken: 18 Mar 1813 - Where taken: Western Islands - Date received: 21 Apr 1813 - From what ship: HMS Bittern - Born: Connecticut - Age: 23 - Sent to Dartmoor on 1 Jul 1813.

Star, G. C. - 2nd Mate - Number: 316 - Prize name: Porcupine - Ship type: MV - How taken: HM Frigate Dryad - When taken: 8 Jan 1813 - Where taken: off Bordeaux - Date received: 21 Jan 1813 - From what ship: HMS Abercrombie - Born: Connecticut - Age: 24 - Sent to Chatham on 29 Mar 1813 on the HMS Braham.

Starbuck, George - Seaman - Number: 2434 - Prize name: US Brig Argus - Ship type: War - How taken: HM Brig Pelican - When taken: 14 Aug 1813 - Where taken: Irish Channel - Date received: 12 Feb 1814 - From what ship: HMS Salvador del Mundo - Born: Nantucket - Age: 25 - Discharged on 10 May 1814 and sent to Dartmoor.

Starbuck, Thaddeus B. - Seaman - Number: 665 - How taken: Impressed at Dublin - When taken: 7 Jan 1813 - Date received: 14 Mar 1813 - From what ship: HMS Neptune - Born: Nantucket - Age: 24 - Sent to Dartmoor on 2 Apr 1813.

Steel, John - Seaman - Number: 771 - Prize name: William Bayard - Ship type: MV - How taken: HM Ship-of-the-Line Warspite - When taken: 3 Mar 1813 - Where taken: Bay of Biscay - Date received: 19 Mar 1813 - From what ship: HMS Warspite - Born: Maryland - Age: 26 - Sent to Dartmoor on 28 Jun 1813.

Stephens, John - Seaman - Number: 2562 - How taken: Impressed at Cork - When taken: 27 Mar 1814 - Date received: 17 Apr 1814 - From what ship: HMS Teazer - Born: Salem - Age: 40 - Discharged on 10 May 1814 and sent to

Stephens, Obadiah - Seaman - Number: 2613 - How taken: MV Alemena - Date received: 16 May 1814 - From what ship: HMS Repulse - Born: Philadelphia - Age: 27 - Discharged on 14 Jun 1814 and sent to Dartmoor.

Stephens, William - Boy - Number: 1489 - Prize name: Paul Jones - Ship type: P - How taken: HM Frigate Leonidas - When taken: 23 May 1813 - Where taken: off Cape Clear - Date received: 26 May 1813 - From what ship: HMS Leonidas - Born: New London, CT - Age: 14 - Discharged on 30 Jun 1813 and sent to Stapleton.

Stevens, Amos - Seaman - Number: 2516 - Prize name: Bunker Hill - Ship type: P - How taken: HM Frigate Pomone & HM Frigate Cydnus - When taken: 4 Mar 1814 - Where taken: Bay of Biscay - Date received: 4 Apr 1814 - From what ship: HMS Virago - Born: Boston - Age: 36 - Discharged on 10 May 1814 and sent to Dartmoor.

Stevens, Hugh - Seaman - Number: 1953 - Prize name: Joel Barlow - Ship type: LM - How taken: HM Frigate Briton - When taken: 3 Jul 1813 - Where taken: off Bordeaux - Date received: 31 Aug 1813 - From what ship: HMS Clarence - Born: Philadelphia - Age: 27 - Discharged on 8 Sep 1813 and sent to Dartmoor.

Stevens, John - Marine - Number: 1862 - Prize name: US Brig Argus - Ship type: War - How taken: HM Brig Pelican - When taken: 14 Aug 1813 - Where taken: Irish Channel - Date received: 23 Aug 1813 - From what ship: HMS Pelican - Born: Delaware - Age: 38 - Discharged on 8 Sep 1813 and sent to Dartmoor.

Stevens, Joseph - Mate - Number: 342 - Prize name: Portsea, prize to the Privateer Thrasher - Ship type: P - How taken: HM Sloop Helena - When taken: 31 Dec 1813 - Where taken: off the Western Islands - Date received: 22 Jan 1813 - From what ship: HMS Helena - Born: Gloucester - Age: 24 - Sent to Chatham on 29 Mar 1813 on the HMS Braham.

Stewart, Adam - Seaman - Number: 2318 - Prize name: Siro - Ship type: LM - How taken: HM Brig Pelican - When taken: 13 Jan 1814 - Where taken: at sea - Date received: 25 Jan 1814 - From what ship: HMS Pelican - Discharged on 14 Jun 1814 and sent to Dartmoor.

Stewart, Isaac - Seaman - Number: 2033 - Prize name: Friendship West, prize of Privateer True Blooded Yankee -

Ship type: P - How taken: HM Schooner Helicon & HM Schooner Whiting - When taken: 25 Oct 1813 - Where taken: Bay of Biscay - Date received: 31 Oct 1813 - From what ship: HMS Whiting - Born: Baltimore - Age: 29 - Race: Black - Discharged on 3 Nov 1813 and sent to Dartmoor.

Stewart, William - Seaman - Number: 1433 - Prize name: Paul Jones - Ship type: P - How taken: HM Frigate Leonidas - When taken: 23 May 1813 - Where taken: off Cape Clear - Date received: 26 May 1813 - From what ship: HMS Leonidas - Born: New York - Age: 19 - Race: Black - Discharged on 30 Jun 1813 and sent to Stapleton.

Stewart, William - Seaman - Number: 1502 - Prize name: Paul Jones - Ship type: P - How taken: HM Frigate Leonidas - When taken: 23 May 1813 - Where taken: off Cape Clear - Date received: 27 May 1813 - From what ship: HMS Leonidas - Born: Manchester - Age: 45 - Discharged on 30 Jun 1813 and sent to Stapleton.

Stewart, William - Seaman - Number: 1744 - Prize name: Paul Jones - Ship type: P - How taken: HM Frigate Leonidas - When taken: 23 May 1813 - Where taken: off Cape Clear - Date received: 13 Jun 1813 - From what ship: Retaken after escape - Born: Manchester - Age: 45 - Discharged on 8 Sep 1813 and sent to Dartmoor.

Stiles, William - 1st Lieutenant - Number: 505 - Prize name: Cashiere - Ship type: LM - How taken: HM Brig Reindeer - When taken: 3 Feb 1813 - Where taken: Bay of Biscay - Date received: 15 Feb 1813 - From what ship: Cashier - Born: Baltimore - Age: 24 - Sent on 18 Feb 1813 to Ashburton on parole.

Stiles, William - 1st Lieutenant - Number: 2317 - Prize name: Siro - Ship type: LM - How taken: HM Brig Pelican - When taken: 13 Jan 1814 - Where taken: at sea - Date received: 25 Jan 1814 - From what ship: HMS Pelican - Born: Baltimore - Age: 25 - Discharged on 25 Jan 1814 and sent to Ashburton on parole.

Still, David - Seaman - Number: 2468 - Prize name: Fair American - Ship type: MV - How taken: HM Frigate Andromache - When taken: 19 Jan 1814 - Where taken: Bay of Biscay - Date received: 22 Feb 1814 - From what ship: HMS York - Born: Wiscasset - Age: 42 - Discharged on 10 May 1814 and sent to Dartmoor.

Stites, Ezra - Seaman - Number: 1786 - Prize name: Jane Barns - Ship type: MV - How taken: HM Frigate Comus - When taken: 14 May 1813 - Where taken: off Lisbon - Date received: 13 Aug 1813 - From what ship: HMS Protection - Born: Connecticut - Age: 22 - Discharged on 8 Sep 1813 and sent to Dartmoor.

Stockman, William B. - Seaman - Number: 1241 - Prize name: Essex - Ship type: MV - How taken: HM Frigate Pyramus - When taken: 2 Apr 1813 - Where taken: Bay of Biscay - Date received: 9 May 1813 - From what ship: HMS Andromache - Born: Massachusetts - Age: 29 - Discharged on 3 Jul 1813 and sent to Stapleton Prison.

Stoddart, Robert - Seaman - Number: 796 - Prize name: Cannoniere - Ship type: P - How taken: HM Ship-of-the-Line Warspite - When taken: 14 Mar 1813 - Where taken: Bay of Biscay - Date received: 19 Mar 1813 - From what ship: HMS Warspite - Born: Long Island - Age: 19 - Sent to Dartmoor on 28 Jun 1813.

Stone, Edmond - Seaman - Number: 2081 - Prize name: captured French frigate - How taken: HM Ship-of-the-Line Rippon, HM Brig Scylla & HM Brig Royalist - When taken: 21 Oct 1813 - Where taken: off Ushant - Date received: 31 Oct 1813 - From what ship: HMS Rippon - Discharged on 29 Nov 1813 and sent to Dartmoor.

Stone, John - 1st Mate - Number: 2624 - Prize name: Ateline - Ship type: MV - How taken: HM Frigate Magiciene - When taken: 14 Mar 1814 - Where taken: off Cape Finisterre, Spain - Date received: 17 May 1814 - From what ship: HMS Tortois - Born: New London - Age: 27 - Discharged on 21 May 1814 and sent to Ashburton on parole.

Stone, Samuel - Seaman - Number: 2662 - Prize name: Indian Lass, prize of the Privateer Grand Turk - Ship type: P - How taken: HM Transport Akbar - When taken: 29 Apr 1814 - Where taken: at sea - Date received: 5 Jun 1814 - From what ship: HMS Gronville - Born: Ipswich - Age: 19 - Discharged on 14 Jun 1814 and sent to Dartmoor.

Stone, William - Seaman - Number: 418 - Prize name: Union - Ship type: MV - How taken: HM Frigate Iris - When taken: 17 Jan 1813 - Where taken: Lat 44 N Long 2.3 W - Date received: 5 Feb 1813 - From what ship: HMS San Josef - Born: New Jersey - Age: 19 - Sent to Chatham on 29 Mar 1813 on the HMS Braham.

Stout, John - Seaman - Number: 2287 - Prize name: Siro - Ship type: LM - How taken: HM Brig Pelican - When taken: 13 Jan 1814 - Where taken: at sea - Date received: 23 Jan 1814 - From what ship: HMS Pelican - Born: Portland - Age: 16 - Discharged on 31 Jan 1814 and sent to Dartmoor.

Stow, Jeremiah - Seaman - Number: 849 - How taken: Delivered himself up from HM Ship-of-the-Line Leyden - Date received: 23 May 1813 - From what ship: HMS Dauntless - Born: Philadelphia - Age: 44 - Sent to Chatham on 8 Jul 1813 on HM Tender Neptune.

Stowell, Peter - Mate - Number: 53 - Prize name: Independence - Ship type: MV - How taken: HM Frigate Medusa - When taken: 9 Nov 1812 - Where taken: San Sebastian - Date received: 27 Nov 1812 - From what ship: HMS Wasp - Born: New York - Age: 27 - Sent on 8 Dec 1812 to Ashburton on parole.

Stowell, John - Master - Number: 74 - Prize name: Independence - Ship type: MV - How taken: HM Frigate Medusa - When taken: 9 Nov 1812 - Where taken: San Sebastian - Date received: 29 Nov 1812 - From what ship: HMS Wasp - Born: Boston - Age: 29 - Sent on 8 Dec 1812 to Ashburton on parole.

Stranton, John - Seaman - Number: 23 - Prize name: Philips Burgh - Ship type: MV - How taken: Taken at Liverpool - When taken: 9 Nov 1812 - Date received: 21 Nov 1812 - From what ship: HMS Salvador del Mundo - Age: 24 - Sent to Portsmouth on 29 Dec 1812 on the HMS Northumberland.

Struby, John - Seaman - Number: 1310 - Prize name: Eliza - Ship type: MV - How taken: HM Frigate Surveillante - When taken: 22 Apr 1813 - Where taken: at sea - Date received: 10 May 1813 - From what ship: HMS Medusa - Born: Massachusetts - Age: 22 - Discharged on 3 Jul 1813 and sent to Stapleton Prison.

Strum, Magdalene - Daughter - Number: 2248 - Prize name: Squirrel - Ship type: MV - How taken: HM Frigate Belle Poule - When taken: 14 Dec 1813 - Where taken: Bay of Biscay - Date received: 15 Jan 1814 - From what ship: HMS Bellona - Born: Switzerland - Liberated on 16 Feb 1814.

Strum, Yves - Merchant & Passenger - Number: 2247 - Prize name: Squirrel - Ship type: MV - How taken: HM Frigate Belle Poule - When taken: 14 Dec 1813 - Where taken: Bay of Biscay - Date received: 15 Jan 1814 - From what ship: HMS Bellona - Born: Switzerland - Liberated on 16 Feb 1814.

Stubbs, James - 1st Mate - Number: 335 - Prize name: Rossie - Ship type: MV - How taken: Rochefort, France - When taken: 1 Jan 1813 - Where taken: Basque Roads, France - Date received: 22 Jan 1813 - From what ship: Rossie - Born: Jamestown - Age: 27 - Sent on 23 Jan 1813 to Ashburton on parole.

Sturges, Bradley - 2nd Mate - Number: 1089 - Prize name: Young Holkar - Ship type: MV - How taken: HM Ship-of-the-Line Superb - When taken: 10 Apr 1813 - Where taken: off Belle Isle - Date received: 22 Apr 1813 - From what ship: HMS Superb - Born: Connecticut - Age: 19 - Sent to Dartmoor on 1 Jul 1813.

Sturges, Bradley - 2nd Mate - Number: 2335 - Prize name: Young Holkar - Ship type: MV - How taken: HM Ship-of-the-Line Superb - When taken: 10 Apr 1813 - Where taken: off Belle Isle - Date received: 29 Jan 1814 - From what ship: Dartmoor prison - Born: Connecticut - Age: 19 - Discharged on 3 Feb 1814 and sent to Dartmoor.

Sturm, Jeannette - Wife - Number: 2249 - Prize name: Squirrel - Ship type: MV - How taken: HM Frigate Belle Poule - When taken: 14 Dec 1813 - Where taken: Bay of Biscay - Date received: 15 Jan 1814 - From what ship: HMS Bellona - Born: Switzerland - Liberated on 16 Feb 1814.

Sullivan, Hampton - Seaman - Number: 704 - Prize name: Star - Ship type: MV - How taken: HM Ship-of-the-Line Superb - When taken: 9 Feb 1813 - Where taken: Bay of Biscay - Date received: 19 Mar 1813 - From what ship: HMS Warspite - Born: Wilmington - Age: 19 - Sent to Dartmoor on 2 Apr 1813.

Summers, Alexander - Chief Mate - Number: 856 - Prize name: William Bayard - Ship type: MV - How taken: HM Ship-of-the-Line Warspite - When taken: 3 Mar 1813 - Where taken: Bay of Biscay - Date received: 25 May 1813 - From what ship: HMS Warspite - Born: New York - Age: 29 - Sent on 26 Mar 1813 to Ashburton on parole.

Summerville, Charles - Boy - Number: 1014 - Prize name: Thrasher - Ship type: P - How taken: HM Frigate Magiciene - When taken: 17 Jan 1813 - Where taken: St. Mary, Western Island - Date received: 20 Apr 1813 - From what ship: HMS Libria - Born: Baltimore - Age: 17 - Sent to Dartmoor on 1 Jul 1813.

Sutter, John - Seaman - Number: 1789 - How taken: Delivered himself up from HM Ship-of-the-Line Scipion - When taken: 4 Nov 1812 - Date received: 13 Aug 1813 - From what ship: HMS Protection - Born: Philadelphia - Age: 33 - Discharged on 8 Sep 1813 and sent to Dartmoor.

Sutton, John - Seaman - Number: 883 - Prize name: Tiger - Ship type: MV - How taken: HM Brig Scylla - When taken: 22 Mar 1813 - Where taken: Bay of Biscay - Date received: 1 Apr 1813 - From what ship: HMS Scylla - Born: Nantucket - Age: 26 - Sent on 21 Jun 1813 to Portsmouth.

Swain, Darius - Seaman - Number: 946 - Prize name: Good Friends - Ship type: MV - How taken: HM Frigate Andromache - When taken: 2 Apr 1813 - Where taken: Bay of Biscay - Date received: 7 Apr 1813 - From what ship: HMS Sea Lark - Born: Caroline - Age: 23 - Sent to Dartmoor on 28 Jun 1813.

Swanston, Jacob - Seaman - Number: 412 - Prize name: Union - Ship type: MV - How taken: HM Frigate Iris - When taken: 17 Jan 1813 - Where taken: Lat 44 N Long 2.3 W - Date received: 5 Feb 1813 - From what ship: HMS San Josef - Born: Christiana - Age: 29 - Sent to Chatham on 29 Mar 1813 on the HMS Braham.

Swasey, John - Seaman - Number: 2511 - Prize name: Bunker Hill - Ship type: P - How taken: HM Frigate Pomone & HM Frigate Cydnus - When taken: 4 Mar 1814 - Where taken: Bay of Biscay - Date received: 4 Apr 1814 - From what ship: HMS Virago - Born: Salem - Age: 20 - Discharged on 10 May 1814 and sent to Dartmoor.

Sweetman, Samuel - Seaman - Number: 380 - Prize name: Orbit - Ship type: MV - How taken: HM Brig Achates - When taken: 29 Jan 1813 - Where taken: Lat 49 N Long 13 W - Date received: 4 Feb 1813 - From what ship: HMS Achates - Born: New York - Age: 25 - Sent to Chatham on 29 Mar 1813 on the HMS Braham.

Swett, David - Seaman - Number: 1298 - Prize name: Price - Ship type: LM - How taken: HM Frigate Medusa - When taken: 13 Apr 1813 - Where taken: Bay of Biscay - Date received: 10 May 1813 - From what ship: HMS Medusa - Born: New York - Age: 22 - Discharged on 3 Jul 1813 and sent to Stapleton Prison.

Swett, Samuel - Chief Mate - Number: 561 - Prize name: Spitfire - Ship type: MV - How taken: HM Brig Achates - When taken: 14 Feb 1814 - Where taken: off Ushant - Date received: 21 Feb 1813 - From what ship: HMS Achates - Born: Marblehead - Age: 27 - Sent on 22 Feb 1813 to Ashburton on parole.

Swinny, Edward - Seaman - Number: 309 - Prize name: Blue Bird - Ship type: MV - How taken: HM Frigate Briton - When taken: 1 Jan 1813 - Where taken: off Bordeaux - Date received: 21 Jan 1813 - From what ship: HMS Abercrombie - Born: Maryland - Age: 21 - Sent to Chatham on 29 Mar 1813 on the HMS Braham.

Symonds, M. John - Prize Master - Number: 2191 - Prize name: General Kempt, prize of the Privateer Grand Turk - Ship type: P - How taken: HM Brig Foxhound - When taken: 18 Dec 1813 - Where taken: Lat 48.4 Long 6 - Date received: 21 Dec 1813 - From what ship: HMS Foxhound - Born: Salem - Age: 23 - Discharged on 31 Jan 1814 and sent to Dartmoor.

Tafe, William - Boy - Number: 1909 - Prize name: US Brig Argus - Ship type: War - How taken: HM Brig Pelican - When taken: 14 Aug 1813 - Where taken: Irish Channel - Date received: 23 Aug 1813 - From what ship: HMS Pelican - Born: New York - Age: 13 - Discharged on 8 Sep 1813 and sent to Dartmoor.

Tamer, John - Seaman - Number: 2696 - Prize name: US Brig Argus - Ship type: War - How taken: HM Brig Pelican - When taken: 14 Aug 1813 - Where taken: Irish Channel - Date received: 17 Jun 1814 - From what ship: Dartmouth - Born: Holland - Age: 32 - Discharged on 20 Jun 1814 and released to the Speedwell Cartel.

Taney, Lewis - Merchant & Passenger - Number: 2237 - Prize name: Squirrel - Ship type: MV - How taken: HM Frigate Belle Poule - When taken: 14 Dec 1813 - Where taken: Bay of Biscay - Date received: 15 Jan 1814 - From what ship: HMS Bellona - Born: France - Age: 21 - Discharged on 29 Jan 1814 and sent to Ashburton on parole.

Tapiau, Vitale - Merchant & Passenger - Number: 2239 - Prize name: Squirrel - Ship type: MV - How taken: HM Frigate Belle Poule - When taken: 14 Dec 1813 - Where taken: Bay of Biscay - Date received: 15 Jan 1814 - From what ship: HMS Bellona - Born: Au Cayes, France - Age: 50 - Discharged on 29 Jan 1814 and sent to Ashburton on parole.

Tarbell, William George - Master - Number: 2349 - Prize name: Hannah - Ship type: MV - How taken: HM Ship-of-the-Line Conquestador - When taken: 14 Jan 1814 - Where taken: Lat 47.3 Long 7 - Date received: 31 Jan 1814 - From what ship: HMS Surveillante - Born: Groton - Age: 28 - Discharged on 31 Jan 1814 and sent to Ashburton on parole.

Tasdeboiz, Louis - Passenger - Number: 1596 - Prize name: Governor Gerry - Ship type: MV - How taken: HM Brig Royalist - When taken: 31 May 1813 - Where taken: Bay of Biscay - Date received: 14 Jun 1813 - From what ship: HMS Royalist - Born: France - Age: 46 - Discharged on 7 Jul 1813 and sent to Ashburton on parole.

Taylor, Archibald - Captain - Number: 1505 - Prize name: Paul Jones - Ship type: P - How taken: HM Frigate Leonidas - When taken: 23 May 1813 - Where taken: off Cape Clear - Date received: 28 May 1813 - From what ship: HMS Leonidas - Born: New York - Age: 26 - Discharged on 30 May 1813 and sent to Ashburton on parole.

Taylor, George - Seaman - Number: 597 - Prize name: St. Martin's Plantation, prize of the Privateer Paul Jones - Ship type: P - How taken: HM Ship-of-the-Line Dublin - When taken: 9 Feb 1815 - Where taken: Lat 43 N Long 33.5 W - Date received: 25 Feb 1813 - From what ship: HMS Dublin - Born: New Jersey - Age: 27 - Sent to Dartmoor on 2 Apr 1813.

Taylor, Jacob - Seaman - Number: 749 - Prize name: Charlotte - Ship type: MV - How taken: HM Ship-of-the-Line Warspite - When taken: 3 Mar 1813 - Where taken: Bay of Biscay - Date received: 19 Mar 1813 - From what ship: HMS Warspite - Born: Philadelphia - Age: 24 - Sent to Dartmoor on 2 Apr 1813.

Taylor, James - Unknown - Number: 916 - How taken: Delivered himself up as a prisoner of war - Date received: 4 Apr 1813 - From what ship: HMS Salvador del Mundo - Born: Philadelphia - Age: 27 - Sent on 21 Jun 1813 to Portsmouth on HMS Prometheus.

Taylor, James - Seaman - Number: 2260 - How taken: Sent to prison by HM Brig Sparrow - Date received: 18 Jan 1814 - From what ship: HMS Sparrow - Born: Philadelphia - Age: 21 - Race: Negro - Discharged on 31 Jan 1814 and sent to Dartmoor.

Taylor, James - Seaman - Number: 2163 - Prize name: Princess - Ship type: MV - How taken: Impressed at Liverpool - When taken: 1 Nov 1813 - Date received: 13 Dec 1813 - From what ship: HMS Bittern - Born: Maryland - Age: 30 - Discharged on 31 Jan 1814 and sent to Dartmoor.

Taylor, John - Seaman - Number: 220 - Prize name: Vengeance - Ship type: LM - How taken: HM Frigate Phoebe - When taken: 1 Jan 1813 - Where taken: Lat 44.4 Long 23 - Date received: 9 Jan 1813 - From what ship: HMS Phoebe - Born: Africa - Age: 30 - Race: Negro - Sent to Portsmouth on 8 Feb 1813 on the HMS Colossus.

Taylor, Samuel E. - Seaman - Number: 1384 - Prize name: Tom - Ship type: LM - How taken: HM Frigate Surveillante - When taken: 27 Apr 1813 - Where taken: Bay of Biscay - Date received: 15 May 1813 - From what ship: HMS Foxhound - Born: Charlestown - Age: 22 - Discharged on 3 Jul 1813 and sent to Stapleton Prison.

Taylor, Thomas - Seaman - Number: 1302 - Prize name: Price - Ship type: LM - How taken: HM Frigate Medusa - When taken: 13 Apr 1813 - Where taken: Bay of Biscay - Date received: 10 May 1813 - From what ship: HMS Medusa - Born: Wilmington - Age: 23 - Discharged on 3 Jul 1813 and sent to Stapleton Prison.

Taylor, Thomas - Seaman - Number: 1668 - Prize name: Gemini - Ship type: MV - How taken: Impressed at Liverpool - When taken: 4 May 1813 - Date received: 17 Jun 1813 - From what ship: HMS Bittern - Born: Boston - Age: 26 - Discharged on 8 Sep 1813 and sent to Dartmoor.

Taylor, William - Seaman - Number: 2604 - How taken: MV Nautilus - When taken: 20 Dec 1812 - Date received: 16 May 1814 - From what ship: HMS Repulse - Born: Wilmington - Age: 42 - Discharged on 14 Jun 1814 and sent to Dartmoor.

Tegget, John - 3rd Mate - Number: 2638 - Prize name: Ateline - Ship type: MV - How taken: HM Frigate Magiciene - When taken: 14 Mar 1814 - Where taken: off Cape Finisterre, Spain - Date received: 17 May 1814 - From what ship: HMS

Tortois - Born: Philadelphia - Age: 32 - Race: Black - Discharged on 14 Jun 1814 and sent to Dartmoor.

Temple, William - Midshipman - Number: 1930 - Prize name: US Brig Argus - Ship type: War - How taken: HM Brig Pelican - When taken: 14 Aug 1813 - Where taken: Irish Channel - Date received: 23 Aug 1813 - From what ship: HMS Pelican - Born: Virginia - Age: 20 - Discharged on 26 Aug 1813 and sent to Ashburton on parole.

Terry, Daniel - Seaman - Number: 109 - Prize name: Otter - Ship type: MV - How taken: HM Ship-Sloop Jalousie - When taken: 1 Dec 1812 - Where taken: off Cape Vincent - Date received: 30 Dec 1812 - From what ship: HMS Leonidas - Born: Connecticut - Age: 20 - Sent to Portsmouth on 4 Jan 1813 on the HMS Revolutionnaire.

Thomas, Abraham - Seaman - Number: 1466 - Prize name: Paul Jones - Ship type: P - How taken: HM Frigate Leonidas - When taken: 23 May 1813 - Where taken: off Cape Clear - Date received: 26 May 1813 - From what ship: HMS Leonidas - Born: New Haven, CT - Age: 30 - Race: Black - Discharged on 30 Jun 1813 and sent to Stapleton.

Thomas, Alexander - Seaman - Number: 213 - Prize name: Hunter - Ship type: P - How taken: HM Frigate Phoebe - When taken: 24 Dec 1812 - Where taken: off Western Islands - Date received: 9 Jan 1813 - From what ship: HMS Phoebe - Born: Portland - Age: 30 - Sent to Chatham on 29 Mar 1813 on the HMS Braham.

Thomas, Andre - Passenger - Number: 1189 - Prize name: Zebra - Ship type: LM - How taken: HM Frigate Pyramus - When taken: 20 Apr 1813 - Where taken: Bay of Biscay - Date received: 9 May 1813 - From what ship: HMS Andromache - Born: Switzerland - Age: 22 - Sent on 11 Jun 1813 to the American ship Hope.

Thomas, Archibald - Carpenter - Number: 632 - Prize name: Criterion - Ship type: MV - How taken: HM Frigate

Belle Poule - When taken: 14 Feb 1813 - Where taken: Bay of Biscay - Date received: 4 Mar 1813 - From what ship: HMS Strenuous - Born: New York - Age: 25 - Sent to Dartmoor on 2 Apr 1813.

Thomas, David - Seaman - Number: 2421 - Prize name: US Brig Argus - Ship type: War - How taken: Sent from HM Ship-of-the-Line Salvador del Mundo - When taken: 14 Aug 1813 - Where taken: Irish Channel - Date received: 12 Feb 1814 - From what ship: HMS Salvador del Mundo - Born: Newport - Age: 32 - Discharged on 10 May 1814 and sent to Dartmoor.

Thomas, David - Seaman - Number: 1893 - Prize name: US Brig Argus - Ship type: War - How taken: HM Brig Pelican - When taken: 14 Aug 1813 - Where taken: Irish Channel - Date received: 23 Aug 1813 - From what ship: HMS Pelican - Born: Newport - Age: 32 - Discharged on 24 Aug 1813 and released to HMS Salvador del Mundo.

Thomas, George - Seaman - Number: 190 - Prize name: Hunter - Ship type: P - How taken: HM Frigate Phoebe - When taken: 24 Dec 1812 - Where taken: off Western Islands - Date received: 9 Jan 1813 - From what ship: HMS Phoebe - Born: Georgetown - Age: 24 - Race: Black - Sent to Portsmouth on 8 Feb 1813 on the HMS Colossus.

Thomas, Henry - Seaman - Number: 462 - Prize name: Dolphin - Ship type: LM - How taken: HM Ship-of-the-Line Colossus, HM Frigate Rhin & HM Brig Goldfinch - When taken: 5 Jan 1813 - Where taken: off Western Islands - Date received: 6 Feb 1813 - From what ship: HMS Rhin - Born: Baltimore - Age: 23 - Race: Negro - Sent to Chatham on 29 Mar 1813 on the HMS Braham.

Thomas, John - Seaman - Number: 1097 - Prize name: Young Holkar - Ship type: MV - How taken: HM Ship-of-the-Line Superb - When taken: 10 Apr 1813 - Where taken: off Belle Isle - Date received: 22 Apr 1813 - From what ship: HMS Superb - Born: New Orleans - Age: 21 - Sent to Dartmoor on 1 Jul 1813.

Thomas, John - Seaman - Number: 1026 - Prize name: Two Brothers - Ship type: MV - How taken: Bootle of Liverpool, letter of marque - When taken: 18 Mar 1813 - Where taken: Western Islands - Date received: 21 Apr 1813 - From what ship: HMS Bittern - Born: Boston - Age: 21 - Sent to Dartmoor on 1 Jul 1813.

Thomas, Spencer - Seaman - Number: 1710 - Prize name: Joseph - Ship type: MV - How taken: HM Frigate Iris - When taken: 8 Jun 1813 - Where taken: Bay of Biscay - Date received: 4 Jul 1813 - From what ship: HMS Iris - Born: Gloucester - Age: 21 - Discharged on 22 Aug 1813 and released to HM Brig Redpole.

Thomas, Stephen - Seaman - Number: 2637 - Prize name: Ateline - Ship type: MV - How taken: HM Frigate Magiciene - When taken: 14 Mar 1814 - Where taken: off Cape Finisterre, Spain - Date received: 17 May 1814 - From what ship: HMS Tortois - Born: Dover - Age: 26 - Discharged on 14 Jun 1814 and sent to Dartmoor.

Thomas, William - Seaman - Number: 135 - Prize name: Nope - Ship type: MV - How taken: HM Schooner Bramble - When taken: 3 Dec 1812 - Where taken: Coruna, Spain - Date received: 7 Jan 1813 - From what ship: Nope - Born: at sea - Age: 56 - Race: Negro - Sent to Portsmouth on 8 Feb 1813 on the HMS Colossus.

Thomas, William - Seaman - Number: 682 - Prize name: Nope - Ship type: MV - How taken: Chance, privateer - When taken: 15 Feb 1813 - Where taken: off Bordeaux - Date received: 15 Mar 1813 - From what ship: Growler - Born: New Orleans - Age: 35 - Sent to Dartmoor on 2 Apr 1813.

Thomas, William - Seaman - Number: 1118 - Prize name: Viper - Ship type: MV - How taken: HM Ship-of-the-Line Superb - When taken: 15 Apr 1813 - Where taken: Bay of Biscay - Date received: 22 Apr 1813 - From what ship: HMS Superb - Born: Delaware - Age: 28 - Sent to Dartmoor on 1 Jul 1813.

Thompson, Andrew - Seaman - Number: 2208 - Prize name: Squirrel - Ship type: MV - How taken: HM Frigate Belle Poule - When taken: 14 Dec 1813 - Where taken: Bay of Biscay - Date received: 27 Dec 1813 - From what ship: HMS Protector - Born: Providence - Age: 20 - Discharged on 31 Jan 1814 and sent to Dartmoor.

Thompson, Charles - Seaman - Number: 1284 - Prize name: Price - Ship type: LM - How taken: HM Frigate Medusa - When taken: 13 Apr 1813 - Where taken: Bay of Biscay - Date received: 10 May 1813 - From what ship: HMS Medusa - Born: New York - Age: 26 - Discharged on 3 Jul 1813 and sent to Stapleton Prison.

Thompson, Courtney - Seaman - Number: 374 - Prize name: Orbit - Ship type: MV - How taken: HM Brig Achates - When taken: 29 Jan 1813 - Where taken: Lat 49 N Long 13 W - Date received: 4 Feb 1813 - From what ship: HMS Achates - Born: New York - Age: 19 - Sent to Chatham on 29 Mar 1813 on the HMS Braham.

Thompson, Henry - Boy - Number: 1484 - Prize name: Paul Jones - Ship type: P - How taken: HM Frigate Leonidas - When taken: 23 May 1813 - Where taken: off Cape Clear - Date received: 26 May 1813 - From what ship: HMS

Leonidas - Born: Connecticut - Age: 11 - Discharged on 8 Sep 1813 and sent to Dartmoor.

Thompson, Henry - Chief Mate - Number: 2220 - Prize name: Zephyr - Ship type: MV - How taken: HM Frigate Pyramus - When taken: 30 Nov 1813 - Where taken: off Lorient - Date received: 3 Jan 1814 - From what ship: HMS Warspite - Born: Philadelphia - Age: 26 - Discharged on 6 Jan 1814 and sent to Ashburton on parole.

Thompson, James - Seaman - Number: 364 - Prize name: Orbit - Ship type: MV - How taken: HM Brig Achates - When taken: 29 Jan 1813 - Where taken: Lat 49 N Long 13 W - Date received: 4 Feb 1813 - From what ship: HMS Achates - Born: New York - Age: 36 - Sent to Chatham on 29 Mar 1813 on the HMS Braham.

Thompson, James - Seaman - Number: 2036 - How taken: Impressed at Liverpool - When taken: 16 Sep 1813 - Date received: 1 Nov 1813 - From what ship: HMS Bittern - Born: Nonith - Age: 28 - Discharged on 3 Nov 1813 and sent to Dartmoor.

Thompson, John - Seaman - Number: 2517 - Prize name: Bunker Hill - Ship type: P - How taken: HM Frigate Pomone & HM Frigate Cydnus - When taken: 4 Mar 1814 - Where taken: Bay of Biscay - Date received: 4 Apr 1814 - From what ship: HMS Virago - Born: Kent County - Age: 30 - Discharged on 10 May 1814 and sent to Dartmoor.

Thompson, Martin - Seaman - Number: 634 - Prize name: Criterion - Ship type: MV - How taken: HM Frigate Belle Poule - When taken: 14 Feb 1813 - Where taken: Bay of Biscay - Date received: 4 Mar 1813 - From what ship: HMS Strenuous - Born: Denmark - Age: 36 - Sent to Dartmoor on 2 Apr 1813.

Thompson, Owen - Seaman - Number: 310 - Prize name: Blue Bird - Ship type: MV - How taken: HM Frigate Briton - When taken: 1 Jan 1813 - Where taken: off Bordeaux - Date received: 21 Jan 1813 - From what ship: HMS Abercrombie - Born: Little York - Age: 31 - Sent to Chatham on 29 Mar 1813 on the HMS Braham.

Thompson, Peter - Seaman - Number: 2251 - Prize name: Squirrel - Ship type: MV - How taken: HM Frigate Belle Poule - When taken: 14 Dec 1813 - Where taken: Bay of Biscay - Date received: 15 Jan 1814 - From what ship: HMS Teazer - Born: Maryland - Age: 26 - Discharged on 31 Jan 1814 and sent to Dartmoor.

Thompson, Robert - Master - Number: 980 - Prize name: Good Friends - Ship type: MV - How taken: HM Frigate Andromache - When taken: 2 Apr 1813 - Where taken: Bay of Biscay - Date received: 12 Apr 1813 - From what ship: Good Friends - Born: Philadelphia - Age: 34 - Sent on 15 Apr 1813 to Ashburton on parole.

Thompson, Samuel - Seaman - Number: 322 - Prize name: Porcupine - Ship type: MV - How taken: HM Frigate Dryad - When taken: 8 Jan 1813 - Where taken: off Bordeaux - Date received: 21 Jan 1813 - From what ship: HMS Abercrombie - Born: New York - Age: 18 - Sent to Chatham on 29 Mar 1813 on the HMS Braham.

Thompson, Thomas - Seaman - Number: 1400 - Prize name: Henry Clements - Ship type: MV - How taken: HM Brig Orestes - When taken: 13 Apr 1813 - Where taken: Bay of Biscay - Date received: 15 May 1813 - From what ship: HMS Salvador del Mundo - Born: Maryland - Age: 39 - Race: Black - Discharged on 23 May 1813.

Thompson, Thomas - Boatswain's Mate - Number: 1824 - Prize name: US Brig Argus - Ship type: War - How taken: HM Brig Pelican - When taken: 14 Aug 1813 - Where taken: Irish Channel - Date received: 17 Aug 1813 - From what ship: USS Argus - Born: Rhode Island - Age: 24 - Discharged on 8 Sep 1813 and sent to Dartmoor.

Thompson, Whitney - Seaman - Number: 595 - Prize name: St. Martin's Plantation, prize of the Privateer Paul Jones - Ship type: P - How taken: HM Ship-of-the-Line Dublin - When taken: 9 Feb 1815 - Where taken: Lat 43 N Long 33.5 W - Date received: 25 Feb 1813 - From what ship: HMS Dublin - Born: Hartford - Age: 23 - Sent to Dartmoor on 2 Apr 1813.

Thompson, William - Seaman - Number: 697 - Prize name: Star - Ship type: MV - How taken: HM Ship-of-the-Line Superb - When taken: 9 Feb 1813 - Where taken: Bay of Biscay - Date received: 19 Mar 1813 - From what ship: HMS Warspite - Born: Copenhagen - Age: 23 - Sent to Dartmoor on 2 Apr 1813.

Thompson, William - Seaman - Number: 411 - Prize name: Union - Ship type: MV - How taken: HM Frigate Iris - When taken: 17 Jan 1813 - Where taken: Lat 44 N Long 2.3 W - Date received: 5 Feb 1813 - From what ship: HMS San Josef - Born: New York - Age: 19 - Sent to Chatham on 29 Mar 1813 on the HMS Braham.

Thompson, William - Seaman - Number: 1102 - Prize name: Viper - Ship type: MV - How taken: HM Ship-of-the-Line Superb - When taken: 15 Apr 1813 - Where taken: Bay of Biscay - Date received: 22 Apr 1813 - From what ship: HMS Superb - Born: Rochester - Age: 19 - Sent to Dartmoor on 1 Jul 1813.

Thompson, William - Mate - Number: 133 - Prize name: Nope - Ship type: MV - How taken: HM Schooner Bramble

- When taken: 3 Dec 1812 - Where taken: Coruna, Spain - Date received: 7 Jan 1813 - From what ship: Nope - Born: Philadelphia - Age: 35 - Sent on 8 Dec 1812 to Ashburton on parole.

Thompson, William - Seaman - Number: 1731 - Prize name: Star - Ship type: MV - How taken: HM Ship-of-the-Line Superb - When taken: 9 Feb 1813 - Where taken: Bay of Biscay - Date received: 10 Jul 1813 - From what ship: Dartmoor Prison - Born: Copenhagen - Age: 23 - Discharged on 10 Jul 1813 and released to HMS Salvador del Mundo.

Thompson, William - Seaman - Number: 1168 - Prize name: Hebe - Ship type: MV - How taken: HM Frigate Stag - When taken: 18 Apr 1813 - Where taken: Bay of Biscay - Date received: 6 May 1813 - From what ship: HMS Stag - Born: York County, PA - Age: 24 - Sent on 3 Jul 1813 to Stapleton prison.

Thompson, William - Cook - Number: 2290 - Prize name: Siro - Ship type: LM - How taken: HM Brig Pelican - When taken: 13 Jan 1814 - Where taken: at sea - Date received: 23 Jan 1814 - From what ship: HMS Pelican - Born: Saint Domingue (Haiti) - Age: 25 - Race: Negro - Discharged on 31 Jan 1814 and sent to Dartmoor.

Thompson, William - Seaman - Number: 2230 - How taken: Delivered himself up from the HM Brig Lyra - Date received: 9 Jan 1814 - From what ship: HMS Lyra - Born: Santee - Age: 32 - Race: Black - Discharged on 31 Jan 1814 and sent to Dartmoor.

Thornson, Joseph - Seaman - Number: 1511 - Prize name: Grand Napoleon - Ship type: MV - How taken: HM Brig Goldfinch - When taken: 17 Apr 1813 - Where taken: Bay of Biscay - Date received: 29 May 1813 - From what ship: HMS Hannibal - Born: Falmouth, MA - Age: 22 - Discharged on 30 Jun 1813 and sent to Stapleton.

Thornton, William - Seaman - Number: 2262 - Prize name: Growler - Ship type: MV - How taken: HM Brig Wolf - When taken: 11 Aug 1813 - Where taken: at sea - Date received: 20 Jan 1814 - From what ship: MV Nero - Born: Richmond - Age: 19 - Discharged on 31 Jan 1814 and sent to Dartmoor.

Thurtleff, William - Chief Mate - Number: 854 - Prize name: Pert - Ship type: MV - How taken: HM Ship-of-the-Line Warspite - When taken: 1 Mar 1813 - Where taken: Bay of Biscay - Date received: 25 May 1813 - From what ship: HMS Warspite - Born: Philadelphia - Age: 23 - Sent on 26 Mar 1813 to Ashburton on parole.

Tibbets, Henry - Prize Master - Number: 1759 - Prize name: Friendship, prize to the Privateer America - Ship type: P - How taken: HM Schooner Whiting - When taken: 15 Jul 1815 - Where taken: Lat 47N Long 8W - Date received: 21 Jul 1813 - From what ship: HMS Whiting - Born: Salem - Age: 28 - Discharged on 8 Sep 1813 and sent to Dartmoor.

Tightham, Peter - Carpenter - Number: 284 - Prize name: Columbia - Ship type: MV - How taken: HM Frigate Briton - When taken: 17 Dec 1812 - Where taken: off Bordeaux - Date received: 21 Jan 1813 - From what ship: HMS Abercrombie - Born: Philadelphia - Age: 22 - Sent to Portsmouth on 8 Feb 1813 on the HMS Colossus.

Tilley, Benjamin - Gunner's Mate - Number: 2576 - How taken: Delivered himself up from the MV Hebrus - Date received: 8 May 1814 - From what ship: MV Hebrus - Born: Philadelphia - Age: 38 - Discharged on 7 May 1814 and sent to Ashburton on parole.

Timmerman, Matthew - Chief Mate - Number: 1674 - Prize name: Tom Thumb - Ship type: MV - How taken: Cutter Lion - When taken: 17 Feb 1813 - Where taken: Bay of Biscay - Date received: 23 Jun 1813 - From what ship: Nimble of Guernsey - Born: New York - Age: 30 - Discharged on 3 Nov 1813 and sent to Dartmoor.

Timmons, John - Seaman - Number: 2664 - Prize name: Traveler, prize of the Privateer Surprize - Ship type: P - How taken: HMS Cawser - When taken: 7 May 1814 - Where taken: off Cape Clear - Date received: 5 Jun 1814 - From what ship: HMS Gronville - Born: Staten Island - Age: 18 - Discharged on 14 Jun 1814 and sent to Dartmoor.

Tipton, Solomon - Seaman - Number: 372 - Prize name: Orbit - Ship type: MV - How taken: HM Brig Achates - When taken: 29 Jan 1813 - Where taken: Lat 49 N Long 13 W - Date received: 4 Feb 1813 - From what ship: HMS Achates - Born: Baltimore - Age: 53 - Sent to Chatham on 29 Mar 1813 on the HMS Braham.

Toby, Elisha - Seaman - Number: 2023 - Prize name: Friendship West, prize of Privateer True Blooded Yankee - Ship type: P - How taken: HM Schooner Helicon & HM Schooner Whiting - When taken: 25 Oct 1813 - Where taken: Bay of Biscay - Date received: 31 Oct 1813 - From what ship: HMS Whiting - Born: Massachusetts - Age: 28 - Discharged on 3 Nov 1813 and sent to Dartmoor.

Toby, Peter - Seaman - Number: 1095 - Prize name: Young Holkar - Ship type: MV - How taken: HM Ship-of-the-

Line Superb - When taken: 10 Apr 1813 - Where taken: off Belle Isle - Date received: 22 Apr 1813 - From what ship: HMS Superb - Born: New Orleans - Age: 30 - Sent to Dartmoor on 1 Jul 1813.

Toby, Peter - Seaman - Number: 2337 - Prize name: Young Holkar - Ship type: MV - How taken: HM Ship-of-the-Line Superb - When taken: 10 Apr 1813 - Where taken: off Belle Isle - Date received: 29 Jan 1814 - From what ship: Dartmoor prison - Born: New Orleans - Age: 30 - Discharged on 3 Feb 1814 and sent to Dartmoor.

Todd, William - Seaman - Number: 295 - Prize name: Columbia - Ship type: MV - How taken: HM Frigate Briton - When taken: 17 Dec 1812 - Where taken: off Bordeaux - Date received: 21 Jan 1813 - From what ship: HMS Abercrombie - Born: Massachusetts - Age: 25 - Sent to Portsmouth on 8 Feb 1813 on the HMS Colossus.

Tolson, Jeremy - Seaman - Number: 71 - Prize name: Elk - Ship type: MV - How taken: Rose, Tender - When taken: 27 Sep 1812 - Where taken: Greenock - Date received: 29 Nov 1812 - From what ship: HMS Salvador del Mundo - Born: Richmond - Age: 22 - Sent to Portsmouth on 29 Dec 1812 on the HMS Northumberland.

Tolvet, Charles - Seaman - Number: 1890 - Prize name: US Brig Argus - Ship type: War - How taken: HM Brig Pelican - When taken: 14 Aug 1813 - Where taken: Irish Channel - Date received: 23 Aug 1813 - From what ship: HMS Pelican - Born: Cohasset, MA - Age: 33 - Discharged on 8 Sep 1813 and sent to Dartmoor.

Thompson, John - Seaman - Number: 2668 - Prize name: Traveler, prize of the Privateer Surprize - Ship type: P - How taken: HMS Cawser - When taken: 7 May 1814 - Where taken: off Cape Clear - Date received: 5 Jun 1814 - From what ship: HMS Gronville - Born: Brandywine - Age: 45 - Discharged on 14 Jun 1814 and sent to Dartmoor.

Thompson, Nathaniel - Seaman - Number: 905 - Prize name: Prompt - Ship type: MV - How taken: Chance, privateer - When taken: 22 Mar 1813 - Where taken: Bay of Biscay - Date received: 3 Apr 1813 - From what ship: Mary - Born: Virginia - Age: 31 - Sent on 21 Jun 1813 to Portsmouth on HMS Prometheus.

Tonver, Michael - Ordinary Seaman - Number: 481 - How taken: Impressed at Greenock - When taken: 8 Jan 1813 - Date received: 10 Feb 1813 - From what ship: HMS Frederick - Born: Hingham - Age: 37 - Sent to Dartmoor on 2 Apr 1813.

Towns, Daniel - Seaman - Number: 199 - Prize name: Hunter - Ship type: P - How taken: HM Frigate Phoebe - When taken: 24 Dec 1812 - Where taken: off Western Islands - Date received: 9 Jan 1813 - From what ship: HMS Phoebe - Born: Danvers - Age: 18 - Sent to Portsmouth on 8 Feb 1813 on the HMS Colossus.

Townsend, Thomas - Seaman - Number: 2579 - How taken: Delivered himself up from the MV Hebrus - Date received: 8 May 1814 - From what ship: MV Hebrus - Born: Biddeford - Age: 36 - Discharged on 7 May 1814 and sent to Ashburton on parole.

Towson, William - 2nd Mate - Number: 957 - Prize name: Courier - Ship type: LM - How taken: HM Frigate Andromache - When taken: 14 Mar 1813 - Where taken: Bay of Biscay - Date received: 7 Apr 1813 - From what ship: HMS Sea Lark - Born: Maryland - Age: 25 - Sent to Dartmoor on 28 Jun 1813.

Treffy, James - Seaman - Number: 1156 - Prize name: Essex - Ship type: MV - How taken: HM Frigate Pyramus - When taken: 2 Apr 1813 - Where taken: Bay of Biscay - Date received: 2 Mar 1813 - From what ship: HMS Rota - Born: Marblehead - Age: 16 - Sent on 3 Jul 1813 to Stapleton prison.

Treffy, Peter - Seaman - Number: 1158 - Prize name: Essex - Ship type: MV - How taken: HM Frigate Pyramus - When taken: 2 Apr 1813 - Where taken: Bay of Biscay - Date received: 2 Mar 1813 - From what ship: HMS Rota - Born: Marblehead - Age: 25 - Sent on 3 Jul 1813 to Stapleton prison.

Trifft, Joseph - Seaman - Number: 1382 - Prize name: Tom - Ship type: LM - How taken: HM Frigate Surveillante - When taken: 27 Apr 1813 - Where taken: Bay of Biscay - Date received: 15 May 1813 - From what ship: HMS Foxhound - Born: Massachusetts - Age: 21 - Discharged on 3 Jul 1813 and sent to Stapleton Prison.

Trout, William - Seaman - Number: 1970 - How taken: Impressed at Liverpool - When taken: 17 Jul 1813 - Date received: 19 Sep 1813 - From what ship: HMS Bittern - Born: Boston - Age: 25 - Discharged on 27 Sep 1813 and sent to Dartmoor.

Trowbridge, Bela - Mate - Number: 10 - Prize name: Anna - Ship type: MV - How taken: Taken at Liverpool - When taken: 17 Oct 1812 - Date received: 21 Nov 1812 - From what ship: HMS Salvador del Mundo - Born: Massachusetts - Age: 35 - Sent on 8 Dec 1812 to Ashburton on parole.

Truman, Alline - Seaman - Number: 32 - Prize name: Catharine - Ship type: MV - How taken: HM Frigate Leonidas

- When taken: 31 Jul 1812 - Where taken: off Ireland - Date received: 23 Nov 1812 - From what ship: HMS Stork - Sent to Portsmouth on 29 Dec 1812 on the HMS Northumberland.

Trusty, Henry - 2nd Mate - Number: 960 - Prize name: Ferox - Ship type: MV - How taken: HM Frigate Medusa & HM Brig Lyra - When taken: 28 Mar 1813 - Where taken: off Cape Ortagle - Date received: 9 Apr 1813 - From what ship: HMS Lyra - Born: Philadelphia - Age: 29 - Sent to Chatham on 8 Jul 1813 on HM Tender Neptune.

Tubbs, Martin - Seaman - Number: 2304 - Prize name: Siro - Ship type: LM - How taken: HM Brig Pelican - When taken: 13 Jan 1814 - Where taken: at sea - Date received: 23 Jan 1814 - From what ship: HMS Pelican - Born: New York - Age: 24 - Discharged on 31 Jan 1814 and sent to Dartmoor.

Tuck, James - 2nd Mate - Number: 1704 - Prize name: Joseph - Ship type: MV - How taken: HM Frigate Iris - When taken: 8 Jun 1813 - Where taken: Bay of Biscay - Date received: 4 Jul 1813 - From what ship: HMS Iris - Born: Manchester - Age: 22 - Discharged on 22 Aug 1813 and released to HM Brig Redpole.

Tucker, Edward - Seaman - Number: 203 - Prize name: Hunter - Ship type: P - How taken: HM Frigate Phoebe - When taken: 24 Dec 1812 - Where taken: off Western Islands - Date received: 9 Jan 1813 - From what ship: HMS Phoebe - Born: Salem - Age: 18 - Sent to Portsmouth on 8 Feb 1813 on the HMS Colossus.

Tucker, George C. - 2nd Mate - Number: 1621 - Prize name: Leo - Ship type: LM - How taken: HM Frigate Magiciene - When taken: 4 Jun 1813 - Where taken: Lat 45 Long 14 - Date received: 14 Jun 1813 - From what ship: HMS Orestes - Born: Portland - Age: 24 - Discharged on 30 Jun 1813 and sent to Stapleton.

Tufts, Zachariah - Seaman - Number: 535 - Prize name: Terrible - Ship type: MV - How taken: HM Brig Foxhound - When taken: 8 Feb 1813 - Where taken: Channel - Date received: 15 Feb 1813 - From what ship: HMS Foxhound - Born: Keene - Age: 28 - Sent to Dartmoor on 2 Apr 1813.

Tufts, Zachariah - Seaman - Number: 1730 - Prize name: Terrible - Ship type: MV - How taken: HM Brig Foxhound - When taken: 8 Feb 1813 - Where taken: Channel - Date received: 10 Jul 1813 - From what ship: Dartmoor Prison - Born: New Hampton - Age: 28 - Discharged on 10 Jul 1813 and released to HMS Salvador del Mundo.

Turner, Benjamin - Seaman - Number: 239 - How taken: Impressed at Liverpool - When taken: 17 Nov 1812 - Date received: 12 Jan 1813 - From what ship: HMS Bittern - Born: Charlestown - Age: 28 - Sent to Portsmouth on 8 Feb 1813 on the HMS Colossus.

Turner, Daniel - Seaman - Number: 2451 - How taken: Taken out of a Russian ship - When taken: 28 Jan 1814 - Where taken: Cove of Cork - Date received: 15 Feb 1814 - From what ship: HMS Zealous - Born: Maryland - Age: 44 - Discharged on 10 May 1814 and sent to Dartmoor.

Turner, Samuel - Master - Number: 7 - Prize name: Purse - Ship type: MV - How taken: HM Frigate Amide - When taken: 29 May 1812 - Where taken: off Bordeaux - Date received: 20 Nov 1812 - From what ship: HMS Salvador del Mundo - Born: New York - Age: 24 - Sent to Portsmouth on 29 Dec 1812 on the HMS Northumberland.

Turner, William - Seaman - Number: 790 - Prize name: Cannoniere - Ship type: P - How taken: HM Ship-of-the-Line Warspite - When taken: 14 Mar 1813 - Where taken: Bay of Biscay - Date received: 19 Mar 1813 - From what ship: HMS Warspite - Born: Philadelphia - Age: 22 - Sent to Dartmoor on 28 Jun 1813.

Turpin, Francis - Seaman - Number: 268 - Prize name: Brutus - Ship type: MV - How taken: HM Frigate Briton - When taken: Jan 1813 - Where taken: Bay of Biscay - Date received: 21 Jan 1813 - From what ship: HMS Briton - Born: New Orleans - Age: 28 - Sent to Portsmouth on 8 Feb 1813 on the HMS Colossus.

Turtle, French - Seaman - Number: 1630 - Prize name: Leo - Ship type: LM - How taken: HM Frigate Magiciene - When taken: 4 Jun 1813 - Where taken: Lat 45 Long 14 - Date received: 14 Jun 1813 - From what ship: HMS Orestes - Born: Falmouth - Age: 23 - Discharged on 30 Jun 1813 and sent to Stapleton.

Tuttle, Joseph - Seaman - Number: 2379 - Prize name: Devon, prize to the Privateer Bunker Hill - Ship type: P - How taken: HM Brig Fly - When taken: 21 Jan 1814 - Where taken: at sea - Date received: 31 Jan 1814 - From what ship: HMS Fly - Born: Freeport - Age: 27 - Discharged on 27 Feb 1814 and sent to Chatham on HMS Haleyon.

Tweed, Caldwell - Chief Mate - Number: 1317 - Prize name: Shadow - Ship type: LM - How taken: HM Brig Reindeer & HM Schooner Helicon - When taken: 6 Apr 1813 - Where taken: Bay of Biscay - Date received: 12 May 1813 - From what ship: HMS Reindeer - Born: Philadelphia - Age: 22 - Race: Black - Discharged on 30 May 1813 and sent to Ashburton on parole.

Tyren, William - Seaman - Number: 1107 - Prize name: Viper - Ship type: MV - How taken: HM Ship-of-the-Line Superb - When taken: 15 Apr 1813 - Where taken: Bay of Biscay - Date received: 22 Apr 1813 - From what ship: HMS Superb - Born: Wendell - Age: 21 - Sent to Dartmoor on 1 Jul 1813.

Underwood, Benjamin - Boatswain - Number: 394 - Prize name: Union - Ship type: MV - How taken: HM Frigate Iris - When taken: 17 Jan 1813 - Where taken: Lat 44 N Long 2.3 W - Date received: 5 Feb 1813 - From what ship: HMS San Josef - Age: 24 - Sent to Chatham on 29 Mar 1813 on the HMS Braham.

Underwood, John - Seaman - Number: 1752 - Prize name: Friendship, prize to the Privateer America - Ship type: P - How taken: HM Schooner Whiting - When taken: 15 Jul 1815 - Where taken: Lat 47N Long 8W - Date received: 20 Jul 1813 - From what ship: HMS Whiting - Born: Massachusetts - Age: 46 - Discharged on 8 Sep 1813 and sent to Dartmoor.

Underwood, John F. - Seaman - Number: 28 - Prize name: Catharine - Ship type: MV - How taken: HM Frigate Leonidas - When taken: 31 Jul 1812 - Where taken: off Ireland - Date received: 23 Nov 1812 - From what ship: HMS Stork - Born: Westport - Age: 21 - Sent to Portsmouth on 29 Dec 1812 on the HMS Northumberland.

Upton, Jeduthun - Captain - Number: 140 - Prize name: Hunter - Ship type: P - How taken: HM Frigate Phoebe - When taken: 24 Dec 1812 - Where taken: off Western Islands - Date received: 9 Jan 1813 - From what ship: HMS Phoebe - Born: Salem - Age: 27 - Sent to Chatham on 29 Mar 1813 on the HMS Braham.

Upton, Samuel - Master's Mate - Number: 152 - Prize name: Hunter - Ship type: P - How taken: HM Frigate Phoebe - When taken: 24 Dec 1812 - Where taken: off Western Islands - Date received: 9 Jan 1813 - From what ship: HMS Phoebe - Born: Salem - Age: 21 - Sent to Portsmouth on 8 Feb 1813 on the HMS Colossus.

Vail, Peter - Seaman - Number: 1223 - Prize name: Grand Napoleon - Ship type: MV - How taken: HM Frigate Belle Poule - When taken: 3 Apr 1813 - Where taken: off Bordeaux - Date received: 9 May 1813 - From what ship: HMS Andromache - Born: Connecticut - Age: 24 - Discharged on 3 Jul 1813 and sent to Stapleton Prison.

Valentine, John - Seaman - Number: 710 - Prize name: Star - Ship type: MV - How taken: HM Ship-of-the-Line Superb - When taken: 9 Feb 1813 - Where taken: Bay of Biscay - Date received: 19 Mar 1813 - From what ship: HMS Warspite - Born: New York - Age: 17 - Sent to Dartmoor on 2 Apr 1813.

Valiant, Richard - Seaman - Number: 2106 - Prize name: Chesapeake - Ship type: LM - How taken: HM Frigate Hotspur & HM Frigate Pyramus - When taken: 26 Oct 1813 - Where taken: Bay of Biscay - Date received: 22 Nov 1813 - From what ship: HMS Pyramus - Born: Maryland - Age: 26 - Discharged on 29 Nov 1813 and sent to Dartmoor.

Van Borgen, Martin - Passenger - Number: 2102 - Prize name: Chesapeake - Ship type: LM - How taken: HM Frigate Hotspur & HM Frigate Pyramus - When taken: 26 Oct 1813 - Where taken: Bay of Biscay - Date received: 22 Nov 1813 - From what ship: HMS Pyramus - Born: New York - Age: 31 - Discharged on 25 Aug 1813 and sent to Ashburton on parole.

Van Dine, Garland - Seaman - Number: 2690 - How taken: Sent into custody from HM Ship-of-the-Line Minden - Date received: 7 Jun 1814 - From what ship: Inap - Born: Philadelphia - Age: 32 - Discharged on 14 Jun 1814 and sent to Dartmoor.

Van Donveer, Peter - Seaman - Number: 63 - Prize name: Independence - Ship type: MV - How taken: HM Frigate Medusa - When taken: 9 Nov 1812 - Where taken: San Sebastian - Date received: 27 Nov 1812 - From what ship: HMS Wasp - Born: New Jersey - Age: 24 - Sent to Portsmouth on 29 Dec 1812 on the HMS Northumberland.

Vandine, Henry - Mate - Number: 1053 - Prize name: Virginia Planter - Ship type: MV - How taken: HM Frigate Pyramus - When taken: 17 Mar 1813 - Where taken: off Nantes - Date received: 22 Apr 1813 - From what ship: HMS Superb - Born: New York - Age: 21 - Sent on 27 Feb 1813 to Ashburton on parole.

Vangosbet, Cato - Seaman - Number: 935 - Prize name: Weasel - Ship type: MV - How taken: HM Brig Foxhound - When taken: 25 Mar 1813 - Where taken: Bay of Biscay - Date received: 6 Apr 1813 - From what ship: HMS Foxhound - Born: New Jersey - Age: 22 - Race: Black - Sent on 21 Jun 1813 to Portsmouth on HMS Prometheus.

Van Kirk, Joseph - Seaman - Number: 525 - Prize name: Phoenix - Ship type: MV - How taken: Detained at Gibraltar - When taken: 8 Aug 1812 - Date received: 15 Feb 1813 - From what ship: HMS Andromeda - Born: Wheelan - Age: 23 - Sent to Dartmoor on 2 Apr 1813.

Vaughan, Nathaniel - Seaman - Number: 819 - Prize name: Decornau - Ship type: MV - How taken: HM Sloop Pheasant - When taken: 15 Mar 1813 - Where taken: Bay of Biscay - Date received: 19 Mar 1813 - From what ship: HMS Pheasant - Born: Newport - Age: 27 - Sent to Dartmoor on 28 Jun 1813.

Vaughan, Thomas - Seaman - Number: 703 - Prize name: Star - Ship type: MV - How taken: HM Ship-of-the-Line Superb - When taken: 9 Feb 1813 - Where taken: Bay of Biscay - Date received: 19 Mar 1813 - From what ship: HMS Warspite - Born: New York - Age: 21 - Sent to Dartmoor on 2 Apr 1813.

Venison, Leven - Boy - Number: 1031 - Prize name: Two Brothers - Ship type: MV - How taken: Bootle of Liverpool, letter of marque - When taken: 18 Mar 1813 - Where taken: Western Islands - Date received: 21 Apr 1813 - From what ship: HMS Bittern - Born: Boston - Age: 18 - Race: Black - Sent to Dartmoor on 1 Jul 1813.

Vezin, Charles - Passenger - Number: 1188 - Prize name: Zebra - Ship type: LM - How taken: HM Frigate Pyramus - When taken: 20 Apr 1813 - Where taken: Bay of Biscay - Date received: 9 May 1813 - From what ship: HMS Andromache - Born: Osnabruck, Germany - Age: 31 - Sent on 11 Jun 1813 to the American ship Hope.

Vicary, Richard - Seaman - Number: 998 - Prize name: Polly - Ship type: MV - How taken: HM Frigate Surveillante - When taken: 27 Mar 1813 - Where taken: Bay of Biscay - Date received: 16 Apr 1813 - From what ship: HMS Fairy - Born: Beverly - Age: 17 - Sent to Chatham on 8 Jul 1813 on HM Tender Neptune.

Vogel, Herman - Seaman - Number: 1554 - Prize name: Good Friends - Ship type: MV - How taken: HM Frigate Andromache - When taken: 2 Apr 1813 - Where taken: Bay of Biscay - Date received: 29 May 1813 - From what ship: HMS Hannibal - Born: Amsterdam - Age: 16 - Discharged on 30 Jun 1813 and sent to Stapleton.

Voorhies, James - Seaman - Number: 928 - Prize name: Weasel - Ship type: MV - How taken: HM Brig Foxhound - When taken: 25 Mar 1813 - Where taken: Bay of Biscay - Date received: 6 Apr 1813 - From what ship: HMS Foxhound - Born: New Jersey - Age: 20 - Sent on 21 Jun 1813 to Portsmouth on HMS Prometheus.

Vetch, William - Seaman - Number: 1440 - Prize name: Paul Jones - Ship type: P - How taken: HM Frigate Leonidas - When taken: 23 May 1813 - Where taken: off Cape Clear - Date received: 26 May 1813 - From what ship: HMS Leonidas - Born: New York - Age: 19 - Discharged on 30 Jun 1813 and sent to Stapleton.

Wadden, Jacob - Seaman - Number: 556 - Prize name: Spitfire - Ship type: MV - How taken: HM Brig Achates - When taken: 14 Feb 1813 - Where taken: off Ushant - Date received: 16 Feb 1813 - From what ship: HMS Achates - Born: Marblehead - Age: 39 - Sent to Dartmoor on 2 Apr 1813.

Wade, John - Seaman - Number: 2107 - Prize name: Chesapeake - Ship type: LM - How taken: HM Frigate Hotspur & HM Frigate Pyramus - When taken: 26 Oct 1813 - Where taken: Bay of Biscay - Date received: 22 Nov 1813 - From what ship: HMS Pyramus - Born: New York - Age: 21 - Discharged on 29 Nov 1813 and sent to Dartmoor.

Wade, Otis - Seaman - Number: 113 - Prize name: Otter - Ship type: MV - How taken: HM Ship-Sloop Jalousie - When taken: 1 Dec 1812 - Where taken: off Cape Vincent - Date received: 30 Dec 1812 - From what ship: HMS Leonidas - Born: Massachusetts - Age: 22 - Sent to Portsmouth on 4 Jan 1813 on the HMS Revolutionnaire.

Wade, Richard - Seaman - Number: 2045 - How taken: Impressed at Liverpool - When taken: 16 Oct 1813 - Date received: 1 Nov 1813 - From what ship: HMS Bittern - Born: Wollish - Age: 22 - Discharged on 3 Nov 1813 and sent to Dartmoor.

Wain, Benjamin - 1st Lieutenant - Number: 141 - Prize name: Hunter - Ship type: P - How taken: HM Frigate Phoebe - When taken: 24 Dec 1812 - Where taken: off Western Islands - Date received: 9 Jan 1813 - From what ship: HMS Phoebe - Born: Boston - Age: 45 - Sent to Chatham on 29 Mar 1813 on the HMS Braham.

Wait, Henry - Seaman - Number: 1239 - Prize name: Essex - Ship type: MV - How taken: HM Frigate Pyramus - When taken: 2 Apr 1813 - Where taken: Bay of Biscay - Date received: 9 May 1813 - From what ship: HMS Andromache - Born: Marblehead - Age: 24 - Discharged on 3 Jul 1813 and sent to Stapleton Prison.

Waite, Abel - Quartermaster - Number: 1821 - Prize name: US Brig Argus - Ship type: War - How taken: HM Brig Pelican - When taken: 14 Aug 1813 - Where taken: Irish Channel - Date received: 17 Aug 1813 - From what ship: USS Argus - Born: Rhode Island - Age: 26 - Discharged on 8 Sep 1813 and sent to Dartmoor.

Wakefield, Nathaniel - Seaman - Number: 2292 - Prize name: Siro - Ship type: LM - How taken: HM Brig Pelican - When taken: 13 Jan 1814 - Where taken: at sea - Date received: 23 Jan 1814 - From what ship: HMS Pelican - Born: Beverly - Age: 52 - Discharged on 31 Jan 1814 and sent to Dartmoor.

Walden, James - Seaman - Number: 139 - Prize name: Nope - Ship type: MV - How taken: HM Sloop Pheasant - When taken: 13 Dec 1812 - Where taken: off Western Islands - Date received: 8 Jan 1813 - From what ship: HMS Pheasant - Born: Virginia - Age: 21 - Sent to Portsmouth on 8 Feb 1813 on the HMS Colossus.

Wilken, James - Prize Master - Number: 2071 - Prize name: Avon, prize of the Privateer True Blooded Yankee - Ship type: P - How taken: HM Frigate Eurotas - When taken: 27 Oct 1813 - Where taken: off Ushant - Date received: 1 Nov 1813 - From what ship: HMS Hannibal - Born: Portland - Age: 26 - Discharged on 3 Nov 1813 and sent to Dartmoor.

Walker, Armstrong - Seaman - Number: 1526 - Prize name: Courier - Ship type: LM - How taken: HM Brig Rover - When taken: 14 Mar 1813 - Where taken: Bay of Biscay - Date received: 29 May 1813 - From what ship: HMS Hannibal - Born: Baltimore - Age: 16 - Discharged on 30 Jun 1813 and sent to Stapleton.

Walker, Francis - Seaman - Number: 1204 - Prize name: Zebra - Ship type: LM - How taken: HM Frigate Pyramus - When taken: 20 Apr 1813 - Where taken: Bay of Biscay - Date received: 9 May 1813 - From what ship: HMS Andromache - Born: Baltimore - Age: 25 - Sent on 3 Jul 1813 to Stapleton prison.

Walker, James - Seaman - Number: 5 - Prize name: Friendship - Ship type: MV - How taken: HM Frigate Rosamond - When taken: Aug 1812 - Where taken: off Halifax - Date received: 24 Oct 1812 - From what ship: HMS Frederick - Born: Boston - Age: 33 - Race: Negro - Sent to Portsmouth on 29 Dec 1812 on the HMS Northumberland.

Walker, Seth - Prize Master - Number: 2352 - Prize name: Zephyr - Ship type: MV - How taken: HM Frigate Surveillante - When taken: 6 Jan 1814 - Where taken: Bay of Biscay - Date received: 31 Jan 1814 - From what ship: HMS Surveillante - Born: Portsmouth - Age: 35 - Discharged on 27 Feb 1814 and sent to Chatham on HMS Haleyon.

Wallace, James - Seaman - Number: 993 - Prize name: Lightning - Ship type: MV - How taken: HM Frigate Medusa - When taken: 2 Apr 1813 - Where taken: Bay of Biscay - Date received: 16 Apr 1813 - From what ship: HMS Fairy - Born: Saint Michael - Age: 31 - Sent to Chatham on 8 Jul 1813 on HM Tender Neptune.

Wallace, William - Seaman - Number: 1405 - How taken: Delivered himself up as a prisoner of war - Date received: 19 May 1813 - From what ship: HMS Salvador del Mundo - Born: New York - Age: 35 - Discharged on 8 Jun 1813.

Walleman, William - Seaman - Number: 1761 - How taken: Impressed at Belfast - When taken: 29 Jun 1813 - Date received: 22 Jul 1813 - From what ship: HMS Elizabeth - Born: Boston - Age: 31 - Discharged on 8 Sep 1813 and sent to Dartmoor.

Ware, William - Seaman - Number: 732 - Prize name: Pert - Ship type: MV - How taken: HM Ship-of-the-Line Warspite - When taken: 1 Mar 1813 - Where taken: Bay of Biscay - Date received: 19 Mar 1813 - From what ship: HMS Warspite - Born: Philadelphia - Age: 38 - Sent to Dartmoor on 2 Apr 1813.

Warner, Charles - Steward's Mate - Number: 159 - Prize name: Hunter - Ship type: P - How taken: HM Frigate Phoebe - When taken: 24 Dec 1812 - Where taken: off Western Islands - Date received: 9 Jan 1813 - From what ship: HMS Phoebe - Born: Brookfield - Age: 26 - Sent to Portsmouth on 8 Feb 1813 on the HMS Colossus.

Warner, George - Seaman - Number: 180 - Prize name: Hunter - Ship type: P - How taken: HM Frigate Phoebe - When taken: 24 Dec 1812 - Where taken: off Western Islands - Date received: 9 Jan 1813 - From what ship: HMS Phoebe - Born: Salem - Age: 22 - Sent to Portsmouth on 8 Feb 1813 on the HMS Colossus.

Warner, Henry - Seaman - Number: 517 - Prize name: Allegany - Ship type: MV - How taken: Detained at Gibraltar - When taken: 8 Aug 1812 - Date received: 15 Feb 1813 - From what ship: HMS Andromeda - Born: New Haven - Age: 28 - Sent to Dartmoor on 2 Apr 1813.

Warner, Samuel - Seaman - Number: 2602 - How taken: MV Furuise - When taken: 23 Sep 1812 - Date received: 16 May 1814 - From what ship: HMS Repulse - Born: Baltimore - Age: 20 - Race: Mulatto - Discharged on 14 Jun 1814 and sent to Dartmoor.

Warner, William - Seaman - Number: 1043 - How taken: Delivered himself up from HM Ship-of-the-Line Ajax - Date received: 22 Apr 1813 - From what ship: HMS Ajax - Born: New York - Age: 28 - Discharged on 7 Jul 1813 and sent on the HMS Salvador del Mundo.

Warrance, John - Seaman - Number: 989 - Prize name: Lightning - Ship type: MV - How taken: HM Frigate Medusa

- When taken: 2 Apr 1813 - Where taken: Bay of Biscay - Date received: 16 Apr 1813 - From what ship: HMS Fairy - Born: Philadelphia - Age: 19 - Sent to Chatham on 8 Jul 1813 on HM Tender Neptune.

Warren, John - Seaman - Number: 344 - How taken: Impressed at Plymouth - When taken: 25 Jan 1813 - Date received: 25 Jan 1813 - From what ship: HMS Salvador del Monde - Born: Baltimore - Age: 40 - Race: Negro - Sent to Chatham on 29 Mar 1813 on the HMS Braham.

Warren, Nathan - Master - Number: 233 - Prize name: Blue Bird - Ship type: MV - How taken: HM Frigate Andromache & HM Frigate Briton - When taken: 23 Dec 1812 - Where taken: at Coruna, Spain - Date received: 11 Jan 1813 - From what ship: Blue Bird - Born: Somersworth - Age: 29 - Sent on 12 Jan 1813 to Ashburton on parole.

Warren, David - Marine - Number: 1478 - Prize name: Paul Jones - Ship type: P - How taken: HM Frigate Leonidas - When taken: 23 May 1813 - Where taken: off Cape Clear - Date received: 26 May 1813 - From what ship: HMS Leonidas - Born: Worcester, MA - Age: 36 - Discharged on 30 Jun 1813 and sent to Stapleton.

Washbuiz, Elbert - Seaman - Number: 2641 - Prize name: Ateline - Ship type: MV - How taken: HM Frigate Magiciene - When taken: 14 Mar 1814 - Where taken: off Cape Finisterre, Spain - Date received: 17 May 1814 - From what ship: HMS Tortois - Born: Portland - Age: 17 - Discharged on 14 Jun 1814 and sent to Dartmoor.

Washburn, Edward - Seaman - Number: 708 - Prize name: Star - Ship type: MV - How taken: HM Ship-of-the-Line Superb - When taken: 9 Feb 1813 - Where taken: Bay of Biscay - Date received: 19 Mar 1813 - From what ship: HMS Warspite - Born: New York - Age: 20 - Sent to Dartmoor on 2 Apr 1813.

Washburn, Edward - Seaman - Number: 1736 - Prize name: Star - Ship type: MV - How taken: HM Ship-of-the-Line Superb - When taken: 9 Feb 1813 - Where taken: Bay of Biscay - Date received: 10 Jul 1813 - From what ship: Dartmoor Prison - Born: New York - Age: 20 - Discharged on 10 Jul 1813 and released to HMS Salvador del Mundo.

Wassen, Joseph Bartram - Master - Number: 1152 - Prize name: Young Holkar - Ship type: MV - How taken: HM Ship-of-the-Line Superb - When taken: 10 Apr 1813 - Where taken: off Belle Isle - Date received: 23 Apr 1813 - From what ship: HMS Plymouth - Born: Connecticut - Age: 31 - Sent on 24 Apr 1813 to Ashburton on parole.

Waters, Abraham - Seaman - Number: 1244 - Prize name: Essex - Ship type: MV - How taken: HM Frigate Pyramus - When taken: 2 Apr 1813 - Where taken: Bay of Biscay - Date received: 9 May 1813 - From what ship: HMS Andromache - Born: Charlestown, MA - Age: 16 - Discharged on 3 Jul 1813 and sent to Stapleton Prison.

Waters, Philip - Seaman - Number: 477 - How taken: Impressed at Greenock - When taken: 27 Nov 1812 - Date received: 10 Feb 1813 - From what ship: HMS Frederick - Born: Baltimore - Age: 29 - Race: Black - Sent on 20 Feb 1813 to Ashburton on parole.

Waters, Robert - Seaman - Number: 1917 - Prize name: US Brig Argus - Ship type: War - How taken: HM Brig Pelican - When taken: 14 Aug 1813 - Where taken: Irish Channel - Date received: 23 Aug 1813 - From what ship: HMS Pelican - Born: Massachusetts - Age: 25 - Race: Black - Discharged on 8 Sep 1813 and sent to Dartmoor.

Wither, John - Seaman - Number: 757 - Prize name: William Bayard - Ship type: MV - How taken: HM Ship-of-the-Line Warspite - When taken: 3 Mar 1813 - Where taken: Bay of Biscay - Date received: 19 Mar 1813 - From what ship: HMS Warspite - Born: Philadelphia - Age: 27 - Sent to Dartmoor on 2 Apr 1813.

Watkins, Fredrick - Seaman - Number: 21 - Prize name: Washington - Ship type: MV - How taken: Taken at Liverpool - When taken: 18 Oct 1812 - Date received: 21 Nov 1812 - From what ship: HMS Salvador del Mundo - Born: New York - Age: 22 - Sent to Portsmouth on 29 Dec 1812 on the HMS Northumberland.

Watkins, Thomas - Seaman - Number: 1607 - Prize name: Revenge - Ship type: LM - How taken: HM Frigate Belle Poule - When taken: 11 May 1813 - Where taken: Bay of Biscay - Date received: 14 Jun 1813 - From what ship: HMS Royalist - Born: Dalton - Age: 38 - Discharged on 30 Jun 1813 and sent to Stapleton.

Watson, Isaac - Seaman - Number: 1338 - Prize name: Fox - Ship type: LM - How taken: HM Sloop Pheasant - When taken: 23 Apr 1813 - Where taken: Bay of Biscay - Date received: 14 May 1813 - From what ship: HMS Pleasant - Born: Charleston - Age: 23 - Race: Black - Discharged on 3 Jul 1813 and sent to Stapleton Prison.

Watson, James - Seaman - Number: 1177 - Prize name: Hebe - Ship type: MV - How taken: HM Frigate Stag - When taken: 18 Apr 1813 - Where taken: Bay of Biscay - Date received: 6 May 1813 - From what ship: HMS Stag -

Born: Troy - Age: 27 - Sent on 3 Jul 1813 to Stapleton prison.

Watson, William - Seaman - Number: 2368 - Prize name: Minerva - Ship type: MV - How taken: HM Ship-of-the-Line Conquestador - When taken: 19 Dec 1814 - Where taken: Bay of Biscay - Date received: 31 Jan 1814 - From what ship: HMS Surveillante - Born: Scarborough - Age: 20 - Discharged on 27 Feb 1814 and sent to Chatham on HMS Haleyon.

Watson, William H. - 1st Lieutenant - Number: 1857 - Prize name: US Brig Argus - Ship type: War - How taken: HM Brig Pelican - When taken: 14 Aug 1813 - Where taken: Irish Channel - Date received: 23 Aug 1813 - From what ship: HMS Pelican - Born: Alexandria - Age: 22 - Discharged on 1 Sep 1813 and sent to Ashburton on parole.

Watterman, John - 2nd Mate - Number: 919 - Prize name: Tiger - Ship type: MV - How taken: HM Brig Scylla - When taken: 22 Mar 1813 - Where taken: Bay of Biscay - Date received: 5 Apr 1813 - From what ship: HMS Scylla - Born: Nantucket - Age: 21 - Sent on 21 Jun 1813 to Portsmouth on HMS Prometheus.

Wattes, Philip - Seaman - Number: 2640 - Prize name: Ateline - Ship type: MV - How taken: HM Frigate Magiciene - When taken: 14 Mar 1814 - Where taken: off Cape Finisterre, Spain - Date received: 17 May 1814 - From what ship: HMS Tortois - Born: Baltimore - Age: 25 - Race: Black - Discharged on 14 Jun 1814 and sent to Dartmoor.

Watts, Herman - Seaman - Number: 798 - Prize name: Cannoniere - Ship type: P - How taken: HM Ship-of-the-Line Warspite - When taken: 14 Mar 1813 - Where taken: Bay of Biscay - Date received: 19 Mar 1813 - From what ship: HMS Warspite - Born: Newburgh - Age: 18 - Sent to Dartmoor on 28 Jun 1813.

Watts, James - 2nd Mate - Number: 867 - Prize name: Pert - Ship type: MV - How taken: HM Ship-of-the-Line Warspite - When taken: 1 Mar 1813 - Where taken: Bay of Biscay - Date received: 28 Mar 1813 - From what ship: HMS Warspite - Born: Philadelphia - Age: 43 - Sent to Dartmoor on 28 Jun 1813.

Way, John - Mate - Number: 1099 - Prize name: Courier - Ship type: MV - How taken: HM Frigate Andromache - When taken: 14 Mar 1813 - Where taken: Bay of Biscay - Date received: 22 Apr 1813 - From what ship: HMS Superb - Born: Baltimore - Age: 26 - Sent on 25 Apr 1813 to Ashburton on parole.

Weaphor, Andrew - Seaman - Number: 807 - Prize name: Decornau - Ship type: MV - How taken: HM Sloop Pheasant - When taken: 15 Mar 1813 - Where taken: Bay of Biscay - Date received: 19 Mar 1813 - From what ship: HMS Pheasant - Born: Philadelphia - Age: 53 - Sent to Dartmoor on 28 Jun 1813.

Webb, Alexander - Seaman - Number: 2044 - How taken: Impressed at Liverpool - When taken: 16 Oct 1813 - Date received: 1 Nov 1813 - From what ship: HMS Bittern - Born: Philadelphia - Age: 20 - Discharged on 3 Nov 1813 and sent to Dartmoor.

Webb, John - Boy - Number: 382 - Prize name: Orbit - Ship type: MV - How taken: HM Brig Achates - When taken: 29 Jan 1813 - Where taken: Lat 49 N Long 13 W - Date received: 4 Feb 1813 - From what ship: HMS Achates - Born: New York - Age: 16 - Sent to Chatham on 29 Mar 1813 on the HMS Braham.

Webb, John - Seaman - Number: 1638 - Prize name: Tickler - Ship type: LM - How taken: HM Frigate Magiciene - When taken: 5 Jun 1813 - Where taken: Lat 47 Long 13 - Date received: 14 Jun 1813 - From what ship: HMS Orestes - Born: Maine - Age: 24 - Discharged on 30 Jun 1813 and sent to Stapleton.

Webb, William - Seaman - Number: 408 - Prize name: Union - Ship type: MV - How taken: HM Frigate Iris - When taken: 17 Jan 1813 - Where taken: Lat 44 N Long 2.3 W - Date received: 5 Feb 1813 - From what ship: HMS San Josef - Born: Philadelphia - Age: 19 - Sent to Chatham on 29 Mar 1813 on the HMS Braham.

Webber, Frederick - Seaman - Number: 2407 - Prize name: Rachel & Ann, prize of the Privateer Prince de Neufchatel - Ship type: P - How taken: HM Frigate Cydnus - When taken: 6 Jan 1814 - Where taken: at sea - Date received: 4 Feb 1814 - From what ship: HMS Cydnus - Born: New Orleans - Age: 28 - Race: Black - Discharged on 10 May 1814 and sent to Dartmoor.

Webber, John - Seaman - Number: 2053 - How taken: Pallas - Where taken: St. Johns, New Brunswick - Date received: 1 Nov 1813 - From what ship: HMS Bittern - Born: Brunswick - Age: 28 - Discharged on 3 Nov 1813 and sent to Dartmoor.

Webster, James - Seaman - Number: 1115 - Prize name: Viper - Ship type: MV - How taken: HM Ship-of-the-Line Superb - When taken: 15 Apr 1813 - Where taken: Bay of Biscay - Date received: 22 Apr 1813 - From what ship: HMS Superb - Born: Edenton - Age: 52 - Sent to Dartmoor on 1 Jul 1813.

Weedon, Anthony - Seaman - Number: 1556 - Prize name: Elvia - Ship type: MV - How taken: HM Frigate Surveillante - When taken: 21 Mar 1813 - Where taken: Bay of Biscay - Date received: 29 May 1813 - From what ship: HMS Hannibal - Born: Rhode Island - Age: 31 - Discharged on 30 Jun 1813 and sent to Stapleton.

Weeks, Benjamin - Seaman - Number: 1746 - How taken: Impressed at Liverpool - When taken: 25 Apr 1813 - Date received: 14 Jun 1813 - From what ship: HMS Bittern - Born: Philadelphia - Age: 30 - Discharged on 8 Sep 1813 and sent to Dartmoor.

Weeks, David - Seaman - Number: 1545 - Prize name: Zebra - Ship type: LM - How taken: HM Frigate Pyramus - When taken: 20 Apr 1813 - Where taken: Bay of Biscay - Date received: 29 May 1813 - From what ship: HMS Hannibal - Born: New Jersey - Age: 21 - Discharged on 30 Jun 1813 and sent to Stapleton.

Weeks, James - Seaman - Number: 2288 - Prize name: Siro - Ship type: LM - How taken: HM Brig Pelican - When taken: 13 Jan 1814 - Where taken: at sea - Date received: 23 Jan 1814 - From what ship: HMS Pelican - Born: Palermo - Age: 21 - Discharged on 31 Jan 1814 and sent to Dartmoor.

Weellid, William - Seaman - Number: 1214 - Prize name: Zebra - Ship type: LM - How taken: HM Frigate Pyramus - When taken: 20 Apr 1813 - Where taken: Bay of Biscay - Date received: 9 May 1813 - From what ship: HMS Andromache - Born: Washington - Age: 28 - Discharged on 2 Sep 1813 and sent to Dartmoor.

Weimain, Lawrence - Private - Number: 2020 - Prize name: 21st U.S. Infantry - Ship type: LF - How taken: British army - When taken: 28 May 1813 - Where taken: Lake Ontario - Date received: 13 Oct 1813 - From what ship: Transport Lord Cathcart - Born: Berlin - Age: 37 - Discharged on 3 Nov 1813 and sent to Dartmoor.

Welch, Benjamin - Seaman - Number: 1229 - Prize name: Essex - Ship type: MV - How taken: HM Frigate Pyramus - When taken: 2 Apr 1813 - Where taken: Bay of Biscay - Date received: 9 May 1813 - From what ship: HMS Andromache - Born: Massachusetts - Age: 22 - Discharged on 3 Jul 1813 and sent to Stapleton Prison.

Wallander, Adam - Seaman - Number: 803 - Prize name: Cannoniere - Ship type: P - How taken: HM Ship-of-the-Line Warspite - When taken: 14 Mar 1813 - Where taken: Bay of Biscay - Date received: 19 Mar 1813 - From what ship: HMS Warspite - Born: Sweden - Age: 28 - Sent to Dartmoor on 28 Jun 1813.

Wells, John - Seaman - Number: 1279 - Prize name: Price - Ship type: LM - How taken: HM Frigate Medusa - When taken: 13 Apr 1813 - Where taken: Bay of Biscay - Date received: 10 May 1813 - From what ship: HMS Medusa - Born: Newburyport - Age: 29 - Discharged on 21 Jun 1813 and released to American ship Mount Hope.

Welsh, William - Seaman - Number: 2291 - Prize name: Siro - Ship type: LM - How taken: HM Brig Pelican - When taken: 13 Jan 1814 - Where taken: at sea - Date received: 23 Jan 1814 - From what ship: HMS Pelican - Born: Port Royal - Age: 17 - Discharged on 31 Jan 1814 and sent to Dartmoor.

Wessel, Samuel - Seaman - Number: 947 - Prize name: Good Friends - Ship type: MV - How taken: HM Frigate Andromache - When taken: 2 Apr 1813 - Where taken: Bay of Biscay - Date received: 7 Apr 1813 - From what ship: HMS Sea Lark - Born: New Jersey - Age: 20 - Sent to Dartmoor on 28 Jun 1813.

Wessels, Robert - Seaman - Number: 1991 - Prize name: Ned - Ship type: LM - How taken: HM Brig Royalist - When taken: 6 Sep 1812 - Where taken: coast of France - Date received: 22 Sep 1813 - From what ship: HMS Royalist - Born: New York - Age: 17 - Discharged on 27 Sep 1813 and sent to Dartmoor.

West, Benjamin - Seaman - Number: 1094 - Prize name: Young Holkar - Ship type: MV - How taken: HM Ship-of-the-Line Superb - When taken: 10 Apr 1813 - Where taken: off Belle Isle - Date received: 22 Apr 1813 - From what ship: HMS Superb - Born: Charlestown - Age: 19 - Sent to Dartmoor on 1 Jul 1813.

West, Dennis - Seaman - Number: 1262 - Prize name: Caroline - Ship type: MV - How taken: HM Frigate Medusa - When taken: 12 Apr 1813 - Where taken: Bay of Biscay - Date received: 10 May 1813 - From what ship: HMS Medusa - Born: Massachusetts - Age: 25 - Discharged on 3 Jul 1813 and sent to Stapleton Prison.

West, John - Seaman - Number: 19 - Prize name: Martin - Ship type: MV - How taken: Taken at Liverpool - When taken: 18 Oct 1812 - Date received: 21 Nov 1812 - From what ship: HMS Salvador del Mundo - Born: New Orleans - Age: 20 - Sent to Portsmouth on 29 Dec 1812 on the HMS Northumberland.

West, John - Seaman - Number: 814 - Prize name: Decornau - Ship type: MV - How taken: HM Sloop Pheasant - When taken: 15 Mar 1813 - Where taken: Bay of Biscay - Date received: 19 Mar 1813 - From what ship: HMS Pheasant - Born: New York - Age: 21 - Sent to Dartmoor on 28 Jun 1813.

West, Nathan - Ordinary Seaman - Number: 352 - How taken: Taken off the HM Frigate Andromache - Date received: 29 Jan 1813 - From what ship: HMS Royal Sovereign - Born: Greenfield - Age: 21 - Sent to Chatham on 29 Mar 1813 on the HMS Braham.

West, Reuben - 2nd Mate - Number: 509 - Prize name: Tyger - Ship type: MV - How taken: Detained at Gibraltar - When taken: 8 Aug 1812 - Date received: 15 Feb 1813 - From what ship: HMS Andromeda - Born: Falmouth - Age: 21 - Sent to Dartmoor on 2 Apr 1813.

West, Simon - Seaman - Number: 1582 - Prize name: Governor Gerry - Ship type: MV - How taken: HM Brig Royalist - When taken: 31 May 1813 - Where taken: Bay of Biscay - Date received: 12 Jun 1813 - From what ship: Governor Gerry - Born: Rhode Island - Age: 23 - Discharged on 30 Jun 1813 and sent to Stapleton.

Wester, Andrew - Seaman - Number: 1258 - Prize name: Caroline - Ship type: MV - How taken: HM Frigate Medusa - When taken: 12 Apr 1813 - Where taken: Bay of Biscay - Date received: 10 May 1813 - From what ship: HMS Medusa - Born: Stockholm, Sweden - Age: 27 - Discharged on 3 Jul 1813 and sent to Stapleton Prison.

Westerbest, William - Seaman - Number: 879 - Prize name: Tiger - Ship type: MV - How taken: HM Brig Scylla - When taken: 22 Mar 1813 - Where taken: Bay of Biscay - Date received: 1 Apr 1813 - From what ship: HMS Scylla - Born: New York - Age: 22 - Sent to Chatham on 8 Jul 1813 on HM Tender Neptune.

Weston, David - Seaman - Number: 1600 - Prize name: Governor Gerry - Ship type: MV - How taken: HM Brig Royalist - When taken: 31 May 1813 - Where taken: Bay of Biscay - Date received: 14 Jun 1813 - From what ship: HMS Royalist - Born: Baltimore - Age: 27 - Discharged on 30 Jun 1813 and sent to Stapleton.

Wheeler, Benjamin - Seaman - Number: 2009 - Prize name: Ned - Ship type: LM - How taken: HM Brig Royalist - When taken: 6 Sep 1812 - Where taken: coast of France - Date received: 24 Sep 181 - From what ship: HMS Rippon - Born: Le Havre - Age: 27 - Discharged on 27 Sep 1813 and sent to Dartmoor.

Wheeler, Charles - Seaman - Number: 717 - Prize name: Criterion - Ship type: MV - How taken: HM Frigate Belle Poule - When taken: 14 Feb 1813 - Where taken: Bay of Biscay - Date received: 19 Mar 1813 - From what ship: HMS Warspite - Born: Smith Town - Age: 34 - Sent to Dartmoor on 2 Apr 1813.

Wheeler, Henry - Seaman - Number: 2061 - Prize name: Betsy, prize to the Privateer True Blooded Yankee - Ship type: P - How taken: HM Frigate Eurotas - When taken: 26 Oct 1813 - Where taken: off Ushant - Date received: 1 Nov 1813 - From what ship: HMS Hannibal - Born: Salem - Age: 22 - Discharged on 3 Nov 1813 and sent to Dartmoor.

Wheeler, John - Seaman - Number: 2202 - Prize name: General Kempt, prize of the Privateer Grand Turk - Ship type: P - How taken: HM Brig Foxhound - When taken: 18 Dec 1813 - Where taken: Lat 48.4 Long 6 - Date received: 21 Dec 1813 - From what ship: HMS Foxhound - Born: Salem - Age: 21 - Discharged on 31 Jan 1814 and sent to Dartmoor.

Wheeler, John W. - Seaman - Number: 1687 - Prize name: Revenge - Ship type: LM - How taken: HM Frigate Belle Poule - When taken: 11 May 1813 - Where taken: Bay of Biscay - Date received: 26 Jun 1813 - From what ship: HMS Duncan - Born: Fairfield - Age: 26 - Discharged on 30 Jun 1813 and sent to Stapleton.

Wheeler, William - Seaman - Number: 371 - Prize name: Orbit - Ship type: MV - How taken: HM Brig Achates - When taken: 29 Jan 1813 - Where taken: Lat 49 N Long 13 W - Date received: 4 Feb 1813 - From what ship: HMS Achates - Born: Newburyport - Age: 20 - Sent to Chatham on 29 Mar 1813 on the HMS Braham.

Wheler, Richard - Cook - Number: 544 - Prize name: Spitfire - Ship type: MV - How taken: HM Brig Achates - When taken: 14 Feb 1813 - Where taken: off Ushant - Date received: 16 Feb 1813 - From what ship: HMS Achates - Born: Marblehead - Age: 23 - Sent to Dartmoor on 2 Apr 1813.

Whetting, Samuel C. - Seaman - Number: 1448 - Prize name: Paul Jones - Ship type: P - How taken: HM Frigate Leonidas - When taken: 23 May 1813 - Where taken: off Cape Clear - Date received: 26 May 1813 - From what ship: HMS Leonidas - Born: Providence, RI - Age: 22 - Discharged on 30 Jun 1813 and sent to Stapleton.

Whipple, Samuel - Seaman - Number: 181 - Prize name: Hunter - Ship type: P - How taken: HM Frigate Phoebe - When taken: 24 Dec 1812 - Where taken: off Western Islands - Date received: 9 Jan 1813 - From what ship: HMS Phoebe - Born: Newport - Age: 25 - Sent to Portsmouth on 8 Feb 1813 on the HMS Colossus.

White, Alden - Seaman - Number: 27 - Prize name: Catharine - Ship type: MV - How taken: HM Frigate Leonidas -

When taken: 31 Jul 1812 - Where taken: off Ireland - Date received: 23 Nov 1812 - From what ship: HMS Stork - Born: New Bedford - Age: 20 - Sent to Portsmouth on 29 Dec 1812 on the HMS Northumberland.

White, Charles - Seaman - Number: 1539 - Prize name: Meteor - Ship type: MV - How taken: HM Frigate Briton - When taken: 12 Mar 1813 - Where taken: Bay of Biscay - Date received: 29 May 1813 - From what ship: HMS Hannibal - Born: Virginia - Age: 28 - Discharged on 30 Jun 1813 and sent to Stapleton.

White, David - Seaman - Number: 1531 - Prize name: Courier - Ship type: LM - How taken: HM Brig Rover - When taken: 14 Mar 1813 - Where taken: Bay of Biscay - Date received: 29 May 1813 - From what ship: HMS Hannibal - Born: Baltimore - Age: 15 - Discharged on 30 Jun 1813 and sent to Stapleton.

White, Henry - Seaman - Number: 720 - Prize name: Criterion - Ship type: MV - How taken: HM Frigate Belle Poule - When taken: 14 Feb 1813 - Where taken: Bay of Biscay - Date received: 19 Mar 1813 - From what ship: HMS Warspite - Born: Maryland - Age: 21 - Sent to Dartmoor on 2 Apr 1813.

White, Isaac - Seaman - Number: 1508 - Prize name: Grand Napoleon - Ship type: MV - How taken: HM Brig Goldfinch - When taken: 17 Apr 1813 - Where taken: Bay of Biscay - Date received: 29 May 1813 - From what ship: HMS Hannibal - Born: Groton, CT - Age: 19 - Discharged on 30 Jun 1813 and sent to Stapleton.

White, James - Seaman - Number: 2070 - Prize name: Betsy, prize to the Privateer True Blooded Yankee - Ship type: P - How taken: HM Frigate Eurotas - When taken: 26 Oct 1813 - Where taken: off Ushant - Date received: 1 Nov 1813 - From what ship: HMS Hannibal - Born: New York - Age: 27 - Discharged on 3 Nov 1813 and sent to Dartmoor.

White, James - Carpenter - Number: 1802 - Prize name: US Brig Argus - Ship type: War - How taken: HM Brig Pelican - When taken: 14 Aug 1813 - Where taken: Irish Channel - Date received: 17 Aug 1813 - From what ship: USS Argus - Born: New York - Age: 21 - Died on 17 Aug 1813 in the Mill Prison Hospital.

White, James - Seaman - Number: 2159 - Prize name: Princess - Ship type: MV - How taken: Impressed at Liverpool - When taken: 2 Nov 1813 - Date received: 13 Dec 1813 - From what ship: HMS Bittern - Born: North Carolina - Age: 23 - Race: Black - Died on in Feb 1814 on the Prison Ship Caton.

White, John - Seaman - Number: 1679 - Prize name: Revenge - Ship type: LM - How taken: HM Frigate Belle Poule - When taken: 11 May 1813 - Where taken: Bay of Biscay - Date received: 26 Jun 1813 - From what ship: HMS Duncan - Born: Boston - Age: 24 - Discharged on 30 Jun 1813 and sent to Stapleton.

White, Lauren - Seaman - Number: 1322 - Prize name: Shadow - Ship type: LM - How taken: HM Brig Reindeer & HM Schooner Helicon - When taken: 6 Apr 1813 - Where taken: Bay of Biscay - Date received: 12 May 1813 - From what ship: HMS Helicon - Born: New Orleans - Age: 14 - Discharged on 7 Jul 183 and released to the Salvador del Mundo.

White, Philip - Seaman - Number: 620 - Prize name: Governor McKean - Ship type: LM - How taken: HM Brig Rover - When taken: 26 Jun 1813 - Where taken: off Bordeaux - Date received: 2 Mar 1813 - From what ship: HMS Insolent - Born: Pennsylvania - Age: 17 - Sent to Dartmoor on 2 Apr 1813.

White, Philip - Prize Master - Number: 2658 - Prize name: Indian Lass, prize of the Privateer Grand Turk - Ship type: P - How taken: HM Transport Akbar - When taken: 29 Apr 1814 - Where taken: at sea - Date received: 5 Jun 1814 - From what ship: HMS Gronville - Born: Marblehead - Age: 30 - Discharged on 14 Jun 1814 and sent to Dartmoor.

White, Sampson - Reserve Officer - Number: 150 - Prize name: Hunter - Ship type: P - How taken: HM Frigate Phoebe - When taken: 24 Dec 1812 - Where taken: off Western Islands - Date received: 9 Jan 1813 - From what ship: HMS Phoebe - Born: Boston - Age: 24 - Sent to Portsmouth on 8 Feb 1813 on the HMS Colossus.

White, William - Seaman - Number: 1180 - How taken: Delivered himself up from HM Ship-of-the-Line Clarence - Date received: 6 May 1813 - From what ship: HMS Stag - Born: New York - Age: 25 - Sent to Chatham on 8 Jul 1813 on HM Tender Neptune.

White, William - Seaman - Number: 1181 - How taken: Delivered himself up from HM Ship-of-the-Line Clarence - Date received: 6 May 1813 - From what ship: HMS Stag - Born: Boston - Age: 23 - Sent to Chatham on 8 Jul 1813 on HM Tender Neptune.

Whitehead, James - Seaman - Number: 2702 - Prize name: Margaret, prize to the Privateer Surprize - Ship type: P - How taken: HM Brig Foxhound - When taken: 27 May 1814 - Where taken: off Scylly - Date received: 16 Jun

1814 - From what ship: HMS Foxhound - Born: Norfolk - Age: 19 - Discharged on 20 Jun 1814 and sent to Dartmoor.

Whitehouse, Asey - Seaman - Number: 666 - How taken: Impressed at Dublin - When taken: 7 Jan 1813 - Date received: 14 Mar 1813 - From what ship: HMS Neptune - Born: Brookfield - Age: 23 - Sent to Dartmoor on 2 Apr 1813.Whiteman, John - Seaman - Number: 2684 - How taken: Sent into custody from HM Frigate Thames - Date received: 6 Jun 1814 - From what ship: Eudam - Discharged on 14 Jun 1814 and sent to Dartmoor.

Whitewood, Charles - Seaman - Number: 728 - Prize name: Mars - Ship type: MV - How taken: HM Ship-of-the-Line Warspite - When taken: 26 Feb 1813 - Where taken: Bay of Biscay - Date received: 19 Mar 1813 - From what ship: HMS Warspite - Born: New York - Age: 28 - Sent to Dartmoor on 2 Apr 1813.

Whitmore, William - Seaman - Number: 120 - How taken: Impressed at Dublin - When taken: 12 Nov 1812 - Date received: 30 Dec 1812 - From what ship: HMS Frederick - Born: American - Age: 24 - Sent to Portsmouth on 4 Jan 1813 on the HMS Revolutionnaire.

Whitney, Joseph - Seaman - Number: 224 - Prize name: Vengeance - Ship type: LM - How taken: HM Frigate Phoebe - When taken: 1 Jan 1813 - Where taken: Lat 44.4 Long 23 - Date received: 9 Jan 1813 - From what ship: HMS Phoebe - Born: Northampton - Age: 20 - Sent to Portsmouth on 8 Feb 1813 on the HMS Colossus.

Whitney, Thomas - Prize Master - Number: 2561 - Prize name: Bunker Hill - Ship type: P - How taken: HM Frigate Pomone & HM Frigate Cydnus - When taken: 4 Mar 1814 - Where taken: Bay of Biscay - Date received: 17 Apr 1814 - From what ship: HMS Teazer - Born: Fairfield - Age: 28 - Discharged on 10 May 1814 and sent to Dartmoor.

Wicks, Littleton - Seaman - Number: 13 - Prize name: Hanna - Ship type: MV - How taken: Taken at Liverpool - When taken: 18 Oct 1812 - Date received: 21 Nov 1812 - From what ship: HMS Salvador del Mundo - Born: Virginia - Age: 27 - Sent to Portsmouth on 29 Dec 1812 on the HMS Northumberland.

Wilcox, Lewis - Seaman - Number: 1517 - Prize name: Grand Napoleon - Ship type: MV - How taken: HM Brig Goldfinch - When taken: 17 Apr 1813 - Where taken: Bay of Biscay - Date received: 29 May 1813 - From what ship: HMS Hannibal - Born: Rhode Island - Age: 32 - Discharged on 30 Jun 1813 and sent to Stapleton.

Wild, Caleb - Merchant & Passenger - Number: 562 - Prize name: Spitfire - Ship type: MV - How taken: HM Brig Achates - When taken: 14 Feb 1814 - Where taken: off Ushant - Date received: 21 Feb 1813 - From what ship: HMS Achates - Born: Boston - Age: 27 - Sent on 22 Feb 1813 to Ashburton on parole.

Wild, Thomas - Seaman - Number: 1445 - Prize name: Paul Jones - Ship type: P - How taken: HM Frigate Leonidas - When taken: 23 May 1813 - Where taken: off Cape Clear - Date received: 26 May 1813 - From what ship: HMS Leonidas - Born: New Castle, DE - Age: 27 - Discharged on 30 Jun 1813 and sent to Stapleton.

Wilkins, William - Seaman - Number: 1343 - Prize name: Fox - Ship type: LM - How taken: HM Sloop Pheasant - When taken: 23 Apr 1813 - Where taken: Bay of Biscay - Date received: 14 May 1813 - From what ship: HMS Pleasant - Born: New Jersey - Age: 21 - Discharged on 3 Jul 1813 and sent to Stapleton Prison.

Wilkson, John - Seaman - Number: 2666 - Prize name: Traveler, prize of the Privateer Surprize - Ship type: P - How taken: HMS Cawser - When taken: 7 May 1814 - Where taken: off Cape Clear - Date received: 5 Jun 1814 - From what ship: HMS Gronville - Born: Baltimore - Age: 23 - Discharged on 14 Jun 1814 and sent to Dartmoor.

Willcox, Caesar - Seaman - Number: 189 - Prize name: Hunter - Ship type: P - How taken: HM Frigate Phoebe - When taken: 24 Dec 1812 - Where taken: off Western Islands - Date received: 9 Jan 1813 - From what ship: HMS Phoebe - Born: Carnatic - Age: 41 - Race: Black - Sent to Portsmouth on 8 Feb 1813 on the HMS Colossus.

Willett, Robert - Ordinary Seaman - Number: 353 - How taken: Taken off the HM Frigate Andromache - Date received: 29 Jan 1813 - From what ship: HMS Royal Sovereign - Born: Newburyport - Age: 21 - Sent to Chatham on 29 Mar 1813 on the HMS Braham.

Willey, Ebenezer - Seaman - Number: 2068 - Prize name: Betsy, prize to the Privateer True Blooded Yankee - Ship type: P - How taken: HM Frigate Eurotas - When taken: 26 Oct 1813 - Where taken: off Ushant - Date received: 1 Nov 1813 - From what ship: HMS Hannibal - Born: New Hampshire - Age: 34 - Discharged on 3 Nov 1813 and sent to Dartmoor.

William, James - Seaman - Number: 1996 - Prize name: Ned - Ship type: LM - How taken: HM Brig Royalist - When taken: 6 Sep 1812 - Where taken: coast of France - Date received: 22 Sep 1813 - From what ship: HMS Royalist - Born: Baltimore - Age: 18 - Race: Black - Discharged on 27 Sep 1813 and sent to Dartmoor.

Williams, Alexander - Seaman - Number: 22 - Prize name: Industry - Ship type: MV - How taken: Taken at Liverpool - When taken: 18 Oct 1812 - Date received: 21 Nov 1812 - From what ship: HMS Salvador del Mundo - Born: Long Island - Age: 21 - Sent to Portsmouth on 29 Dec 1812 on the HMS Northumberland.

Williams, Alexander - Seaman - Number: 2409 - Prize name: Rachel & Ann, prize of the Privateer Prince de Neufchatel - Ship type: P - How taken: HM Frigate Cydnus - When taken: 6 Jan 1814 - Where taken: at sea - Date received: 4 Feb 1814 - From what ship: HMS Cydnus - Born: New York - Age: 22 - Race: Black - Discharged on 10 May 1814 and sent to Dartmoor.

Williams, Ambrose - Seaman - Number: 1675 - Prize name: Rolla - Ship type: MV - How taken: HM Frigate Surveillante - When taken: 11 Feb 1813 - Where taken: Bay of Biscay - Date received: 24 Jun 1813 - From what ship: Dartmoor Prison - Born: Salem - Age: 19 - Discharged on 27 Sep 1813 and sent to Dartmoor.

Williams, Ambrose - Seaman - Number: 564 - Prize name: Rolla - Ship type: MV - How taken: HM Frigate Surveillante - When taken: 11 Feb 1813 - Where taken: Bay of Biscay - Date received: 23 Feb 1813 - From what ship: HMS Surveillante - Born: Salem - Age: 19 - Sent to Dartmoor on 2 Apr 1813.

Williams, David - Chief Mate - Number: 828 - Prize name: Star - Ship type: MV - How taken: HM Ship-of-the-Line Superb - When taken: 9 Feb 1813 - Where taken: Bay of Biscay - Date received: 22 Mar 1813 - From what ship: HMS Warspite - Born: New London - Age: 27 - Sent on 23 Mar 1813 to Ashburton on parole.

Williams, Edward - Seaman - Number: 2447 - How taken: Taken out of a Russian ship - When taken: 28 Jan 1814 - Where taken: Cove of Cork - Date received: 15 Feb 1814 - From what ship: HMS Zealous - Born: Virginia - Age: 22 - Discharged on 10 May 1814 and sent to Dartmoor.

Williams, George - Seaman - Number: 752 - Prize name: Charlotte - Ship type: MV - How taken: HM Ship-of-the-Line Warspite - When taken: 3 Mar 1813 - Where taken: Bay of Biscay - Date received: 19 Mar 1813 - From what ship: HMS Warspite - Born: Maryland - Age: 24 - Race: Black - Sent to Dartmoor on 2 Apr 1813.

Williams, George - Seaman - Number: 1737 - Prize name: Charlotte - Ship type: MV - How taken: HM Ship-of-the-Line Warspite - When taken: 3 Mar 1813 - Where taken: Bay of Biscay - Date received: 10 Jul 1813 - From what ship: Dartmoor Prison - Born: Maryland - Age: 24 - Discharged on 10 Jul 1813 and released to HMS Salvador del Mundo.

Williams, George - Seaman - Number: 29 - Prize name: Catharine - Ship type: MV - How taken: HM Frigate Leonidas - When taken: 31 Jul 1812 - Where taken: off Ireland - Date received: 23 Nov 1812 - From what ship: HMS Stork - Born: Queen Anne's county - Age: 23 - Race: Black - Sent to Portsmouth on 29 Dec 1812 on the HMS Northumberland.

Williams, George - Seaman - Number: 30 - Prize name: Catharine - Ship type: MV - How taken: HM Frigate Leonidas - When taken: 31 Jul 1812 - Where taken: off Ireland - Date received: 23 Nov 1812 - From what ship: HMS Stork - Born: Baltimore - Age: 56 - Race: Black - Sent to Portsmouth on 29 Dec 1812 on the HMS Northumberland.

Williams, Hendrick - Seaman - Number: 2127 - Prize name: Chesapeake - Ship type: LM - How taken: HM Frigate Hotspur & HM Frigate Pyramus - When taken: 26 Oct 1813 - Where taken: Bay of Biscay - Date received: 22 Nov 1813 - From what ship: HMS Pyramus - Born: Maryland - Age: 16 - Discharged on 29 Nov 1813 and sent to Dartmoor.

Williams, Henry - Seaman - Number: 714 - Prize name: Star - Ship type: MV - How taken: HM Ship-of-the-Line Superb - When taken: 9 Feb 1813 - Where taken: Bay of Biscay - Date received: 19 Mar 1813 - From what ship: HMS Warspite - Born: New Orleans - Age: 36 - Sent to Dartmoor on 2 Apr 1813.

Williams, James - Seaman - Number: 1259 - Prize name: Caroline - Ship type: MV - How taken: HM Frigate Medusa - When taken: 12 Apr 1813 - Where taken: Bay of Biscay - Date received: 10 May 1813 - From what ship: HMS Medusa - Born: Connecticut - Age: 23 - Discharged on 3 Jul 1813 and sent to Stapleton Prison.

Williams, James - Seaman - Number: 323 - Prize name: Porcupine - Ship type: MV - How taken: HM Frigate Dryad - When taken: 8 Jan 1813 - Where taken: off Bordeaux - Date received: 21 Jan 1813 - From what ship: HMS Abercrombie - Born: New York - Age: 28 - Sent to Chatham on 29 Mar 1813 on the HMS Braham.

Williams, James - Seaman - Number: 2464 - Prize name: Fair American - Ship type: MV - How taken: HM Frigate Andromache - When taken: 19 Jan 1814 - Where taken: Bay of Biscay - Date received: 22 Feb 1814 - From what ship: HMS York - Born: Garcia, Italy - Age: 30 - Discharged on 10 May 1814 and sent to Dartmoor.

Williams, John - Seaman - Number: 1108 - Prize name: Viper - Ship type: MV - How taken: HM Ship-of-the-Line Superb - When taken: 15 Apr 1813 - Where taken: Bay of Biscay - Date received: 22 Apr 1813 - From what ship: HMS Superb - Born: New Orleans - Age: 22 - Sent to Dartmoor on 1 Jul 1813.

Williams, John - Seaman - Number: 1106 - Prize name: Viper - Ship type: MV - How taken: HM Ship-of-the-Line Superb - When taken: 15 Apr 1813 - Where taken: Bay of Biscay - Date received: 22 Apr 1813 - From what ship: HMS Superb - Born: New Orleans - Age: 21 - Sent to Dartmoor on 1 Jul 1813.

Williams, John - Seaman - Number: 1305 - Prize name: Eliza - Ship type: MV - How taken: HM Frigate Surveillante - When taken: 22 Apr 1813 - Where taken: at sea - Date received: 10 May 1813 - From what ship: HMS Medusa - Born: Rhode Island - Age: 29 - Discharged on 3 Jul 1813 and sent to Stapleton Prison.

Williams, John - Seaman - Number: 1169 - Prize name: Hebe - Ship type: MV - How taken: HM Frigate Stag - When taken: 18 Apr 1813 - Where taken: Bay of Biscay - Date received: 6 May 1813 - From what ship: HMS Stag - Born: Amsterdam - Age: 31 - Sent on 8 Sep 1813 to Dartmoor.

Williams, John - Seaman - Number: 1395 - Prize name: Henry Clements - Ship type: MV - How taken: HM Brig Orestes - When taken: 13 Apr 1813 - Where taken: Bay of Biscay - Date received: 15 May 1813 - From what ship: HMS Orestes - Born: Newburyport - Age: 22 - Race: Black - Discharged on 3 Jul 1813 and sent to Stapleton Prison.

Williams, John - Seaman - Number: 763 - Prize name: William Bayard - Ship type: MV - How taken: HM Ship-of-the-Line Warspite - When taken: 3 Mar 1813 - Where taken: Bay of Biscay - Date received: 19 Mar 1813 - From what ship: HMS Warspite - Born: Baltimore - Age: 25 - Sent to Dartmoor on 28 Jun 1813.

Williams, John - Seaman - Number: 2490 - Prize name: Sherbrook - Ship type: MV - How taken: Impressed at Falmouth - When taken: 20 Mar 1814 - Date received: 26 Mar 1814 - From what ship: HMS Prince Frederick - Born: Massachusetts - Age: 30 - Race: Mulatto - Discharged on 10 May 1814 and sent to Dartmoor.

Williams, John - Seaman - Number: 1998 - Prize name: Ned - Ship type: LM - How taken: HM Brig Royalist - When taken: 6 Sep 1812 - Where taken: coast of France - Date received: 22 Sep 1813 - From what ship: HMS Royalist - Born: Baltimore - Age: 27 - Discharged on 27 Sep 1813 and sent to Dartmoor.

Williams, John - Seaman - Number: 1901 - Prize name: US Brig Argus - Ship type: War - How taken: HM Brig Pelican - When taken: 14 Aug 1813 - Where taken: Irish Channel - Date received: 23 Aug 1813 - From what ship: HMS Pelican - Born: New York - Age: 22 - Race: Black - Discharged on 8 Sep 1813 and sent to Dartmoor.

Williams, John - Seaman - Number: 2132 - Prize name: Amiable - Ship type: LM - How taken: HM Ship-of-the-Line Magnificant - When taken: 30 Oct 1813 - Where taken: off Lorient - Date received: 29 Nov 1813 - From what ship: HMS Dublin - Born: New Jersey - Age: 45 - Discharged on 4 Dec 1813 and sent to Dartmoor.

Williams, John - Seaman - Number: 2475 - Prize name: US Brig Argus - Ship type: War - How taken: HM Brig Pelican - When taken: 14 Aug 1813 - Where taken: Irish Channel - Date received: 28 Feb 1814 - From what ship: Dartmoor prison - Born: Memel - Age: 20 - Discharged on 4 Mar 1814 and sent to Harwich for HMS Rosemond.

Williams, John - Seaman - Number: 2682 - How taken: Sent into custody from HM Ship-of-the-Line Minden - Date received: 5 Jun 1814 - From what ship: HMS Gronville - Born: Virginia - Age: 33 - Race: Black - Discharged on 14 Jun 1814 and sent to Dartmoor.

Williams, John - Seaman - Number: 1883 - Prize name: US Brig Argus - Ship type: War - How taken: HM Brig Pelican - When taken: 14 Aug 1813 - Where taken: Irish Channel - Date received: 23 Aug 1813 - From what ship: HMS Pelican - Born: New York - Age: 20 - Discharged on 8 Sep 1813 and sent to Dartmoor.

Williams, Joseph - Seaman - Number: 1722 - Prize name: Hannah & Eliza - Ship type: MV - How taken: HM Brig Lyra - When taken: 29 May 1813 - Where taken: off north coast of Spain - Date received: 4 Jul 1813 - From what ship: HMS Iris - Born: New London - Age: 19 - Discharged on 22 Aug 1813 and released to HM Brig Redpole.

Williams, Joshua - Chief Mate - Number: 1673 - Prize name: Tom Thumb - Ship type: MV - How taken: Cutter Lion - When taken: 17 Feb 1813 - Where taken: Bay of Biscay - Date received: 23 Jun 1813 - From what ship: Nimble of Guernsey - Born: Nova Scotia - Age: 30 - Discharged on 9 Jul 1813 and sent to Ashburton on parole.

Williams, Peter - Seaman - Number: 427 - Prize name: Union - Ship type: MV - How taken: HM Frigate Iris - When taken: 17 Jan 1813 - Where taken: Lat 44 N Long 2.3 W - Date received: 5 Feb 1813 - From what ship: HMS San Josef - Born: Philadelphia - Age: 24 - Race: Black - Sent to Chatham on 29 Mar 1813 on the HMS Braham.

Williams, Peter - Seaman - Number: 1148 - Prize name: Magdalen - Ship type: MV - How taken: HM Ship-of-the-Line Superb - When taken: 15 Apr 1813 - Where taken: off Belle Isle - Date received: 22 Apr 1813 - From what ship: HMS Superb - Born: New Orleans - Age: 33 - Sent to Dartmoor on 1 Jul 1813.

Williams, R. N. - Seaman - Number: 2160 - Prize name: Princess - Ship type: MV - How taken: Impressed at Liverpool - When taken: 13 Nov 1813 - Date received: 13 Dec 1813 - From what ship: HMS Bittern - Born: New Jersey - Age: 20 - Race: Black - Discharged on 10 Feb 1814 and sent to Dartmoor.

Williams, Thomas - Seaman - Number: 1399 - Prize name: Henry Clements - Ship type: MV - How taken: HM Brig Orestes - When taken: 13 Apr 1813 - Where taken: Bay of Biscay - Date received: 15 May 1813 - From what ship: HMS Orestes - Born: Charlestown - Age: 24 - Discharged on 8 Sep 1813 and sent to Dartmoor.

Williams, Thomas - Seaman - Number: 1235 - Prize name: Essex - Ship type: MV - How taken: HM Frigate Pyramus - When taken: 2 Apr 1813 - Where taken: Bay of Biscay - Date received: 9 May 1813 - From what ship: HMS Andromache - Born: New York - Age: 22 - Race: Black - Discharged on 3 Jul 1813 and sent to Stapleton Prison.

Williams, Thomas - Seaman - Number: 1109 - Prize name: Viper - Ship type: MV - How taken: HM Ship-of-the-Line Superb - When taken: 15 Apr 1813 - Where taken: Bay of Biscay - Date received: 22 Apr 1813 - From what ship: HMS Superb - Born: Connecticut - Age: 23 - Sent to Dartmoor on 1 Jul 1813.

Williams, William - Seaman - Number: 1096 - Prize name: Young Holkar - Ship type: MV - How taken: HM Ship-of-the-Line Superb - When taken: 10 Apr 1813 - Where taken: off Belle Isle - Date received: 22 Apr 1813 - From what ship: HMS Superb - Born: Virginia - Age: 25 - Sent to Dartmoor on 1 Jul 1813.

Williams, William - Seaman - Number: 1969 - How taken: Impressed at Liverpool - When taken: 10 Sep 1813 - Date received: 19 Sep 1813 - From what ship: HMS Bittern - Born: New York - Age: 29 - Discharged on 27 Sep 1813 and sent to Dartmoor.

Williamson, Charles - Seaman - Number: 93 - Prize name: Experiment - Ship type: MV - How taken: HM Brig Rover - When taken: 21 Oct 1812 - Where taken: off Bordeaux - Date received: 25 Dec 1812 - From what ship: HMS Northumberland - Born: Maryland - Age: 27 - Sent to Portsmouth on 29 Dec 1812 on the HMS Northumberland.

Williamson, David - Seaman - Number: 2356 - Prize name: Zephyr - Ship type: MV - How taken: HM Frigate Surveillante - When taken: 6 Jan 1814 - Where taken: Bay of Biscay - Date received: 31 Jan 1814 - From what ship: HMS Surveillante - Born: Philadelphia - Age: 25 - Discharged on 27 Feb 1814 and sent to Chatham on HMS Haleyon.

Williamson, John - Seaman - Number: 1042 - How taken: Delivered himself up from HM Ship-of-the-Line Ajax - Date received: 22 Apr 1813 - From what ship: HMS Ajax - Born: Germantown - Age: 20 - Sent to Chatham on 8 Jul 1813 on HM Tender Neptune.

Williamson, William - Boy - Number: 539 - Prize name: Terrible - Ship type: MV - How taken: HM Brig Foxhound - When taken: 8 Feb 1813 - Where taken: Channel - Date received: 15 Feb 1813 - From what ship: HMS Foxhound - Born: Bridgeport - Age: 15 - Sent to Dartmoor on 2 Apr 1813.

Williams, Elge - Seaman - Number: 1321 - Prize name: Shadow - Ship type: LM - How taken: HM Brig Reindeer & HM Schooner Helicon - When taken: 6 Apr 1813 - Where taken: Bay of Biscay - Date received: 12 May 1813 - From what ship: HMS Helicon - Born: Delaware - Age: 34 - Race: Black - Discharged on 3 Jul 1813 and sent to Stapleton Prison.

Willingsworth, Jeffery - Seaman - Number: 95 - Prize name: Experiment - Ship type: MV - How taken: HM Brig Rover - When taken: 21 Oct 1812 - Where taken: off Bordeaux - Date received: 25 Dec 1812 - From what ship: HMS Northumberland - Born: Lancaster - Age: 40 - Race: Negro - Sent to Portsmouth on 29 Dec 1812 on the HMS Northumberland.

Wills Jr., John - Master - Number: 103 - Prize name: Leader - Ship type: MV - How taken: HM Frigate Briton - When taken: 10 Dec 1812 - Where taken: off Bordeaux - Date received: 28 Dec 1812 - From what ship: MV Leader -

Born: Philadelphia - Age: 28 - Sent on 8 Dec 1812 to Ashburton on parole.

Wills, John - Seaman - Number: 1269 - Prize name: Caroline - Ship type: MV - How taken: HM Frigate Medusa - When taken: 12 Apr 1813 - Where taken: Bay of Biscay - Date received: 10 May 1813 - From what ship: HMS Medusa - Born: Lancaster - Age: 47 - Discharged on 3 Jul 1813 and sent to Stapleton Prison.

Wills, Peter - Seaman - Number: 601 - Prize name: St. Martin's Plantation, prize of the Privateer Paul Jones - Ship type: P - How taken: HM Ship-of-the-Line Dublin - When taken: 9 Feb 1815 - Where taken: Lat 43 N Long 33.5 W - Date received: 25 Feb 1813 - From what ship: HMS Dublin - Born: Norwich - Age: 29 - Race: Black - Sent to Dartmoor on 2 Apr 1813.

Willson, Charles - Seaman - Number: 2578 - How taken: Delivered himself up from the MV Hebrus - Date received: 8 May 1814 - From what ship: MV Hebrus - Born: Rhode Island - Age: 46 - Race: Black - Discharged on 7 May 1814 and sent to Ashburton on parole.

Willson, George - Seaman - Number: 1840 - Prize name: US Brig Argus - Ship type: War - How taken: HM Brig Pelican - When taken: 14 Aug 1813 - Where taken: Irish Channel - Date received: 17 Aug 1813 - From what ship: USS Argus - Born: New York - Age: 14 - Discharged on 8 Sep 1813 and sent to Dartmoor.

Willson, John - Seaman - Number: 1714 - Prize name: Orders in Council - Ship type: LM - How taken: HM Frigate Surveillante - When taken: 1 Jun 1813 - Where taken: Bay of Biscay - Date received: 4 Jul 1813 - From what ship: HMS Iris - Born: Charlestown - Age: 23 - Discharged on 22 Aug 1813 and released to HM Brig Redpole.

Willson, John - Seaman - Number: 1375 - Prize name: Tom - Ship type: LM - How taken: HM Frigate Surveillante - When taken: 27 Apr 1813 - Where taken: Bay of Biscay - Date received: 15 May 1813 - From what ship: HMS Foxhound - Born: New York - Age: 19 - Discharged on 3 Jul 1813 and sent to Stapleton Prison.

Willson, Thomas - Seaman - Number: 2622 - How taken: HM Ship-of-the-Line Edinburgh - When taken: 28 Oct 1812 - Date received: 16 May 1814 - From what ship: HMS Repulse - Born: Baltimore - Age: 35 - Race: Black - Discharged on 14 Jun 1814 and sent to Dartmoor.

Wilson, Charles - Seaman - Number: 2311 - Prize name: Siro - Ship type: LM - How taken: HM Brig Pelican - When taken: 13 Jan 1814 - Where taken: at sea - Date received: 23 Jan 1814 - From what ship: HMS Pelican - Born: Baltimore - Age: 19 - Race: Negro - Discharged on 31 Jan 1814 and sent to Dartmoor.

Wilson, Francis - Seaman - Number: 1011 - Prize name: Amphitrite - Ship type: MV - How taken: HM Ketch Gleaner - When taken: 27 Feb 1813 - Where taken: Bay of Biscay - Date received: 20 Apr 1813 - From what ship: HMS Libria - Born: New York - Age: 24 - Sent to Dartmoor on 1 Jul 1813.

Wilson, George - Captain - Number: 504 - Prize name: Cashiere - Ship type: LM - How taken: HM Brig Reindeer - When taken: 3 Feb 1813 - Where taken: Bay of Biscay - Date received: 15 Feb 1813 - From what ship: Cashier - Born: Wiscasset, MA - Age: 34 - Sent on 18 Feb 1813 to Ashburton on parole.

Wilson, James - Seaman - Number: 1434 - Prize name: Paul Jones - Ship type: P - How taken: HM Frigate Leonidas - When taken: 23 May 1813 - Where taken: off Cape Clear - Date received: 26 May 1813 - From what ship: HMS Leonidas - Born: Philadelphia - Age: 30 - Race: Black - Discharged on 30 Jun 1813 and sent to Stapleton.

Wilson, John - Seaman - Number: 1581 - Prize name: Governor Gerry - Ship type: MV - How taken: HM Brig Royalist - When taken: 31 May 1813 - Where taken: Bay of Biscay - Date received: 12 Jun 1813 - From what ship: Governor Gerry - Born: Wilkes-Barre - Age: 20 - Discharged on 30 Jun 1813 and sent to Stapleton.

Wilson, John - Seaman - Number: 1034 - How taken: Impressed at Liverpool - When taken: 29 Mar 1813 - Date received: 21 Apr 1813 - From what ship: HMS Bittern - Born: Norfolk - Age: 25 - Sent to Dartmoor on 1 Jul 1813.

Wilson, John - Seaman - Number: 1729 - Prize name: Governor Gerry - Ship type: MV - How taken: HM Brig Royalist - When taken: 31 May 1813 - Where taken: Bay of Biscay - Date received: 7 Jul 1813 - From what ship: Escaped and retaken - Born: Wilkes-Barre - Age: 20 - Discharged on 1 Aug 1813 and released to HMS Salvador del Mundo.

Wilson, John - Seaman - Number: 2090 - How taken: Delivered himself up from HM Frigate Fortunee - Date received: 2 Nov 1813 - From what ship: HMS Bacchus - Born: Bath - Age: 22 - Discharged on 3 Nov 1813 and sent to Dartmoor.

Wilson, John - Seaman - Number: 2227 - Prize name: Zephyr - Ship type: MV - How taken: HM Frigate Pyramus - When taken: 30 Nov 1813 - Where taken: off Lorient - Date received: 3 Jan 1814 - From what ship: HMS Warspite

- Born: New York - Age: 22 - Discharged on 31 Jan 1814 and sent to Dartmoor.

Wilson, Peter - Seaman - Number: 59 - Prize name: Independence - Ship type: MV - How taken: HM Frigate Medusa - When taken: 9 Nov 1812 - Where taken: San Sebastian - Date received: 27 Nov 1812 - From what ship: HMS Wasp - Born: New York - Age: 42 - Sent to Portsmouth on 29 Dec 1812 on the HMS Northumberland.

Wilson, Peter - Seaman - Number: 1056 - Prize name: Virginia Planter - Ship type: MV - How taken: HM Frigate Pyramus - When taken: 17 Mar 1813 - Where taken: off Nantes - Date received: 22 Apr 1813 - From what ship: HMS Superb - Born: New York - Age: 25 - Sent to Dartmoor on 1 Jul 1813.

Wilson, Thomas - Unknown - Number: 915 - How taken: Delivered himself up as a prisoner of war - Date received: 4 Apr 1813 - From what ship: HMS Salvador del Mundo - Born: Alexandria - Age: 34 - Sent on 21 Jun 1813 to Portsmouth on HMS Prometheus.

Wilson, William - Seaman - Number: 57 - Prize name: Independence - Ship type: MV - How taken: HM Frigate Medusa - When taken: 9 Nov 1812 - Where taken: San Sebastian - Date received: 27 Nov 1812 - From what ship: HMS Wasp - Born: Richmond - Age: 21 - Sent to Portsmouth on 29 Dec 1812 on the HMS Northumberland.

Wilson, William - Seaman - Number: 1992 - Prize name: Ned - Ship type: LM - How taken: HM Brig Royalist - When taken: 6 Sep 1812 - Where taken: coast of France - Date received: 22 Sep 1813 - From what ship: HMS Royalist - Born: Newport - Age: 23 - Discharged on 27 Sep 1813 and sent to Dartmoor.

Winant, Peter - Seaman - Number: 1280 - Prize name: Price - Ship type: LM - How taken: HM Frigate Medusa - When taken: 13 Apr 1813 - Where taken: Bay of Biscay - Date received: 10 May 1813 - From what ship: HMS Medusa - Born: New York - Age: 24 - Discharged on 21 Jun 1813 and released to American ship Mount Hope.

Wendell, Isaac - Seaman - Number: 834 - Prize name: Pallas - Ship type: MV - How taken: HM Brig Rebuff - When taken: 23 Dec 1812 - Where taken: off Cadiz - Date received: 22 Mar 1813 - From what ship: HMS Dauntless - Born: Boston - Age: 35 - Sent to Dartmoor on 28 Jun 1813.

Windham, John - Seaman - Number: 2011 - Prize name: Ned - Ship type: LM - How taken: HM Brig Royalist - When taken: 6 Sep 1812 - Where taken: coast of France - Date received: 24 Sep 181 - From what ship: HMS Rippon - Born: New York - Age: 27 - Discharged on 27 Sep 1813 and sent to Dartmoor.

Wing, John - Seaman - Number: 1723 - Prize name: Hannah & Eliza - Ship type: MV - How taken: HM Brig Lyra - When taken: 29 May 1813 - Where taken: off north coast of Spain - Date received: 4 Jul 1813 - From what ship: HMS Iris - Born: Sandwich - Age: 24 - Discharged on 22 Aug 1813 and released to HM Brig Redpole.

Wing, Joseph - Seaman - Number: 1020 - Prize name: Two Brothers - Ship type: MV - How taken: Bootle of Liverpool, letter of marque - When taken: 18 Mar 1813 - Where taken: Western Islands - Date received: 21 Apr 1813 - From what ship: HMS Bittern - Born: Boston - Age: 22 - Sent to Dartmoor on 1 Jul 1813.

Winkley, John - 2nd Lieutenant - Number: 340 - Prize name: Portsea, prize to the Privateer Thrasher - Ship type: P - How taken: HM Sloop Helena - When taken: 31 Dec 1813 - Where taken: off the Western Islands - Date received: 22 Jan 1813 - From what ship: HMS Helena - Born: Kittery - Age: 21 - Sent to Chatham on 29 Mar 1813 on the HMS Braham.

Winslow, Elijah - Seaman - Number: 1101 - Prize name: Viper - Ship type: MV - How taken: HM Ship-of-the-Line Superb - When taken: 15 Apr 1813 - Where taken: Bay of Biscay - Date received: 22 Apr 1813 - From what ship: HMS Superb - Born: Massachusetts - Age: 18 - Sent to Dartmoor on 1 Jul 1813.

Winter, Peter - Seaman - Number: 621 - Prize name: Governor McKean - Ship type: LM - How taken: HM Brig Rover - When taken: 26 Jun 1813 - Where taken: off Bordeaux - Date received: 2 Mar 1813 - From what ship: HMS Insolent - Born: Newcastle - Age: 25 - Race: Negro - Sent to Dartmoor on 2 Apr 1813.

Winter, William - Mate - Number: 2094 - Prize name: Fire Fly - Ship type: MV - How taken: HM Frigate Revolutionnaire - When taken: 20 Oct 1813 - Where taken: off Cape Ortagle - Date received: 6 Nov 1813 - From what ship: Fire Fly - Born: Cape Ann - Age: 38 - Discharged on 9 Nov 1813 and sent to Ashburton on parole.

Wintory, Anthony - Seaman - Number: 1113 - Prize name: Viper - Ship type: MV - How taken: HM Ship-of-the-Line Superb - When taken: 15 Apr 1813 - Where taken: Bay of Biscay - Date received: 22 Apr 1813 - From what ship: HMS Superb - Born: New Orleans - Age: 28 - Sent to Dartmoor on 1 Jul 1813.

Witheram, Nicholas - Seaman - Number: 552 - Prize name: Spitfire - Ship type: MV - How taken: HM Brig Achates

- When taken: 14 Feb 1813 - Where taken: off Ushant - Date received: 16 Feb 1813 - From what ship: HMS Achates - Born: Marblehead - Age: 27 - Sent to Dartmoor on 2 Apr 1813.

Witherett, Charles - Seaman - Number: 2324 - Prize name: Amity, prize of the Privateer Prince de Neufchatel - Ship type: P - How taken: HM Brig Achates - When taken: 22 Dec 1814 - Where taken: Bay of Biscay - Date received: 25 Jan 1814 - From what ship: HMS Conflict - Born: Massachusetts - Age: 28 - Discharged on 27 Feb 1814 and sent to Chatham on HMS Haleyon.

Wittington, Joshua - Seaman - Number: 1140 - Prize name: Magdalen - Ship type: MV - How taken: HM Ship-of-the-Line Superb - When taken: 15 Apr 1813 - Where taken: off Belle Isle - Date received: 22 Apr 1813 - From what ship: HMS Superb - Born: Baltimore - Age: 33 - Sent on 21 Jun 1813 to the American ship Hope.

Wolfe, William - Seaman - Number: 1999 - Prize name: Ned - Ship type: LM - How taken: HM Brig Royalist - When taken: 6 Sep 1812 - Where taken: coast of France - Date received: 22 Sep 1813 - From what ship: HMS Royalist - Born: Baltimore - Age: 20 - Discharged on 27 Sep 1813 and sent to Dartmoor.

Wood, John - Seaman - Number: 1535 - Prize name: Courier - Ship type: LM - How taken: HM Brig Rover - When taken: 14 Mar 1813 - Where taken: Bay of Biscay - Date received: 29 May 1813 - From what ship: HMS Hannibal - Born: Virginia - Age: 22 - Discharged on 30 Jun 1813 and sent to Stapleton.

Wood, Peleg - Boy - Number: 804 - Prize name: Cannoniere - Ship type: P - How taken: HM Ship-of-the-Line Warspite - When taken: 14 Mar 1813 - Where taken: Bay of Biscay - Date received: 19 Mar 1813 - From what ship: HMS Warspite - Born: Newport - Age: 15 - Sent on 8 Apr 1813 to Ashburton on parole.

Wood, Sylvanus - Seaman - Number: 1628 - Prize name: Leo - Ship type: LM - How taken: HM Frigate Magiciene - When taken: 4 Jun 1813 - Where taken: Lat 45 Long 14 - Date received: 14 Jun 1813 - From what ship: HMS Orestes - Born: New York - Age: 23 - Discharged on 30 Jun 1813 and sent to Stapleton.

Wood, Thomas - Seaman - Number: 491 - Prize name: Cashiere - Ship type: LM - How taken: HM Brig Reindeer - When taken: 3 Feb 1813 - Where taken: Bay of Biscay - Date received: 12 Feb 1813 - From what ship: HMS Reindeer - Born: Prince Georges County - Age: 21 - Sent to Dartmoor on 2 Apr 1813.

Wood, William - Seaman - Number: 2648 - How taken: Delivered himself up from HM Brig Scylla - Date received: 22 May 1814 - From what ship: HMS Scylla - Born: Philadelphia - Age: 50 - Race: Black - Discharged on 14 Jun 1814 and sent to Dartmoor.

Woodard, Elijah - Seaman - Number: 82 - Prize name: Argus - Ship type: MV - How taken: Fancy, Cutter - When taken: 19 Dec 1812 - Where taken: Bay of Biscay - Date received: 24 Dec 1812 - From what ship: Fancy, Cutter - Born: Massachusetts - Age: 24 - Sent to Portsmouth on 29 Dec 1812 on the HMS Northumberland.

Woodberry, Caleb - Prize Master - Number: 355 - Prize name: Louisa, prize of the Privateer Decatur - Ship type: P - How taken: HM Frigate Andromache - When taken: 11 Jan 1813 - Where taken: off Bordeaux - Date received: 4 Feb 1813 - From what ship: HMS Cornwall - Born: Gloucester - Age: 38 - Sent to Chatham on 29 Mar 1813 on the HMS Braham.

Woodbury, John - Seaman - Number: 1708 - Prize name: Joseph - Ship type: MV - How taken: HM Frigate Iris - When taken: 8 Jun 1813 - Where taken: Bay of Biscay - Date received: 4 Jul 1813 - From what ship: HMS Iris - Born: Gloucester - Age: 24 - Discharged on 22 Aug 1813 and released to HM Brig Redpole.

Woodcraft, John - Prize Master - Number: 1751 - Prize name: Fox Packet, prize of the Privateer Fox - Ship type: P - How taken: Superior, letter of marque - When taken: 25 Jun 1813 - Where taken: Lat 50N Long 21W - Date received: 19 Jul 1813 - From what ship: HMS Bittern - Born: Great Island, Canada - Age: 18 - Discharged on 8 Sep 1813 and sent to Dartmoor.

Woods, Edward - Seaman - Number: 2344 - Prize name: Apparencen, prize of the Privateer Bunker Hill - Ship type: P - How taken: Cartdian - When taken: 27 Jan 1814 - Where taken: off Ushant - Date received: 30 Jan 1814 - From what ship: Cardian - Born: Gloucester - Age: 21 - Discharged on 31 Jan 1814 and sent to Dartmoor.

Woodward, Anthony - Seaman - Number: 588 - Prize name: St. Martin's Plantation, prize of the Privateer Paul Jones - Ship type: P - How taken: HM Ship-of-the-Line Dublin - When taken: 9 Feb 1815 - Where taken: Lat 43 N Long 33.5 W - Date received: 25 Feb 1813 - From what ship: HMS Dublin - Born: Philadelphia - Age: 17 - Sent to Dartmoor on 2 Apr 1813.

Wooldridge, Thomas - Seaman - Number: 549 - Prize name: Spitfire - Ship type: MV - How taken: HM Brig Achates - When taken: 14 Feb 1813 - Where taken: off Ushant - Date received: 16 Feb 1813 - From what ship: HMS Achates - Born: Marblehead - Age: 19 - Sent to Dartmoor on 2 Apr 1813.

Wores, Robert - Seaman - Number: 1334 - Prize name: Fox - Ship type: LM - How taken: HM Sloop Pheasant - When taken: 23 Apr 1813 - Where taken: Bay of Biscay - Date received: 14 May 1813 - From what ship: HMS Pleasant - Born: Philadelphia - Age: 28 - Race: Negro - Discharged on 3 Jul 1813 and sent to Stapleton Prison.

Works, Alford - Seaman - Number: 2659 - Prize name: Indian Lass, prize of the Privateer Grand Turk - Ship type: P - How taken: HM Transport Akbar - When taken: 29 Apr 1814 - Where taken: at sea - Date received: 5 Jun 1814 - From what ship: HMS Gronville - Born: Stafford - Age: 21 - Discharged on 14 Jun 1814 and sent to Dartmoor.

Wright, Amos - Seaman - Number: 2440 - Prize name: Mary, prize of the Privateer Prince de Neufchatel - Ship type: P - How taken: Retaken by the crew of the Mary - Date received: 14 Feb 1814 - From what ship: HMS Halcyon - Born: Philadelphia - Age: 28 - Discharged on 10 May 1814 and sent to Dartmoor.

Wright, George - Seaman - Number: 1252 - Prize name: Caroline - Ship type: MV - How taken: HM Frigate Medusa - When taken: 12 Apr 1813 - Where taken: Bay of Biscay - Date received: 10 May 1813 - From what ship: HMS Medusa - Born: Delaware - Age: 29 - Race: Negro - Discharged on 3 Jul 1813 and sent to Stapleton Prison.

Wright, Isaac - Yeoman - Number: 1038 - How taken: Delivered himself up from HM Ship-of-the-Line Ajax - Date received: 22 Apr 1813 - From what ship: HMS Ajax - Born: Philadelphia - Age: 33 - Sent to Chatham on 8 Jul 1813 on HM Tender Neptune.

Wright, John - Seaman - Number: 1283 - Prize name: Price - Ship type: LM - How taken: HM Frigate Medusa - When taken: 13 Apr 1813 - Where taken: Bay of Biscay - Date received: 10 May 1813 - From what ship: HMS Medusa - Born: New York - Age: 21 - Discharged on 3 Jul 1813 and sent to Stapleton Prison.

Wright, William - Seaman - Number: 2537 - Prize name: Lyon - Ship type: MV - How taken: Brilliant, privateer - When taken: 1 Jan 1814 - Date received: 13 Apr 1814 - From what ship: HMS Bittern - Born: Portland - Age: 23 - Discharged on 10 May 1814 and sent to Dartmoor.

Wright, William - Seaman - Number: 2283 - Prize name: Siro - Ship type: LM - How taken: HM Brig Pelican - When taken: 13 Jan 1814 - Where taken: at sea - Date received: 23 Jan 1814 - From what ship: HMS Pelican - Born: Baltimore - Age: 37 - Discharged on 31 Jan 1814 and sent to Dartmoor.

Wyer, Joseph - Seaman - Number: 2197 - Prize name: General Kempt, prize of the Privateer Grand Turk - Ship type: P - How taken: HM Brig Foxhound - When taken: 18 Dec 1813 - Where taken: Lat 48.4 Long 6 - Date received: 21 Dec 1813 - From what ship: HMS Foxhound - Born: Beverly - Age: 17 - Discharged on 31 Jan 1814 and sent to Dartmoor.

Yale, Nathaniel - Cook - Number: 128 - Prize name: Nope - Ship type: MV - How taken: HM Sloop Pheasant - When taken: 13 Dec 1812 - Where taken: off Oporto - Date received: 2 Jan 1813 - From what ship: Nope - Born: Maryland - Age: 27 - Sent to Portsmouth on 4 Jan 1813 on the HMS Revolutionnaire.

Yard, William - Seaman - Number: 1566 - Prize name: Miranda, prize of the Privateer Paul Jones - Ship type: P - How taken: HM Frigate Unicorn - When taken: 21 Mar 1813 - Where taken: off Ushant - Date received: 9 Jun 1813 - From what ship: HMS Conquestador - Born: Trenton, NJ - Age: 26 - Discharged on 30 Jun 1813 and sent to Stapleton.

Yilstone, George - Seaman - Number: 2120 - Prize name: Chesapeake - Ship type: LM - How taken: HM Frigate Hotspur & HM Frigate Pyramus - When taken: 26 Oct 1813 - Where taken: Bay of Biscay - Date received: 22 Nov 1813 - From what ship: HMS Pyramus - Born: New Jersey - Age: 32 - Discharged on 29 Nov 1813 and sent to Dartmoor.

Young, Abubakar - Chief Mate - Number: 1575 - Prize name: Orders in Council - Ship type: LM - How taken: HM Frigate Surveillante - When taken: 1 Jun 1813 - Where taken: Bay of Biscay - Date received: 12 Jun 1813 - From what ship: Cutter Earl Wellington - Born: Providence - Age: 26 - Discharged on 30 Jun 1813 and sent to Stapleton.

Young, John - Seaman - Number: 1103 - Prize name: Viper - Ship type: MV - How taken: HM Ship-of-the-Line Superb - When taken: 15 Apr 1813 - Where taken: Bay of Biscay - Date received: 22 Apr 1813 - From what ship: HMS Superb - Born: Kent County - Age: 36 - Sent to Dartmoor on 1 Jul 1813.

Young, John - Quartermaster - Number: 1803 - Prize name: US Brig Argus - Ship type: War - How taken: HM Brig Pelican - When taken: 14 Aug 1813 - Where taken: Irish Channel - Date received: 17 Aug 1813 - From what ship: USS Argus - Discharged on 14 June 1813 and sent to Dartmoor.

Young, John - Seaman - Number: 1793 - How taken: Delivered himself up from HM Ship-of-the-Line Tremendous - When taken: 1 Feb 1813 - Date received: 13 Aug 1813 - From what ship: HMS Protection - Born: Albany, NY - Age: 34 - Discharged on 8 Sep 1813 and sent to Dartmoor.

Young, John - Chief Mate - Number: 2478 - How taken: Sent to Mill Prison by the major of Plymouth - Date received: 5 Mar 1814 - From what ship: Plymouth - Born: Virginia - Age: 28 - Discharged on 10 May 1814 and sent to Dartmoor.

Young, Joseph - Seaman - Number: 1755 - Prize name: Friendship, prize to the Privateer America - Ship type: P - How taken: HM Schooner Whiting - When taken: 15 Jul 1815 - Where taken: Lat 47N Long 8W - Date received: 20 Jul 1813 - From what ship: HMS Whiting - Born: Plymouth - Age: 28 - Discharged on 8 Sep 1813 and sent to Dartmoor.

Young, Rufus - Seaman - Number: 1982 - Prize name: Ned - Ship type: LM - How taken: HM Brig Royalist - When taken: 6 Sep 1812 - Where taken: coast of France - Date received: 22 Sep 1813 - From what ship: HMS Royalist - Born: Providence - Age: 18 - Discharged on 27 Sep 1813 and sent to Dartmoor.

Young, Thomas - Seaman - Number: 1892 - Prize name: US Brig Argus - Ship type: War - How taken: HM Brig Pelican - When taken: 14 Aug 1813 - Where taken: Irish Channel - Date received: 23 Aug 1813 - From what ship: HMS Pelican - Born: Providence - Age: 35 - Discharged on 8 Sep 1813 and sent to Dartmoor.

Young, Thomas - Seaman - Number: 2183 - Prize name: Charlotte - Ship type: MV - How taken: Cutter Dwarf - When taken: 4 Nov 1812 - Where taken: off Bordeaux - Date received: 18 Dec 1813 - From what ship: HMS Conquistador - Born: Wilmington, NC - Age: 24 - Discharged on 31 Jan 1814 and sent to Dartmoor.

Young, William - Seaman - Number: 1290 - Prize name: Price - Ship type: LM - How taken: HM Frigate Medusa - When taken: 13 Apr 1813 - Where taken: Bay of Biscay - Date received: 10 May 1813 - From what ship: HMS Medusa - Born: New York - Age: 21 - Discharged on 3 Jul 1813 and sent to Stapleton Prison.

Young, William - Seaman - Number: 115 - Prize name: Otter - Ship type: MV - How taken: HM Ship-Sloop Jalousie - When taken: 1 Dec 1812 - Where taken: off Cape Vincent - Date received: 30 Dec 1812 - From what ship: HMS Leonidas - Born: Richmond - Age: 23 - Sent to Portsmouth on 4 Jan 1813 on the HMS Revolutionnaire.

Young, William - Seaman - Number: 1021 - Prize name: Two Brothers - Ship type: MV - How taken: Bootle of Liverpool, letter of marque - When taken: 18 Mar 1813 - Where taken: Western Islands - Date received: 21 Apr 1813 - From what ship: HMS Bittern - Born: New York - Age: 22 - Sent to Dartmoor on 1 Jul 1813.

Young, William - Seaman - Number: 1884 - Prize name: Betsey, prize to the US Brig Argus - Ship type: War - How taken: HM Frigate Leonidas - When taken: 12 Aug 1813 - Where taken: Channel - Date received: 23 Aug 1813 - From what ship: HMS Pelican - Born: Portsmouth - Age: 38 - Discharged on 27 Sep 1813 and sent to Dartmoor.

Young, William - Seaman - Number: 1795 - How taken: Impressed at Liverpool - When taken: 12 Jul 1813 - Date received: 16 Aug 1813 - From what ship: HMS Bittern - Born: Portsmouth, NH - Age: 32 - Race: Black - Discharged on 8 Sep 1813 and sent to Dartmoor.

Younger, William - Seaman - Number: 2649 - How taken: Delivered himself up from HM Brig Scylla - Date received: 22 May 1814 - From what ship: HMS Scylla - Born: New York - Age: 36 - Discharged on 14 Jun 1814 and sent to Dartmoor.

Numeric listing by prisoner number
October 1812 through June 1814

1	Gardiner, Charles	59	Wilson, Peter	117	Smith, Thomas
2	Bramblecome, David	60	Farrell, John	118	Carr, Laurence
3	Pearson, Samuel	61	Lewis, John	119	Hitchi, John
4	Knapp, Walker	62	Garrison, John	120	Whitmore, William
5	Walker, James	63	Van Donveer, Peter	121	Coleman, John
6	Codsifershall, Charles	64	Manuel, John	122	Humphries, Thomas
7	Turner, Samuel	65	Francois, James	123	Dixon, Michael
8	Dasing, Caesar	66	Davis, John	124	Douglas, Thomas
9	Latish, Joseph	67	Harris, George	125	Castor, Charles
10	Trowbridge, Bela	68	Brown, John	126	Cobb, Samuel
11	Scott, James	69	Carr, Richard	127	Oleily, Thomas
12	Evans, John	70	Kerban, Thomas	128	Yale, Nathaniel
13	Wicks, Littleton	71	Tolson, Jeremy	129	Michael, Peter
14	Beckett, William	72	Johnson, Robert	130	Labery, Pierre Joseph
15	Osbourne, Peter	73	Murray, John	131	Gadet, Martin
16	Johnstone, Henry	74	Stowell, John	132	Evans, Thomas L.
17	Scannell, Cornell	75	Richard, John	133	Thompson, William
18	Bateman, Charles	76	Mesey, William	134	Dominic, John
19	West, John	77	Ducassau, Charles A.	135	Thomas, William
20	Higgins, William	78	Mills, John	136	Peter, John
21	Watkins, Fredrick	79	Packard, William	137	Glenn, Robert
22	Williams, Alexander	80	Smith, Jacob	138	Kennedy, John
23	Stranton, John	81	Post, Anthony Oswald	139	Walden, James
24	Pattingale, Enoch	82	Woodard, Elijah	140	Upton, Jeduthun
25	Brightman, Joseph	83	Howey, Artemas	141	Wain, Benjamin
26	Gifford, Barry	84	Green, Samuel	142	Hatches, Levi
27	White, Alden	85	Gibbons, Andrew	143	Leach, Charles
28	Underwood, John F.	86	Arthur, Alexander	144	Obrion, Jeuth.
29	Williams, George (1)	87	Joseph, Michael	145	Carter, Moses
30	Williams, George (2)	88	Ryder, Philip	146	Hutchinson, Henry P.
31	Allen, Elihu	89	Everett, Francois	147	Hall, William
32	Truman, Alhine	90	Cobb, Samuel	148	Scribner, Elijah
33	Hitch, Ebenezer	91	Fate, Thomas	149	Poland, David
34	Neil, James	92	Anthony, John	150	White, Sampson
35	Green, John	93	Williamson, Charles	151	Barchman, John
36	McKinzey, John	94	Benner, Lewis	152	Upton, Samuel
37	Cook, Joseph B.	95	Willingsworth, Jeffery	153	Bickford, Ebenezer
38	Peterson, John	96	Ellis, John	154	Leveatt, William
39	Jane, Joseph	97	Lynch, William	155	Dean, Daniel
40	Joseph, John	98	Jewell, Samuel	156	Holden, Charles
41	Hazard, Thomas	99	Harris, James	157	Rowell, James
42	Manuel, Joseph	100	Dunn, Hezekiah	158	Cogswell, Edward
43	Hathaway, William N.	101	Hilliard, Robert B.	159	Warner, Charles
44	Slayter, John	102	Drybourgh, James	160	Hosmer, Joseph
45	Riley, William	103	Wills Jr., John	161	Fletcher, Henry
46	Pearce, Edward	104	Hartford, James	162	Peterson, James
47	Cunningham, John	105	Dawson, John	163	Card, Nathaniel
48	Floyd, James	106	Brown, Thomas	164	Gardner, Samuel
49	Pardoe, John	107	Gisby, John	165	Cloutman, Samuel
50	Lendison, Henry	108	Brown, John	166	Nye, Charles N.
51	Conway, Samuel	109	Terry, Daniel	167	Schroder, Henrick
52	Bears, Moses	110	Shepherd, Daniel	168	Moor, Thomas
53	Stowell, Peter	111	Bentsten, John	169	Grunlief, Timothy
54	McFaul, James	112	Howland, Samuel	170	Parsons, Samuel
55	Beachman, George	113	Wade, Otis	171	Jenkins, Richard
56	James, Peter	114	Hook, Aaron	172	Anthony, Abraham
57	Wilson, William	115	Young, William	173	Hammet, John
58	Fountain, Isaac	116	Evans, Thomas	174	Huma, John Lewis

175	Ross, George	237	Morris, William	298	Skiddy, William
176	Knox, Thomas	238	Scot, Benjamin	299	Harris, William
177	Peterson, Nicholas	239	Turner, Benjamin	300	Gardner, Anthony
178	Hobert, George	240	Gibby, John	301	Benny, Malloc
179	Signard, Samuel Francis	241	Moss, William	302	Bailey, Joseph
180	Warner, George	242	Martin, Ephraim	303	Graham, George
181	Whipple, Samuel	243	Coffin, Frederick Henry	304	Dickson, Richard
182	Levis, Raymond	244	Davis, John	305	Robinson, James
183	Dolores, John	245	Gadson, John	306	Quan, George
184	Carves, John	246	Haywood, Simon	307	Porgan, Theodore
185	Lee, Nathaniel	247	Marchley, William	308	Marshall, John
186	Boyd, Edward	248	Ranter, Thomas	309	Swinny, Edward
187	Ingalls, Edmond	249	Dowling, Anthony	310	Thompson, Owen
188	Foster, Asa	250	Golfin, John	311	Deshell, Alexander
189	Willcox, Caesar	251	Arnold, James	312	Jones, John
190	Thomas, George	252	Bribion, Madam	313	Fuller, Zachariah
191	Glasgow, John		(first name not listed)	314	Spryon, John
192	Mitchell, John	253	Bribion - 1st child	315	Can, Moses
193	Fannol, Augustus	254	Bribion - 2nd child	316	Star, G. C.
194	Carrol, Robert	255	Bribion - 3rd child	317	Cochran, Peter
195	Robins, Jeremiah	256	Lyons, Henry	318	Henry, John
196	Anderson, James	257	Anderson, William	319	Prutty, Henry
197	Neal, Henry	258	Pigott, James	320	Lyons, Peter
198	Peters, Jacob	259	Ramblett, Charles	321	Bunker, Peter
199	Towns, Daniel	260	Nartigue, John	322	Thompson, Samuel
200	Child, Samuel	261	Scott, Andrew	323	Williams, James
201	Rowe, Stephen	262	Rich, Francis	324	Myers, Frederick
202	Michell, James	263	Gomez, Manuel	325	Duffy, Nathaniel
203	Tucker, Edward	264	Perez, Joseph	326	Melville, John
204	Monett, Samuel	265	Slate, Henry	327	Kinley, Joseph
205	Brown, John	266	Lewis, John	328	Parr, James
206	Brown, Jesse	267	Parker, Samuel	329	Carver, Abraham
207	Bowdler, Thomas	268	Turpin, Francis	330	Evans, Hale
208	Bunoth, Mansfield	269	Johns, Bellona	331	Harris, William
209	Main, Henry	270	Coperris, Nicholas	332	Lyons, Samuel
210	Sodoburgh, John	271	Roy, Charles	333	Abraham, William
211	Henry, Edward	272	Capron, William	334	Daniels, J. D.
212	Johnstone, Samuel	273	Calkings, Zero	335	Stubbs, James
213	Thomas, Alexander	274	Bell, George	336	Gould, Thomas
214	Dowdell, George R.	275	Hatch, Abraham	337	Hurtall, Peter
215	Adams, Theophilus	276	Brewer, James	338	Coffin, Alexander
216	Bowen, Lewis	277	Keith, James	339	Kneeland, Ebenezer
217	Lynch, Joseph	278	Anderson, William	340	Winkley, John
218	Couet, John	279	Gardner, James	341	Davids, Thomas
219	Simpson, William	280	Richardson, Daniel	342	Stevens, Joseph
220	Taylor, John	281	Roach, Nicholas	343	Mains, John
221	Birnent, Edward	282	Cousor, Adam	344	Warren, John
222	Aulajo, Thomas	283	Neaver, Thomas	345	Moneys, Robert
223	Green, Horace	284	Tightham, Peter	346	Paddock, Benjamin M.
224	Whitney, Joseph	285	Henderson, David	347	Ross, Richard
225	Cochran, Stephen	286	Dickenson, Francis	348	Bagot, Lewis
226	Bevers, Clement	287	Douchney, Hiram	349	Richardson, John
227	Robinson, Michael	288	Hartford, John	350	Clark, Elisha
228	Law, John	289	Garthon, Willey	351	Kingley, Benjamin
229	Smith, Chester	290	Kain, Peter	352	West, Nathan
230	Harper, Robert	291	Canada, Prince	353	Willett, Robert
231	Menard, Nicholas	292	Barton, Nathan	354	Dempsey, John
232	Buchanan, James	293	Robinson, Benjamin	355	Woodberry, Caleb
233	Warren, Nathan	294	Belford, Isaac	356	Jones, John
234	Peterson, John	295	Todd, William	357	Boss, Thomas
235	Brandy, Francis	296	Helm, Charles	358	Jenin, Samuel
236	Carson, Robert	297	Jackson, John	359	Chestly, Amos

360	Cox Miles	422	Morris, Thomas	484	Smith, Richard
361	Huntress, Robert	423	Graham, William	485	Spear, Samuel
362	Brown, Wheeler	424	Finch, William	486	Mullan, John
363	Stanley, Joseph	425	Robinson, David	487	Deal, William
364	Thompson, James	426	Parker, David	488	Foster, Joseph
365	Frees, James	427	Williams, Peter	489	Halfpenny, Robert
366	Rape, Nicholas	428	Mainers, Benjamin	490	Schoeman, Frederick
367	Kennedy, Peter	429	Irwin, Magnus	491	Wood, Thomas
368	Albert, Hezekiah	430	Douglas, Thomas	492	Galloway, Joseph
369	Cleaveland, Davis	431	Cutis, George	493	Birch, Andrew
370	Dow, John	432	Length, Francis	494	Jentle, Joseph
371	Wheeler, William	433	Knight, Isaac D.	495	Ricco, Joachim
372	Tipton, Solomon	434	Burch, James	496	Blackstone, Edward
373	Moore, Benjamin	435	Chapel, William	497	Gilbert, Thomas
374	Thompson, Courtney	436	Jordon, David	498	Drake, Henry
375	Greenleaf, Thomas	437	De Colville, Laur	499	Hoage, Rufus
376	Johnston, James	438	Gibb, James	500	Evans, Henry
377	Lewis, George	439	Dempsey, William	501	Morell, Thomas
378	Johnson, Jacob	440	Ross, John	502	Kimmins, John
379	Larrabee, William	441	Gronard, Peter	503	Madden, Fredrick
380	Sweetman, Samuel	442	Rice, Thomas	504	Wilson, George
381	Jackson, Thomas	443	Rice, John	505	Stiles, William
382	Webb, John	444	Johnston, Edward	506	Hudson, William
383	Ashton, Joseph	445	Daly, John	507	Bartell, William
384	Kirby, Benjamin	446	Ruliff, London	508	Snow, Joseph
385	Sparks, Henry	447	Lawrence, David	509	West, Reuben
386	Chekes, William	448	Lapavous, Philip	510	Newell, John
387	Denckle, Christian	449	Fryer, John	511	Atkins, Uriah
388	Barradaile, Thomas	450	Fisher, Richard D.	512	Hubbard, John
389	Legrand, Claudius Francis	451	Richardson, Samuel	513	Leeds, Leonard
390	Martin, John	452	Myer, John	514	Blanchard, Craven
391	Goetz, Lewis	453	Carney, Edward	515	Hough, Ebenezer
392	Seimandy, Gasper	454	Low, Thomas	516	Oden, John
393	Fairweather, Robert	455	Pannell, Hugh	517	Warner, Henry
394	Underwood, Benjamin	456	Carman, Francis	518	Berry, Brook
395	Johnston, Samuel	457	Spencer, Samuel	519	Conway, William
396	Crofts, William	458	Hamilton, John	520	Higgins, George
397	McElroy, William	459	Smasher, Allen	521	Mires, John
398	Hassam, John	460	Coffin, George	522	Glover, John
399	Lownsburg, Carpenter	461	Nargney, James	523	Hovey, Joseph
400	Mercer, Chaumont	462	Thomas, Henry	524	Foster, Joseph
401	Johnston, William	463	James John	525	Van Kirk, Joseph
402	Harrison, John	464	Chew, Joseph	526	Peterson, Hance
403	Elliot, Francis	465	Lovering, William	527	Barnes, William
404	Reeves, Joseph	466	Robinson, John	528	Barnes, Robert
405	Allen, Jacob	467	-- Missing Name --	529	Lovet, John
406	Gilligan, William	468	Olmstead, Moss	530	Smith, John
407	McDonald, John	469	Grisse, William	531	Martin, Peter
408	Webb, William	470	Barnard, Timothy	532	Kromhout, Barney
409	Morris, John	471	Allen, Eleazer	533	Root, Eliakim F.
410	Schyder, Jacob Knapp	472	Cowell, Slater	534	Smith, Eldridge
411	Thompson, William	473	Ramsdell, John	535	Tufts, Zachariah
412	Swanston, Jacob	474	Forbes, James	536	Leach, William
413	Carnes, Richard	475	Reid, Joseph	537	Smith, William
414	Seibold, John	476	Jackson, Thomas	538	Cook, James
415	Jones, John	477	Waters, Philip	539	Williamson, William
416	Hill, William	478	Scott, Ezekiel	540	Freeman, Prince
417	Armstrong, Nicholas	479	Gallo, Joseph	541	Lampson, Charles
418	Stone, William	480	Crowell, Uriel	542	Carton, Thomas
419	Randal, Frederick	481	Tonver, Michael	543	Bishop, William
421	Godshall, John	482	Dixey, John	544	Wheeler, Richard
421	Mason, Francis	483	Grubb, James Julian	545	Bartlett, Thomas

546	Jones, Francis	608	Parsons, Ignatius	670	Rock, Oliver
547	Jones, Francis Vickery	609	Grey, Thomas	671	Allen, John
548	Jervis, John	610	Nones, A. B.	672	Check, Stephen
549	Wooldridge, Thomas	611	Sanders, Joseph	673	Dingell, George
550	Dodd, Samuel	612	Buffs, James	674	Ripley, William
551	Bridge, Francis	613	Baptist, John	675	Eddy, John
552	Witheram, Nicholas	614	Likens, Andrew	676	Gilbert, Thomas
553	Jutt, Benjamin	615	Capewell, Bartholomew	677	Flood, David
554	Dolliver, Francis	616	Regens, Jonathan	678	Allen, William
555	Lovett, William	617	Hook, John	679	Rosset, Samuel
556	Wadden, Jacob	618	Farrell, Francis	680	Peterson, Alexander
557	Fletcher, William B.	619	Hayes, Simon	681	Rhodrick, Joseph
558	Lewis, Angus T.	620	White, Philip	682	Thomas, William
559	Cole, Richard	621	Winter, Peter	683	Blue, Peter
560	Burnham, Francis A.	622	Humphries, James	684	Crete, John Nicholas
561	Swett, Samuel	623	Augustus, Amos	685	Smith, Charles
562	Wild, Caleb	624	Davis, Benjamin S.	686	Reed, Hartwell
563	Miller, Henry	625	Sargant, Samuel Allen	687	Holts, Daniel
564	Williams, Ambrose	626	Millet, James	688	Mallery, Michael
565	Johnson, John	627	Andrews, Thomas	689	Hughes, John
566	Morris, John	628	Dobbins, John	690	Right, Joshua
567	Ryan, William	629	Dye, William	691	Ervin, William
568	Ireland, James	630	Bronston, Job Ellis	692	Dennison, Jedidiah
569	Armstrong, William	631	Johnson, John	693	Duston, Peter
570	Dwight, Samuel	632	Thomas, Archibald	694	Romain, Samuel
571	Rotner, James	633	Miller, John	695	Jolyon, Robert
572	Reeves, James	634	Thompson, Martin	696	Johnston, Gershom
573	Augustin, Anthony	635	Royal, John	697	Thompson, William
574	Johnson, Edward	636	Marble, Jabez	698	Gordon, Thomas
575	Lassalle, Francisco	637	Griffiths, Thomas	699	Richman, Joshua
576	Colcocha, Anthony	638	Dickson, Charles	700	Isaac, Moses
577	McCauley, George M.	639	Chambers, Henry	701	Clerk, William
578	Bloom, Joseph	640	Neel, David	702	Borgin, Gabriel
579	Longworthy, John	641	Schultz, John	703	Vaughan, Thomas
580	Harris, John	642	Manatt, Richard	704	Sullivan, Hampton
581	Ingbriton, Nicholas	643	Barron, William	705	Frash, James
582	Bannister, George	644	Bedson, Thomas	706	Clements, John C.
583	Cooper, Daniel	645	Sage, Moses	707	Joseph, Francis
584	Pack, Abraham	646	DeWitt, John	708	Washburn, Edward
585	Shammton, Samuel	647	Reeves, William	709	Louring, Henry
586	Jones, William P.	648	Brown, Henry	710	Valentine, John
587	Grey, John	649	Ludlow, Charles	711	Mix, William A.
588	Woodward, Anthony	650	Brereton, Benjamin	712	Bisley, Horace
589	Ellis, William	651	Cartwright, Alexander J.	713	Cox, John
590	Covell, Isaac	652	Harry, John	714	Williams, Henry
591	Allen, John L.	653	Fawcett, William	715	Lewis, John
592	Mains, Darius	654	Fossender, William	716	Johnston, Thomas
593	Russell, Isaac	655	McClelland, William	717	Wheeler, Charles
594	Hulet, Michael	656	Statman, James	718	Monks, John
595	Thompson, Whitney	657	Mathews, Joseph	719	Porter, William
596	Merritt, Almond	658	Foster, Cato	720	White, Henry
597	Taylor, George	659	Pierce, Amos	721	Henderson, Alexander
598	Brown, John	660	Coleman, Charles	722	Lawson, James
599	Gilbert, John	661	Johnson, William	723	Jones, Benjamin
600	Morrison, David	662	Ray, Nathaniel	724	Parker, George
601	Wills, Peter	663	Morris, Patterson	725	Barret, Anthony
602	Spofford, Jacob	664	Brown, John	726	Baker, John
603	Marshall, David	665	Starbuck, Thaddeus B.	727	Caldwell, James
604	Athroun, Samuel	666	Whitehouse, Asley	728	Whitewood, Charles
605	Ostin, John	667	Smith, John	729	Allen, Henry
606	Anderson, Robert	668	Brown, George	730	Saunders, William
607	Evans, James	669	Fox, Washington	731	Smith, William

732	Ware, William	794	Colville, John	856	Summers, Alexander
733	Freddle, John	795	Gault, William	857	Malbony, William T.
734	Lawson, Thomas	796	Stoddart, Robert	858	Mather, Allyn
735	Gellens, William	797	Divers, Charles	859	Hazard, William
736	Court, Robert	798	Watts, Herman	860	Lellifer, Jacob
737	Null, William	799	Neath, Henry	861	Hale, Horace
738	Bellinger, William	800	Criger, Namon	863	Davis, Henry
739	Stoddard, Reuben	801	LeRoy, Alexander	863	Doolittle, Isaac
740	Norman, Peter	802	Richardson, William	864	Everly, John
741	Brown, William	803	Wallander, Adam	865	Crandall, John
742	Holms, Andres	804	Wood, Peleg	866	Glass, Reuben
743	Lammers, Henry	805	Jackson, John W.	867	Watts, James
744	Forbes, William	806	Philen, Richard	868	Blake, Philip
745	Christy, Alexander	807	Weaphor, Andrew	869	Mann, John
746	Davis, Charles	808	Bennet, James	870	Folger, Shubert
747	Goff, James	809	Norton, George	871	Sherman, Reuben
748	Pike, John	810	Lemercier, Peter	872	Brown, Jesse
749	Taylor, Jacob	811	Robinson, Samuel	873	Easterly, George
750	Mitchell, Ezekiel	812	Brownell, Richard	874	Goodday, James
751	Akins, William	813	Hurd, William	875	Clark, Joseph Wanton
752	Williams, George	814	West, John	876	Bartell, Robert
753	Cerk, Frederick	815	Haydon, Eli	877	Conklin, William
754	Blanchard, George	816	Brown, Benjamin	878	Sproson, James
755	Roe, John	817	Garret, James	879	Westerbest, William
756	Averell, Loring	818	Brown, Lodwick	880	Spencer, Leonard
757	Wither, John	819	Vaughan, Nathaniel	881	Calder, John H.
758	Barton, Robert	820	Brown, Samuel	882	Churchill, Timothy
759	Lewis, John	821	Chauvet, Frederic George	883	Sutton, John
760	Senet, John	822	Audibert, Joseph	884	Morslender, Reuben
761	Jones, Benjamin	823	Evans, Jacob	885	Quackenbush, William
762	Gerling, George C.	824	Cooper, James	886	Kershon, Abraham
763	Williams, John	825	Ferret, George	887	Mars, George
764	Bunker, James	826	Burdge, Samuel	888	Hill, Joseph
765	Jackson, John	827	McCormick, Simon	889	Ludlow, Reuben
766	Sherwood, John	828	Williams, David	890	Ryer, Peter
767	Kellum, John C.	829	Henshaw, Jacob S.	891	Caldwell, Samuel
768	Hoskins, James	830	Hudleback, George	892	Bartholf, Nicholas
769	Hall, James	831	Dolinson, Andrew	893	Small, George D.
770	Francis, Benjamin	832	Caesar, Joseph	894	Nicholas, John
771	Steel, John	833	Robinson, Ebenezer	895	Sampson, Jacob
772	Leach, William	834	Wendell, Isaac	896	Hill, Ephraim
773	Riley, Michael	835	Ratoon, Thomas	897	Jones, James
774	Harens, William	836	Colfax, William	898	Davis, Robert
775	Davis, James	837	Edgerly, George	899	Farrell, Jeffery
776	Dorrell, John	838	Robinson, William	900	Archer, Joseph
777	Rollands, Joseph	839	Burton, John	901	Donaldson, Joseph
778	Jones, John	840	Reardon, Anthony	902	Bishop, Edward
779	Richardson, Francis	841	Brown, James	903	Hubbard, Alfred
780	Elm, John	842	Douarty, Angelo	904	Cole, William
781	Jennings, Nathaniel	843	Smith, George	905	Thompson, Nathaniel
782	Coffin, Samuel	844	Hockman, William	906	Lopaus, William
783	Landerman, Joseph	845	Edward, Prince	907	Bartis, John
784	Blacklee, Barnard	846	Moor, Lawrence	908	Chase, Nathaniel
785	Grant, Peter	847	Lewis, Edward	909	Atwood, Thomas
786	Greenwood, Thales	848	Hawking, John	910	Bucher, William Palmer
787	Eaton, George	849	Stow, Jeremiah	911	Copland, James
788	Hock, J. Nicholas	850	Ferrald, John	912	Pain, James
789	Jones, William	851	Meigs, John	913	Miller, William
790	Turner, William	852	Davis, Ezra	914	Sillock, Amos
791	Holmes, Caleb K.	853	Cleveland, Lawrence	915	Wilson, Thomas
792	Erlstroom, John	854	Thurtleff, William	916	Taylor, James
793	Palmer, John	855	Croft, George	917	Coffin, William P.

918	Coffin, George P.	980	Thompson, Robert	1042	Williamson, John
919	Waterman, John	981	Enderson, James	1043	Warner, William
920	Morris, James	982	Sims, Oliver	1044	Nelson, Richard
921	Beecher, Thaddeus	983	Bowman, Benjamin	1045	Hayes, Elias Warner
922	Smith, Henry	984	Collison, Joseph	1046	Bonie, James
923	De Forest, John H.	985	Ruddick, William	1047	Griffin, William
924	Allen, John	986	Peterson, Peter	1048	Howell, Thomas
925	Lemmon, Henry	987	Jordan, Peter	1049	Robinson, Joseph
926	Gibbs, Daniel	988	Garthy, James	1050	Hodgins, Daniel
927	Hull, Edward	989	Warrantee, John	1051	Harvey, Joseph
928	Voorhies, James	990	Crawford, Nelson	1052	Marsh, Jesse
929	Ashfield, Henry	991	Jones, Richard	1053	Vandine, Henry
930	Chapel, John	992	Anderson, James	1054	Rice, Francis
931	King, John	993	Wallace, James	1055	Irving, William
932	Blackman, Moses	994	Bertine, John	1056	Wilson, Peter
933	Dunn, David	995	Adams, Samuel Reed	1057	Andrews, Charles
934	Bailey, John	996	Harris, William	1058	Combs, William
935	Vangosbet, Cato	997	Rogers, Nathaniel	1059	Harding, John
936	Jiffs, Joseph	998	Vicary, Richard	1060	Harter, Henry
937	Brill, John	999	Bryant, James	1061	Dougherty, Michael
938	Macomb, James	1000	Goss, Joshua	1062	Bundaberg, Nicholas
939	Longfield, Thomas	1001	Fairchild, Robert	1063	Latimore, Matthew
940	Brevoort, William	1002	Handel, Henry	1064	Forester, Francis
941	Cortu, John	1003	Dunchller, Isaac	1065	Perney, William
942	Layton, William	1004	Casey, Henry	1067	Soderstrum, Ola
943	Gifford, Thomas	1005	Alley, Samuel	1068	Kennedy, John
944	Adgate, William	1006	Keen, Nathaniel	1069	Hill, Leonard
945	Hogabets, John	1007	Dover, Charles	1070	Snate, George
946	Swain, Darius	1008	McKenzie, Alexander	1071	Burn, Reuben
947	Wessel, Samuel	1009	Hunt, David	1072	Perne, George
948	Morris, John	1010	Shairs, Samuel	1073	Anderson, Samuel
949	Moss, John	1011	Wilson, Francis	1074	Fayola, Eugene
950	Johnson, Samuel	1012	Smith, William	1075	Laquerene, Peter Louis
951	Morris, George	1013	Davison, Henry	1076	Baas, John Thomas
952	Morse, Henry	1014	Summerville, Charles	1077	Carnes, Joseph
953	Dowell, Isaac	1015	Eldridge, Nathaniel	1078	Joseph, Thomas
954	Hall, David	1016	Delaney, Matthew	1079	Miller, Stephen
955	Hall, Richard	1017	Rich, William	1080	Griffin, John
956	Powers, James	1018	Mazeb, James	1081	Rogers, James
957	Towson, William	1019	Stanton, Henry	1082	Mumford, Thomas
958	Law, William	1020	Wing, Joseph	1083	Cammon, Robert
959	Kalm, John	1021	Young, William	1084	Higgins, James
960	Trusty, Henry	1022	Leas, Anthony	1085	Howard, Samuel
961	Shaw, Jacob	1023	Sheppard, William	1086	Montierne, Francois
962	Bailey, Samuel	1024	Paulin, Nicholas	1087	Russell, John
963	Holts, Peter	1025	Atkinson, William	1088	Blackwood, Bailey
964	Courtney, James	1026	Thomas, John	1089	Sturges, Bradley
965	Lee, George	1027	Bridges, John	1090	Le Baith, John J.
966	Myers, David	1028	Deer, Andrew	1091	Burr, Francis L.
967	Deham, Charles	1029	Pickering, George	1092	James, Thomas
968	Jones, William	1030	Jackson, Daniel	1093	Blake, Alexander
969	Deal, John	1031	Venison, Leven	1094	West, Benjamin
970	Crandall, John	1032	Giggord, Robert	1095	Toby, Peter
971	Aural, Leonard	1033	Bagley, William	1096	Williams, William
972	Freeman, Charles	1034	Wilson, John	1097	Thomas, John
973	Caldwell, James	1035	Ervin, Ely	1098	Cowen, Robert
974	Gilbert, George	1036	Bristol, Nehemiah	1099	Way, John
975	Charles, Paul	1037	James, Sacket	1100	Morgan, Joseph
976	Bradford, Henry	1038	Wright, Isaac	1101	Winslow, Elijah
977	Rerette, John Francis	1039	Jackson, Daniel	1102	Thompson, William
978	Moffat, John	1040	Scott, John	1103	Young, John
979	Reynolds, Frederick	1041	Kennedy, William	1104	Allen, Thomas

1105	Foster, David	1167	Mode, David	1229	Welch, Benjamin
1106	Williams, John	1168	Thompson, William	1230	Blodget, Caleb
1107	Tyren, William	1169	Williams, John	1231	Ripley, Eden
1108	Williams, John	1170	Hanson, Christian	1232	Dilus, Benjamin
1109	Williams, Thomas	1171	Brown, William	1233	Chandler, Simon
1110	Smally, Thomas	1172	Page, Thomas	1234	Parker, John A.
1111	Hinkley, Aaron	1173	Fish, James	1235	Williams, Thomas
1112	Churchill, Stephen	1174	Cannoway, Charles	1236	Pomeril, Robert
1113	Wintory, Anthony	1175	Chisselsine, John	1237	Shovel, John
1114	Farrell, John	1176	Denham, John	1238	Davis, James
1115	Webster, James	1177	Watson, James	1239	Wait, Henry
1116	McNelly, Thomas	1178	Jones, Peter	1240	Moss, Thomas
1117	Baker, John	1179	Bansal, Lewis	1241	Stockman, William B.
1118	Thomas, William	1180	White, William (1)	1242	Burnham, Enoch
1119	Bradley, Hugh	1181	White, William (2)	1243	Libby, Moses
1120	Hunter, George	1182	Jones, William	1244	Waters, Abraham
1121	Duncan, Peter	1183	Roach, Reuben	1245	Lowe, John
1122	Arnold, Alfred	1184	Patton, Robert	1246	Mathews, Richard
1123	Robinson, Charles	1185	Farrell, Andrew	1247	Gorham, John
1124	Small, Richard	1186	Cosevin, Pierre Mathrie	1248	Morris, Robert
1125	Paine, Joseph	1187	Dunlap, James Alexander	1249	Southcombie, Plummer
1126	Hay, Cornelius	1188	Vezin, Charles	1250	Spafford, Samuel
1127	Jones, George	1189	Thomas, Andre	1251	Jones, Thomas
1128	Crouch, Richard	1190	Le Graet, Peter	1252	Wright, George
1129	Campbell, John	1191	Delbos, Felix	1253	Plumere, Joseph
1130	Belfast, Richard	1192	Phillips, John	1254	Edwards, David
1131	Barkman, Henry	1193	Manson, Jeremiah	1255	Martin, Anthony
1132	Morrey, Henry	1194	Harrington, Simon	1256	Anderson, David
1133	Leverage, William	1195	Jackson, John	1257	Fritts, Joseph
1134	Foster, Thomas	1196	Cheney, Amos	1258	Wester, Andrew
1135	Harre, John	1197	Payer, Walter	1259	Williams, James
1136	Kennedy, John	1198	Mills, William	1260	Evans, Moses
1137	Holmes, James	1199	Garcia, Francois	1261	Beriston, Peter
1138	Lewis, Francis B.	1200	Jones, Charles	1262	West, Dennis
1139	Jackson, Thomas	1201	Minugh, John	1263	Burns, Charles
1140	Wittington, Joshua	1202	Mercer, Benjamin	1264	Lambert, Joseph
1141	Norris, Lament	1203	Brooks, Hayden T.	1265	Mitchel, Jacob
1142	Adams, John	1204	Walker, Francis	1266	Peregrine, Taggert
1143	Reynolds, Owen	1205	Aris, James	1267	Johnson, James
1144	Harts, William	1206	Lyon, Charles	1268	Metley, Thomas
1145	Prince, Benjamin	1207	Lawrence, John	1269	Wills, John
1146	Johnston, John	1208	Fay, Salmon	1270	Broadwater, Samuel
1147	Johnston, Peter	1209	Membelly, William	1271	Ingle, John
1148	Williams, Peter	1210	Brant, Thomas	1272	Saff, Francis
1149	Fairchild, Hamlet	1211	Phipps, Stephen	1273	Morgan, John
1150	Howard, John	1212	Davis, William	1274	Simpson, Thomas
1151	Fitzpatrick, William	1213	Stag, John B.	1275	Albertson, John N.
1152	Wassen, Joseph Bartram	1214	Weellid, William	1276	Murry, James
1153	Ryan, Thomas	1215	Pomp, William	1277	Hile, Shadrick
1154	Eastland, James	1216	Hanford, William	1278	Murray, John
1155	Bancroft, Samuel	1217	Barker, Reuben	1279	Wells, John
1156	Treffy, James	1218	Morgan, Henry	1280	Winant, Peter
1157	Dolliver, William	1219	Bradford, Charles	1281	Barrett, James
1158	Treffy, Peter	1220	Everill, Daniel	1282	Jones, Gardner
1159	Clothey, Thomas	1221	Sailes, James	1283	Wright, John
1160	Merrett, Thomas	1222	Gray, Morehouse	1284	Thompson, Charles
1161	Russell, Patten	1223	Vail, Peter	1285	Bennett, Charles
1162	Fisher, Lewis	1224	Pratt, Benjamin	1286	May, Walter
1163	Main, William	1225	Hopson, Ebenezer	1287	Robertson, James
1164	Elwell, Thomas	1226	Clapp, Abraham	1288	Dillin, Pierre
1165	Grimsby, George	1227	Ammerson, Charles	1289	Price, Jacob
1166	Francis, Joseph Butler	1228	Brown, Samuel	1290	Young, William

1291	Cooke, Samuel	1353	Campbell, Reynold	1415	Lippert, Thomas D.
1292	Hickman, Joseph	1354	Johnston, Simon	1416	Chadwick, M. T.
1293	Ingerson, Michael	1355	Esdale, Thomas	1417	Lockwood, Rufus
1294	Francis, John	1356	Innes, John	1418	Keven, Edward
1295	Hunter, William	1357	Mullen, Edward	1419	Smith, Thomas
1296	Francis, James	1358	McMaken, John	1420	Edwards, John
1297	Hatterson, Nathan	1359	Fletcher, James	1421	Friday, John
1298	Swett, David	1360	Bright, George	1422	King, Joseph
1299	Dean, Jonas	1361	Otis, Charles	1423	Sanford, William
1300	Brown, John	1362	Ford, George L.	1424	Fink, Johannes
1301	Blanchard, Simon	1363	Russell, George	1425	Roberson, John
1302	Taylor, Thomas	1364	Hollinger, William	1426	Allman, John
1303	Sparrow, James	1365	Rowley, Henry	1427	Freeman, John
1304	Hays, Moses	1366	Cantrell, Norvell	1428	Lacoen, John B.
1305	Williams, John	1367	Brown, William	1429	Miller, Thomas
1306	Murray, Jacob	1368	Davis, John	1430	Joseph, Francis
1307	Longford, Samuel	1369	Brown, John William	1431	Martin, John
1308	Ferrite, Francis	1370	Frazer, John	1432	Jackson, James
1309	Hammond, Joseph	1371	Gee, Thomas	1433	Stewart, William
1310	Struby, John	1372	Lerves, Robert	1434	Wilson, James
1311	Austin, Jonathan	1373	Layfield, Littleton	1435	Bayard, Joseph
1312	Farley, John	1374	Martian, John B.	1436	Mully, Etienne
1313	McBride, James	1375	Wilson, John	1437	Godfry, William
1314	Leron, Alexander	1376	Beckworth, Benjamin	1438	Louis, Nicholas
1315	Doolittle, Isaac	1377	McIntire, Samuel	1439	Peck, James
1316	Davis, Ezra	1378	Smith, Andrew	1440	Vetch, William
1317	Tweed, Caldwell	1379	Nelley, Richard John	1441	Sandford, John W.
1318	Phillips, John	1380	Pilton, Alexander	1442	Smith, John
1319	Marshall, Alexander	1381	Batman, John	1443	Gibbs, Henry
1320	Hansell, John	1382	Trifft, Joseph	1444	Fardy, Anthony
1321	Williams, Elge	1383	Mansfield, James	1445	Wild, Thomas
1322	White, Laurence	1384	Taylor, Samuel E.	1446	Smith, John
1323	Mingle, William	1385	Gilmore, William H.	1447	Jackson, Lambert
1324	Huggins, Daniel	1386	Sheffers, Steuben	1448	Whetting, Samuel C.
1325	Lilley, Simon	1387	Davis, John	1449	Newell, Isaac
1326	Merrill, Mark	1388	Rowe, John	1450	Peterson, Peter
1327	Bazin, Francois Gaston	1389	Penman, Richard	1451	Jasseciu, Louis
1328	Sketchly, William	1390	Doolittle, Henry	1452	Anderson, James
1329	Harris, William	1391	Cummings, James	1453	Cato, John
1330	Howard, William	1392	Heard, Thomas	1454	Edwards, John
1331	Rawle, Benjamin	1393	Jenkins, Nathaniel	1455	Cooper, Andrew A.
1332	Perry, Charles	1394	Dean, Nathaniel B.	1456	Cramstead, James
1333	Dunn, John	1395	Williams, John	1457	Prince, George
1334	Wores, Robert	1396	Beek, Steward	1458	Rosings, Cornelius
1335	Mingle, Thomas	1397	Gallibrandt, Bernard	1459	Senile, Francis
1336	Rennabin, Benjamin	1398	Smith, Thomas	1460	Martin, Manuel
1337	Moore, Richard	1399	Williams, Thomas	1461	Riddlefield, James
1338	Watson, Isaac	1400	Thompson, Thomas	1462	Hitre, Dempsey
1339	Harris, John	1401	Payne, Joseph Smith	1463	Johnston, Joseph
1340	Slowin, Abraham	1402	Hazel, George	1464	Cooke, William
1341	James, Daniel	1403	Cotterill, James	1465	Ruffield, Samuel
1342	Baldwin, John	1404	Richards, James	1466	Thomas, Abraham
1343	Wilkins, William	1405	Wallace, William	1467	Brown, Charles
1344	McCoy, James A.	1406	Harper, Nicholas	1468	Green, William
1345	Cainton, Joseph	1407	Russell, William	1469	Cook, Charles H.
1346	Calhoun, Richard	1408	Horsydise, William	1470	Allen, John
1347	Beard, Francis	1409	Singleton, Samuel	1471	Dibble, Reuben
1348	Andres, David	1410	Lamoure, Peter	1472	Roley, John
1349	Mack, Theron	1411	Roberts, John	1473	Risings, John
1350	Irvin, Mathew	1412	Johnson, John	1474	Parsons, Samuel D.
1351	Duran, Pierre	1413	Grover, Edmond	1475	Ligan, Pierre
1352	Price, John	1414	Jackson, Henry	1476	Guillard, Peter

1477	Bursted, John	1539	White, Charles	1600	Weston, David
1478	Warren, David	1540	Durand, John	1601	Cudworth, Henry
1479	Barnard, John	1541	Carter, Edward	1602	Smith, John
1480	Samsuillen, John	1542	Carter, Daniel	1603	Smith, Richard
1481	Mooney, Peter	1543	Gellie, Thomas	1604	Cross, Oliver
1482	Guillard, Louis	1544	Barber, William	1605	Hudson, Thomas
1483	Martin, Isaac	1545	Weeks, David	1606	Lemond, John
1484	Thompson, Henry	1546	Pitts, William	1607	Watkins, Thomas
1485	Hamilton, Anthony M.	1547	Miller, Charles	1608	English, Edward
1486	Irvin, Anthony	1548	Roff, Isaac	1609	Gland, Benjamin
1487	Grosse, William	1549	Reynolds, James	1610	McKenny, John
1488	Coleman, David	1550	Edson, John	1611	Eveleth, William
1489	Stephens, William	1551	Martin, Henry	1612	Flynn, Abraham
1490	Campbell, John	1552	Boriesa, John	1613	Price, Samuel
1491	Freeman, Prince	1553	Lerma, John	1614	Page, John
1492	Anderson, Robert	1554	Vogel, Herman	1615	Lothrop, James
1493	Evans, James	1555	Ludson, Daniel	1616	Holland, Richard
1494	Parsons, Ignatius	1556	Weedon, Anthony	1617	Holmstrom, Lorentz
1495	Harvey, John	1557	Smith, Chris	1618	Holmstrom, Alexander
1496	Burbank, G. W.	1558	Edgerly, William	1619	Riddle, Thomas
1497	Long, R. S.	1559	Gardner, Peter	1620	Miller, William
1498	Little, George	1560	Bergman, John	1621	Tucker, George C.
1499	Askwith, William Vickery	1561	Anthony, Stephen	1622	Bartlett, Caleb
1500	Colton, Walter	1562	Neal, Dennis	1623	Hume, George
1501	Barnes, Nathaniel	1563	Hilaire, Gasper	1624	Codman, Richard
1502	Stewart, William	1564	Brown, George	1625	Manson, Nathaniel
1503	Jamison, Peter	1565	Fox, Washington	1626	Davis, John
1504	Gandell, Epine	1566	Yard, William	1627	Peters, John
1505	Taylor, Archibald	1567	Parnell, Benjamin	1628	Wood, Sylvanus
1506	Rozier, William	1568	Robertson, Robert	1629	Doughty, Jesse
1507	Brown, Arn	1569	Baransan, John	1630	Turtle, French
1508	White, Isaac	1570	Littefield, Rufus	1631	Gove, William
1509	Hutchins, Henry	1571	Boyd, John	1632	Foss, Edmond
1510	Lane, James	1572	Allen, William	1633	Anderson, Daniel
1511	Thornson, Joseph	1573	More, Francis	1634	Cary, John
1512	Pote, Jeremiah	1574	Laurencau, Jean	1635	Foss, Joseph
1513	Poland, Joshua	1575	Young, Aboukir	1636	Foster, John Thomas
1514	Redding, John	1576	Kenny, Harris	1637	Gray, William
1515	Parish, Samuel	1577	Champy, Peter Felix	1638	Webb, John
1516	Haught, John	1578	De la Batiet, Francois	1639	Christie, James
1517	Wilcox, Lewis	1579	Madey, Gustave Charles	1640	Brandage, John
1518	Cornwall, Arthur		D'Eocouthant	1641	Dinsmore, John
1519	Cooper, Charles	1580	Meilleur, Simon	1642	Hokinson, Nathaniel
1520	Nicholas, Stephen	1581	Wilson, John	1643	Dougal, Thomas
1521	Ostand, Michael	1582	West, Simon	1644	Best, Robert
1522	Atkins, James	1583	Godwin, William	1645	Jules, Robert
1523	Mart, James	1584	Butler, George	1646	Pickrage, George
1524	Manson, William	1585	Doughty, Levi	1647	Monroe, John
1525	Hutchins, Edward	1586	Huskill, Robert	1648	Hacking, Robert
1526	Walker, Armstrong	1587	Robinson, Elias	1649	Butman, Charles
1527	Shaw, Richard	1588	Nicherson, Josiah	1650	Atkins, Uriah
1528	Biss, Daniel W.	1589	Mouls, Benjamin	1651	Hubbard, John
1529	Manson, William	1590	Braley, George	1652	Fossender, William
1530	Mullins, James	1591	Hitch, Joshua	1653	Dennison, Judah
1531	White, David	1592	Miller, John	1654	Schultz, John
1532	Roles, John	1593	Gannett, Felix	1655	Chambers, Henry
1533	Logan, William	1594	Deshays, Jean Francois	1656	Blanchard, George
1534	Dickinson, Chester	1595	Adam, Joseph	1657	Barnes, Robert
1535	Wood, John	1596	Tasdeboiz, Louis	1658	Allen, Henry
1536	Mills, William	1597	Merrett, Robert	1659	Armstrong, William
1537	Hopkins, Daniel	1598	Gage, Isaac	1660	Sennett, John
1538	Bower, Joseph	1599	Charter, Samuel	1661	Holtz, Daniel

1662	Smith, William	1724	Fry, John	1786	Stites, Ezra
1663	Lovett, Thomas L.	1725	Baday, John	1787	Manuel, Diego
1664	Rich, Ebenezer	1726	Roman, John	1788	Molineux, William
1665	Russell, Morris	1727	Lund, Peter M.	1789	Sutter, John
1666	Ingles, David	1728	Boggs, James	1790	Hartfield, James
1667	Armstrong, James	1729	Wilson, John	1791	Johnson, Samuel
1668	Taylor, Thomas	1730	Tufts, Zachariah	1792	Hubbard, John
1669	Norris, Augusta George	1731	Thompson, William	1793	Young, John
1670	Jones, Thomas	1732	Borgin, Gabriel	1794	Marshall, Thomas
1671	Hopkins, Elisha	1733	Blackstone, Edward	1795	Young, William
1672	Scott, John	1734	Brown, Henry	1796	Freely, Henry
1673	Williams, Joshua	1735	Joseph, Francis	1797	Shepherd, William
1674	Timmerman, Matthew	1736	Washburn, Edward	1798	Brown, Francis
1675	Williams, Ambrose	1737	Williams, George	1799	Peckham, Henry
1676	Reeves, James	1738	Augustin, Anthony	1800	Pratt, Paris
1677	Geyer, Joshua	1739	Lassalle, Francisco	1801	McLeod, Collin
1678	Putman, Charles	1740	Bishop, William	1802	White, James
1679	White, John	1741	Cox, John	1803	Young, John
1680	Kellinger, John	1742	Schoeman, Frederick	1804	Eggert, Francis
1681	Mesick, Elihu	1743	Hughes, John	1805	Nugent, John
1682	Healy, John	1744	Stewart, William	1806	Baxter, Charles
1683	Brown, Benjamin	1745	Page, John	1807	Kellam, James
1684	Bowen, John	1746	Weeks, Benjamin	1808	Hovrington, William
1685	Merritt, Jonathan	1747	Jenkins, Joseph	1809	Hall, James
1686	Burrett, Benjamin	1748	Brown, Thomas	1810	Jamison, Robert
1687	Wheeler, John W.	1749	Jackson, William	1811	Famer, John
1688	Gardner, Joseph	1750	Dymoss, Peter	1812	Freeman, John
1689	Gabriel, Joshua	1751	Woodcraft, John	1813	Jordan, Joseph
1690	Keller, John	1752	Underwood, John	1814	Delphi, Richard
1691	Harris, James	1753	Green, Moses	1815	Conklin, Robert
1692	Smith, John	1754	Currien, Stephen	1816	Clerk, George
1693	Basset, David	1755	Young, Joseph	1817	Fleming, John Joseph
1694	Morrie, Joseph	1756	Courris, Francis	1818	Day, Syles C.
1695	Linsey, Alexander	1757	Courtis, Harry	1819	Cooper, James
1696	Hill, John	1758	Giles, Edward	1820	Smith, John
1697	Hooper, Benjamin E.	1759	Tibbet, Henry	1821	Waite, Abel
1698	Armstrong, William	1760	Martin, William	1822	Salley, James
1699	Jones, Thomas	1761	Walleman, William	1823	Place, John
1700	Smith, John	1762	D'Olivera, Manuel	1824	Thompson, Thomas
1701	Clawe, Maurice	1763	Grace, Allen	1825	Sears, Abram
1702	Hamill, William	1764	Denham, Cornelius	1826	Curtis, Joseph
1703	Carpenter, Henry	1765	Robinson, John	1827	Phillips, Benjamin
1704	Tuck, James	1766	Roberts, Hugh	1828	Hopkins, William
1705	Davis, Morris	1767	Gray, Isaac	1829	Allister, Isaac
1706	Bartlett, John	1768	Bell, James	1830	Lee, Ely
1707	Parsons, Daniel	1769	Drew, William	1831	Hennery, James
1708	Woodbury, John	1770	Curtis, John	1832	Gavel, Oliver
1709	Rogers, Gorham	1771	Hall, John	1833	Barlow, John
1710	Thomas, Spencer	1772	Little, John	1834	Roberson, John
1711	Cogan, John	1773	Fardel, Thomas	1835	Myatt, Frederick
1712	Dolliver, John	1774	Gibbs, William	1836	Nagel, George
1713	Michael, Car.	1775	Eskindson, George	1837	Scott, Robert
1714	Wilson, John	1776	Brown, William	1838	Graham, David
1715	Barttist, John	1777	Andrews, Charles	1839	Bogart, Killian William
1716	Gardner, William	1778	Harton, Henry	1840	Wilson, George
1717	Slocum, William	1779	Hall, James	1841	Carbenett, John
1718	Allen, John	1780	Russell, James	1842	Smith, James
1719	Martin, William	1781	Abbott, Timothy	1843	Bladen, John
1720	Fox, Richard	1782	Amos, Cheney	1844	Leach, Daniel
1721	Billings, John	1783	Jordan, Richard	1845	Jones, Charles
1722	Williams, Joseph	1784	Patten, James	1846	Garner, James
1723	Wing, John	1785	Malbone, Evan	1847	Allen, William Henry

1848	Juderwicks, James	1910	Griffith, Benjamin	1971	McFall, William
1849	Dennison, Henry	1911	Henry, John	1972	Brown, William
1850	Bacus, David	1912	Ayres, Robert	1973	Newby, Lamb
1851	Snelson, Robert Lewis	1913	Coombes, James	1974	Brown, William
1852	Hudson, John	1914	Fadden, John	1975	May, Henry
1853	Baron, Thomas	1915	Godfry, Edward	1976	Brannigan, John
1854	Laurnsberry, Benjamin	1916	Noney, Peter	1977	Jackson, David
1855	Lovett, Henry	1917	Waters, Robert	1978	Bradford, James
1856	Appine	1918	Hunter, Charles	1979	Booth, Stephen
1857	Watson, William H.	1919	Collins, Andrew	1980	Marshall, John
1858	Allen, William Howard	1920	Croker, Nathaniel	1981	Ransom, Joseph
1859	Levi, Uriah Phillips	1921	Henry, William	1982	Young, Rufus
1860	Hunt, Charles	1922	Butnet, William	1983	Smith, Elisha
1861	Bradt, Francis	1923	Ash, Samson	1984	Robinson, Nicholas
1862	Stevens, John	1924	Daniels, John	1985	Moore, William
1863	Himes, Walter	1925	Scott, John	1986	Neil, Henry
1864	Benedict, William	1926	Sniffen, John	1987	Buchanan, John
1865	Blackledge, John	1927	Scott, John	1988	Duncan, George
1866	Leary, Daniel	1928	Pettinger, William	1989	Lewis, Thomas
1867	Richards, George	1929	Jamison, William	1990	Mathis, Hendrick
1868	Chuglar, John	1930	Temple, William	1991	Wessels, Robert
1869	Dillon, William	1931	Adams, James	1992	Wilson, William
1870	Crawford, Robert	1932	Guillard, Joseph	1993	Sickless, James M.
1871	Addigo, Henry	1933	Bennett, Charles	1994	Snowdon, Jacob
1872	Brown, Alexander	1934	Rodwin, Walter	1995	Norris, Robert
1873	Dunham, Abinesar	1935	Silva, Francis	1996	William, James
1874	Groger, Henry	1936	Allen, Philip	1997	Ashmore, Edward
1875	Barry, James	1937	Fedie, John	1998	Williams, John
1876	Preston, Jonas	1938	More, Henry	1999	Wolfe, William
1877	Ferguson, Henry	1939	Hart, Joshua	2000	Mines, Artemas
1878	Russell, Frederick	1940	Roberson, Henry	2001	Morweek, Andre
1879	Couret, Francis	1941	Montel, Pierre	2002	Meiniffe, Charles
1880	Nash, Manuel	1942	Sanis, Lewis	2003	Coffin, John
1881	Allen, Joseph	1943	Charlies, James	2004	Lettington, James
1882	Benson, James	1944	Atkins, Martyn	2005	Hebron, John
1883	Williams, John	1945	Menard, Augustus	2006	Spires, Thomas
1884	Young, William	1946	Booth, William	2007	Dalloway, John
1885	Blisset, James	1947	Barstow, Samuel	2008	Churchill, Manuel
1886	Rosah, Samuel	1948	Johnson, Robert	2009	Wheeler, Benjamin
1887	Clark, John	1949	Buckhannon, Edwin	2010	Anderson, Mathew
1888	Smith, Adam	1950	Coder, Antoine	2011	Windham, John
1889	Noyes, Charles	1951	Peters, Aaron	2012	McNab, James
1890	Tolvet, Charles	1952	Allen, James	2013	Baker, Israel
1891	Price, William	1953	Stevens, Hugh	2014	Sanderhin, Daniel
1892	Young, Thomas	1954	Hart, Bartholomew	2015	Spangle, Frederick
1893	Thomas, David	1955	Palmer, Pero	2016	Hacket, William
1894	Leach, Josem	1956	Bell, James	2017	Stansbury, James
1895	Bevin, William	1957	Griffin, William	2018	Gundy, James
1896	Collin, William	1958	Morell, Benjamin	2019	Johnson, John
1897	McDonald, John	1959	Smith, William	2020	Weimain, Lawrence
1898	Paris, William	1960	Palmer, Robert	2021	Oakley, Jacob
1899	Johns, Thomas	1961	Burdick, Simon	2022	Anderson, James
1900	Davis, William	1962	Chipman, Christopher	2023	Toby, Elisha
1901	Williams, John	1963	Johnson, William	2024	Lindsborough, Charles
1902	Brown, John	1964	Black, George	2025	Livermore, Arthur
1903	Bowne, Asher	1965	Durvolf, Stephen	2026	Jones, Walter
1904	Ginkins, William	1966	Brown, Ebenezer	2027	Luce, Prispery
1905	Conley, Samuel	1966	Scholle, John	2028	Dennison, Thomas
1906	Randol, Thomas	1967	Jeffery, Job	2029	Garrot, William
1907	Smith, William	1968	Jackson, James	2030	Humphries, Warren
1908	Hanscom, Moses	1969	Williams, William	2031	Holden, George
1909	Tafe, William	1970	Trout, William	2032	Hennay, Jacob

2033	Stewart, Isaac	2095	Haskell, Thomas	2157	Bowman, Leon
2034	Moore, George	2096	Lane, James	2158	Kennisen, William
2035	Ogden, James	2097	Franks, Francis	2159	White, James
2036	Thompson, James	2098	Nicholas, John Bruss	2160	Williams, R. N.
2037	Marsh, Beverly	2099	Pass, Samuel William	2161	Cartwright, George
2038	Davis, George	2100	Pearce, Alexander	2162	Smith, James
2039	Montgomery, John	2101	Shaw, Edward	2163	Taylor, James
2040	Shipley, Daniel	2102	Van Borgen, Martin	2164	Redman, William
2041	Blasdell, Jonathan	2103	Oudin, Pierre Francois	2165	Cloutman, Robert
2042	Carter, James W.	2104	Babi, Joseph	2166	Phippen, John
2043	Peters, Peter	2105	Gustave, Peter	2167	Clough, Isaac
2044	Webb, Alexander	2106	Valiant, Richard	2168	Grand, William
2045	Wade, Richard	2107	Wade, John	2169	Stacey, Benjamin
2046	Jansen, Knut	2108	Rommel, Francis	2170	Andrews, Joseph
2047	Nicholson, James	2109	Bassinet, Charles	2171	Follet, William
2048	Bannister, Joshua	2110	Pearer, John	2172	Bass, Prince
2049	Pynne, James R.	2111	Goldsberry, William	2173	Easton Joseph
2050	Baughton, Glover	2112	Johnson, John	2174	Alexander, Richard
2051	Sandburn, John	2113	Dusheets, Arthur	2175	Knapp, Samuel
2052	Muggin, John	2114	McDonald, John	2176	Stacey, Samuel
2053	Webber, John	2115	Palmer, John	2177	Chevers, Joseph
2054	Mason, John B.	2116	Moore, John	2178	Carnes, John
2055	Holmes, Charles	2117	Cornish, Charles	2179	Quinan, Stephen
2056	Nicholson, James	2118	Oakley, Joseph	2180	Price, David
2057	Sears, Charles	2119	Cook, Benjamin	2181	Lyon, Charles
2058	Elliott, Benjamin	2120	Yilstone, George	2182	Lymones, Procter
2059	Lansford, John	2121	Hamell, Christopher	2183	Young, Thomas
2060	Bunker, Isiah	2122	Barton, Eseek	2184	Creping, Thomas
2061	Wheeler, Henry	2123	Marshall, Emanuel	2185	Booth, Charles
2062	Nugent, John	2124	Crawford, William	2186	Ganson, Richard
2063	Selston, Isaac	2125	Martin, Stephen	2187	Gallen, John
2064	Skilder, Samuel	2126	Fletcher, John	2188	Castera, Jean S.
2065	Green, Reuben	2127	Williams, Hendrick	2189	Lactortier, Charles F.
2066	Harrington, John	2128	Ewing, James	2190	Conroy, William M.
2067	Bienfaux, Allen	2129	Schew, Richard	2191	Symonds, M. John
2068	Willey, Ebenezer	2130	Green, Peter	2192	Archer, Joseph
2069	Meden, William	2131	Creeps, Warren	2193	Fisher, John
2070	White, James	2132	Williams, John	2194	Mone, Denis
2071	Walken, James	2133	Smith, John	2195	Parley, William
2072	Morris, John	2134	Crosby, George	2196	Dolliver, Richard
2073	Carland, Lewis	2135	Perkins, Thomas	2197	Wyer, Joseph
2074	Dyer, Johnathan	2136	Nelson, Thomas	2198	Spur, Elijah
2075	Smith, John	2137	Astrop, Hans Christian	2199	Rogers, Robert
2076	McDowall, John	2138	Peterson, Laurence	2200	Perkins, Elijah
2077	Olsen, Elias	2139	Campbell, John	2201	Diemen, John
2078	Johnson, Elias	2140	Peane, Henry	2202	Wheeler, John
2079	Limonden, William	2141	Rodyman, John	2203	Smith, Richard
2080	Campbell, Abraham	2142	Kitchen, Richard	2204	Andrews, Benjamin
2081	Stone, Edmond	2143	Prymroy, Richard	2205	McNeil, John
2082	Allen, Elissa	2144	Blake, Thomas	2206	Hughes, John
2083	Prior, Essa	2145	Newell, Nathaniel	2207	Gregory, George
2084	Small, Enoch	2146	Jackson, William	2208	Thompson, Andrew
2085	Ross, Philip	2147	Bowers, Jonathan	2209	Gilchrist, John
2086	Dubois, Alexander	2148	Rogers, Henry	2210	Chain, John
2087	Michell, Fairly	2149	Bowers, Mrs.	2211	Road, John
2088	Andrews, George	2150	Peirce, Joseph	2212	Martin, John
2089	Horn, Abner	2151	Lehart, Jacob	2213	McFarland, John
2090	Wilson, John	2152	Jackson, Curtis	2214	Long, Charles
2091	Bymer, George	2153	Adams, John	2215	Pagechin, Henry
2092	Edgar, William	2154	Hopkin, William	2216	Gable, John
2093	Brown, Richard	2155	Hillman, John	2217	Fleur, Francois Xavier
2094	Winter, William	2156	Ingles, Daniel	2218	Pierson, Elijah

2219	Roper, Thomas William	2281	Hiler, George	2343	Bunkerson, John
2220	Thompson, Henry	2282	Perkins, John	2344	Woods, Edward
2221	Postlock, William	2283	Wright, William	2345	Raquinis, Francois
2222	Hancock, William	2284	Patterson, Andrew	2346	Peridis, Pierre
2223	Graham, James	2285	Sheardon, Ellison	2347	Castania, John Baptiste
2224	Coffin, Alexander	2286	Harris, Bradley	2348	Campbell, Alexander
2225	Mathews, Joseph	2287	Stout, John	2349	Tarbell, William George
2226	Brown, David	2288	Weeks, James	2350	Bawnham, Francis A.
2227	Wilson, John	2289	Prosper, John	2351	Coutie, Thomas
2228	Jones, Thomas	2290	Thompson, William	2352	Walker, Seth
2229	Scheen, John	2291	Welsh, William	2353	Plumer, William Reed
2230	Thompson, William	2292	Wakefield, Nathaniel	2354	Donelson, Joseph
2231	Dixon, Benjamin	2293	Pinkham, Amos	2355	Modre, Joseph
2232	Mathews, Henry	2294	Pain, George	2356	Williamson, David
2233	Cummings, Samuel	2295	Bodfish, William	2357	Keen, Joseph
2234	Baldwin, Pierson	2296	Nye, Cornelius	2358	Carr, James
2235	Lapierre, Etienne	2297	Hanson, Henry	2359	Harris, Simon
2236	Frenarye, Pierre	2298	Hanson, Peter	2360	Golandre, John
2237	Taney, Lewis	2299	Soderback, Andrew	2361	Pritchard, Israel
2238	Dumas, Raymond	2300	Patterson, Joshua	2362	Selman, John
2239	Tapiau, Vitale	2301	Samuel, Nicholas	2363	Baird, David
2240	Montaindon, Pierre Hans	2302	Brown, Sandy	2364	Harvey, Joseph
2241	Montaindon, Julien	2303	Masson, William	2365	Orne, Oliver P.
2242	Montaindon, Aimee	2304	Tubbs, Martin	2366	Merrill, Enoch
2243	Montaindon, Henry	2305	Coffer, William	2367	Harris, Ebenezer
2244	Montaindon, Marianne	2306	Bottelio, Antony	2368	Watson, William
2245	Montaindon, Julie	2307	Man, Tabez	2369	Lepberg, John P.
2246	Montaindon, Louise	2308	Ringgold, Peregrine	2370	Conner, Michael
2247	Strum, Yves	2309	Miller, William	2371	Hill, Daniel
2248	Strum, Magdalene	2310	Moore, Alexander	2372	Campbell, William
2249	Sturm, Jeannette	2311	Wilson, Charles	2373	Debaize, Francois Jean
2250	Plan, Madame G.	2312	Cooper, John	2374	Perott, Jean Francois
2251	Thompson, Peter	2313	Guillemot, Richard	2375	Gravely, Joseph
2252	Laborde, Peter	2314	Smiles, John	2376	Chauvel, Thomas
2253	Cross, Peter	2315	Jackson, Samuel	2377	Damerie, Etienne
2254	Barker, Charles G.	2316	Matthews, Leonard	2378	Grossetti, Jean Maurice
2255	Conrad, Godfred	2317	Stiles, William	2379	Tuttle, Joseph
2256	Anthony, Joseph	2318	Stewart, Adam	2380	Carter, Enoch
2257	Lingrin, Peter	2319	Furlong, William	2381	Bloomdose, John
2258	Campbell, James	2320	Hobert, William	2382	Dorsey, John
2259	Clark, Jacob	2321	Popal, Richard	2383	Hubbell, James
2260	Taylor, James	2322	David, William	2384	Ewell, Edward
2261	Roberts, Joel	2323	Duff, James	2385	Gardner, Jerry
2262	Thornton, William	2324	Witherett, Charles	2386	Conklin, Smith
2263	Dunn, William	2325	Moore, John	2387	Harris, John
2264	Lambert, William	2326	Leger, Andrew	2388	Mains, John
2265	McFadden, James	2327	Lovely, Henry	2389	Sprunwell, Peter
2266	Hoselquist, John	2328	Davis, Charles S.	2390	Edwards, John
2267	Hayes, Benjamin	2329	Ramsden, John	2391	Johnson, Samuel B.
2268	Cheney, Daniel	2330	Knight, John	2392	Holden, Andrew
2269	Abbey, Obadiah	2331	Charles, John	2393	Murphy, William
2270	Sereder, Martin	2332	Poland, Thomas	2394	Port, John
2271	Demetra, George	2333	Le Core, Peter Murry	2395	Carter, Henry
2272	Routemen, Bernard	2334	Relavoine, Louis	2396	Riva, Antonio
2273	Drinkwater, Peter	2335	Sturges, Bradley	2397	Morisau, Jean Baptiste
2274	Brown, Stephen	2336	Lebarth, John	2398	Rouger, Desire
2275	Neal, John	2337	Toby, Peter	2399	Grenaux, Jean Yves
2276	Denison, George	2338	Burr, Isaac	2400	Robin, John
2277	Priestly, James	2339	Pelham, Robert	2401	Brutus, Mario
2278	Hutchins, William	2340	Anderson, Edward	2402	Deparvier, John
2279	Gordnow, Saal B.	2341	Neall, Daniel	2403	Creighton, Manuel
2280	Gall, Michael	2342	Barbadoes, Robert	2404	Rogers, Henry

2405	Smith, John	2467	Downe, William	2529	Nooney, William
2406	Morgan, John	2468	Still, David	2530	Cotterill, Henry
2407	Webber, Frederick	2469	Gerard, William	2531	Chesbrough, Benjamin F.
2408	Ferris, Joseph	2470	Sparks, Robert	2532	Bedson, Jacob
2409	Williams, Alexander	2471	Morgan, Jonathan	2533	Page, Josiah Clark
2410	Brown, Ambrose James	2472	Hinkley, Thomas	2534	Smith, John
2411	Bredey, Jason	2473	Peterson, Andre	2535	Coleman, William
2412	Glover, John	2474	Patterson, John	2536	Jackson, James
2413	Hammond, William	2475	Williams, John	2537	Wright, William
2414	Gardner, Edward	2476	Schrader, Martin	2538	Henson, J. A.
2415	Stacy, William	2477	Long, Charles	2539	Hatch, Samuel
2416	Besson, Phillipe	2478	Young, John	2540	Parker, Peter
2417	Chambers, Elias	2479	Nicholls, John	2541	Fisher, James
2418	Ober, Ezra	2480	Freeman, David	2542	Grant, James
2419	Crawford, Robert	2481	Elliot, Robert	2543	Lerwick, John
2420	Conklin, Robert	2482	Phipping, Nathaniel	2544	Dean, Moses
2421	Thomas, David	2483	Bird, Comfort	2545	Neel, William
2422	Smith, John	2484	Carrot, Charles	2546	Rey, David
2423	Cooper, James	2485	Cassam, Michael	2547	Damon, William
2424	Price, William	2486	Davey, Charles	2548	Spark, Samuel
2425	Robinson, James	2487	Cooper, Tannic	2549	Rideout, James
2426	Leary, Daniel	2488	Rodden, Charles	2550	Davis, John
2427	Gilbert, Jurdonet	2489	Simons, William	2551	Bowman, Isaac
2428	Been, James	2490	Williams, John	2552	Christian, John
2429	Shaw, William	2491	Lewis, John	2553	Brooks, Russell
2430	Smith, William	2492	Brooks, Thomas	2554	Fisher, John
2431	Hill, Thomas	2493	Handfield, Robert	2555	Day, Benjamin
2432	Kennedy, Richard	2494	McConnell, John	2556	Borddell, Justice
2433	Anderson, George	2495	Briggs, William	2557	Drew, Charles
2434	Starbuck, George	2496	Moore, James	2558	Joseph, Francis
2435	Olcott, Jedediah	2497	Phillips, James	2559	Conner, Edward
2436	Gisenado, Samuel H.	2498	Fletcher, John William	2560	Dagger, Robert
2437	Ham, John	2499	Mecoh, David	2561	Whitney, Thomas
2438	Berto, John	2500	Potter, John	2562	Stephens, John
2439	Pascrell, John	2501	Atwick, Thomas C.	2563	Robinson, John
2440	Wright, Amos	2502	Crowder, James	2564	Catwood, Zenas
2441	Hudson, James	2503	Armistead, Edward	2565	Nesbitt, Richard
2442	Castor, Thomas	2504	Autest, Jean	2566	Soreby, Robert
2443	Frederick, Charles	2505	Cottle, William	2567	Minois, Pierre
2444	Gassy, Raymond	2506	Orne, William B.	2568	Fuller, Enoch
2445	Lunberge, Berg	2507	Leggett, William	2569	Paulin, Nicholas
2446	Graham, John	2508	Miles, William	2570	Montierne, Francois
2447	Williams, Edward	2509	Hedenburg, Jacob	2571	Crosby, John
2448	Boyd, William	2510	Gardner, Joseph	2572	Emmerton, James
2449	Benstead, John	2511	Swasey, John	2573	Jackson, Thomas
2450	Norcross, Nehemiah	2512	Graves, Ebenezer	2574	Adams, Peter William
2451	Turner, Daniel	2513	Hulbert, Henry	2575	Johnson, Edward
2452	Stafford, John	2514	Gatewood, James	2576	Tilley, Benjamin
2453	Brown, John	2515	Channing, John	2577	Shepherd, William
2454	Hinkley, Joseph	2516	Stevens, Amos	2578	Wilson, Charles
2455	Lyon, John	2517	Thompson, John	2579	Townsend, Thomas
2456	Gardner, Edward	2518	Brightman, John	2580	Neeve, Thomas
2457	Butts, Joseph	2519	Smith, John D.	2581	Henderson, Benjamin
2458	Richardson, Cheney	2520	Gillet, John Francis	2582	Norgrave, Jeremiah
2459	Clapp, George	2521	Elves, Manuel	2583	Ide, John Henry
2460	Mayor, James	2522	Elves, Joseph	2584	Rice, John
2461	Myers, Daniel	2523	Mitchell, John	2585	Murray, Nathaniel
2462	Abbott, Ephraim	2524	Drake, John	2586	Short, Clement
2463	Baldwin, John	2525	Polease, Jean M.	2587	Ley, John
2464	Williams, James	2526	Lewis, James	2588	Adams, John
2465	Mathews, William	2527	Garcia, Anthony	2589	Snell, David
2466	Homes, Michael	2528	Lopez, Joseph	2590	Greenland, Stephen

2591	Arnold, Welcome	2630	Conklin, Edward	2669	Long, Joseph
2592	Potter, Richard H.	2631	Lorenza, Thomas	2670	Harker, John
2593	Arnold, Thomas	2632	Bertol, Samuel	2671	McCurdie, Alexander
2594	Chadwick, John	2633	Flatt, Robert	2672	Bitters, John
2595	Holsten, Peter	2634	Sole, Edward	2673	Phillips, John
2596	Akerman, William	2635	Holbrook, Elias	2674	Simpson, William
2597	Holstade, Joseph	2636	Sole, Elias	2675	Sone, John
2598	Brown, Thomas	2637	Thomas, Stephen	2676	Farmer, George
2599	Jarvis, James	2638	Tegget, John	2677	Johnson, Henry
2600	Brown, Jesse	2639	Sole, Thomas	2678	Skinner, Tilman
2601	Musick, Joseph	2640	Wattes, Philip	2679	Kimball, Nathaniel
2602	Warner, Samuel	2641	Washbuiz, Elbert	2680	Day, James
2603	Rose, William	2642	Roberts, Moses	2681	Bonnell, James
2604	Taylor, William	2643	Benny, David	2682	Williams, John
2605	Smith, William	2644	Jackson, William	2683	Caldwell, James
2606	McCarthy, John	2645	Atwood, John	2684	Whiteman, John
2607	Green, Solomon	2646	Lapierre, Etienne	2685	McQuillan, James
2608	Hartwell, Barry	2647	Johnson, Peter	2686	Norman, Michael
2609	Ford, Philip	2648	Wood, William	2687	Lechner, Anthony
2610	Johnson, Andrew	2649	Younger, William	2688	Parker, Edward
2611	Ayers, John	2650	Gensler, George	2689	Gregory, Elijah
2612	Craig, William	2651	Gilpin, William	2690	Van Dine, Garland
2613	Stephens, Obadiah	2652	Alton, Peter	2691	Snyder, William
2614	Day, John	2653	Armstrong, Joseph	2692	Roulemmon, Bernard
2615	Marchant, William	2654	Knox John	2693	Payechin, Henry
2616	Daniel, Robert	2655	Barford, William	2694	Road, John
2617	Hooper, William	2656	Hoyt, James	2695	Scholle, John
2618	Cox, John	2657	Chane, Daniel	2696	Tamer, John
2619	Holberg, Emanuel	2658	White, Philip	2697	Gerling, George C.
2620	Driver, John	2659	Works, Elfrod	2698	Nuganes, Daniel
2621	Chase, Joseph	2660	Bassett, Gorham	2699	Helenburg, James
2622	Wilson, Thomas	2661	Holland, James	2700	May, Henry
2623	Ford, Charles	2662	Stone, Samuel	2701	Donelson, Joseph
2624	Stone, John	2663	Merrill, John	2702	Whitehead, James
2625	Morris, Isaac	2664	Timmons, John	2703	Armstrong, Daniel
2626	Johnson, Edward	2665	Church, William	2704	Grubb, Andrews
2627	Cole, John	2666	Wilkson, John	2705	Banks, Perry
2628	Newman, Daniel	2667	Flynn, Pierre	2706	Jones, Hensley
2629	Pottigal, John	2668	Thompson, John	2707	Howe, Phineas

Crew listing by ship
October 1812 through June 1814

<u>Unknown</u>

Adams, Peter William
Akerman, William
Allen, John
Alton, Peter
Anderson, James
Anderson, Robert
Armstrong, James
Armstrong, Joseph
Arnold, James
Atwood, John
Austin, Jonathan
Ayers, John
Bagley, William
Bannister, Joshua
Bansal, Lewis
Barford, William
Barker, Charles G.
Baughton, Glover
Benstead, John
Bird, Comfort
Blasdell, Jonathan
Bonnell, James
Bowman, Benjamin
Boyd, William
Bready, Jason
Briggs, William
Brightman, Joseph
Brooks, Thomas
Brown, Jesse
Brown, Thomas
Brown, William
Caldwell, James
Campbell, John
Carpenter, Henry
Carr, Richard
Carrot, Charles
Carter, James W.
Cassam, Michael
Catwood, Zenas
Chane, Daniel
Charles, John
Chase, Joseph
Check, Stephen
Clark, Elisha
Clark, Jacob
Clawe, Maurice
Coffin, Frederick Henry
Coleman, John
Conklin, Smith
Conrad, Godfred
Conway, Samuel
Cooper, Tannic
Cotterill, James
Cox, John
Craig, William
Crosby, John
Crowell, Uriel

Cunningham, John
Daniel, Robert
Davey, Charles
Davis, George
Dawson, John
Day, James
Day, John
De Colville, Lair
Delaney, Matthew
Denham, Cornelius
Dingell, George
Dixon, Benjamin
Dobbins, John
D'Olivera, Manuel
Driver, John
Eddey, John
Edgerly, William
Edwards, John
Eldridge, Nathaniel
Elliott, Benjamin
Enderson, James
Ervin, Ely
Evans, James
Farrell, Andrew
Fawcett, William
Farley, John
Ferrald, John
Fisher, James
Flood, David
Floyd, James
Ford, Charles
Ford, Philip
Fessenden, William
Foster, Cato
Freely, Henry
Freeman, David
Freeman, Prince
Fuller, Enoch
Gallo, Joseph
Gandell, Epine
Ganson, Richard
Gensler, George
Gardner, Jerry
Gardner, Peter
Garner, James
Gibb, James
Gibbs, William
Gibby, John
Giggord, Robert
Gilbert, Thomas
Gilpin, William
Golfin, John
Grace, Allen
Graham, John
Grant, James
Green, Solomon
Gregory, Elijah
Gronard, Peter

Hamill, William
Handfield, Robert
Harris, John
Harry, John
Hartfield, James
Hartford, James
Hartwell, Barry
Harvey, John
Hawking, John
Holberg, Emanuel
Holmes, Charles
Holsien, Peter
Holstade, Joseph
Hooper, William
Horn, Abner
Howe, Phineas
Hoyt, James
Hubbard, John
Hunt, David
Ingles, David
Jackson, Daniel
Jackson, Thomas
Jackson, William
James, Sacket
Jamison, Peter
Jansen, Knut
Jarvis, James
Johnson, Andrew
Johnson, Edward
Johnson, Peter
Johnson, Samuel
Johnson, Samuel B.
Johnson, William
Jones, Charles
Jones, Peter
Jones, William
Kennedy, William
Kerban, Thomas
Kimball, Nathaniel
Kingley, Benjamin
Knox John
Langoth, Francis
Leach, Daniel
Lechler, Anthony
Lendison, Henry
Leron, Alexander
Mains, John
Mains, John
Marchant, William
Marsh, Beverly
Marshall, John
Marshall, Thomas
Martian, Ephraim
Martin, William
Mason, John B.
Mathews, Henry
Mathews, Joseph
Mathews, Richard

Mazeb, James
McBride, James
McCarthy, John
McConnell, John
McFall, William
McClelland, William
McNeil, John
Miller, William
Mills, John
Montgomery, John
Moore, James
Morris, William
Moss, William
Muggin, John
Musick, Joseph
Neeve, Thomas
Nelson, Richard
Newby, Lamb
Nicholls, John
Nicholson, James
Norcross, Nehemiah
Norman, Michael
Oakley, Jacob
Ogden, James
Ostin, John
Packard, William
Pain, James
Pardoe, John
Parker, Edward
Parsons, Ignatius
Patton, Robert
Pearce, Edward
Peters, Peter
Peterson, John
Peterson, Alexander
Phillips, James
Pierce, Amos
Poland, Thomas
Potter, John
Pynne, James R.
Reid, Joseph
Rich, William
Richards, James
Richardson, John
Rideout, James
Riley, William
Ripley, William
Roach, Reuben
Road, John
Roberts, John
Rose, William
Rosset, Samuel
Ross, John
Russell, Morris
Ryan, Thomas
Sandburn, John
Scott, Ezekiel
Scott, John
Scott, John
Sears, Charles
Shepherd, William
Shipley, Daniel

Sillock, Amos
Sims, Oliver
Smith, Charles
Smith, John
Smith, William
Snyder, William
Sprunwell, Peter
Stafford, John
Staitman, James
Starbuck, Thaddeus B.
Stephens, John
Stephens, Obadiah
Stow, Jeremiah
Sutter, John
Taylor, James
Taylor, William
Thompson, James
Thompson, William
Tilley, Benjamin
Tonver, Michael
Townsend, Thomas
Trout, William
Turner, Benjamin
Turner, Daniel
Van Dine, Garland
Wade, Richard
Wallace, William
Walleman, William
Warner, Samuel
Warner, William
Warren, John
Waters, Philip
Webb, Alexander
Webber, John
Weeks, Benjamin
West, Nathan
White, William (1)
White, William (2)
Whitehouse, Asey
Whiteman, John
Whitmore, William
Willett, Robert
Williams, Edward
Williams, John
Williams, William
Williamson, John
Wilson, Charles
Wilson, Thomas
Wilson, John
Wilson, John
Wilson, Thomas
Wood, William
Wright, Isaac
Young, John
Young, William
Younger, William

14th U.S. Infantry
Johnson, John
21st U.S. Infantry
Weinman, Lawrence

Agnes

Adams, John

Conroy, William M.
Hillman, John
Hopping, William
Jackson, Curtis
Lehar, Jacob
Peirce, Joseph

Allegany

Berry, Brook

Allegany

Conway, William
Dolinson, Andrew
Higgins, George
Hough, Ebenezer
Mires, John
Oden, John
Warner, Henry

Alligator

Emmerton, James

Amiable

Astrop, Hans Christian
Blake, Thomas
Campbell, John
Creeps, Warren
Crosby, George
Ewing, James
Franks, Francis
Green, Peter
Jackson, William
Kitchen, Richard
Lane, James
Nelson, Thomas
Newell, Nathaniel
Peane, Henry
Perkins, Thomas
Peterson, Laurence
Prymroy, Richard
Rodyman, John
Schew, Richard
Smith, John
Williams, John

Amity

David, William
Duff, James
Hobert, William
Leger, Andrew
Moore, John
Popal, Richard
Witherett, Charles

Amphitrite

Bristol, Nehemiah
Cartwright, Alexander J.
Shairs, Samuel
Wilson, Francis

Ann

Jackson, David

Anna

Trowbridge, Bela

Antonette

Scheen, John

Apparencen

Anderson, Edward
Barbadoes, Robert

Bunkerson, John
Neall, Daniel
Pelham, Robert
Peridis, Pierre
Raquinis, Francois
Woods, Edward

Argus

Arthur, Alexander
Drybourgh, James
Gibbons, Andrew
Green, Samuel
Hilliard, Robert B.
Howey, Artemas
Joseph, Michael
Paddock, Benjamin Mead
Post, Anthony Oswald
Woodard, Elijah

Asia

Gardiner, Charles

Ateline

Benny, David
Bertol, Samuel
Cole, John
Conklin, Edward
Flatt, Robert
Holbrook, Elias
Johnson, Edward
Lorenze, Thomas
Morris, Isaac
Newman, Daniel
Pottigal, John
Roberts, Moses
Sole, Edward
Sole, Elias
Sole, Thomas
Stone, John
Tegget, John
Thomas, Stephen
Washbuiz, Elbert
Wattes, Philip

Avon

Carland, Lewis
Dyer, Johnathan
Johnson, Elias
McDowall, John
Morris, John
Olsen, Elias
Smith, John
Walker, James

Betsey

Bevin, William
Couret, Francis
Croker, Nathaniel
Daniels, John
Noyes, Charles
Paris, William
Preston, Jonas
Randol, Thomas
Young, William

Betsy

Bienfaux, Allen
Bunker, Isiah

Green, Reuben
Harrington, John
Landford, John
Meden, William
Nugent, John
Selston, Isaac
Skilder, Samuel
Wheeler, Henry
White, James
Willey, Ebenezer

Blue Bird

Dishele, Alexander
Fuller, Zachariah
Jones, John
Marshall, John
Spryon, John
Swinny, Edward
Thompson, Owen
Warren, Nathan

Brizeland

Allen, William

Brutus

Calkings, Zero
Capron, William
Copperas, Nicholas
Gomez, Manuel
Hurtall, Peter
Johns, Bellona
Lewis, John
Nartigue, John
Parker, Samuel
Perez, Joseph
Ramblett, Charles
Rich, Francis
Roy, Charles
Scott, Andrew
Slate, Henry
Turpin, Francis

Bunker Hill

Borddell, Justice
Bowman, Isaac
Brightman, John
Brooks, Russell
Channing, John
Christian, John
Conner, Edward
Cottle, William
Dagger, Robert
Davis, John
Day, Benjamin
Drew, Charles
Fisher, John
Gardner, Joseph
Gatewood, James
Graves, Ebenezer
Hedenburg, Jacob
Helenburg, James
Hulbert, Henry
Joseph, Francis
Leggett, William
Miles, William
Nooney, William

Orne, William B.
Rodden, Charles
Stevens, Amos
Swasey, John
Thompson, John
Whitney, Thomas

Cannoniere

Blacklee, Barnard
Coffin, Samuel
Colville, John
Criger, Namon
Divers, Charles
Eaton, George
Erlstroom, John
Gault, William
Grant, Peter
Greenwood, Thales
Hazard, William
Hock, J. Nicholas
Holmes, Caleb K.
Jackson, John W.
Jennings, Nathaniel
Jones, William
Linderman, Joseph
LeRoy, Alexander
Malbony, William T.
Mann, John
Neath, Henry
Palmer, John
Richardson, William
Stoddart, Robert
Turner, William
Watts, Herman
Wallander, Adam
Wood, Peleg

A captured French frigate

Campbell, Abraham
Limonden, William
Stone, Edmond

Carantine

Minois, Pierre

Caroline (1)

Albertson, John Nathaniel
Anderson, David
Beriston, Peter
Broadwater, Samuel
Burns, Charles
Edwards, David
Evans, Moses
Fritts, Joseph
Hile, Shadrick
Ingle, John
Johnson, James
Jones, Thomas
Lambert, Joseph
Martin, Anthony
Motley, Thomas
Mitchel, Jacob
Morgan, John
Murray, John
Murry, James
Peregrine, Taggert

Plumeria, Joseph
Saff, Francis
Simpson, Thomas
Spafford, Samuel
West, Dennis
Wester, Andrew
Williams, James
Wills, John
Wright, George

Caroline (2)

Hatch, Samuel
Honson, J. A.

Caroline (3)

Rice, John
Rice, Thomas

Cashiere

Bannister, George
Birch, Andrew
Blackstone, Edward
Bloom, Joseph
Cooper, Daniel
Deal, William
Drake, Henry
Evans, Henry
Foster, Joseph
Galloway, Joseph
Gilbert, Thomas
Halfpenny, Robert
Harris, John
Hoage, Rufus
Hudson, William
Ingbriton, Nicholas
Jentle, Joseph
Kimmins, John
Longworthy, John
Madden, Fredrick
Morell, Thomas
Mullan, John
Pack, Abraham
Ricco, Joachim
Schoeman, Frederick
Shammton, Samuel
Smith, Richard
Spear, Samuel
Stiles, William
Wilson, George
Wood, Thomas

Catharine

Allen, Elihu
Gifford, Barry
Truman, Alhine
Underwood, John F.
White, Alden
Williams, George (1)
Williams, George (2)

Ceres

Johnson, Robert

Charles

Dempsey, John
Dempsey, William
Forbes, James
Higgins, William

Pearson, Samuel

Charlotte (1)

Booth, Charles
Bowers, Jonathan
Bowers, Mrs.
Bowman, Leon
Castera, Jean S.
Creping, Thomas
Gallin, John
Lactortier, Charles Felid
Rogers, Henry
Young, Thomas

Charlotte (2)

Akins, William
Blake, Philip
Brown, William
Cerk, Frederick
Christy, Alexander
Clark, Joseph Wanton
Croft, George
Davis, Charles
Forbes, William
Goff, James
Holms, Andres
Lammers, Henry
Mitchell, Ezekiel
Norman, Peter
Pike, John
Stoddert, Reuben
Taylor, Jacob
Williams, George

Chesapeake

Babi, Joseph
Barton, Eseek
Bassinet, Charles
Cook, Benjamin
Cornish, Charles
Crawford, William
Dusheets, Arthur
Fletcher, John
Goldsberry, William
Gustave, Peter
Hamell, Christopher
Johnson, John
Marshall, Emanuel
Martin, Stephen
McDonald, John
Moore, John
Oakley, Joseph
Oudin, Pierre Francois
Palmer, John
Pearer, John
Rommel, Francis
Shaw, Edward
Valiant, Richard
Van Borgen, Martin
Wade, John
Williams, Hendrick
Yilstone, George

Colin West

Allen, Elissa
Prior, Essa

Columbia (1)

Barton, Nathan
Belford, Isaac
Canada, Prince
Castor, Charles
Courser, Adam
Dickenson, Francis
Dixon, Michael
Douchney, Hiram
Douglas, Thomas
Garthon, Willey
Hartford, John
Helm, Charles
Henderson, David
Jackson, John
Kain, Peter
Neaver, Thomas
Roach, Nicholas
Robinson, Benjamin
Tightham, Peter
Todd, William

Columbia (2)

Merrill, John

Commerce

Norris, Augusta George

Cornelia

Bramblecome, David

Courier

Anthony, Stephen
Atkins, James
Biss, Daniel W.
Bower, Joseph
Bergman, John
Davis, Robert
Dickinson, Chester
Dover, Charles
Dowell, Isaac
Hall, David
Hall, Richard
Hilaire, Gasper
Hopkins, Daniel
Hutchins, Edward
Keen, Nathaniel
Logan, William
Lowe, John
Manson, William
Mart, James
McKenzie, Alexander
Mills, William
Mullins, James
Neal, Dennis
Powers, James
Roles, John
Shaw, Richard
Towson, William
Walker, Armstrong
Way, John
White, David
Wood, John

Coxerien

Crandall, John

Criterion

Barron, William
Bidson, Thomas
Brereton, Benjamin
Brown, Henry
Chambers, Henry
DeWitt, John
Dickson, Charles
Griffiths, Thomas
Henderson, Alexander
Henshaw, Jacob S.
Johnson, John
Johnston, Thomas
Ludlow, Charles
Manatt, Richard
Marble, Jabez
Miller, John
Monks, John
Neel, David
Porter, William
Reeves, William
Royal, John
Sage, Moses
Schultz, John
Sherman, Reuben
Thomas, Archibald
Thompson, Martin
Wheeler, Charles
White, Henry

Dart

Nicholas, John Bruss
Pass, Samuel William
Pearce, Alexander

Deborah

Ober, Ezra

Decornau

Audibert, Joseph
Bennet, James
Brown, Benjamin
Brown, Lodwick
Brown, Samuel
Brownell, Richard
Burdge, Samuel
Chauvet, Frederic
George
Davis, Henry
Doolittle, Isaac
Garret, James
Hale, Horace
Haydon, Eli
Hurd, William
Lemercier, Peter
McCormick, Simon
Norton, George
Philon, Richard
Robinson, Samuel
Vaughan, Nathaniel
Weaphor, Andrew
West, John

Devon

Bloomdose, John
Carter, Enoch
Chauvet, Thomas

Damerie, Etienne
Debaize, Francois Jean
Dorsey, John
Ewell, Edward
Gravely, Joseph
Grosette, Jean Maurice
Hubbell, James
Perrott, Jean Francois
Tuttle, Joseph

Diamond

Armistead, Edward
Atwick, Thomas C.
Autist, Jean
Crowder, James

Dick

Archer, Joseph
Bishop, Edward
Donaldson, Joseph
Lellifer, Jacob

Dolphin

Carman, Francis
Carney, Edward
Chew, Joseph
Coffin, George
Fisher, Richard D.
Fryer, John
Hamilton, John
James John
Lovering, William
Low, Thomas
Myer, John
Nargney, James
Pannell, Hugh
Richardson, Samuel
Robinson, John
Smasher, Allen
Spencer, Samuel
Thomas, Henry

Dorotha

Nesbitt, Richard
Soreby, Robert

Ducornau

Doolittle, Isaac

Eliza

Alley, Samuel
Casey, Henry
Dunchller, Isaac
Fairchild, Hamlet
Fairchild, Robert
Faritte, Francis
Hammond, Joseph
Handel, Henry
Longford, Samuel
Murray, Jacob
Struby, John
Williams, John

Elizabeth

Scott, James

Elk

Tolson, Jeremy

Ellen & Emeline

Adams, James

Booth, William
Guillard, Joseph

Elvia

Ludson, Daniel
Smith, Chris
Weedon, Anthony

Empress

Buchanan, James
Gadet, Martin
Labery, Pierre Joseph
Menard, Nicholas

Enterprize

Davis, Ezra

Essex

Ammerson, Charles
Bancroft, Samuel
Blodget, Caleb
Brown, Samuel
Burnham, Enoch
Chandler, Simon
Clothey, Thomas
Davis, James
Dilus, Benjamin
Dolliver, William
Eastland, James
Fisher, Lewis
Fitzpatrick, William
Libby, Moses
Main, William
Merrett, Thomas
Moss, Thomas
Parker, John A.
Pomeril, Robert
Ripley, Eden
Russell, Patten
Shovel, John
Stockman, William B.
Treffy, James
Treffy, Peter
Wait, Henry
Waters, Abraham
Welch, Benjamin
Williams, Thomas

Eunice

Brandy, Francis

Expectation

Evans, Thomas L.

Experiment (1)

Anthony, John
Benner, Lewis
Cobb, Samuel
Dunn, Hezekiah
Ellis, John
Evelett, Francois
Fate, Thomas
Harris, James
Jewell, Samuel
Lynch, William
Ryder, Philip
Smith, Jacob
Williamson, Charles
Willingsworth, Jeffery

Experiment (2)
Murray, John

Fair American
Abbott, Ephraim
Baldwin, John
Butts, Joseph
Clapp, George
Downe, William
Gardner, Edward
Gerard, William
Hinkley, Joseph
Hinkley, Thomas
Homes, Michael
Lyon, John
Mathews, William
Mayor, James
Morgan, Jonathan
Myers, Daniel
Richardson, Cheney
Sparks, Robert
Still, David
Williams, James

Fanny
Abbey, Obadiah
Cheney, Daniel
Dematra, George
Hayes, Benjamin
Hoselquist, John
Lambert, William
McFadden, James
Roulemmon, Bernard
Schrader, Martin
Sereder, Martin

Favorite
Arnold, Thomas

Ferox
Aurel, Leonard
Bailey, Samuel
Caldwell, James
Courtney, James
Crandall, John
Deal, John
Deham, Charles
Freeman, Charles
Gilbert, George
Holts, Peter
Jones, William
Kalm, John
Lee, George
Myers, David
Shaw, Jacob
Trusty, Henry

Fire Fly
Haskell, Thomas
Winter, William

Flash
Lovett, Thomas L.
Rich, Ebenezer

Fortune
Kennisen, William

Fox
Andres, David

Baldwin, John
Basin, Francois Gaston
Beard, Francis
Bright, George
Cainton, Joseph
Calhoun, Richard
Dunn, John
Duran, Pierre
Fletcher, James
Harris, John
Harris, William
Howard, William
Irvin, Mathew
James, Daniel
Lamoree, Peter
Mack, Theron
McCoy, James A.
McMaken, John
Mingle, Thomas
Moore, Richard
Mullen, Edward
Perry, Charles
Rawle, Benjamin
Rennabin, Benjamin
Singleton, Samuel
Slowin, Abraham
Watson, Isaac
Wilkins, William
Wores, Robert

Fox Packet
Brown, Thomas
Dymoss, Peter
Jackson, William
Jenkins, Joseph
Woodcraft, John

Francis & Ann
Coffin, John
Hebron, John
Lettington, James

Frederick & Augusta
Booth, Stephen

Friendship
Ross, Richard
Walker, James

Friendship West
Dennison, Thomas
Garrot, William
Hennay, Jacob
Hoden, George
Humphries, Warren
Jones, Walter
Lindsborough, Charles
Livermore, Arthur
Luce, Prispery
Moore, George
Stewart, Isaac
Toby, Elisha
Courris, Francis
Courtis, Harry
Currien, Stephen
Giles, Edward
Green, Moses

Tibbet, Henry
Underwood, John
Young, Joseph

Gemini
Taylor, Thomas

General Kempt
Andrews, Benjamin
Archer, Joseph
Dieman, John
Dolliver, Richard
Fisher, John
Mone, Denis
Parley, William
Perkins, Elijah
Rogers, Robert
Smith, Richard
Spurr, Elijah
Symonds, M. John
Wheeler, John
Wyer, Joseph

Gleamer
Brevoort, William
Cortu, John
Laytin, William

Godfrey & Mary
Brown, Francis
Peckham, Henry
Pratt, Paris
Shepherd, William

Gold Coiner
Meigs, John

Goldfinch
Brown, John

Good Friends
Adgate, William
Boriesa, John
Gifford, Thomas
Hogabets, John
Johnson, Samuel
Lerna, John
Martin, Henry
Morris, George
Morris, John
Morse, Henry
Moss, John
Swain, Darius
Thompson, Robert
Vogel, Herman
Wessel, Samuel

Good Intent (1)
Jones, Thomas

Good Intent (2)
Andrews, Thomas
Davis, Benjamin S.
Elwell, Thomas
Millet, James
Sargant, Samuel Allen

Governor Gerry
Adam, Joseph
Allen, William
Armstrong, William
Basset, David

Braley, George
Charter, Samuel
Cross, Oliver
Cudworth, Henry
Deshays, Jean Francoise
Gabriel, Joshua
Gage, Isaac
Gannett, Felix
Harris, James
Hill, John
Hitch, Joshua
Hooper, Benjamin E.
Jones, Thomas
Keller, John
Linsey, Alexander
Merrett, Robert
Miller, John
Morie, Joseph
Smith, John
Smith, Richard
Tasdeboiz, Louis
West, Simon
Weston, David
Wilson, John

Governor McKean
Augustus, Amos
Baptist, John
Boffs, James
Boggs, James
Capewell, Bartholomew
Daly, John
Farrell, Francis
Hayes, Simon
Hook, John
Humphries, James
Likens, Andrew
Nones, A. B.
Regens, Jonathan
Sanders, Joseph
White, Philip
Winter, Peter

Grand Napoleon
Brown, Arn
Cooper, Charles
Cornwall, Arthur
Haught, John
Hutchins, Henry
Lane, James
Nicholas, Stephen
Ostand, Michael
Parish, Samuel
Poland, Joshua
Pote, Jeremiah
Redding, John
Rozier, William
Thornson, Joseph
White, Isaac
Wilcox, Lewis
Bradford, Charles
Bradford, Henry
Clapp, Abraham
Everill, Daniel

Gray, Morehouse
Hopson, Ebenezer
Howard, John
Moffit, John
Morgan, Henry
Pratt, Benjamin
Rerette, John Francis
Reynolds, Frederick
Sailes, James
Vail, Peter

Growler
Dunn, William
Roberts, Joel
Thornton, William

Hall

Hanna
Castania, John Baptiste

Hannah
Johnstone, Henry
Wicks, Littleton

Baird, David
Bawnham, Francis Aortas
Harvey, Joseph
Knight, John
Orne, Oliver P.
Pritchard, Israel
Ramsden, John
Selman, John
Tarbell, William George

Hannah & Eliza
Baday, John
Billings, John
Boyd, John
Fox, Richard
Fry, John
Laurencau, Jean
Lund, Peter M.
More, Francis
Roman, John
Williams, Joseph
Wing, John

Harvest
Carter, Henry
Coutie, Thomas
Grenaux, Jean Yves
Holden, Andrew
Morisau, Jean Baptiste
Murphy, William
Port, John
Riva, Antonio
Rouger, Desire

Hebe
Brown, William
Cannoway, Charles
Chisselsine, John
Denham, John
Fish, James
Francis, Joseph Butler
Gramsby, George
Hanson, Christian
Mode, David
Page, Thomas

Thompson, William
Watson, James
Williams, John

Henry Clements
Beek, Steward
Gallibrandt, Bernard
Harper, Nicholas
Horsydise, William
Russell, William
Smith, Thomas
Thompson, Thomas
Williams, John
Williams, Thomas

Hepa
Bymer, George
Edgar, William

Hero
Bradford, James

Hibernia
Beckett, William

Hope (1)
Harper, Robert

Hope (2)
Drake, John
Elves, Joseph
Elves, Manuel
Gillet, John Francis
Mitchell, John
Polease, Jean M.
Smith, John D.

Howard
Barnes, Robert
Barnes, William

Hunter
Anderson, James
Anthony, Abraham
Barchman, John
Bickford, Ebenezer
Bowdler, Thomas
Boyd, Edward
Brown, Jesse
Brown, John
Bunoth, Mansfield
Card, Nathaniel
Carrol, Robert
Carter, Moses
Carves, John
Child, Samuel
Cloutman, Samuel
Cogswell, Edward
Dean, Daniel
Dolorer, John
Fannol, Augustus
Fletcher, Henry
Foster, Asa
Gardner, Samuel
Glasgow, John
Grunlief, Timothy
Hall, William
Hammet, John
Hatchess, Levi
Henry, Edward

Hobert, George
Holden, Charles
Hosmer, Joseph
Huma, John Lewis
Hutchinson, Henry P.
Ingalls, Edmond
Jenkins, Richard
Johnstone, Samuel
Knox, Thomas
Leach, Charles
Lee, Nathaniel
Leveatt, William
Levis, Raymond
Main, Henry
Michell, James
Mitchell, John
Monett, Samuel
Moor, Thomas
Neal, Henry
Nye, Charles N.
Obrion, Jeuth.
Parsons, Samuel
Peters, Jacob
Peterson, James
Peterson, Nicholas
Poland, David
Robins, Jeremiah
Ross, George
Rowe, Stephen
Rowell, James
Ruliff, London
Schroder, Henrick
Scribner, Elijah
Signard, Samuel Francis
Sodoburgh, John
Thomas, Alexander
Thomas, George
Towns, Daniel
Tucker, Edward
Upton, Jeduthun
Upton, Samuel
Wain, Benjamin
Warner, Charles
Warner, George
Whipple, Samuel
White, Sampson
Willcox, Caesar

Independence (1)
Beachman, George
Bears, Moses
Davis, John
Ducassau, Charles
Arnold
Farrell, John
Fountain, Isaac
Francois, James
Garrison, John
Harris, George
James, Peter
Lewis, John
Manuel, John
McFuel, James

Mesey, William
Richard, John
Stowell, Peter
Stowell, John
Van Donveer, Peter
Wilson, Peter
Wilson, William

Independence (2)
Bonie, James
Griffin, William
Harvey, Joseph
Hayes, Elias Warner
Hodgins, Daniel
Howell, Thomas
Law, William
Marsh, Jesse
Robinson, Joseph

Indian
Brown, Richard

Indian Lass
Bassett, Gorham
Holland, James
Stone, Samuel
White, Philip
Works, Alfred

Indostan
Lerwick, John

Industry (1)
Williams, Alexander

Industry (2)
Lawrence, David

James
Henderson, Benjamin
Ide, John Henry
Ley, John
Murray, Nathaniel
Norgrave, Jeremiah
Rice, John
Short, Clement

Jane Barns
Malborne, Evan
Manuel, Diego
Molineux, William
Stites, Ezra

Jenny
Bateman, Charles

Joel Barlow
Allen, James
Bell, James
Berdick, Simon
Black, George
Brown, Ebenezer
Buckhannon, Edwin
Chipman, Christopher
Durvolf, Stephen
Griffin, William
Hart, Bartholomew
Jeffery, Job
Johnson, William
Morell, Benjamin
Palmer, Pero
Palmer, Robert

Peters, Aaron
Smith, William
Stevens, Hugh

John (1)
Scannell, Cornell

John (2)
Bitters, John
Farmer, George
Johnson, Henry
McKerbie, Alexander
Phillips, John
Simpson, William
Skinner, Tilman
Sone, John

John & Frances
Anderson, Samuel
Baas, John Thomas
Cammon, Robert
Carnes, Joseph
Easterly, George
Fayolle, Eugene
Griffin, John
Higgins, James
Howard, Samuel
Joseph, Thomas
Laquerene, Peter Louis
Miller, Stephen
Montierne, Francois
Mumford, Thomas
Rogers, James
Russell, John

John Barnes
Carson, Robert
Scot, Benjamin

Joseph (1)
Bartlett, John
Cogan, John
Davis, Morris
Dolliver, John
Michael, Car.
Parsons, Daniel
Rogers, Gorham
Thomas, Spencer
Tuck, James
Woodbury, John

Joseph (2)
Besson, Phillipe
Brown, Ambrose James
Chambers, Elias
Gardner, Edward
Glover, John
Hammond, William
Stacy, William

Janice & Lydia
Davis, John
Gedson, John
Haywood, Simon

King George
Cooper, James
Evans, Jacob
Farret, George

Leader

Anderson, William
Bell, George
Brewer, James
Gardner, James
Hatch, Abraham
Keith, James
Richardson, Daniel
Wills Jr., John

Leo

Anderson, Daniel
Bartlett, Caleb
Cary, John
Codman, Richard
Davis, John
Doughty, Jesse
Foss, Edmond
Foss, Joseph
Gove, William
Hume, George
Manson, Nathaniel
Miller, William
Peters, John
Tucker, George C.
Turtle, French
Wood, Sylvanus

Lightning

Anderson, James
Bertine, John
Collison, Joseph
Crawford, Nelson
Garthy, James
Jones, Richard
Jordan, Peter
Morris, Robert
Peterson, Peter
Ruddick, William
Wallace, James
Warrance, John

Louisa

Boss, Thomas
Chestly, Amos
Jennin, Samuel
Jones, John
Woodberry, Caleb

Lyon

Coleman, William
Jackson, James
Wright, William

Magdalen

Adams, John
Barkman, Henry
Foster, Thomas
Harre, John
Harts, William
Holmes, James
Jackson, Thomas
Johnston, John
Johnston, Peter
Kennedy, John
Leverage, William
Lewis, Francis B.
Morrey, Henry

Norris, Lament
Prince, Benjamin
Reynolds, Owen
Sketchly, William
Williams, Peter
Wittington, Joshua

Manilla

Coleman, Charles

Margaret (1)

Brown, George
Fox, Washington
Rock, Oliver

Margaret (2)

Armstrong, Daniel
Banks, Perry
Grubb, Andrews
Jones, Hensey
McQuillan, James
Whitehead, James

Maria Christiana

Garcia, Anthony
Lewis, James
Lopez, Joseph

Mariner

Abraham, William

Marmion

Allen, Philip
Atkins, Martyn
Barstow, Samuel
Bennett, Charles
Charlies, James
Coader, Antoine
Fedie, John
Hart, Joshua
Johnson, Robert
Menard, Augustus
Montel, Pierre
More, Henry
Roberson, Henry
Rodwin, Walter
Sanis, Lewis
Silva, Francis

Mars

Allen, Henry
Baker, John
Barret, Anthony
Bronston, Job Ellis
Caldwell, James
Cleveland, Lawrence
Dye, William
Glass, Reuben
Hudleback, George
Jones, Benjamin
Lawson, James
Parker, George
Saunders, William
Whitewood, Charles

Martin

West, John

Mary (1)

Castor, Thomas
Frederick, Charles

Gassy, Raymond
Hudson, James
Lunberge, Berg
Wright, Amos

Mary (2)

Fletcher, John William
Mecoh, David

Mary (3)

Dean, Moses
Diamon, William
Neel, William
Rey, David
Spark, Samuel

Matilda

Brown, William
Eskindson, George

Messenger

Farrell, Jeffery
Gorham, John

Meteor

Bartell, Robert
Bunaberg, Nicholas
Burn, Reuben
Dougherty, Michael
Forester, Francis
Hill, Leonard
Kennedy, John
Latimore, Matthew
Perne, George
Perney, William
Scholle, John
Snate, George
Soderstrum, Ola
White, Charles

Minerva (1)

Campbell, Alexander

Minerva (2)

Fardel, Thomas
Hall, John
Little, John

Minerva (3)

Campbell, William
Conner, Michael
Harris, Ebenezer
Hill, Daniel
Lepberg, John P.
Merrill, Enoch
Watson, William

Miranda

Baransan, John
Grover, Edmond
Littlefield, Rufus
Parmell, Benjamin
Robertson, Robert
Yard, William

Monticello

Greenland, Stephen
Snell, David

Montgomery

Brannigan, John
May, Henry
May, Henry

Narius

    Le Core, Peter Murry
    Relavoine, Louis

Ned

    Anderson, Mathew
    Ashmore, Edward
    Baker, Israel
    Buchanan, John
    Churchill, Manuel
    Dalloway, John
    Duncan, George
    Gundy, James
    Hacket, William
    Lewis, Thomas
    Mathis, Hendrick
    McNab, James
    Meiniffe, Charles
    Mines, Artemas
    Moore, William
    Morweek, Andre
    Neil, Henry
    Norris, Robert
    Ransom, Joseph
    Robinson, Nicholas
    Sanderhin, Daniel
    Sickless, James M.
    Smith, Elisha
    Snowdon, Jacob
    Spangle, Frederick
    Spires, Thomas
    Stansbury, James
    Wessels, Robert
    Wheeler, Benjamin
    Williams, James
    Williams, John
    Wilson, William
    Windham, John
    Wolfe, William
    Young, Rufus

Nope (1)

    Dominic, John
    Peter, John
    Thomas, William
    Thompson, William

Nope (2)

    Glenn, Robert
    Kennedy, John
    Michael, Peter
    Oleily, Thomas
    Walden, James
    Yale, Nathaniel

Nope (3)

    Blue, Peter
    Brown, John
    Crete, John Nicholas
    Morris, Patterson
    Rodrick, Joseph
    Thomas, William

Ocean

    Anderson, William
    Lyons, Henry
    Pigott, James

Omer

    Hopkins, Elisha

Orbit

    Albert, Hezekiah
    Allen, Eleazer
    Barnard, Timothy
    Brown, Wheeler
    Cleaveland, Davis
    Cox Miles
    Dow, John
    Frees, James
    Greenleaf, Thomas
    Grubb, James Julian
    Huntress, Robert
    Jackson, Thomas
    Johnson, Jacob
    Johnston, James
    Kennedy, Peter
    Larrabee, William
    Lewis, George
    Moore, Benjamin
    Rape, Nicholas
    Stanley, Joseph
    Sweetman, Samuel
    Thompson, Courtney
    Thompson, James
    Tipton, Solomon
    Webb, John
    Wheeler, William

Orders in Council

    Abbott, Timothy
    Allen, John
    Amos, Cheney
    Barttist, John
    Butler, George
    Champy, Peter Felix
    De la Batiet, Francois
    Gabriel
    Doughty, Levi
    Gardner, William
    Godwin, William
    Hall, James
    Haskell, Robert
    Jordan, Richard
    Kenny, Harris
    Madey, Gustave Charles
    D'Eocouthant
    Martin, William
    Meilleur, Simon
    Mouls, Benjamin
    Nickerson, Josiah
    Patten, James
    Robinson, Elias
    Russell, James
    Slocum, William
    Wilson, John
    Young, Abiakar

Otter

    Bentsten, John
    Brown, John
    Brown, Thomas
    Carr, Laurence

    Evans, Thomas
    Gisby, John
    Hitchi, John
    Hook, Aaron
    Howland, Samuel
    Shepherd, Daniel
    Smith, Thomas
    Terry, Daniel
    Wade, Otis
    Young, William

Pallas

    Brown, James
    Burton, John
    Colfax, William
    Douarty, Angelo
    Edgerly, George
    Edward, Prince
    Hockman, William
    Lewis, Edward
    Moor, Lawrence
    Ratoon, Thomas
    Reardon, Anthony
    Robinson, Ebenezer
    Robinson, William
    Smith, George
    Wendell, Isaac

Paul Jones

    Allen, John
    Allman, John
    Anderson, James
    Askwith, William Vickery
    Barnard, John
    Barnes, Nathaniel
    Bayard, Joseph
    Brown, Charles
    Burbank, G. W.
    Bursted, John
    Cato, John
    Chadwick, M. T.
    Coleman, David
    Colton, Walter
    Cook, Charles H.
    Cooke, William
    Cooper, Andrew A.
    Cramstead, James
    Dibble, Reuben
    Edwards, John
    Fardy, Anthony
    Fink, Johannes
    Freeman, John
    Friday, John
    Gibbs, Henry
    Godfry, William
    Green, William
    Grosse, William
    Guillard, Louis
    Guillard, Peter
    Hamilton, Anthony M.
    Hitre, Dempsey
    Irvin, Anthony
    Jackson, Henry
    Jackson, James

Jackson, Lambert
Jasseciu, Louis
Johnston, Joseph
Joseph, Francis
Keven, Edward
King, Joseph
Lacoen, John B.
Ligan, Pierre
Lippert, Thomas D.
Little, George
Lockwood, Rufus
Long, R. S.
Louis, Nicholas
Martin, Isaac
Martin, John
Martin, Manuel
Miller, Thomas
Mooney, Peter
Mully, Etienne
Newell, Isaac
Parsons, Samuel D.
Peck, James
Peterson, Peter
Prince, George
Riddlefield, James
Risings, John
Roberson, John
Roley, John
Rosings, Cornelius
Ruffield, Samuel
Samsuillen, John
Sandford, John W.
Sanford, William
Sentille, Francis
Smith, John
Smith, Thomas
Stephens, William
Stewart, William
Taylor, Archibald
Thomas, Abraham
Thompson, Henry
Vutch, William
Warren, David
Whetting, Samuel C.
Wild, Thomas
Wilson, James

Paulina
Caesar, Joseph
Perseverance
Hathaway, William N.
Pert
Bellinger, William
Court, Robert
Everly, John
Freddle, John
Gellens, William
Lawson, Thomas
Null, William
Ray, Nathaniel
Smith, William
Shurtleff, William
Ware, William

Watts, James
Philips Burgh
Pattingale, Enoch
Stranton, John
Phoebe & Jane
Moneys, Robert
Phoenix
Foster, Joseph
Glover, John
Hovey, Joseph
Peterson, Hance
Van Kirk, Joseph
Point
Lapavous, Philip
Polly
Adams, Samuel Reed
Bryant, James
Goss, Joshua
Harris, William
Rogers, Nathaniel
Vicary, Richard
Porcupine (1)
Bunker, Peter
Can, Moses
Carver, Abraham
Cochran, Peter
Coffin, Alexander
Duffy, Nathaniel
Evans, Hale
Harris, William
Henry, John
Kinlay, Joseph
Kneeland, Ebenezer
Lyons, Peter
Lyons, Samuel
Melville, John
Myers, Frederick
Parr, James
Prutty, Henry
Star, G. C.
Thompson, Samuel
Williams, James
Porcupine (2)
Cross, Peter
Postsea
Davids, Thomas
Stevens, Joseph
Winkley, John
Price
Barrett, James
Bennett, Charles
Blanchard, Simon
Brown, John
Cooke, Samuel
Dean, Jonas
Dillin, Pierre
Francis, James
Francis, John
Hatterson, Nathan
Hays, Moses
Hickman, Joseph
Hunter, William

Ingerson, Michael
Jones, Gardner
May, Walter
Price, Jacob
Robertson, James
Southcombie, Plummer
Sparrow, James
Swett, David
Taylor, Thomas
Thompson, Charles
Wells, John
Winant, Peter
Wright, John
Young, William
Prince of Wales
Berto, John
Gisenado, Samuel H.
Ham, John
Posnrill, John
Princess
Cartwright, George
Redman, William
Smith, James
Taylor, James
White, James
Williams, R. N.
Print
Cowell, Slater
Dixey, John
Ramsdell, John
Prompt
Allen, John
Atwood, Thomas
Bartis, John
Beecher, Thaddeus
Bucher, William Palmer
Chase, Nathaniel
Cole, William
De Forest, John H.
Hubbard, Alfred
Lopaus, William
Smith, Henry
Thompson, Nathaniel
Purse
Dasing, Caesar
Latish, Joseph
Turner, Samuel
Rachel & Ann
Brutus, Mario
Creighton, Manuel
Deparvier, John
Ferris, Joseph
Morgan, John
Robin, John
Rogers, Henry
Smith, John
Webber, Frederick
Williams, Alexander
Rambler
Adams, John
Resolution
Burch, James

Chapel, William
Johnston, Edward
Jordon, David

Revenge

Bowen, John
Brown, Benjamin
Barrett, Benjamin
English, Edward
Eveleth, William
Flynn, Abraham
Gardner, Joseph
Geyer, Joshua
Gland, Benjamin
Hazel, George
Healy, John
Holland, Richard
Holmstrom, Alexander
Holmstrom, Lorentz
Hudson, Thomas
Kellinger, John
Lemond, John
Lothrop, James
McKenny, John
Merritt, Jonathan
Mesick, Elihu
Page, John
Price, Samuel
Putman, Charles
Riddle, Thomas
Watkins, Thomas
Wheeler, John W.
White, John

Rhode & Betsey

Knapp, Walker

Rising States

Slayter, John

Rising Sun

Dowling, Anthony
Marchley, William
Osbourne, Peter
Ranter, Thomas

Rolla

Armstrong, William
Athroun, Samuel
Augustin, Anthony
Colcocha, Anthony
Dwight, Samuel
Grey, Thomas
Ireland, James
Johnson, Edward
Johnson, John
Lassalle, Francisco
Marshall, David
McCauley, George M.
Miller, Henry
Morris, John
Reeves, James
Ratner, James
Ryan, William
Spofford, Jacob
Williams, Ambrose

Rossie

Daniels, J. D.
Gould, Thomas
Stubbs, James

Sally

Page, Josiah Clark

Science

Codsifershall, Charles

Shadow

Campbell, Reynold
Esdaile, Thomas
Hansell, John
Huggins, Daniel
Innes, John
Johnston, Simon
Lilley, Simon
Marshall, Alexander
Mingle, William
Mural, Mark
Nuganes, Daniel
Phillips, John
Price, John
Tweed, Caldwell
White, Laurence
Williams, Elge

Sherbrook

Lewis, John
Simons, William
Williams, John

Siro

Bodfish, William
Botello, Antony
Brown, Stephen
Brown, Sandy
Coffer, William
Cooper, John
Davis, Charles S.
Denison, George
Drinkwater, Peter
Furlong, William
Gall, Michael
Goodnow, Saul B.
Guillot, Richard
Hanson, Henry
Hanson, Peter
Harris, Bradley
Hiler, George
Hutchins, William
Jackson, Samuel
Lovely, Henry
Man, Tabez
Masson, William
Matthews, Leonard
Miller, William
Moore, Alexander
Neal, John
Nye, Cornelius
Pain, George
Patterson, Joshua
Patterson, Andrew
Patterson, John
Perkins, John
Peterson, Andre

Pinkham, Amos
Priestly, James
Prosper, John
Ringgold, Peregrine
Samuel, Nicholas
Sheardon, Ellison
Smiles, John
Soderback, Andrew
Stewart, Adam
Stiles, William
Stout, John
Thompson, William
Tubbs, Martin
Wakefield, Nathaniel
Weeks, James
Welsh, William
Wilson, Charles
Wright, William

Spitfire

Bartlett, Thomas
Bishop, William
Bridge, Francis
Burnham, Francis A.
Carton, Thomas
Dodd, Samuel
Dolliver, Francis
Fletcher, William B.
Jervis, John
Jones, Francis
Jones, Francis Vickery
Jutt, Benjamin
Lampson, Charles
Lovett, William
Swett, Samuel
Wadden, Jacob
Wheler, Richard
Wild, Caleb
Witheram, Nicholas
Wooldridge, Thomas

Squirrel

Baldwin, Pierson
Campbell, James
Chain, John
Cummings, Samuel
Dumas, Raymond
Frenarye, Pierre
Gable, John
Gilchrist, John
Gregory, George
Hughes, John
Lapierre, Etienne
Long, Charles
Long, Charles
Martin, John
McFarland, John
Montaindon, Julie
Montaindon, Louise
Montaindon, Aimee
Montaindon, Henry
Montaindon, Julien
Montaindon, Marianne
Montaindon, Pierre Hans

Pagechin, Henry
Payechin, Henry
Plan, Madame G.
Road, John
Strum, Magdalene
Strum, Yves
Sturm, Jeannette
Taney, Lewis
Tapiau, Vitale
Thompson, Andrew
Thompson, Peter

St. Martin's Plantation
Allen, John L.
Brown, John
Covell, Isaac
Ellis, William
Gilbert, John
Grey, John
Hulet, Michael
Jones, William P.
Mains, Darius
Merritt, Almond
Morrison, David
Russell, Isaac
Taylor, George
Thompson, Whitney
Wills, Peter
Woodward, Anthony

Star
Bisley, Horace
Borgin, Gabriel
Brown, Jesse
Clements, John C.
Clerk, William
Cox, John
Dennison, Jedidiah
Dennison, Judah
Duston, Peter
Ervin, William
Frash, James
Gordon, Thomas
Holts, Daniel
Hughes, John
Isaac, Moses
Johnston, Gershom
Joseph, Francis
Joulyon, Robert
Lewis, John
Louring, Henry
Mallery, Michael
Mix, William A.
Reed, Hartwell
Richman, Joshua
Right, Joshua
Romain, Samuel
Sullivan, Hampton
Thompson, William
Valentine, John
Vaughan, Thomas
Washburn, Edward
Williams, David
Williams, Henry

Stephen
Bailey, Joseph
Benny, Malloc
Dickson, Richard
Gardner, Anthony
Graham, George
Harris, William
Humphries, Thomas
Porgan, Theodore
Quan, George
Robinson, James
Skiddy, William

Sybella
Andrews, George
Dubois, Alexander
Michell, Fairly
Ross, Philip
Small, Enoch

Terrible
Cole, Richard
Cook, James
Kromhout, Barney
Leach, William
Lewis, Angus T.
Lovet, John
Martian, Peter
Root, Eliakim F.
Smith, Eldridge
Smith, John
Smith, William
Tufts, Zachariah
Williamson, William

Thomas Wilson
Evans, John

Thrasher
Davison, Henry
Summerville, Charles

Three Brothers
Elliot, Robert
Phipping, Nathaniel

Tickler
Best, Robert
Brandage, John
Butman, Charles
Christie, James
Dinsmore, John
Dougal, Thomas
Foster, John Thomas
Gray, William
Hacking, Robert
Hokinson, Nathaniel
Joles, Robert
Monroe, John
Pickrage, George
Webb, John

Tiger
Bartholf, Nicholas
Calder, John H.
Caldwell, Samuel
Carles, Paul
Churchill, Timothy
Coffin, George P.

Coffin, William P.
Conklin, William
Copland, James
Hill, Ephraim
Hill, Joseph
Jones, James
Kershon, Abraham
Ludlow, Reuben
Mars, George
Morris, James
Morslender, Reuben
Nicholas, John
Quackenbush, William
Ryer, Peter
Sampson, Jacob
Small, George D.
Spencer, Leonard
Sproson, James
Sutton, John
Waterman, John
Westerbest, William

Tom
Batman, John
Beckworth, Benjamin
Brown, John William
Brown, William
Cantrell, Norvell
Cummings, James
Davis, John
Davis, John
Dean, Nathaniel B.
Doolittle, Henry
Ford, George L.
Frazer, John
Gee, Thomas
Gilmore, William H.
Heard, Thomas
Hollinger, William
Jenkins, Nathaniel
Layfield, Littleton
Lerves, Robert
Mansfield, James
Martian, John B.
McIntire, Samuel
Nelley, Richard John
Otis, Charles
Payne, Joseph Smith
Penman, Richard
Pilton, Alexander
Rowe, John
Rowley, Henry
Russell, George
Sheffers, Steuben
Smith, Andrew
Taylor, Samuel E.
Trifft, Joseph
Wilson, John

Tom Thumb
Timmerman, Matthew
Williams, Joshua

Traveler
Church, William

Flynn, Pierre
Harker, John
Long, Joseph
Timmons, John
Thompson, John
Wilkson, John

Two Brothers

Atkinson, William
Bridges, John
Deer, Andrew
Jackson, Daniel
Leas, Anthony
Paulin, Nicholas
Pickering, George
Sheppard, William
Stanton, Henry
Thomas, John
Venison, Leven
Wing, Joseph
Young, William

Tyger

Atkins, Uriah
Bartell, William
Blanchard, Carvan
Hubbard, John
Leeds, Leonard
Newell, John
Snow, Joseph
West, Reuben

Union (1)

Allen, Jacob
Armstrong, Nicholas
Ashton, Joseph
Barradaile, Thomas
Carnes, Richard
Chekes, William
Crofts, William
Custis, George
Denckle, Christian
Douglas, Thomas
Elliot, Francis
Fairweather, Robert
Finch, William
Gilligan, William
Godshall, John
Goetz, Lewis
Graham, William
Grisse, William
Harrison, John
Hasem, John
Hill, William
Irwin, Magnus
Johnston, Samuel
Johnston, William
Jones, John
Kirby, Benjamin
Knight, Isaac D.
Legrand, Claudius F.
Lownsburg, Carpenter
Maniers, Benjamin
Martin, John
Mason, Francis

McDonald, John
McElroy, William
Mercer, Chaumont
Morris, John
Morris, Thomas
Olmstead, Moss
Parker, David
Randale, Frederick
Reeves, Joseph
Robinson, David
Schyder, Jacob Knapp
Seibold, John
Seimandy, Gasper
Sparks, Henry
Stone, William
Swanston, Jacob
Thompson, William
Underwood, Benjamin
Webb, William
Williams, Peter

Union (2)

Bell, James
Curtis, John
Drew, William
Gray, Isaac
Roberts, Hugh
Robinson, John

US Brig Argus

Addigo, Henry
Allen, Joseph
Allen, William Henry
Allen, William Howard
Allister, Isaac
Anderson, George
Appine
Ash, Samson
Ayres, Robert
Bacus, David
Barlow, John
Baron, Thomas
Barry, James
Baxter, Charles
Been, James
Benedict, William
Benson, James
Blackledge, John
Bladen, John
Blisset, James
Bogart, Killian William
Bowne, Asher
Bradt, Francis
Brown, Alexander
Brown, John
Butnet, William
Carbenett, John
Chuglar, John
Clark, John
Clerk, George
Collin, William
Collins, Andrew
Conklin, Robert
Conley, Samuel

Coombes, James
Cooper, James
Crawford, Robert
Curtis, Joseph
Davis, William
Day, Syles C.
Delphy, Richard
Dennison, Henry
Dillon, William
Donham, Abinesar
Eggert, Francis
Fadden, John
Famer, John
Fleming, John Joseph
Freeman, John
Ferguson, Henry
Gilbert, Jurdonet
Ginkins, William
Givel, Oliver
Godfry, Edward
Graham, David
Griffith, Benjamin
Groger, Henry
Hall, James
Hanscom, Moses
Hennery, James
Henry, John
Henry, William
Hill, Thomas
Himes, Walter
Hopkins, William
Hovrington, William
Hudson, John
Hunt, Charles
Hunter, Charles
Jamison, Robert
Jamison, William
Johns, Thomas
Jordan, Joseph
Juderwicks, James
Kellam, James
Kennedy, Richard
Lounsbury, Benjamin
Leach, Josem
Leary, Daniel
Lee, Ely
Levi, Uriah Phillips
Lovett, Henry
McDonald, John
McLeod, Collin
Myatt, Frederick
Nagel, George
Nash, Manuel
Noney, Peter
Nugent, John
Phillips, Benjamin
Place, John
Pottenger, William
Price, William
Price, William
Richards, George
Roberson, John

Robinson, James
Robinson, John
Rosah, Samuel
Russell, Frederick
Salley, James
Scott, John
Scott, Robert
Sears, Abram
Shaw, William
Smith, Adam
Smith, James
Smith, John
Smith, William
Snelson, Robert Lewis
Sniffen, John
Starbuck, George
Stevens, John
Tafe, William
Tamer, John
Temple, William
Thomas, David
Thompson, Thomas
Tolvet, Charles
Waite, Abel
Waters, Robert
Watson, William H.
White, James
Williams, John
Wilson, George
Young, John
Young, Thomas

US Frigate Chesapeake
Anthony, Joseph
Lingrin, Peter

Valentine
Arnold, Welcome
Chadwick, John
Potter, Richard H.

Vengeance
Adams, Theophilus
Aulajo, Thomas
Bagot, Lewis
Bevers, Clement
Birnent, Edward
Bowen, Lewis
Bribion - 1st child
Bribion - 2nd child
Bribion - 3rd child
Bribion, Madam
   (first name not listed)
Cochran, Stephen
Couet, John
Dowdell, George R.
Green, Horace
Law, John
Lynch, Joseph
Robinson, Michael
Simpson, William
Smith, Chester
Taylor, John
Whitney, Joseph

Vengeance (French ship)

Parker, Peter

Viper
Allen, Thomas
Arnold, Alfred
Baker, John
Belfast, Richard
Bradley, Hugh
Campbell, John
Churchill, Stephen
Crouch, Richard
Duncan, Peter
Farrell, John
Foster, David
Hay, Cornelius
Hinkley, Aaron
Hunter, George
Jones, George
McNelly, Thomas
Morgan, Joseph
Paine, Joseph
Robinson, Charles
Small, Richard
Smally, Thomas
Thomas, William
Thompson, William
Tyren, William
Webster, James
Williams, John
Williams, Thomas
Winslow, Elijah
Wintory, Anthony
Young, John

Virginia Planter
Andrews, Charles
Combs, William
Goodday, James
Harding, John
Harter, Henry
Horton, Henry
Irving, William
Rice, Francis
Vandine, Henry
Wilson, Peter

Volante
Laborde, Peter

Warren
Cook, Joseph B.
Hazard, Thomas
Jane, Joseph
Joseph, John
Manuel, Joseph
Peterson, John

Washington
Watkins, Fredrick

Wasp
Green, John
Hitch, Ebenezer
McKinzey, John
Neil, James

Weasel
Ashfield, Henry
Bailey, John

Blackman, Moses
Brill, John
Chapple, John
Dunn, David
Gibbs, Daniel
Hull, Edward
Jiffs, Joseph
King, John
Lemmon, Henry
Longfield, Thomas
Macomb, James
Vangosbet, Cato
Voorhies, James

William Bayard
Averell, Loring
Barton, Robert
Blanchard, George
Bunker, James
Davis, James
Dorrell, John
Elm, John
Folger, Shubert
Francis, Benjamin
Gorling, George C.
Hall, James
Harens, William
Hoskins, James
Jackson, John
Jones, Benjamin
Jones, John
Kellum, John C.
Leach, William
Lewis, John
Mather, Allyn
Richardson, Francis
Riley, Michael
Roe, John
Rollands, Joseph
Sennet, John
Sherwood, John
Steel, John
Summers, Alexander
Wather, John
Williams, John

Wolf
Alexander, Richard
Andrews, Joseph
Bass, Prince
Carnes, John
Chevers, Joseph
Clough, Isaac
Cloutman, Robert
Easton Joseph
Follet, William
Grand, William
Knapp, Samuel
Lymones, Procter
Lyon, Charles
Phippen, John
Price, David
Quinen, Stephen
Stacey, Benjamin

        Stacey, Samuel

Young Dixon

        Badson, Jacob
        Chesbrough, Benjamin F.
        Cotterill, Henry

Young Holkar

        Blackwood, Bailey
        Blake, Alexander
        Burr, Francis L.
        Burr, Isaac
        Cowen, Robert
        James, Thomas
        Le Baith, John J.
        Lebarth, John
        Sturges, Bradley
        Thomas, John
        Toby, Peter
        Wassen, Joseph Bartram
        West, Benjamin
        Williams, William

Zebra

        Aris, James
        Barber, William
        Barker, Reuben
        Brant, Thomas
        Brooks, Hayden
        Theophilus
        Carter, Daniel
        Carter, Edward
        Cheney, Amos

        Cosevin, Pierre Mathrie
        Davis, William
        Delbos, Felix
        Dunlap, James A.
        Durand, John
        Edsom, John
        Fay, Salmon
        Garcia, Francois
        Gellie, Thomas
        Hanford, William
        Harrington, Simon
        Jackson, John
        Jones, Charles
        Lawrence, John
        Le Graet, Peter
        Lyon, Charles
        Manson, Jeremiah
        Membelly, William
        Merser, Benjamin
        Miller, Charles
        Mills, William
        Minugh, John
        Payer, Walter
        Phillips, John
        Phipps, Stephen
        Pitts, William
        Pomp, William
        Reynolds, James
        Roff, Isaac
        Stag, John B.

        Thomas, Andre
        Vezin, Charles
        Walker, Francis
        Weeks, David
        Weellid, William

Zephyr (1)

        Brown, David
        Coffin, Alexander
        Fleur, Francois Xavier
        Graham, James
        Hancock, William
        Ingles, Daniel
        Jones, Thomas
        Mathews, Joseph
        Olcott, Jedediah
        Pierson, Elijah
        Postlock, William
        Roper, Thomas William
        Thompson, Henry
        Wilson, John

Zephyr (2)

        Carr, James
        Donelson, Joseph
        Golandre, John
        Harris, Simon
        Keen, Joseph
        Modre, Joseph
        Plumer, William Reed
        Walker, Seth
        Williamson, David

Abbott, David - Seaman - Number: 671 - Prize name: Avon - Ship type: P - How taken: HM Frigate Barbadoes - When taken: 8 Mar 1815 - Where taken: West Indies - Date received: 24 Apr 1815 - From what ship: HMS Bellerophon - Born: New Hampshire - Age: 25 - Discharged on 2 Jun 1815 and released to the Cartel Shakespeare.

Abbott, Francis - Seaman - Number: 724 - Prize name: Sine Qua Non - Ship type: P - How taken: HM Brig Elk - When taken: 20 Feb 1815 - Where taken: off Madeira - Date received: 29 Aug 1815 - From what ship: HMS Tonnant - Born: Massachusetts - Age: 33 - Discharged on 2 Jun 1815 and released to the Cartel Shakespeare.

Adams, Jean B. - Seaman - Number: 469 - Prize name: Fox - Ship type: P - How taken: HM Frigate Barbadoes - When taken: 11 Jan 1815 - Where taken: off Amelia Island, Florida - Date received: 16 Apr 1815 - From what ship: HMS Swifsure - Born: Saint-Domingue (Haiti) - Age: 20 - Race: Negro - Discharged on 8 May 1815.

Adams, Nathaniel - Master's Mate - Number: 859 - Prize name: US Brig Syren - Ship type: War - How taken: HM Ship-of-the-Line Medway - When taken: 12 Jul 1814 - Where taken: off Cape of Good Hope - Date received: 11 Jul 1815 - From what ship: from Portsmouth - Born: Not listed - Discharged on 11 Jul 1815 and released to the Cartel Wooddrop Sims.

Adams, Robert - Seaman - Number: 56 - How taken: Taken out of HM Sloop Volentaire - Date received: 15 Feb 1815 - From what ship: HMS Myrmidon - Born: Massachusetts - Age: 24 - Discharged on 17 Feb 1815 and sent to Dartmoor.

Adore, William - Mate - Number: 453 - Prize name: Sophie - Ship type: MV - How taken: HMS Ister - When taken: 2 Dec 1814 - Where taken: off Amelia Island, Florida - Date received: 16 Apr 1815 - From what ship: HMS Swifsure - Age: 28 - Discharged on 2 Jun 1815 and released to the Cartel Sovereign.

Ager, Thomas - Seaman - Number: 516 - Prize name: Hope, prize to US Sloop-of-War Wasp - Ship type: War - How taken: HM Sloop Fairy - When taken: 6 Jan 1815 - Where taken: Unknown - Date received: 16 Apr 1815 - From what ship: HMS Swifsure - Born: New Jersey - Age: 31 - Discharged on 2 Jun 1815 and released to the Cartel Sovereign.

Aide, John L. - Quarter Gunner - Number: 145 - Prize name: US Brig Syren - Ship type: War - How taken: HM Ship-of-the-Line Medway - When taken: 12 Jul 1814 - Where taken: off Cape of Good Hope - Date received: 21 Feb 1815 - From what ship: HMS Slaney - Discharged on 24 Feb 1815 and sent to Dartmoor.

Alden, Henry - Seaman - Number: 321 - Prize name: Thomas, prize to Privateer Scourge - Ship type: P - How taken: HM Frigate Aquilon - When taken: 11 Mar 1815 - Where taken: off Cape Finisterre, Spain - Date received: 25 Apr 1815 - From what ship: HMS Opossum - Discharged on 27 Apr 1815 and released to the Cartel Minework.

Alexander, George - Seaman - Number: 313 - Prize name: Transit - Ship type: MV - How taken: Impressed at North Shields - When taken: 24 Nov 1814 - Where taken: North Shields - Date received: 4 Mar 1814 - From what ship: HM Cutter Surley - Race: Mulatto - Discharged on 7 Mar 1815 and sent to Dartmoor.

Alexander, James - Seaman - Number: 95 - Prize name: Prince de Neufchatel - Ship type: P - How taken: HM Ship-of-the-Line Leander, HM Ship-of-the-Line Newcastle & HM Frigate Acasta - When taken: 28 Dec 1814 - Where taken: Lat 35N Long 52W - Date received: 18 Feb 1815 - From what ship: HMS Sybille - Born: Portland - Age: 27 - Discharged on 19 Feb 1815 and sent to Dartmoor.

Alexander, Philip - Seaman - Number: 503 - Prize name: Fox - Ship type: P - How taken: HM Frigate Barbadoes - When taken: 11 Jan 1815 - Where taken: off Amelia Island, Florida - Date received: 16 Apr 1815 - From what ship: HMS Swifsure - Born: Not listed - Discharged on 2 Jun 1815 and released to the Cartel Sovereign.

Alfonso, (-----) - Seaman - Number: 493 - Prize name: Fox - Ship type: P - How taken: HM Frigate Barbadoes - When taken: 11 Jan 1815 - Where taken: off Amelia Island, Florida - Date received: 16 Apr 1815 - From what ship: HMS Swifsure - Born: New Orleans - Age: 24 - Race: Negro - Discharged on 24 Apr 1815.

Alfords, John - Seaman - Number: 536 - Prize name: St. Francis - Ship type: MV - How taken: HMS Ister - When taken: 8 Dec 1814 - Where taken: Amelia Island, Florida - Date received: 16 Apr 1815 - From what ship: HMS

Swifsure - Born: Massachusetts - Age: 20 - Discharged on 2 Jun 1815 and released to the Cartel Sovereign.

Atlantis, Aug. - Seaman - Number: 491 - Prize name: Fox - Ship type: P - How taken: HM Frigate Barbadoes - When taken: 11 Jan 1815 - Where taken: off Amelia Island, Florida - Date received: 16 Apr 1815 - From what ship: HMS Swifsure - Born: Savannah - Age: 18 - Race: Negro - Discharged on 2 Jun 1815 and released to the Cartel Sovereign.

Ames, William - Seaman - Number: 119 - Prize name: Prince de Neufchatel - Ship type: P - How taken: HM Ship-of-the-Line Leander, HM Ship-of-the-Line Newcastle & HM Frigate Acasta - When taken: 28 Dec 1814 - Where taken: Lat 35N Long 52W - Date received: 18 Feb 1815 - From what ship: HMS Sybille - Age: 30 - Discharged on 19 Feb 1815 and sent to Dartmoor.

Anderson, John - Seaman - Number: 233 - Prize name: Decatur - Ship type: P - How taken: HM Frigate Rhin - When taken: 4 Jun 1814 - Where taken: at sea - Date received: 2 Mar 1815 - From what ship: HMS Dannemark - Born: Charleston - Age: 20 - Discharged on 3 Mar 1815 and sent to Dartmoor.

Anderson, John - Seaman - Number: 134 - How taken: Taken off the HM Frigate Iris - Date received: 21 Feb 1815 - From what ship: HMS Slaney - Born: Charlestown - Age: 32 - Race: Mulatto - Discharged on 24 Feb 1815 and sent to Dartmoor.

Andrews, Ebenezer - Seaman - Number: 669 - Prize name: Avon - Ship type: P - How taken: HM Frigate Barbadoes - When taken: 8 Mar 1815 - Where taken: West Indies - Date received: 24 Apr 1815 - From what ship: HMS Bellerophon - Born: Massachusetts - Age: 35 - Discharged on 2 Jun 1815 and released to the Cartel Shakespeare.

Andrews, Joseph - Seaman - Number: 368 - Prize name: Leo - Ship type: P - How taken: HM Frigate Tiber - When taken: 11 Mar 1815 - Where taken: Lat 45.24N Long 12.3W - Date received: 29 Mar 1814 - From what ship: HMS Tiber - Born: Massachusetts - Age: 26 - Discharged on 2 May 1815 and released to the Cartel Ariel.

Andrews, Samuel - Ordinary Seaman - Number: 138 - Prize name: US Brig Syren - Ship type: War - How taken: HM Ship-of-the-Line Medway - When taken: 12 Jul 1814 - Where taken: off Cape of Good Hope - Date received: 21 Feb 1815 - From what ship: HMS Slaney - Born: Norway - Age: 21 - Discharged on 24 Feb 1815 and sent to Dartmoor.

Anthony, Joseph - Seaman - Number: 630 - Prize name: Helene, prize of Privateer Morgianna - Ship type: P - How taken: HM Frigate Pique - When taken: 7 Mar 1815 - Where taken: coast of America - Date received: 20 Apr 1815 - From what ship: HMS Musquito - Age: 20 - Discharged on 27 Apr 1815.

Arden, Peter - Seaman - Number: 558 - Prize name: Mary - Ship type: MV - How taken: HMS Muros - When taken: 8 Dec 1814 - Where taken: West Indies - Date received: 16 Apr 1815 - From what ship: HMS Swifsure - Born: Pennsylvania - Age: 26 - Discharged on 2 Jun 1815 and released to the Cartel Sovereign.

Argues, Francis - Seaman - Number: 331 - Prize name: Leo - Ship type: P - How taken: HM Frigate Tiber - When taken: 11 Mar 1815 - Where taken: Lat 45.24N Long 12.3W - Date received: 28 Mar 1815 - From what ship: HMS Tiber - Discharged on 27 Apr 1815 and released to the Cartel Minework.

Armstrong, Andrew - Seaman - Number: 579 - Prize name: Sorin, prize to Privateer Prince of Neufchatel - Ship type: P - How taken: HM Ship-of-the-Line Medway - When taken: 12 Jul 1814 - Where taken: Unknown - Date received: 16 Apr 1815 - From what ship: HMS Swifsure - Born: New York - Age: 22 - Discharged on 2 Jun 1815 and released to the Cartel Sovereign.

Armstrong, Charles - 2nd Mate - Number: 186 - Prize name: Chance - Ship type: P - How taken: HMS Statire - When taken: 1 Apr 1814 - Where taken: at sea - Date received: 2 Mar 1815 - From what ship: HMS Dannemark - Born: Hampton - Age: 34 - Discharged on 3 Mar 1815 and sent to Dartmoor.

Ash, Oliver - Seaman - Number: 287 - How taken: Taken off the HM Ship-of-the-Line Severn and other ships - Date received: 2 Mar 1815 - From what ship: HMS Dannemark - Born: Norfolk - Age: 23 - Race: Mulatto - Discharged on 3 Mar 1815 and sent to Dartmoor.

Ashford, Graves - Seaman - Number: 769 - Prize name: Sine Qua Non - Ship type: P - How taken: HM Brig Elk - When taken: 20 Feb 1815 - Where taken: off Madeira - Date received: 29 Aug 1815 - From what ship: HMS Tonnant - Age: 25 - Race: Negro - Discharged on 2 Jun 1815 and released to the Cartel Shakespeare.

Autel, Stephen - Seaman - Number: 29 - Prize name: Plutarch - Ship type: MV - How taken: HM Schooner Helicon

- When taken: 5 Feb 1815 - Where taken: Lat 45.5N Long 7W - Date received: 8 Feb 1815 - From what ship: HMS Helicon - Age: 28 - Discharged on 10 Feb 1815 and sent to Dartmoor.

Ayres, John - Purser - Number: 639 - Prize name: Avon - Ship type: P - How taken: HM Frigate Barbadoes - When taken: 8 Mar 1815 - Where taken: West Indies - Date received: 24 Apr 1815 - From what ship: HMS Bellerophon - Age: 20 - Discharged on 26 Apr 1815.

Bailey, Samuel - Passenger - Number: 256 - Prize name: Wolfe - Ship type: P - How taken: Not known - Date received: 2 Mar 1815 - From what ship: HMS Dannemark - Discharged on 3 Mar 1815 and sent to Dartmoor.

Bain, John - Seaman - Number: 794 - Prize name: US Brig Syren - Ship type: War - How taken: HM Ship-of-the-Line Medway - When taken: 12 Jul 1814 - Where taken: off Cape of Good Hope - Date received: 11 Jul 1815 - From what ship: HMS Royal Sovereign - Born: Not listed - Discharged on 11 Jul 1815 and released to the Cartel Wooddrop Sims.

Baker, Joseph - Seaman - Number: 511 - Prize name: Hope, prize to US Sloop-of-War Wasp - Ship type: War - How taken: HM Sloop Fairy - When taken: 6 Jan 1815 - Where taken: Unknown - Date received: 16 Apr 1815 - From what ship: HMS Swifsure - Born: New York - Age: 27 - Discharged on 2 Jun 1815 and released to the Cartel Sovereign.

Baker, Thomas - Captain - Number: 265 - Prize name: Netterville - Ship type: MV - How taken: HM Brig Onyx - When taken: 25 Dec 1814 - Where taken: at sea - Date received: 2 Mar 1815 - From what ship: HMS Dannemark - Born: Rhode Island - Age: 26 - Discharged on 3 Mar 1815 and sent to Dartmoor.

Baldura, Theophilus - Seaman - Number: 279 - Prize name: Tartan - Ship type: MV - How taken: Given up - Date received: 2 Mar 1815 - From what ship: HMS Dannemark - Discharged on 3 Mar 1815 and sent to Dartmoor.

Bales, John - Seaman - Number: 624 - Prize name: Helene, prize of Privateer Morgianna - Ship type: P - How taken: HM Frigate Pique - When taken: 7 Mar 1815 - Where taken: coast of America - Date received: 20 Apr 1815 - From what ship: HMS Musquito - Born: New York - Age: 24 - Discharged on 27 Apr 1815.

Bangs, George - 5th Lieutenant - Number: 65 - Prize name: Prince de Neufchatel - Ship type: P - How taken: HM Ship-of-the-Line Leander, HM Ship-of-the-Line Newcastle & HM Frigate Acasta - When taken: 28 Dec 1814 - Where taken: Lat 35N Long 52W - Date received: 18 Feb 1815 - From what ship: HMS Sybille - Born: Boston - Age: 21 - Discharged on 19 Feb 1815 and sent to Dartmoor.

Bans, William - Seaman - Number: 690 - Prize name: George Little - Ship type: P - How taken: HM Frigate Granicus - When taken: 20 Jan 1815 - Where taken: off Cape Finisterre, Spain - Date received: 16 Jul 1815 - From what ship: HMS Hope - Born: New Orleans - Race: Negro - Discharged on 2 Jun 1815 and released to the Cartel Shakespeare.

Baptiste, Jean - Seaman - Number: 475 - Prize name: Fox - Ship type: P - How taken: HM Frigate Barbadoes - When taken: 11 Jan 1815 - Where taken: off Amelia Island, Florida - Date received: 16 Apr 1815 - From what ship: HMS Swifsure - Born: Guadeloupe - Age: 30 - Race: Negro - Discharged on 2 Jun 1815 and released to the Cartel Sovereign.

Baptiste, John - Seaman - Number: 270 - Prize name: Netterville - Ship type: MV - How taken: HM Brig Onyx - When taken: 25 Dec 1814 - Where taken: at sea - Date received: 2 Mar 1815 - From what ship: HMS Dannemark - Born: New Orleans - Age: 27 - Discharged on 3 Mar 1815 and sent to Dartmoor.

Barbadoes, Joseph - Seaman - Number: 526 - Prize name: Dolphin - Ship type: MV - How taken: HM Frigate Barbadoes - When taken: 4 Dec 1814 - Where taken: off St. Bartholomew - Date received: 16 Apr 1815 - From what ship: HMS Swifsure - Born: Boston - Age: 21 - Race: Negro - Discharged on 2 Jun 1815 and released to the Cartel Sovereign.

Barbeca, Jacque - Seaman - Number: 550 - Prize name: Hera - Ship type: MV - How taken: HM Frigate Pique - When taken: 16 Dec 1814 - Where taken: off Porto Rico - Date received: 16 Apr 1815 - From what ship: HMS Swifsure - Age: 26 - Race: Negro - Discharged on 2 Jun 1815 and released to the Cartel Sovereign.

Barker, Thomas - Seaman - Number: 108 - Prize name: Prince de Neufchatel - Ship type: P - How taken: HM Ship-of-the-Line Leander, HM Ship-of-the-Line Newcastle & HM Frigate Acasta - When taken: 28 Dec 1814 - Where taken: Lat 35N Long 52W - Date received: 18 Feb 1815 - From what ship: HMS Sybille - Born: Boston - Age: 21 - Discharged on 19 Feb 1815 and sent to Dartmoor.

Barney, O. C. - Seaman - Number: 707 - Prize name: Sine Qua Non - Ship type: P - How taken: HM Brig Elk - When taken: 20 Feb 1815 - Where taken: off Madeira - Date received: 16 Jul 1815 - From what ship: HMS Hope - Age: 22 - Discharged on 2 Jun 1815 and released to the Cartel Shakespeare.

Bartlett, E. - Seaman - Number: 596 - Prize name: George Little - Ship type: P - How taken: HM Frigate Granicus - When taken: 20 Jan 1815 - Where taken: off Cape Finisterre, Spain - Date received: 18 Apr 1815 - From what ship: HMS Euryalus - Born: Massachusetts - Age: 68 - Discharged on 2 Jun 1815 and released to the Cartel Sovereign.

Bartlett, Nathaniel - Seaman - Number: 363 - Prize name: Leo - Ship type: P - How taken: HM Frigate Tiber - When taken: 11 Mar 1815 - Where taken: Lat 45.24N Long 12.3W - Date received: 29 Mar 1814 - From what ship: HMS Tiber - Born: Massachusetts - Age: 24 - Discharged on 2 May 1815 and released to the Cartel Ariel.

Bartlett, Peekwood - Seaman - Number: 733 - Prize name: George Little - Ship type: P - How taken: HM Frigate Granicus - When taken: 20 Jan 1815 - Where taken: off Cape Finisterre, Spain - Date received: 29 Aug 1815 - From what ship: HMS Tonnant - Born: Massachusetts - Age: 18 - Discharged on 2 Jun 1815 and released to the Cartel Shakespeare.

Barton, Mathew - Seaman - Number: 60 - How taken: Taken out of HM Brig Sabine - Where taken: Portsmouth - Date received: 15 Feb 1815 - From what ship: HMS Myrmidon - Born: Kent County - Age: 39 - Race: Mulatto - Discharged on 17 Feb 1815 and sent to Dartmoor.

Baxter, Charles - Seaman - Number: 670 - Prize name: Avon - Ship type: P - How taken: HM Frigate Barbadoes - When taken: 8 Mar 1815 - Where taken: West Indies - Date received: 24 Apr 1815 - From what ship: HMS Bellerophon - Born: Massachusetts - Age: 28 - Discharged on 2 Jun 1815 and released to the Cartel Shakespeare.

Beck, Francis - Seaman - Number: 232 - Prize name: Decatur - Ship type: P - How taken: HM Frigate Rhin - When taken: 4 Jun 1814 - Where taken: at sea - Date received: 2 Mar 1815 - From what ship: HMS Dannemark - Born: Charleston - Age: 20 - Race: Negro - Discharged on 3 Mar 1815 and sent to Dartmoor.

Belt, W. L. - Midshipman - Number: 855 - Prize name: US Brig Syren - Ship type: War - How taken: HM Ship-of-the-Line Medway - When taken: 12 Jul 1814 - Where taken: off Cape of Good Hope - Date received: 11 Jul 1815 - From what ship: from Portsmouth - Born: Not listed - Discharged on 11 Jul 1815 and released to the Cartel Wooddrop Sims.

Bennett, Horsham - Seaman - Number: 837 - Prize name: US Brig Syren - Ship type: War - How taken: HM Ship-of-the-Line Medway - When taken: 12 Jul 1814 - Where taken: off Cape of Good Hope - Date received: 11 Jul 1815 - From what ship: HMS Royal Sovereign - Born: Not listed - Discharged on 11 Jul 1815 and released to the Cartel Wooddrop Sims.

Bennister, George - Boy - Number: 400 - Prize name: Leo - Ship type: P - How taken: HM Frigate Tiber - When taken: 11 Mar 1815 - Where taken: Lat 45.24N Long 12.3W - Date received: 29 Mar 1815 - From what ship: HMS Tiber - Born: Massachusetts - Age: 13 - Discharged on 30 May 1815 and released to the Cartel Atlas.

Benson, John - Seaman - Number: 283 - How taken: Taken off the HM Ship-of-the-Line Severn and other ships - Date received: 2 Mar 1815 - From what ship: HMS Dannemark - Born: New York - Age: 22 - Race: Negro - Discharged on 3 Mar 1815 and sent to Dartmoor.

Benton, Samuel - Seaman - Number: 177 - Prize name: Farmer's Daughter - Ship type: MV - How taken: HM Ship-of-the-Line Leviathan - When taken: 20 Mar 1814 - Where taken: at sea - Date received: 2 Mar 1815 - From what ship: HMS Dannemark - Born: Kent County - Age: 20 - Discharged on 3 Mar 1815 and sent to Dartmoor.

Bertham, William Henry - Master - Number: 35 - Prize name: Plutarch - Ship type: MV - How taken: HM Schooner Helicon - When taken: 5 Feb 1815 - Where taken: Lat 45.5N Long 7W - Date received: 10 Feb 1815 - From what ship: HMS Helicon - Born: Boston - Age: 27 - Discharged on 10 Feb 1815 and sent to Ashburton.

Bertram, John - Seaman - Number: 560 - Prize name: Mary - Ship type: MV - How taken: HMS Muros - When taken: 8 Dec 1814 - Where taken: West Indies - Date received: 16 Apr 1815 - From what ship: HMS Swifsure - Born: Massachusetts - Age: 29 - Discharged on 2 Jun 1815 and released to the Cartel Sovereign.

Betarsl, Samuel - Seaman - Number: 247 - Prize name: John - Ship type: MV - How taken: HM Brig Zenobia - When taken: 11 Sep 1814 - Where taken: at sea - Date received: 2 Mar 1815 - From what ship: HMS Dannemark - Discharged on 3 Mar 1815 and sent to Dartmoor.

Bissett, Robert - Seaman - Number: 172 - Prize name: Enterprise - Ship type: P - How taken: Not known - Where taken: at sea - Date received: 2 Mar 1815 - From what ship: HMS Dannemark - Born: Albany - Age: 25 - Discharged on 3 Mar 1815 and sent to Dartmoor.

Blair, Samuel - Seaman - Number: 823 - Prize name: US Brig Syren - Ship type: War - How taken: HM Ship-of-the-Line Medway - When taken: 12 Jul 1814 - Where taken: off Cape of Good Hope - Date received: 11 Jul 1815 - From what ship: HMS Royal Sovereign - Born: Not listed - Discharged on 11 Jul 1815 and released to the Cartel Wooddrop Sims.

Blake, D. - Seaman - Number: 543 - How taken: Gave himself up from HMS Harmony - When taken: 4 Sep 1814 - Where taken: Barbados - Date received: 16 Apr 1815 - From what ship: HMS Swifsure - Born: Massachusetts - Age: 19 - Discharged on 2 Jun 1815 and released to the Cartel Sovereign.

Blanchard, C. - Purser - Number: 614 - Prize name: Sine Qua Non - Ship type: P - How taken: HM Brig Elk - When taken: 20 Feb 1815 - Where taken: off Madeira - Date received: 18 Apr 1815 - From what ship: HMS Euryalus - Born: Massachusetts - Age: 23 - Discharged on 2 Jun 1815 and released to the Cartel Sovereign.

Blodget, Phineas - Boy - Number: 397 - Prize name: Leo - Ship type: P - How taken: HM Frigate Tiber - When taken: 11 Mar 1815 - Where taken: Lat 45.24N Long 12.3W - Date received: 29 Mar 1814 - From what ship: HMS Tiber - Born: Massachusetts - Age: 16 - Discharged on 30 May 1815 and released to the Cartel Atlas.

Boniface, Anthony - 2nd Lieutenant - Number: 219 - Prize name: Decatur - Ship type: P - How taken: HM Frigate Rhin - When taken: 4 Jun 1814 - Where taken: at sea - Date received: 2 Mar 1815 - From what ship: HMS Dannemark - Born: Charleston - Age: 30 - Discharged on 3 Mar 1815 and sent to Dartmoor.

Bonner, John - Seaman - Number: 167 - How taken: Impressed from MV Rolla - Date received: 1 Mar 1815 - From what ship: Royal Naval Hospital Plymouth - Age: 20 - Discharged on 3 Mar 1815 and sent to Dartmoor.

Boston, George - Seaman - Number: 684 - Prize name: George Little - Ship type: P - How taken: HM Frigate Granicus - When taken: 20 Jan 1815 - Where taken: off Cape Finisterre, Spain - Date received: 16 Jul 1815 - From what ship: HMS Hope - Age: 34 - Discharged on 2 Jun 1815 and released to the Cartel Shakespeare.

Bottes, John - Cook - Number: 131 - Prize name: St. Joanne - Ship type: MV - How taken: HM Brig Sabine - When taken: 25 Oct 1814 - Where taken: Lat 44 - Date received: 21 Feb 1815 - From what ship: HMS Slaney - Race: Black - Discharged on 24 Feb 1815 and sent to Dartmoor.

Boucher, Louis - Seaman - Number: 505 - Prize name: Fox - Ship type: P - How taken: HM Frigate Barbadoes - When taken: 11 Jan 1815 - Where taken: off Amelia Island, Florida - Date received: 16 Apr 1815 - From what ship: HMS Swifsure - Born: Saint-Domingue (Haiti) - Age: 20 - Discharged on 2 Jun 1815 and released to the Cartel Sovereign.

Bourdon, Amond - Passenger - Number: 39 - Prize name: Leo - Ship type: P - How taken: HM Frigate Granicus - When taken: 2 Dec 1814 - Where taken: off Lorient, France - Date received: 12 Feb 1815 - From what ship: HMS Impregnable - Born: Lorient, France - Age: 25 - Discharged on 17 Feb 1815 and sent to Dartmoor.

Bourtell, Caleb - Surgeon - Number: 583 - Prize name: George Little - Ship type: P - How taken: HM Frigate Granicus - When taken: 20 Jan 1815 - Where taken: off Cape Finisterre, Spain - Date received: 18 Apr 1815 - From what ship: HMS Euryalus - Born: Massachusetts - Age: 30 - Discharged on 24 Apr 1815.

Bowder, Jacob - Seaman - Number: 848 - Prize name: US Brig Syren - Ship type: War - How taken: HM Ship-of-the-Line Medway - When taken: 12 Jul 1814 - Where taken: off Cape of Good Hope - Date received: 11 Jul 1815 - From what ship: from Portsmouth - Born: Not listed - Discharged on 11 Jul 1815 and released to the Cartel Wooddrop Sims.

Bowen, John - Seaman - Number: 235 - Prize name: Decatur - Ship type: P - How taken: HM Frigate Rhin - When taken: 4 Jun 1814 - Where taken: at sea - Date received: 2 Mar 1815 - From what ship: HMS Dannemark - Born: Warren - Age: 34 - Discharged on 3 Mar 1815 and sent to Dartmoor.

Boyd, James - Seaman - Number: 504 - Prize name: Fox - Ship type: P - How taken: HM Frigate Barbadoes - When taken: 11 Jan 1815 - Where taken: off Amelia Island, Florida - Date received: 16 Apr 1815 - From what ship: HMS Swifsure - Born: Philadelphia - Age: 24 - Discharged on 2 Jun 1815 and released to the Cartel Sovereign.

Boyent, William - Seaman - Number: 54 - How taken: Christiane of Petersburgh - Where taken: Leith - Date received: 15 Feb 1815 - From what ship: HMS Myrmidon - Age: 25 - Discharged on 17 Feb 1815 and sent to Dartmoor.

Boyer, David - Seaman - Number: 328 - Prize name: Leo - Ship type: P - How taken: HM Frigate Tiber - When taken: 11 Mar 1815 - Where taken: Lat 45.24N Long 12.3W - Date received: 28 Mar 1815 - From what ship: HMS Tiber - Discharged on 27 Apr 1815 and released to the Cartel Minework.

Boylston, Zebadiah - 4th Lieutenant - Number: 64 - Prize name: Prince de Neufchatel - Ship type: P - How taken: HM Ship-of-the-Line Leander, HM Ship-of-the-Line Newcastle & HM Frigate Acasta - When taken: 28 Dec 1814 - Where taken: Lat 35N Long 52W - Date received: 18 Feb 1815 - From what ship: HMS Sybille - Born: Springfield - Age: 31 - Discharged on 19 Feb 1815 and sent to Dartmoor.

Brackett, James - Prize Master - Number: 156 - Prize name: Nancy, prize of the Privateer United States - Ship type: P - How taken: HM Sloop Papillon - When taken: 27 Dec 1814 - Where taken: Lat 30 Long 10 - Date received: 27 Feb 1815 - From what ship: HMS Nimble - Born: New Hampshire - Age: 25 - Discharged on 3 Mar 1815 and sent to Dartmoor.

Braun, James - Seaman - Number: 825 - Prize name: US Brig Syren - Ship type: War - How taken: HM Ship-of-the-Line Medway - When taken: 12 Jul 1814 - Where taken: off Cape of Good Hope - Date received: 11 Jul 1815 - From what ship: HMS Royal Sovereign - Born: Not listed - Discharged on 11 Jul 1815 and released to the Cartel Wooddrop Sims.

Bray, George - Seaman - Number: 820 - Prize name: US Brig Syren - Ship type: War - How taken: HM Ship-of-the-Line Medway - When taken: 12 Jul 1814 - Where taken: off Cape of Good Hope - Date received: 11 Jul 1815 - From what ship: HMS Royal Sovereign - Born: Not listed - Discharged on 11 Jul 1815 and released to the Cartel Wooddrop Sims.

Brewers, Nathaniel - Merchant - Number: 441 - Prize name: Perveance, cartel - How taken: Detained at Barbados - When taken: 23 Oct 1814 - Date received: 16 Apr 1815 - From what ship: HMS Swifsure - Age: 33 - Discharged on 24 Apr 1815.

Brezur, Benjamin - Seaman - Number: 343 - Prize name: Leo - Ship type: P - How taken: HM Frigate Tiber - When taken: 11 Mar 1815 - Where taken: Lat 45.24N Long 12.3W - Date received: 28 Mar 1815 - From what ship: HMS Tiber - Born: Boston - Age: 46 - Discharged on 27 Apr 1815 and released to the Cartel Minework.

Briggs, William - Seaman - Number: 545 - How taken: Gave himself up from HMS Harmony - When taken: 4 Sep 1814 - Where taken: Barbados - Date received: 16 Apr 1815 - From what ship: HMS Swifsure - Born: Massachusetts - Age: 20 - Discharged on 2 Jun 1815 and released to the Cartel Sovereign.

Brinnie, Das. - Seaman - Number: 351 - Prize name: Leo - Ship type: P - How taken: HM Frigate Tiber - When taken: 11 Mar 1815 - Where taken: Lat 45.24N Long 12.3W - Date received: 29 Mar 1814 - From what ship: HMS Tiber - Born: Massachusetts - Age: 27 - Discharged on 27 Apr 1815 and released to the Cartel Minework.

Brown, Andrew - Seaman - Number: 735 - Prize name: Sine Qua Non - Ship type: P - How taken: HM Brig Elk - When taken: 20 Feb 1815 - Where taken: off Madeira - Date received: 29 Aug 1815 - From what ship: HMS Tonnant - Born: Massachusetts - Age: 37 - Discharged on 2 Jun 1815 and released to the Cartel Shakespeare.

Brown, Henry - Seaman - Number: 295 - How taken: Taken off the HM Ship-of-the-Line Severn and other ships - Date received: 2 Mar 1815 - From what ship: HMS Dannemark - Born: Huntingdon - Age: 22 - Race: Black - Discharged on 3 Mar 1815 and sent to Dartmoor.

Brown, James - Seaman - Number: 530 - Prize name: Gallant - Ship type: MV - How taken: HM Frigate Barbadoes - When taken: 6 Dec 1815 - Where taken: off St. Bartholomew - Date received: 16 Apr 1815 - From what ship: HMS Swifsure - Born: Maryland - Age: 40 - Race: Mulatto - Discharged on 2 Jun 1815 and released to the Cartel Sovereign.

Brown, Jesse J. - Captain - Number: 455 - Prize name: Fox - Ship type: P - How taken: HM Frigate Barbadoes - When taken: 11 Jan 1815 - Where taken: off Amelia Island, Florida - Date received: 16 Apr 1815 - From what ship: HMS Swifsure - Born: New Orleans - Age: 42 - Discharged on 22 Apr 1815.

Brown, John - Seaman - Number: 549 - Prize name: Hera - Ship type: MV - How taken: HM Frigate Pique - When taken: 16 Dec 1814 - Where taken: off Porto Rico - Date received: 16 Apr 1815 - From what ship: HMS Swifsure - Age: 25 - Discharged on 2 Jun 1815 and released to the Cartel Sovereign.

Brown, John - Seaman - Number: 532 - Prize name: Gallant - Ship type: MV - How taken: HM Frigate Barbadoes - When taken: 6 Dec 1815 - Where taken: off St. Bartholomew - Date received: 16 Apr 1815 - From what ship: HMS Swifsure - Born: Maryland - Age: 25 - Discharged on 2 Jun 1815 and released to the Cartel Sovereign.

Brown, Robert - Seaman - Number: 539 - Prize name: Engineer - Ship type: MV - How taken: HMS Muros - When taken: 21 Sep 1814 - Where taken: off Porto Rico - Date received: 16 Apr 1815 - From what ship: HMS Swifsure - Born: New York - Age: 27 - Discharged on 2 Jun 1815 and released to the Cartel Sovereign.

Brown, Samuel - Seaman - Number: 382 - Prize name: Leo - Ship type: P - How taken: HM Frigate Tiber - When taken: 11 Mar 1815 - Where taken: Lat 45.24N Long 12.3W - Date received: 29 Mar 1814 - From what ship: HMS Tiber - Born: Massachusetts - Age: 20 - Discharged on 30 May 1815 and released to the Cartel Atlas.

Brown, William - Seaman - Number: 243 - Prize name: Mary, recaptured prize - Ship type: P - How taken: HM Brig Sophie - When taken: 19 Jun 1814 - Where taken: at sea - Date received: 2 Mar 1815 - From what ship: HMS Dannemark - Discharged on 3 Mar 1815 and sent to Dartmoor.

Brown, William - Seaman - Number: 555 - Prize name: Mary - Ship type: MV - How taken: HMS Muros - When taken: 8 Dec 1814 - Where taken: West Indies - Date received: 16 Apr 1815 - From what ship: HMS Swifsure - Born: Rhode Island - Age: 23 - Discharged on 8 May 1815.

Brown, Zach. - Seaman - Number: 625 - Prize name: Helene, prize of Privateer Morgianna - Ship type: P - How taken: HM Frigate Pique - When taken: 7 Mar 1815 - Where taken: coast of America - Date received: 20 Apr 1815 - From what ship: HMS Musquito - Born: Rhode Island - Age: 20 - Discharged on 27 Apr 1815.

Bubier, John - Midshipman - Number: 857 - Prize name: US Brig Syren - Ship type: War - How taken: HM Ship-of-the-Line Medway - When taken: 12 Jul 1814 - Where taken: off Cape of Good Hope - Date received: 11 Jul 1815 - From what ship: from Portsmouth - Born: Not listed - Discharged on 11 Jul 1815 and released to the Cartel Wooddrop Sims.

Burnham, Abraham - Surgeon - Number: 63 - Prize name: Prince de Neufchatel - Ship type: P - How taken: HM Ship-of-the-Line Leander, HM Ship-of-the-Line Newcastle & HM Frigate Acasta - When taken: 28 Dec 1814 - Where taken: Lat 35N Long 52W - Date received: 18 Feb 1815 - From what ship: HMS Sybille - Born: Ipswich - Age: 30 - Discharged on 19 Feb 1815 and sent to Dartmoor.

Burnham, Henry - Seaman - Number: 752 - Prize name: Melvina - Ship type: MV - How taken: HM Frigate Andromeda - When taken: 28 Feb 1815 - Where taken: off Lisbon - Date received: 29 Aug 1815 - From what ship: HMS Tonnant - Age: 28 - Discharged on 2 Jun 1815 and released to the Cartel Shakespeare.

Burns, John - Seaman - Number: 446 - Prize name: High Flyer - Ship type: LM - How taken: HMS Muros - When taken: 14 Nov 1814 - Where taken: off St. Bartholomew - Date received: 16 Apr 1815 - From what ship: HMS Swifsure - Age: 23 - Discharged on 2 Jun 1815 and released to the Cartel Sovereign.

Butcher, Jacob - Seaman - Number: 173 - Prize name: Enterprise - Ship type: P - How taken: Not known - Where taken: at sea - Date received: 2 Mar 1815 - From what ship: HMS Dannemark - Born: Massachusetts - Age: 23 - Race: Mulatto - Discharged on 3 Mar 1815 and sent to Dartmoor.

Butler, David - Seaman - Number: 188 - Prize name: Chance - Ship type: P - How taken: HMS Statire - When taken: 1 Apr 1814 - Where taken: at sea - Date received: 2 Mar 1815 - From what ship: HMS Dannemark - Born: Boston - Age: 41 - Discharged on 3 Mar 1815 and sent to Dartmoor.

Callaghan, James - Seaman - Number: 21 - Prize name: Mary Ann - Ship type: MV - How taken: Impressed - When taken: 20 Dec 1814 - Date received: 3 Feb 1815 - From what ship: HMS Bittern - Born: Pennsylvania - Age: 23 - Race: Mulatto - Discharged on 4 Feb 1815 and sent to Dartmoor.

Campbell, Henry - Seaman - Number: 237 - How taken: Gave himself up - Date received: 2 Mar 1815 - From what ship: HMS Dannemark - Born: Delaware - Age: 28 - Discharged on 3 Mar 1815 and sent to Dartmoor.

Canot, Lawrence - Seaman - Number: 651 - Prize name: Avon - Ship type: P - How taken: HM Frigate Barbadoes - When taken: 8 Mar 1815 - Where taken: West Indies - Date received: 24 Apr 1815 - From what ship: HMS Bellerophon - Born: Massachusetts - Age: 43 - Discharged on 2 Jun 1815 and released to the Cartel Shakespeare.

Card, Jacob - Seaman - Number: 685 - Prize name: George Little - Ship type: P - How taken: HM Frigate Granicus - When taken: 20 Jan 1815 - Where taken: off Cape Finisterre, Spain - Date received: 16 Jul 1815 - From what ship: HMS Hope - Born: New York - Age: 31 - Discharged on 2 Jun 1815 and released to the Cartel Shakespeare.

Carle, James - Seaman - Number: 182 - Prize name: Farmer's Daughter - Ship type: MV - How taken: HM Ship-of-the-Line Leviathan - When taken: 20 Mar 1814 - Where taken: at sea - Date received: 2 Mar 1815 - From what ship: HMS Dannemark - Born: Boston - Age: 32 - Race: Black - Discharged on 3 Mar 1815 and sent to Dartmoor.

Carlo, William - Seaman - Number: 464 - Prize name: Fox - Ship type: P - How taken: HM Frigate Barbadoes - When taken: 11 Jan 1815 - Where taken: off Amelia Island, Florida - Date received: 16 Apr 1815 - From what ship: HMS Swifsure - Born: Maryland - Age: 22 - Discharged on 2 Jun 1815 and released to the Cartel Sovereign.

Carmere, (-----) - Seaman - Number: 507 - Prize name: Fox - Ship type: P - How taken: HM Frigate Barbadoes - When taken: 11 Jan 1815 - Where taken: off Amelia Island, Florida - Date received: 16 Apr 1815 - From what ship: HMS Swifsure - Born: New Orleans - Age: 26 - Race: Negro - Discharged on 24 Apr 1815.

Carnes, John - 1st Lieutenant - Number: 412 - Prize name: Leo - Ship type: P - How taken: HM Frigate Tiber - When taken: 11 Mar 1815 - Where taken: Lat 45.24N Long 12.3W - Date received: 2 Apr 1815 - From what ship: HMS Tiber - Born: Boston - Age: 26 - Discharged on 2 Jun 1815 and released to the Cartel Sovereign.

Carney, John - Seaman - Number: 484 - Prize name: Fox - Ship type: P - How taken: HM Frigate Barbadoes - When taken: 11 Jan 1815 - Where taken: off Amelia Island, Florida - Date received: 16 Apr 1815 - From what ship: HMS Swifsure - Born: Maryland - Age: 22 - Discharged on 2 Jun 1815 and released to the Cartel Sovereign.

Carpenter, Jacob - Seaman - Number: 804 - Prize name: US Brig Syren - Ship type: War - How taken: HM Ship-of-the-Line Medway - When taken: 12 Jul 1814 - Where taken: off Cape of Good Hope - Date received: 11 Jul 1815 - From what ship: HMS Royal Sovereign - Born: Not listed - Discharged on 11 Jul 1815 and released to the Cartel Wooddrop Sims.

Carpenter, William - Seaman - Number: 813 - Prize name: US Brig Syren - Ship type: War - How taken: HM Ship-of-the-Line Medway - When taken: 12 Jul 1814 - Where taken: off Cape of Good Hope - Date received: 11 Jul 1815 - From what ship: HMS Royal Sovereign - Born: Not listed - Discharged on 11 Jul 1815 and released to the Cartel Wooddrop Sims.

Carter, John - Seaman - Number: 667 - Prize name: Avon - Ship type: P - How taken: HM Frigate Barbadoes - When taken: 8 Mar 1815 - Where taken: West Indies - Date received: 24 Apr 1815 - From what ship: HMS Bellerophon - Born: Massachusetts - Age: 30 - Race: Negro - Discharged on 2 Jun 1815 and released to the Cartel Shakespeare.

Casper, William - Boy - Number: 146 - Prize name: US Brig Syren - Ship type: War - How taken: HM Ship-of-the-Line Medway - When taken: 12 Jul 1814 - Where taken: off Cape of Good Hope - Date received: 21 Feb 1815 - From what ship: HMS Slaney - Born: Charlestown - Age: 19 - Discharged on 24 Feb 1815 and sent to Dartmoor.

Chase, B. - Prize Master - Number: 676 - Prize name: Theodore, prize to Privateer Chasseur - Ship type: P - How taken: HM Ship-of-the-Line Saturn - When taken: 7 Nov 1814 - Where taken: Lat 41N Long 69W - Date received: 24 Apr 1815 - From what ship: Royal Naval Hospital Plymouth - Born: Massachusetts - Age: 39 - Discharged on 3 May 1815.

Chateau, Point - Seaman - Number: 479 - Prize name: Fox - Ship type: P - How taken: HM Frigate Barbadoes - When taken: 11 Jan 1815 - Where taken: off Amelia Island, Florida - Date received: 16 Apr 1815 - From what ship: HMS Swifsure - Born: Saint-Domingue (Haiti) - Age: 40 - Race: Mulatto - Discharged on 2 Jun 1815 and released to the Cartel Sovereign.

Chick, Moses - Seaman - Number: 136 - Prize name: US Brig Syren - Ship type: War - How taken: HM Ship-of-the-Line Medway - When taken: 12 Jul 1814 - Where taken: off Cape of Good Hope - Date received: 21 Feb 1815 - From what ship: HMS Slaney - Born: Wells - Age: 38 - Discharged on 24 Feb 1815 and sent to Dartmoor.

Chilles, John - Seaman - Number: 713 - Prize name: Reindeer - Ship type: P - How taken: HM Sloop Calypso - When taken: 20 Feb 1815 - Where taken: off Lisbon - Date received: 16 Jul 1815 - From what ship: HMS Hope - Born: Massachusetts - Age: 28 - Discharged on 2 Jun 1815 and released to the Cartel Shakespeare.

Church, Henry - Seaman - Number: 369 - Prize name: Leo - Ship type: P - How taken: HM Frigate Tiber - When taken: 11 Mar 1815 - Where taken: Lat 45.24N Long 12.3W - Date received: 29 Mar 1814 - From what ship: HMS Tiber - Born: Massachusetts - Age: 31 - Race: Negro - Discharged on 2 May 1815 and released to the Cartel Ariel.

Churchill, Joseph - Seaman - Number: 118 - Prize name: Prince de Neufchatel - Ship type: P - How taken: HM Ship-of-the-Line Leander, HM Ship-of-the-Line Newcastle & HM Frigate Acasta - When taken: 28 Dec 1814 - Where taken: Lat 35N Long 52W - Date received: 18 Feb 1815 - From what ship: HMS Sybille - Born: Plymouth - Age: 21 - Discharged on 19 Feb 1815 and sent to Dartmoor.

Churchill, Joseph - Seaman - Number: 605 - Prize name: George Little - Ship type: P - How taken: HM Frigate Granicus - When taken: 20 Jan 1815 - Where taken: off Cape Finisterre, Spain - Date received: 18 Apr 1815 - From what ship: HMS Euryalus - Born: Massachusetts - Age: 25 - Discharged on 2 Jun 1815 and released to the Cartel Sovereign.

Clark, Francis Marie - Passenger - Number: 40 - Prize name: Leo - Ship type: P - How taken: HM Frigate Granicus - When taken: 2 Dec 1814 - Where taken: off Lorient, France - Date received: 12 Feb 1815 - From what ship: HMS Impregnable - Born: Lorient, France - Age: 23 - Discharged on 17 Feb 1815 and sent to Dartmoor.

Clark, James - Seaman - Number: 206 - Prize name: Sennett, Swedish merchant vessel - Ship type: MV - How taken: HM Frigate Rhin - When taken: Not known - Date received: 2 Mar 1815 - From what ship: HMS Dannemark - Discharged on 3 Mar 1815 and sent to Dartmoor.

Clarke, B. - Seaman - Number: 565 - Prize name: Mary - Ship type: MV - How taken: HMS Muros - When taken: 8 Dec 1814 - Where taken: West Indies - Date received: 16 Apr 1815 - From what ship: HMS Swifsure - Born: North Caroline - Age: 27 - Discharged on 2 Jun 1815 and released to the Cartel Sovereign.

Clarke, Benjamin - Seaman - Number: 732 - Prize name: Harpy - Ship type: P - How taken: HM Brig Zenobia - When taken: 5 Jan 1815 - Where taken: off Lisbon - Date received: 29 Aug 1815 - From what ship: HMS Tonnant - Born: Massachusetts - Age: 21 - Discharged on 2 Jun 1815 and released to the Cartel Shakespeare.

Clarke, George - Seaman - Number: 675 - Prize name: Leo - Ship type: LM - How taken: HM Frigate Granicus - When taken: 26 Nov 1814 - Where taken: off Lisbon - Date received: 24 Apr 1815 - From what ship: Royal Naval Hospital Plymouth - Born: New York - Age: 30 - Discharged on 3 May 1815.

Clarke, James - Seaman - Number: 562 - Prize name: Mary - Ship type: MV - How taken: HMS Muros - When taken: 8 Dec 1814 - Where taken: West Indies - Date received: 16 Apr 1815 - From what ship: HMS Swifsure - Born: North Caroline - Age: 16 - Discharged on 2 Jun 1815 and released to the Cartel Sovereign.

Clarke, Reuben - Seaman - Number: 559 - Prize name: Mary - Ship type: MV - How taken: HMS Muros - When taken: 8 Dec 1814 - Where taken: West Indies - Date received: 16 Apr 1815 - From what ship: HMS Swifsure - Born: North Caroline - Age: 41 - Discharged on 2 Jun 1815 and released to the Cartel Sovereign.

Cleland, Thomas J. - Seaman - Number: 405 - Prize name: Leo - Ship type: P - How taken: HM Frigate Tiber - When taken: 11 Mar 1815 - Where taken: Lat 45.24N Long 12.3W - Date received: 29 Mar 1815 - From what ship: HMS Tiber - Born: Massachusetts - Age: 20 - Discharged on 30 May 1815 and released to the Cartel Atlas.

Clerge, N. - Seaman - Number: 472 - Prize name: Fox - Ship type: P - How taken: HM Frigate Barbadoes - When taken: 11 Jan 1815 - Where taken: off Amelia Island, Florida - Date received: 16 Apr 1815 - From what ship: HMS Swifsure - Born: Saint-Domingue (Haiti) - Age: 20 - Race: Negro - Discharged on 2 Jun 1815 and released to the Cartel Sovereign.

Clerk, James - Seaman - Number: 22 - Prize name: Jane - Ship type: MV - How taken: Impressed - When taken: 3 Dec 1814 - Date received: 3 Feb 1815 - From what ship: HMS Bittern - Born: Bristol - Age: 23 - Discharged on 4 Feb 1815 and sent to Dartmoor.

Clestair, (-----) - Seaman - Number: 478 - Prize name: Fox - Ship type: P - How taken: HM Frigate Barbadoes - When taken: 11 Jan 1815 - Where taken: off Amelia Island, Florida - Date received: 16 Apr 1815 - From what ship: HMS Swifsure - Born: Saint-Domingue (Haiti) - Age: 24 - Race: Mulatto - Discharged on 2 Jun 1815 and released to the Cartel Sovereign.

Cloughman, Joseph - Seaman - Number: 693 - Prize name: Sine Qua Non - Ship type: P - How taken: HM Brig Elk - When taken: 20 Feb 1815 - Where taken: off Madeira - Date received: 16 Jul 1815 - From what ship: HMS Hope - Age: 25 - Discharged on 2 Jun 1815 and released to the Cartel Shakespeare.

Cloutman, Ephraim - Seaman - Number: 137 - Prize name: US Brig Syren - Ship type: War - How taken: HM Ship-of-the-Line Medway - When taken: 12 Jul 1814 - Where taken: off Cape of Good Hope - Date received: 21 Feb 1815 - From what ship: HMS Slaney - Born: Salem - Age: 32 - Discharged on 24 Feb 1815 and sent to Dartmoor.

Coates, Samuel M. - Boatswain - Number: 142 - Prize name: US Brig Syren - Ship type: War - How taken: HM Ship-of-the-Line Medway - When taken: 12 Jul 1814 - Where taken: off Cape of Good Hope - Date received: 21 Feb 1815 - From what ship: HMS Slaney - Born: Baltimore - Age: 33 - Discharged on 24 Feb 1815 and sent to Dartmoor.

Coffee, Jacob - Seaman - Number: 49 - How taken: Taken out of HMS Groster - Date received: 15 Feb 1815 - From what ship: HMS Myrmidon - Born: Boston - Age: 22 - Race: Mulatto - Discharged on 17 Feb 1815 and sent to Dartmoor.

Coffin, Robert S. - Seaman - Number: 527 - Prize name: Gallant - Ship type: MV - How taken: HM Frigate Barbadoes - When taken: 6 Dec 1815 - Where taken: off St. Bartholomew - Date received: 16 Apr 1815 - From what ship: HMS Swifsure - Born: Massachusetts - Age: 20 - Discharged on 2 Jun 1815 and released to the Cartel Sovereign.

Coffis, Theodore - Seaman - Number: 121 - Prize name: Prince de Neufchatel - Ship type: P - How taken: HM Ship-of-the-Line Leander, HM Ship-of-the-Line Newcastle & HM Frigate Acasta - When taken: 28 Dec 1814 - Where taken: Lat 35N Long 52W - Date received: 18 Feb 1815 - From what ship: HMS Sybille - Age: 26 - Discharged on 19 Feb 1815 and sent to Dartmoor.

Cole, Isaac - Seaman - Number: 599 - Prize name: George Little - Ship type: P - How taken: HM Frigate Granicus - When taken: 20 Jan 1815 - Where taken: off Cape Finisterre, Spain - Date received: 18 Apr 1815 - From what ship: HMS Euryalus - Born: Boston - Age: 28 - Discharged on 2 Jun 1815 and released to the Cartel Sovereign.

Collins, John - Seaman - Number: 664 - Prize name: Avon - Ship type: P - How taken: HM Frigate Barbadoes - When taken: 8 Mar 1815 - Where taken: West Indies - Date received: 24 Apr 1815 - From what ship: HMS Bellerophon - Born: Massachusetts - Age: 20 - Discharged on 2 Jun 1815 and released to the Cartel Shakespeare.

Colmar, Samuel - Seaman - Number: 709 - Prize name: Sine Qua Non - Ship type: P - How taken: HM Brig Elk - When taken: 20 Feb 1815 - Where taken: off Madeira - Date received: 16 Jul 1815 - From what ship: HMS Hope - Born: Massachusetts - Age: 24 - Discharged on 2 Jun 1815 and released to the Cartel Shakespeare.

Colson, Christopher - Seaman - Number: 102 - Prize name: Prince de Neufchatel - Ship type: P - How taken: HM Ship-of-the-Line Leander, HM Ship-of-the-Line Newcastle & HM Frigate Acasta - When taken: 28 Dec 1814 - Where taken: Lat 35N Long 52W - Date received: 18 Feb 1815 - From what ship: HMS Sybille - Age: 25 - Discharged on 19 Feb 1815 and sent to Dartmoor.

Combes, John - Seaman - Number: 231 - Prize name: Decatur - Ship type: P - How taken: HM Frigate Rhin - When taken: 4 Jun 1814 - Where taken: at sea - Date received: 2 Mar 1815 - From what ship: HMS Dannemark - Born: Norfolk - Age: 22 - Race: Negro - Discharged on 3 Mar 1815 and sent to Dartmoor.

Comer, John - Seaman - Number: 672 - Prize name: Avon - Ship type: P - How taken: HM Frigate Barbadoes - When taken: 8 Mar 1815 - Where taken: West Indies - Date received: 24 Apr 1815 - From what ship: HMS Bellerophon - Born: New Orleans - Age: 38 - Race: Negro - Discharged on 2 Jun 1815 and released to the Cartel Shakespeare.

Condell, B. - Seaman - Number: 498 - Prize name: Fox - Ship type: P - How taken: HM Frigate Barbadoes - When taken: 11 Jan 1815 - Where taken: off Amelia Island, Florida - Date received: 16 Apr 1815 - From what ship: HMS Swifsure - Born: Guadeloupe - Age: 48 - Race: Negro - Discharged on 2 Jun 1815 and released to the Cartel Sovereign.

Conner, Henry - Seaman - Number: 419 - Prize name: Mary - Ship type: LM - How taken: HM Frigate Erne - When taken: 7 Jul 1814 - Where taken: off Lezard - Date received: 16 Apr 1815 - From what ship: HMS Swifsure - Born: New York - Age: 20 - Discharged on 2 Jun 1815 and released to the Cartel Sovereign.

Conner, Peter - Prize Master - Number: 782 - Prize name: Eliza, prize of the Privateer Yankee - Ship type: P - How taken: HM Schooner Shelbourne - When taken: 28 Aug 1813 - Where taken: off Amelia Island, Florida - Date received: 11 Jul 1815 - From what ship: Royal Naval Hospital Plymouth - Born: Philadelphia - Age: 35 -

Discharged on 11 Jul 1815 and released to the Cartel Wooddrop Sims.

Connor, Peter - Seaman - Number: 520 - Prize name: High Flyer - Ship type: LM - How taken: HMS Muros - When taken: 14 Nov 1814 - Where taken: off St. Bartholomew - Date received: 16 Apr 1815 - From what ship: HMS Swifsure - Born: Philadelphia - Age: 29 - Discharged on 8 May 1815.

Conrad, John L. - Seaman - Number: 790 - Prize name: US Brig Syren - Ship type: War - How taken: HM Ship-of-the-Line Medway - When taken: 12 Jul 1814 - Where taken: off Cape of Good Hope - Date received: 11 Jul 1815 - From what ship: HMS Royal Sovereign - Born: Not listed - Discharged on 11 Jul 1815 and released to the Cartel Wooddrop Sims.

Constant, William - Seaman - Number: 101 - Prize name: Prince de Neufchatel - Ship type: P - How taken: HM Ship-of-the-Line Leander, HM Ship-of-the-Line Newcastle & HM Frigate Acasta - When taken: 28 Dec 1814 - Where taken: Lat 35N Long 52W - Date received: 18 Feb 1815 - From what ship: HMS Sybille - Born: Modena - Age: 29 - Discharged on 19 Feb 1815 and sent to Dartmoor.

Conton, Philip - Seaman - Number: 47 - How taken: HM Ship-of-the-Line Adamant - When taken: 16 Jan 1815 - Where taken: Portsmouth - Date received: 15 Feb 1815 - From what ship: HMS Myrmidon - Born: New York - Age: 28 - Discharged on 17 Feb 1815 and sent to Dartmoor.

Cook, C. - Master's Mate - Number: 641 - Prize name: Avon - Ship type: P - How taken: HM Frigate Barbadoes - When taken: 8 Mar 1815 - Where taken: West Indies - Date received: 24 Apr 1815 - From what ship: HMS Bellerophon - Age: 45 - Discharged on 2 Jun 1815 and released to the Cartel Shakespeare.

Cook, Francis - Seaman - Number: 566 - Prize name: Mary - Ship type: MV - How taken: HMS Muros - When taken: 8 Dec 1814 - Where taken: West Indies - Date received: 16 Apr 1815 - From what ship: HMS Swifsure - Born: Massachusetts - Age: 23 - Discharged on 2 Jun 1815 and released to the Cartel Sovereign.

Cooper, Peter - Seaman - Number: 274 - Prize name: Netterville - Ship type: MV - How taken: HM Brig Onyx - When taken: 25 Dec 1814 - Where taken: at sea - Date received: 2 Mar 1815 - From what ship: HMS Dannemark - Born: Baltimore - Age: 23 - Race: Mulatto - Discharged on 3 Mar 1815 and sent to Dartmoor.

Cooper, Thomas - Seaman - Number: 25 - Prize name: Plutarch - Ship type: MV - How taken: HM Schooner Helicon - When taken: 5 Feb 1815 - Where taken: Lat 45.5N Long 7W - Date received: 8 Feb 1815 - From what ship: HMS Helicon - Born: Boston - Age: 26 - Discharged on 10 Feb 1815 and sent to Dartmoor.

Coreen, Hugh - Seaman - Number: 20 - Prize name: Brothers - Ship type: MV - How taken: Impressed at Liverpool - When taken: 14 Jan 1815 - Date received: 3 Feb 1815 - From what ship: HMS Bittern - Born: Maryland - Age: 24 - Discharged on 4 Feb 1815 and sent to Dartmoor.

Corvill, Nathaniel - Seaman - Number: 31 - Prize name: Plutarch - Ship type: MV - How taken: HM Schooner Helicon - When taken: 5 Feb 1815 - Where taken: Lat 45.5N Long 7W - Date received: 8 Feb 1815 - From what ship: HMS Helicon - Born: Chatham - Age: 19 - Discharged on 10 Feb 1815 and sent to Dartmoor.

Cossioty, John - Seaman - Number: 650 - Prize name: Avon - Ship type: P - How taken: HM Frigate Barbadoes - When taken: 8 Mar 1815 - Where taken: West Indies - Date received: 24 Apr 1815 - From what ship: HMS Bellerophon - Age: 44 - Discharged on 2 Jun 1815 and released to the Cartel Shakespeare.

Cowett, Jesse - Seaman - Number: 408 - Prize name: Leo - Ship type: P - How taken: HM Frigate Tiber - When taken: 11 Mar 1815 - Where taken: Lat 45.24N Long 12.3W - Date received: 29 Mar 1815 - From what ship: HMS Tiber - Born: Massachusetts - Age: 25 - Discharged on 2 Jun 1815 and released to the Cartel Sovereign.

Craft, John - Seaman - Number: 761 - Prize name: Harpy - Ship type: P - How taken: HM Brig Zenobia - When taken: 5 Jan 1815 - Where taken: off Lisbon - Date received: 29 Aug 1815 - From what ship: HMS Tonnant - Born: Massachusetts - Age: 20 - Discharged on 2 Jun 1815 and released to the Cartel Shakespeare.

Crandall, Joseph - Seaman - Number: 697 - Prize name: Sine Qua Non - Ship type: P - How taken: HM Brig Elk - When taken: 20 Feb 1815 - Where taken: off Madeira - Date received: 16 Jul 1815 - From what ship: HMS Hope - Age: 26 - Discharged on 2 Jun 1815 and released to the Cartel Shakespeare.

Crawford, J. G. - Seaman - Number: 601 - Prize name: George Little - Ship type: P - How taken: HM Frigate Granicus - When taken: 20 Jan 1815 - Where taken: off Cape Finisterre, Spain - Date received: 18 Apr 1815 - From what ship: HMS Euryalus - Born: Massachusetts - Age: 27 - Discharged on 2 Jun 1815 and released to the

Cartel Sovereign.

Crawford, John - Seaman - Number: 513 - Prize name: Hope, prize to US Sloop-of-War Wasp - Ship type: War - How taken: HM Sloop Fairy - When taken: 6 Jan 1815 - Where taken: Unknown - Date received: 16 Apr 1815 - From what ship: HMS Swifsure - Born: Philadelphia - Age: 65 - Discharged on 2 Jun 1815 and released to the Cartel Sovereign.

Creker, Edward - Seaman - Number: 467 - Prize name: Fox - Ship type: P - How taken: HM Frigate Barbadoes - When taken: 11 Jan 1815 - Where taken: off Amelia Island, Florida - Date received: 16 Apr 1815 - From what ship: HMS Swifsure - Age: 22 - Discharged on 2 Jun 1815.

Crosby, J. F. - Seaman - Number: 712 - Prize name: Reindeer - Ship type: P - How taken: HM Sloop Calypso - When taken: 20 Feb 1815 - Where taken: off Lisbon - Date received: 16 Jul 1815 - From what ship: HMS Hope - Born: Massachusetts - Age: 21 - Discharged on 2 Jun 1815 and released to the Cartel Shakespeare.

Cross, George - Seaman - Number: 704 - Prize name: Sine Qua Non - Ship type: P - How taken: HM Brig Elk - When taken: 20 Feb 1815 - Where taken: off Madeira - Date received: 16 Jul 1815 - From what ship: HMS Hope - Born: Massachusetts - Age: 21 - Discharged on 2 Jun 1815 and released to the Cartel Shakespeare.

Cross, Thomas - Boy - Number: 378 - Prize name: Leo - Ship type: P - How taken: HM Frigate Tiber - When taken: 11 Mar 1815 - Where taken: Lat 45.24N Long 12.3W - Date received: 29 Mar 1814 - From what ship: HMS Tiber - Born: Massachusetts - Age: 18 - Discharged on 30 May 1815 and released to the Cartel Atlas.

Crouise, John L. - Seaman - Number: 492 - Prize name: Fox - Ship type: P - How taken: HM Frigate Barbadoes - When taken: 11 Jan 1815 - Where taken: off Amelia Island, Florida - Date received: 16 Apr 1815 - From what ship: HMS Swifsure - Born: New Orleans - Age: 19 - Race: Negro - Discharged on 24 Apr 1815.

Culber, Richard - Seaman - Number: 580 - Prize name: Sorin, prize to Privateer Prince of Neufchatel - Ship type: P - How taken: HM Ship-of-the-Line Medway - When taken: 12 Jul 1814 - Where taken: Unknown - Date received: 16 Apr 1815 - From what ship: HMS Swifsure - Born: New London - Age: 42 - Discharged on 2 Jun 1815 and released to the Cartel Sovereign.

Cullin, Philip - Seaman - Number: 418 - Prize name: Mary - Ship type: LM - How taken: HM Frigate Erne - When taken: 7 Jul 1814 - Where taken: off Lezard - Date received: 16 Apr 1815 - From what ship: HMS Swifsure - Age: 37 - Discharged on 2 Jun 1815 and released to the Cartel Sovereign.

Cumrie, Horrico - Seaman - Number: 428 - Prize name: Engineer - Ship type: MV - How taken: HMS Murros - When taken: 21 Sep 1814 - Where taken: off Porto Rico - Date received: 16 Apr 1815 - From what ship: HMS Swifsure - Born: Rhode Island - Age: 21 - Discharged on 2 Jun 1815 and released to the Cartel Sovereign.

Daird, Samuel - Passenger - Number: 437 - Prize name: Acquilar - Ship type: P - How taken: HM Frigate Pique - When taken: 4 Sep 1814 - Where taken: off Newfoundland - Date received: 16 Apr 1815 - From what ship: HMS Swifsure - Born: Massachusetts - Age: 32 - Discharged on 2 Jun 1815 and released to the Cartel Sovereign.

Daniels, Henry - Seaman - Number: 303 - Prize name: US Brig Syren - Ship type: War - How taken: HM Ship-of-the-Line Medway - When taken: 12 Jul 1814 - Where taken: off Cape of Good Hope - Date received: 4 Mar 1814 - From what ship: HM Cutter Surley - Born: New Jersey - Age: 25 - Discharged on 7 Mar 1815 and sent to Dartmoor.

Darvit, Anthony - Marine Captain - Number: 462 - Prize name: Fox - Ship type: P - How taken: HM Frigate Barbadoes - When taken: 11 Jan 1815 - Where taken: off Amelia Island, Florida - Date received: 16 Apr 1815 - From what ship: HMS Swifsure - Born: New Orleans - Age: 28 - Discharged on 22 Apr 1815.

David, Samuel - Marine - Number: 72 - Prize name: Prince de Neufchatel - Ship type: P - How taken: HM Ship-of-the-Line Leander, HM Ship-of-the-Line Newcastle & HM Frigate Acasta - When taken: 28 Dec 1814 - Where taken: Lat 35N Long 52W - Date received: 18 Feb 1815 - From what ship: HMS Sybille - Born: New York - Age: 18 - Discharged on 19 Feb 1815 and sent to Dartmoor.

Davie, William - Seaman - Number: 365 - Prize name: Leo - Ship type: P - How taken: HM Frigate Tiber - When taken: 11 Mar 1815 - Where taken: Lat 45.24N Long 12.3W - Date received: 29 Mar 1814 - From what ship: HMS Tiber - Born: Massachusetts - Age: 43 - Discharged on 2 May 1815 and released to the Cartel Ariel.

Davies, Thomas - Seaman - Number: 570 - Prize name: Heron - Ship type: MV - How taken: HM Frigate Pique - When taken: 16 Dec 1814 - Where taken: off Porto Rico - Date received: 16 Apr 1815 - From what ship: HMS Swifsure - Born: Philadelphia - Age: 49 - Race: Negro - Discharged on 27 May 1815.

Davis, B. - Prize Master - Number: 637 - Prize name: Avon - Ship type: P - How taken: HM Frigate Barbadoes - When taken: 8 Mar 1815 - Where taken: West Indies - Date received: 24 Apr 1815 - From what ship: HMS Bellerophon - Age: 24 - Discharged on 2 Jun 1815 and released to the Cartel Shakespeare.

Davis, Elias S. - Seaman - Number: 154 - Prize name: US Brig Syren - Ship type: War - How taken: HM Ship-of-the-Line Medway - When taken: 12 Jul 1814 - Where taken: off Cape of Good Hope - Date received: 21 Feb 1815 - From what ship: HMS Slaney - Born: Vermont - Age: 22 - Discharged on 24 Feb 1815 and sent to Dartmoor.

Davis, Frederick - Marine - Number: 73 - Prize name: Prince de Neufchatel - Ship type: P - How taken: HM Ship-of-the-Line Leander, HM Ship-of-the-Line Newcastle & HM Frigate Acasta - When taken: 28 Dec 1814 - Where taken: Lat 35N Long 52W - Date received: 18 Feb 1815 - From what ship: HMS Sybille - Born: Boston - Age: 22 - Race: Mulatto - Discharged on 19 Feb 1815 and sent to Dartmoor.

Davis, James - Seaman - Number: 290 - How taken: Taken off the HM Ship-of-the-Line Severn and other ships - Date received: 2 Mar 1815 - From what ship: HMS Dannemark - Born: Raymond - Age: 33 - Discharged on 3 Mar 1815 and sent to Dartmoor.

Davis, John - Seaman - Number: 590 - Prize name: George Little - Ship type: P - How taken: HM Frigate Granicus - When taken: 20 Jan 1815 - Where taken: off Cape Finisterre, Spain - Date received: 18 Apr 1815 - From what ship: HMS Euryalus - Born: New Hampshire - Age: 32 - Discharged on 2 Jun 1815 and released to the Cartel Sovereign.

Davis, Richard - Seaman - Number: 281 - Prize name: Tartan - Ship type: MV - How taken: Given up - Date received: 2 Mar 1815 - From what ship: HMS Dannemark - Discharged on 3 Mar 1815 and sent to Dartmoor.

Davis, W. B. - Seaman - Number: 542 - Prize name: Engineer - Ship type: MV - How taken: HMS Muros - When taken: 21 Sep 1814 - Where taken: off Porto Rico - Date received: 16 Apr 1815 - From what ship: HMS Swifsure - Age: 25 - Discharged on 8 May 1815.

Dawson, John - Seaman - Number: 178 - Prize name: Farmer's Daughter - Ship type: MV - How taken: HM Ship-of-the-Line Leviathan - When taken: 20 Mar 1814 - Where taken: at sea - Date received: 2 Mar 1815 - From what ship: HMS Dannemark - Born: Fredericksburg - Age: 49 - Discharged on 3 Mar 1815 and sent to Dartmoor.

Day, Samuel - Seaman - Number: 254 - Prize name: Decatur - Ship type: P - How taken: HM Frigate Rhin - When taken: 4 Jun 1814 - Where taken: at sea - Date received: 2 Mar 1815 - From what ship: HMS Dannemark - Born: Massachusetts - Age: 22 - Discharged on 3 Mar 1815 and sent to Dartmoor.

De Four, Victor - Seaman - Number: 496 - Prize name: Fox - Ship type: P - How taken: HM Frigate Barbadoes - When taken: 11 Jan 1815 - Where taken: off Amelia Island, Florida - Date received: 16 Apr 1815 - From what ship: HMS Swifsure - Born: Saint-Domingue (Haiti) - Age: 22 - Race: Negro - Discharged on 2 Jun 1815 and released to the Cartel Sovereign.

De Lage, Jean - Seaman - Number: 494 - Prize name: Fox - Ship type: P - How taken: HM Frigate Barbadoes - When taken: 11 Jan 1815 - Where taken: off Amelia Island, Florida - Date received: 16 Apr 1815 - From what ship: HMS Swifsure - Born: Italy - Age: 37 - Discharged on 2 Jun 1815 and released to the Cartel Sovereign.

Deal, John - Seaman - Number: 340 - Prize name: Leo - Ship type: P - How taken: HM Frigate Tiber - When taken: 11 Mar 1815 - Where taken: Lat 45.24N Long 12.3W - Date received: 28 Mar 1815 - From what ship: HMS Tiber - Race: Negro - Discharged on 27 Apr 1815 and released to the Cartel Minework.

Deck, Lewis - Seaman - Number: 791 - Prize name: US Brig Syren - Ship type: War - How taken: HM Ship-of-the-Line Medway - When taken: 12 Jul 1814 - Where taken: off Cape of Good Hope - Date received: 11 Jul 1815 - From what ship: HMS Royal Sovereign - Born: Not listed - Discharged on 11 Jul 1815 and released to the Cartel Wooddrop Sims.

Dede Veer, Eppcough - Seaman - Number: 207 - Prize name: Sennett, Swedish merchant vessel - Ship type: MV - How taken: HM Frigate Rhin - When taken: Not known - Date received: 2 Mar 1815 - From what ship: HMS Dannemark - Discharged on 3 Mar 1815 and sent to Dartmoor.

Delaver, Samuel - Seaman - Number: 772 - Prize name: Sine Qua Non - Ship type: P - How taken: HM Brig Elk - When taken: 20 Feb 1815 - Where taken: off Madeira - Date received: 29 Aug 1815 - From what ship: HMS Tonnant - Age: 22 - Discharged on 2 Jun 1815 and released to the Cartel Shakespeare.

Delaware, John - Boy - Number: 76 - Prize name: Prince de Neufchatel - Ship type: P - How taken: HM Ship-of-the-Line Leander, HM Ship-of-the-Line Newcastle & HM Frigate Acasta - When taken: 28 Dec 1814 - Where taken: Lat 35N Long 52W - Date received: 18 Feb 1815 - From what ship: HMS Sybille - Born: Boston - Age: 17 - Discharged on 19 Feb 1815 and sent to Dartmoor.

Demedorff, John - Seaman - Number: 98 - Prize name: Prince de Neufchatel - Ship type: P - How taken: HM Ship-of-the-Line Leander, HM Ship-of-the-Line Newcastle & HM Frigate Acasta - When taken: 28 Dec 1814 - Where taken: Lat 35N Long 52W - Date received: 18 Feb 1815 - From what ship: HMS Sybille - Born: Petersburg - Age: 42 - Discharged on 19 Feb 1815 and sent to Dartmoor.

Desomon, Thomas - Seaman - Number: 302 - Prize name: US Brig Syren - Ship type: War - How taken: HM Ship-of-the-Line Medway - When taken: 12 Jul 1814 - Where taken: off Cape of Good Hope - Date received: 4 Mar 1814 - From what ship: HM Cutter Surley - Discharged on 7 Mar 1815 and sent to Dartmoor.

Detende, Joseph - Seaman - Number: 259 - Prize name: Dorothy - Ship type: MV - How taken: HM Frigate North Star - When taken: 15 Nov 1814 - Where taken: at sea - Date received: 2 Mar 1815 - From what ship: HMS Dannemark - Discharged on 3 Mar 1815 and sent to Dartmoor.

Dill, Jobes - Seaman - Number: 648 - Prize name: Avon - Ship type: P - How taken: HM Frigate Barbadoes - When taken: 8 Mar 1815 - Where taken: West Indies - Date received: 24 Apr 1815 - From what ship: HMS Bellerophon - Born: Massachusetts - Age: 25 - Discharged on 2 Jun 1815 and released to the Cartel Shakespeare.

Dine, William - Seaman - Number: 53 - How taken: Taken out of HM Ship-of-the-Line Namur - Where taken: Nore - Date received: 15 Feb 1815 - From what ship: HMS Myrmidon - Born: Philadelphia - Age: 36 - Discharged on 17 Feb 1815 and sent to Dartmoor.

Doans, William - Seaman - Number: 374 - Prize name: Leo - Ship type: P - How taken: HM Frigate Tiber - When taken: 11 Mar 1815 - Where taken: Lat 45.24N Long 12.3W - Date received: 29 Mar 1814 - From what ship: HMS Tiber - Age: 53 - Discharged on 2 May 1815 and released to the Cartel Ariel.

Doing, Denis O. - Seaman - Number: 171 - Prize name: Enterprise - Ship type: P - How taken: Not known - Where taken: at sea - Date received: 2 Mar 1815 - From what ship: HMS Dannemark - Born: Massachusetts - Age: 23 - Discharged on 3 Mar 1815 and sent to Dartmoor.

Dolliver, William - Seaman - Number: 862 - Prize name: US Brig Syren - Ship type: War - How taken: HM Ship-of-the-Line Medway - When taken: 12 Jul 1814 - Where taken: off Cape of Good Hope - Date received: 11 Jul 1815 - From what ship: from Portsmouth - Born: Not listed - Discharged on 11 Jul 1815 and released to the Cartel Wooddrop Sims.

Dolphin, Joseph - Seaman - Number: 159 - Prize name: Nancy, prize of the Privateer United States - Ship type: P - How taken: HM Sloop Papillon - When taken: 27 Dec 1814 - Where taken: Lat 30 Long 10 - Date received: 27 Feb 1815 - From what ship: HMS Nimble - Born: Exeter - Age: 22 - Discharged on 3 Mar 1815 and sent to Dartmoor.

Domingo, (-----) - Seaman - Number: 557 - Prize name: Mary - Ship type: MV - How taken: HMS Muros - When taken: 8 Dec 1814 - Where taken: West Indies - Date received: 16 Apr 1815 - From what ship: HMS Swifsure - Age: 23 - Discharged on 2 Jun 1815 and released to the Cartel Sovereign.

Dominico, John - Cook - Number: 196 - Prize name: Chance - Ship type: P - How taken: HMS Statire - When taken: 1 Apr 1814 - Where taken: at sea - Date received: 2 Mar 1815 - From what ship: HMS Dannemark - Born: Maryland - Age: 25 - Race: Negro - Discharged on 3 Mar 1815 and sent to Dartmoor.

Donald, Joseph - Seaman - Number: 538 - Prize name: Engineer - Ship type: MV - How taken: HMS Muros - When taken: 21 Sep 1814 - Where taken: off Porto Rico - Date received: 16 Apr 1815 - From what ship: HMS Swifsure - Born: Massachusetts - Age: 22 - Discharged on 2 Jun 1815 and released to the Cartel Sovereign.

Donald, Michael - Seaman - Number: 758 - Prize name: George Little - Ship type: P - How taken: HM Frigate Granicus - When taken: 20 Jan 1815 - Where taken: off Cape Finisterre, Spain - Date received: 29 Aug 1815 - From what ship: HMS Tonnant - Born: Massachusetts - Age: 22 - Discharged on 2 Jun 1815 and released to the

Cartel Shakespeare.

Door, Ebenezer - 6th Lieutenant - Number: 66 - Prize name: Prince de Neufchatel - Ship type: P - How taken: HM Ship-of-the-Line Leander, HM Ship-of-the-Line Newcastle & HM Frigate Acasta - When taken: 28 Dec 1814 - Where taken: Lat 35N Long 52W - Date received: 18 Feb 1815 - From what ship: HMS Sybille - Born: Roxborough - Age: 26 - Discharged on 19 Feb 1815 and sent to Dartmoor.

Dotor, Samuel - Lieutenant - Number: 582 - Prize name: George Little - Ship type: P - How taken: HM Frigate Granicus - When taken: 20 Jan 1815 - Where taken: off Cape Finisterre, Spain - Date received: 18 Apr 1815 - From what ship: HMS Euryalus - Born: Massachusetts - Age: 32 - Discharged on 24 Apr 1815.

Douglas, Mathew - Seaman - Number: 9 - Prize name: Albion - Ship type: MV - How taken: HM Schooner Helicon - When taken: 17 Jan 1815 - Where taken: Lat 20N Long 52W - Date received: 3 Feb 1815 - From what ship: HMS Bittern - Born: Maryland - Age: 24 - Discharged on 4 Feb 1815 and sent to Dartmoor.

Down, J. - Seaman - Number: 746 - Prize name: Sine Qua Non - Ship type: P - How taken: HM Brig Elk - When taken: 20 Feb 1815 - Where taken: off Madeira - Date received: 29 Aug 1815 - From what ship: HMS Tonnant - Born: Massachusetts - Age: 36 - Discharged on 2 Jun 1815 and released to the Cartel Shakespeare.

Downing, B. - Boy - Number: 679 - Prize name: Snap Dragon - Ship type: P - How taken: HMS Mentor - When taken: 10 Jul 1814 - Where taken: off Halifax - Date received: 14 May 1815 - From what ship: from the police - Age: 16 - Discharged on 2 Jun 1815 and released to the Cartel Shakespeare.

Downs, F. - Sailing Master - Number: 852 - Prize name: US Brig Syren - Ship type: War - How taken: HM Ship-of-the-Line Medway - When taken: 12 Jul 1814 - Where taken: off Cape of Good Hope - Date received: 11 Jul 1815 - From what ship: from Portsmouth - Born: Not listed - Discharged on 11 Jul 1815 and released to the Cartel Wooddrop Sims.

Downs, Jesse - Marine - Number: 80 - Prize name: Prince de Neufchatel - Ship type: P - How taken: HM Ship-of-the-Line Leander, HM Ship-of-the-Line Newcastle & HM Frigate Acasta - When taken: 28 Dec 1814 - Where taken: Lat 35N Long 52W - Date received: 18 Feb 1815 - From what ship: HMS Sybille - Born: Boston - Age: 20 - Discharged on 19 Feb 1815 and sent to Dartmoor.

Downs, John - Seaman - Number: 447 - Prize name: High Flyer - Ship type: LM - How taken: HMS Muros - When taken: 14 Nov 1814 - Where taken: off St. Bartholomew - Date received: 16 Apr 1815 - From what ship: HMS Swifsure - Born: Massachusetts - Age: 26 - Discharged on 2 Jun 1815 and released to the Cartel Sovereign.

Drott, Henry D. - Master's Mate - Number: 458 - Prize name: Fox - Ship type: P - How taken: HM Frigate Barbadoes - When taken: 11 Jan 1815 - Where taken: off Amelia Island, Florida - Date received: 16 Apr 1815 - From what ship: HMS Swifsure - Born: New Jersey - Age: 20 - Discharged on 2 Jun 1815 and released to the Cartel Sovereign.

Drummond, Alexander - Prize Master - Number: 623 - Prize name: Helene, prize of Privateer Morgianna - Ship type: P - How taken: HM Frigate Pique - When taken: 7 Mar 1815 - Where taken: coast of America - Date received: 20 Apr 1815 - From what ship: HMS Musquito - Born: New York - Age: 26 - Discharged on 28 Apr 1815.

Dufires, Simon - Master's Mate - Number: 459 - Prize name: Fox - Ship type: P - How taken: HM Frigate Barbadoes - When taken: 11 Jan 1815 - Where taken: off Amelia Island, Florida - Date received: 16 Apr 1815 - From what ship: HMS Swifsure - Born: New Orleans - Age: 38 - Discharged on 22 Apr 1815.

Dunham, William - Ordinary Seaman - Number: 139 - Prize name: US Brig Syren - Ship type: War - How taken: HM Ship-of-the-Line Medway - When taken: 12 Jul 1814 - Where taken: off Cape of Good Hope - Date received: 21 Feb 1815 - From what ship: HMS Slaney - Born: Lyme - Age: 20 - Discharged on 24 Feb 1815 and sent to Dartmoor.

Dunklin, Jesse - Seaman - Number: 127 - Prize name: Prince de Neufchatel - Ship type: P - How taken: HM Ship-of-the-Line Leander, HM Ship-of-the-Line Newcastle & HM Frigate Acasta - When taken: 28 Dec 1814 - Where taken: Lat 35N Long 52W - Date received: 18 Feb 1815 - From what ship: HMS Sybille - Born: New Hampshire - Age: 31 - Discharged on 19 Feb 1815 and sent to Dartmoor.

Dupoent, Julius - Seaman - Number: 500 - Prize name: Fox - Ship type: P - How taken: HM Frigate Barbadoes - When taken: 11 Jan 1815 - Where taken: off Amelia Island, Florida - Date received: 16 Apr 1815 - From what ship: HMS Swifsure - Born: New Orleans - Age: 24 - Discharged on 2 Jun 1815 and released to the Cartel Sovereign.

Dwerell, John - Seaman - Number: 203 - Prize name: Sennett, Swedish merchant vessel - Ship type: MV - How taken: HM Frigate Rhin - When taken: Not known - Date received: 2 Mar 1815 - From what ship: HMS Dannemark - Discharged on 3 Mar 1815 and sent to Dartmoor.

Dyer, Charles - Seaman - Number: 214 - Prize name: Sennett, Swedish merchant vessel - Ship type: MV - How taken: HM Frigate Rhin - When taken: Not known - Date received: 2 Mar 1815 - From what ship: HMS Dannemark - Discharged on 3 Mar 1815 and sent to Dartmoor.

Dyk, Es. - Seaman - Number: 763 - Prize name: George Little - Ship type: P - How taken: HM Frigate Granicus - When taken: 20 Jan 1815 - Where taken: off Cape Finisterre, Spain - Date received: 29 Aug 1815 - From what ship: HMS Tonnant - Born: Massachusetts - Age: 25 - Discharged on 2 Jun 1815 and released to the Cartel Shakespeare.

Edwards, William - Seaman - Number: 179 - Prize name: Farmer's Daughter - Ship type: MV - How taken: HM Ship-of-the-Line Leviathan - When taken: 20 Mar 1814 - Where taken: at sea - Date received: 2 Mar 1815 - From what ship: HMS Dannemark - Born: Accomack - Age: 20 - Discharged on 3 Mar 1815 and sent to Dartmoor.

Elles, Edward - Seaman - Number: 381 - Prize name: Leo - Ship type: P - How taken: HM Frigate Tiber - When taken: 11 Mar 1815 - Where taken: Lat 45.24N Long 12.3W - Date received: 29 Mar 1814 - From what ship: HMS Tiber - Born: Massachusetts - Age: 27 - Discharged on 30 May 1815 and released to the Cartel Atlas.

Ellingwood, Herbert - Seaman - Number: 354 - Prize name: Leo - Ship type: P - How taken: HM Frigate Tiber - When taken: 11 Mar 1815 - Where taken: Lat 45.24N Long 12.3W - Date received: 29 Mar 1814 - From what ship: HMS Tiber - Born: Massachusetts - Age: 26 - Discharged on 27 Apr 1815 and released to the Cartel Minework.

Emery, John - Seaman - Number: 764 - Prize name: Sine Qua Non - Ship type: P - How taken: HM Brig Elk - When taken: 20 Feb 1815 - Where taken: off Madeira - Date received: 29 Aug 1815 - From what ship: HMS Tonnant - Born: Massachusetts - Age: 20 - Discharged on 2 Jun 1815 and released to the Cartel Shakespeare.

Endley, James - Seaman - Number: 425 - Prize name: Engineer - Ship type: MV - How taken: HMS Murros - When taken: 21 Sep 1814 - Where taken: off Porto Rico - Date received: 16 Apr 1815 - From what ship: HMS Swifsure - Born: Massachusetts - Age: 19 - Discharged on 2 Jun 1815 and released to the Cartel Sovereign.

Eskill, John - Seaman - Number: 743 - Prize name: Sine Qua Non - Ship type: P - How taken: HM Brig Elk - When taken: 20 Feb 1815 - Where taken: off Madeira - Date received: 29 Aug 1815 - From what ship: HMS Tonnant - Born: Massachusetts - Age: 25 - Discharged on 2 Jun 1815 and released to the Cartel Shakespeare.

Evans, Thomas - Seaman - Number: 46 - How taken: Taken out of HM Ship-of-the-Line Prince of Wales - Where taken: Spithead - Date received: 15 Feb 1815 - From what ship: HMS Myrmidon - Born: Virginia - Age: 33 - Discharged on 17 Feb 1815 and sent to Dartmoor.

Falcomb, William - Seaman - Number: 525 - Prize name: Dolphin - Ship type: MV - How taken: HM Frigate Barbadoes - When taken: 4 Dec 1814 - Where taken: off St. Bartholomew - Date received: 16 Apr 1815 - From what ship: HMS Swifsure - Born: Newburyport - Age: 23 - Discharged on 2 Jun 1815 and released to the Cartel Sovereign.

Farr, William - Seaman - Number: 96 - Prize name: Prince de Neufchatel - Ship type: P - How taken: HM Ship-of-the-Line Leander, HM Ship-of-the-Line Newcastle & HM Frigate Acasta - When taken: 28 Dec 1814 - Where taken: Lat 35N Long 52W - Date received: 18 Feb 1815 - From what ship: HMS Sybille - Age: 24 - Race: Black - Discharged on 19 Feb 1815 and sent to Dartmoor.

Fathan, Jacob - Seaman - Number: 341 - Prize name: Leo - Ship type: P - How taken: HM Frigate Tiber - When taken: 11 Mar 1815 - Where taken: Lat 45.24N Long 12.3W - Date received: 28 Mar 1815 - From what ship: HMS Tiber - Discharged on 27 Apr 1815 and released to the Cartel Minework.

Fernandez, George - Seaman - Number: 110 - Prize name: Prince de Neufchatel - Ship type: P - How taken: HM Ship-of-the-Line Leander, HM Ship-of-the-Line Newcastle & HM Frigate Acasta - When taken: 28 Dec 1814 - Where taken: Lat 35N Long 52W - Date received: 18 Feb 1815 - From what ship: HMS Sybille - Born: Brazil -

Age: 24 - Race: Black - Discharged on 19 Feb 1815 and sent to Dartmoor.

Fielding, John - Seaman - Number: 348 - Prize name: Leo - Ship type: P - How taken: HM Frigate Tiber - When taken: 11 Mar 1815 - Where taken: Lat 45.24N Long 12.3W - Date received: 28 Mar 1815 - From what ship: HMS Tiber - Born: Massachusetts - Age: 20 - Discharged on 27 Apr 1815 and released to the Cartel Minework.

Finch, Abraham - Seaman - Number: 252 - How taken: Taken up at Kingston - Date received: 2 Mar 1815 - From what ship: HMS Dannemark - Born: New Hampshire - Age: 26 - Discharged on 3 Mar 1815 and sent to Dartmoor.

Fincho, Etienne - Seaman - Number: 262 - Prize name: Dorothy - Ship type: MV - How taken: HM Frigate North Star - When taken: 15 Nov 1814 - Where taken: at sea - Date received: 2 Mar 1815 - From what ship: HMS Dannemark - Discharged on 3 Mar 1815 and sent to Dartmoor.

Finder, John - Seaman - Number: 289 - How taken: Taken off the HM Ship-of-the-Line Severn and other ships - Date received: 2 Mar 1815 - From what ship: HMS Dannemark - Race: Mulatto - Discharged on 3 Mar 1815 and sent to Dartmoor.

Finney, James - Seaman - Number: 51 - How taken: Taken out of MV Essex - Date received: 15 Feb 1815 - From what ship: HMS Myrmidon - Born: Sandwich - Age: 31 - Discharged on 17 Feb 1815 and sent to Dartmoor.

Fisher, James - Seaman - Number: 315 - Prize name: Transit - Ship type: MV - How taken: Impressed at North Shields - When taken: 24 Nov 1814 - Where taken: North Shields - Date received: 4 Mar 1814 - From what ship: HM Cutter Surley - Discharged on 7 Mar 1815 and sent to Dartmoor.

Fisher, John - Seaman - Number: 334 - Prize name: Leo - Ship type: P - How taken: HM Frigate Tiber - When taken: 11 Mar 1815 - Where taken: Lat 45.24N Long 12.3W - Date received: 28 Mar 1815 - From what ship: HMS Tiber - Discharged on 27 Apr 1815 and released to the Cartel Minework.

Fisher, Robert - Seaman - Number: 104 - Prize name: Prince de Neufchatel - Ship type: P - How taken: HM Ship-of-the-Line Leander, HM Ship-of-the-Line Newcastle & HM Frigate Acasta - When taken: 28 Dec 1814 - Where taken: Lat 35N Long 52W - Date received: 18 Feb 1815 - From what ship: HMS Sybille - Age: 26 - Discharged on 19 Feb 1815 and sent to Dartmoor.

Fontaine, (-----) - Seaman - Number: 480 - Prize name: Fox - Ship type: P - How taken: HM Frigate Barbadoes - When taken: 11 Jan 1815 - Where taken: off Amelia Island, Florida - Date received: 16 Apr 1815 - From what ship: HMS Swifsure - Born: Guadeloupe - Age: 30 - Race: Negro - Discharged on 2 Jun 1815 and released to the Cartel Sovereign.

Ford, John - Seaman - Number: 518 - Prize name: Mary Ann - Ship type: MV - How taken: HM Ship-of-the-Line Elizabeth - When taken: 28 Feb 1815 - Where taken: West Indies - Date received: 16 Apr 1815 - From what ship: HMS Swifsure - Born: St. Marys - Age: 30 - Discharged on 11 Jul 1815 and released to the Cartel Wooddrop Sims.

Foreman, John - Seaman - Number: 100 - Prize name: Prince de Neufchatel - Ship type: P - How taken: HM Ship-of-the-Line Leander, HM Ship-of-the-Line Newcastle & HM Frigate Acasta - When taken: 28 Dec 1814 - Where taken: Lat 35N Long 52W - Date received: 18 Feb 1815 - From what ship: HMS Sybille - Age: 22 - Race: Black - Discharged on 19 Feb 1815 and sent to Dartmoor.

Forestte, (-----) - Seaman - Number: 502 - Prize name: Fox - Ship type: P - How taken: HM Frigate Barbadoes - When taken: 11 Jan 1815 - Where taken: off Amelia Island, Florida - Date received: 16 Apr 1815 - From what ship: HMS Swifsure - Born: Saint-Domingue (Haiti) - Age: 29 - Race: Mulatto - Discharged on 2 Jun 1815 and released to the Cartel Sovereign.

Foster, B. C. - Seaman - Number: 699 - Prize name: Sine Qua Non - Ship type: P - How taken: HM Brig Elk - When taken: 20 Feb 1815 - Where taken: off Madeira - Date received: 16 Jul 1815 - From what ship: HMS Hope - Born: Massachusetts - Age: 23 - Discharged on 2 Jun 1815 and released to the Cartel Shakespeare.

Foster, William - Seaman - Number: 673 - Prize name: Avon - Ship type: P - How taken: HM Frigate Barbadoes - When taken: 8 Mar 1815 - Where taken: West Indies - Date received: 24 Apr 1815 - From what ship: HMS Bellerophon - Born: Massachusetts - Age: 25 - Discharged on 2 Jun 1815 and released to the Cartel Shakespeare.

Fouche, Jean - Seaman - Number: 517 - Prize name: Hope, prize to US Sloop-of-War Wasp - Ship type: War - How taken: HM Sloop Fairy - When taken: 6 Jan 1815 - Where taken: Unknown - Date received: 16 Apr 1815 - From what ship: HMS Swifsure - Born: Saint-Domingue (Haiti) - Age: 57 - Discharged on 2 Jun 1815 and released to the Cartel Sovereign.

Foulton, John - Seaman - Number: 573 - Prize name: Heron - Ship type: MV - How taken: HM Frigate Pique - When taken: 16 Dec 1814 - Where taken: off Porto Rico - Date received: 16 Apr 1815 - From what ship: HMS Swifsure - Born: Rhode Island - Age: 18 - Discharged on 27 May 1815.

Fox, Edward - Master - Number: 257 - How taken: Taken up at Kingston after breaking his parole at Nassau, New Providence - Date received: 2 Mar 1815 - From what ship: HMS Dannemark - Discharged on 3 Mar 1815 and sent to Dartmoor.

Foy, Peter - Prize Master - Number: 613 - Prize name: Sine Qua Non - Ship type: P - How taken: HM Brig Elk - When taken: 20 Feb 1815 - Where taken: off Madeira - Date received: 18 Apr 1815 - From what ship: HMS Euryalus - Born: Massachusetts - Age: 31 - Discharged on 23 May 1815.

Fray, Samuel - Prize Master - Number: 155 - Prize name: Nancy, prize of the Privateer United States - Ship type: P - How taken: HM Sloop Papillon - When taken: 27 Dec 1814 - Where taken: Lat 30 Long 10 - Date received: 27 Feb 1815 - From what ship: HMS Nimble - Born: Massachusetts - Age: 25 - Discharged on 3 Mar 1815 and sent to Dartmoor.

Frazier, Charles - 2nd Mate - Number: 37 - Prize name: Plutarch - Ship type: MV - How taken: HM Schooner Helicon - When taken: 5 Feb 1815 - Where taken: Lat 45.5N Long 7W - Date received: 10 Feb 1815 - From what ship: HMS Helicon - Born: New York - Age: 44 - Discharged on 17 Feb 1815 and sent to Dartmoor.

Frederick, C. M. - Boy - Number: 688 - Prize name: George Little - Ship type: P - How taken: HM Frigate Granicus - When taken: 20 Jan 1815 - Where taken: off Cape Finisterre, Spain - Date received: 16 Jul 1815 - From what ship: HMS Hope - Born: Massachusetts - Age: 17 - Discharged on 2 Jun 1815 and released to the Cartel Shakespeare.

Free, Isaac - Seaman - Number: 700 - Prize name: Sine Qua Non - Ship type: P - How taken: HM Brig Elk - When taken: 20 Feb 1815 - Where taken: off Madeira - Date received: 16 Jul 1815 - From what ship: HMS Hope - Born: Massachusetts - Age: 33 - Discharged on 2 Jun 1815 and released to the Cartel Shakespeare.

Fremont, Lawrence - Seaman - Number: 91 - Prize name: Prince de Neufchatel - Ship type: P - How taken: HM Ship-of-the-Line Leander, HM Ship-of-the-Line Newcastle & HM Frigate Acasta - When taken: 28 Dec 1814 - Where taken: Lat 35N Long 52W - Date received: 18 Feb 1815 - From what ship: HMS Sybille - Race: Black - Discharged on 19 Feb 1815 and sent to Dartmoor.

Frost, Thomas Bell - Seaman - Number: 57 - How taken: Nancy, merchant vessel - Date received: 15 Feb 1815 - From what ship: HMS Myrmidon - Born: Newcastle - Age: 30 - Discharged on 17 Feb 1815 and sent to Dartmoor.

Frovemcs, Ford. - Seaman - Number: 298 - Prize name: US Brig Syren - Ship type: War - How taken: HM Ship-of-the-Line Medway - When taken: 12 Jul 1814 - Where taken: off Cape of Good Hope - Date received: 4 Mar 1814 - From what ship: HM Cutter Surley - Discharged on 7 Mar 1815 and sent to Dartmoor.

Fuller, Benjamin - Seaman - Number: 309 - Prize name: Transit - Ship type: MV - How taken: Impressed at North Shields - When taken: 24 Nov 1814 - Where taken: North Shields - Date received: 4 Mar 1814 - From what ship: HM Cutter Surley - Discharged on 7 Mar 1815 and sent to Dartmoor.

Fuller, Moses - Seaman - Number: 124 - Prize name: Prince de Neufchatel - Ship type: P - How taken: HM Ship-of-the-Line Leander, HM Ship-of-the-Line Newcastle & HM Frigate Acasta - When taken: 28 Dec 1814 - Where taken: Lat 35N Long 52W - Date received: 18 Feb 1815 - From what ship: HMS Sybille - Age: 17 - Discharged on 19 Feb 1815 and sent to Dartmoor.

Furdge, Henry - Seaman - Number: 439 - Prize name: Nancy - Ship type: P - How taken: HM Frigate Amelia - When taken: 6 Sep 1814 - Where taken: off Newfoundland - Date received: 16 Apr 1815 - From what ship: HMS Swifsure - Born: Pennsylvania - Age: 20 - Discharged on 2 Jun 1815 and released to the Cartel Sovereign.

Furrel, John - Boy - Number: 248 - Prize name: John - Ship type: MV - How taken: HM Brig Zenobia - When taken: 11 Sep 1814 - Where taken: at sea - Date received: 2 Mar 1815 - From what ship: HMS Dannemark - Born: New London - Age: 16 - Discharged on 3 Mar 1815 and sent to Dartmoor.

Gale, Russel - Seaman - Number: 194 - Prize name: Chance - Ship type: P - How taken: HMS Statire - When taken: 1 Apr 1814 - Where taken: at sea - Date received: 2 Mar 1815 - From what ship: HMS Dannemark - Born: North Carolina - Age: 18 - Discharged on 3 Mar 1815 and sent to Dartmoor.

Gallen, Samuel - Seaman - Number: 657 - Prize name: Avon - Ship type: P - How taken: HM Frigate Barbadoes - When taken: 8 Mar 1815 - Where taken: West Indies - Date received: 24 Apr 1815 - From what ship: HMS Bellerophon - Age: 28 - Discharged on 1 Jun 1815.

Gammel, Joseph - Seaman - Number: 793 - Prize name: US Brig Syren - Ship type: War - How taken: HM Ship-of-the-Line Medway - When taken: 12 Jul 1814 - Where taken: off Cape of Good Hope - Date received: 11 Jul 1815 - From what ship: HMS Royal Sovereign - Born: Not listed - Discharged on 11 Jul 1815 and released to the Cartel Wooddrop Sims.

Garbrierre, John - Mate - Number: 227 - Prize name: Decatur - Ship type: P - How taken: HM Frigate Rhin - When taken: 4 Jun 1814 - Where taken: at sea - Date received: 2 Mar 1815 - From what ship: HMS Dannemark - Born: Charleston - Age: 20 - Race: Mulatto - Discharged on 3 Mar 1815 and sent to Dartmoor.

Garrish, George - Seaman - Number: 589 - Prize name: George Little - Ship type: P - How taken: HM Frigate Granicus - When taken: 20 Jan 1815 - Where taken: off Cape Finisterre, Spain - Date received: 18 Apr 1815 - From what ship: HMS Euryalus - Born: New Hampshire - Age: 25 - Discharged on 2 Jun 1815 and released to the Cartel Sovereign.

Gregory, Cornelius - Seaman - Number: 448 - Prize name: High Flyer - Ship type: LM - How taken: HMS Muros - When taken: 14 Nov 1814 - Where taken: off St. Bartholomew - Date received: 16 Apr 1815 - From what ship: HMS Swifsure - Born: North Caroline - Age: 26 - Discharged on 2 Jun 1815 and released to the Cartel Sovereign.

Gerrish, Samuel - Seaman - Number: 810 - Prize name: US Brig Syren - Ship type: War - How taken: HM Ship-of-the-Line Medway - When taken: 12 Jul 1814 - Where taken: off Cape of Good Hope - Date received: 11 Jul 1815 - From what ship: HMS Royal Sovereign - Born: Not listed - Discharged on 11 Jul 1815 and released to the Cartel Wooddrop Sims.

Gerry, Joseph - Seaman - Number: 785 - Prize name: US Brig Syren - Ship type: War - How taken: HM Ship-of-the-Line Medway - When taken: 12 Jul 1814 - Where taken: off Cape of Good Hope - Date received: 11 Jul 1815 - From what ship: HMS Royal Sovereign - Born: Not listed - Discharged on 11 Jul 1815 and released to the Cartel Wooddrop Sims.

Getcher, Samuel - Seaman - Number: 842 - Prize name: US Brig Syren - Ship type: War - How taken: HM Ship-of-the-Line Medway - When taken: 12 Jul 1814 - Where taken: off Cape of Good Hope - Date received: 11 Jul 1815 - From what ship: HMS Royal Sovereign - Born: Not listed - Discharged on 11 Jul 1815 and released to the Cartel Wooddrop Sims.

Gibbs, Moses - Prize Master - Number: 635 - Prize name: Avon - Ship type: P - How taken: HM Frigate Barbadoes - When taken: 8 Mar 1815 - Where taken: West Indies - Date received: 24 Apr 1815 - From what ship: HMS Bellerophon - Age: 20 - Discharged on 23 May 1815.

Gibson, Francis - Seaman - Number: 317 - How taken: Impressed off HM Ship-of-the-Line Scepter - When taken: 15 Feb 1815 - Where taken: Cork - Date received: 5 Mar 1815 - From what ship: HMS Tartarus - Born: Pennsylvania - Age: 42 - Race: Negro - Discharged on 7 Mar 1815 and sent to Dartmoor.

Gibson, Samuel - Seaman - Number: 386 - Prize name: Leo - Ship type: P - How taken: HM Frigate Tiber - When taken: 11 Mar 1815 - Where taken: Lat 45.24N Long 12.3W - Date received: 29 Mar 1814 - From what ship: HMS Tiber - Born: Massachusetts - Age: 19 - Discharged on 30 May 1815 and released to the Cartel Atlas.

Glover, William - Seaman - Number: 123 - Prize name: Prince de Neufchatel - Ship type: P - How taken: HM Ship-of-the-Line Leander, HM Ship-of-the-Line Newcastle & HM Frigate Acasta - When taken: 28 Dec 1814 - Where taken: Lat 35N Long 52W - Date received: 18 Feb 1815 - From what ship: HMS Sybille - Born: Salem - Age: 18 - Discharged on 19 Feb 1815 and sent to Dartmoor.

Goodhall, Joseph - Seaman - Number: 107 - Prize name: Prince de Neufchatel - Ship type: P - How taken: HM Ship-of-the-Line Leander, HM Ship-of-the-Line Newcastle & HM Frigate Acasta - When taken: 28 Dec 1814 - Where taken: Lat 35N Long 52W - Date received: 18 Feb 1815 - From what ship: HMS Sybille - Born: Bordeaux - Age:

33 - Discharged on 19 Feb 1815 and sent to Dartmoor.

Gordon, Elijah - Seaman - Number: 775 - Prize name: Aurora - Ship type: MV - How taken: HM Frigate Andromeda - When taken: 28 Feb 1815 - Where taken: off Lisbon - Date received: 29 Aug 1815 - From what ship: HMS Tonnant - Age: 19 - Discharged on 2 Jun 1815 and released to the Cartel Shakespeare.

Gordon, Emanuel - Seaman - Number: 751 - Prize name: Sine Qua Non - Ship type: P - How taken: HM Brig Elk - When taken: 20 Feb 1815 - Where taken: off Madeira - Date received: 29 Aug 1815 - From what ship: HMS Tonnant - Age: 18 - Discharged on 2 Jun 1815 and released to the Cartel Shakespeare.

Gordon, John - Seaman - Number: 750 - Prize name: Sine Qua Non - Ship type: P - How taken: HM Brig Elk - When taken: 20 Feb 1815 - Where taken: off Madeira - Date received: 29 Aug 1815 - From what ship: HMS Tonnant - Age: 20 - Discharged on 2 Jun 1815 and released to the Cartel Shakespeare.

Gordon, William L. - Lieutenant - Number: 851 - Prize name: US Brig Syren - Ship type: War - How taken: HM Ship-of-the-Line Medway - When taken: 12 Jul 1814 - Where taken: off Cape of Good Hope - Date received: 11 Jul 1815 - From what ship: from Portsmouth - Born: Not listed - Discharged on 11 Jul 1815 and released to the Cartel Wooddrop Sims.

Grant, Christian - Seaman - Number: 766 - Prize name: George Little - Ship type: P - How taken: HM Frigate Granicus - When taken: 20 Jan 1815 - Where taken: off Cape Finisterre, Spain - Date received: 29 Aug 1815 - From what ship: HMS Tonnant - Born: Massachusetts - Age: 24 - Discharged on 2 Jun 1815 and released to the Cartel Shakespeare.

Greavier, Gabriel - Seaman - Number: 509 - Prize name: Fox - Ship type: P - How taken: HM Frigate Barbadoes - When taken: 11 Jan 1815 - Where taken: off Amelia Island, Florida - Date received: 16 Apr 1815 - From what ship: HMS Swifsure - Born: Saint-Domingue (Haiti) - Age: 35 - Race: Mulatto - Discharged on 2 Jun 1815 and released to the Cartel Sovereign.

Green, Peter - Seaman - Number: 11 - Prize name: Albion - Ship type: MV - How taken: HM Schooner Helicon - When taken: 17 Jan 1815 - Where taken: Lat 20N Long 52W - Date received: 3 Feb 1815 - From what ship: HMS Bittern - Born: Salem - Age: 23 - Race: Black - Discharged on 4 Feb 1815 and sent to Dartmoor.

Greenlaw, Jeremiah - 3rd Lieutenant - Number: 62 - Prize name: Prince de Neufchatel - Ship type: P - How taken: HM Ship-of-the-Line Leander, HM Ship-of-the-Line Newcastle & HM Frigate Acasta - When taken: 28 Dec 1814 - Where taken: Lat 35N Long 52W - Date received: 18 Feb 1815 - From what ship: HMS Sybille - Born: Virginia - Age: 21 - Discharged on 19 Feb 1815 and sent to Dartmoor.

Greenough, William - Seaman - Number: 435 - Prize name: Mars, prize of the Privateer David Porter - Ship type: P - How taken: HM Frigate Pique - When taken: 12 Aug 1814 - Where taken: off Newfoundland - Date received: 16 Apr 1815 - From what ship: HMS Swifsure - Born: New Hampshire - Age: 19 - Discharged on 8 May 1815.

Griffin, Heathcote - Seaman - Number: 184 - Prize name: Farmer's Daughter - Ship type: MV - How taken: HM Ship-of-the-Line Leviathan - When taken: 20 Mar 1814 - Where taken: at sea - Date received: 2 Mar 1815 - From what ship: HMS Dannemark - Born: New Haven - Age: 20 - Discharged on 3 Mar 1815 and sent to Dartmoor.

Groce, Samuel - 2nd Lieutenant - Number: 413 - Prize name: Leo - Ship type: P - How taken: HM Frigate Tiber - When taken: 11 Mar 1815 - Where taken: Lat 45.24N Long 12.3W - Date received: 2 Apr 1815 - From what ship: HMS Tiber - Born: Boston - Age: 28 - Discharged on 2 Jun 1815 and released to the Cartel Sovereign.

Grove, Thomas - Seaman - Number: 740 - Prize name: Sine Qua Non - Ship type: P - How taken: HM Brig Elk - When taken: 20 Feb 1815 - Where taken: off Madeira - Date received: 29 Aug 1815 - From what ship: HMS Tonnant - Born: Massachusetts - Age: 26 - Discharged on 2 Jun 1815 and released to the Cartel Shakespeare.

Gunmar, Lewis - Lieutenant - Number: 850 - Prize name: US Brig Syren - Ship type: War - How taken: HM Ship-of-the-Line Medway - When taken: 12 Jul 1814 - Where taken: off Cape of Good Hope - Date received: 11 Jul 1815 - From what ship: from Portsmouth - Born: Not listed - Discharged on 11 Jul 1815 and released to the Cartel Wooddrop Sims.

Hackley, Walter - Prize Master - Number: 245 - Prize name: Mary, recaptured prize - Ship type: P - How taken: HM Brig Sophie - When taken: 19 Jun 1814 - Where taken: at sea - Date received: 2 Mar 1815 - From what ship: HMS Dannemark - Discharged on 3 Mar 1815 and sent to Dartmoor.

Hadley, William - Seaman - Number: 841 - Prize name: US Brig Syren - Ship type: War - How taken: HM Ship-of-the-Line Medway - When taken: 12 Jul 1814 - Where taken: off Cape of Good Hope - Date received: 11 Jul 1815 - From what ship: HMS Royal Sovereign - Born: Not listed - Discharged on 11 Jul 1815 and released to the Cartel Wooddrop Sims.

Hall, George - Seaman - Number: 581 - Prize name: Sorin, prize to Privateer Prince of Neufchatel - Ship type: P - How taken: HM Ship-of-the-Line Medway - When taken: 12 Jul 1814 - Where taken: Unknown - Date received: 16 Apr 1815 - From what ship: HMS Swifsure - Born: New York - Age: 25 - Discharged on 2 Jun 1815 and released to the Cartel Sovereign.

Hall, John - Seaman - Number: 779 - Prize name: Sine Qua Non - Ship type: P - How taken: HM Brig Elk - When taken: 20 Feb 1815 - Where taken: off Madeira - Date received: 29 Aug 1815 - From what ship: HMS Tonnant - Age: 45 - Discharged on 2 Jun 1815 and released to the Cartel Shakespeare.

Hall, John - Marine - Number: 97 - Prize name: Prince de Neufchatel - Ship type: P - How taken: HM Ship-of-the-Line Leander, HM Ship-of-the-Line Newcastle & HM Frigate Acasta - When taken: 28 Dec 1814 - Where taken: Lat 35N Long 52W - Date received: 18 Feb 1815 - From what ship: HMS Sybille - Born: Methuen - Age: 20 - Discharged on 19 Feb 1815 and sent to Dartmoor.

Hall, Noah - Seaman - Number: 696 - Prize name: Sine Qua Non - Ship type: P - How taken: HM Brig Elk - When taken: 20 Feb 1815 - Where taken: off Madeira - Date received: 16 Jul 1815 - From what ship: HMS Hope - Born: New Jersey - Age: 45 - Discharged on 2 Jun 1815 and released to the Cartel Shakespeare.

Hall, Samuel - Seaman - Number: 723 - Prize name: Sine Qua Non - Ship type: P - How taken: HM Brig Elk - When taken: 20 Feb 1815 - Where taken: off Madeira - Date received: 29 Aug 1815 - From what ship: HMS Tonnant - Born: Massachusetts - Age: 46 - Discharged on 2 Jun 1815 and released to the Cartel Shakespeare.

Hallick, Edward - Seaman - Number: 522 - Prize name: High Flyer - Ship type: LM - How taken: HMS Muros - When taken: 14 Nov 1814 - Where taken: off St. Bartholomew - Date received: 16 Apr 1815 - From what ship: HMS Swifsure - Born: Massachusetts - Age: 25 - Race: Negro - Discharged on 2 Jun 1815 and released to the Cartel Sovereign.

Hammett, C. H - Purser - Number: 585 - Prize name: George Little - Ship type: P - How taken: HM Frigate Granicus - When taken: 20 Jan 1815 - Where taken: off Cape Finisterre, Spain - Date received: 18 Apr 1815 - From what ship: HMS Euryalus - Born: Massachusetts - Age: 24 - Discharged on 24 Apr 1815.

Hamond, David - Boy - Number: 393 - Prize name: Leo - Ship type: P - How taken: HM Frigate Tiber - When taken: 11 Mar 1815 - Where taken: Lat 45.24N Long 12.3W - Date received: 29 Mar 1814 - From what ship: HMS Tiber - Born: Massachusetts - Age: 16 - Discharged on 30 May 1815 and released to the Cartel Atlas.

Hannah, Samuel - Seaman - Number: 748 - Prize name: Harpy - Ship type: P - How taken: HM Brig Zenobia - When taken: 5 Jan 1815 - Where taken: off Lisbon - Date received: 29 Aug 1815 - From what ship: HMS Tonnant - Age: 31 - Discharged on 2 Jun 1815 and released to the Cartel Shakespeare.

Hannes, George R. - Seaman - Number: 358 - Prize name: Leo - Ship type: P - How taken: HM Frigate Tiber - When taken: 11 Mar 1815 - Where taken: Lat 45.24N Long 12.3W - Date received: 29 Mar 1814 - From what ship: HMS Tiber - Born: Massachusetts - Age: 42 - Discharged on 27 Apr 1815 and released to the Cartel Minework.

Hansford, Stephen - Seaman - Number: 488 - Prize name: Fox - Ship type: P - How taken: HM Frigate Barbadoes - When taken: 11 Jan 1815 - Where taken: off Amelia Island, Florida - Date received: 16 Apr 1815 - From what ship: HMS Swifsure - Born: Virginia - Age: 28 - Discharged on 2 Jun 1815 and released to the Cartel Sovereign.

Hanstable, William - Seaman - Number: 165 - How taken: Impressed from MV Rolla - Date received: 27 Feb 1815 - From what ship: HMS Nimble - Age: 21 - Discharged on 3 Mar 1815 and sent to Dartmoor.

Harding, Joseph - Master's Mate - Number: 860 - Prize name: US Brig Syren - Ship type: War - How taken: HM Ship-of-the-Line Medway - When taken: 12 Jul 1814 - Where taken: off Cape of Good Hope - Date received: 11 Jul 1815 - From what ship: from Portsmouth - Born: Not listed - Discharged on 11 Jul 1815 and released to the Cartel Wooddrop Sims.

Harker, John - Seaman - Number: 682 - Prize name: William, prize to Privateer Surprize - Ship type: P - How taken: HMS Ister - When taken: 8 Jun 1814 - Date received: 14 May 1815 - From what ship: from the police - Age: 35 -

Discharged on 2 Jun 1815 and released to the Cartel Shakespeare.

Harris, Richard - Boy - Number: 396 - Prize name: Leo - Ship type: P - How taken: HM Frigate Tiber - When taken: 11 Mar 1815 - Where taken: Lat 45.24N Long 12.3W - Date received: 29 Mar 1814 - From what ship: HMS Tiber - Born: Boston - Age: 19 - Discharged on 30 May 1815 and released to the Cartel Atlas.

Harvey, Frederick - Master's Mate - Number: 460 - Prize name: Fox - Ship type: P - How taken: HM Frigate Barbadoes - When taken: 11 Jan 1815 - Where taken: off Amelia Island, Florida - Date received: 16 Apr 1815 - From what ship: HMS Swifsure - Age: 35 - Discharged on 2 Jun 1815 and released to the Cartel Sovereign.

Harvey, James - Seaman - Number: 268 - Prize name: Netterville - Ship type: MV - How taken: HM Brig Onyx - When taken: 25 Dec 1814 - Where taken: at sea - Date received: 2 Mar 1815 - From what ship: HMS Dannemark - Born: North Carolina - Age: 32 - Discharged on 3 Mar 1815 and sent to Dartmoor.

Harwood, Edward - Seaman - Number: 338 - Prize name: Leo - Ship type: P - How taken: HM Frigate Tiber - When taken: 11 Mar 1815 - Where taken: Lat 45.24N Long 12.3W - Date received: 28 Mar 1815 - From what ship: HMS Tiber - Discharged on 27 Apr 1815 and released to the Cartel Minework.

Harwood, Morocco - Servant - Number: 622 - Prize name: Sine Qua Non - Ship type: P - How taken: HM Brig Elk - When taken: 20 Feb 1815 - Where taken: off Madeira - Date received: 18 Apr 1815 - From what ship: HMS Euryalus - Age: 24 - Race: Negro - Discharged on 22 Apr 1815.

Hawdy, Joseph - Seaman - Number: 280 - Prize name: Tartan - Ship type: MV - How taken: Given up - Date received: 2 Mar 1815 - From what ship: HMS Dannemark - Discharged on 3 Mar 1815 and sent to Dartmoor.

Hawes, J. P. - Seaman - Number: 335 - Prize name: Leo - Ship type: P - How taken: HM Frigate Tiber - When taken: 11 Mar 1815 - Where taken: Lat 45.24N Long 12.3W - Date received: 28 Mar 1815 - From what ship: HMS Tiber - Discharged on 27 Apr 1815 and released to the Cartel Minework.

Hawkins, John - Seaman - Number: 114 - Prize name: Prince de Neufchatel - Ship type: P - How taken: HM Ship-of-the-Line Leander, HM Ship-of-the-Line Newcastle & HM Frigate Acasta - When taken: 28 Dec 1814 - Where taken: Lat 35N Long 52W - Date received: 18 Feb 1815 - From what ship: HMS Sybille - Discharged on 19 Feb 1815 and sent to Dartmoor.

Haycock, Joseph - Quarter Gunner - Number: 149 - Prize name: US Brig Syren - Ship type: War - How taken: HM Ship-of-the-Line Medway - When taken: 12 Jul 1814 - Where taken: off Cape of Good Hope - Date received: 21 Feb 1815 - From what ship: HMS Slaney - Born: Portland - Age: 55 - Discharged on 24 Feb 1815 and sent to Dartmoor.

Haywood, Abraham - Merchant - Number: 442 - Prize name: Perveance, cartel - How taken: Detained at Barbados - When taken: 23 Oct 1814 - Date received: 16 Apr 1815 - From what ship: HMS Swifsure - Age: 25 - Discharged on 2 Jun 1815 and released to the Cartel Sovereign.

Haywood, Samuel - Seaman - Number: 94 - Prize name: Prince de Neufchatel - Ship type: P - How taken: HM Ship-of-the-Line Leander, HM Ship-of-the-Line Newcastle & HM Frigate Acasta - When taken: 28 Dec 1814 - Where taken: Lat 35N Long 52W - Date received: 18 Feb 1815 - From what ship: HMS Sybille - Born: Roxborough - Age: 26 - Discharged on 19 Feb 1815 and sent to Dartmoor.

Hazlet, Henry - Merchant Clerk - Number: 510 - Prize name: Fox - Ship type: P - How taken: HM Frigate Barbadoes - When taken: 11 Jan 1815 - Where taken: off Amelia Island, Florida - Date received: 16 Apr 1815 - From what ship: HMS Swifsure - Age: 29 - Discharged on 22 Apr 1815.

Healy, Samuel - Seaman - Number: 84 - Prize name: Prince de Neufchatel - Ship type: P - How taken: HM Ship-of-the-Line Leander, HM Ship-of-the-Line Newcastle & HM Frigate Acasta - When taken: 28 Dec 1814 - Where taken: Lat 35N Long 52W - Date received: 18 Feb 1815 - From what ship: HMS Sybille - Age: 18 - Discharged on 19 Feb 1815 and sent to Dartmoor.

Heasey, John - Seaman - Number: 311 - Prize name: Transit - Ship type: MV - How taken: Impressed at North Shields - When taken: 24 Nov 1814 - Where taken: North Shields - Date received: 4 Mar 1814 - From what ship: HM Cutter Surley - Discharged on 7 Mar 1815 and sent to Dartmoor.

Hebley, John - Seaman - Number: 832 - Prize name: US Brig Syren - Ship type: War - How taken: HM Ship-of-the-Line Medway - When taken: 12 Jul 1814 - Where taken: off Cape of Good Hope - Date received: 11 Jul 1815 -

From what ship: HMS Royal Sovereign - Born: Not listed - Discharged on 11 Jul 1815 and released to the Cartel Wooddrop Sims.

Hedley, Andrew - Seaman - Number: 836 - Prize name: US Brig Syren - Ship type: War - How taken: HM Ship-of-the-Line Medway - When taken: 12 Jul 1814 - Where taken: off Cape of Good Hope - Date received: 11 Jul 1815 - From what ship: HMS Royal Sovereign - Born: Not listed - Discharged on 11 Jul 1815 and released to the Cartel Wooddrop Sims.

Hedrick, Jan - Seaman - Number: 454 - Prize name: San Francisco - Ship type: MV - How taken: HMS Ister - When taken: 2 Dec 1814 - Where taken: off Amelia Island, Florida - Date received: 16 Apr 1815 - From what ship: HMS Swifsure - Age: 36 - Discharged on 2 Jun 1815 and released to the Cartel Sovereign.

Henderson, Joseph - Seaman - Number: 246 - Prize name: John - Ship type: MV - How taken: HM Brig Zenobia - When taken: 11 Aug 1814 - Where taken: at sea - Date received: 2 Mar 1815 - From what ship: HMS Dannemark - Born: New London - Age: 21 - Discharged on 3 Mar 1815 and sent to Dartmoor.

Hendry, William - Seaman - Number: 238 - How taken: Gave himself up - Date received: 2 Mar 1815 - From what ship: HMS Dannemark - Born: South Carolina - Age: 49 - Discharged on 3 Mar 1815 and sent to Dartmoor.

Henry, Jean B. - Seaman - Number: 470 - Prize name: Fox - Ship type: P - How taken: HM Frigate Barbadoes - When taken: 11 Jan 1815 - Where taken: off Amelia Island, Florida - Date received: 16 Apr 1815 - From what ship: HMS Swifsure - Born: Saint-Domingue (Haiti) - Age: 25 - Race: Negro - Discharged on 2 Jun 1815 and released to the Cartel Sovereign.

Henry, Joshua - Seaman - Number: 780 - Prize name: George Little - Ship type: P - How taken: HM Frigate Granicus - When taken: 20 Jan 1815 - Where taken: off Cape Finisterre, Spain - Date received: 29 Aug 1815 - From what ship: HMS Tonnant - Born: Massachusetts - Age: 18 - Discharged on 2 Jun 1815 and released to the Cartel Shakespeare.

Henshaw, Jacob - Prize Master - Number: 409 - Prize name: Thomas, prize to Privateer Scourge - Ship type: P - How taken: HM Frigate Aquilon - When taken: 11 Mar 1815 - Where taken: off Cape Finisterre, Spain - Date received: 29 Mar 1815 - From what ship: HMS Icarus - Born: Massachusetts - Age: 29 - Discharged on 2 Jun 1815 and released to the Cartel Sovereign.

Henson, Peter - Boatswain - Number: 187 - Prize name: Chance - Ship type: P - How taken: HMS Statire - When taken: 1 Apr 1814 - Where taken: at sea - Date received: 2 Mar 1815 - From what ship: HMS Dannemark - Born: Not legible - Age: 50 - Discharged on 3 Mar 1815 and sent to Dartmoor.

Hepburn, James - Seaman - Number: 38 - Prize name: Plutarch - Ship type: MV - How taken: HM Schooner Helicon - When taken: 5 Feb 1815 - Where taken: Lat 45.5N Long 7W - Date received: 10 Feb 1815 - From what ship: HMS Helicon - Born: Charleston - Age: 30 - Race: Mulatto - Discharged on 17 Feb 1815 and sent to Dartmoor.

Herkwright, John - Seaman - Number: 463 - Prize name: Fox - Ship type: P - How taken: HM Frigate Barbadoes - When taken: 11 Jan 1815 - Where taken: off Amelia Island, Florida - Date received: 16 Apr 1815 - From what ship: HMS Swifsure - Born: Maryland - Age: 24 - Discharged on 2 Jun 1815 and released to the Cartel Sovereign.

Hessings, Lot - Seaman - Number: 633 - How taken: Impressed by HM Ship-of-the-Line Namur - When taken: 7 Nov 1814 - Date received: 24 Apr 1815 - From what ship: HMS Bellerophon - Age: 18 - Discharged on 2 Jun 1815 and released to the Cartel Shakespeare.

Hicks, (-----) - Seaman - Number: 240 - Prize name: Mary, recaptured prize - Ship type: P - How taken: HM Brig Sophie - When taken: 19 Jun 1814 - Where taken: at sea - Date received: 2 Mar 1815 - From what ship: HMS Dannemark - Discharged on 3 Mar 1815 and sent to Dartmoor.

Hiels, John L. - Master - Number: 457 - Prize name: Fox - Ship type: P - How taken: HM Frigate Barbadoes - When taken: 11 Jan 1815 - Where taken: off Amelia Island, Florida - Date received: 16 Apr 1815 - From what ship: HMS Swifsure - Born: Connecticut - Age: 26 - Discharged on 2 Jun 1815 and released to the Cartel Sovereign.

Hill, Francis - Seaman - Number: 195 - Prize name: Chance - Ship type: P - How taken: HMS Statire - When taken: 1 Apr 1814 - Where taken: at sea - Date received: 2 Mar 1815 - From what ship: HMS Dannemark - Born: Virginia - Age: 18 - Discharged on 3 Mar 1815 and sent to Dartmoor.

Hill, Stephen - Master - Number: 1 - Prize name: Albion - Ship type: MV - How taken: HM Schooner Helicon - When taken: 17 Jan 1815 - Where taken: Lat 20N Long 52W - Date received: 3 Feb 1815 - From what ship: HMS Bittern - Born: Delaware - Age: 26 - Discharged on 4 Feb 1815 and sent to Dartmoor.

Hinchman, George - Captain - Number: 250 - Prize name: John - Ship type: MV - How taken: HM Brig Zenobia - When taken: 11 Sep 1814 - Where taken: at sea - Date received: 2 Mar 1815 - From what ship: HMS Dannemark - Born: Boston - Age: 21 - Discharged on 3 Mar 1815 and sent to Dartmoor.

Hinchmes, Louis - Master's Mate - Number: 858 - Prize name: US Brig Syren - Ship type: War - How taken: HM Ship-of-the-Line Medway - When taken: 12 Jul 1814 - Where taken: off Cape of Good Hope - Date received: 11 Jul 1815 - From what ship: from Portsmouth - Born: Not listed - Discharged on 11 Jul 1815 and released to the Cartel Wooddrop Sims.

Hinckley, Richard - Purser - Number: 324 - Prize name: Leo - Ship type: P - How taken: HM Frigate Tiber - When taken: 11 Mar 1815 - Where taken: Lat 45.24N Long 12.3W - Date received: 28 Mar 1815 - From what ship: HMS Tiber - Discharged on 9 Apr 1815.

Hindman, John - Seaman - Number: 269 - Prize name: Netterville - Ship type: MV - How taken: HM Brig Onyx - When taken: 25 Dec 1814 - Where taken: at sea - Date received: 2 Mar 1815 - From what ship: HMS Dannemark - Born: Annapolis - Age: 41 - Race: Negro - Discharged on 3 Mar 1815 and sent to Dartmoor.

Hingston, Richard - 1st Mate - Number: 2 - Prize name: Albion - Ship type: MV - How taken: HM Schooner Helicon - When taken: 17 Jan 1815 - Where taken: Lat 20N Long 52W - Date received: 3 Feb 1815 - From what ship: HMS Bittern - Born: Vienna - Age: 21 - Discharged on 4 Feb 1815 and sent to Dartmoor.

Histoss, Mathew - Seaman - Number: 487 - Prize name: Fox - Ship type: P - How taken: HM Frigate Barbadoes - When taken: 11 Jan 1815 - Where taken: off Amelia Island, Florida - Date received: 16 Apr 1815 - From what ship: HMS Swifsure - Born: South Carolina - Age: 19 - Discharged on 2 Jun 1815 and released to the Cartel Sovereign.

Hitchburn, Alexander - Seaman - Number: 370 - Prize name: Leo - Ship type: P - How taken: HM Frigate Tiber - When taken: 11 Mar 1815 - Where taken: Lat 45.24N Long 12.3W - Date received: 29 Mar 1814 - From what ship: HMS Tiber - Born: Massachusetts - Age: 21 - Discharged on 2 May 1815 and released to the Cartel Ariel.

Hitchin, George - Seaman - Number: 148 - Prize name: US Brig Syren - Ship type: War - How taken: HM Ship-of-the-Line Medway - When taken: 12 Jul 1814 - Where taken: off Cape of Good Hope - Date received: 21 Feb 1815 - From what ship: HMS Slaney - Discharged on 24 Feb 1815 and sent to Dartmoor.

Hodges, Edward - Ordinary Seaman - Number: 144 - Prize name: US Brig Syren - Ship type: War - How taken: HM Ship-of-the-Line Medway - When taken: 12 Jul 1814 - Where taken: off Cape of Good Hope - Date received: 21 Feb 1815 - From what ship: HMS Slaney - Born: Boston - Age: 26 - Discharged on 24 Feb 1815 and sent to Dartmoor.

Hodges, William - Master - Number: 612 - Prize name: Sine Qua Non - Ship type: P - How taken: HM Brig Elk - When taken: 20 Feb 1815 - Where taken: off Madeira - Date received: 18 Apr 1815 - From what ship: HMS Euryalus - Born: Massachusetts - Age: 26 - Discharged on 2 Jun 1815 and released to the Cartel Sovereign.

Holbrook, Gideon - Seaman - Number: 364 - Prize name: Leo - Ship type: P - How taken: HM Frigate Tiber - When taken: 11 Mar 1815 - Where taken: Lat 45.24N Long 12.3W - Date received: 29 Mar 1814 - From what ship: HMS Tiber - Born: Massachusetts - Age: 28 - Discharged on 2 May 1815 and released to the Cartel Ariel.

Holding, John - Seaman - Number: 631 - Prize name: Helene, prize of Privateer Morgianna - Ship type: P - How taken: HM Frigate Pique - When taken: 7 Mar 1815 - Where taken: coast of America - Date received: 20 Apr 1815 - From what ship: HMS Musquito - Age: 17 - Discharged on 2 Jun 1815 and released to the Cartel Shakespeare.

Holding, Lawrence - Seaman - Number: 574 - Prize name: Sorin, prize to Privateer Prince of Neufchatel - Ship type: P - How taken: HM Ship-of-the-Line Medway - When taken: 12 Jul 1814 - Where taken: Unknown - Date received: 16 Apr 1815 - From what ship: HMS Swifsure - Born: Rhode Island - Age: 22 - Race: Negro - Discharged on 2 Jun 1815 and released to the Cartel Sovereign.

Holland, Thomas - Seaman - Number: 629 - Prize name: Helene, prize of Privateer Morgianna - Ship type: P - How taken: HM Frigate Pique - When taken: 7 Mar 1815 - Where taken: coast of America - Date received: 20 Apr

1815 - From what ship: HMS Musquito - Age: 24 - Race: Negro - Discharged on 27 Apr 1815.

Holmes, Samuel D. - Seaman - Number: 356 - Prize name: Leo - Ship type: P - How taken: HM Frigate Tiber - When taken: 11 Mar 1815 - Where taken: Lat 45.24N Long 12.3W - Date received: 29 Mar 1814 - From what ship: HMS Tiber - Born: Massachusetts - Age: 22 - Discharged on 27 Apr 1815 and released to the Cartel Minework.

Holmes, William - Seaman - Number: 662 - Prize name: Avon - Ship type: P - How taken: HM Frigate Barbadoes - When taken: 8 Mar 1815 - Where taken: West Indies - Date received: 24 Apr 1815 - From what ship: HMS Bellerophon - Born: Massachusetts - Age: 52 - Discharged on 2 Jun 1815 and released to the Cartel Shakespeare.

Holmes, William - Seaman - Number: 344 - Prize name: Leo - Ship type: P - How taken: HM Frigate Tiber - When taken: 11 Mar 1815 - Where taken: Lat 45.24N Long 12.3W - Date received: 28 Mar 1815 - From what ship: HMS Tiber - Born: Massachusetts - Age: 24 - Discharged on 27 Apr 1815 and released to the Cartel Minework.

Holmes, Zachus - Seaman - Number: 603 - Prize name: George Little - Ship type: P - How taken: HM Frigate Granicus - When taken: 20 Jan 1815 - Where taken: off Cape Finisterre, Spain - Date received: 18 Apr 1815 - From what ship: HMS Euryalus - Born: Massachusetts - Age: 53 - Discharged on 2 Jun 1815 and released to the Cartel Sovereign.

Homer, James - Seaman - Number: 796 - Prize name: US Brig Syren - Ship type: War - How taken: HM Ship-of-the-Line Medway - When taken: 12 Jul 1814 - Where taken: off Cape of Good Hope - Date received: 11 Jul 1815 - From what ship: HMS Royal Sovereign - Born: Not listed - Discharged on 11 Jul 1815 and released to the Cartel Wooddrop Sims.

Hood, Daniel - Seaman - Number: 819 - Prize name: US Brig Syren - Ship type: War - How taken: HM Ship-of-the-Line Medway - When taken: 12 Jul 1814 - Where taken: off Cape of Good Hope - Date received: 11 Jul 1815 - From what ship: HMS Royal Sovereign - Born: Not listed - Discharged on 11 Jul 1815 and released to the Cartel Wooddrop Sims.

Hopping, Edward - Seaman - Number: 571 - Prize name: Heron - Ship type: MV - How taken: HM Frigate Pique - When taken: 16 Dec 1814 - Where taken: off Porto Rico - Date received: 16 Apr 1815 - From what ship: HMS Swifsure - Born: Providence - Age: 17 - Discharged on 27 May 1815.

Hose, Richard - Carpenter - Number: 113 - Prize name: Prince de Neufchatel - Ship type: P - How taken: HM Ship-of-the-Line Leander, HM Ship-of-the-Line Newcastle & HM Frigate Acasta - When taken: 28 Dec 1814 - Where taken: Lat 35N Long 52W - Date received: 18 Feb 1815 - From what ship: HMS Sybille - Born: Boston - Discharged on 19 Feb 1815 and sent to Dartmoor.

Hoty, Charles - 2nd Lieutenant - Number: 456 - Prize name: Fox - Ship type: P - How taken: HM Frigate Barbadoes - When taken: 11 Jan 1815 - Where taken: off Amelia Island, Florida - Date received: 16 Apr 1815 - From what ship: HMS Swifsure - Born: New York - Age: 46 - Discharged on 22 Apr 1815.

Hours, Samuel - Seaman - Number: 430 - Prize name: Engineer - Ship type: MV - How taken: HMS Murros - When taken: 21 Sep 1814 - Where taken: off Porto Rico - Date received: 16 Apr 1815 - From what ship: HMS Swifsure - Born: Massachusetts - Age: 24 - Discharged on 8 May 1815.

Howard, Samuel - Sailing Master - Number: 584 - Prize name: George Little - Ship type: P - How taken: HM Frigate Granicus - When taken: 20 Jan 1815 - Where taken: off Cape Finisterre, Spain - Date received: 18 Apr 1815 - From what ship: HMS Euryalus - Born: Massachusetts - Age: 34 - Discharged on 24 Apr 1815.

Howes, Caleb - Seaman - Number: 81 - Prize name: Prince de Neufchatel - Ship type: P - How taken: HM Ship-of-the-Line Leander, HM Ship-of-the-Line Newcastle & HM Frigate Acasta - When taken: 28 Dec 1814 - Where taken: Lat 35N Long 52W - Date received: 18 Feb 1815 - From what ship: HMS Sybille - Born: Boston - Age: 22 - Discharged on 19 Feb 1815 and sent to Dartmoor.

Hoy, James - Seaman - Number: 846 - Prize name: US Brig Syren - Ship type: War - How taken: HM Ship-of-the-Line Medway - When taken: 12 Jul 1814 - Where taken: off Cape of Good Hope - Date received: 11 Jul 1815 - From what ship: HMS Royal Sovereign - Born: Not listed - Discharged on 11 Jul 1815 and released to the Cartel Wooddrop Sims.

Hoyt, Samuel - Seaman - Number: 333 - Prize name: Leo - Ship type: P - How taken: HM Frigate Tiber - When taken: 11 Mar 1815 - Where taken: Lat 45.24N Long 12.3W - Date received: 28 Mar 1815 - From what ship:

HMS Tiber - Discharged on 27 Apr 1815 and released to the Cartel Minework.

Hubbard, John - Seaman - Number: 618 - Prize name: Sine Qua Non - Ship type: P - How taken: HM Brig Elk - When taken: 20 Feb 1815 - Where taken: off Madeira - Date received: 18 Apr 1815 - From what ship: HMS Euryalus - Age: 40 - Race: Negro - Discharged on 2 Jun 1815 and released to the Cartel Sovereign.

Hughes, John - Seaman - Number: 755 - Prize name: Melvina - Ship type: MV - How taken: HM Frigate Andromeda - When taken: 28 Feb 1815 - Where taken: off Lisbon - Date received: 29 Aug 1815 - From what ship: HMS Tonnant - Born: Massachusetts - Age: 40 - Discharged on 2 Jun 1815 and released to the Cartel Shakespeare.

Hulse, J. L. - Seaman - Number: 217 - Prize name: Adamant - Ship type: MV - Date received: 2 Mar 1815 - From what ship: HMS Dannemark - Discharged on 3 Mar 1815 and sent to Dartmoor.

Hunt, James - Seaman - Number: 176 - Prize name: Farmer's Daughter - Ship type: MV - How taken: HM Ship-of-the-Line Leviathan - When taken: 20 Mar 1814 - Where taken: at sea - Date received: 2 Mar 1815 - From what ship: HMS Dannemark - Born: Baltimore - Age: 24 - Race: Black - Discharged on 3 Mar 1815 and sent to Dartmoor.

Huntingdon, E. - 2nd Mate - Number: 422 - Prize name: Engineer - Ship type: MV - How taken: HMS Murros - When taken: 21 Sep 1814 - Where taken: off Porto Rico - Date received: 16 Apr 1815 - From what ship: HMS Swifsure - Born: Connecticut - Age: 23 - Discharged on 2 Jun 1815 and released to the Cartel Sovereign.

Hunty, Jacob - Seaman - Number: 211 - Prize name: Sennett, Swedish merchant vessel - Ship type: MV - How taken: HM Frigate Rhin - When taken: Not known - Date received: 2 Mar 1815 - From what ship: HMS Dannemark - Discharged on 3 Mar 1815 and sent to Dartmoor.

Huston, Samuel - Seaman - Number: 661 - Prize name: Avon - Ship type: P - How taken: HM Frigate Barbadoes - When taken: 8 Mar 1815 - Where taken: West Indies - Date received: 24 Apr 1815 - From what ship: HMS Bellerophon - Born: Massachusetts - Age: 24 - Discharged on 2 Jun 1815 and released to the Cartel Shakespeare.

Huston, William - Seaman - Number: 52 - How taken: Taken out of HMS Bellejoux - Where taken: Chatham - Date received: 15 Feb 1815 - From what ship: HMS Myrmidon - Age: 55 - Discharged on 17 Feb 1815 and sent to Dartmoor.

Hutchinson, Robert - Seaman - Number: 346 - Prize name: Leo - Ship type: P - How taken: HM Frigate Tiber - When taken: 11 Mar 1815 - Where taken: Lat 45.24N Long 12.3W - Date received: 28 Mar 1815 - From what ship: HMS Tiber - Born: Massachusetts - Age: 26 - Discharged on 27 Apr 1815 and released to the Cartel Minework.

Hutckins, William - Seaman - Number: 830 - Prize name: US Brig Syren - Ship type: War - How taken: HM Ship-of-the-Line Medway - When taken: 12 Jul 1814 - Where taken: off Cape of Good Hope - Date received: 11 Jul 1815 - From what ship: HMS Royal Sovereign - Born: Not listed - Discharged on 11 Jul 1815 and released to the Cartel Wooddrop Sims.

Hutson, William - Seaman - Number: 786 - Prize name: US Brig Syren - Ship type: War - How taken: HM Ship-of-the-Line Medway - When taken: 12 Jul 1814 - Where taken: off Cape of Good Hope - Date received: 11 Jul 1815 - From what ship: HMS Royal Sovereign - Born: Not listed - Discharged on 11 Jul 1815 and released to the Cartel Wooddrop Sims.

Iney, Samuel - Seaman - Number: 544 - How taken: Gave himself up from HMS Harmony - When taken: 4 Sep 1814 - Where taken: Barbados - Date received: 16 Apr 1815 - From what ship: HMS Swifsure - Born: Massachusetts - Age: 23 - Discharged on 2 Jun 1815 and released to the Cartel Sovereign.

Ingusal, Calvin - Boy - Number: 390 - Prize name: Leo - Ship type: P - How taken: HM Frigate Tiber - When taken: 11 Mar 1815 - Where taken: Lat 45.24N Long 12.3W - Date received: 29 Mar 1814 - From what ship: HMS Tiber - Born: Massachusetts - Age: 15 - Discharged on 30 May 1815 and released to the Cartel Atlas.

Inns, John - Seaman - Number: 781 - Prize name: Sine Qua Non - Ship type: P - How taken: HM Brig Elk - When taken: 20 Feb 1815 - Where taken: off Madeira - Date received: 29 Aug 1815 - From what ship: HMS Tonnant - Born: Massachusetts - Age: 30 - Discharged on 2 Jun 1815 and released to the Cartel Shakespeare.

Jack, John - Seaman - Number: 652 - Prize name: Avon - Ship type: P - How taken: HM Frigate Barbadoes - When taken: 8 Mar 1815 - Where taken: West Indies - Date received: 24 Apr 1815 - From what ship: HMS Bellerophon - Born: Massachusetts - Age: 27 - Discharged on 2 Jun 1815 and released to the Cartel Shakespeare.

Jackson, Ebenezer - Boy - Number: 125 - Prize name: Prince de Neufchatel - Ship type: P - How taken: HM Ship-of-the-Line Leander, HM Ship-of-the-Line Newcastle & HM Frigate Acasta - When taken: 28 Dec 1814 - Where taken: Lat 35N Long 52W - Date received: 18 Feb 1815 - From what ship: HMS Sybille - Born: Charlestown - Age: 17 - Discharged on 19 Feb 1815 and sent to Dartmoor.

Jackson, John - Seaman - Number: 175 - Prize name: Enterprise - Ship type: P - How taken: Not known - Where taken: at sea - Date received: 2 Mar 1815 - From what ship: HMS Dannemark - Born: New York - Age: 21 - Race: Black - Discharged on 3 Mar 1815 and sent to Dartmoor.

Jackson, John - Seaman - Number: 285 - How taken: Taken off the HM Ship-of-the-Line Severn and other ships - Date received: 2 Mar 1815 - From what ship: HMS Dannemark - Race: Negro - Discharged on 3 Mar 1815 and sent to Dartmoor.

Jackson, John - Seaman - Number: 332 - Prize name: Leo - Ship type: P - How taken: HM Frigate Tiber - When taken: 11 Mar 1815 - Where taken: Lat 45.24N Long 12.3W - Date received: 28 Mar 1815 - From what ship: HMS Tiber - Discharged on 27 Apr 1815 and released to the Cartel Minework.

Jackson, Joseph - Seaman - Number: 326 - Prize name: Leo - Ship type: P - How taken: HM Frigate Tiber - When taken: 11 Mar 1815 - Where taken: Lat 45.24N Long 12.3W - Date received: 28 Mar 1815 - From what ship: HMS Tiber - Discharged on 27 Apr 1815 and released to the Cartel Minework.

Jackson, Thomas - Boy - Number: 291 - How taken: Taken off the HM Ship-of-the-Line Severn and other ships - Date received: 2 Mar 1815 - From what ship: HMS Dannemark - Born: New York - Age: 14 - Race: Negro - Discharged on 3 Mar 1815 and sent to Dartmoor.

James, William - Seaman - Number: 30 - Prize name: Plutarch - Ship type: MV - How taken: HM Schooner Helicon - When taken: 5 Feb 1815 - Where taken: Lat 45.5N Long 7W - Date received: 8 Feb 1815 - From what ship: HMS Helicon - Born: New York - Age: 27 - Discharged on 10 Feb 1815 and sent to Dartmoor.

Jarrett, Abraham - Seaman - Number: 483 - Prize name: Fox - Ship type: P - How taken: HM Frigate Barbadoes - When taken: 11 Jan 1815 - Where taken: off Amelia Island, Florida - Date received: 16 Apr 1815 - From what ship: HMS Swifsure - Born: Maryland - Age: 50 - Discharged on 2 Jun 1815 and released to the Cartel Sovereign.

Jeanason, Isaac - Seaman - Number: 388 - Prize name: Leo - Ship type: P - How taken: HM Frigate Tiber - When taken: 11 Mar 1815 - Where taken: Lat 45.24N Long 12.3W - Date received: 29 Mar 1814 - From what ship: HMS Tiber - Born: Massachusetts - Age: 29 - Discharged on 30 May 1815 and released to the Cartel Atlas.

Jenkins, Thomas - 2nd Mate - Number: 267 - Prize name: Netterville - Ship type: MV - How taken: HM Brig Onyx - When taken: 25 Dec 1814 - Where taken: at sea - Date received: 2 Mar 1815 - From what ship: HMS Dannemark - Born: Virginia - Age: 23 - Discharged on 3 Mar 1815 and sent to Dartmoor.

Jenkins, Walter - Seaman - Number: 534 - Prize name: St. Francis - Ship type: MV - How taken: HMS Ister - When taken: 8 Dec 1814 - Where taken: Amelia Island, Florida - Date received: 16 Apr 1815 - From what ship: HMS Swifsure - Age: 23 - Discharged on 2 Jun 1815 and released to the Cartel Sovereign.

Jennings, Benjamin - Seaman - Number: 745 - Prize name: Sine Qua Non - Ship type: P - How taken: HM Brig Elk - When taken: 20 Feb 1815 - Where taken: off Madeira - Date received: 29 Aug 1815 - From what ship: HMS Tonnant - Born: Boston - Age: 36 - Discharged on 2 Jun 1815 and released to the Cartel Shakespeare.

Jennings, Francis - Seaman - Number: 174 - Prize name: Enterprise - Ship type: P - How taken: Not known - Where taken: at sea - Date received: 2 Mar 1815 - From what ship: HMS Dannemark - Born: New York - Age: 30 - Discharged on 3 Mar 1815 and sent to Dartmoor.

Jerring, Josiah - Clerk - Number: 640 - Prize name: Avon - Ship type: P - How taken: HM Frigate Barbadoes - When taken: 8 Mar 1815 - Where taken: West Indies - Date received: 24 Apr 1815 - From what ship: HMS Bellerophon - Age: 25 - Discharged on 2 Jun 1815 and released to the Cartel Shakespeare.

Jervis, Peter - Seaman - Number: 70 - Prize name: Prince de Neufchatel - Ship type: P - How taken: HM Ship-of-the-Line Leander, HM Ship-of-the-Line Newcastle & HM Frigate Acasta - When taken: 28 Dec 1814 - Where taken: Lat 35N Long 52W - Date received: 18 Feb 1815 - From what ship: HMS Sybille - Born: Saint Malo - Age: 25 - Discharged on 19 Feb 1815 and sent to Dartmoor.

Jewell, Philip - Seaman - Number: 655 - Prize name: Avon - Ship type: P - How taken: HM Frigate Barbadoes - When taken: 8 Mar 1815 - Where taken: West Indies - Date received: 24 Apr 1815 - From what ship: HMS Bellerophon - Born: Massachusetts - Age: 28 - Discharged on 2 Jun 1815 and released to the Cartel Shakespeare.

Johannes, Logan - Seaman - Number: 278 - Prize name: Netterville - Ship type: MV - How taken: HM Brig Onyx - When taken: 25 Dec 1814 - Where taken: at sea - Date received: 2 Mar 1815 - From what ship: HMS Dannemark - Born: Calcutta - Age: 26 - Discharged on 3 Mar 1815 and sent to Dartmoor.

Johnson, Abraham - Seaman - Number: 216 - Prize name: Romes - Ship type: MV - How taken: Impressed - Date received: 2 Mar 1815 - From what ship: HMS Dannemark - Discharged on 3 Mar 1815 and sent to Dartmoor.

Johnson, Charles - Seaman - Number: 229 - Prize name: Decatur - Ship type: P - How taken: HM Frigate Rhin - When taken: 4 Jun 1814 - Where taken: at sea - Date received: 2 Mar 1815 - From what ship: HMS Dannemark - Born: New York - Age: 42 - Race: Negro - Discharged on 3 Mar 1815 and sent to Dartmoor.

Johnson, Jesse - Seaman - Number: 10 - Prize name: Albion - Ship type: MV - How taken: HM Schooner Helicon - When taken: 17 Jan 1815 - Where taken: Lat 20N Long 52W - Date received: 3 Feb 1815 - From what ship: HMS Bittern - Born: New York - Age: 20 - Race: Mulatto - Discharged on 4 Feb 1815 and sent to Dartmoor.

Johnson, Moses - Seaman - Number: 843 - Prize name: US Brig Syren - Ship type: War - How taken: HM Ship-of-the-Line Medway - When taken: 12 Jul 1814 - Where taken: off Cape of Good Hope - Date received: 11 Jul 1815 - From what ship: HMS Royal Sovereign - Born: Not listed - Discharged on 11 Jul 1815 and released to the Cartel Wooddrop Sims.

Johnson, Samuel - Seaman - Number: 803 - Prize name: US Brig Syren - Ship type: War - How taken: HM Ship-of-the-Line Medway - When taken: 12 Jul 1814 - Where taken: off Cape of Good Hope - Date received: 11 Jul 1815 - From what ship: HMS Royal Sovereign - Born: Not listed - Discharged on 11 Jul 1815 and released to the Cartel Wooddrop Sims.

Johnson, Samuel J, - Seaman - Number: 320 - Prize name: Thomas, prize to Privateer Scourge - Ship type: P - How taken: HM Frigate Aquilon - When taken: 11 Mar 1815 - Where taken: off Cape Finisterre, Spain - Date received: 25 Apr 1815 - From what ship: HMS Opossum - Discharged on 27 Apr 1815 and released to the Cartel Minework.

Johnson, Thomas - Seaman - Number: 181 - Prize name: Farmer's Daughter - Ship type: MV - How taken: HM Ship-of-the-Line Leviathan - When taken: 20 Mar 1814 - Where taken: at sea - Date received: 2 Mar 1815 - From what ship: HMS Dannemark - Born: Taunton - Age: 25 - Discharged on 3 Mar 1815 and sent to Dartmoor.

Johnson, William - Seaman - Number: 230 - Prize name: Decatur - Ship type: P - How taken: HM Frigate Rhin - When taken: 4 Jun 1814 - Where taken: at sea - Date received: 2 Mar 1815 - From what ship: HMS Dannemark - Born: New York - Age: 23 - Race: Black - Discharged on 3 Mar 1815 and sent to Dartmoor.

Johnson, William - Seaman - Number: 818 - Prize name: US Brig Syren - Ship type: War - How taken: HM Ship-of-the-Line Medway - When taken: 12 Jul 1814 - Where taken: off Cape of Good Hope - Date received: 11 Jul 1815 - From what ship: HMS Royal Sovereign - Born: Not listed - Discharged on 11 Jul 1815 and released to the Cartel Wooddrop Sims.

Johnston, Frederick - Seaman - Number: 432 - Prize name: Engineer - Ship type: MV - How taken: HMS Murros - When taken: 21 Sep 1814 - Where taken: off Porto Rico - Date received: 16 Apr 1815 - From what ship: HMS Swifsure - Born: New Orleans - Age: 33 - Discharged on 2 Jun 1815 and released to the Cartel Sovereign.

Johnston, George - Seaman - Number: 429 - Prize name: Engineer - Ship type: MV - How taken: HMS Murros - When taken: 21 Sep 1814 - Where taken: off Porto Rico - Date received: 16 Apr 1815 - From what ship: HMS Swifsure - Born: New York - Age: 20 - Discharged on 2 Jun 1815 and released to the Cartel Sovereign.

Johnston, Perry - Seaman - Number: 4 - Prize name: Albion - Ship type: MV - How taken: HM Schooner Helicon - When taken: 17 Jan 1815 - Where taken: Lat 20N Long 52W - Date received: 3 Feb 1815 - From what ship: HMS Bittern - Born: Maryland - Age: 26 - Race: Black - Discharged on 4 Feb 1815 and sent to Dartmoor.

Johnston, Samuel - Seaman - Number: 529 - Prize name: Gallant - Ship type: MV - How taken: HM Frigate

Barbadoes - When taken: 6 Dec 1815 - Where taken: off St. Bartholomew - Date received: 16 Apr 1815 - From what ship: HMS Swifsure - Born: Baltimore - Age: 25 - Race: Negro - Discharged on 2 Jun 1815 and released to the Cartel Sovereign.

Jolmes, Isaac - Seaman - Number: 205 - Prize name: Sennett, Swedish merchant vessel - Ship type: MV - How taken: HM Frigate Rhin - When taken: Not known - Date received: 2 Mar 1815 - From what ship: HMS Dannemark - Race: Negro - Discharged on 3 Mar 1815 and sent to Dartmoor.

Jones, Calvin - Seaman - Number: 162 - Prize name: Nancy, prize of the Privateer United States - Ship type: P - How taken: HM Sloop Papillon - When taken: 27 Dec 1814 - Where taken: Lat 30 Long 10 - Date received: 27 Feb 1815 - From what ship: HMS Nimble - Born: Allentown - Age: 19 - Discharged on 3 Mar 1815 and sent to Dartmoor.

Jones, Edward - Marine - Number: 77 - Prize name: Prince de Neufchatel - Ship type: P - How taken: HM Ship-of-the-Line Leander, HM Ship-of-the-Line Newcastle & HM Frigate Acasta - When taken: 28 Dec 1814 - Where taken: Lat 35N Long 52W - Date received: 18 Feb 1815 - From what ship: HMS Sybille - Born: Roxborough - Age: 18 - Discharged on 19 Feb 1815 and sent to Dartmoor.

Jones, Ezekiel - Seaman - Number: 829 - Prize name: US Brig Syren - Ship type: War - How taken: HM Ship-of-the-Line Medway - When taken: 12 Jul 1814 - Where taken: off Cape of Good Hope - Date received: 11 Jul 1815 - From what ship: HMS Royal Sovereign - Born: Not listed - Discharged on 11 Jul 1815 and released to the Cartel Wooddrop Sims.

Jones, Henry - Seaman - Number: 263 - Prize name: Dorothy - Ship type: MV - How taken: HM Frigate North Star - When taken: 15 Nov 1814 - Where taken: at sea - Date received: 2 Mar 1815 - From what ship: HMS Dannemark - Discharged on 3 Mar 1815 and sent to Dartmoor.

Jones, James - Seaman - Number: 276 - Prize name: Netterville - Ship type: MV - How taken: HM Brig Onyx - When taken: 25 Dec 1814 - Where taken: at sea - Date received: 2 Mar 1815 - From what ship: HMS Dannemark - Born: Baltimore - Age: 28 - Discharged on 3 Mar 1815 and sent to Dartmoor.

Jones, John - Seaman - Number: 563 - Prize name: Mary - Ship type: MV - How taken: HMS Muros - When taken: 8 Dec 1814 - Where taken: West Indies - Date received: 16 Apr 1815 - From what ship: HMS Swifsure - Born: North Caroline - Age: 21 - Discharged on 2 Jun 1815 and released to the Cartel Sovereign.

Jones, John - Seaman - Number: 161 - Prize name: Nancy, prize of the Privateer United States - Ship type: P - How taken: HM Sloop Papillon - When taken: 27 Dec 1814 - Where taken: Lat 30 Long 10 - Date received: 27 Feb 1815 - From what ship: HMS Nimble - Born: Portsmouth - Age: 26 - Discharged on 3 Mar 1815 and sent to Dartmoor.

Jory, Elisha - Seaman - Number: 742 - Prize name: George Little - Ship type: P - How taken: HM Frigate Granicus - When taken: 20 Jan 1815 - Where taken: off Cape Finisterre, Spain - Date received: 29 Aug 1815 - From what ship: HMS Tonnant - Born: Massachusetts - Age: 18 - Discharged on 2 Jun 1815 and released to the Cartel Shakespeare.

Jourdan, Conrad - Seaman - Number: 535 - Prize name: St. Francis - Ship type: MV - How taken: HMS Ister - When taken: 8 Dec 1814 - Where taken: Amelia Island, Florida - Date received: 16 Apr 1815 - From what ship: HMS Swifsure - Age: 23 - Discharged on 2 Jun 1815 and released to the Cartel Sovereign.

Judah, David - Seaman - Number: 58 - How taken: Taken out of the West Indiaman Alfred - Where taken: London - Date received: 15 Feb 1815 - From what ship: HMS Myrmidon - Born: Connecticut - Age: 27 - Discharged on 17 Feb 1815 and sent to Dartmoor.

Jürgen, Morgan - Seaman - Number: 512 - Prize name: Hope, prize to US Sloop-of-War Wasp - Ship type: War - How taken: HM Sloop Fairy - When taken: 6 Jan 1815 - Where taken: Unknown - Date received: 16 Apr 1815 - From what ship: HMS Swifsure - Born: Pennsylvania - Age: 30 - Discharged on 8 May 1815.

Kecting, John - Seaman - Number: 627 - Prize name: Helene, prize of Privateer Morgianna - Ship type: P - How taken: HM Frigate Pique - When taken: 7 Mar 1815 - Where taken: coast of America - Date received: 20 Apr 1815 - From what ship: HMS Musquito - Age: 25 - Race: Negro - Discharged on 27 Apr 1815.

Keisoner, Samuel - Seaman - Number: 521 - Prize name: High Flyer - Ship type: LM - How taken: HMS Muros - When taken: 14 Nov 1814 - Where taken: off St. Bartholomew - Date received: 16 Apr 1815 - From what ship:

HMS Swifsure - Born: New Jersey - Age: 23 - Race: Negro - Discharged on 2 Jun 1815 and released to the Cartel Sovereign.

Kell, Francis - Seaman - Number: 271 - Prize name: Netterville - Ship type: MV - How taken: HM Brig Onyx - When taken: 25 Dec 1814 - Where taken: at sea - Date received: 2 Mar 1815 - From what ship: HMS Dannemark - Born: Baltimore - Age: 22 - Discharged on 3 Mar 1815 and sent to Dartmoor.

Kellock, Stephen - Seaman - Number: 299 - Prize name: US Brig Syren - Ship type: War - How taken: HM Ship-of-the-Line Medway - When taken: 12 Jul 1814 - Where taken: off Cape of Good Hope - Date received: 4 Mar 1814 - From what ship: HM Cutter Surley - Born: Wilmington - Age: 33 - Discharged on 7 Mar 1815 and sent to Dartmoor.

Kelly, Samuel - Seaman - Number: 445 - Prize name: High Flyer - Ship type: LM - How taken: HMS Muros - When taken: 14 Nov 1814 - Where taken: off St. Bartholomew - Date received: 16 Apr 1815 - From what ship: HMS Swifsure - Born: Massachusetts - Age: 25 - Discharged on 2 Jun 1815 and released to the Cartel Sovereign.

Kemenes, Hocao - Seaman - Number: 501 - Prize name: Fox - Ship type: P - How taken: HM Frigate Barbadoes - When taken: 11 Jan 1815 - Where taken: off Amelia Island, Florida - Date received: 16 Apr 1815 - From what ship: HMS Swifsure - Born: Kadir - Age: 40 - Discharged on 2 Jun 1815 and released to the Cartel Sovereign.

Kemp, Nathaniel - Prize Master - Number: 636 - Prize name: Avon - Ship type: P - How taken: HM Frigate Barbadoes - When taken: 8 Mar 1815 - Where taken: West Indies - Date received: 24 Apr 1815 - From what ship: HMS Bellerophon - Age: 31 - Discharged on 2 Jun 1815 and released to the Cartel Shakespeare.

Kempton, Mathew - Seaman - Number: 361 - Prize name: Leo - Ship type: P - How taken: HM Frigate Tiber - When taken: 11 Mar 1815 - Where taken: Lat 45.24N Long 12.3W - Date received: 29 Mar 1814 - From what ship: HMS Tiber - Born: Massachusetts - Age: 52 - Discharged on 2 May 1815 and released to the Cartel Ariel.

Kendall, David - Seaman - Number: 551 - Prize name: Hera - Ship type: MV - How taken: HM Frigate Pique - When taken: 16 Dec 1814 - Where taken: off Porto Rico - Date received: 16 Apr 1815 - From what ship: HMS Swifsure - Born: Saint-Domingue (Haiti) - Age: 24 - Race: Mulatto - Discharged on 2 Jun 1815 and released to the Cartel Sovereign.

Kendell, Franklin - Boy - Number: 12 - Prize name: Albion - Ship type: MV - How taken: HM Schooner Helicon - When taken: 17 Jan 1815 - Where taken: Lat 20N Long 52W - Date received: 3 Feb 1815 - From what ship: HMS Bittern - Born: Leominster - Age: 16 - Discharged on 4 Feb 1815 and sent to Dartmoor.

Kennedy, Henry - Seaman - Number: 554 - Prize name: Hera - Ship type: MV - How taken: HM Frigate Pique - When taken: 16 Dec 1814 - Where taken: off Porto Rico - Date received: 16 Apr 1815 - From what ship: HMS Swifsure - Born: Boston - Age: 29 - Discharged on 2 Jun 1815 and released to the Cartel Sovereign.

Kent, John - Boy - Number: 384 - Prize name: Leo - Ship type: P - How taken: HM Frigate Tiber - When taken: 11 Mar 1815 - Where taken: Lat 45.24N Long 12.3W - Date received: 29 Mar 1814 - From what ship: HMS Tiber - Born: Massachusetts - Age: 16 - Discharged on 30 May 1815 and released to the Cartel Atlas.

Ketchell, George - Seaman - Number: 241 - Prize name: Mary, recaptured prize - Ship type: P - How taken: HM Brig Sophie - When taken: 19 Jun 1814 - Where taken: at sea - Date received: 2 Mar 1815 - From what ship: HMS Dannemark - Discharged on 3 Mar 1815 and sent to Dartmoor.

Kethan, George - Seaman - Number: 718 - Prize name: Melvina - Ship type: MV - How taken: HM Frigate Andromeda - When taken: 28 Feb 1815 - Where taken: off Lisbon - Date received: 16 Jul 1815 - From what ship: HMS Hope - Born: Massachusetts - Age: 33 - Discharged on 2 Jun 1815 and released to the Cartel Shakespeare.

Kingdom, John - Seaman - Number: 180 - Prize name: Farmer's Daughter - Ship type: MV - How taken: HM Ship-of-the-Line Leviathan - When taken: 20 Mar 1814 - Where taken: at sea - Date received: 2 Mar 1815 - From what ship: HMS Dannemark - Born: Maryland - Age: 18 - Discharged on 3 Mar 1815 and sent to Dartmoor.

Kirby, Abel - Seaman - Number: 540 - Prize name: Engineer - Ship type: MV - How taken: HMS Muros - When taken: 21 Sep 1814 - Where taken: off Porto Rico - Date received: 16 Apr 1815 - From what ship: HMS Swifsure - Born: Connecticut - Age: 21 - Discharged on 2 Jun 1815 and released to the Cartel Sovereign.

Kirby, Robert - Boy - Number: 150 - Prize name: US Brig Syren - Ship type: War - How taken: HM Ship-of-the-

Line Medway - When taken: 12 Jul 1814 - Where taken: off Cape of Good Hope - Date received: 21 Feb 1815 - From what ship: HMS Slaney - Born: Boston - Age: 17 - Discharged on 24 Feb 1815 and sent to Dartmoor.

Kitton, Abraham - Boy - Number: 272 - Prize name: Netterville - Ship type: MV - How taken: HM Brig Onyx - When taken: 25 Dec 1814 - Where taken: at sea - Date received: 2 Mar 1815 - From what ship: HMS Dannemark - Born: Baltimore - Age: 14 - Discharged on 3 Mar 1815 and sent to Dartmoor.

Knowles, David - Seaman - Number: 644 - Prize name: Avon - Ship type: P - How taken: HM Frigate Barbadoes - When taken: 8 Mar 1815 - Where taken: West Indies - Date received: 24 Apr 1815 - From what ship: HMS Bellerophon - Age: 21 - Discharged on 2 Jun 1815 and released to the Cartel Shakespeare.

Knowles, William - Seaman - Number: 668 - Prize name: Avon - Ship type: P - How taken: HM Frigate Barbadoes - When taken: 8 Mar 1815 - Where taken: West Indies - Date received: 24 Apr 1815 - From what ship: HMS Bellerophon - Born: Massachusetts - Age: 26 - Discharged on 23 May 1815.

Kumar, Samuel - Boy - Number: 392 - Prize name: Leo - Ship type: P - How taken: HM Frigate Tiber - When taken: 11 Mar 1815 - Where taken: Lat 45.24N Long 12.3W - Date received: 29 Mar 1814 - From what ship: HMS Tiber - Born: Massachusetts - Age: 16 - Race: Mulatto - Discharged on 30 May 1815 and released to the Cartel Atlas.

Lacey, Zachariah - Pilot - Number: 226 - Prize name: Decatur - Ship type: P - How taken: HM Frigate Rhin - When taken: 4 Jun 1814 - Where taken: at sea - Date received: 2 Mar 1815 - From what ship: HMS Dannemark - Born: Charleston - Age: 31 - Discharged on 3 Mar 1815 and sent to Dartmoor.

Lahay, Jean B. - Seaman - Number: 499 - Prize name: Fox - Ship type: P - How taken: HM Frigate Barbadoes - When taken: 11 Jan 1815 - Where taken: off Amelia Island, Florida - Date received: 16 Apr 1815 - From what ship: HMS Swifsure - Born: New Orleans - Age: 32 - Discharged on 2 Jun 1815 and released to the Cartel Sovereign.

Lake, William - Seaman & Passenger - Number: 415 - Prize name: W. S. Many - Ship type: MV - How taken: HM Brig Reynard - When taken: 4 Feb 1814 - Where taken: off Cadiz - Date received: 2 Apr 1815 - From what ship: HMS Basilisk - Age: 22 - Discharged on 2 Jun 1815 and released to the Cartel Sovereign.

Lakes, Leonard - Seaman - Number: 647 - Prize name: Avon - Ship type: P - How taken: HM Frigate Barbadoes - When taken: 8 Mar 1815 - Where taken: West Indies - Date received: 24 Apr 1815 - From what ship: HMS Bellerophon - Born: Massachusetts - Age: 28 - Discharged on 2 Jun 1815 and released to the Cartel Shakespeare.

Lambert, Andrew - Seaman - Number: 26 - Prize name: Plutarch - Ship type: MV - How taken: HM Schooner Helicon - When taken: 5 Feb 1815 - Where taken: Lat 45.5N Long 7W - Date received: 8 Feb 1815 - From what ship: HMS Helicon - Born: Hanover - Age: 26 - Discharged on 10 Feb 1815 and sent to Dartmoor.

Lambert, John - Seaman - Number: 815 - Prize name: US Brig Syren - Ship type: War - How taken: HM Ship-of-the-Line Medway - When taken: 12 Jul 1814 - Where taken: off Cape of Good Hope - Date received: 11 Jul 1815 - From what ship: HMS Royal Sovereign - Born: Not listed - Discharged on 11 Jul 1815 and released to the Cartel Wooddrop Sims.

Lamon, John - Seaman - Number: 691 - Prize name: Lion - Ship type: P - How taken: HM Frigate Granicus - When taken: 26 Jun 1814 - Where taken: off Cape Finisterre, Spain - Date received: 16 Jul 1815 - From what ship: HMS Hope - Age: 25 - Discharged on 2 Jun 1815 and released to the Cartel Shakespeare.

Lane, Emui - Seaman - Number: 737 - Prize name: Sine Qua Non - Ship type: P - How taken: HM Brig Elk - When taken: 20 Feb 1815 - Where taken: off Madeira - Date received: 29 Aug 1815 - From what ship: HMS Tonnant - Born: Rhode Island - Age: 24 - Discharged on 2 Jun 1815 and released to the Cartel Shakespeare.

Lane, John - Seaman - Number: 362 - Prize name: Leo - Ship type: P - How taken: HM Frigate Tiber - When taken: 11 Mar 1815 - Where taken: Lat 45.24N Long 12.3W - Date received: 29 Mar 1814 - From what ship: HMS Tiber - Born: Massachusetts - Age: 25 - Discharged on 2 May 1815 and released to the Cartel Ariel.

Lavenio, Louis - Seaman - Number: 497 - Prize name: Fox - Ship type: P - How taken: HM Frigate Barbadoes - When taken: 11 Jan 1815 - Where taken: off Amelia Island, Florida - Date received: 16 Apr 1815 - From what ship: HMS Swifsure - Born: Saint-Domingue (Haiti) - Age: 20 - Discharged on 30 May 1815 and released to the Cartel Sovereign.

Lawrence, Trent - Seaman - Number: 258 - Prize name: Dorothy - Ship type: MV - How taken: HM Frigate North Star - When taken: 15 Nov 1814 - Where taken: at sea - Date received: 2 Mar 1815 - From what ship: HMS Dannemark - Discharged on 3 Mar 1815 and sent to Dartmoor.

Leach, Samuel - Seaman - Number: 789 - Prize name: US Brig Syren - Ship type: War - How taken: HM Ship-of-the-Line Medway - When taken: 12 Jul 1814 - Where taken: off Cape of Good Hope - Date received: 11 Jul 1815 - From what ship: HMS Royal Sovereign - Born: Not listed - Discharged on 11 Jul 1815 and released to the Cartel Wooddrop Sims.

Lee, Abraham - Seaman - Number: 5 - Prize name: Albion - Ship type: MV - How taken: HM Schooner Helicon - When taken: 17 Jan 1815 - Where taken: Lat 20N Long 52W - Date received: 3 Feb 1815 - From what ship: HMS Bittern - Born: New York - Age: 16 - Discharged on 4 Feb 1815 and sent to Dartmoor.

Lee, John - Seaman - Number: 546 - How taken: Gave himself up from HMS Harmony - When taken: 4 Sep 1814 - Where taken: Barbados - Date received: 16 Apr 1815 - From what ship: HMS Swifsure - Age: 29 - Discharged on 2 Jun 1815 and released to the Cartel Sovereign.

Legiois, William - Seaman - Number: 770 - Prize name: Aurora - Ship type: MV - How taken: HM Frigate Andromeda - When taken: 28 Feb 1815 - Where taken: off Lisbon - Date received: 29 Aug 1815 - From what ship: HMS Tonnant - Age: 22 - Discharged on 2 Jun 1815 and released to the Cartel Shakespeare.

Lehay, George - Seaman - Number: 568 - Prize name: Heron - Ship type: MV - How taken: HM Frigate Pique - When taken: 16 Dec 1814 - Where taken: off Porto Rico - Date received: 16 Apr 1815 - From what ship: HMS Swifsure - Born: New York - Age: 22 - Discharged on 2 Jun 1815 and released to the Cartel Sovereign.

Lemon, John - Seaman - Number: 169 - How taken: Impressed from MV Rolla - Date received: 1 Mar 1815 - From what ship: Royal Naval Hospital Plymouth - Age: 36 - Race: Negro - Discharged on 3 Mar 1815 and sent to Dartmoor.

Leonard, Thomas - Seaman - Number: 99 - Prize name: Prince de Neufchatel - Ship type: P - How taken: HM Ship-of-the-Line Leander, HM Ship-of-the-Line Newcastle & HM Frigate Acasta - When taken: 28 Dec 1814 - Where taken: Lat 35N Long 52W - Date received: 18 Feb 1815 - From what ship: HMS Sybille - Born: Belgium - Age: 24 - Discharged on 19 Feb 1815 and sent to Dartmoor.

Levitt, Caleb - Seaman - Number: 83 - Prize name: Prince de Neufchatel - Ship type: P - How taken: HM Ship-of-the-Line Leander, HM Ship-of-the-Line Newcastle & HM Frigate Acasta - When taken: 28 Dec 1814 - Where taken: Lat 35N Long 52W - Date received: 18 Feb 1815 - From what ship: HMS Sybille - Born: Boston - Age: 18 - Discharged on 19 Feb 1815 and sent to Dartmoor.

Lewis, Charles - Seaman - Number: 767 - Prize name: Aurora - Ship type: MV - How taken: HM Frigate Andromeda - When taken: 28 Feb 1815 - Where taken: off Lisbon - Date received: 29 Aug 1815 - From what ship: HMS Tonnant - Born: Massachusetts - Age: 20 - Discharged on 2 Jun 1815 and released to the Cartel Shakespeare.

Lewis, Jesse L. - Marine - Number: 86 - Prize name: Prince de Neufchatel - Ship type: P - How taken: HM Ship-of-the-Line Leander, HM Ship-of-the-Line Newcastle & HM Frigate Acasta - When taken: 28 Dec 1814 - Where taken: Lat 35N Long 52W - Date received: 18 Feb 1815 - From what ship: HMS Sybille - Born: Connecticut - Age: 22 - Discharged on 19 Feb 1815 and sent to Dartmoor.

Lewis, M. - Seaman - Number: 619 - Prize name: Sine Qua Non - Ship type: P - How taken: HM Brig Elk - When taken: 20 Feb 1815 - Where taken: off Madeira - Date received: 18 Apr 1815 - From what ship: HMS Euryalus - Born: New York - Age: 25 - Race: Negro - Discharged on 2 Jun 1815 and released to the Cartel Sovereign.

Lewis, Raymond - Seaman - Number: 792 - Prize name: US Brig Syren - Ship type: War - How taken: HM Ship-of-the-Line Medway - When taken: 12 Jul 1814 - Where taken: off Cape of Good Hope - Date received: 11 Jul 1815 - From what ship: HMS Royal Sovereign - Born: Not listed - Discharged on 11 Jul 1815 and released to the Cartel Wooddrop Sims.

Lewis, Winslow - Seaman - Number: 152 - Prize name: US Brig Syren - Ship type: War - How taken: HM Ship-of-the-Line Medway - When taken: 12 Jul 1814 - Where taken: off Cape of Good Hope - Date received: 21 Feb 1815 - From what ship: HMS Slaney - Born: Boston - Age: 25 - Discharged on 24 Feb 1815 and sent to Dartmoor.

Lickerany, William - Seaman - Number: 548 - Prize name: Hera - Ship type: MV - How taken: HM Frigate Pique - When taken: 16 Dec 1814 - Where taken: off Porto Rico - Date received: 16 Apr 1815 - From what ship: HMS Swifsure - Born: Boston - Age: 25 - Discharged on 2 Jun 1815 and released to the Cartel Sovereign.

Lidings, John - Marine - Number: 79 - Prize name: Prince de Neufchatel - Ship type: P - How taken: HM Ship-of-the-Line Leander, HM Ship-of-the-Line Newcastle & HM Frigate Acasta - When taken: 28 Dec 1814 - Where taken: Lat 35N Long 52W - Date received: 18 Feb 1815 - From what ship: HMS Sybille - Age: 19 - Discharged on 19 Feb 1815 and sent to Dartmoor.

Linciere, Drury - Seaman - Number: 845 - Prize name: US Brig Syren - Ship type: War - How taken: HM Ship-of-the-Line Medway - When taken: 12 Jul 1814 - Where taken: off Cape of Good Hope - Date received: 11 Jul 1815 - From what ship: HMS Royal Sovereign - Born: Not listed - Discharged on 11 Jul 1815 and released to the Cartel Wooddrop Sims.

Lincoln, Robert - Seaman - Number: 799 - Prize name: US Brig Syren - Ship type: War - How taken: HM Ship-of-the-Line Medway - When taken: 12 Jul 1814 - Where taken: off Cape of Good Hope - Date received: 11 Jul 1815 - From what ship: HMS Royal Sovereign - Born: Not listed - Discharged on 11 Jul 1815 and released to the Cartel Wooddrop Sims.

Limbourg, Charles - Seaman - Number: 19 - Prize name: King of Prussia - Ship type: MV - How taken: Impressed at Liverpool - When taken: 31 Dec 1814 - Date received: 3 Feb 1815 - From what ship: HMS Bittern - Born: Savannah - Age: 37 - Discharged on 4 Feb 1815 and sent to Dartmoor.

Lingham, Nates - Seaman - Number: 714 - Prize name: Reindeer - Ship type: P - How taken: HM Sloop Calypso - When taken: 20 Feb 1815 - Where taken: off Lisbon - Date received: 16 Jul 1815 - From what ship: HMS Hope - Born: Massachusetts - Age: 19 - Discharged on 2 Jun 1815 and released to the Cartel Shakespeare.

Lion, Anthony - Seaman - Number: 654 - Prize name: Avon - Ship type: P - How taken: HM Frigate Barbadoes - When taken: 8 Mar 1815 - Where taken: West Indies - Date received: 24 Apr 1815 - From what ship: HMS Bellerophon - Born: Massachusetts - Age: 19 - Discharged on 2 Jun 1815 and released to the Cartel Shakespeare.

Little, Charles - Seaman - Number: 264 - Prize name: Dorothy - Ship type: MV - How taken: HM Frigate North Star - When taken: 15 Nov 1814 - Where taken: at sea - Date received: 2 Mar 1815 - From what ship: HMS Dannemark - Discharged on 3 Mar 1815 and sent to Dartmoor.

Little, John - Seaman - Number: 477 - Prize name: Fox - Ship type: P - How taken: HM Frigate Barbadoes - When taken: 11 Jan 1815 - Where taken: off Amelia Island, Florida - Date received: 16 Apr 1815 - From what ship: HMS Swifsure - Born: Pennsylvania - Age: 34 - Discharged on 2 Jun 1815 and released to the Cartel Sovereign.

Little, John S. - Boy - Number: 414 - Prize name: Leo - Ship type: P - How taken: HM Frigate Tiber - When taken: 11 Mar 1815 - Where taken: Lat 45.24N Long 12.3W - Date received: 2 Apr 1815 - From what ship: HMS Tiber - Born: Boston - Age: 18 - Discharged on 2 Jun 1815 and released to the Cartel Sovereign.

Little, William - Seaman - Number: 189 - Prize name: Chance - Ship type: P - How taken: HMS Statire - When taken: 1 Apr 1814 - Where taken: at sea - Date received: 2 Mar 1815 - From what ship: HMS Dannemark - Born: New York - Age: 24 - Discharged on 3 Mar 1815 and sent to Dartmoor.

Lloyd, Samuel - Seaman - Number: 591 - Prize name: George Little - Ship type: P - How taken: HM Frigate Granicus - When taken: 20 Jan 1815 - Where taken: off Cape Finisterre, Spain - Date received: 18 Apr 1815 - From what ship: HMS Euryalus - Born: New Hampshire - Age: 32 - Discharged on 8 May 1815.

Longiain, Samuel - Seaman - Number: 593 - Prize name: George Little - Ship type: P - How taken: HM Frigate Granicus - When taken: 20 Jan 1815 - Where taken: off Cape Finisterre, Spain - Date received: 18 Apr 1815 - From what ship: HMS Euryalus - Age: 43 - Discharged on 2 Jun 1815 and released to the Cartel Sovereign.

Lord, John - Prize Master - Number: 318 - Prize name: Maid of the Mill, prized of the Privateer President - Ship type: P - How taken: HM Sloop Calypso - When taken: 25 Feb 1815 - Where taken: off Lisbon - Date received: 24 Mar 1815 - From what ship: HMS Hyperion - Discharged on 27 Apr 1815 and released to the Cartel Minework.

Loss, Anthony - Seaman - Number: 617 - Prize name: Sine Qua Non - Ship type: P - How taken: HM Brig Elk - When taken: 20 Feb 1815 - Where taken: off Madeira - Date received: 18 Apr 1815 - From what ship: HMS Euryalus - Age: 27 - Discharged on 2 Jun 1815 and released to the Cartel Sovereign.

Louis, Joseph - Seaman - Number: 15 - Prize name: Philbes - Ship type: MV - How taken: Impressed at Greenock - When taken: 26 Dec 1814 - Date received: 3 Feb 1815 - From what ship: HMS Bittern - Born: Baltimore - Age: 21 - Discharged on 4 Feb 1815 and sent to Dartmoor.

Love, Mark - Seaman - Number: 515 - Prize name: Hope, prize to US Sloop-of-War Wasp - Ship type: War - How taken: HM Sloop Fairy - When taken: 6 Jan 1815 - Where taken: Unknown - Date received: 16 Apr 1815 - From what ship: HMS Swifsure - Born: Philadelphia - Age: 30 - Discharged on 2 Jun 1815 and released to the Cartel Sovereign.

Lovell, William - Seaman - Number: 656 - Prize name: Avon - Ship type: P - How taken: HM Frigate Barbadoes - When taken: 8 Mar 1815 - Where taken: West Indies - Date received: 24 Apr 1815 - From what ship: HMS Bellerophon - Born: Massachusetts - Age: 21 - Discharged on 2 Jun 1815 and released to the Cartel Shakespeare.

Lowe, Isaac - Seaman - Number: 3 - Prize name: Albion - Ship type: MV - How taken: HM Schooner Helicon - When taken: 17 Jan 1815 - Where taken: Lat 20N Long 52W - Date received: 3 Feb 1815 - From what ship: HMS Bittern - Born: Maryland - Age: 34 - Race: Black - Discharged on 4 Feb 1815 and sent to Dartmoor.

Lowner, Joseph - Seaman - Number: 209 - Prize name: Sennett, Swedish merchant vessel - Ship type: MV - How taken: HM Frigate Rhin - When taken: Not known - Date received: 2 Mar 1815 - From what ship: HMS Dannemark - Discharged on 3 Mar 1815 and sent to Dartmoor.

Lukins, Thomas - Seaman - Number: 212 - Prize name: Sennett, Swedish merchant vessel - Ship type: MV - How taken: HM Frigate Rhin - When taken: Not known - Date received: 2 Mar 1815 - From what ship: HMS Dannemark - Discharged on 3 Mar 1815 and sent to Dartmoor.

Lumey, John - Boy - Number: 126 - Prize name: Prince de Neufchatel - Ship type: P - How taken: HM Ship-of-the-Line Leander, HM Ship-of-the-Line Newcastle & HM Frigate Acasta - When taken: 28 Dec 1814 - Where taken: Lat 35N Long 52W - Date received: 18 Feb 1815 - From what ship: HMS Sybille - Discharged on 19 Feb 1815 and sent to Dartmoor.

Lumsby, Frederick - Seaman - Number: 242 - Prize name: Mary, recaptured prize - Ship type: P - How taken: HM Brig Sophie - When taken: 19 Jun 1814 - Where taken: at sea - Date received: 2 Mar 1815 - From what ship: HMS Dannemark - Discharged on 3 Mar 1815 and sent to Dartmoor.

Lyon, John - Marine Captain - Number: 116 - Prize name: Prince de Neufchatel - Ship type: P - How taken: HM Ship-of-the-Line Leander, HM Ship-of-the-Line Newcastle & HM Frigate Acasta - When taken: 28 Dec 1814 - Where taken: Lat 35N Long 52W - Date received: 18 Feb 1815 - From what ship: HMS Sybille - Born: Virginia - Age: 34 - Discharged on 19 Feb 1815 and sent to Dartmoor.

Maceman, Thomas - Seaman - Number: 192 - Prize name: Chance - Ship type: P - How taken: HMS Statire - When taken: 1 Apr 1814 - Where taken: at sea - Date received: 2 Mar 1815 - From what ship: HMS Dannemark - Born: New York - Age: 30 - Discharged on 3 Mar 1815 and sent to Dartmoor.

Mack, William F. - Mate - Number: 130 - Prize name: St. Joanne - Ship type: MV - How taken: HM Brig Sabine - When taken: 25 Oct 1814 - Where taken: Lat 44 - Date received: 21 Feb 1815 - From what ship: HMS Slaney - Discharged on 24 Feb 1815 and sent to Dartmoor.

Mackey, Charles - Seaman - Number: 812 - Prize name: US Brig Syren - Ship type: War - How taken: HM Ship-of-the-Line Medway - When taken: 12 Jul 1814 - Where taken: off Cape of Good Hope - Date received: 11 Jul 1815 - From what ship: HMS Royal Sovereign - Born: Not listed - Discharged on 11 Jul 1815 and released to the Cartel Wooddrop Sims.

Mackford, John - Seaman - Number: 304 - Prize name: US Brig Syren - Ship type: War - How taken: HM Ship-of-the-Line Medway - When taken: 12 Jul 1814 - Where taken: off Cape of Good Hope - Date received: 4 Mar 1814 - From what ship: HM Cutter Surley - Discharged on 7 Mar 1815 and sent to Dartmoor.

Macklin, John - Seaman - Number: 193 - Prize name: Chance - Ship type: P - How taken: HMS Statire - When taken: 1 Apr 1814 - Where taken: at sea - Date received: 2 Mar 1815 - From what ship: HMS Dannemark - Born: Georgetown - Age: 23 - Discharged on 3 Mar 1815 and sent to Dartmoor.

Mackster, William - Seaman - Number: 273 - Prize name: Netterville - Ship type: MV - How taken: HM Brig Onyx - When taken: 25 Dec 1814 - Where taken: at sea - Date received: 2 Mar 1815 - From what ship: HMS Dannemark - Race: Mulatto - Discharged on 3 Mar 1815 and sent to Dartmoor.

Madden, John - Seaman - Number: 275 - Prize name: Netterville - Ship type: MV - How taken: HM Brig Onyx - When taken: 25 Dec 1814 - Where taken: at sea - Date received: 2 Mar 1815 - From what ship: HMS Dannemark - Born: Talbot County - Age: 26 - Race: Negro - Discharged on 3 Mar 1815 and sent to Dartmoor.

Magown, Joseph - Seaman - Number: 731 - Prize name: Harpy - Ship type: P - How taken: HM Brig Zenobia - When taken: 5 Jan 1815 - Where taken: off Lisbon - Date received: 29 Aug 1815 - From what ship: HMS Tonnant - Born: New Hampshire - Age: 28 - Discharged on 2 Jun 1815 and released to the Cartel Shakespeare.

Maker, James - Seaman - Number: 739 - Prize name: Sine Qua Non - Ship type: P - How taken: HM Brig Elk - When taken: 20 Feb 1815 - Where taken: off Madeira - Date received: 29 Aug 1815 - From what ship: HMS Tonnant - Born: Massachusetts - Age: 30 - Discharged on 2 Jun 1815 and released to the Cartel Shakespeare.

Manchard, Louis - Seaman - Number: 508 - Prize name: Fox - Ship type: P - How taken: HM Frigate Barbadoes - When taken: 11 Jan 1815 - Where taken: off Amelia Island, Florida - Date received: 16 Apr 1815 - From what ship: HMS Swifsure - Born: New Orleans - Age: 25 - Discharged on 2 Jun 1815 and released to the Cartel Sovereign.

Manning, John - Seaman - Number: 760 - Prize name: Harpy - Ship type: P - How taken: HM Brig Zenobia - When taken: 5 Jan 1815 - Where taken: off Lisbon - Date received: 29 Aug 1815 - From what ship: HMS Tonnant - Age: 26 - Discharged on 2 Jun 1815 and released to the Cartel Shakespeare.

Mansfield, Stod. - Boy - Number: 398 - Prize name: Leo - Ship type: P - How taken: HM Frigate Tiber - When taken: 11 Mar 1815 - Where taken: Lat 45.24N Long 12.3W - Date received: 29 Mar 1814 - From what ship: HMS Tiber - Born: Massachusetts - Age: 15 - Discharged on 30 May 1815 and released to the Cartel Atlas.

Manuel, Joseph - Seaman - Number: 817 - Prize name: US Brig Syren - Ship type: War - How taken: HM Ship-of-the-Line Medway - When taken: 12 Jul 1814 - Where taken: off Cape of Good Hope - Date received: 11 Jul 1815 - From what ship: HMS Royal Sovereign - Born: Not listed - Discharged on 11 Jul 1815 and released to the Cartel Wooddrop Sims.

Marcey, W. R. - Seaman - Number: 687 - Prize name: George Little - Ship type: P - How taken: HM Frigate Granicus - When taken: 20 Jan 1815 - Where taken: off Cape Finisterre, Spain - Date received: 16 Jul 1815 - From what ship: HMS Hope - Born: Massachusetts - Age: 22 - Discharged on 2 Jun 1815 and released to the Cartel Shakespeare.

Marole, Richard - Seaman - Number: 749 - Prize name: Harpy - Ship type: P - How taken: HM Brig Zenobia - When taken: 5 Jan 1815 - Where taken: off Lisbon - Date received: 29 Aug 1815 - From what ship: HMS Tonnant - Age: 20 - Discharged on 2 Jun 1815 and released to the Cartel Shakespeare.

Marshall, Anthony - Seaman - Number: 34 - How taken: Taken up at Plymouth - When taken: 1 Feb 1815 - Date received: 8 Feb 1815 - From what ship: HMS Impregnable - Born: New Orleans - Age: 28 - Race: Mulatto - Discharged on 10 Feb 1815 and sent to Dartmoor.

Marshall, William - Boy - Number: 132 - Prize name: St. Joanne - Ship type: MV - How taken: HM Brig Sabine - When taken: 25 Oct 1814 - Where taken: Lat 44 - Date received: 21 Feb 1815 - From what ship: HMS Slaney - Discharged on 24 Feb 1815 and sent to Dartmoor.

Martin, John - 2nd Lieutenant - Number: 67 - Prize name: Prince de Neufchatel - Ship type: P - How taken: HM Ship-of-the-Line Leander, HM Ship-of-the-Line Newcastle & HM Frigate Acasta - When taken: 28 Dec 1814 - Where taken: Lat 35N Long 52W - Date received: 18 Feb 1815 - From what ship: HMS Sybille - Born: Boston - Age: 25 - Discharged on 19 Feb 1815 and sent to Dartmoor.

Martin, John - Seaman - Number: 129 - Prize name: Prince de Neufchatel - Ship type: P - How taken: HM Ship-of-the-Line Leander, HM Ship-of-the-Line Newcastle & HM Frigate Acasta - When taken: 28 Dec 1814 - Where taken: Lat 35N Long 52W - Date received: 18 Feb 1815 - From what ship: HMS Sybille - Born: New Orleans - Age: 35 - Discharged on 19 Feb 1815 and sent to Dartmoor.

Martin, John - Seaman - Number: 166 - How taken: Impressed from MV Rolla - Date received: 1 Mar 1815 - From what ship: Royal Naval Hospital Plymouth - Age: 20 - Discharged on 3 Mar 1815 and sent to Dartmoor.

Martyn, John G. - Seaman - Number: 719 - Prize name: Harpy - Ship type: P - How taken: HM Brig Zenobia - When taken: 5 Jan 1815 - Where taken: off Lisbon - Date received: 16 Jul 1815 - From what ship: HMS Hope - Age: 28 - Discharged on 2 Jun 1815 and released to the Cartel Shakespeare.

Mason, Samuel - Boy - Number: 391 - Prize name: Leo - Ship type: P - How taken: HM Frigate Tiber - When taken: 11 Mar 1815 - Where taken: Lat 45.24N Long 12.3W - Date received: 29 Mar 1814 - From what ship: HMS Tiber - Born: Massachusetts - Age: 16 - Discharged on 30 May 1815 and released to the Cartel Atlas.

Mason, William - Seaman - Number: 588 - Prize name: George Little - Ship type: P - How taken: HM Frigate Granicus - When taken: 20 Jan 1815 - Where taken: off Cape Finisterre, Spain - Date received: 18 Apr 1815 - From what ship: HMS Euryalus - Born: Rhode Island - Age: 44 - Discharged on 2 Jun 1815 and released to the Cartel Sovereign.

Massabi, Pierre - Master's Mate - Number: 461 - Prize name: Fox - Ship type: P - How taken: HM Frigate Barbadoes - When taken: 11 Jan 1815 - Where taken: off Amelia Island, Florida - Date received: 16 Apr 1815 - From what ship: HMS Swifsure - Born: New Orleans - Age: 19 - Discharged on 24 Apr 1815.

Matemes, Enu - Seaman - Number: 450 - Prize name: High Flyer - Ship type: LM - How taken: HMS Muros - When taken: 14 Nov 1814 - Where taken: off St. Bartholomew - Date received: 16 Apr 1815 - From what ship: HMS Swifsure - Born: New Jersey - Age: 24 - Discharged on 2 Jun 1815 and released to the Cartel Sovereign.

Mathers, Allen - Boy - Number: 703 - Prize name: Sine Qua Non - Ship type: P - How taken: HM Brig Elk - When taken: 20 Feb 1815 - Where taken: off Madeira - Date received: 16 Jul 1815 - From what ship: HMS Hope - Born: Massachusetts - Age: 16 - Discharged on 2 Jun 1815 and released to the Cartel Shakespeare.

Maxan, Donrill - Seaman - Number: 474 - Prize name: Fox - Ship type: P - How taken: HM Frigate Barbadoes - When taken: 11 Jan 1815 - Where taken: off Amelia Island, Florida - Date received: 16 Apr 1815 - From what ship: HMS Swifsure - Born: Saint-Domingue (Haiti) - Age: 19 - Discharged on 2 Jun 1815 and released to the Cartel Sovereign.

May, William - Seaman - Number: 632 - How taken: Impressed by HM Ship-of-the-Line Namur - When taken: 7 Nov 1814 - Date received: 24 Apr 1815 - From what ship: HMS Bellerophon - Age: 30 - Discharged on 2 Jun 1815 and released to the Cartel Shakespeare.

Mazely, William - Seaman - Number: 678 - Prize name: Snap Dragon - Ship type: P - How taken: HMS Mentor - When taken: 10 Jul 1814 - Where taken: off Halifax - Date received: 14 May 1815 - From what ship: from the police - Age: 27 - Discharged on 2 Jun 1815 and released to the Cartel Shakespeare.

McCally, John - Seaman - Number: 839 - Prize name: US Brig Syren - Ship type: War - How taken: HM Ship-of-the-Line Medway - When taken: 12 Jul 1814 - Where taken: off Cape of Good Hope - Date received: 11 Jul 1815 - From what ship: HMS Royal Sovereign - Born: Not listed - Discharged on 11 Jul 1815 and released to the Cartel Wooddrop Sims.

McCarter, M. - Seaman - Number: 350 - Prize name: Leo - Ship type: P - How taken: HM Frigate Tiber - When taken: 11 Mar 1815 - Where taken: Lat 45.24N Long 12.3W - Date received: 29 Mar 1814 - From what ship: HMS Tiber - Born: Massachusetts - Age: 25 - Discharged on 27 Apr 1815 and released to the Cartel Minework.

McCarthy, Samuel - Seaman - Number: 261 - Prize name: Dorothy - Ship type: MV - How taken: HM Frigate North Star - When taken: 15 Nov 1814 - Where taken: at sea - Date received: 2 Mar 1815 - From what ship: HMS Dannemark - Discharged on 3 Mar 1815 and sent to Dartmoor.

McCaslin, James - Seaman - Number: 449 - Prize name: High Flyer - Ship type: LM - How taken: HMS Muros - When taken: 14 Nov 1814 - Where taken: off St. Bartholomew - Date received: 16 Apr 1815 - From what ship: HMS Swifsure - Born: Pennsylvania - Age: 28 - Discharged on 2 Jun 1815 and released to the Cartel Sovereign.

McCormick, James - Seaman - Number: 200 - Prize name: Saucy Jack - Ship type: MV - How taken: Unknown - Date received: 2 Mar 1815 - From what ship: HMS Dannemark - Discharged on 3 Mar 1815 and sent to Dartmoor.

McDaniel, John - Seaman - Number: 48 - How taken: Impressed from HM Frigate Phoenix - When taken: 14 Jan 1815 - Date received: 15 Feb 1815 - From what ship: HMS Myrmidon - Born: New Orleans - Age: 24 - Race: Black - Discharged on 17 Feb 1815 and sent to Dartmoor.

McDermott, A. - Seaman - Number: 553 - Prize name: Hera - Ship type: MV - How taken: HM Frigate Pique - When taken: 16 Dec 1814 - Where taken: off Porto Rico - Date received: 16 Apr 1815 - From what ship: HMS Swifsure - Born: Pennsylvania - Age: 23 - Discharged on 2 Jun 1815 and released to the Cartel Sovereign.

McHam, Daniel - 2nd Mate - Number: 444 - Prize name: High Flyer - Ship type: LM - How taken: HMS Muros - When taken: 14 Nov 1814 - Where taken: off St. Bartholomew - Date received: 16 Apr 1815 - From what ship: HMS Swifsure - Born: Pennsylvania - Age: 23 - Discharged on 2 Jun 1815 and released to the Cartel Sovereign.

McIntyre, John (alias F. M. Intro) - Seaman - Number: 323 - Prize name: Thomas, prize to Privateer Scourge - Ship type: P - How taken: HM Frigate Aquilon - When taken: 11 Mar 1815 - Where taken: off Cape Finisterre, Spain - Date received: 25 Apr 1815 - From what ship: HMS Opossum - Discharged on 27 Apr 1815 and released to the Cartel Minework.

McLachlan, John - Seaman - Number: 628 - Prize name: Helene, prize of Privateer Morgianna - Ship type: P - How taken: HM Frigate Pique - When taken: 7 Mar 1815 - Where taken: coast of America - Date received: 20 Apr 1815 - From what ship: HMS Musquito - Age: 25 - Discharged on 27 Apr 1815.

McLean, Judes - Seaman - Number: 506 - Prize name: Fox - Ship type: P - How taken: HM Frigate Barbadoes - When taken: 11 Jan 1815 - Where taken: off Amelia Island, Florida - Date received: 16 Apr 1815 - From what ship: HMS Swifsure - Born: Saint-Domingue (Haiti) - Age: 24 - Race: Negro - Discharged on 2 Jun 1815 and released to the Cartel Sovereign.

McNeil, Dennis - Seaman - Number: 223 - Prize name: Decatur - Ship type: P - How taken: HM Frigate Rhin - When taken: 4 Jun 1814 - Where taken: at sea - Date received: 2 Mar 1815 - From what ship: HMS Dannemark - Born: Boston - Age: 32 - Discharged on 3 Mar 1815 and sent to Dartmoor.

Medford, James - Seaman - Number: 701 - Prize name: Sine Qua Non - Ship type: P - How taken: HM Brig Elk - When taken: 20 Feb 1815 - Where taken: off Madeira - Date received: 16 Jul 1815 - From what ship: HMS Hope - Born: South Carolina - Age: 35 - Discharged on 2 Jun 1815 and released to the Cartel Shakespeare.

Medford, Joseph - Seaman - Number: 727 - Prize name: George Little - Ship type: P - How taken: HM Frigate Granicus - When taken: 20 Jan 1815 - Where taken: off Cape Finisterre, Spain - Date received: 29 Aug 1815 - From what ship: HMS Tonnant - Born: Massachusetts - Age: 32 - Discharged on 2 Jun 1815 and released to the Cartel Shakespeare.

Meddler, John - Seaman - Number: 547 - How taken: Gave himself up from HMS Harmony - When taken: 4 Sep 1814 - Where taken: Barbados - Date received: 16 Apr 1815 - From what ship: HMS Swifsure - Born: Maryland - Age: 40 - Discharged on 2 Jun 1815 and released to the Cartel Sovereign.

Medley, Samuel - Seaman - Number: 339 - Prize name: Leo - Ship type: P - How taken: HM Frigate Tiber - When taken: 11 Mar 1815 - Where taken: Lat 45.24N Long 12.3W - Date received: 28 Mar 1815 - From what ship: HMS Tiber - Discharged on 27 Apr 1815 and released to the Cartel Minework.

Menzer, James - Seaman - Number: 308 - Prize name: US Brig Syren - Ship type: War - How taken: HM Ship-of-the-Line Medway - When taken: 12 Jul 1814 - Where taken: off Cape of Good Hope - Date received: 4 Mar 1814 - From what ship: HM Cutter Surley - Discharged on 7 Mar 1815 and sent to Dartmoor.

Merchant, Joseph - Seaman - Number: 666 - Prize name: Avon - Ship type: P - How taken: HM Frigate Barbadoes - When taken: 8 Mar 1815 - Where taken: West Indies - Date received: 24 Apr 1815 - From what ship: HMS Bellerophon - Born: Massachusetts - Age: 34 - Discharged on 2 Jun 1815 and released to the Cartel Shakespeare.

Merrick, William - Seaman - Number: 705 - Prize name: Sine Qua Non - Ship type: P - How taken: HM Brig Elk - When taken: 20 Feb 1815 - Where taken: off Madeira - Date received: 16 Jul 1815 - From what ship: HMS Hope - Age: 23 - Discharged on 29 May 1815.

Michael, Samuel M. - Seaman - Number: 756 - Prize name: George Little - Ship type: P - How taken: HM Frigate Granicus - When taken: 20 Jan 1815 - Where taken: off Cape Finisterre, Spain - Date received: 29 Aug 1815 - From what ship: HMS Tonnant - Born: Massachusetts - Age: 18 - Discharged on 2 Jun 1815 and released to the Cartel Shakespeare.

Micker, Charles - Seaman - Number: 305 - Prize name: US Brig Syren - Ship type: War - How taken: HM Ship-of-the-Line Medway - When taken: 12 Jul 1814 - Where taken: off Cape of Good Hope - Date received: 4 Mar 1814 - From what ship: HM Cutter Surley - Discharged on 7 Mar 1815 and sent to Dartmoor.

Miller, Isaac - Seaman - Number: 329 - Prize name: Leo - Ship type: P - How taken: HM Frigate Tiber - When taken: 11 Mar 1815 - Where taken: Lat 45.24N Long 12.3W - Date received: 28 Mar 1815 - From what ship: HMS Tiber - Discharged on 27 Apr 1815 and released to the Cartel Minework.

Miller, James - Seaman - Number: 59 - How taken: Taken out of HM Ship-of-the-Line Montague - Where taken: Portsmouth - Date received: 15 Feb 1815 - From what ship: HMS Myrmidon - Age: 50 - Discharged on 17 Feb 1815 and sent to Dartmoor.

Miller, John - Seaman - Number: 122 - Prize name: Prince de Neufchatel - Ship type: P - How taken: HM Ship-of-the-Line Leander, HM Ship-of-the-Line Newcastle & HM Frigate Acasta - When taken: 28 Dec 1814 - Where taken: Lat 35N Long 52W - Date received: 18 Feb 1815 - From what ship: HMS Sybille - Born: Baltimore - Age: 39 - Discharged on 19 Feb 1815 and sent to Dartmoor.

Miller, Peter - Seaman - Number: 402 - Prize name: Leo - Ship type: P - How taken: HM Frigate Tiber - When taken: 11 Mar 1815 - Where taken: Lat 45.24N Long 12.3W - Date received: 29 Mar 1815 - From what ship: HMS Tiber - Born: Massachusetts - Age: 32 - Discharged on 24 Apr 1815.

Miller, Samuel - Seaman - Number: 821 - Prize name: US Brig Syren - Ship type: War - How taken: HM Ship-of-the-Line Medway - When taken: 12 Jul 1814 - Where taken: off Cape of Good Hope - Date received: 11 Jul 1815 - From what ship: HMS Royal Sovereign - Born: Not listed - Discharged on 11 Jul 1815 and released to the Cartel Wooddrop Sims.

Millow, John - Seaman - Number: 147 - Prize name: US Brig Syren - Ship type: War - How taken: HM Ship-of-the-Line Medway - When taken: 12 Jul 1814 - Where taken: off Cape of Good Hope - Date received: 21 Feb 1815 - From what ship: HMS Slaney - Born: Alexandria - Age: 25 - Discharged on 24 Feb 1815 and sent to Dartmoor.

Mills, Samuel - Boy - Number: 385 - Prize name: Leo - Ship type: P - How taken: HM Frigate Tiber - When taken: 11 Mar 1815 - Where taken: Lat 45.24N Long 12.3W - Date received: 29 Mar 1814 - From what ship: HMS Tiber - Born: Massachusetts - Age: 16 - Discharged on 30 May 1815 and released to the Cartel Atlas.

Minor, Charles - Seaman - Number: 410 - Prize name: Thomas, prize to Privateer Scourge - Ship type: P - How taken: HM Frigate Aquilon - When taken: 11 Mar 1815 - Where taken: off Cape Finisterre, Spain - Date received: 29 Mar 1815 - From what ship: HMS Icarus - Born: Massachusetts - Age: 20 - Race: Negro - Discharged on 2 Jun 1815 and released to the Cartel Sovereign.

Mitchell, R. J. - Clerk - Number: 615 - Prize name: Sine Qua Non - Ship type: P - How taken: HM Brig Elk - When taken: 20 Feb 1815 - Where taken: off Madeira - Date received: 18 Apr 1815 - From what ship: HMS Euryalus - Age: 26 - Discharged on 24 Apr 1815.

Monk, John - Seaman - Number: 519 - Prize name: High Flyer - Ship type: LM - How taken: HMS Muros - When taken: 14 Nov 1814 - Where taken: off St. Bartholomew - Date received: 16 Apr 1815 - From what ship: HMS Swifsure - Born: Rhode Island - Age: 28 - Discharged on 2 Jun 1815 and released to the Cartel Sovereign.

Monk, Philip - Seaman - Number: 111 - Prize name: Prince de Neufchatel - Ship type: P - How taken: HM Ship-of-the-Line Leander, HM Ship-of-the-Line Newcastle & HM Frigate Acasta - When taken: 28 Dec 1814 - Where taken: Lat 35N Long 52W - Date received: 18 Feb 1815 - From what ship: HMS Sybille - Discharged on 19 Feb 1815 and sent to Dartmoor.

Moore, John - Seaman - Number: 717 - Prize name: Melvina - Ship type: MV - How taken: HM Frigate Andromeda - When taken: 28 Feb 1815 - Where taken: off Lisbon - Date received: 16 Jul 1815 - From what ship: HMS Hope - Born: Massachusetts - Age: 29 - Discharged on 2 Jun 1815 and released to the Cartel Shakespeare.

Moore, Joseph H. - Seaman - Number: 379 - Prize name: Leo - Ship type: P - How taken: HM Frigate Tiber - When taken: 11 Mar 1815 - Where taken: Lat 45.24N Long 12.3W - Date received: 29 Mar 1814 - From what ship: HMS Tiber - Born: Massachusetts - Age: 17 - Discharged on 30 May 1815 and released to the Cartel Atlas.

Moore, Joshua - Marine - Number: 82 - Prize name: Prince de Neufchatel - Ship type: P - How taken: HM Ship-of-the-Line Leander, HM Ship-of-the-Line Newcastle & HM Frigate Acasta - When taken: 28 Dec 1814 - Where taken: Lat 35N Long 52W - Date received: 18 Feb 1815 - From what ship: HMS Sybille - Born: Portsmouth - Age: 21 - Discharged on 19 Feb 1815 and sent to Dartmoor.

Moore, Robert - Seaman - Number: 163 - Prize name: Nancy, prize of the Privateer United States - Ship type: P - How taken: HM Sloop Papillon - When taken: 27 Dec 1814 - Where taken: Lat 30 Long 10 - Date received: 27 Feb 1815 - From what ship: HMS Nimble - Born: Kittery - Age: 20 - Discharged on 3 Mar 1815 and sent to Dartmoor.

Moore, Seth H. - Seaman - Number: 389 - Prize name: Leo - Ship type: P - How taken: HM Frigate Tiber - When taken: 11 Mar 1815 - Where taken: Lat 45.24N Long 12.3W - Date received: 29 Mar 1814 - From what ship: HMS Tiber - Born: Massachusetts - Age: 18 - Discharged on 30 May 1815 and released to the Cartel Atlas.

Moore, Warren - Seaman - Number: 105 - Prize name: Prince de Neufchatel - Ship type: P - How taken: HM Ship-of-the-Line Leander, HM Ship-of-the-Line Newcastle & HM Frigate Acasta - When taken: 28 Dec 1814 - Where taken: Lat 35N Long 52W - Date received: 18 Feb 1815 - From what ship: HMS Sybille - Born: Massachusetts - Age: 22 - Discharged on 19 Feb 1815 and sent to Dartmoor.

Moore, William - Seaman - Number: 239 - Prize name: Mary, recaptured prize - Ship type: P - How taken: HM Brig Sophie - When taken: 19 Jun 1814 - Where taken: at sea - Date received: 2 Mar 1815 - From what ship: HMS Dannemark - Discharged on 3 Mar 1815 and sent to Dartmoor.

Morey, Silvis - Seaman - Number: 375 - Prize name: Leo - Ship type: P - How taken: HM Frigate Tiber - When taken: 11 Mar 1815 - Where taken: Lat 45.24N Long 12.3W - Date received: 29 Mar 1814 - From what ship: HMS Tiber - Born: Massachusetts - Age: 28 - Discharged on 2 May 1815 and released to the Cartel Ariel.

Morris, John - Boatswain's Mate - Number: 135 - Prize name: US Brig Syren - Ship type: War - How taken: HM Ship-of-the-Line Medway - When taken: 12 Jul 1814 - Where taken: off Cape of Good Hope - Date received: 21 Feb 1815 - From what ship: HMS Slaney - Born: Marblehead - Age: 27 - Discharged on 24 Feb 1815 and sent to Dartmoor.

Morton, J. J. - Seaman - Number: 606 - Prize name: George Little - Ship type: P - How taken: HM Frigate Granicus - When taken: 20 Jan 1815 - Where taken: off Cape Finisterre, Spain - Date received: 18 Apr 1815 - From what ship: HMS Euryalus - Born: Massachusetts - Age: 27 - Discharged on 2 Jun 1815 and released to the Cartel Sovereign.

Morton, Levi - Seaman - Number: 608 - Prize name: George Little - Ship type: P - How taken: HM Frigate Granicus - When taken: 20 Jan 1815 - Where taken: off Cape Finisterre, Spain - Date received: 18 Apr 1815 - From what ship: HMS Euryalus - Born: Massachusetts - Age: 27 - Discharged on 2 Jun 1815 and released to the Cartel Sovereign.

Morton, P. - Surgeon - Number: 616 - Prize name: Sine Qua Non - Ship type: P - How taken: HM Brig Elk - When taken: 20 Feb 1815 - Where taken: off Madeira - Date received: 18 Apr 1815 - From what ship: HMS Euryalus - Age: 26 - Discharged on 24 Apr 1815.

Morton, Samuel N. - Lieutenant - Number: 674 - Prize name: Leo - Ship type: LM - How taken: HM Frigate Granicus - When taken: 26 Nov 1814 - Where taken: off Lisbon - Date received: 24 Apr 1815 - From what ship: Royal Naval Hospital Plymouth - Age: 34 - Discharged on 3 May 1815.

Moses, John - Seaman - Number: 831 - Prize name: US Brig Syren - Ship type: War - How taken: HM Ship-of-the-Line Medway - When taken: 12 Jul 1814 - Where taken: off Cape of Good Hope - Date received: 11 Jul 1815 - From what ship: HMS Royal Sovereign - Born: Not listed - Discharged on 11 Jul 1815 and released to the Cartel Wooddrop Sims.

Moss, Richard - Seaman - Number: 296 - How taken: Taken off the HM Ship-of-the-Line Severn and other ships - Date received: 2 Mar 1815 - From what ship: HMS Dannemark - Born: Bridgewater - Age: 24 - Discharged on 3 Mar 1815 and sent to Dartmoor.

Moulston, Nathaniel - Seaman - Number: 284 - How taken: Taken off the HM Ship-of-the-Line Severn and other ships - Date received: 2 Mar 1815 - From what ship: HMS Dannemark - Born: Newburyport - Age: 18 - Discharged on 3 Mar 1815 and sent to Dartmoor.

Mount, John - Seaman - Number: 6 - Prize name: Albion - Ship type: MV - How taken: HM Schooner Helicon - When taken: 17 Jan 1815 - Where taken: Lat 20N Long 52W - Date received: 3 Feb 1815 - From what ship: HMS Bittern - Born: New Jersey - Age: 24 - Discharged on 4 Feb 1815 and sent to Dartmoor.

Mown, Hitcher - Seaman - Number: 327 - Prize name: Leo - Ship type: P - How taken: HM Frigate Tiber - When taken: 11 Mar 1815 - Where taken: Lat 45.24N Long 12.3W - Date received: 28 Mar 1815 - From what ship: HMS Tiber - Discharged on 27 Apr 1815 and released to the Cartel Minework.

Murphy, William - Seaman - Number: 753 - Prize name: Melvina - Ship type: MV - How taken: HM Frigate Andromeda - When taken: 28 Feb 1815 - Where taken: off Lisbon - Date received: 29 Aug 1815 - From what ship: HMS Tonnant - Age: 24 - Discharged on 2 Jun 1815 and released to the Cartel Shakespeare.

Murray, M. - Seaman - Number: 726 - Prize name: Sine Qua Non - Ship type: P - How taken: HM Brig Elk - When taken: 20 Feb 1815 - Where taken: off Madeira - Date received: 29 Aug 1815 - From what ship: HMS Tonnant - Born: New Orleans - Age: 24 - Discharged on 2 Jun 1815 and released to the Cartel Shakespeare.

Myrick, John - Seaman - Number: 716 - Prize name: Reindeer - Ship type: P - How taken: HM Sloop Calypso - When taken: 20 Feb 1815 - Where taken: off Lisbon - Date received: 16 Jul 1815 - From what ship: HMS Hope - Born: Massachusetts - Age: 25 - Discharged on 2 Jun 1815 and released to the Cartel Shakespeare.

Names, Nathaniel - Seaman - Number: 367 - Prize name: Leo - Ship type: P - How taken: HM Frigate Tiber - When taken: 11 Mar 1815 - Where taken: Lat 45.24N Long 12.3W - Date received: 29 Mar 1814 - From what ship: HMS Tiber - Born: Massachusetts - Age: 22 - Discharged on 2 May 1815 and released to the Cartel Ariel.

Names, Samuel - Seaman - Number: 355 - Prize name: Leo - Ship type: P - How taken: HM Frigate Tiber - When taken: 11 Mar 1815 - Where taken: Lat 45.24N Long 12.3W - Date received: 29 Mar 1814 - From what ship: HMS Tiber - Born: Massachusetts - Age: 30 - Discharged on 27 Apr 1815 and released to the Cartel Minework.

Namiss, Peter - Seaman - Number: 55 - How taken: Taken out of HMS Necover - Where taken: Sternness - Date received: 15 Feb 1815 - From what ship: HMS Myrmidon - Age: 32 - Race: Black - Discharged on 17 Feb 1815 and sent to Dartmoor.

Nantes, John - 2nd Lieutenant - Number: 720 - Prize name: Hussar - Ship type: P - How taken: HM Brig Zenobia - When taken: 5 Jan 1815 - Where taken: off Lisbon - Date received: 29 Aug 1815 - From what ship: HMS Tonnant - Age: 25 - Discharged on 2 Jun 1815 and released to the Cartel Shakespeare.

Napp, George - Seaman - Number: 572 - Prize name: Heron - Ship type: MV - How taken: HM Frigate Pique - When taken: 16 Dec 1814 - Where taken: off Porto Rico - Date received: 16 Apr 1815 - From what ship: HMS Swifsure - Born: Massachusetts - Age: 23 - Discharged on 2 Jun 1815 and released to the Cartel Sovereign.

Nash, Alexander - Seaman - Number: 112 - Prize name: Prince de Neufchatel - Ship type: P - How taken: HM Ship-of-the-Line Leander, HM Ship-of-the-Line Newcastle & HM Frigate Acasta - When taken: 28 Dec 1814 - Where taken: Lat 35N Long 52W - Date received: 18 Feb 1815 - From what ship: HMS Sybille - Born: Boston - Age: 36 - Discharged on 19 Feb 1815 and sent to Dartmoor.

Natcheder, George - Seaman - Number: 352 - Prize name: Leo - Ship type: P - How taken: HM Frigate Tiber - When taken: 11 Mar 1815 - Where taken: Lat 45.24N Long 12.3W - Date received: 29 Mar 1814 - From what ship: HMS Tiber - Born: Massachusetts - Age: 24 - Discharged on 27 Apr 1815 and released to the Cartel Minework.

Nettle, Joseph - Seaman - Number: 251 - How taken: Taken up at Kingston - Date received: 2 Mar 1815 - From what ship: HMS Dannemark - Discharged on 3 Mar 1815 and sent to Dartmoor.

Newall, John B. L. - Marine - Number: 87 - Prize name: Prince de Neufchatel - Ship type: P - How taken: HM Ship-of-the-Line Leander, HM Ship-of-the-Line Newcastle & HM Frigate Acasta - When taken: 28 Dec 1814 - Where taken: Lat 35N Long 52W - Date received: 18 Feb 1815 - From what ship: HMS Sybille - Born: Lynn - Age: 20 - Discharged on 19 Feb 1815 and sent to Dartmoor.

Newell, Thomas - Marine Captain - Number: 638 - Prize name: Avon - Ship type: P - How taken: HM Frigate Barbadoes - When taken: 8 Mar 1815 - Where taken: West Indies - Date received: 24 Apr 1815 - From what ship: HMS Bellerophon - Age: 24 - Discharged on 23 May 1815.

Nichelle, Pierre - Passenger - Number: 41 - Prize name: Leo - Ship type: P - How taken: HM Frigate Granicus - When taken: 2 Dec 1814 - Where taken: off Lorient, France - Date received: 12 Feb 1815 - From what ship: HMS Impregnable - Born: Lorient, France - Age: 34 - Discharged on 17 Feb 1815 and sent to Dartmoor.

Nicholas, Jacob - Seaman - Number: 301 - Prize name: US Brig Syren - Ship type: War - How taken: HM Ship-of-the-Line Medway - When taken: 12 Jul 1814 - Where taken: off Cape of Good Hope - Date received: 4 Mar 1814 - From what ship: HM Cutter Surley - Born: Not legible - Discharged on 7 Mar 1815 and sent to Dartmoor.

Nichols, N. D. - Lieutenant - Number: 849 - Prize name: US Brig Syren - Ship type: War - How taken: HM Ship-of-the-Line Medway - When taken: 12 Jul 1814 - Where taken: off Cape of Good Hope - Date received: 11 Jul 1815 - From what ship: from Portsmouth - Born: Not listed - Discharged on 11 Jul 1815 and released to the Cartel Wooddrop Sims.

Nicon, John M. - Master's Mate - Number: 587 - Prize name: George Little - Ship type: P - How taken: HM Frigate Granicus - When taken: 20 Jan 1815 - Where taken: off Cape Finisterre, Spain - Date received: 18 Apr 1815 - From what ship: HMS Euryalus - Born: Rhode Island - Age: 31 - Discharged on 2 Jun 1815 and released to the Cartel Sovereign.

Ninner, George - Seaman - Number: 403 - Prize name: Leo - Ship type: P - How taken: HM Frigate Tiber - When taken: 11 Mar 1815 - Where taken: Lat 45.24N Long 12.3W - Date received: 29 Mar 1815 - From what ship: HMS Tiber - Born: Massachusetts - Age: 18 - Discharged on 30 May 1815 and released to the Cartel Atlas.

Noles, E. D. - Seaman - Number: 715 - Prize name: Reindeer - Ship type: P - How taken: HM Sloop Calypso - When taken: 20 Feb 1815 - Where taken: off Lisbon - Date received: 16 Jul 1815 - From what ship: HMS Hope - Born: Massachusetts - Age: 22 - Discharged on 2 Jun 1815 and released to the Cartel Shakespeare.

Nolan, Bartholomew - Seaman - Number: 778 - Prize name: Sine Qua Non - Ship type: P - How taken: HM Brig Elk - When taken: 20 Feb 1815 - Where taken: off Madeira - Date received: 29 Aug 1815 - From what ship: HMS Tenant - Age: 23 - Race: Negro - Discharged on 2 Jun 1815 and released to the Cartel Shakespeare.

Norman, William - Seaman - Number: 288 - How taken: Taken off the HM Ship-of-the-Line Severn and other ships - Date received: 2 Mar 1815 - From what ship: HMS Dannemark - Born: Boston - Age: 26 - Discharged on 3 Mar 1815 and sent to Dartmoor.

Norris, Benjamin - Seaman - Number: 157 - Prize name: Nancy, prize of the Privateer United States - Ship type: P - How taken: HM Sloop Papillon - When taken: 27 Dec 1814 - Where taken: Lat 30 Long 10 - Date received: 27 Feb 1815 - From what ship: HMS Nimble - Born: Not legible - Age: 20 - Discharged on 3 Mar 1815 and sent to Dartmoor.

Norris, Thomas - Seaman - Number: 725 - Prize name: Sine Qua Non - Ship type: P - How taken: HM Brig Elk - When taken: 20 Feb 1815 - Where taken: off Madeira - Date received: 29 Aug 1815 - From what ship: HMS Tonnant - Born: Massachusetts - Age: 32 - Discharged on 2 Jun 1815 and released to the Cartel Shakespeare.

Noy, Jolly - Seaman - Number: 490 - Prize name: Fox - Ship type: P - How taken: HM Frigate Barbadoes - When taken: 11 Jan 1815 - Where taken: off Amelia Island, Florida - Date received: 16 Apr 1815 - From what ship: HMS Swifsure - Born: New Orleans - Age: 26 - Race: Negro - Discharged on 24 Apr 1815.

Nudd, Stephen - Seaman - Number: 747 - Prize name: Harpy - Ship type: P - How taken: HM Brig Zenobia - When taken: 5 Jan 1815 - Where taken: off Lisbon - Date received: 29 Aug 1815 - From what ship: HMS Tonnant - Age: 21 - Discharged on 2 Jun 1815 and released to the Cartel Shakespeare.

Nutmas, William - Seaman - Number: 376 - Prize name: Leo - Ship type: P - How taken: HM Frigate Tiber - When taken: 11 Mar 1815 - Where taken: Lat 45.24N Long 12.3W - Date received: 29 Mar 1814 - From what ship: HMS Tiber - Born: Massachusetts - Age: 31 - Discharged on 2 May 1815 and released to the Cartel Ariel.

Nutter, Henry - Seaman - Number: 185 - Prize name: Farmer's Daughter - Ship type: MV - How taken: HM Ship-of-the-Line Leviathan - When taken: 20 Mar 1814 - Where taken: at sea - Date received: 2 Mar 1815 - From what ship: HMS Dannemark - Age: 35 - Race: Negro - Discharged on 3 Mar 1815 and sent to Dartmoor.

Nutter, Henry - Seaman - Number: 738 - Prize name: Harpy - Ship type: P - How taken: HM Brig Zenobia - When taken: 5 Jan 1815 - Where taken: off Lisbon - Date received: 29 Aug 1815 - From what ship: HMS Tonnant - Age: 21 - Discharged on 2 Jun 1815 and released to the Cartel Shakespeare.

Oatham, Henry - Seaman - Number: 759 - Prize name: Sine Qua Non - Ship type: P - How taken: HM Brig Elk - When taken: 20 Feb 1815 - Where taken: off Madeira - Date received: 29 Aug 1815 - From what ship: HMS Tonnant - Born: Massachusetts - Age: 25 - Discharged on 2 Jun 1815 and released to the Cartel Shakespeare.

Ocklin, Joseph - Seaman - Number: 847 - Prize name: US Brig Syren - Ship type: War - How taken: HM Ship-of-the-Line Medway - When taken: 12 Jul 1814 - Where taken: off Cape of Good Hope - Date received: 11 Jul 1815 - From what ship: HMS Royal Sovereign - Born: Not listed - Discharged on 11 Jul 1815 and released to the Cartel Wooddrop Sims.

Offa, Torivio - Seaman - Number: 473 - Prize name: Fox - Ship type: P - How taken: HM Frigate Barbadoes - When taken: 11 Jan 1815 - Where taken: off Amelia Island, Florida - Date received: 16 Apr 1815 - From what ship: HMS Swifsure - Born: Spain - Age: 17 - Race: Negro - Discharged on 8 May 1815.

Oliver, Mathew - Seaman - Number: 14 - Prize name: Philbes - Ship type: MV - How taken: Taken out of a Spanish schooner - When taken: 25 Nov 1814 - Where taken: Greenock - Date received: 3 Feb 1815 - From what ship: HMS Bittern - Born: Bath - Age: 26 - Discharged on 4 Feb 1815 and sent to Dartmoor.

Orn, Joseph - Lieutenant - Number: 610 - Prize name: Sine Qua Non - Ship type: P - How taken: HM Brig Elk - When taken: 20 Feb 1815 - Where taken: off Madeira - Date received: 18 Apr 1815 - From what ship: HMS Euryalus - Age: 27 - Discharged on 23 May 1815.

Osborne, John L. - Seaman - Number: 158 - Prize name: Nancy, prize of the Privateer United States - Ship type: P - How taken: HM Sloop Papillon - When taken: 27 Dec 1814 - Where taken: Lat 30 Long 10 - Date received: 27 Feb 1815 - From what ship: HMS Nimble - Born: Newburyport - Age: 18 - Discharged on 3 Mar 1815 and sent to Dartmoor.

Owen, Thomas - Seaman - Number: 838 - Prize name: US Brig Syren - Ship type: War - How taken: HM Ship-of-the-Line Medway - When taken: 12 Jul 1814 - Where taken: off Cape of Good Hope - Date received: 11 Jul 1815 - From what ship: HMS Royal Sovereign - Born: Not listed - Discharged on 11 Jul 1815 and released to the Cartel Wooddrop Sims.

Pachum, John - Seaman - Number: 482 - Prize name: Fox - Ship type: P - How taken: HM Frigate Barbadoes - When taken: 11 Jan 1815 - Where taken: off Amelia Island, Florida - Date received: 16 Apr 1815 - From what ship: HMS Swifsure - Born: Lorient, France - Age: 22 - Discharged on 2 Jun 1815 and released to the Cartel Sovereign.

Parker, Samuel - Seaman - Number: 665 - Prize name: Avon - Ship type: P - How taken: HM Frigate Barbadoes - When taken: 8 Mar 1815 - Where taken: West Indies - Date received: 24 Apr 1815 - From what ship: HMS Bellerophon - Born: Massachusetts - Age: 22 - Discharged on 2 Jun 1815 and released to the Cartel Shakespeare.

Parnet, Henry - Seaman - Number: 306 - Prize name: US Brig Syren - Ship type: War - How taken: HM Ship-of-the-Line Medway - When taken: 12 Jul 1814 - Where taken: off Cape of Good Hope - Date received: 4 Mar 1814 - From what ship: HM Cutter Surley - Discharged on 7 Mar 1815 and sent to Dartmoor.

Parsons, James - Seaman - Number: 253 - How taken: Taken up at Kingston - Date received: 2 Mar 1815 - From what ship: HMS Dannemark - Born: Long Island - Age: 27 - Discharged on 3 Mar 1815 and sent to Dartmoor.

Parsons, Thomas - Chief Mate - Number: 36 - Prize name: Plutarch - Ship type: MV - How taken: HM Schooner Helicon - When taken: 5 Feb 1815 - Where taken: Lat 45.5N Long 7W - Date received: 10 Feb 1815 - From what ship: HMS Helicon - Discharged on 10 Feb 1815 and sent to Ashburton.

Partex, Samuel - Seaman - Number: 85 - Prize name: Prince de Neufchatel - Ship type: P - How taken: HM Ship-of-the-Line Leander, HM Ship-of-the-Line Newcastle & HM Frigate Acasta - When taken: 28 Dec 1814 - Where taken: Lat 35N Long 52W - Date received: 18 Feb 1815 - From what ship: HMS Sybille - Age: 20 - Discharged on 19 Feb 1815 and sent to Dartmoor.

Patterson, Lucas - Seaman - Number: 598 - Prize name: George Little - Ship type: P - How taken: HM Frigate Granicus - When taken: 20 Jan 1815 - Where taken: off Cape Finisterre, Spain - Date received: 18 Apr 1815 - From what ship: HMS Euryalus - Born: Boston - Age: 42 - Discharged on 2 Jun 1815 and released to the Cartel Sovereign.

Patton, William - Seaman - Number: 757 - Prize name: Aurora - Ship type: MV - How taken: HM Frigate Andromeda - When taken: 28 Feb 1815 - Where taken: off Lisbon - Date received: 29 Aug 1815 - From what ship: HMS Tonnant - Age: 39 - Discharged on 2 Jun 1815 and released to the Cartel Shakespeare.

Pavish, John - Seaman - Number: 244 - Prize name: Mary, recaptured prize - Ship type: P - How taken: HM Brig Sophie - When taken: 19 Jun 1814 - Where taken: at sea - Date received: 2 Mar 1815 - From what ship: HMS Dannemark - Discharged on 3 Mar 1815 and sent to Dartmoor.

Payne, James - Seaman - Number: 42 - How taken: Taken out of HM Brig Havock - Where taken: Chatham - Date received: 15 Feb 1815 - From what ship: HMS Myrmidon - Born: New York - Age: 36 - Race: Black - Discharged on 17 Feb 1815 and sent to Dartmoor.

Payne, John J. - Seaman - Number: 741 - Prize name: Sine Qua Non - Ship type: P - How taken: HM Brig Elk - When taken: 20 Feb 1815 - Where taken: off Madeira - Date received: 29 Aug 1815 - From what ship: HMS Tonnant - Age: 21 - Discharged on 2 Jun 1815 and released to the Cartel Shakespeare.

Payton, James - Prize Master - Number: 128 - Prize name: Prince de Neufchatel - Ship type: P - How taken: HM Ship-of-the-Line Leander, HM Ship-of-the-Line Newcastle & HM Frigate Acasta - When taken: 28 Dec 1814 -

Where taken: Lat 35N Long 52W - Date received: 18 Feb 1815 - From what ship: HMS Sybille - Born: Charleston - Age: 26 - Discharged on 19 Feb 1815 and sent to Dartmoor.

Pearson, Thomas - Seaman - Number: 325 - Prize name: Leo - Ship type: P - How taken: HM Frigate Tiber - When taken: 11 Mar 1815 - Where taken: Lat 45.24N Long 12.3W - Date received: 28 Mar 1815 - From what ship: HMS Tiber - Discharged on 27 Apr 1815 and released to the Cartel Minework.

Pease, Richard - Seaman - Number: 722 - Prize name: Sine Qua Non - Ship type: P - How taken: HM Brig Elk - When taken: 20 Feb 1815 - Where taken: off Madeira - Date received: 29 Aug 1815 - From what ship: HMS Tonnant - Born: Massachusetts - Age: 40 - Discharged on 2 Jun 1815 and released to the Cartel Shakespeare.

Pendall, James - Seaman - Number: 426 - Prize name: Engineer - Ship type: MV - How taken: HMS Murros - When taken: 21 Sep 1814 - Where taken: off Porto Rico - Date received: 16 Apr 1815 - From what ship: HMS Swifsure - Born: Massachusetts - Age: 22 - Discharged on 2 Jun 1815 and released to the Cartel Sovereign.

Perkins, Clement - Seaman - Number: 160 - Prize name: Nancy, prize of the Privateer United States - Ship type: P - How taken: HM Sloop Papillon - When taken: 27 Dec 1814 - Where taken: Lat 30 Long 10 - Date received: 27 Feb 1815 - From what ship: HMS Nimble - Born: Massachusetts - Age: 21 - Discharged on 3 Mar 1815 and sent to Dartmoor.

Perkins, Lewis - Seaman - Number: 728 - Prize name: George Little - Ship type: P - How taken: HM Frigate Granicus - When taken: 20 Jan 1815 - Where taken: off Cape Finisterre, Spain - Date received: 29 Aug 1815 - From what ship: HMS Tonnant - Born: New Orleans - Age: 36 - Discharged on 2 Jun 1815 and released to the Cartel Shakespeare.

Perry, Ebenezer - Marine - Number: 74 - Prize name: Prince de Neufchatel - Ship type: P - How taken: HM Ship-of-the-Line Leander, HM Ship-of-the-Line Newcastle & HM Frigate Acasta - When taken: 28 Dec 1814 - Where taken: Lat 35N Long 52W - Date received: 18 Feb 1815 - From what ship: HMS Sybille - Born: Cambridge - Age: 19 - Discharged on 19 Feb 1815 and sent to Dartmoor.

Perry, John - Mate - Number: 754 - Prize name: Aurora - Ship type: MV - How taken: HM Frigate Andromeda - When taken: 28 Feb 1815 - Where taken: off Lisbon - Date received: 29 Aug 1815 - From what ship: HMS Tonnant - Age: 26 - Discharged on 2 Jun 1815 and released to the Cartel Shakespeare.

Perry, Miles - Seaman - Number: 406 - Prize name: Leo - Ship type: P - How taken: HM Frigate Tiber - When taken: 11 Mar 1815 - Where taken: Lat 45.24N Long 12.3W - Date received: 29 Mar 1815 - From what ship: HMS Tiber - Born: Massachusetts - Age: 25 - Discharged on 30 May 1815 and released to the Cartel Atlas.

Perez, Emanuel - Seaman - Number: 677 - Prize name: Eliza of Liverpool, prize of the Privateer Zealous - Ship type: P - How taken: Sent by HM Frigate President to the hospital - Where taken: coast of America - Date received: 2 May 1815 - From what ship: Royal Naval Hospital Plymouth - Age: 20 - Discharged on 2 Jun 1815 and released to the Cartel Shakespeare.

Peters, John - Seaman - Number: 286 - How taken: Taken off the HM Ship-of-the-Line Severn and other ships - Date received: 2 Mar 1815 - From what ship: HMS Dannemark - Race: Mulatto - Discharged on 3 Mar 1815 and sent to Dartmoor.

Peters, Rolls - Seaman - Number: 734 - Prize name: Sine Qua Non - Ship type: P - How taken: HM Brig Elk - When taken: 20 Feb 1815 - Where taken: off Madeira - Date received: 29 Aug 1815 - From what ship: HMS Tonnant - Born: Massachusetts - Age: 58 - Discharged on 2 Jun 1815 and released to the Cartel Shakespeare.

Peterson, Alexander - Seaman - Number: 32 - How taken: Impressed at Greenock - Date received: 8 Feb 1815 - From what ship: HMS Impregnable - Born: Maryland - Age: 25 - Race: Black - Discharged on 10 Feb 1815 and sent to Dartmoor.

Peterson, David - Seaman - Number: 103 - Prize name: Prince de Neufchatel - Ship type: P - How taken: HM Ship-of-the-Line Leander, HM Ship-of-the-Line Newcastle & HM Frigate Acasta - When taken: 28 Dec 1814 - Where taken: Lat 35N Long 52W - Date received: 18 Feb 1815 - From what ship: HMS Sybille - Born: Wiscasset - Age: 21 - Discharged on 19 Feb 1815 and sent to Dartmoor.

Peterson, John - Seaman - Number: 33 - How taken: Taken up at Plymouth - When taken: 1 Feb 1815 - Date received: 8 Feb 1815 - From what ship: HMS Impregnable - Born: Albany - Age: 31 - Discharged on 10 Feb 1815 and sent to Dartmoor.

Peterson, Thomas - Seaman - Number: 293 - How taken: Taken off the HM Ship-of-the-Line Severn and other ships - Date received: 2 Mar 1815 - From what ship: HMS Dannemark - Born: Norfolk - Age: 25 - Race: Black - Discharged on 3 Mar 1815 and sent to Dartmoor.

Phillips, Jacob - Seaman - Number: 575 - Prize name: Sorin, prize to Privateer Prince of Neufchatel - Ship type: P - How taken: HM Ship-of-the-Line Medway - When taken: 12 Jul 1814 - Where taken: Unknown - Date received: 16 Apr 1815 - From what ship: HMS Swifsure - Born: New York - Age: 17 - Discharged on 27 May 1815.

Phillips, John - Seaman - Number: 236 - How taken: Gave himself up - Date received: 2 Mar 1815 - From what ship: HMS Dannemark - Born: Connecticut - Age: 22 - Discharged on 3 Mar 1815 and sent to Dartmoor.

Phillips, Mathew - Seaman - Number: 788 - Prize name: US Brig Syren - Ship type: War - How taken: HM Ship-of-the-Line Medway - When taken: 12 Jul 1814 - Where taken: off Cape of Good Hope - Date received: 11 Jul 1815 - From what ship: HMS Royal Sovereign - Born: Not listed - Discharged on 11 Jul 1815 and released to the Cartel Wooddrop Sims.

Phillips, William - Pilot - Number: 224 - Prize name: Decatur - Ship type: P - How taken: HM Frigate Rhin - When taken: 4 Jun 1814 - Where taken: at sea - Date received: 2 Mar 1815 - From what ship: HMS Dannemark - Born: Philadelphia - Age: 21 - Discharged on 3 Mar 1815 and sent to Dartmoor.

Pike, John - Seaman - Number: 706 - Prize name: Sine Qua Non - Ship type: P - How taken: HM Brig Elk - When taken: 20 Feb 1815 - Where taken: off Madeira - Date received: 16 Jul 1815 - From what ship: HMS Hope - Age: 22 - Discharged on 2 Jun 1815 and released to the Cartel Shakespeare.

Pinchmas, John - Seaman - Number: 576 - Prize name: Sorin, prize to Privateer Prince of Neufchatel - Ship type: P - How taken: HM Ship-of-the-Line Medway - When taken: 12 Jul 1814 - Where taken: Unknown - Date received: 16 Apr 1815 - From what ship: HMS Swifsure - Born: Boston - Age: 15 - Discharged on 2 Jun 1815 and released to the Cartel Sovereign.

Pendleton, C. L. - Seaman - Number: 621 - Prize name: Sine Qua Non - Ship type: P - How taken: HM Brig Elk - When taken: 20 Feb 1815 - Where taken: off Madeira - Date received: 18 Apr 1815 - From what ship: HMS Euryalus - Age: 26 - Discharged on 2 Jun 1815 and released to the Cartel Sovereign.

Pitts, Francis - Seaman - Number: 17 - Prize name: Steven Getard - Ship type: MV - How taken: Providence, privateer - When taken: 18 Dec 1814 - Date received: 3 Feb 1815 - From what ship: HMS Bittern - Born: Philadelphia - Age: 24 - Race: Black - Discharged on 4 Feb 1815 and sent to Dartmoor.

Pitts, George - Seaman - Number: 24 - Prize name: Plutarch - Ship type: MV - How taken: HM Schooner Helicon - When taken: 5 Feb 1815 - Where taken: Lat 45.5N Long 7W - Date received: 8 Feb 1815 - From what ship: HMS Helicon - Born: Newport - Age: 35 - Discharged on 10 Feb 1815 and sent to Dartmoor.

Plenighan, Thomas - Seaman - Number: 776 - Prize name: George Little - Ship type: P - How taken: HM Frigate Granicus - When taken: 20 Jan 1815 - Where taken: off Cape Finisterre, Spain - Date received: 29 Aug 1815 - From what ship: HMS Tonnant - Age: 38 - Discharged on 2 Jun 1815 and released to the Cartel Shakespeare.

Plester, Stephen - Seaman - Number: 528 - Prize name: Gallant - Ship type: MV - How taken: HM Frigate Barbadoes - When taken: 6 Dec 1815 - Where taken: off St. Bartholomew - Date received: 16 Apr 1815 - From what ship: HMS Swifsure - Born: Baltimore - Age: 25 - Discharged on 2 Jun 1815 and released to the Cartel Sovereign.

Plumber, James - Seaman - Number: 164 - Prize name: Nancy, prize of the Privateer United States - Ship type: P - How taken: HM Sloop Papillon - When taken: 27 Dec 1814 - Where taken: Lat 30 Long 10 - Date received: 27 Feb 1815 - From what ship: HMS Nimble - Born: Bristol - Age: 26 - Discharged on 3 Mar 1815 and sent to Dartmoor.

Pockmitt, Jackness - Seaman - Number: 202 - Prize name: Sennett, Swedish merchant vessel - Ship type: MV - How taken: HM Frigate Rhin - When taken: Not known - Date received: 2 Mar 1815 - From what ship: HMS Dannemark - Race: Negro - Discharged on 3 Mar 1815 and sent to Dartmoor.

Poen, William - Seaman - Number: 834 - Prize name: US Brig Syren - Ship type: War - How taken: HM Ship-of-the-Line Medway - When taken: 12 Jul 1814 - Where taken: off Cape of Good Hope - Date received: 11 Jul 1815 - From what ship: HMS Royal Sovereign - Born: Not listed - Discharged on 11 Jul 1815 and released to the Cartel Wooddrop Sims.

Pollard, John - Seaman - Number: 359 - Prize name: Leo - Ship type: P - How taken: HM Frigate Tiber - When taken: 11 Mar 1815 - Where taken: Lat 45.24N Long 12.3W - Date received: 29 Mar 1814 - From what ship: HMS Tiber - Born: Massachusetts - Age: 39 - Discharged on 2 May 1815 and released to the Cartel Ariel.

Pope, Alexander - Cook - Number: 151 - Prize name: US Brig Syren - Ship type: War - How taken: HM Ship-of-the-Line Medway - When taken: 12 Jul 1814 - Where taken: off Cape of Good Hope - Date received: 21 Feb 1815 - From what ship: HMS Slaney - Discharged on 24 Feb 1815 and sent to Dartmoor.

Porce, John - Seaman - Number: 797 - Prize name: US Brig Syren - Ship type: War - How taken: HM Ship-of-the-Line Medway - When taken: 12 Jul 1814 - Where taken: off Cape of Good Hope - Date received: 11 Jul 1815 - From what ship: HMS Royal Sovereign - Born: Not listed - Discharged on 11 Jul 1815 and released to the Cartel Wooddrop Sims.

Potter, Isaac - Seaman - Number: 595 - Prize name: George Little - Ship type: P - How taken: HM Frigate Granicus - When taken: 20 Jan 1815 - Where taken: off Cape Finisterre, Spain - Date received: 18 Apr 1815 - From what ship: HMS Euryalus - Born: New Hampshire - Age: 21 - Race: Mulatto - Discharged on 2 Jun 1815 and released to the Cartel Sovereign.

Potter, James - Seaman - Number: 537 - Prize name: Engineer - Ship type: MV - How taken: HMS Muros - When taken: 21 Sep 1814 - Where taken: off Porto Rico - Date received: 16 Apr 1815 - From what ship: HMS Swifsure - Age: 26 - Discharged on 2 Jun 1815 and released to the Cartel Sovereign.

Pressey, Joseph - Seaman - Number: 387 - Prize name: Leo - Ship type: P - How taken: HM Frigate Tiber - When taken: 11 Mar 1815 - Where taken: Lat 45.24N Long 12.3W - Date received: 29 Mar 1814 - From what ship: HMS Tiber - Born: Massachusetts - Age: 27 - Discharged on 30 May 1815 and released to the Cartel Atlas.

Price, Peter - Seaman - Number: 314 - Prize name: Transit - Ship type: MV - How taken: Impressed at North Shields - When taken: 24 Nov 1814 - Where taken: North Shields - Date received: 4 Mar 1814 - From what ship: HM Cutter Surley - Discharged on 7 Mar 1815 and sent to Dartmoor.

Prince, Samuel - Seaman - Number: 377 - Prize name: Leo - Ship type: P - How taken: HM Frigate Tiber - When taken: 11 Mar 1815 - Where taken: Lat 45.24N Long 12.3W - Date received: 29 Mar 1814 - From what ship: HMS Tiber - Born: Massachusetts - Age: 22 - Discharged on 2 May 1815 and released to the Cartel Ariel.

Princess, George - Seaman - Number: 694 - Prize name: Sine Qua Non - Ship type: P - How taken: HM Brig Elk - When taken: 20 Feb 1815 - Where taken: off Madeira - Date received: 16 Jul 1815 - From what ship: HMS Hope - Born: Rhode Island - Age: 42 - Discharged on 2 Jun 1815 and released to the Cartel Shakespeare.

Putman, Joseph - Seaman - Number: 822 - Prize name: US Brig Syren - Ship type: War - How taken: HM Ship-of-the-Line Medway - When taken: 12 Jul 1814 - Where taken: off Cape of Good Hope - Date received: 11 Jul 1815 - From what ship: HMS Royal Sovereign - Born: Not listed - Discharged on 11 Jul 1815 and released to the Cartel Wooddrop Sims.

Quague, Samuel - Servant - Number: 863 - Prize name: US Brig Syren - Ship type: War - How taken: HM Ship-of-the-Line Medway - When taken: 12 Jul 1814 - Where taken: off Cape of Good Hope - Date received: 11 Jul 1815 - From what ship: from Portsmouth - Born: Not listed - Discharged on 11 Jul 1815 and released to the Cartel Wooddrop Sims.

Quinn, Patrick - Seaman - Number: 824 - Prize name: US Brig Syren - Ship type: War - How taken: HM Ship-of-the-Line Medway - When taken: 12 Jul 1814 - Where taken: off Cape of Good Hope - Date received: 11 Jul 1815 - From what ship: HMS Royal Sovereign - Born: Not listed - Discharged on 11 Jul 1815 and released to the Cartel Wooddrop Sims.

Rake, Martin - Seaman - Number: 168 - How taken: Impressed from MV Rolla - Date received: 1 Mar 1815 - From what ship: Royal Naval Hospital Plymouth - Age: 30 - Discharged on 3 Mar 1815 and sent to Dartmoor.

Ramsdell, Charles - Seaman - Number: 260 - Prize name: Dorothy - Ship type: MV - How taken: HM Frigate North Star - When taken: 15 Nov 1814 - Where taken: at sea - Date received: 2 Mar 1815 - From what ship: HMS

Dannemark - Discharged on 3 Mar 1815 and sent to Dartmoor.

Ranbre, Mathew - Seaman - Number: 71 - Prize name: Prince de Neufchatel - Ship type: P - How taken: HM Ship-of-the-Line Leander, HM Ship-of-the-Line Newcastle & HM Frigate Acasta - When taken: 28 Dec 1814 - Where taken: Lat 35N Long 52W - Date received: 18 Feb 1815 - From what ship: HMS Sybille - Age: 25 - Discharged on 19 Feb 1815 and sent to Dartmoor.

Randell, Henry - Seaman - Number: 814 - Prize name: US Brig Syren - Ship type: War - How taken: HM Ship-of-the-Line Medway - When taken: 12 Jul 1814 - Where taken: off Cape of Good Hope - Date received: 11 Jul 1815 - From what ship: HMS Royal Sovereign - Born: Not listed - Discharged on 11 Jul 1815 and released to the Cartel Wooddrop Sims.

Randell, Thomas - Seaman - Number: 170 - Prize name: Enterprise - Ship type: P - How taken: Not known - Where taken: at sea - Date received: 2 Mar 1815 - From what ship: HMS Dannemark - Born: Rochester - Age: 23 - Discharged on 3 Mar 1815 and sent to Dartmoor.

Randolph, Henry - Seaman - Number: 564 - Prize name: Mary - Ship type: MV - How taken: HMS Muros - When taken: 8 Dec 1814 - Where taken: West Indies - Date received: 16 Apr 1815 - From what ship: HMS Swifsure - Born: Baltimore - Age: 24 - Discharged on 2 Jun 1815 and released to the Cartel Sovereign.

Randolph, Samuel - Seaman - Number: 802 - Prize name: US Brig Syren - Ship type: War - How taken: HM Ship-of-the-Line Medway - When taken: 12 Jul 1814 - Where taken: off Cape of Good Hope - Date received: 11 Jul 1815 - From what ship: HMS Royal Sovereign - Born: Not listed - Discharged on 11 Jul 1815 and released to the Cartel Wooddrop Sims.

Rankins, William - Seaman - Number: 805 - Prize name: US Brig Syren - Ship type: War - How taken: HM Ship-of-the-Line Medway - When taken: 12 Jul 1814 - Where taken: off Cape of Good Hope - Date received: 11 Jul 1815 - From what ship: HMS Royal Sovereign - Born: Not listed - Discharged on 11 Jul 1815 and released to the Cartel Wooddrop Sims.

Ray, Gideon - Seaman - Number: 330 - Prize name: Leo - Ship type: P - How taken: HM Frigate Tiber - When taken: 11 Mar 1815 - Where taken: Lat 45.24N Long 12.3W - Date received: 28 Mar 1815 - From what ship: HMS Tiber - Discharged on 27 Apr 1815 and released to the Cartel Minework.

Ray, Richard - Seaman - Number: 23 - Prize name: Plutarch - Ship type: MV - How taken: HM Schooner Helicon - When taken: 5 Feb 1815 - Where taken: Lat 45.5N Long 7W - Date received: 8 Feb 1815 - From what ship: HMS Helicon - Born: Newland - Age: 30 - Discharged on 10 Feb 1815 and sent to Dartmoor.

Raymond, John - Cook - Number: 78 - Prize name: Prince de Neufchatel - Ship type: P - How taken: HM Ship-of-the-Line Leander, HM Ship-of-the-Line Newcastle & HM Frigate Acasta - When taken: 28 Dec 1814 - Where taken: Lat 35N Long 52W - Date received: 18 Feb 1815 - From what ship: HMS Sybille - Born: New Orleans - Age: 24 - Race: Black - Discharged on 19 Feb 1815 and sent to Dartmoor.

Reed, Aug. - Seaman - Number: 380 - Prize name: Leo - Ship type: P - How taken: HM Frigate Tiber - When taken: 11 Mar 1815 - Where taken: Lat 45.24N Long 12.3W - Date received: 29 Mar 1814 - From what ship: HMS Tiber - Born: Massachusetts - Age: 26 - Discharged on 30 May 1815 and released to the Cartel Atlas.

Reed, George - Seaman - Number: 816 - Prize name: US Brig Syren - Ship type: War - How taken: HM Ship-of-the-Line Medway - When taken: 12 Jul 1814 - Where taken: off Cape of Good Hope - Date received: 11 Jul 1815 - From what ship: HMS Royal Sovereign - Born: Not listed - Discharged on 11 Jul 1815 and released to the Cartel Wooddrop Sims.

Reed, Lewis - Seaman - Number: 626 - Prize name: Helene, prize of Privateer Morgianna - Ship type: P - How taken: HM Frigate Pique - When taken: 7 Mar 1815 - Where taken: coast of America - Date received: 20 Apr 1815 - From what ship: HMS Musquito - Born: Rhode Island - Age: 17 - Discharged on 2 Jun 1815 and released to the Cartel Sovereign.

Rhodes, Charles H. - Seaman - Number: 811 - Prize name: US Brig Syren - Ship type: War - How taken: HM Ship-of-the-Line Medway - When taken: 12 Jul 1814 - Where taken: off Cape of Good Hope - Date received: 11 Jul 1815 - From what ship: HMS Royal Sovereign - Born: Not listed - Discharged on 11 Jul 1815 and released to the Cartel Wooddrop Sims.

Rich, John - Lieutenant - Number: 634 - Prize name: Avon - Ship type: P - How taken: HM Frigate Barbadoes -

When taken: 8 Mar 1815 - Where taken: West Indies - Date received: 24 Apr 1815 - From what ship: HMS Bellerophon - Age: 26 - Discharged on 2 Jun 1815 and released to the Cartel Shakespeare.

Richmond, Alpheus - Seaman - Number: 600 - Prize name: George Little - Ship type: P - How taken: HM Frigate Granicus - When taken: 20 Jan 1815 - Where taken: off Cape Finisterre, Spain - Date received: 18 Apr 1815 - From what ship: HMS Euryalus - Born: Massachusetts - Age: 33 - Discharged on 2 Jun 1815 and released to the Cartel Sovereign.

Richmond, Caleb - Seaman - Number: 294 - How taken: Taken off the HM Ship-of-the-Line Severn and other ships - Date received: 2 Mar 1815 - From what ship: HMS Dannemark - Born: Philadelphia - Age: 34 - Discharged on 3 Mar 1815 and sent to Dartmoor.

Rickmes, Rub. - Seaman - Number: 420 - Prize name: Mary - Ship type: LM - How taken: HM Frigate Erne - When taken: 7 Jul 1814 - Where taken: off Lezard - Date received: 16 Apr 1815 - From what ship: HMS Swifsure - Born: Pennsylvania - Age: 24 - Discharged on 8 May 1815.

Ridgeway, Ebenezer - Midshipman - Number: 856 - Prize name: US Brig Syren - Ship type: War - How taken: HM Ship-of-the-Line Medway - When taken: 12 Jul 1814 - Where taken: off Cape of Good Hope - Date received: 11 Jul 1815 - From what ship: from Portsmouth - Born: Not listed - Discharged on 11 Jul 1815 and released to the Cartel Wooddrop Sims.

Ripley, Daniel - Prize Master - Number: 586 - Prize name: George Little - Ship type: P - How taken: HM Frigate Granicus - When taken: 20 Jan 1815 - Where taken: off Cape Finisterre, Spain - Date received: 18 Apr 1815 - From what ship: HMS Euryalus - Born: Massachusetts - Age: 33 - Discharged on 2 Jun 1815 and released to the Cartel Sovereign.

Roberts, James - Seaman - Number: 282 - How taken: Taken off the HM Ship-of-the-Line Severn and other ships - Date received: 2 Mar 1815 - From what ship: HMS Dannemark - Born: Wilmington - Age: 34 - Race: Negro - Discharged on 3 Mar 1815 and sent to Dartmoor.

Robins, Anselm - Seaman - Number: 357 - Prize name: Leo - Ship type: P - How taken: HM Frigate Tiber - When taken: 11 Mar 1815 - Where taken: Lat 45.24N Long 12.3W - Date received: 29 Mar 1814 - From what ship: HMS Tiber - Born: Massachusetts - Age: 23 - Discharged on 27 Apr 1815 and released to the Cartel Minework.

Robins, John - Seaman - Number: 729 - Prize name: George Little - Ship type: P - How taken: HM Frigate Granicus - When taken: 20 Jan 1815 - Where taken: off Cape Finisterre, Spain - Date received: 29 Aug 1815 - From what ship: HMS Tonnant - Born: New Orleans - Age: 20 - Discharged on 2 Jun 1815 and released to the Cartel Shakespeare.

Robinson, William - Seaman - Number: 27 - Prize name: Plutarch - Ship type: MV - How taken: HM Schooner Helicon - When taken: 5 Feb 1815 - Where taken: Lat 45.5N Long 7W - Date received: 8 Feb 1815 - From what ship: HMS Helicon - Born: Philadelphia - Age: 40 - Discharged on 10 Feb 1815 and sent to Dartmoor.

Robinson, William - Seaman & Passenger - Number: 416 - Prize name: W. S. Many - Ship type: MV - How taken: HM Brig Reynard - When taken: 4 Feb 1814 - Where taken: off Cadiz - Date received: 2 Apr 1815 - From what ship: HMS Basilisk - Age: 20 - Discharged on 2 Jun 1815 and released to the Cartel Sovereign.

Robinson, William - Seaman - Number: 431 - Prize name: Engineer - Ship type: MV - How taken: HMS Murros - When taken: 21 Sep 1814 - Where taken: off Porto Rico - Date received: 16 Apr 1815 - From what ship: HMS Swifsure - Born: New York - Age: 27 - Discharged on 2 Jun 1815 and released to the Cartel Shakespeare.

Rodmond, Moses - Seaman - Number: 569 - Prize name: Heron - Ship type: MV - How taken: HM Frigate Pique - When taken: 16 Dec 1814 - Where taken: off Porto Rico - Date received: 16 Apr 1815 - From what ship: HMS Swifsure - Born: Rhode Island - Age: 17 - Race: Negro - Discharged on 2 Jun 1815 and released to the Cartel Sovereign.

Rogers, Smith - Seaman - Number: 427 - Prize name: Engineer - Ship type: MV - How taken: HMS Murros - When taken: 21 Sep 1814 - Where taken: off Porto Rico - Date received: 16 Apr 1815 - From what ship: HMS Swifsure - Born: Massachusetts - Age: 23 - Discharged on 2 Jun 1815 and released to the Cartel Sovereign.

Rondel, N. - Seaman - Number: 440 - Prize name: Nancy - Ship type: P - How taken: HM Frigate Amelia - When taken: 6 Sep 1814 - Where taken: off Newfoundland - Date received: 16 Apr 1815 - From what ship: HMS Swifsure - Born: New Orleans - Age: 50 - Discharged on 24 Apr 1815.

Roper, Nathaniel - Seaman - Number: 835 - Prize name: US Brig Syren - Ship type: War - How taken: HM Ship-of-the-Line Medway - When taken: 12 Jul 1814 - Where taken: off Cape of Good Hope - Date received: 11 Jul 1815 - From what ship: HMS Royal Sovereign - Born: Not listed - Discharged on 11 Jul 1815 and released to the Cartel Wooddrop Sims.

Rosenburg, Charles - Seaman - Number: 310 - Prize name: Transit - Ship type: MV - How taken: Impressed at North Shields - When taken: 24 Nov 1814 - Where taken: North Shields - Date received: 4 Mar 1814 - From what ship: HM Cutter Surley - Discharged on 7 Mar 1815 and sent to Dartmoor.

Ross, George - Seaman - Number: 730 - Prize name: Sine Qua Non - Ship type: P - How taken: HM Brig Elk - When taken: 20 Feb 1815 - Where taken: off Madeira - Date received: 29 Aug 1815 - From what ship: HMS Tonnant - Born: New Orleans - Age: 64 - Discharged on 2 Jun 1815 and released to the Cartel Shakespeare.

Roundy, John - Seaman - Number: 808 - Prize name: US Brig Syren - Ship type: War - How taken: HM Ship-of-the-Line Medway - When taken: 12 Jul 1814 - Where taken: off Cape of Good Hope - Date received: 11 Jul 1815 - From what ship: HMS Royal Sovereign - Born: Not listed - Discharged on 11 Jul 1815 and released to the Cartel Wooddrop Sims.

Rue, Samuel - Pilot - Number: 421 - Prize name: Engineer - Ship type: MV - How taken: HMS Murros - When taken: 21 Sep 1814 - Where taken: off Porto Rico - Date received: 16 Apr 1815 - From what ship: HMS Swifsure - Born: New York - Age: 24 - Discharged on 2 Jun 1815 and released to the Cartel Sovereign.

Rundell, William - Seaman - Number: 607 - Prize name: George Little - Ship type: P - How taken: HM Frigate Granicus - When taken: 20 Jan 1815 - Where taken: off Cape Finisterre, Spain - Date received: 18 Apr 1815 - From what ship: HMS Euryalus - Born: Massachusetts - Age: 41 - Discharged on 2 Jun 1815 and released to the Cartel Sovereign.

Rupey, Luther - Seaman - Number: 531 - Prize name: Gallant - Ship type: MV - How taken: HM Frigate Barbadoes - When taken: 6 Dec 1815 - Where taken: off St. Bartholomew - Date received: 16 Apr 1815 - From what ship: HMS Swifsure - Born: Massachusetts - Age: 34 - Discharged on 8 May 1815.

Rust, John - Seaman - Number: 319 - Prize name: Maid of the Mill, prized of the Privateer President - Ship type: P - How taken: HM Sloop Calypso - When taken: 25 Feb 1815 - Where taken: off Lisbon - Date received: 24 Mar 1815 - From what ship: HMS Hyperion - Discharged on 27 Apr 1815 and released to the Cartel Minework.

Rust, Zebulon - Seaman - Number: 153 - Prize name: US Brig Syren - Ship type: War - How taken: HM Ship-of-the-Line Medway - When taken: 12 Jul 1814 - Where taken: off Cape of Good Hope - Date received: 21 Feb 1815 - From what ship: HMS Slaney - Born: Massachusetts - Age: 29 - Discharged on 24 Feb 1815 and sent to Dartmoor.

Rutledge, James - Seaman - Number: 533 - Prize name: St. Francis - Ship type: MV - How taken: HMS Ister - When taken: 8 Dec 1814 - Where taken: Amelia Island, Florida - Date received: 16 Apr 1815 - From what ship: HMS Swifsure - Born: New Hampshire - Age: 30 - Discharged on 2 Jun 1815 and released to the Cartel Sovereign.

Ryder, M. - Seaman - Number: 620 - Prize name: Sine Qua Non - Ship type: P - How taken: HM Brig Elk - When taken: 20 Feb 1815 - Where taken: off Madeira - Date received: 18 Apr 1815 - From what ship: HMS Euryalus - Born: North Caroline - Age: 23 - Discharged on 2 Jun 1815 and released to the Cartel Sovereign.

Salisbury, James - Seaman - Number: 201 - Prize name: Sennett, Swedish merchant vessel - Ship type: MV - How taken: HM Frigate Rhin - When taken: Not known - Date received: 2 Mar 1815 - From what ship: HMS Dannemark - Discharged on 3 Mar 1815 and sent to Dartmoor.

Samson, Capes - Seaman - Number: 373 - Prize name: Leo - Ship type: P - How taken: HM Frigate Tiber - When taken: 11 Mar 1815 - Where taken: Lat 45.24N Long 12.3W - Date received: 29 Mar 1814 - From what ship: HMS Tiber - Age: 21 - Discharged on 2 May 1815 and released to the Cartel Ariel.

Samuels, Samuel - Seaman - Number: 109 - Prize name: Prince de Neufchatel - Ship type: P - How taken: HM Ship-of-the-Line Leander, HM Ship-of-the-Line Newcastle & HM Frigate Acasta - When taken: 28 Dec 1814 - Where taken: Lat 35N Long 52W - Date received: 18 Feb 1815 - From what ship: HMS Sybille - Race: Black - Discharged on 19 Feb 1815 and sent to Dartmoor.

Sanman, Jan - Seaman - Number: 604 - Prize name: George Little - Ship type: P - How taken: HM Frigate Granicus - When taken: 20 Jan 1815 - Where taken: off Cape Finisterre, Spain - Date received: 18 Apr 1815 - From what ship: HMS Euryalus - Born: Massachusetts - Age: 27 - Discharged on 2 Jun 1815 and released to the Cartel Sovereign.

Sanster, John - Seaman - Number: 90 - Prize name: Prince de Neufchatel - Ship type: P - How taken: HM Ship-of-the-Line Leander, HM Ship-of-the-Line Newcastle & HM Frigate Acasta - When taken: 28 Dec 1814 - Where taken: Lat 35N Long 52W - Date received: 18 Feb 1815 - From what ship: HMS Sybille - Discharged on 19 Feb 1815 and sent to Dartmoor.

Scargle, William - Seaman - Number: 541 - Prize name: Engineer - Ship type: MV - How taken: HMS Muros - When taken: 21 Sep 1814 - Where taken: off Porto Rico - Date received: 16 Apr 1815 - From what ship: HMS Swifsure - Born: Philadelphia - Age: 26 - Discharged on 28 Jul 1815.

Scott, Edward - Boy & Passenger - Number: 417 - Prize name: W. S. Many - Ship type: MV - How taken: HM Brig Reynard - When taken: 4 Feb 1814 - Where taken: off Cadiz - Date received: 2 Apr 1815 - From what ship: HMS Basilisk - Age: 16 - Discharged on 2 Jun 1815 and released to the Cartel Sovereign.

Seaves, Louis - Seaman - Number: 434 - Prize name: Mars, prize of the Privateer David Porter - Ship type: P - How taken: HM Frigate Pique - When taken: 12 Aug 1814 - Where taken: off Newfoundland - Date received: 16 Apr 1815 - From what ship: HMS Swifsure - Born: Massachusetts - Age: 23 - Discharged on 2 Jun 1815 and released to the Cartel Sovereign.

Sewell, John - Seaman - Number: 312 - Prize name: Transit - Ship type: MV - How taken: Impressed at North Shields - When taken: 24 Nov 1814 - Where taken: North Shields - Date received: 4 Mar 1814 - From what ship: HM Cutter Surley - Discharged on 7 Mar 1815 and sent to Dartmoor.

Shawley, Lewis - Seaman - Number: 801 - Prize name: US Brig Syren - Ship type: War - How taken: HM Ship-of-the-Line Medway - When taken: 12 Jul 1814 - Where taken: off Cape of Good Hope - Date received: 11 Jul 1815 - From what ship: HMS Royal Sovereign - Born: Not listed - Discharged on 11 Jul 1815 and released to the Cartel Wooddrop Sims.

Shelton, C. J. - Seaman - Number: 486 - Prize name: Fox - Ship type: P - How taken: HM Frigate Barbadoes - When taken: 11 Jan 1815 - Where taken: off Amelia Island, Florida - Date received: 16 Apr 1815 - From what ship: HMS Swifsure - Born: Virginia - Age: 27 - Discharged on 2 Jun 1815 and released to the Cartel Sovereign.

Shepherd, John - Seaman - Number: 649 - Prize name: Avon - Ship type: P - How taken: HM Frigate Barbadoes - When taken: 8 Mar 1815 - Where taken: West Indies - Date received: 24 Apr 1815 - From what ship: HMS Bellerophon - Born: Massachusetts - Age: 22 - Discharged on 1 Jun 1815.

Shepherd, John - Seaman - Number: 800 - Prize name: US Brig Syren - Ship type: War - How taken: HM Ship-of-the-Line Medway - When taken: 12 Jul 1814 - Where taken: off Cape of Good Hope - Date received: 11 Jul 1815 - From what ship: HMS Royal Sovereign - Born: Not listed - Discharged on 11 Jul 1815 and released to the Cartel Wooddrop Sims.

Sidney, Stephen - Seaman - Number: 777 - Prize name: George Little - Ship type: P - How taken: HM Frigate Granicus - When taken: 20 Jan 1815 - Where taken: off Cape Finisterre, Spain - Date received: 29 Aug 1815 - From what ship: HMS Tonnant - Born: Massachusetts - Age: 19 - Discharged on 2 Jun 1815 and released to the Cartel Shakespeare.

Simon, Nes. - Seaman - Number: 489 - Prize name: Fox - Ship type: P - How taken: HM Frigate Barbadoes - When taken: 11 Jan 1815 - Where taken: off Amelia Island, Florida - Date received: 16 Apr 1815 - From what ship: HMS Swifsure - Born: New York - Age: 22 - Race: Negro - Discharged on 2 Jun 1815 and released to the Cartel Sovereign.

Skipper, John - Seaman - Number: 208 - Prize name: Sennett, Swedish merchant vessel - Ship type: MV - How taken: HM Frigate Rhin - When taken: Not known - Date received: 2 Mar 1815 - From what ship: HMS Dannemark - Discharged on 3 Mar 1815 and sent to Dartmoor.

Small, John - Boy - Number: 407 - Prize name: Leo - Ship type: P - How taken: HM Frigate Tiber - When taken: 11 Mar 1815 - Where taken: Lat 45.24N Long 12.3W - Date received: 29 Mar 1815 - From what ship: HMS Tiber - Born: Massachusetts - Age: 13 - Discharged on 30 May 1815 and released to the Cartel Atlas.

Smet, John - Seaman - Number: 50 - How taken: Taken out of HM Ship-of-the-Line Cornwallis - Date received: 15 Feb 1815 - From what ship: HMS Myrmidon - Born: New York - Age: 21 - Race: Black - Discharged on 17 Feb

1815 and sent to Dartmoor.

Smith, Abraham - Seaman - Number: 686 - Prize name: George Little - Ship type: P - How taken: HM Frigate Granicus - When taken: 20 Jan 1815 - Where taken: off Cape Finisterre, Spain - Date received: 16 Jul 1815 - From what ship: HMS Hope - Born: Massachusetts - Age: 33 - Discharged on 2 Jun 1815 and released to the Cartel Shakespeare.

Smith, B. - Seaman - Number: 552 - Prize name: Hera - Ship type: MV - How taken: HM Frigate Pique - When taken: 16 Dec 1814 - Where taken: off Porto Rico - Date received: 16 Apr 1815 - From what ship: HMS Swifsure - Born: Baltimore - Age: 20 - Discharged on 2 Jun 1815 and released to the Cartel Sovereign.

Smith, Charles - Seaman - Number: 592 - Prize name: George Little - Ship type: P - How taken: HM Frigate Granicus - When taken: 20 Jan 1815 - Where taken: off Cape Finisterre, Spain - Date received: 18 Apr 1815 - From what ship: HMS Euryalus - Born: New Hampshire - Age: 19 - Discharged on 2 Jun 1815 and released to the Cartel Sovereign.

Smith, David - Seaman - Number: 698 - Prize name: Sine Qua Non - Ship type: P - How taken: HM Brig Elk - When taken: 20 Feb 1815 - Where taken: off Madeira - Date received: 16 Jul 1815 - From what ship: HMS Hope - Born: Massachusetts - Age: 21 - Discharged on 2 Jun 1815 and released to the Cartel Shakespeare.

Smith, John - Seaman - Number: 481 - Prize name: Fox - Ship type: P - How taken: HM Frigate Barbadoes - When taken: 11 Jan 1815 - Where taken: off Amelia Island, Florida - Date received: 16 Apr 1815 - From what ship: HMS Swifsure - Born: Maryland - Age: 24 - Discharged on 2 Jun 1815 and released to the Cartel Sovereign.

Smith, John - Seaman - Number: 199 - Prize name: Saucy Jack - Ship type: MV - How taken: Unknown - Date received: 2 Mar 1815 - From what ship: HMS Dannemark - Discharged on 3 Mar 1815 and sent to Dartmoor.

Smith, John - Seaman - Number: 372 - Prize name: Leo - Ship type: P - How taken: HM Frigate Tiber - When taken: 11 Mar 1815 - Where taken: Lat 45.24N Long 12.3W - Date received: 29 Mar 1814 - From what ship: HMS Tiber - Born: New Hampshire - Age: 45 - Discharged on 2 May 1815 and released to the Cartel Ariel.

Smith, John - Seaman - Number: 702 - Prize name: Sine Qua Non - Ship type: P - How taken: HM Brig Elk - When taken: 20 Feb 1815 - Where taken: off Madeira - Date received: 16 Jul 1815 - From what ship: HMS Hope - Born: Massachusetts - Age: 20 - Discharged on 2 Jun 1815 and released to the Cartel Shakespeare.

Smith, John - Seaman - Number: 443 - Prize name: Olio - Ship type: MV - How taken: HMS Muros - When taken: 13 Nov 1814 - Where taken: coast of America - Date received: 16 Apr 1815 - From what ship: HMS Swifsure - Age: 22 - Discharged on 24 Apr 1815.

Smith, John - Seaman - Number: 660 - Prize name: Avon - Ship type: P - How taken: HM Frigate Barbadoes - When taken: 8 Mar 1815 - Where taken: West Indies - Date received: 24 Apr 1815 - From what ship: HMS Bellerophon - Age: 29 - Discharged on 2 Jun 1815 and released to the Cartel Shakespeare.

Smith, Philip - Seaman - Number: 642 - Prize name: Avon - Ship type: P - How taken: HM Frigate Barbadoes - When taken: 8 Mar 1815 - Where taken: West Indies - Date received: 24 Apr 1815 - From what ship: HMS Bellerophon - Age: 28 - Discharged on 2 Jun 1815 and released to the Cartel Shakespeare.

Smith, Robert - Seaman - Number: 411 - Prize name: Thomas, prize to Privateer Scourge - Ship type: P - How taken: HM Frigate Aquilon - When taken: 11 Mar 1815 - Where taken: off Cape Finisterre, Spain - Date received: 29 Mar 1815 - From what ship: HMS Icarus - Born: Massachusetts - Age: 25 - Race: Negro - Discharged on 2 Jun 1815 and released to the Cartel Sovereign.

Smith, Samuel - Seaman - Number: 191 - Prize name: Chance - Ship type: P - How taken: HMS Statire - When taken: 1 Apr 1814 - Where taken: at sea - Date received: 2 Mar 1815 - From what ship: HMS Dannemark - Born: Norfolk - Age: 20 - Discharged on 3 Mar 1815 and sent to Dartmoor.

Smith, Samuel - Seaman - Number: 643 - Prize name: Avon - Ship type: P - How taken: HM Frigate Barbadoes - When taken: 8 Mar 1815 - Where taken: West Indies - Date received: 24 Apr 1815 - From what ship: HMS Bellerophon - Age: 27 - Discharged on 2 Jun 1815 and released to the Cartel Shakespeare.

Smith, William - Seaman - Number: 371 - Prize name: Leo - Ship type: P - How taken: HM Frigate Tiber - When taken: 11 Mar 1815 - Where taken: Lat 45.24N Long 12.3W - Date received: 29 Mar 1814 - From what ship: HMS Tiber - Born: Massachusetts - Age: 28 - Discharged on 2 May 1815 and released to the Cartel Ariel.

Snider, Enos - Seaman - Number: 809 - Prize name: US Brig Syren - Ship type: War - How taken: HM Ship-of-the-Line Medway - When taken: 12 Jul 1814 - Where taken: off Cape of Good Hope - Date received: 11 Jul 1815 - From what ship: HMS Royal Sovereign - Born: Not listed - Discharged on 11 Jul 1815 and released to the Cartel Wooddrop Sims.

Snow, Daniel - Marine - Number: 88 - Prize name: Prince de Neufchatel - Ship type: P - How taken: HM Ship-of-the-Line Leander, HM Ship-of-the-Line Newcastle & HM Frigate Acasta - When taken: 28 Dec 1814 - Where taken: Lat 35N Long 52W - Date received: 18 Feb 1815 - From what ship: HMS Sybille - Born: Litchfield - Age: 19 - Discharged on 19 Feb 1815 and sent to Dartmoor.

Snow, Godfrey - Seaman - Number: 695 - Prize name: Sine Qua Non - Ship type: P - How taken: HM Brig Elk - When taken: 20 Feb 1815 - Where taken: off Madeira - Date received: 16 Jul 1815 - From what ship: HMS Hope - Born: Massachusetts - Age: 19 - Discharged on 2 Jun 1815 and released to the Cartel Shakespeare.

Southcombe, Peter - 1st Mate - Number: 266 - Prize name: Netterville - Ship type: MV - How taken: HM Brig Onyx - When taken: 25 Dec 1814 - Where taken: at sea - Date received: 2 Mar 1815 - From what ship: HMS Dannemark - Born: Philadelphia - Age: 27 - Discharged on 3 Mar 1815 and sent to Dartmoor.

Sparkes, John - Gunner - Number: 141 - Prize name: US Brig Syren - Ship type: War - How taken: HM Ship-of-the-Line Medway - When taken: 12 Jul 1814 - Where taken: off Cape of Good Hope - Date received: 21 Feb 1815 - From what ship: HMS Slaney - Born: Philadelphia - Age: 36 - Discharged on 24 Feb 1815 and sent to Dartmoor.

Spear, Joseph - Seaman - Number: 307 - Prize name: US Brig Syren - Ship type: War - How taken: HM Ship-of-the-Line Medway - When taken: 12 Jul 1814 - Where taken: off Cape of Good Hope - Date received: 4 Mar 1814 - From what ship: HM Cutter Surley - Born: Boston - Age: 22 - Discharged on 7 Mar 1815 and sent to Dartmoor.

Spencer, Robert - Seaman - Number: 710 - Prize name: Sine Qua Non - Ship type: P - How taken: HM Brig Elk - When taken: 20 Feb 1815 - Where taken: off Madeira - Date received: 16 Jul 1815 - From what ship: HMS Hope - Age: 22 - Discharged on 2 Jun 1815 and released to the Cartel Shakespeare.

Sponarts, James - Seaman - Number: 807 - Prize name: US Brig Syren - Ship type: War - How taken: HM Ship-of-the-Line Medway - When taken: 12 Jul 1814 - Where taken: off Cape of Good Hope - Date received: 11 Jul 1815 - From what ship: HMS Royal Sovereign - Born: Not listed - Discharged on 11 Jul 1815 and released to the Cartel Wooddrop Sims.

Sprague, Benjamin - Lieutenant - Number: 611 - Prize name: Sine Qua Non - Ship type: P - How taken: HM Brig Elk - When taken: 20 Feb 1815 - Where taken: off Madeira - Date received: 18 Apr 1815 - From what ship: HMS Euryalus - Born: Massachusetts - Age: 24 - Discharged on 23 May 1815.

Spring, Friends - Seaman - Number: 744 - Prize name: Sine Qua Non - Ship type: P - How taken: HM Brig Elk - When taken: 20 Feb 1815 - Where taken: off Madeira - Date received: 29 Aug 1815 - From what ship: HMS Tonnant - Born: Massachusetts - Age: 22 - Discharged on 2 Jun 1815 and released to the Cartel Shakespeare.

Stacey, William - Seaman - Number: 771 - Prize name: George Little - Ship type: P - How taken: HM Frigate Granicus - When taken: 20 Jan 1815 - Where taken: off Cape Finisterre, Spain - Date received: 29 Aug 1815 - From what ship: HMS Tonnant - Age: 25 - Discharged on 2 Jun 1815 and released to the Cartel Shakespeare.

Stackpole, Nathaniel - Seaman - Number: 383 - Prize name: Leo - Ship type: P - How taken: HM Frigate Tiber - When taken: 11 Mar 1815 - Where taken: Lat 45.24N Long 12.3W - Date received: 29 Mar 1814 - From what ship: HMS Tiber - Born: Massachusetts - Age: 19 - Discharged on 30 May 1815 and released to the Cartel Atlas.

Stanton, William - Seaman - Number: 438 - Prize name: Acquilar - Ship type: P - How taken: HM Frigate Pique - When taken: 4 Sep 1814 - Where taken: off Newfoundland - Date received: 16 Apr 1815 - From what ship: HMS Swifsure - Born: Lyme - Age: 18 - Discharged on 2 Jun 1815 and released to the Cartel Sovereign.

Starks, Jand. - Seaman - Number: 451 - Prize name: High Flyer - Ship type: LM - How taken: HMS Muros - When taken: 14 Nov 1814 - Where taken: off St. Bartholomew - Date received: 16 Apr 1815 - From what ship: HMS Swifsure - Age: 32 - Discharged on 2 Jun 1815 and released to the Cartel Sovereign.

Stickney, E. F. - Prize Master - Number: 433 - Prize name: Mars, prize of the Privateer David Porter - Ship type: P - How taken: HM Frigate Pique - When taken: 12 Aug 1814 - Where taken: off Newfoundland - Date received: 16 Apr 1815 - From what ship: HMS Swifsure - Born: Salem - Age: 33 - Discharged on 2 Jun 1815 and released to the Cartel Sovereign.

Steel, John - Seaman - Number: 221 - Prize name: Decatur - Ship type: P - How taken: HM Frigate Rhin - When taken: 4 Jun 1814 - Where taken: at sea - Date received: 2 Mar 1815 - From what ship: HMS Dannemark - Born: Hertford - Age: 32 - Discharged on 3 Mar 1815 and sent to Dartmoor.

Stenson, David - Seaman - Number: 13 - Prize name: Philbes - Ship type: MV - How taken: Taken out of a Spanish schooner - When taken: 25 Nov 1814 - Where taken: Greenock - Date received: 3 Feb 1815 - From what ship: HMS Bittern - Born: Georgetown - Age: 24 - Discharged on 4 Feb 1815 and sent to Dartmoor.

Stevens, John - Seaman - Number: 653 - Prize name: Avon - Ship type: P - How taken: HM Frigate Barbadoes - When taken: 8 Mar 1815 - Where taken: West Indies - Date received: 24 Apr 1815 - From what ship: HMS Bellerophon - Born: Massachusetts - Age: 22 - Discharged on 2 Jun 1815 and released to the Cartel Shakespeare.

Stevens, John - Seaman - Number: 297 - Prize name: US Brig Syren - Ship type: War - How taken: HM Ship-of-the-Line Medway - When taken: 12 Jul 1814 - Where taken: off Cape of Good Hope - Date received: 4 Mar 1814 - From what ship: HM Cutter Surley - Born: Ipswich - Age: 25 - Discharged on 7 Mar 1815 and sent to Dartmoor.

Stevens, William - Seaman - Number: 18 - Prize name: Jane - Ship type: MV - How taken: Impressed at Liverpool - When taken: 25 Nov 1814 - Date received: 3 Feb 1815 - From what ship: HMS Bittern - Born: Newburyport - Age: 22 - Discharged on 4 Feb 1815 and sent to Dartmoor.

Stickney, Benjamin - Seaman - Number: 353 - Prize name: Leo - Ship type: P - How taken: HM Frigate Tiber - When taken: 11 Mar 1815 - Where taken: Lat 45.24N Long 12.3W - Date received: 29 Mar 1814 - From what ship: HMS Tiber - Born: Massachusetts - Age: 20 - Discharged on 27 Apr 1815 and released to the Cartel Minework.

Stock, Caleb - Seaman - Number: 783 - Prize name: Herald (New York0 - Ship type: P - How taken: HM Frigate Endymion - When taken: 15 Aug 1814 - Where taken: coast of America - Date received: 11 Jul 1815 - From what ship: Royal Naval Hospital Plymouth - Born: New London - Age: 25 - Discharged on 11 Jul 1815 and released to the Cartel Wooddrop Sims.

Stone, Benjamin - Seaman - Number: 773 - Prize name: George Little - Ship type: P - How taken: HM Frigate Granicus - When taken: 20 Jan 1815 - Where taken: off Cape Finisterre, Spain - Date received: 29 Aug 1815 - From what ship: HMS Tonnant - Age: 22 - Discharged on 2 Jun 1815 and released to the Cartel Shakespeare.

Stone, Jerry - Seaman - Number: 7 - Prize name: Albion - Ship type: MV - How taken: HM Schooner Helicon - When taken: 17 Jan 1815 - Where taken: Lat 20N Long 52W - Date received: 3 Feb 1815 - From what ship: HMS Bittern - Born: New Orleans - Age: 19 - Discharged on 4 Feb 1815 and sent to Dartmoor.

Stone, John - Seaman - Number: 300 - Prize name: US Brig Syren - Ship type: War - How taken: HM Ship-of-the-Line Medway - When taken: 12 Jul 1814 - Where taken: off Cape of Good Hope - Date received: 4 Mar 1814 - From what ship: HM Cutter Surley - Born: Gloucester - Age: 33 - Discharged on 7 Mar 1815 and sent to Dartmoor.

Story, Elias - Seaman - Number: 689 - Prize name: George Little - Ship type: P - How taken: HM Frigate Granicus - When taken: 20 Jan 1815 - Where taken: off Cape Finisterre, Spain - Date received: 16 Jul 1815 - From what ship: HMS Hope - Born: Massachusetts - Age: 31 - Discharged on 2 Jun 1815 and released to the Cartel Shakespeare.

Studley, Warren - Marine Corporal - Number: 69 - Prize name: Prince de Neufchatel - Ship type: P - How taken: HM Ship-of-the-Line Leander, HM Ship-of-the-Line Newcastle & HM Frigate Acasta - When taken: 28 Dec 1814 - Where taken: Lat 35N Long 52W - Date received: 18 Feb 1815 - From what ship: HMS Sybille - Born: Not legible - Age: 22 - Discharged on 19 Feb 1815 and sent to Dartmoor.

Summers, William - Seaman - Number: 784 - Prize name: US Brig Syren - Ship type: War - How taken: HM Ship-of-the-Line Medway - When taken: 12 Jul 1814 - Where taken: off Cape of Good Hope - Date received: 11 Jul 1815 - From what ship: HMS Royal Sovereign - Born: Not listed - Discharged on 11 Jul 1815 and released to the Cartel Wooddrop Sims.

Swady, William - Seaman - Number: 567 - Prize name: Mary - Ship type: MV - How taken: HMS Muros - When taken: 8 Dec 1814 - Where taken: West Indies - Date received: 16 Apr 1815 - From what ship: HMS Swifsure - Born: North Caroline - Age: 40 - Discharged on 2 Jun 1815 and released to the Cartel Sovereign.

Swasey, William - Seaman - Number: 8 - Prize name: Albion - Ship type: MV - How taken: HM Schooner Helicon -

When taken: 17 Jan 1815 - Where taken: Lat 20N Long 52W - Date received: 3 Feb 1815 - From what ship: HMS Bittern - Born: Salem - Age: 19 - Discharged on 4 Feb 1815 and sent to Dartmoor.

Swift, William - Surgeon - Number: 853 - Prize name: US Brig Syren - Ship type: War - How taken: HM Ship-of-the-Line Medway - When taken: 12 Jul 1814 - Where taken: off Cape of Good Hope - Date received: 11 Jul 1815 - From what ship: from Portsmouth - Born: Not listed - Discharged on 11 Jul 1815 and released to the Cartel Wooddrop Sims.

Swinney, William - Seaman - Number: 663 - Prize name: Avon - Ship type: P - How taken: HM Frigate Barbadoes - When taken: 8 Mar 1815 - Where taken: West Indies - Date received: 24 Apr 1815 - From what ship: HMS Bellerophon - Born: Massachusetts - Age: 22 - Discharged on 2 Jun 1815 and released to the Cartel Shakespeare.

Symons, Moses - Seaman - Number: 115 - Prize name: Prince de Neufchatel - Ship type: P - How taken: HM Ship-of-the-Line Leander, HM Ship-of-the-Line Newcastle & HM Frigate Acasta - When taken: 28 Dec 1814 - Where taken: Lat 35N Long 52W - Date received: 18 Feb 1815 - From what ship: HMS Sybille - Born: New York - Age: 27 - Race: Black - Discharged on 19 Feb 1815 and sent to Dartmoor.

Tatey, Ephraim - Seaman - Number: 347 - Prize name: Leo - Ship type: P - How taken: HM Frigate Tiber - When taken: 11 Mar 1815 - Where taken: Lat 45.24N Long 12.3W - Date received: 28 Mar 1815 - From what ship: HMS Tiber - Born: Massachusetts - Age: 32 - Discharged on 27 Apr 1815 and released to the Cartel Minework.

Taulton, Elias - Seaman - Number: 762 - Prize name: Harpy - Ship type: P - How taken: HM Brig Zenobia - When taken: 5 Jan 1815 - Where taken: off Lisbon - Date received: 29 Aug 1815 - From what ship: HMS Tonnant - Age: 20 - Discharged on 2 Jun 1815 and released to the Cartel Shakespeare.

Taylor, John - Seaman - Number: 826 - Prize name: US Brig Syren - Ship type: War - How taken: HM Ship-of-the-Line Medway - When taken: 12 Jul 1814 - Where taken: off Cape of Good Hope - Date received: 11 Jul 1815 - From what ship: HMS Royal Sovereign - Born: Not listed - Discharged on 11 Jul 1815 and released to the Cartel Wooddrop Sims.

Taylor, John - Seaman - Number: 133 - How taken: Impressed at Portsmouth - Date received: 21 Feb 1815 - From what ship: HMS Slaney - Born: Annapolis - Age: 29 - Discharged on 24 Feb 1815 and sent to Dartmoor.

Taylor, William - Seaman - Number: 404 - Prize name: Leo - Ship type: P - How taken: HM Frigate Tiber - When taken: 11 Mar 1815 - Where taken: Lat 45.24N Long 12.3W - Date received: 29 Mar 1815 - From what ship: HMS Tiber - Born: Massachusetts - Age: 35 - Discharged on 30 May 1815 and released to the Cartel Atlas.

Taylor, William - Seaman - Number: 222 - Prize name: Decatur - Ship type: P - How taken: HM Frigate Rhin - When taken: 4 Jun 1814 - Where taken: at sea - Date received: 2 Mar 1815 - From what ship: HMS Dannemark - Born: Charleston - Age: 23 - Discharged on 3 Mar 1815 and sent to Dartmoor.

Tessendes, Moses - Seaman - Number: 342 - Prize name: Leo - Ship type: P - How taken: HM Frigate Tiber - When taken: 11 Mar 1815 - Where taken: Lat 45.24N Long 12.3W - Date received: 28 Mar 1815 - From what ship: HMS Tiber - Discharged on 27 Apr 1815 and released to the Cartel Minework.

Thachster, John - Seaman - Number: 721 - Prize name: Sine Qua Non - Ship type: P - How taken: HM Brig Elk - When taken: 20 Feb 1815 - Where taken: off Madeira - Date received: 29 Aug 1815 - From what ship: HMS Tonnant - Age: 19 - Discharged on 2 Jun 1815 and released to the Cartel Shakespeare.

Thomas, Charles - Prize Master - Number: 117 - Prize name: Prince de Neufchatel - Ship type: P - How taken: HM Ship-of-the-Line Leander, HM Ship-of-the-Line Newcastle & HM Frigate Acasta - When taken: 28 Dec 1814 - Where taken: Lat 35N Long 52W - Date received: 18 Feb 1815 - From what ship: HMS Sybille - Born: Massachusetts - Age: 30 - Discharged on 19 Feb 1815 and sent to Dartmoor.

Thomas, John - Master at Arms - Number: 143 - Prize name: US Brig Syren - Ship type: War - How taken: HM Ship-of-the-Line Medway - When taken: 12 Jul 1814 - Where taken: off Cape of Good Hope - Date received: 21 Feb 1815 - From what ship: HMS Slaney - Born: South Carolina - Age: 42 - Discharged on 24 Feb 1815 and sent to Dartmoor.

Thomas, Theodore - Seaman - Number: 106 - Prize name: Prince de Neufchatel - Ship type: P - How taken: HM Ship-of-the-Line Leander, HM Ship-of-the-Line Newcastle & HM Frigate Acasta - When taken: 28 Dec 1814 - Where taken: Lat 35N Long 52W - Date received: 18 Feb 1815 - From what ship: HMS Sybille - Born: Bordeaux - Age: 27 - Race: Black - Discharged on 19 Feb 1815 and sent to Dartmoor.

Thompson, Ephraim - Seaman - Number: 609 - Prize name: George Little - Ship type: P - How taken: HM Frigate Granicus - When taken: 20 Jan 1815 - Where taken: off Cape Finisterre, Spain - Date received: 18 Apr 1815 - From what ship: HMS Euryalus - Born: Massachusetts - Age: 23 - Discharged on 2 Jun 1815 and released to the Cartel Sovereign.

Thompson, John - Seaman - Number: 468 - Prize name: Fox - Ship type: P - How taken: HM Frigate Barbadoes - When taken: 11 Jan 1815 - Where taken: off Amelia Island, Florida - Date received: 16 Apr 1815 - From what ship: HMS Swifsure - Born: Massachusetts - Age: 27 - Discharged on 2 Jun 1815 and released to the Cartel Sovereign.

Thompson, John - Seaman - Number: 711 - Prize name: Sine Qua Non - Ship type: P - How taken: HM Brig Elk - When taken: 20 Feb 1815 - Where taken: off Madeira - Date received: 16 Jul 1815 - From what ship: HMS Hope - Age: 25 - Discharged on 2 Jun 1815 and released to the Cartel Shakespeare.

Thompson, Joseph (alias Robert Taylor) - Seaman - Number: 316 - How taken: Impressed off HM Ship-of-the-Line Scepter - When taken: 15 Feb 1815 - Where taken: Cork - Date received: 5 Mar 1815 - From what ship: HMS Tartarus - Born: Connecticut - Age: 30 - Discharged on 7 Mar 1815 and sent to Dartmoor.

Thompson, William - Seaman - Number: 43 - How taken: Taken out of HM Ship-of-the-Line Minden - Where taken: East Indies - Date received: 15 Feb 1815 - From what ship: HMS Myrmidon - Born: New York - Age: 27 - Race: Mulatto - Discharged on 17 Feb 1815 and sent to Dartmoor.

Tibble, Joseph - Seaman - Number: 337 - Prize name: Leo - Ship type: P - How taken: HM Frigate Tiber - When taken: 11 Mar 1815 - Where taken: Lat 45.24N Long 12.3W - Date received: 28 Mar 1815 - From what ship: HMS Tiber - Discharged on 27 Apr 1815 and released to the Cartel Minework.

Todd, Robert - Seaman - Number: 681 - Prize name: Frolick - Ship type: P - How taken: HM Frigate Orpheus - When taken: 6 Apr 1814 - Date received: 14 May 1815 - From what ship: from the police - Age: 53 - Discharged on 2 Jun 1815 and released to the Cartel Shakespeare.

Todd, William - Seaman - Number: 523 - Prize name: Dolphin - Ship type: MV - How taken: HM Frigate Barbadoes - When taken: 4 Dec 1814 - Where taken: off St. Bartholomew - Date received: 16 Apr 1815 - From what ship: HMS Swifsure - Born: Newburyport - Age: 27 - Discharged on 2 Jun 1815 and released to the Cartel Sovereign.

Torre, Vincente - Seaman - Number: 476 - Prize name: Fox - Ship type: P - How taken: HM Frigate Barbadoes - When taken: 11 Jan 1815 - Where taken: off Amelia Island, Florida - Date received: 16 Apr 1815 - From what ship: HMS Swifsure - Age: 20 - Race: Indian - Discharged on 2 Jun 1815 and released to the Cartel Sovereign.

Torrey, Isaac - Seaman - Number: 401 - Prize name: Leo - Ship type: P - How taken: HM Frigate Tiber - When taken: 11 Mar 1815 - Where taken: Lat 45.24N Long 12.3W - Date received: 29 Mar 1815 - From what ship: HMS Tiber - Born: Massachusetts - Age: 20 - Discharged on 30 May 1815 and released to the Cartel Atlas.

Touler, Louis - Seaman - Number: 471 - Prize name: Fox - Ship type: P - How taken: HM Frigate Barbadoes - When taken: 11 Jan 1815 - Where taken: off Amelia Island, Florida - Date received: 16 Apr 1815 - From what ship: HMS Swifsure - Born: Saint-Domingue (Haiti) - Age: 20 - Discharged on 2 Jun 1815 and released to the Cartel Sovereign.

Towley, Joshua - Seaman - Number: 861 - Prize name: US Brig Syren - Ship type: War - How taken: HM Ship-of-the-Line Medway - When taken: 12 Jul 1814 - Where taken: off Cape of Good Hope - Date received: 11 Jul 1815 - From what ship: from Portsmouth - Born: Not listed - Discharged on 11 Jul 1815 and released to the Cartel Wooddrop Sims.

Train, Joseph - Cook - Number: 89 - Prize name: Prince de Neufchatel - Ship type: P - How taken: HM Ship-of-the-Line Leander, HM Ship-of-the-Line Newcastle & HM Frigate Acasta - When taken: 28 Dec 1814 - Where taken: Lat 35N Long 52W - Date received: 18 Feb 1815 - From what ship: HMS Sybille - Race: Black - Discharged on 19 Feb 1815 and sent to Dartmoor.

Treffick, Charles - Seaman - Number: 659 - Prize name: Avon - Ship type: P - How taken: HM Frigate Barbadoes - When taken: 8 Mar 1815 - Where taken: West Indies - Date received: 24 Apr 1815 - From what ship: HMS Bellerophon - Age: 26 - Discharged on 2 Jun 1815 and released to the Cartel Shakespeare.

Trammel, Henry - Seaman - Number: 795 - Prize name: US Brig Syren - Ship type: War - How taken: HM Ship-of-the-Line Medway - When taken: 12 Jul 1814 - Where taken: off Cape of Good Hope - Date received: 11 Jul

1815 - From what ship: HMS Royal Sovereign - Born: Not listed - Discharged on 11 Jul 1815 and released to the Cartel Wooddrop Sims.

Tubill, William - Seaman - Number: 597 - Prize name: George Little - Ship type: P - How taken: HM Frigate Granicus - When taken: 20 Jan 1815 - Where taken: off Cape Finisterre, Spain - Date received: 18 Apr 1815 - From what ship: HMS Euryalus - Born: Massachusetts - Age: 28 - Discharged on 2 Jun 1815 and released to the Cartel Sovereign.

Tune, Samuel - Seaman - Number: 61 - How taken: Taken out of HM Brig Sabine - Where taken: Portsmouth - Date received: 15 Feb 1815 - From what ship: HMS Myrmidon - Age: 28 - Discharged on 17 Feb 1815 and sent to Dartmoor.

Truelove, John - Seaman - Number: 292 - How taken: Taken off the HM Ship-of-the-Line Severn and other ships - Date received: 2 Mar 1815 - From what ship: HMS Dannemark - Discharged on 3 Mar 1815 and sent to Dartmoor.

Turner, Joshua - Seaman - Number: 215 - Prize name: Sennett, Swedish merchant vessel - Ship type: MV - How taken: HM Frigate Rhin - When taken: Not known - Date received: 2 Mar 1815 - From what ship: HMS Dannemark - Discharged on 3 Mar 1815 and sent to Dartmoor.

Twine, William - Seaman - Number: 561 - Prize name: Mary - Ship type: MV - How taken: HMS Muros - When taken: 8 Dec 1814 - Where taken: West Indies - Date received: 16 Apr 1815 - From what ship: HMS Swifsure - Born: North Caroline - Age: 25 - Discharged on 2 Jun 1815 and released to the Cartel Sovereign.

Vans, William - Seaman - Number: 394 - Prize name: Leo - Ship type: P - How taken: HM Frigate Tiber - When taken: 11 Mar 1815 - Where taken: Lat 45.24N Long 12.3W - Date received: 29 Mar 1814 - From what ship: HMS Tiber - Born: Massachusetts - Age: 21 - Discharged on 30 May 1815 and released to the Cartel Atlas.

Viels, Jean - Seaman - Number: 495 - Prize name: Fox - Ship type: P - How taken: HM Frigate Barbadoes - When taken: 11 Jan 1815 - Where taken: off Amelia Island, Florida - Date received: 16 Apr 1815 - From what ship: HMS Swifsure - Born: Saint-Domingue (Haiti) - Age: 29 - Race: Negro - Discharged on 2 Jun 1815 and released to the Cartel Sovereign.

Ving, William - Seaman - Number: 768 - Prize name: Aurora - Ship type: MV - How taken: HM Frigate Andromeda - When taken: 28 Feb 1815 - Where taken: off Lisbon - Date received: 29 Aug 1815 - From what ship: HMS Tonnant - Age: 47 - Discharged on 2 Jun 1815 and released to the Cartel Shakespeare.

Wales, N. - Seaman - Number: 692 - Prize name: Sine Qua Non - Ship type: P - How taken: HM Brig Elk - When taken: 20 Feb 1815 - Where taken: off Madeira - Date received: 16 Jul 1815 - From what ship: HMS Hope - Age: 36 - Discharged on 2 Jun 1815 and released to the Cartel Shakespeare.

Walker, Daniel - Steward - Number: 423 - Prize name: Engineer - Ship type: MV - How taken: HMS Murros - When taken: 21 Sep 1814 - Where taken: off Porto Rico - Date received: 16 Apr 1815 - From what ship: HMS Swifsure - Born: Boston - Age: 16 - Discharged on 2 Jun 1815 and released to the Cartel Sovereign.

Walker, John - Seaman - Number: 436 - Prize name: Mars, prize of the Privateer David Porter - Ship type: P - How taken: HM Frigate Pique - When taken: 12 Aug 1814 - Where taken: off Newfoundland - Date received: 16 Apr 1815 - From what ship: HMS Swifsure - Born: Massachusetts - Age: 19 - Discharged on 2 Jun 1815 and released to the Cartel Sovereign.

Wanett, William - Seaman - Number: 806 - Prize name: US Brig Syren - Ship type: War - How taken: HM Ship-of-the-Line Medway - When taken: 12 Jul 1814 - Where taken: off Cape of Good Hope - Date received: 11 Jul 1815 - From what ship: HMS Royal Sovereign - Born: Not listed - Discharged on 11 Jul 1815 and released to the Cartel Wooddrop Sims.

Washburn, George - Seaman - Number: 360 - Prize name: Leo - Ship type: P - How taken: HM Frigate Tiber - When taken: 11 Mar 1815 - Where taken: Lat 45.24N Long 12.3W - Date received: 29 Mar 1814 - From what ship: HMS Tiber - Born: Massachusetts - Age: 24 - Discharged on 2 May 1815 and released to the Cartel Ariel.

Waters, Louis - Steward - Number: 197 - Prize name: Chance - Ship type: P - How taken: HMS Statire - When taken: 1 Apr 1814 - Where taken: at sea - Date received: 2 Mar 1815 - From what ship: HMS Dannemark - Born: Maryland - Age: 22 - Race: Black - Discharged on 3 Mar 1815 and sent to Dartmoor.

Watson, George - Seaman - Number: 827 - Prize name: US Brig Syren - Ship type: War - How taken: HM Ship-of-the-Line Medway - When taken: 12 Jul 1814 - Where taken: off Cape of Good Hope - Date received: 11 Jul 1815 - From what ship: HMS Royal Sovereign - Born: Not listed - Discharged on 11 Jul 1815 and released to the Cartel Wooddrop Sims.

Watt, Samuel - Seaman - Number: 277 - Prize name: Netterville - Ship type: MV - How taken: HM Brig Onyx - When taken: 25 Dec 1814 - Where taken: at sea - Date received: 2 Mar 1815 - From what ship: HMS Dannemark - Born: Delaware - Age: 22 - Discharged on 3 Mar 1815 and sent to Dartmoor.

Watton, John - Seaman - Number: 524 - Prize name: Dolphin - Ship type: MV - How taken: HM Frigate Barbadoes - When taken: 4 Dec 1814 - Where taken: off St. Bartholomew - Date received: 16 Apr 1815 - From what ship: HMS Swifsure - Born: Newburyport - Age: 36 - Discharged on 2 Jun 1815 and released to the Cartel Sovereign.

Weaver, W. A. - Midshipman - Number: 854 - Prize name: US Brig Syren - Ship type: War - How taken: HM Ship-of-the-Line Medway - When taken: 12 Jul 1814 - Where taken: off Cape of Good Hope - Date received: 11 Jul 1815 - From what ship: from Portsmouth - Born: Not listed - Discharged on 11 Jul 1815 and released to the Cartel Wooddrop Sims.

Webster, Benjamin - Seaman - Number: 345 - Prize name: Leo - Ship type: P - How taken: HM Frigate Tiber - When taken: 11 Mar 1815 - Where taken: Lat 45.24N Long 12.3W - Date received: 28 Mar 1815 - From what ship: HMS Tiber - Born: Massachusetts - Age: 28 - Discharged on 27 Apr 1815 and released to the Cartel Minework.

Webster, Michael - Seaman - Number: 183 - Prize name: Farmer's Daughter - Ship type: MV - How taken: HM Ship-of-the-Line Leviathan - When taken: 20 Mar 1814 - Where taken: at sea - Date received: 2 Mar 1815 - From what ship: HMS Dannemark - Born: Annapolis - Age: 18 - Discharged on 3 Mar 1815 and sent to Dartmoor.

Weeks, William - Seaman - Number: 514 - Prize name: Hope, prize to US Sloop-of-War Wasp - Ship type: War - How taken: HM Sloop Fairy - When taken: 6 Jan 1815 - Where taken: Unknown - Date received: 16 Apr 1815 - From what ship: HMS Swifsure - Born: Massachusetts - Age: 29 - Discharged on 2 Jun 1815 and released to the Cartel Sovereign.

Weldon, Daniel - Seaman - Number: 485 - Prize name: Fox - Ship type: P - How taken: HM Frigate Barbadoes - When taken: 11 Jan 1815 - Where taken: off Amelia Island, Florida - Date received: 16 Apr 1815 - From what ship: HMS Swifsure - Born: Rhode Island - Age: 25 - Discharged on 2 Jun 1815 and released to the Cartel Sovereign.

Welbourg, David - Seaman - Number: 213 - Prize name: Sennett, Swedish merchant vessel - Ship type: MV - How taken: HM Frigate Rhin - When taken: Not known - Date received: 2 Mar 1815 - From what ship: HMS Dannemark - Race: Negro - Discharged on 3 Mar 1815 and sent to Dartmoor.

Welch, John - Seaman - Number: 120 - Prize name: Prince de Neufchatel - Ship type: P - How taken: HM Ship-of-the-Line Leander, HM Ship-of-the-Line Newcastle & HM Frigate Acasta - When taken: 28 Dec 1814 - Where taken: Lat 35N Long 52W - Date received: 18 Feb 1815 - From what ship: HMS Sybille - Born: Boston - Age: 36 - Discharged on 19 Feb 1815 and sent to Dartmoor.

Welker, John - Seaman - Number: 204 - Prize name: Sennett, Swedish merchant vessel - Ship type: MV - How taken: HM Frigate Rhin - When taken: Not known - Date received: 2 Mar 1815 - From what ship: HMS Dannemark - Race: Negro - Discharged on 3 Mar 1815 and sent to Dartmoor.

Wells, Francis - Seaman - Number: 645 - Prize name: Avon - Ship type: P - How taken: HM Frigate Barbadoes - When taken: 8 Mar 1815 - Where taken: West Indies - Date received: 24 Apr 1815 - From what ship: HMS Bellerophon - Age: 22 - Discharged on 2 Jun 1815 and released to the Cartel Shakespeare.

Wells, N. - Seaman - Number: 578 - Prize name: Sorin, prize to Privateer Prince of Neufchatel - Ship type: P - How taken: HM Ship-of-the-Line Medway - When taken: 12 Jul 1814 - Where taken: Unknown - Date received: 16 Apr 1815 - From what ship: HMS Swifsure - Born: Boston - Age: 29 - Discharged on 2 Jun 1815 and released to the Cartel Sovereign.

Welsh, Ebenezer - Seaman - Number: 399 - Prize name: Leo - Ship type: P - How taken: HM Frigate Tiber - When taken: 11 Mar 1815 - Where taken: Lat 45.24N Long 12.3W - Date received: 29 Mar 1814 - From what ship:

HMS Tiber - Born: Massachusetts - Age: 22 - Discharged on 30 May 1815 and released to the Cartel Atlas.

Welsh, Richard - Seaman - Number: 228 - Prize name: Decatur - Ship type: P - How taken: HM Frigate Rhin - When taken: 4 Jun 1814 - Where taken: at sea - Date received: 2 Mar 1815 - From what ship: HMS Dannemark - Born: New York - Age: 24 - Discharged on 3 Mar 1815 and sent to Dartmoor.

Wilson, Michael - Seaman - Number: 708 - Prize name: Sine Qua Non - Ship type: P - How taken: HM Brig Elk - When taken: 20 Feb 1815 - Where taken: off Madeira - Date received: 16 Jul 1815 - From what ship: HMS Hope - Age: 40 - Discharged on 2 Jun 1815 and released to the Cartel Shakespeare.

Wentworth, John - Marine - Number: 75 - Prize name: Prince de Neufchatel - Ship type: P - How taken: HM Ship-of-the-Line Leander, HM Ship-of-the-Line Newcastle & HM Frigate Acasta - When taken: 28 Dec 1814 - Where taken: Lat 35N Long 52W - Date received: 18 Feb 1815 - From what ship: HMS Sybille - Born: Canton, China - Age: 20 - Discharged on 19 Feb 1815 and sent to Dartmoor.

West, Jacob - Seaman - Number: 140 - Prize name: US Brig Syren - Ship type: War - How taken: HM Ship-of-the-Line Medway - When taken: 12 Jul 1814 - Where taken: off Cape of Good Hope - Date received: 21 Feb 1815 - From what ship: HMS Slaney - Born: Albany - Age: 30 - Discharged on 24 Feb 1815 and sent to Dartmoor.

West, Stephen - Seaman - Number: 424 - Prize name: Engineer - Ship type: MV - How taken: HMS Murros - When taken: 21 Sep 1814 - Where taken: off Porto Rico - Date received: 16 Apr 1815 - From what ship: HMS Swifsure - Born: Massachusetts - Age: 21 - Discharged on 2 Jun 1815 and released to the Cartel Sovereign.

Westcott, William - 1st Lieutenant - Number: 218 - Prize name: Decatur - Ship type: P - How taken: HM Frigate Rhin - When taken: 4 Jun 1814 - Where taken: at sea - Date received: 2 Mar 1815 - From what ship: HMS Dannemark - Born: Baltimore - Age: 45 - Discharged on 3 Mar 1815 and sent to Dartmoor.

Wester, John - Seaman - Number: 349 - Prize name: Leo - Ship type: P - How taken: HM Frigate Tiber - When taken: 11 Mar 1815 - Where taken: Lat 45.24N Long 12.3W - Date received: 28 Mar 1815 - From what ship: HMS Tiber - Born: Massachusetts - Age: 26 - Discharged on 27 Apr 1815 and released to the Cartel Minework.

Western, Henry - Seaman - Number: 602 - Prize name: George Little - Ship type: P - How taken: HM Frigate Granicus - When taken: 20 Jan 1815 - Where taken: off Cape Finisterre, Spain - Date received: 18 Apr 1815 - From what ship: HMS Euryalus - Born: Massachusetts - Age: 34 - Discharged on 8 May 1815.

Westar, Henry - Seaman - Number: 366 - Prize name: Leo - Ship type: P - How taken: HM Frigate Tiber - When taken: 11 Mar 1815 - Where taken: Lat 45.24N Long 12.3W - Date received: 29 Mar 1814 - From what ship: HMS Tiber - Born: Massachusetts - Age: 23 - Discharged on 2 May 1815 and released to the Cartel Ariel.

Wheeler, John - Seaman - Number: 577 - Prize name: Sorin, prize to Privateer Prince of Neufchatel - Ship type: P - How taken: HM Ship-of-the-Line Medway - When taken: 12 Jul 1814 - Where taken: Unknown - Date received: 16 Apr 1815 - From what ship: HMS Swifsure - Born: Boston - Age: 19 - Discharged on 2 Jun 1815 and released to the Cartel Sovereign.

Whiney, Josias - Pilot - Number: 225 - Prize name: Decatur - Ship type: P - How taken: HM Frigate Rhin - When taken: 4 Jun 1814 - Where taken: at sea - Date received: 2 Mar 1815 - From what ship: HMS Dannemark - Discharged on 3 Mar 1815 and sent to Dartmoor.

Whipple, John - Seaman - Number: 190 - Prize name: Chance - Ship type: P - How taken: HMS Statire - When taken: 1 Apr 1814 - Where taken: at sea - Date received: 2 Mar 1815 - From what ship: HMS Dannemark - Born: New London - Age: 19 - Discharged on 3 Mar 1815 and sent to Dartmoor.

White, James - Seaman - Number: 249 - Prize name: John - Ship type: MV - How taken: HM Brig Zenobia - When taken: 11 Sep 1814 - Where taken: at sea - Date received: 2 Mar 1815 - From what ship: HMS Dannemark - Born: Charleston - Age: 32 - Discharged on 3 Mar 1815 and sent to Dartmoor.

White, William - Seaman - Number: 844 - Prize name: US Brig Syren - Ship type: War - How taken: HM Ship-of-the-Line Medway - When taken: 12 Jul 1814 - Where taken: off Cape of Good Hope - Date received: 11 Jul 1815 - From what ship: HMS Royal Sovereign - Born: Not listed - Discharged on 11 Jul 1815 and released to the Cartel Wooddrop Sims.

White, William - Seaman - Number: 828 - Prize name: US Brig Syren - Ship type: War - How taken: HM Ship-of-the-Line Medway - When taken: 12 Jul 1814 - Where taken: off Cape of Good Hope - Date received: 11 Jul

1815 - From what ship: HMS Royal Sovereign - Born: Not listed - Discharged on 11 Jul 1815 and released to the Cartel Wooddrop Sims.

Whitehead, John - Seaman - Number: 198 - Prize name: Saucy Jack - Ship type: MV - How taken: Unknown - Date received: 2 Mar 1815 - From what ship: HMS Dannemark - Discharged on 3 Mar 1815 and sent to Dartmoor.

Whiter, Amos - Seaman - Number: 322 - Prize name: Thomas, prize to Privateer Scourge - Ship type: P - How taken: HM Frigate Aquilon - When taken: 11 Mar 1815 - Where taken: off Cape Finisterre, Spain - Date received: 25 Apr 1815 - From what ship: HMS Opossum - Discharged on 27 Apr 1815 and released to the Cartel Minework.

Whiting, Benjamin - Boy - Number: 395 - Prize name: Leo - Ship type: P - How taken: HM Frigate Tiber - When taken: 11 Mar 1815 - Where taken: Lat 45.24N Long 12.3W - Date received: 29 Mar 1814 - From what ship: HMS Tiber - Born: Massachusetts - Age: 16 - Discharged on 30 May 1815 and released to the Cartel Atlas.

Whiting, Samuel - Seaman - Number: 833 - Prize name: US Brig Syren - Ship type: War - How taken: HM Ship-of-the-Line Medway - When taken: 12 Jul 1814 - Where taken: off Cape of Good Hope - Date received: 11 Jul 1815 - From what ship: HMS Royal Sovereign - Born: Not listed - Discharged on 11 Jul 1815 and released to the Cartel Wooddrop Sims.

Whitney, William - Seaman - Number: 210 - Prize name: Sennett, Swedish merchant vessel - Ship type: MV - How taken: HM Frigate Rhin - When taken: Not known - Date received: 2 Mar 1815 - From what ship: HMS Dannemark - Discharged on 3 Mar 1815 and sent to Dartmoor.

Whittington, Samuel - Seaman - Number: 93 - Prize name: Prince de Neufchatel - Ship type: P - How taken: HM Ship-of-the-Line Leander, HM Ship-of-the-Line Newcastle & HM Frigate Acasta - When taken: 28 Dec 1814 - Where taken: Lat 35N Long 52W - Date received: 18 Feb 1815 - From what ship: HMS Sybille - Born: Providence - Age: 20 - Discharged on 19 Feb 1815 and sent to Dartmoor.

Wicks, Ebenezer - Marine Sergeant - Number: 68 - Prize name: Prince de Neufchatel - Ship type: P - How taken: HM Ship-of-the-Line Leander, HM Ship-of-the-Line Newcastle & HM Frigate Acasta - When taken: 28 Dec 1814 - Where taken: Lat 35N Long 52W - Date received: 18 Feb 1815 - From what ship: HMS Sybille - Age: 29 - Discharged on 19 Feb 1815 and sent to Dartmoor.

Widbee, Joseph - Prize Master - Number: 220 - Prize name: Decatur - Ship type: P - How taken: HM Frigate Rhin - When taken: 4 Jun 1814 - Where taken: at sea - Date received: 2 Mar 1815 - From what ship: HMS Dannemark - Discharged on 3 Mar 1815 and sent to Dartmoor.

Willey, Fredrick - Seaman - Number: 840 - Prize name: US Brig Syren - Ship type: War - How taken: HM Ship-of-the-Line Medway - When taken: 12 Jul 1814 - Where taken: off Cape of Good Hope - Date received: 11 Jul 1815 - From what ship: HMS Royal Sovereign - Born: Not listed - Discharged on 11 Jul 1815 and released to the Cartel Wooddrop Sims.

Williams, Alexander - Seaman - Number: 465 - Prize name: Fox - Ship type: P - How taken: HM Frigate Barbadoes - When taken: 11 Jan 1815 - Where taken: off Amelia Island, Florida - Date received: 16 Apr 1815 - From what ship: HMS Swifsure - Born: Kentucky - Age: 34 - Discharged on 2 Jun 1815 and released to the Cartel Sovereign.

Williams, Alexander - Seaman - Number: 683 - How taken: Gave himself up from HM Ship-of-the-Line Majestic - Date received: 14 May 1815 - From what ship: from the police - Age: 44 - Race: Negro - Discharged on 2 Jun 1815 and released to the Cartel Shakespeare.

Williams, George - Seaman - Number: 44 - How taken: Taken out of HM Ship-of-the-Line Gloucester - Where taken: Sternness - Date received: 15 Feb 1815 - From what ship: HMS Myrmidon - Born: Philadelphia - Age: 29 - Race: Black - Discharged on 17 Feb 1815 and sent to Dartmoor.

Williams, James - Seaman - Number: 16 - Prize name: Sophie - Ship type: MV - How taken: Impressed at Greenock - When taken: 10 Dec 1814 - Date received: 3 Feb 1815 - From what ship: HMS Bittern - Born: Staten Island - Age: 21 - Race: Black - Discharged on 4 Feb 1815 and sent to Dartmoor.

Williams, John - Seaman - Number: 255 - Prize name: Wolfe - Ship type: P - How taken: Not known - Date received: 2 Mar 1815 - From what ship: HMS Dannemark - Born: Norfolk - Age: 17 - Discharged on 3 Mar 1815 and sent to Dartmoor.

Williams, John - Seaman - Number: 234 - Prize name: Decatur - Ship type: P - How taken: HM Frigate Rhin - When taken: 4 Jun 1814 - Where taken: at sea - Date received: 2 Mar 1815 - From what ship: HMS Dannemark - Born: Roxborough - Age: 42 - Discharged on 3 Mar 1815 and sent to Dartmoor.

Williams, Lloyd - Seaman - Number: 556 - Prize name: Mary - Ship type: MV - How taken: HMS Muros - When taken: 8 Dec 1814 - Where taken: West Indies - Date received: 16 Apr 1815 - From what ship: HMS Swifsure - Born: Massachusetts - Age: 24 - Discharged on 2 Jun 1815 and released to the Cartel Sovereign.

Williams, Pompey - Seaman - Number: 765 - Prize name: George Little - Ship type: P - How taken: HM Frigate Granicus - When taken: 20 Jan 1815 - Where taken: off Cape Finisterre, Spain - Date received: 29 Aug 1815 - From what ship: HMS Tonnant - Age: 25 - Race: Negro - Discharged on 2 Jun 1815 and released to the Cartel Shakespeare.

Williams, Stephen - Seaman - Number: 594 - Prize name: George Little - Ship type: P - How taken: HM Frigate Granicus - When taken: 20 Jan 1815 - Where taken: off Cape Finisterre, Spain - Date received: 18 Apr 1815 - From what ship: HMS Euryalus - Born: New Hampshire - Age: 21 - Discharged on 2 Jun 1815 and released to the Cartel Sovereign.

Wilson, Samuel - Seaman - Number: 680 - Prize name: Frolick - Ship type: P - How taken: HM Frigate Orpheus - When taken: 6 Apr 1814 - Date received: 14 May 1815 - From what ship: from the police - Age: 24 - Discharged on 2 Jun 1815 and released to the Cartel Shakespeare.

Wilson, Thomas - Seaman - Number: 774 - Prize name: Sine Qua Non - Ship type: P - How taken: HM Brig Elk - When taken: 20 Feb 1815 - Where taken: off Madeira - Date received: 29 Aug 1815 - From what ship: HMS Tonnant - Age: 24 - Discharged on 2 Jun 1815 and released to the Cartel Shakespeare.

Winkler, Mathew - Seaman - Number: 646 - Prize name: Avon - Ship type: P - How taken: HM Frigate Barbadoes – When taken: 8 Mar 1815 - Where taken: West Indies - Date received: 24 Apr 1815 - From what ship: HMS Bellerophon - Age: 22 - Discharged on 2 Jun 1815 and released to the Cartel Shakespeare.

Wood, R. - Seaman - Number: 336 - Prize name: Leo - Ship type: P - How taken: HM Frigate Tiber - When taken: 11 Mar 1815 - Where taken: Lat 45.24N Long 12.3W - Date received: 28 Mar 1815 - From what ship: HMS Tiber - Discharged on 27 Apr 1815 and released to the Cartel Minework.

Wood, Robert - Seaman - Number: 45 - How taken: Taken out of HM Frigate Iris - Where taken: Spithead - Date received: 15 Feb 1815 - From what ship: HMS Myrmidon - Born: New York - Age: 28 - Race: Black - Discharged on 17 Feb 1815 and sent to Dartmoor.

Woodford, Joseph - Seaman - Number: 798 - Prize name: US Brig Syren - Ship type: War - How taken: HM Ship-of-the-Line Medway - When taken: 12 Jul 1814 - Where taken: off Cape of Good Hope - Date received: 11 Jul 1815 - From what ship: HMS Royal Sovereign - Born: Not listed - Discharged on 11 Jul 1815 and released to the Cartel Wooddrop Sims.

Woods, Samuel - Seaman - Number: 452 - Prize name: High Flyer - Ship type: LM - How taken: HMS Muros - When taken: 14 Nov 1814 - Where taken: off St. Bartholomew - Date received: 16 Apr 1815 - From what ship: HMS Swifsure - Born: Massachusetts - Age: 26 - Discharged on 2 Jun 1815 and released to the Cartel Sovereign.

Woodward, John - Seaman - Number: 658 - Prize name: Avon - Ship type: P - How taken: HM Frigate Barbadoes - When taken: 8 Mar 1815 - Where taken: West Indies - Date received: 24 Apr 1815 - From what ship: HMS Bellerophon - Born: Boston - Age: 20 - Discharged on 2 Jun 1815 and released to the Cartel Shakespeare.

Woodward, Samuel - Seaman - Number: 736 - Prize name: Sine Qua Non - Ship type: P - How taken: HM Brig Elk - When taken: 20 Feb 1815 - Where taken: off Madeira - Date received: 29 Aug 1815 - From what ship: HMS Tonnant - Born: Massachusetts - Age: 34 - Discharged on 2 Jun 1815 and released to the Cartel Shakespeare.

Wright, Joseph - Seaman - Number: 92 - Prize name: Prince de Neufchatel - Ship type: P - How taken: HM Ship-of-the-Line Leander, HM Ship-of-the-Line Newcastle & HM Frigate Acasta - When taken: 28 Dec 1814 - Where taken: Lat 35N Long 52W - Date received: 18 Feb 1815 - From what ship: HMS Sybille - Age: 19 - Discharged on 19 Feb 1815 and sent to Dartmoor.

Young, John - Seaman - Number: 28 - Prize name: Plutarch - Ship type: MV - How taken: HM Schooner Helicon - When taken: 5 Feb 1815 - Where taken: Lat 45.5N Long 7W - Date received: 8 Feb 1815 - From what ship:

HMS Helicon - Born: Wiscasset - Age: 22 - Discharged on 10 Feb 1815 and sent to Dartmoor.

Young, Samuel - Seaman - Number: 466 - Prize name: Fox - Ship type: P - How taken: HM Frigate Barbadoes - When taken: 11 Jan 1815 - Where taken: off Amelia Island, Florida - Date received: 16 Apr 1815 - From what ship: HMS Swifsure - Born: Maryland - Age: 23 - Discharged on 2 Jun 1815 and released to the Cartel Sovereign.

Young, William - Seaman - Number: 787 - Prize name: US Brig Syren - Ship type: War - How taken: HM Ship-of-the-Line Medway - When taken: 12 Jul 1814 - Where taken: off Cape of Good Hope - Date received: 11 Jul 1815 - From what ship: HMS Royal Sovereign - Born: Not listed - Discharged on 11 Jul 1815 and released to the Cartel Wooddrop Sims.

Numeric listing by prisoner number
February 1815 through June 1815

1 Hill, Stephen
2 Hingston, Richard
3 Lowe, Isaac
4 Johnston, Perry
5 Lee, Abraham
6 Mount, John
7 Stone, Jerry
8 Swasey, William
9 Douglas, Mathew
10 Johnson, Jesse
11 Green, Peter
12 Kendell, Franklin
13 Stenson, David
14 Oliver, Mathew
15 Louis, Joseph
16 Williams, James
17 Pitts, Francis
18 Stevens, William
19 Lindborg, Charles
20 Coren, Hugh
21 Callaghan, James
22 Clerk, James
23 Ray, Richard
24 Pitts, George
25 Cooper, Thomas
26 Lambert, Andrew
27 Robinson, William
28 Young, John
29 Autel, Stephen
30 James, William
31 Covell, Nathaniel
32 Peterson, Alexander
33 Peterson, John
34 Marshall, Anthony
35 Bertram, William Henry
36 Parsons, Thomas
37 Frazier, Charles
38 Hepburn, James
39 Bourdon, Amend
40 Clark, Francis Marie
41 Nichelle, Pierre
42 Payne, James
43 Thompson, William
44 Williams, George
45 Wood, Robert
46 Evans, Thomas
47 Conton, Philip
48 McDaniel, John
49 Coffee, Jacob
50 Smet, John
51 Finney, James
52 Huston, William
53 Dine, William
54 Boyent, William
55 Namiss, Peter
56 Adams, Robert
57 Frost, Thomas Bell

58 Judah, David
59 Miller, James
60 Barton, Mathew
61 Tune, Samuel
62 Greenlaw, Jeremiah
63 Burnham, Abraham
64 Boylston, Zebadiah
65 Bangs, George
66 Door, Ebenezer
67 Martin, John
68 Wicks, Ebenezer
69 Studley, Warren
70 Jervis, Peter
71 Ranbre, Mathew
72 David, Samuel
73 Davis, Frederick
74 Perry, Ebenezer
75 Wentworth, John
76 Delaware, John
77 Jones, Edward
78 Raymond, John
79 Lidings, John
80 Downs, Jesse
81 Howes, Caleb
82 Moore, Joshua
83 Levitt, Caleb
84 Healy, Samuel
85 Partex, Samuel
86 Lewis, Jesse L.
87 Newall, John B. L.
88 Snow, Daniel
89 Train, Joseph
90 Sanster, John
91 Fremont, Lawrence
92 Wright, Joseph
93 Whittington, Samuel
94 Haywood, Samuel
95 Alexander, James
96 Farr, William
97 Hall, John
98 Dimitroff, John
99 Leonard, Thomas
100 Foreman, John
101 Constant, William
102 Colson, Christopher
103 Peterson, David
104 Fisher, Robert
105 Moore, Warren
106 Thomas, Theodore
107 Goodhall, Joseph
108 Barker, Thomas
109 Samuels, Samuel
110 Fernandez, George
111 Monk, Philip
112 Nash, Alexander
113 Hose, Richard
114 Hawkins, John

115 Symons, Moses
116 Lyon, John
117 Thomas, Charles
118 Churchill, Joseph
119 Ames, William
120 Welch, John
121 Coffis, Theodore
122 Miller, John
123 Glover, William
124 Fuller, Moses
125 Jackson, Ebenezer
126 Lumey, John
127 Dunklin, Jesse
128 Payton, James
129 Martin, John
130 Mack, William F.
131 Bottes, John
132 Marshall, William
133 Taylor, John
134 Anderson, John
135 Morris, John
136 Chick, Moses
137 Cloutman, Ephraim
138 Andrews, Samuel
139 Dunham, William
140 West, Jacob
141 Sparkes, John
142 Coates, Samuel M.
143 Thomas, John
144 Hodges, Edward
145 Aide, John L.
146 Casper, William
147 Millow, John
148 Hitchens, George
149 Haycock, Joseph
150 Kirby, Robert
151 Pope, Alexander
152 Lewis, Winslow
153 Rust, Zebulon
154 Davis, Elias S.
155 Fray, Samuel
156 Brackett, James
157 Norris, Benjamin
158 Osborne, John L.
159 Dolphin, Joseph
160 Perkins, Clement
161 Jones, John
162 Jones, Calvin
163 Moore, Robert
164 Plumber, James
165 Hanstable, William
166 Martin, John
167 Bonner, John
168 Rake, Martin
169 Lemon, John
170 Randell, Thomas
171 Doing, Denis O.

172 Bissett, Robert	234 Williams, John	296 Moss, Richard
173 Butcher, Jacob	235 Bowen, John	297 Stevens, John
174 Jennings, Francis	236 Phillips, John	298 Frovemcs, Ford.
175 Jackson, John	237 Campbell, Henry	299 Kellock, Stephen
176 Hunt, James	238 Hendry, William	300 Stone, John
177 Benton, Samuel	239 Moore, William	301 Nicholas, Jacob
178 Dawson, John	240 Hicks, (-----)	302 Desomon, Thomas
179 Edwards, William	241 Ketchell, George	303 Daniels, Henry
180 Kingdom, John	242 Lumsby, Frederick	304 Mackford, John
181 Johnson, Thomas	243 Brown, William	305 Micker, Charles
182 Carle, James	244 Pavish, John	306 Parnet, Henry
183 Webster, Michael	245 Hackley, Walter	307 Spear, Joseph
184 Griffin, Heathcote	246 Henderson, Joseph	308 Menzer, James
185 Nutter, Henry	247 Betarsl, Samuel	309 Fuller, Benjamin
186 Armstrong, Charles	248 Farrell, John	310 Rosenburg, Charles
187 Henson, Peter	249 White, James	311 Heasey, John
188 Butler, David	250 Henchman, George	312 Sewell, John
189 Little, William	251 Nettle, Joseph	313 Alexander, George
190 Whipple, John	252 Finch, Abraham	314 Price, Peter
191 Smith, Samuel	253 Parsons, James	315 Fisher, James
192 Maceman, Thomas	254 Day, Samuel	316 Thompson, Joseph
193 Macklin, John	255 Williams, John	317 Gibson, Francis
194 Gale, Russel	256 Bailey, Samuel	318 Lord, John
195 Hill, Francis	257 Fox, Edward	319 Rust, John
196 Dominico, John	258 Lawrence, Trent	320 Johnson, Samuel J.
197 Waters, Louis	259 Detende, Joseph	321 Alden, Henry
198 Whitehead, John	260 Ramsdell, Charles	322 Whiter, Amos
199 Smith, John	261 McCarthy, Samuel	323 McIntyre, John
200 McCormick, James	262 Fincho, Etienne	324 Hinckley, Richard
201 Salisbury, James	263 Jones, Henry	325 Pearson, Thomas
202 Pockmitt, Jackness	264 Little, Charles	326 Jackson, Joseph
203 Dwerell, John	265 Baker, Thomas	327 Mown, Hitcher
204 Welker, John	266 Southcombe, Peter	328 Boyer, David
205 Jolmes, Isaac	267 Jenkins, Thomas	329 Miller, Isaac
206 Clark, James	268 Harvey, James	330 Ray, Gideon
207 Dede Veer, Eppcough	269 Hindman, John	331 Argues, Francis
208 Skipper, John	270 Baptiste, John	332 Jackson, John
209 Lowner, Joseph	271 Kell, Francis	333 Hoyt, Samuel
210 Whitney, William	272 Kitton, Abraham	334 Fisher, John
211 Hunty, Jacob	273 Mackster, William	335 Hawes, J. P.
212 Lukins, Thomas	274 Cooper, Peter	336 Wood, R.
213 Welbourg, David	275 Madden, John	337 Tibble, Joseph
214 Dyer, Charles	276 Jones, James	338 Harwood, Edward
215 Turner, Joshua	277 Watt, Samuel	339 Medley, Samuel
216 Johnson, Abraham	278 Johannes, Logan	340 Deal, John
217 Hulse, J. L.	279 Baldura, Theophilus	341 Farhan, Jacob
218 Westcott, William	280 Howdy, Joseph	342 Tessendes, Moses
219 Bonfonce, Anthony	281 Davis, Richard	343 Brezur, Benjamin
220 Widbee, Joseph	282 Roberts, James	344 Holmes, William
221 Steel, John	283 Benson, John	345 Webster, Benjamin
222 Taylor, William	284 Moulston, Nathaniel	346 Hutchinson, Robert
223 McNeil, Dennis	285 Jackson, John	347 Tatey, Ephraim
224 Phillips, William	286 Peters, John	348 Fielden, John
225 Whiney, Josias	287 Ash, Oliver	349 Wester, John
226 Lacey, Zachariah	288 Norman, William	350 McCarter, M.
227 Garbrierre, John	289 Finder, John	351 Brinnie, Das.
228 Welsh, Richard	290 Davis, James	352 Natcheder, George
229 Johnson, Charles	291 Jackson, Thomas	353 Stickney, Benjamin
230 Johnson, William	292 Truelove, John	354 Ellingwood, Herbert
231 Combes, John	293 Peterson, Thomas	355 Names, Samuel
232 Beck, Francis	294 Richmond, Caleb	356 Holmes, Samuel D.
233 Anderson, John	295 Brown, Henry	357 Robins, Anselm

358 Hannes, George R.
359 Pollard, John
360 Washburn, George
361 Kempton, Mathew
362 Lane, John
363 Bartlett, Nathaniel
364 Holbrook, Gideon
365 Davie, William
366 Westar, Henry
367 Names, Nathaniel
368 Andrews, Joseph
369 Church, Henry
370 Hitchburn, Alexander
371 Smith, William
372 Smith, John
373 Samson, Capes
374 Doans, William
375 Morey, Silves
376 Nutmas, William
377 Prince, Samuel
378 Cross, Thomas
379 Moore, Joseph H.
380 Reed, Aug.
381 Ellis, Edward
382 Brown, Samuel
383 Stackpole, Nathaniel
384 Kent, John
385 Mills, Samuel
386 Gibson, Samuel
387 Pressey, Joseph
388 Jeanason, Isaac
389 Moore, Seth H.
390 Ingusal, Calvin
391 Mason, Samuel
392 Kumer, Samuel
393 Hamond, David
394 Vans, William
395 Whiting, Benjamin
396 Harris, Richard
397 Blodget, Phineas
398 Mansfield, Stod.
399 Welsh, Ebenezer
400 Bennister, George
401 Torrey, Isaac
402 Miller, Peter
403 Ninner, George
404 Taylor, William
405 Cleland, Thomas J.
406 Perry, Miles
407 Small, John
408 Cowett, Jesse
409 Henshaw, Jacob
410 Minor, Charles
411 Smith, Robert
412 Carnes, John
413 Groce, Samuel
414 Little, John S.
415 Lake, William
416 Robinson, William
417 Scott, Edward
418 Cullin, Philip
419 Conner, Henry

420 Rickmes, Rub.
421 Rue, Samuel
422 Huntingdon, E.
423 Walker, Daniel
424 West, Stephen
425 Endley, James
426 Pendall, James
427 Rogers, Smith
428 Cumrie, Horrico
429 Johnston, George
430 Hours, Samuel
431 Robinson, William
432 Johnston, Frederick
433 Stickney, E. F.
434 Seaves, Louis
435 Greenough, William
436 Walker, John
437 Daird, Samuel
438 Stanton, William
439 Furdge, Henry
440 Rondel, N.
441 Brewers, Nathaniel
442 Haywood, Abraham
443 Smith, John
444 McHam, Daniel
445 Kelly, Samuel
446 Burns, John
447 Downs, John
448 Georgory, Cornelius
449 McCaslin, James
450 Matemes, Enu
451 Starks, Jand.
452 Woods, Samuel
453 Adroe, William
454 Hedrick, Jan
455 Brown, Jesse J.
456 Hoty, Charles
457 Hiels, John L.
458 Drott, Henry D.
459 Dumfries, Simon
460 Harvey, Frederick
461 Massabi, Pierre
462 Davit, Anthony
463 Herkwright, John
464 Carlo, William
465 Williams, Alexander
466 Young, Samuel
467 Creker, Edward
468 Thompson, John
469 Adams, Jean B.
470 Henry, Jean B.
471 Touler, Louis
472 Clerge, N.
473 Offa, Torrivi
474 Maxan, Donrill
475 Baptiste, Jean
476 Torre, Vincente
477 Little, John
478 Clestair, (-----)
479 Chateau, Point
480 Fontaine, (-----)
481 Smith, John

482 Pachum, John
483 Jarrett, Abraham
484 Carney, John
485 Weeden, Daniel
486 Shelton, C. J.
487 Histoss, Mathew
488 Hansford, Stephen
489 Simon, Nes.
490 Noy, Jolly
491 Atlantis, Aug.
492 Crouise, John L.
493 Alfonso, (-----)
494 De Lage, Jean
495 Viels, Jean
496 De Four, Victor
497 Lavenio, Louis
498 Condell, B.
499 Lahay, Jean B.
500 Dupoent, Julius
501 Kemenes, Hocao
502 Forsett, (-----)
503 Alexander, Philip
504 Boyd, James
505 Boucher, Louis
506 McLean, Judes
507 Carmere, (-----)
508 Manchard, Louis
509 Greavier, Gabriel
510 Hazlet, Henry
511 Baker, Joseph
512 Jürgen, Morgan
513 Crawford, John
514 Weeks, William
515 Love, Mark
516 Agerm Thomas
517 Fouche, Jean
518 Ford, John
519 Monk, John
520 Connor, Peter
521 Keisoner, Samuel
522 Halleck, Edward
523 Todd, William
524 Watton, John
525 Falcomb, William
526 Barbadoes, Joseph
527 Coffin, Robert S.
528 Plester, Stephen
529 Johnston, Samuel
530 Brown, James
531 Rupey, Luther
532 Brown, John
533 Rutledge, James
534 Jenkins, Walter
535 Jourdan, Conrad
536 Alfords, John
537 Potter, James
538 Donald, Joseph
539 Brown, Robert
540 Kirby, Abel
541 Scargle, William
542 Davis, W. B.
543 Blake, D.

544 Iney, Samuel
545 Briggs, William
546 Lee, John
547 Meddler, John
548 Lickerany, William
549 Brown, John
550 Barbeca, Jacque
551 Kendall, David
552 Smith, B.
553 McDermott, A.
554 Kennedy, Henry
555 Brown, William
556 Williams, Lloyd
557 Domingo, (-----)
558 Arden, Peter
559 Clarke, Reuben
560 Bertram, John
561 Twine, William
562 Clarke, James
563 Jones, John
564 Randolph, Henry
565 Clarke, B.
566 Cook, Francis
567 Swady, William
568 Lehay, George
569 Redmond, Moses
570 Davies, Thomas
571 Hopping, Edward
572 Napp, George
573 Fulton, John
574 Holding, Lawrence
575 Phillips, Jacob
576 Pinchmas, John
577 Wheeler, John
578 Wells, N.
579 Armstrong, Andrew
580 Culver, Richard
581 Hall, George
582 Dotor, Samuel
583 Bourtell, Caleb
584 Howard, Samuel
585 Hammett, C. H
586 Ripley, Daniel
587 Nicon, John M.
588 Mason, William
589 Garish, George
590 Davis, John
591 Lloyd, Samuel
592 Smith, Charles
593 Longiain, Samuel
594 Williams, Stephen
595 Potter, Isaac
596 Bartlett, E.
597 Tubill, William
598 Patterson, Lucas
599 Cole, Isaac
600 Richmond, Alpheus
601 Crawford, J. G.
602 Western, Henry
603 Holmes, Zachus
604 Sandman, Jan
605 Churchill, Joseph

606 Morton, J. J.
607 Rondell, William
608 Morton, Levi
609 Thompson, Ephraim
610 Orn, Joseph
611 Sprague, Benjamin
612 Hodges, William
613 Foy, Peter
614 Blanchard, C.
615 Mitchell, R. J.
616 Morton, P.
617 Loss, Anthony
618 Hubbard, John
619 Lewis, M.
620 Rydor, M.
621 Pendleton, C. L.
622 Harwood, Morocco
623 Drummond, Alexander
624 Bales, John
625 Brown, Zach.
626 Reed, Lewis
627 Keating, John
628 McLachlan, John
629 Holland, Thomas
630 Anthony, Joseph
631 Holding, John
632 May, William
633 Hissings, Lot
634 Rich, John
635 Gibbs, Moses
636 Kemp, Nathaniel
637 Davis, B.
638 Newell, Thomas
639 Ayres, John
640 Jarring, Josiah
641 Cook, C.
642 Smith, Philip
643 Smith, Samuel
644 Knowles, David
645 Wells, Francis
646 Winkler, Mathew
647 Lakes, Leonard
648 Dill, Jobs
649 Shepherd, John
650 Cossioty, John
651 Canot, Lawrence
652 Jack, John
653 Stevens, John
654 Lion, Anthony
655 Jewell, Philip
656 Lovell, William
657 Gallen, Samuel
658 Woodward, John
659 Treffick, Charles
660 Smith, John
661 Huston, Samuel
662 Holmes, William
663 Swinney, William
664 Collins, John
665 Parker, Samuel
666 Merchant, Joseph
667 Carter, John

668 Knowles, William
669 Andrews, Ebenezer
670 Baxter, Charles
671 Abbott, David
672 Comer, John
673 Foster, William
674 Morton, Samuel N.
675 Clarke, George
676 Chase, B.
677 Perez, Emanuel
678 Mazely, William
679 Downing, B.
680 Wilson, Samuel
681 Todd, Robert
682 Harker, John
683 Williams, Alexander
684 Boston, George
685 Card, Jacob
686 Smith, Abraham
687 Marcey, W. R.
688 Frederick, C. M.
689 Story, Elias
690 Bans, William
691 Lamon, John
692 Wales, N.
693 Cloughman, Joseph
694 Princess, George
695 Snow, Godfrey
696 Hall, Noah
697 Crandall, Joseph
698 Smith, David
699 Foster, B. C.
700 Free, Isaac
701 Medford, James
702 Smith, John
703 Mathers, Allen
704 Cross, George
705 Merrick, William
706 Pike, John
707 Barney, O. C.
708 Welson, Michael
709 Colmar, Samuel
710 Spencer, Robert
711 Thompson, John
712 Crosby, J. F.
713 Chilles, John
714 Lingham, Nates
715 Noles, E. D.
716 Myrick, John
717 Moore, John
718 Kethan, George
719 Martyn, John G.
720 Nantes, John
721 Thachster, John
722 Pease, Richard
723 Hall, Samuel
724 Abbott, Francis
725 Norris, Thomas
726 Murray, M.
727 Medford, Joseph
728 Perkins, Lewis
729 Robins, John

730 Ross, George
731 Magown, Joseph
732 Clarke, Benjamin
733 Bartlett, Peekwood
734 Peters, Rolls
735 Brown, Andrew
736 Woodward, Samuel
737 Lane, Emil
738 Nutter, Henry
739 Maker, James
740 Grove, Thomas
741 Payne, John J.
742 Jory, Elisha
743 Eskill, John
744 Spring, Friends
745 Jennings, Benjamin
746 Down, J.
747 Nudd, Stephen
748 Hannah, Samuel
749 Marole, Richard
750 Gordon, John
751 Gordon, Emanuel
752 Burnham, Henry
753 Murphy, William
754 Perry, John
755 Hughes, John
756 Michael, Samuel M.
757 Patton, William
758 Donald, Michael
759 Oatham, Henry
760 Manning, John
761 Craft, John
762 Taulton, Elias
763 Dyk, Es.
764 Emery, John
765 Williams, Pompey
766 Grant, Christian
767 Lewis, Charles
768 Ving, William
769 Ashford, Graves
770 Legiois, William
771 Stacey, William
772 Delaver, Samuel
773 Stone, Benjamin
774 Wilson, Thomas

775 Gordon, Elijah
776 Plenighan, Thomas
777 Sidney, Stephen
778 Nolan, Bartholomew
779 Hall, John
780 Henry, Joshua
781 Inns, John
782 Conner, Peter
783 Stock, Caleb
784 Summers, William
785 Gerry, Joseph
786 Hutson, William
787 Young, William
788 Phillips, Mathew
789 Leach, Samuel
790 Conrad, John L.
791 Deck, Lewis
792 Lewis, Raymond
793 Gimmel, Joseph
794 Bain, John
795 Trammel, Henry
796 Homer, James
797 Porce, John
798 Woodford, Joseph
799 Lincoln, Robert
800 Shepherd, John
801 Shawley, Lewis
802 Randolph, Samuel
803 Johnson, Samuel
804 Carpenter, Jacob
805 Rankins, William
806 Wanett, William
807 Sponarts, James
808 Roundy, John
809 Snider, Enos
810 Gerrich, Samuel
811 Rhodes, Charles H.
812 Mackey, Charles
813 Carpenter, William
814 Randell, Henry
815 Lambert, John
816 Reed, George
817 Manuel, Joseph
818 Johnson, William
819 Hood, Daniel

820 Bray, George
821 Miller, Samuel
822 Putman, Joseph
823 Blair, Samuel
824 Quinn, Patrick
825 Braun, James
826 Taylor, John
827 Watson, George
828 White, William
829 Jones, Ezekiel
830 Hutchins, William
831 Moses, John
832 Hebley, John
833 Whiting, Samuel
834 Poen, William
835 Roper, Nathaniel
836 Hedley, Andrew
837 Bennett, Horsham
838 Owen, Thomas
839 McCally, John
840 Willey, Fredrick
841 Hadley, William
842 Getcher, Samuel
843 Johnson, Moses
844 White, William
845 Linciere, Drury
846 Hoy, James
847 Ocklin, Joseph
848 Bowder, Jacob
849 Nichols, N. D.
850 Gummer, Lewis
851 Gordon, William L.
852 Downs, F.
853 Swift, William
854 Weaver, W. A.
855 Belt, W. L.
856 Ridgeway, Ebenezer
857 Bubier, John
858 Hinchmes, Louis
859 Adams, Nathaniel
860 Harding, Joseph
861 Towley, Joshua
862 Dolliver, William
863 Quogue, Samuel

Crew listing by ship
February 1815 through June 1815

Unknown

Adams, Robert
Anderson, John
Ash, Oliver
Barton, Mathew
Benson, John
Blake, D.   Acquilar
Bonner, John
Boyent, William   Daird, Samuel
Briggs, William   Adamant   Stanton, William
Brown, Henry
Campbell, Henry   Albion   Hulse, J. L.
Coffee, Jacob
Conton, Philip   Douglas, Mathew
Davis, James   Green, Peter
Dine, William   Hill, Stephen
Evans, Thomas   Hingston, Richard   Brothers
Finch, Abraham   Johnson, Jesse
Finder, John   Johnston, Perry   Chance
Finney, James   Kendell, Franklin
Fox, Edward   Lee, Abraham
Frost, Thomas Bell   Lowe, Isaac
Gibson, Francis   Mount, John
Hanstable, William   Stone, Jerry
Hendry, William   Aurora   Swazi, William
Hissings, Lot
Huston, William   Gordon, Elijah
Iney, Samuel   Legiois, William
Jackson, John   Lewis, Charles
Jackson, Thomas   Patton, William
Judah, David   Perry, John
Lee, John   Avon   Ving, William   Decatur
Lemon, John
Marshall, Anthony   Abbott, David
Martin, John   Andrews, Ebenezer
May, William   Ayres, John
McDaniel, John   Baxter, Charles
Meddler, John   Canot, Lawrence
Miller, James   Carter, John
Moss, Richard   Collins, John
Moulston, Nathaniel   Comer, John
Namiss, Peter   Cook, C.
Nettle, Joseph   Cossioty, John
Norman, William   Davis, B.
Parsons, James   Dill, Jobes
Payne, James   Foster, William
Peters, John   Gallen, Samuel
Peterson, Alexander   Gibbs, Moses
Peterson, John   Holmes, William
Peterson, Thomas   Huston, Samuel
Phillips, John   Jack, John
Rake, Martin   Jerring, Josiah   Dolphin
Richmond, Caleb   Jewell, Philip
Roberts, James   Kemp, Nathaniel
Smett, John   Knowles, David
Taylor, John   Knowles, William
Thompson, Joseph   Lakes, Leonard   Dorothy
                                   Lion, Anthony

Thompson, William   Lovell, William
Tune, Samuel   Merchant, Joseph
Truelove, John   Newell, Thomas
Williams, Alexander   Parker, Samuel
Williams, George   Rich, John
Wood, Robert   Shepherd, John
                                   Smith, John
                                   Smith, Philip
                                   Smith, Samuel
                                   Stevens, John
                                   Swiney, William
                                   Treffick, Charles
                                   Wells, Francis
                                   Winkler, Mathew
                                   Woodward, John

                                   Coren, Hugh

                                   Armstrong, Charles
                                   Butler, David
                                   Dominico, John
                                   Gale, Russel
                                   Henson, Peter
                                   Hill, Francis
                                   Little, William
                                   Maceman, Thomas
                                   Macklin, John
                                   Smith, Samuel
                                   Waters, Louis
                                   Whipple, John

                                   Anderson, John
                                   Beck, Francis
                                   Bonfonce, Anthony
                                   Bowen, John
                                   Combes, John
                                   Day, Samuel
                                   Garbrierre, John
                                   Johnson, Charles
                                   Johnson, William
                                   Lacey, Zachariah
                                   McNeil, Dennis
                                   Phillips, William
                                   Steel, John
                                   Taylor, William
                                   Welsh, Richard
                                   Westcott, William
                                   Whiney, Josias
                                   Widbee, Joseph
                                   Williams, John

                                   Barbadoes, Joseph
                                   Falcomb, William
                                   Todd, William
                                   Watton, John

                                   Detende, Joseph

Fincho, Etienne
Jones, Henry
Lawrence, Trent
Little, Charles
McCarthy, Samuel
Ramsdell, Charles

Eliza (1)

Perez, Emanuel

Eliza (2)

Conner, Peter

Engineer

Brown, Robert
Cumrie, Horrico
Davis, W. B.
Donald, Joseph
Endley, James
Hours, Samuel
Huntingdon, E.
Johnston, Frederick
Johnston, George
Kirby, Abel
Pendall, James
Potter, James
Robinson, William
Rogers, Smith
Rue, Samuel
Scargle, William
Walker, Daniel
West, Stephen

Enterprise

Bissett, Robert
Butcher, Jacob
Doing, Denis O.
Jackson, John
Jennings, Francis
Randell, Thomas

Farmer's Daughter

Benton, Samuel
Carle, James
Dawson, John
Edwards, William
Griffin, Heathcote
Hunt, James
Johnson, Thomas
Kingdom, John
Nutter, Henry
Webster, Michael

Fox

Adams, Jean B.
Alexander, Philip
Alfonso, (-----)
Atlantis, Aug.
Baptiste, Jean
Boucher, Louis
Boyd, James
Brown, Jesse J.
Carlo, William
Carmere, (-----)
Carney, John
Chateau, Point
Clerge, N.
Clestair, (-----)

Condell, B.
Creker, Edward
Crouise, John L.
Davit, Anthony
De Four, Victor
De Lage, Jean
Drott, Henry D.
Dumfries, Simon
Dupoent, Julius
Fontaine, (-----)
Forsett, (-----)
Greavier, Gabriel
Hansford, Stephen
Harvey, Frederick
Hazlet, Henry
Henry, Jean B.
Herkwright, John
Hiels, John L.
Histoss, Mathew
Hoty, Charles
Jarrett, Abraham
Kemenes, Hocao
Lahay, Jean B.
Lavenio, Louis
Little, John
Manchard, Louis
Massabi, Pierre
Maxan, Donrill
McLean, Judes
Noy, Jolly
Offa, Torrivi
Pachum, John
Shelton, C. J.
Simon, Nes.
Smith, John
Thompson, John
Torre, Vincente
Touler, Louis
Viels, Jean
Weidon, Daniel
Williams, Alexander
Young, Samuel

Frolick

Todd, Robert
Wilson, Samuel

Gallant

Brown, James
Brown, John
Coffin, Robert S.
Johnston, Samuel
Plester, Stephen
Rupey, Luther

George Little

Bans, William
Bartlett, E.
Bartlett, Peekwood
Boston, George
Bourtell, Caleb
Card, Jacob
Churchill, Joseph
Cole, Isaac
Crawford, J. G.

Davis, John
Donald, Michael
Dotor, Samuel
Dyk, Es.
Frederick, C. M.
Garrish, George
Grant, Christian
Hammett, C. H
Henry, Joshua
Holmes, Zachus
Howard, Samuel
Jory, Elisha
Lloyd, Samuel
Longiain, Samuel
Marcey, W. R.
Mason, William
Medford, Joseph
Michael, Samuel M.
Morton, J. J.
Morton, Levi
Nicon, John M.
Patterson, Lucas
Perkins, Lewis
Plenighan, Thomas
Potter, Isaac
Richmond, Alpheus
Ripley, Daniel
Robins, John
Rondell, William
Sandman, Jan
Sidney, Stephen
Smith, Abraham
Smith, Charles
Stacey, William
Stone, Benjamin
Story, Elias
Thompson, Ephraim
Tubill, William
Western, Henry
Williams, Pompey
Williams, Stephen

Harpy

Clarke, Benjamin
Craft, John
Hannah, Samuel
Magown, Joseph
Manning, John
Marole, Richard
Martyn, John G.
Nudd, Stephen
Nutter, Henry
Taulton, Elias

Helene

Anthony, Joseph
Bales, John
Brown, Zach.
Drummond,
Alexander
Holding, John
Holland, Thomas
Kecting, John
McLachlan, John

Hera

Reed, Lewis

Barbeca, Jacque
Brown, John
Kendall, David
Kennedy, Henry
Lickerany, William
McDermott, A.
Smith, B.

Herald

Stock, Caleb

Heron

Davies, Thomas
Foulton, John
Hopping, Edward
Lehay, George
Napp, George
Rodmond, Moses

High Flyer

Burns, John
Connor, Peter
Downs, John
Georgory, Cornelius
Halleck, Edward
Keisoner, Samuel
Kelly, Samuel
Matemes, Enu
McCaslin, James
McHam, Daniel
Monk, John
Starks, Jand.
Woods, Samuel

Hope

Agerm Thomas
Baker, Joseph
Crawford, John
Fouche, Jean
Jurgin, Morgan
Love, Mark
Weeks, William

Hussar

Nantes, John

Jane

Clerk, James
Stevens, William

John

Betarsl, Samuel
Farrell, John
Henderson, Joseph
Hinchman, George
White, James

King of Prussia

Lindborg, Charles

Leo (1)

Andrews, Joseph
Argues, Francis
Bartlett, Nathaniel
Bennister, George
Blodget, Phineas
Boyer, David
Brezur, Benjamin
Brinnie, Das.

Brown, Samuel
Carnes, John
Church, Henry
Clarke, George
Cleland, Thomas J.
Cowett, Jesse
Cross, Thomas
Davie, William
Deal, John
Doans, William
Ellis, Edward
Ellingwood, Herbert
Fathan, Jacob
Fielden, John
Fisher, John
Gibson, Samuel
Groce, Samuel
Hamond, David
Hannes, George R.
Harris, Richard
Harwood, Edward
Hawes, J. P.
Hinckley, Richard
Hitchburn, Alexander
Holbrook, Gideon
Holmes, Samuel D.
Holmes, William
Hoyt, Samuel
Hutchinson, Robert
Ingusal, Calvin
Jackson, John
Jackson, Joseph
Jeanason, Isaac
Kempton, Mathew
Kent, John
Kumer, Samuel
Lane, John
Little, John S.
Mansfield, Stod.
Mason, Samuel
McCarter, M.
Medley, Samuel
Miller, Isaac
Miller, Peter
Mills, Samuel
Moore, Joseph H.
Moore, Seth H.
Morey, Silves
Morton, Samuel N.
Mown, Hitcher
Names, Nathaniel
Names, Samuel
Natcheder, George
Ninner, George
Nutmas, William
Pearson, Thomas
Perry, Miles
Pollard, John
Pressey, Joseph
Prince, Samuel
Ray, Gideon
Reed, Aug.

Robins, Anselm
Samson, Capis
Small, John
Smith, John
Smith, William
Stackpole, Nathaniel
Stickney, Benjamin
Tatey, Ephraim
Taylor, William
Tessendes, Moses
Tibble, Joseph
Torrey, Isaac
Vans, William
Washburn, George
Webster, Benjamin
Welsh, Ebenezer
Wester, John
Wester, Henry
Whiting, Benjamin
Wood, R.

Leo (2)

Bourdon, Amond
Clark, Francis Marie
Nichelle, Pierre

Lion

Lamon, John

Maid of the Mill

Lord, John
Rust, John

Mars

Greenough, William
Seaves, Louis
Stickney, E. F.
Walker, John

Mary (1)

Arden, Peter
Bertram, John
Brown, William
Clarke, B.
Clarke, James
Clarke, Reuben
Cook, Francis
Domingo, (-----)
Jones, John
Randolph, Henry
Swady, William
Twine, William
Williams, Lloyd

Mary (2)

Conner, Henry
Cullen, Philip
Rickmes, Rub.

Mary (3)

Brown, William
Hackley, Walter
Hicks, (-----)
Ketchell, George
Lumsby, Frederick
Moore, William
Pavish, John

Mary Ann

Callaghan, James

Ford, John

Melvina

Burnham, Henry
Hughes, John
Kethan, George
Moore, John
Murphy, William

Nancy (1)

Furdge, Henry
Rondel, N.

Nancy (2)

Brackett, James
Dolphin, Joseph
Fray, Samuel
Jones, Calvin
Jones, John
Moore, Robert
Norris, Benjamin
Osborne, John L.
Perkins, Clement
Plumber, James

Netterville

Baker, Thomas
Baptiste, John
Cooper, Peter
Harvey, James
Hindman, John
Jenkins, Thomas
Johannes, Logan
Jones, James
Kell, Francis
Kitton, Abraham
Mackster, William
Madden, John
Southcombe, Peter
Watt, Samuel

Olio

Smith, John

Perveranace

Brewers, Nathaniel
Haywood, Abraham

Philbes

Louis, Joseph
Oliver, Mathew
Stenson, David

Plutarch

Autel, Stephen
Bertram, William
Henry
Cooper, Thomas
Covell, Nathaniel
Frazier, Charles
Hepburn, James
James, William
Lambert, Andrew
Parsons, Thomas
Pitts, George
Ray, Richard
Robinson, William
Young, John

Prince de Neufchatel

Alexander, James

Ames, William
Bangs, George
Barker, Thomas
Boylston, Zebadiah
Burnham, Abraham
Churchill, Joseph
Coffis, Theodore
Colson, Christopher
Constant, William
David, Samuel
Davis, Frederick
Delaware, John
Demedorff, John
Door, Ebenezer
Downs, Jesse
Dunklin, Jesse
Farr, William
Fernandez, George
Fisher, Robert
Foreman, John
Fremont, Lawrence
Fuller, Moses
Glover, William
Goodall, Joseph
Greenlaw, Jeremiah
Hall, John
Hawkins, John
Haywood, Samuel
Healy, Samuel
Hose, Richard
Howes, Caleb
Jackson, Ebenezer
Jervis, Peter
Jones, Edward
Leonard, Thomas
Levitt, Caleb
Lewis, Jesse L.
Lidings, John
Lumey, John
Lyon, John
Martin, John
Miller, John
Monk, Philip
Moore, Joshua
Moore, Warren
Nash, Alexander
Newall, John B. L.
Partex, Samuel
Payton, James
Perry, Ebenezer
Peterson, David
Ranbre, Mathew
Raymond, John
Samuels, Samuel
Sanster, John
Snow, Daniel
Studley, Warren
Symons, Moses
Thomas, Charles
Thomas, Theodore
Train, Joseph
Welch, John

Wentworth, John
Whittington, Samuel
Wicks, Ebenezer
Wright, Joseph

Reindeer

Chilles, John
Crosby, J. F.
Lingham, Nates
Myrick, John
Noles, E. D.

Romes

Johnson, Abraham

San Francisco

Hedrick, Jan

Saucy Jack

McCormick, James
Smith, John
Whitehead, John

Sennett, Swedish merchant vessel

Clark, James
Dede Veer,
Eppcough
Dwerell, John
Dyer, Charles
Hunty, Jacob
Jolmes, Isaac
Lowner, Joseph
Lukins, Thomas
Pockmitt, Jackness
Salisbury, James
Skipper, John
Turner, Joshua
Welbourn, David
Welker, John
Whitney, William

Sine Qua Non

Abbott, Francis
Ashford, Graves
Barney, O. C.
Blanchard, C.
Brown, Andrew
Cloughman, Joseph
Colmer, Samuel
Crandall, Joseph
Cross, George
Delaver, Samuel
Down, J.
Emery, John
Eskill, John
Foster, B. C.
Foy, Peter
Free, Isaac
Gordon, Emanuel
Gordon, John
Grove, Thomas
Hall, John
Hall, Noah
Hall, Samuel
Harwood, Morocco
Hodges, William
Hubbard, John
Inns, John

Jennings, Benjamin
Lane, Emil
Lewis, M.
Loss, Anthony
Maker, James
Mathers, Allen
Medford, James
Merrick, William
Mitchell, R. J.
Morton, P.
Murray, M.
Nolan, Bartholomew
Norris, Thomas
Oatham, Henry
Orn, Joseph
Payne, John J.
Pease, Richard
Peters, Rolls
Pike, John
Pendleton, C. L.
Princess, George
Ross, George
Ryder, M.
Smith, David
Smith, John
Snow, Godfrey
Spencer, Robert
Sprague, Benjamin
Spring, Friends
Thachster, John
Thompson, John
Wales, N.
Welson, Michael
Wilson, Thomas
Woodward, Samuel

Snap Dragon

Downing, B.
Mazely, William

Sophie

Adroe, William
Williams, James

Sorin

Armstrong, Andrew
Culber, Richard
Hall, George
Holding, Lawrence
Phillips, Jacob
Pinchmas, John
Wells, N.
Wheeler, John

St. Francis

Alfords, John
Jenkins, Walter
Jourdan, Conrad
Rutledge, James

St. Joanne

Bottes, John
Mack, William F.
Marshall, William

Steven Getard

Pitts, Francis

Tartan

Baldura, Theophilus
Davis, Richard
Hawdy, Joseph

Theodore

Chase, B.
Alden, Henry
Henshaw, Jacob
Johnson, Samuel J,
McIntyre, John
Minor, Charles
Smith, Robert
Whiter, Amos

Transit

Alexander, George
Fisher, James
Fuller, Benjamin
Heasey, John
Price, Peter
Rosenburg, Charles
Sewell, John

US Brig Syren

Adams, Nathaniel
Aide, John L.
Andrews, Samuel
Bain, John
Belt, W. L.
Bennett, Horsham
Blair, Samuel
Bowder, Jacob
Braun, James
Bray, George
Bubier, John
Carpenter, Jacob
Carpenter, William
Casper, William
Chick, Moses
Cloutman, Ephraim
Coates, Samuel M.
Conrad, John L.
Daniels, Henry
Davis, Elias S.
Deck, Lewis
Desomon, Thomas
Dolliver, William
Downs, F.
Dunham, William
Frovemcs, Ford.
Gimmel, Joseph
Gerrich, Samuel
Gerry, Joseph
Getcher, Samuel
Gordon, William L.
Gunmar, Lewis
Hadley, William
Harding, Joseph
Haycock, Joseph
Hebley, John
Hedley, Andrew
Hinchmes, Louis
Hitchens, George
Hodges, Edward
Homer, James

Hood, Daniel
Hoy, James
Hutchins, William
Hutson, William
Johnson, Moses
Johnson, Samuel
Johnson, William
Jones, Ezekiel
Kellock, Stephen
Kirby, Robert
Lambert, John
Leach, Samuel
Lewis, Raymond
Lewis, Winslow
Linciere, Drury
Lincoln, Robert
Mackey, Charles
Mackford, John
Manuel, Joseph
McCally, John
Menzer, James
Micker, Charles
Miller, Samuel
Millow, John
Morris, John
Moses, John
Nicholas, Jacob
Nichols, N. D.
Ocklin, Joseph
Owen, Thomas
Parnet, Henry
Phillips, Mathew
Poen, William
Pope, Alexander
Porce, John
Putman, Joseph
Quogue, Samuel
Quinn, Patrick
Randell, Henry
Randolph, Samuel
Rankins, William
Reed, George
Rhodes, Charles H.
Ridgeway, Ebenezer
Roper, Nathaniel
Roundy, John
Rust, Zebulon
Shawley, Lewis
Shepherd, John
Snider, Enos
Sparkes, John
Spear, Joseph
Sponarts, James
Stevens, John
Stone, John
Summers, William
Swift, William
Taylor, John
Thomas, John
Towley, Joshua
Trummel, Henry
Wanett, William

Watson, George
Weaver, W. A.
West, Jacob   W. S. Many
White, William
Whiting, Samuel
Willey, Fredrick

Woodford, Joseph   William
Young, William   Harker, John

Lake, William   Wolfe   Bailey, Samuel
Robinson, William   Williams, John
Scott, Edward